T0181621

 Birkhäuser

George Grätzer

# Lattice Theory: Foundation

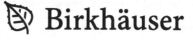 Birkhäuser

George Grätzer
Department of Mathematics
University of Manitoba
Winnipeg, Manitoba R3T 2N2
Canada
gratzer@me.com

2010 Mathematics Subject Classification 06-01, 06-02

ISBN 978-3-0348-0017-4        e-ISBN 978-3-0348-0018-1
DOI 10.1007/978-3-0348-0018-1

Library of Congress Control Number: 2011921250

*Cover design*: deblik, Berlin

Printed on acid-free paper

Springer Basel AG is part of Springer Science+Business Media

www.birkhauser-science.com

To Cheryl and David,
for the support they gave me,
when it was most needed ...

# Short Contents

# Contents

# *Preface*

My book, *General Lattice Theory,* was published in 1978. Its goal: "to discuss in depth the basics of general lattice theory". Each chapter concluded with a section, *Further Topics and References,* providing brief outlines of, and references to, related topics. Each chapter contained a long list of open problems.

The *second edition* appeared twenty years later, in 1998. It included the material of the first edition, and a series of appendices. The first, *Retrospective,* reviewed developments of the 20 years between the two editions, especially, solutions of the open problems proposed in the first edition. The other seven appendices surveyed new fields. They were written by the best experts available. Obviously, I could no longer command an overview of all of lattice theory. The book provided foundation, the appendices surveyed contemporary research.

The explosive growth of the field continued. While the nineteen sixties provided under 1,500 papers and books, the seventies 2,700, the eighties over 3,200, the nineties almost 3,600, and the first decade of this century about 4,000. As a result, it became almost inevitable that we split the book into two volumes.

This book, *Lattice Theory: Foundation,* lays the foundations of the field. There are no *Retrospectives* and no lists of open problems. Its companion volume, *Lattice Theory: Special Topics and Applications,* completes the picture; it is written by experts in the various topics covered.

To help the readers of this book to acquire a wider view, almost a thousand exercises are provided. And there are over forty *diamond sections,* brief sections marked by the symbol $\Diamond$, that provide brief glimpses into research fields beyond the horizon of this book.

## Contributors

The following mathematicians contributed diamond sections:

- Kira Adaricheva (Sections VI.2.7 and VII.2.7);
- Gábor Czédli (Section V.1.9);
- Brian A. Davey and Miroslav Haviar (Section VI.2.8);
- Kalle Kaarli (Section II.2.5);
- Jimmie D. Lawson (Section I.3.16);
- Joseph P. S. Kung (Section V.5.13);
- Tibor Katriňák (Section V.1.8);
- J. B. Nation (Section IV.4.4);
- Hilary A. Priestley (Section II.5.6);
- Aleš Pultr (Section II.5.7);
- Manfred Stern (Section V.2.6);
- Friedrich Wehrung (Sections III.4.4, IV.5.5, V.5.11, and V.5.12).

I am deeply appreciative to all of them.

## Acknowledgements

To keep this *Preface* short, I put the history of this book and the very extensive acknowledgements into the *Afterword*. But let me repeat one point made there. I started writing this book in 1968. In the forty plus years of this endeavor, I received help from hundreds of mathematicians. I am forever grateful.

# Foreword

In the first half of the nineteenth century, George Boole's attempt to formalize propositional logic led to the concept of boolean algebras. While investigating the axiomatics of boolean algebras at the end of the nineteenth century, Charles S. Pierce and Ernst Schröder found it useful to introduce the lattice concept. Independently, Richard Dedekind's research on ideals of algebraic numbers led to the same discovery. In fact, Dedekind also introduced modularity, a weakened form of distributivity. Although some of the early results of these mathematicians and of Edward V. Huntington are very elegant and far from trivial, they did not attract the attention of the mathematical community.

It was Garrett Birkhoff's work in the mid-1930s that started the general development of lattice theory. In a brilliant series of papers, he demonstrated the importance of lattice theory and showed that it provides a unifying framework for hitherto unrelated developments in many mathematical disciplines. Birkhoff himself, Valère Glivenko, Karl Menger, John von Neumann, Oystein Ore, and others had developed enough of this new field for Birkhoff to attempt to "sell" it to the general mathematical community, which he did with astonishing success in the first edition of his *Lattice Theory*. The further early development of the subject matter can best be followed by comparing the first, second, and third editions of his book: G. Birkhoff [65] (1940), [70] (1948), and [71] (1967).

The goal of the present volume can be stated very simply: to discuss in depth the foundation of lattice theory. I tried to include the most important results and research methods that form the basis of all the work in this field.

Special topics and applications of lattice theory are presented in the companion volume. As I mentioned in the *Preface,* over forty *diamond sections* whet the appetite of the reader by providing brief glimpses into areas not covered in this volume.

In my view, distributive lattices have played a many-faceted role in the development of lattice theory. Historically, lattice theory started with (boolean) distributive lattices; as a result, the theory of distributive lattices is one of the most extensive and most satisfying chapters of lattice theory. Distributive lattices have provided the motivation for many results in general lattice theory. Several conditions on lattices and on elements and ideals of lattices are weakened forms of distributivity. Therefore, a thorough understanding of distributive lattices is indispensable for work in lattice theory.

This viewpoint moved me to break with the traditional approach to lattice theory, which proceeds from orders to general lattices, semimodular lattices, modular lattices, and, finally, to distributive lattices. My approach has the added advantage that the reader reaches interesting and deep results early in the book.

**Chapter I** develops the basic concepts of orders and lattices. Diagrams are emphasized because I believe that an important part of learning lattice theory is the acquisition of skill in drawing diagrams. This point of view is stressed throughout the book by about 130 diagrams (heeding Alice's advice: "and what is the use of a book without pictures", L. Carroll [1865]); the reader would be well advised to draw lots more while reading the book.

A special feature of this chapter is a detailed development of free lattices generated by a partial lattice over an arbitrary variety; this is one of the most important research tools of lattice theory.

*Diamond section* topics include tolerances, continuous lattices, the characterization theorem of congruence lattices of universal algebras, finitely presented lattices, and various axiom systems for lattices.

**Chapter II** develops distributive lattices including representation theorems, congruences, boolean algebras, and topological representations. The last section is a brief introduction to the theory of distributive lattices with pseudocomplementation. While the theory of distributive lattices is developed in detail, the reader should keep in mind that the purpose of this chapter is, basically, to serve as a model for the rest of lattice theory.

*Diamond section* topics include polynomial completeness, Priestley spaces, frames (a lattice theoretic approach to topology), and generalizations of Stone algebras.

In **Chapter III**, we discuss congruences and ideals of general lattices. The various types of ideals discussed all imitate to some extent the behavior of ideals in distributive lattices.

There is only one *diamond section*, discussing infinite direct decompositions of complete lattices.

Lattice constructions play a central role in lattice theory. **Chapter IV** discusses a construction of old: gluing (1941) and the newer One-Point Extension (1992), the crucial chopped lattices (from the 1970s), and the newest construction (1999): boolean triples.

*Diamond section* topics include generalized gluing constructions, congruence lattices of (i) finite lattices, (ii) finite lattices in special classes, (iii) more than one finite lattice, (iv) general lattices, (v) complete lattices; furthermore, independence results, tensor products, and congruence-permutable, congruence-preserving extensions.

After presenting the basic facts concerning modular and semimodular lattices, **Chapter V** investigates in detail the connection between lattice theory and geometry. We develop the theory of geometric lattices, in particular, direct decompositions and geometric lattices arising out of geometries and graphs. As an important example, we investigate partition lattices. The last section deals with complemented modular lattices and projective geometries, including the Coordinatization Theorem and Frink's Embedding Theorem.

*Diamond section* topics include pseudocomplementation in modular lattices, identities of submodule lattices, consistency (a generalization of modularity different from semimodularity), type 2 and 3 congruence lattices for universal algebras, special topics on partition lattices, coordinatization results of sectionally complemented modular lattices, the dimension monoid of a lattice (a precursor of the congruence lattice), and Dilworth's covering theorem.

**Chapter VI** deals with varieties of lattices. It covers the basic properties, including Jónsson's Lemma, the lattice of varieties of lattices, equational bases, and the Amalgamation Property.

*Diamond section* topics include products of varieties, lattices of (quasi-) equational theories, and modified Priestley dualities.

**Chapter VII** presents free products of lattices, including the Structure Theorem, the Common Refinement Property, sublattices of a free lattice, reduced free products, and hopfian lattices.

*Diamond section* topics include amalgamated free products, the word problem for modular lattices, transferable lattices and finite sublattices of a free lattice, semidistributive lattices, and Dean's Lemma.

The exercises form an integral part of the book; do not leave a section without doing a good number of them.

The Bibliography contains about 700 entries; it is not a comprehensive bibliography of this field. With a few exceptions, it contains only items referred to in the text. To find the references for a topic, use the AMS online database, *MathSciNet*, or turn to *Zentralblatt*.

A very detailed *Index* and the *Glossary of Notation* should help the reader in finding where a concept or notation is first introduced. For names and concepts, such as "Jónsson, B." and "Priestley space", use the *Index*; symbols, such as $\operatorname{Con} L$, $\operatorname{rank}(p)$, should be looked up in the *Glossary*.

I assume a rudimentary knowledge of basic set theory and algebra.

## Notation

More difficult exercises are marked by *. Theorems (lemmas) presented without proofs are marked by the diamond symbol $\Diamond$.

Section 5 refers to a section in the chapter you are reading, whereas Section II.5 refers to a section in Chapter II. Exercise 5.2 refers to the second exercise in Section 5 of the chapter you are in, while Exercise V.5.2 refers to the second exercise in Section 5 of Chapter V. Finally, Lemma 403(ii) refers to

the second statement of Lemma 403 and Definition 41(i) to the first condition
of Definition 41.

If you are curious how the mathematical notational system used in this
book developed, consult the *Afterword*.

Winnipeg, Manitoba
November 2010

*George Grätzer*

gratzer@me.com

# Glossary of Notation

xxiii

| Symbol | Explanation | Page |
|---|---|---|
| Cov $P$ | covering graph of order $P$ | 7 |
| **D** | class (variety) of distributive lattices | 15, 75 |
| $\dim(P)$ | order-dimension of an order $P$ | 9 |
| Distr $L$ | set of all distributive elements of $L$ | 223 |
| Distr$^\delta$ $L$ | set of all dually distributive elements of $L$ | 224 |
| Dns $L$ | dense set of $L$ | 101, 194 |
| Down $P$ | order of down-sets of the order $P$ | 7 |
| ext | for $K \le L$, extension map: $\alpha \mapsto \mathrm{con}_L(\alpha)$ | 216 |
| End $L$ | endomorphism monoid of $L$ | 122, 515 |
| End$_{\{0,1\}}$ $L$ | $\{0,1\}$-endomorphism monoid of $L$ | 122, 515 |
| Equ $A$ | lattice of all equivalences on $A$ | 3 |
| (Ex) | condition for the existence of a free lattice | 77 |
| fil$(a)$ | filter generated by the element $a$ | 34 |
| fil$(H)$ | filter generated by the set $H$ | 34 |
| (Fil) | a condition for partial lattices | 88 |
| Fil $L$ | filter lattice of a lattice $L$ | 34 |
| Fil$_0$ $L$ | augmented filter lattice of a lattice $L$ | 34, 88 |
| Free$(\mathfrak{m})$ | free lattice on $\mathfrak{m}$ generators | 76 |
| Free $P$ | free lattice over the order $P$ | 76 |
| Free $\mathfrak{A}$ | free lattice over the partial lattice $\mathfrak{A}$ | 90 |
| Free$_{\mathbf{D}}(3)$ | free distributive lattice on three generators | 82 |
| Free$_{\mathbf{K}}$ $\mathfrak{A}$ | free lattice over $\mathfrak{A}$ in a variety $\mathbf{K}$ | 89 |
| Free$_{\mathbf{K}}$ $P$ | free lattice over the order $P$ in a variety $\mathbf{K}$ | 76 |
| Free$_{\mathbf{M}}(3)$ | free modular lattice on three generators | 83 |
| Free$(P; \mathcal{J}, \mathcal{M})$ | free lattice in Dean's Lemma | 517 |
| (GC) | compactness condition on open sets | 170 |
| $(G; E)$ | graph on set $G$ with edges $E$ | 7 |
| height$(a)$ | height of an element | 4 |
| **H(K)** | class of homomorphic images of members of **K** | 413 |
| id$(a)$ | principal ideal generated by the element $a$ | 32, 88 |
| id$(H)$ | ideal generated by the set $H$ | 32 |
| **I(K)** | class of isomorphic copies of members of **K** | 413 |
| Id $L$ | ideal lattice of $L$ | 33, 52, 270 |
| Id$_0$ $L$ | augmented ideal lattice of $L$ | 33, 88 |
| (Id) | for ideals in chopped lattices | 270 |
| (Idem) | idempotency condition for a binary operation | 10 |
| Iden$(\mathbf{K})$ | set of identities holding in the class **K** | 409 |
| (Idl) | a condition for partial lattices | 88 |
| $\inf H, \bigwedge H$ | greatest lower bound of $H$ | 5 |
| Ji $L$ | order of join-irreducible elements of $L$ | 102 |
| (JID) | Join Infinite Distributive Identity | 154 |

| Symbol | Explanation | Page |
|---|---|---|
| $\mathrm{Ker}(\varphi)$ | congruence kernel of $\varphi$ | 41 |
| **L** | class (variety) of all lattices | 75 |
| **Λ** | lattice of all varieties of lattices | 423 |
| $\mathrm{Lat}\,G$ | lattice representation of the graph $G$ | 515 |
| $\mathit{L}^{\mathrm{alg}}$ | the (order) lattice $L$ as an algebra | 13 |
| $\mathit{L}^{\mathrm{ord}}$ | the (algebra) lattice $L$ as an order | 13 |
| $\mathrm{len}(P)$ | length of a finite order $P$ | 4 |
| (Lin) | linearity condition for binary relations | 1 |
| **M** | class (variety) of modular lattices | 15, 75 |
| $\mathsf{M}_3$ | five-element modular nondistributive lattice | 23 |
| $\mathbf{M}_3$ | variety generated by $\mathsf{M}_3$ | 425 |
| $\mathsf{M}_4$ | a modular lattice with four atoms | 425 |
| $\mathbf{M}_4$ | variety generated by $\mathsf{M}_4$ | 425 |
| $\mathsf{M}_{3,3}$ | two copies of $\mathsf{M}_3$ glued together | 425 |
| $\mathbf{M}_{3,3}$ | variety generated by $\mathsf{M}_{3,3}$ | 425 |
| $\mathrm{Max}(M)$ | maximal elements of a chopped lattice | 270 |
| $\mathrm{Merge}(C, D)$ | merging of lattices $C$ and $D$ | 269 |
| $\mathrm{Mi}\,L$ | order of meet-irreducible elements of a lattice $L$ | 102 |
| (MID) | Meet Infinite Distributive Identity | 154 |
| $\mathbf{Mod}(\Sigma)$ | class of all lattice models of $\Sigma$ | 409 |
| $\mathsf{M}_3[L]$ | order of boolean triples of the lattice $L$ | 295 |
| $\mathsf{M}_3[\boldsymbol{\alpha}]$ | congruence on $\mathsf{M}_3[L]$ | 297 |
| $\mathsf{N}_5$ | five-element nonmodular lattice | 23 |
| $\mathbf{N}_5$ | the variety generated by $\mathsf{N}_5$ | 423 |
| $\mathsf{N}_6 = \mathsf{N}(p, q)$ | six-element nonmodular lattice | 277 |
| $\mathrm{Neutr}\,L$ | set of all neutral elements of $L$ | 226 |
| (OP) | condition for One-Point Extension | 256 |
| $\mathfrak{p}, \mathfrak{q}$ | prime intervals | 35 |
| $P^\delta$ | dual of $P$ | 5 |
| $P^{\mathrm{max}}$ | a partial lattice formed from the order $P$ | 90 |
| $P^{\mathrm{min}}$ | a partial lattice formed from the order $P$ | 90 |
| $\mathrm{Part}\,A$ | partition lattice of $A$ | 3, 359 |
| $\mathrm{Part}_{\mathrm{fin}}\,A$ | set of finite partitions | 361 |
| $\mathrm{PG}(D, \mathfrak{m})$ | $\mathfrak{m}$-dimensional projective geometry over $D$ | 378 |
| $\mathrm{Pow}\,X$ | power set lattice of $X$ | 4 |
| $\mathbf{P}(\mathbf{K})$ | class of direct products of members of $\mathbf{K}$ | 413 |
| $\mathbf{P_s}(\mathbf{K})$ | class of subdirect products of members of $\mathbf{K}$ | 414 |
| $\mathbf{P_u}(\mathbf{K})$ | class of ultraproducts of members of $\mathbf{K}$ | 416 |
| $\mathrm{PrInt}(L)$ | set of prime intervals of $L$ | 213 |
| $\mathbb{Q}$ | chain of rational numbers | 158 |

| Symbol | Explanation | Page |
|---|---|---|
| rank($p$) | rank of a term $p$ | 68 |
| re | for $K \leq L$, restriction map: $\alpha \mapsto \alpha \rceil K$ | 214 |
| rep($p$) | terms equivalent to $p$ in a free product | 474 |
| (Refl) | reflexivity condition for binary relations | 1 |
| sub($H$) | sublattice generated by $H$ | 31 |
| sup $H, \bigvee H$ | least upper bound of $H$ | 5 |
| spec($a$) | spectrum of an element $a$ | 112, 118 |
| (SD$_\vee$) | join-semidistributive law | 479 |
| (SD$_\wedge$) | meet-semidistributive law | 479 |
| **S(K)** | class of subalgebras of members of **K** | 413 |
| **Si(K)** | class of subdirectly irreducible members of **K** | 418 |
| S$_1$ | three-element Stone algebra | 201 |
| $\mathcal{S}^{\text{B}}$ | booleanization of the Stone space $\mathcal{S}$ | 176 |
| Spec $L$ | spectrum of a distributive lattice $L$ | 116 |
| Skel $L$ | skeleton of $L$ | 99 |
| (SP$_\vee$) | join-substitution property | 36, 43, 269 |
| (SP$_\wedge$) | meet-substitution property | 36, 43, 269 |
| Sub $\mathfrak{A}$ | subalgebra lattice of an algebra $\mathfrak{A}$ | 57 |
| Stand $L$ | set of all standard elements of $L$ | 224 |
| (Stone1)–(Stone3) | Stone conditions on a topological space | 171, 174 |
| Sub $L$ | sublattice lattice (including $\varnothing$) of a lattice $L$ | 51 |
| **T** | class (variety) of trivial lattices | 92, 423 |
| T$_n$ | Tamari lattice | 27 |
| Term($n$) | $n$-ary lattice terms | 66 |
| Term$_{\text{B}}(n)$ | $n$-ary lattice terms in **B** | 129 |
| Term$_{\text{D}}(n)$ | $n$-ary lattice terms in **D** | 126 |
| tran($\varrho$) | transitive closure for binary relation $\varrho$ | 3 |
| (Trans) | transitivity condition for binary relations | 1 |
| (W) | Whitman condition for free product | 479 |
| **Var(K)** | smallest variety containing the class **K** | 414 |
| $W_i$ | Whitney number | 353 |

| Symbol | Explanation | Page |
|---|---|---|

### Relations and Congruences

| | | |
|---|---|---|
| $A^2$ | set of ordered pairs of $A$ | 2 |
| $\varepsilon, \varrho, \tau, \pi, \ldots$ ` | binary relations | 2 |
| $\mathrm{equ}(\pi)$ | binary relation from a partition $\pi$ | 3 |
| $\boldsymbol{\alpha}, \boldsymbol{\beta}, \ldots, \boldsymbol{\theta}$ | congruences | 36 |
| $\boldsymbol{0}, \boldsymbol{1}$ | zero and unit of Part $A$ and Con $L$ | 36 |
| $(a, b) \in \varepsilon$ | $a$ and $b$ are in relation $\varepsilon$ | 3 |
| $a \, \varepsilon \, b$ | $a$ and $b$ are in relation $\varepsilon$ | 3 |
| $a \equiv b \pmod{\varepsilon}$ | $a$ and $b$ are in relation $\varepsilon$ | 3 |
| $\mathfrak{p} \in \boldsymbol{\alpha}$ | prime interval $\mathfrak{p}$ collapsed by $\boldsymbol{\alpha}$ | 213 |
| $a/\pi$ | block containing $a$ | 3, 36 |
| $A/\pi$ | set of all block of $\pi$ | 3, 40 |
| $L/\boldsymbol{\alpha}$ | quotient lattice | 40 |
| $\boldsymbol{\beta}/\boldsymbol{\alpha}$ | quotient congruence, tolerance | 42, 43, 198 |
| $\boldsymbol{\alpha}\rceil_K$ | restriction of $\boldsymbol{\alpha}$ to the sublattice $K$ | 41 |
| $\pi_i$ | projection map | 46 |
| $\boldsymbol{\alpha} \times \boldsymbol{\beta}$ | direct product of congruences | 46 |

### Orders

| | | |
|---|---|---|
| $\leq, <$ | ordering | 2 |
| $\geq, >$ | ordering, inverse notation | 2 |
| $K \leq L$ | $K$ is a sublattice of $L$ | 31 |
| $L \geq K$ | $L$ is an extension of $K$ | 31 |
| $\leq_Q$ | ordering of $P$ restricted to a subset $Q$ | 4 |
| $a \parallel b$ | $a$ incomparable with $b$ | 4 |
| $a \prec b$ | $a$ is covered by $b$ | 6 |
| $a \preceq b$ | $a \prec b$ or $a = b$ | 6 |
| $b \succ a$ | $b$ covers $a$ | 6 |
| $b \succeq a$ | $b \succ a$ or $b = a$ | 6 |
| $0, 1$ | zero and unit of order | 5 |
| $a \vee b$ | join operation | 10 |
| $\bigvee H$ | least upper bound of $H$ | 5 |
| $a \wedge b$ | meet operation | 10 |
| $\bigwedge H$ | greatest lower bound of $H$ | 5 |
| $[a, b]$ | interval | 35 |
| $\mathrm{down}(H)$ | down-set generated by $H$ | 7 |
| $\mathrm{down}(a), \downarrow a$ | down-set generated by $a$ | 7 |
| $P \cong Q$ | order (lattice) $P$ isomorphic to $Q$ | 4, 12, 28 |

| Symbol | Explanation | Page |
|---|---|---|

**Miscellaneous**

| | | |
|---|---|---|
| $a^*$ | pseudocomplement of $a$ | 99 |
| $a_*$ | unique lower cover of $a$ | 102 |
| $a'$ | complement of $a$ | 99 |
| $\overline{x}, \overline{X}$ | closure of $x$ and $X$ | 47, 49 |
| $\varnothing$ | empty set | 5 |
| $\equiv_{\mathbf{B}}$ | equivalence of boolean terms | 128 |
| $p/\mathbf{B}$ | block of boolean terms | 129 |
| $\equiv_{\mathbf{D}}$ | equivalence of distributive lattice terms | 126 |
| $p/\mathbf{D}$ | block of distributive lattice terms | 126 |
| $x + y$ | symmetric difference | 132 |
| $\alpha \circ \beta$ | product of binary relations | 2 |
| $\mathbf{V} \circ \mathbf{W}$ | product of the varieties $\mathbf{V}$ and $\mathbf{W}$ | 430 |
| $a \, M \, b$ | modular pair | 335 |
| $p^{(i)}, p_{(i)}$ | upper and lower $i$-cover of $p$ in a free product | 471 |
| $\overline{p}, \underline{p}$ | upper and lower cover of $p$ in $P$ | 517 |
| $A^b$ | lattice $A$ with two new bounds | 471 |

# Chapter

# I

# First Concepts

## 1. Two Definitions of Lattices

### 1.1 Orders

Whereas the arithmetical properties of the set of reals $\mathbb{R}$ can be expressed in terms of addition and multiplication, the order theoretic, and thus the topological, properties are expressed in terms of the ordering $\leq$. The basic properties of this relation are as follows.

For all $a, b, c \in \mathbb{R}$, the following rules hold for the ordering:

| (Refl) | Reflexivity: | $a \leq a$. |
|---|---|---|
| (ASym) | Antisymmetry: | $a \leq b$ and $b \leq a$ imply that $a = b$. |
| (Trans) | Transitivity: | $a \leq b$ and $b \leq c$ imply that $a \leq c$. |
| (Lin) | Linearity: | $a \leq b$ or $b \leq a$. |

There are many examples of binary relations sharing these properties with the ordering of reals, and there are even more enjoying the first three properties. This fact, by itself, would not justify the introduction of a new concept. However, it has been observed that many basic concepts and results about the reals depend only on the first three properties, and these can be profitably used whenever we have a relation satisfying them. A relation satisfying the properties: reflexivity, antisymmetry, and transitivity (the conditions: (Refl), (ASym), and (Trans)) is called an *ordering*; a nonempty set equipped with such a relation is called an *ordered set* or an *order* (or a *partially ordered set* or a *poset*).

To make the definitions formal, let us start with two sets $A$ and $B$ and form the set $A \times B$ of all ordered pairs $(a, b)$ with $a \in A$ and $b \in B$. If $A = B$,

G. Grätzer, *Lattice Theory: Foundation*, DOI 10.1007/978-3-0348-0018-1_1,
© Springer Basel AG 2011

we write $A^2$ for $A \times A$. Then a *binary relation* $\varrho$ on $A$ is a subset of $A^2$. The elements $a, b \in A$ are *in relation* with respect to $\varrho$ if $(a, b) \in \varrho$, for case we shall also use the notation $a \varrho b$. Binary relations will be denoted by small Greek letters or by special symbols.

Compare this formal definition with the intuitive one: The binary relation $\varrho$ on $A$ is a "rule" that decides whether or not $a \varrho b$ for any given pair $a, b$ of elements of $A$. Of course, any such rule will determine the set

$$\{ (a, b) \in A^2 \mid a \varrho b \},$$

and this set determines $\varrho$, so we might as well regard $\varrho$ as being the same as this set.

For the binary relations $\varrho$ and $\sigma$ on $A$, we introduce the *product* $\varrho \circ \sigma$, a binary relation on $A$ defined as follows: $(a, b) \in \varrho \circ \sigma$ if there is an element $c \in A$ satisfying $(a, c) \in \varrho$ and $(c, b) \in \sigma$. So the binary relation $\varrho$ is transitive if $\varrho \circ \varrho \subseteq \varrho$.

An *order* $(A; \varrho)$ consists of a nonempty set $A$ and a binary relation $\varrho$ on $A$ such that $\varrho$ satisfies properties reflexivity, antisymmetry, and transitivity. Note that these can be restated as follows:

For all $a, b, c \in A$,

$(a, a) \in \varrho$;

$(a, b), (b, a) \in \varrho$ imply that $a = b$;

$(a, b), (b, c) \in \varrho$ imply that $(a, c) \in \varrho$.

If $\varrho$ satisfies the properties reflexivity, antisymmetry, and transitivity, then $\varrho$ is an *ordering* (or *partial ordering relation*), and will usually be denoted by $\leq$. If $a \leq b$, sometimes we say that $a$ is *majorized by* $b$ or $b$ *majorizes* $a$. Also,

$a \geq b$ means that $b \leq a$;

$a < b$ means that $a \leq b$ and $a \neq b$;

$a > b$ means that $b < a$.

Most of the time we shall say that $A$ (rather than $(A; \leq)$) is an order, meaning that the ordering is understood. This is an ambiguous, although widely accepted, practice.

## 1.2  Equivalence relations and preorderings

There is another important property—almost the opposite of (ASym)—a binary relation $\varepsilon$ can have:

(Sym) Symmetry:   $(a, b) \in \varepsilon$ implies that $(b, a) \in \varepsilon$.

A binary relation $\varepsilon$ on the nonempty set $A$ satisfying the three properties: reflexivity, symmetry, and transitivity, that is the conditions: (Refl), (Sym), and (Trans), is called an *equivalence relation*. If $\varepsilon$ is an equivalence relation, the relation $(a, b) \in \varepsilon$ is often denoted by $a \equiv b \pmod{\varepsilon}$.

For an equivalence relation $\varepsilon$ on the nonempty set $A$ and for an $a \in A$, we define *the block of $\varepsilon$ containing $a$* (often called, the *equivalence class* of $\varepsilon$ containing $a$) as follows:

$$a/\varepsilon = \{\, b \in A \mid (a, b) \in \varepsilon \,\}.$$

Note that if $b \in A/\varepsilon$, then $a/\varepsilon = b/\varepsilon$.

Let $A/\varepsilon$ denote the set of all blocks of $\varepsilon$; this set forms a *partition* of $A$, that is, a family of pairwise disjoint nonempty subsets of $A$ whose union is $A$.

Conversely, if $\pi$ is a partition of $A$ and we define a binary relation $\mathrm{equ}(\pi)$ on $A$ as follows:

$$(a, b) \in \mathrm{equ}(\pi) \quad \text{if} \quad a, b \in X \text{ for some } X \in \pi,$$

then $\mathrm{equ}(\pi)$ is an equivalence relation. So equivalence relations and partitions are two ways of describing the same situation.

We denote by $\mathrm{Equ}\, A$ the *set of all equivalence relations* and by $\mathrm{Part}\, A$ the *set of all partitions* on the nonempty set $A$. Clearly, set inclusion makes $\mathrm{Equ}\, A$ an order. So for the partitions $\pi_1$ and $\pi_2$ of $A$, we can define

$$\pi_1 \leq \pi_2 \text{ if } \mathrm{equ}(\pi_1) \leq \mathrm{equ}(\pi_2) \text{ in } \mathrm{Equ}\, A,$$

that is, if $\mathrm{equ}(\pi_1) \subseteq \mathrm{equ}(\pi_2)$. For further observations and results on $\mathrm{Equ}\, A$ and $\mathrm{Part}\, A$, see the Exercises and all of Section V.4.

Sometimes, we start with a binary relation $\varrho$ on $A$ which is reflexive and transitive, but not necessarily antisymmetric; we call such relations *preorderings* (often called *quasiorderings*). If $\varrho$ is a preordering on $A$, then we can define an equivalence relation $\mathrm{sym}(\varrho)$ on $A$: the elements $a$ and $b$ are related under $\mathrm{sym}(\varrho)$ if $(a, b), (b, a) \in \varrho$. Define a relation $\hat{\varrho}$ on the set, $A/\mathrm{sym}(\varrho)$, of all blocks of $\mathrm{sym}(\varrho)$: $(X, Y) \in \hat{\varrho}$ if $(x, y) \in \varrho$ for any/all $x \in X$ and $y \in Y$. It is easy to see that $A/\mathrm{sym}(\varrho)$ under $\hat{\varrho}$ is an order. We call $\hat{\varrho}$ the *ordering associated with the preordering* $\varrho$.

Again, we may start with a reflexive binary relation $\varrho$ on $A$ satisfying the condition that $\varrho$ has no cycles; a *cycle* is a sequence of elements $x_0, x_1, \ldots, x_n$, for some $n \geq 1$, of elements of $A$ such that

$$(x_0, x_1), \ldots, (x_{n-1}, x_n), (x_n, x_0) \in \varrho.$$

Then we can define on $A$ an ordering, the *transitive closure* $\mathrm{tran}(\varrho)$ of $\varrho$:

$(a, b) \in \mathrm{tran}(\varrho)$ if there is a sequence $a = x_0, x_1, \ldots, x_n = b$, for $n \geq 1$, such that

$$(x_0, x_1), \ldots, (x_{n-1}, x_n) \in \varrho.$$

It is easy to see that $\mathrm{tran}(\varrho)$ is an ordering on $A$. Reflexivity was assumed; antisymmetry follows from the absence of loops, and transitivity is built in.

## 1.3  Basic order concepts

The orders $(A_0; \leq)$ and $(A_1; \leq)$ *are isomorphic*, in symbols,

$$(A_0; \leq) \cong (A_1; \leq),$$

and the map $\varphi \colon L_0 \to L_1$ is an *isomorphism* if $\varphi$ is a bijection and

$$a \leq b \text{ in } (L_0; \leq) \quad \text{iff} \quad \varphi(a) \leq \varphi(b) \text{ in } (L_1; \leq).$$

As a rule, we study orders up to isomorphism.

Let $(A; \leq)$ be an order. The elements $a, b \in A$ are *comparable* if $a \leq b$ or $b \leq a$. Otherwise, $a$ and $b$ are *incomparable*, in notation, $a \parallel b$. (A few authors, for instance, F. Maeda and S. Maeda [521], use $a \parallel b$ for the lattice theoretic generalizations of geometric parallelism.)

An order $(A; \leq)$ satisfying (Lin) is called a *chain* (also called a *totally ordered set* and a *linearly ordered set*). A chain is, therefore, an order in which there are no pairs of incomparable elements. On the other hand, we call $(A; \leq)$ an *unordered set* or an *antichain* if $a \parallel b$ for all $a \neq b$.

Let $\mathsf{C}_n$ denote the set $\{0, \dots, n-1\}$ ordered by $0 < 1 < 2 < \cdots < n-1$. Then $\mathsf{C}_n$ is an $n$-element chain. Clearly, any $n$-element chain is isomorphic to $\mathsf{C}_n$.

Let $(P; \leq)$ be an order and let $Q$ be a nonempty subset of $P$. Then there is a natural ordering $\leq_Q$ on $Q$ induced by the order $\leq$ on $P$: for the elements $a, b \in Q$, we define $a \leq_Q b$ if $a \leq b$ in $P$; we call $(Q; \leq_Q)$ (or simply, $(Q; \leq)$ or $Q$) a *suborder* of $(P; \leq)$.

A *chain $C$ in an order $P$* is a nonempty subset, which, as a suborder, is a chain; $C$ is a *maximal chain* in $P$ if $C$ is a chain in $P$ and whenever $C \subseteq D \subseteq P$ and $D$ is a chain in $P$, then $C = D$.

An *antichain $C$ in an order $P$* is a nonempty subset which is unordered.

The *length*, $\operatorname{len}(C)$, of a finite chain $C$ is $|C|-1$; the number of "jumps"—not the number of elements. ($|C|$ is the cardinality of $C$.) Note that $\operatorname{len}(\mathsf{C}_n) = n-1$. An order $P$ is said to be *of length $n$* (in symbols, $\operatorname{len}(P) = n$), where $n$ is a natural number, if there is a chain in $P$ of length $n$ and all chains in $P$ are of length $\leq n$. An *order $P$ is of finite length* if it is of length $n$ for some natural number $n$. The *height* of an element $a \in P$, denoted by $\operatorname{height}(a)$, is the length of the order $\{x \in P \mid x \leq a\}$.

Let $\operatorname{Pow} X$ (also $P(X)$, $\mathcal{P}(X)$, $2^X$, and $\operatorname{Exp} X$ in the literature) denote the set of all subsets of a set $X$, ordered by $\subseteq$. Observe that if $X$ has $n$ elements, then $\operatorname{len}(\operatorname{Pow} X) = n$.

Let $n$ be a natural number. The *width* of an order $P$ is $n$ if there is an antichain in $P$ of $n$ elements and all antichains in $P$ have $\leq n$ elements; in formula, $\operatorname{width}(P) = n$.

Next we shall define inf and sup in an arbitrary order $P$ (that is, $(P; \leq)$), as it is usually done for infinite sets of real numbers.

Let $H \subseteq P$ and $a \in P$. Then $a$ is an *upper bound* of $H$ if $a$ majorizes all $h \in H$. An upper bound $a$ of $H$ is the *least upper bound* of $H$ or *supremum* of $H$ if $a$ is majorized by all upper bounds of $H$. We shall write $a = \sup H$ or $a = \bigvee H$. (The notation $a = \mathrm{l.\,u.\,b.}\,H$ and $a = \sum H$ is also common in the literature.)

To show the uniqueness of the supremum when it exists, let $a_0$ and $a_1$ be both suprema of $H$; then $a_0 \leq a_1$, since $a_1$ is an upper bound and $a_0$ is a supremum. Similarly, $a_1 \leq a_0$; thus $a_0 = a_1$ by (ASym).

Let $\varnothing$ be the empty set. Let us assume that $a = \sup \varnothing$ exists. Every $b \in P$ is an upper bound of $\varnothing$ since $b$ majorizes all $h \in \varnothing$ (there is no such $h$). Thus $a$ is majorized by all $b \in P$. We conclude that $\sup \varnothing$ exists iff $P$ has a smallest element. We call $\sup \varnothing$ the *zero* of $P$ and denote it by $0_P$, or $0$ if $P$ is understood. In some papers, the notation $\perp$ is used for $0$.

The concepts of *lower bound* and *greatest lower bound* or *infimum* are similarly defined; the latter is denoted by $\inf H$ or $\bigwedge H$. (The notation $\mathrm{g.\,l.\,b.}\,H$ and $\prod H$ is also used in the literature.) The uniqueness is proved as in the last but one paragraph. Observe that $\inf \varnothing$ exists iff $P$ has a largest element. We call $\inf \varnothing$ the *unit* (or *identity*) and denote it by $1_P$, or $1$ if $P$ is understood. In some papers, the notation $\top$ is used for $1$.

A *bounded* order is one that has both a zero and a unit.

The adverb "similarly" in the paragraph introducing infima can be given a very concrete meaning. Let $(P; \leq)$ be an order. The relation $a \geq b$ can also be regarded as a definition of a binary relation on $P$. This binary relation $\geq$ satisfies (Refl), (ASym), and (Trans); as an example, let us check (ASym). If $a \geq b$ and $b \geq a$, then $b \leq a$ and $a \leq b$ hold, by the definition of $\geq$; using (ASym) for $\leq$, we conclude that $a = b$. (Refl) and (Trans) are equally trivial. Thus $(P; \geq)$ is also an order, called the *dual* of $(P; \leq)$; we shall denote it by $P^\delta$ (in the literature, $\tilde{P}, \overline{P}, P^d$, and so on). Now if $\Phi$ is a "statement" about orders, and if in $\Phi$ we replace all occurrences of $\leq$ by $\geq$, we get the *dual* of $\Phi$, in notation, $\Phi^\delta$.

**Duality Principle.** *If a statement $\Phi$ is true in all orders, then its dual, $\Phi^\delta$, is also true in all orders.*

This is true simply because $\Phi$ holds for $(P; \leq)$ iff $\Phi^\delta$ holds for $(P; \geq)$, which also ranges over all orders.

As an example, take for $\Phi$ the statement: "If $\sup H$ exists, then it is unique." We get as its dual: "If $\inf H$ exists, then it is unique." The dual of "$(P; \leq)$ has a zero" is "$(P; \geq)$ has a unit".

It is hard to imagine that anything as trivial as the Duality Principle could yield anything profound—it does not. But it can save a lot of work.

## 1.4   Ordering and covers

Consider the order $\mathrm{Pow}\,X$, where $X = \{u, v\}$; we use the notation $0 = \varnothing$, $a = \{u\}, b = \{v\}, 1 = \{u, v\}$. We can describe the ordering on $\mathrm{Pow}\,X$ by listing all pairs $(x, y)$ with $x \leq y$:

$$\{(0,0), (0,a), (0,b), (0,1), (a,a), (a,1), (b,b), (b,1), (1,1)\}.$$

All pairs of the form $(x, x)$ can be omitted from the list, since we know that $x \leq x$. Also, if $x \leq y$ and $y \leq z$, then $x \leq z$ by transitivity. For instance, if we know that $0 \leq a$ and $a \leq 1$, we do not have to be told that $0 \leq 1$.

Let us make this idea more precise. In the order $(P; \leq)$, *the element $x$ is covered by the element $y$* if $x < y$ but $x < z < y$ for no $z \in P$; we use the notation $x \prec y$. We also say that $y$ *covers* $x$, in notation, $y \succ x$.

Let us write $a \preceq b$ for $a \prec b$ or $a = b$ and $a \succeq b$ for $a \succ b$ or $a = b$.

The covering relation $\prec$ of the preceding example is the much smaller set

$$\{(0,a), (0,b), (a,1), (b,1)\}.$$

Does the covering relation determine the ordering? The following lemma shows that in the finite case it does.

**Lemma 1.** *Let $(P; \leq)$ be a finite order. Then $a \leq b$ if $a = b$ or if there exists a finite sequence of elements $x_0, \ldots, x_{n-1}$ such that $x_0 = a$, $x_{n-1} = b$, and $x_i \prec x_{i+1}$ for all $0 \leq i < n - 1$.*

*Proof.* If there is such a finite sequence, then $a = x_0 \leq x_1 \leq \cdots \leq x_{n-1} = b$, and a trivial induction on $n$ yields that $a \leq b$. Thus it suffices to prove that if $a < b$, then there is such a sequence. Fix $a, b \in P$ with $a < b$, and take all subsets $H$ of $P$ such that $H$ is a chain in $P$, the element $a$ is the smallest in $H$, and the element $b$ is the largest in $H$. There are such subsets, $\{a, b\}$, for example. Choose such an $H$ with the largest possible number of elements, say with $m$ elements. (Recall that $P$ is finite.) Then $H = \{x_0, \ldots, x_{m-1}\}$, and we can assume that $x_0 < x_1 < \cdots < x_{m-1}$. We claim that

$$a = x_0 \prec x_1 \prec \cdots \prec x_{m-1} = b$$

in $(P; \leq)$. Indeed, $x_i < x_{i+1}$ by assumption. Thus if $x_i \prec x_{i+1}$ does not hold, then $x_i < x < x_{i+1}$ for some $x \in P$, and $H \cup \{x\}$ will be a chain of $m + 1$ elements between $a$ and $b$, contrary to the maximality of the number of elements of $H$. $\qquad\square$

## 1.5  Order diagrams

The *diagram* of a finite order $(P; \leq)$ represents the elements by dots (in the figures of this book, small circles, $\circ$). The circles representing the elements $x$ and $y$ are connected by a straight line if one covers the other; if $x$ covers $y$, then the circle representing $x$ is higher than the circle representing $y$. Figure 1 shows three diagrams of the same order $P$.

By Lemma 1, the diagram of a finite order determines the order up to isomorphism.

In a diagram the intersection of two line segments does not necessarily mark an element. A diagram is *planar* if no two line segments intersect. An order $P$ is *planar* if it has a diagram that is planar. Since the third diagram in Figure 1 is planar, the order $P$ with diagrams in Figure 1 is planar.

For aesthetic reasons, intersecting line segments are sometimes drawn as though one passes in front of another, as in Figure 7; this has no impact on the order structure.

Sometimes the "generalized diagram" of an infinite order is drawn; see, for instance, Figure 121 on page 490. Such diagrams are always accompanied by explanations in the text.

The diagram of an order can be viewed as a *graph* $(G; E)$: a set $G$ with a fixed set $E$ (the *edges*) of two-element subsets of $G$, namely, the sets $\{a, b\}$ satisfying $a \prec b$. This graph is called the *covering graph*, Cov $P$, of the order $P$. Of course, the diagram of $P$ contains much more information than the covering graph.

## 1.6   Order constructions

For an order $P$, a subset $A \subseteq P$ is called *down-set* if $x \in A$ and $y \leq x$ imply that $y \in A$. Note that, by definition, the empty set is a down-set. For $H \subseteq P$, let $\downarrow_P H$ be the down-set generated by $H$ in $P$, that is,

$$\downarrow_P H = \{\, x \in P \mid x \leq y \text{ for some } y \in H \,\},$$

and we write $\downarrow H$ if the order $P$ is understood. For $H = \{h\}$, let $\downarrow h = \downarrow \{h\}$.

The order Down $P$ is the set of all down-sets of $P$ ordered under set inclusion. Again, note that $\varnothing \in$ Down $P$.

The dual of a down-set is an *up-set*: a subset $A \subseteq P$ such that if $x \in A$ and $y \geq x$, then $y \in A$. We use the notation $\uparrow_P H$, $\uparrow H$, $\uparrow h$.

A nonempty down-set (dually, an up-set) of an order is an order.

Given the orders $P$ and $Q$, we can form the *direct product* $P \times Q$, consisting of all ordered pairs $(x_1, x_2)$ with $x_1 \in P$ and $x_2 \in Q$, ordered componentwise, that is, $(x_1, x_2) \leq (y_1, y_2)$ if $x_1 \leq y_1$ in $P$ and $x_2 \leq y_2$ in $Q$. If $P = Q$, then we write $P^2$ for $P \times Q$. Similarly, we use the notation $P^n$ for $P^{n-1} \times P$ for every $n > 2$. Figure 2 shows a diagram of $C_2 \times P$, where $P$ is the order with diagrams in Figure 1. (Recall that $C_2$ is the two-element chain, see Section 1.3.)

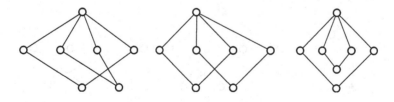

Figure 1. Three diagrams of an order $P$

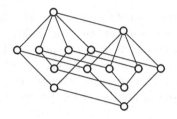

Figure 2. A diagram of $C_2 \times P$

Another often used construction is the (ordinal) *sum* $P + Q$ of $P$ and $Q$, defined on the (disjoint) union $P \cup Q$ ordered as follows: for the elements $x, y \in P \cup Q$, define $x \leq y$ if one of the following conditions holds:

(i) $x, y \in P$ and $x \leq_P y$;
(ii) $x, y \in Q$ and $x \leq_Q y$;
(iii) $x \in P$ and $y \in Q$.

Figure 3 shows diagrams of $C_2 + P$ and $P + C_2$, where $P$ is the order with diagrams in Figure 1. In both diagrams, the elements of $C_2$ are black-filled.

A variant of this is the *glued sum*, $P \mathbin{\dot{+}} Q$, applied to an order $P$ with unit, $1_P$, and an order $Q$ with zero, $0_Q$; then $P \mathbin{\dot{+}} Q$ is obtained from $P + Q$ by identifying $1_P$ and $0_Q$, that is, $1_P = 0_Q$ in $P \mathbin{\dot{+}} Q$. The third diagram in Figure 3 is a diagram of $P \mathbin{\dot{+}} C_2$.

### 1.7   Two more numeric invariants

We have already discussed a number of numeric invariants of a (finite) order such as length and width. Two more numeric invariants are in common use.

Let $n$ be a positive integer. We say that an order $P$ has *breadth at most $n$* if for all elements $x_0, x_1, \ldots, x_n, y_0, y_1, \ldots, y_n$ in $P$, if $x_i \leq y_j$ for all $i \neq j$ in

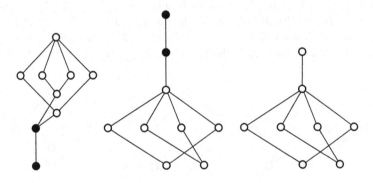

Figure 3. Diagrams of two sums and a glued sum

$\{0, 1, \ldots, n\}$, then there exists $i \in \{0, 1, \ldots, n\}$ such that $x_i \leq y_i$. The *breadth* of $P$, in notation, breadth$(P)$, is the least positive integer $n$ such that $P$ has breadth at most $n$ if such an $n$ exists.

Observe that this definition of breadth is selfdual. For an equivalent definition for a join-semilattice (or for a lattice), see Exercise 1.19.

Define the *order-dimension*, $\dim(P)$, of an order $P$ as the smallest cardinal $\mathfrak{m}$ such that $P$ is a suborder of a product of $\mathfrak{m}$ chains. For a finite lattice $L$, planarity is equivalent to $\dim(L) \leq 2$. For infinite lattices, "order-dimension at most 2" substitutes for planarity, see, for instance, G. Grätzer and R. W. Quackenbush [327]. See also Exercises 1.20–1.23 and 6.23.

## 1.8  Lattices as orders

An order $(L; \leq)$ is a *lattice* if $\sup\{a, b\}$ and $\inf\{a, b\}$ exist for all $a, b \in L$.

In other words, lattice theory singles out a special type of order for detailed investigation. To make such a definition worthwhile, it must be shown that this class of orders is a very useful class, that there are many such orders in various branches of mathematics (analysis, topology, logic, algebra, geometry, and so on), and that a general study of these orders will lead to a better understanding of the behavior of the examples. This was done around 1934–1940 and published in the first edition of G. Birkhoff's *Lattice Theory* [65].

As we go along, we shall see many examples of lattices, most of them in the Exercises and, of course, in the companion volume of this book. Many of the orders discussed in this section are lattices, chains, for example. The order Equ $A$ is a lattice; the inf is set intersection but the sup requires some work, see Exercise 1.16; this computation is similar to that in Theorem 12 and is discussed in more detail (in terms of partitions) in Section V.4.1.

For a general survey of lattices in mathematics, see M. K. Bennett [54], G. Birkhoff [71], [72], [73], H. H. Crapo and G.-C. Rota [102], and T. S. Fofanova [178], [179].

Our definition of lattices in terms of suprema and infima of two-element sets may seem arbitrary; but in fact, it is equivalent to a very natural condition:

**Lemma 2.** *An order $(L; \leq)$ is a lattice if $\sup H$ and $\inf H$ exist for every finite nonempty subset $H$ of $L$.*

*Proof.* It is enough to prove that the original definition implies the new one. So let $(L; \leq)$ satisfy the original definition and let $H \subseteq L$ be nonempty and finite. If $H = \{a\}$, then $\sup H = \inf H = a$. If $H = \{a_0, \ldots, a_{n-1}\}$ for some $n \geq 1$, then

$$\sup\{\ldots \sup\{\sup\{a_0, a_1\}, a_2\}, \ldots, a_{n-1}\} = \sup H,$$

by an easy inductive argument. For example, if $H = \{a, b, c\}$, then set $d = \sup\{a, b\}$ and $e = \sup\{c, d\}$; we claim that $e = \sup H$. First, $a, b \leq d$ and

$c, d \leq e$; therefore (by transitivity), $x \leq e$ for all $x \in H$. Second, if $f$ is an upper bound of $H$, then $a, b \leq f$ and thus $d \leq f$; also $c \leq f$, so that $c, d \leq f$, therefore, $e \leq f$, since $e = \sup\{c, d\}$. Thus $e$ is the sup of $H$.

By duality (in other words, by applying the Duality Principle), we conclude that $\inf H$ exists. $\qquad\qquad\qquad\qquad\qquad\qquad\qquad\qquad\qquad\qquad\qquad\qquad\square$

A lattice need not have a zero or a unit, so $\sup \varnothing$ and $\inf \varnothing$ may not exist. A *bounded* lattice has both a zero and a unit. Every finite lattice is bounded.

The simple proof of Lemma 2 can be varied to yield a large number of equally trivial statements about lattices and orders. Some of these will be stated as exercises and used later. To make use of the Duality Principle legitimate for lattices, note:

If $(P; \leq)$ is a lattice, so is its dual $P^\delta = (P; \geq)$.

Thus the Duality Principle applies to lattices.

We shall use the notation

$$a \vee b = \sup\{a, b\},$$
$$a \wedge b = \inf\{a, b\},$$

and call $\vee$ the *join*, and $\wedge$ the *meet*. In lattices, they are both *binary operations*, which means that they can be applied to a pair of elements $a, b$ of $L$ to yield again an element of $L$. Thus $\vee$ is a map of $L^2$ into $L$ and so is $\wedge$, a remark that might fail to be very illuminating at this point.

The previous proof yields that

$$(\ldots((a_1 \vee a_2) \vee a_3)\ldots) \vee a_n = \sup\{a_1, \ldots, a_n\},$$

and there is a similar formula for inf. Now observe that the right-hand side does not depend on the way the elements $a_i$ are listed. Thus $\vee$ and $\wedge$ are idempotent, commutative, and associative—that is, they satisfy the following conditions:

(Idem) Idempotency: $\qquad\qquad\qquad\quad a \vee a = a,$

$\qquad\qquad\qquad\qquad\qquad\qquad\qquad\quad a \wedge a = a.$

(Comm) Commutativity: $\qquad\qquad\qquad a \vee b = b \vee a,$

$\qquad\qquad\qquad\qquad\qquad\qquad\qquad\quad a \wedge b = b \wedge a.$

(Assoc) Associativity: $\qquad\qquad (a \vee b) \vee c = a \vee (b \vee c),$

$\qquad\qquad\qquad\qquad\qquad\qquad (a \wedge b) \wedge c = a \wedge (b \wedge c).$

These properties of the operations are also called the *idempotent identities, commutative identities*, and *associative identities*, respectively. (We introduce identities, in general, in Section 4.2.)

As is always the case when we have an associative operation, we may write iterated operations without parentheses, for instance,

$$a_1 \vee a_2 \vee \cdots \vee a_n$$

(and the same for $\wedge$).

There is another pair of rules that connect $\vee$ and $\wedge$. To derive them, note that if $a \leq b$, then $\sup\{a, b\} = b$; that is, $a \vee b = b$, and conversely. Thus

$$a \leq b \quad \text{iff} \quad a \vee b = b.$$

By duality (and by interchanging $a$ and $b$),

$$a \leq b \quad \text{iff} \quad a \wedge b = a$$

holds. Applying the "only if" part of the first rule to $a$ and $a \wedge b$, and that of the second rule to $a \vee b$ and $a$, we get the *Absorption identities*:

(Absorp) Absorption:    $a \vee (a \wedge b) = a,$

$a \wedge (a \vee b) = a.$

Now we are faced with the crucial question: Do we know enough about $\vee$ and $\wedge$ so that lattices can be characterized purely in terms of the properties of these two operations?

Two comments are in order. It is obvious that $\leq$ can be characterized by $\vee$ and $\wedge$ (in fact, by either of them); therefore, obtaining such a characterization is only a matter of persistence. More importantly, why should we try to get such a characterization? To rephrase the question, why should we want to characterize $(L; \leq)$ as $(L; \vee, \wedge)$, which is an algebra—that is, a set equipped with operations (in this case, two binary operations)? Note that $\leq$ is a subset of $L^2$, whereas $\vee$ and $\wedge$ are maps from $L^2$ into $L$.

The answer is simple. We want such a characterization because if we can treat lattices as algebras, then all the concepts and methods of universal algebra will become applicable. The usefulness of treating lattices as algebras will soon become clear.

## 1.9  Algebras

In this section, we started out with orders and then introduced lattices as orders. Now we introduce (universal) algebras, and then introduce lattices as algebras.

Let $A$ be a nonempty set. An *n-ary operation* $f$ on a set $A$ is a map from $A^n$ into $A$; in other words, if $a_1, \ldots, a_n \in A$, then $f(a_1, \ldots, a_n) \in A$. If $n = 1$, the operation $f$ is called *unary*; if $n = 2$, the operation $f$ is called *binary*. Since $A^0 = \{\varnothing\}$, a *nullary operation* $(n = 0)$ is determined by $f(\varnothing) \in A$, and $f$ is usually identified with $f(\varnothing)$. A *universal algebra*, or simply *algebra*, consists of a nonempty set $A$ and a set $F$ of operations; each $f \in F$ is an $n$-ary operation for some $n$ (depending on $f$). We denote this algebra by $\mathfrak{A}$ or $(A; F)$. (In universal algebra, German uppercase characters such as $\mathfrak{A}$ are often used to denote algebras; we use this convention for lattices and orders regularly for the next couple of pages—and occasionally after that—to avoid ambiguity.)

A *type* $\tau$ of algebras is a sequence $(n_0, n_1, \ldots, n_\gamma, \ldots)$ of nonnegative integers, $\gamma < o(\tau)$, where $o(\tau)$ is an ordinal called the *order* of $\tau$. An algebra $\mathfrak{A}$ of *type* $\tau$ is an ordered pair $(A; F)$, where $A$ is a nonempty set and $F$ is a sequence $(f_0, \ldots, f_\gamma, \ldots)$, where $f_\gamma$ is an $n_\gamma$-ary operation on $A$ for $\gamma < o(\tau)$. If $o(\tau)$ is finite, $o(\tau) = n$, then we may write $(A; f_0, \ldots, f_{n-1})$ for $(A; F)$.

Let $(A; F)$ and $(B; F)$ be algebras of the same type $\tau$; we denote by $f_\gamma$ the $\gamma$-th operation in both algebras. We say that $(A; F)$ and $(B; F)$ are *isomorphic*, and $\varphi$ is an *isomorphism*, if $\varphi$ is a one-to-one and onto map from $A$ to $B$ and

$$\varphi(f_\gamma(a_1, \ldots, a_{n_\gamma})) = f_\gamma(\varphi(a_1), \ldots, \varphi(a_{n_\gamma}))$$

for all $a_1, \ldots, a_{n_\gamma} \in A$.

All isomorphism concepts involve a one-to-one and onto map; so we introduce a special name for them: *bijections*.

Many of the concepts and results discussed in this chapter are special cases of universal algebraic concepts and results. Some of the results of Sections 3–5 (some of them easy, some nontrivial) can be formulated and proved for arbitrary universal algebras. See Exercises 3.78–3.85, 4.25–4.27, and the discussion of the Second Isomorphism Theorem and Birkhoff's Subdirect Representation Theorem in Section II.6.5.

Algebras naturally generalize to *infinitary universal algebras*. In the definition of an $n$-ary operation, replace $n$ by an ordinal $\delta$, and in the type, replace $n_\gamma$ by the ordinal $\delta_\gamma$. Infinitary universal algebras behave very differently from their finitary counterparts.

For more details, see Chapters 1–4 of the author's book [263].

### 1.10   Lattices as algebras

Let $(A; \circ)$ be an algebra with one binary operation $\circ$. The algebra $(A; \circ)$ is a *semilattice* if $\circ$ is idempotent, commutative, and associative.

An algebra $(L; \vee, \wedge)$ (of type $(2, 2)$) is called a *lattice* if $L$ is a nonempty set, $(L; \vee)$ and $(L; \wedge)$ are semilattices, and the two absorption identities are satisfied. The following theorem states that a lattice as an algebra and a lattice as an order are "equivalent" concepts. (The word "equivalent" will not be defined.)

**Theorem 3.**

(i) *Let the order* $\mathfrak{L} = (L; \leq)$ *be a lattice. Set*

$$a \vee b = \sup\{a, b\},$$
$$a \wedge b = \inf\{a, b\}.$$

*Then the algebra $\mathfrak{L}^{\mathrm{alg}} = (L; \vee, \wedge)$ is a lattice.*

(ii) *Let the algebra $\mathfrak{L} = (L; \vee, \wedge)$ be a lattice. Set*

$$a \leq b \quad \text{if} \quad a \vee b = b.$$

*Then $\mathfrak{L}^{\mathrm{ord}} = (L; \leq)$ is an order, and the order $\mathfrak{L}^{\mathrm{ord}}$ is a lattice.*

(iii) *Let the order $\mathfrak{L} = (L; \leq)$ be a lattice. Then*

$$(\mathfrak{L}^{\mathrm{alg}})^{\mathrm{ord}} = \mathfrak{L}.$$

(iv) *Let the algebra $\mathfrak{L} = (L; \vee, \wedge)$ be a lattice. Then*

$$(\mathfrak{L}^{\mathrm{ord}})^{\mathrm{alg}} = \mathfrak{L}.$$

*Remark.* (i) and (ii) describe the way in which we pass from an order to an algebra and back, whereas (iii) and (iv) state that going there and back returns us to where we started. See also Lemma 4.

*Proof.*

(i) This has already been proved.

(ii) First, we set $a \leq b$ to mean that $a \vee b = b$. Now the relation $\leq$ is reflexive since $\vee$ is idempotent; the relation $\leq$ is antisymmetric since $a \leq b$ and $b \leq a$ mean that $a \vee b = b$ and $b \vee a = a$, which, by the commutativity of $\vee$, imply that $a = a \vee b = b \vee a = b$; the relation $\leq$ is transitive, since if $a \leq b$ and $b \leq c$, then $b = a \vee b$ and $c = b \vee c$, and so

$$\begin{aligned} c = b \vee c &= (a \vee b) \vee c \qquad &&(\vee \text{ is associative}) \\ &= a \vee (b \vee c) = a \vee c, \end{aligned}$$

that is, $a \leq c$. Thus $(L; \leq)$ is an order. To prove that $(L; \leq)$ is a lattice, we shall verify that $a \vee b = \sup\{a, b\}$ and $a \wedge b = \inf\{a, b\}$ (these are not definitions here). Indeed, $a \leq a \vee b$, since

$$a \vee (a \vee b) = (a \vee a) \vee b = a \vee b,$$

using the associativity and idempotency of $\vee$; similarly, $b \leq a \vee b$. So $a \vee b$ is an upper bound of $a$ and $b$. Now if $c$ is any upper bound of $a$ and $b$, then $a \leq c$ and $b \leq c$, that is, $a \vee c = c$ and $b \vee c = c$, then

$$(a \vee b) \vee c = a \vee (b \vee c) = a \vee c = c;$$

thus $a \vee b \leq c$, proving that $a \vee b = \sup\{a, b\}$.

Second, $a \wedge b \leq a$ and $a \wedge b \leq b$, because $(a \wedge b) \vee a = a$ and $(a \wedge b) \vee b = a$ by the first absorption identity. If $c \leq a$ and $c \leq b$, that is, if $a = a \vee c$ and

$b = b \vee c$, then $a \wedge c = (a \vee c) \wedge c = c$ and $b \wedge c = c$ (by the second absorption identity). Thus

$$
\begin{aligned}
(a \wedge b) \vee c &= (a \wedge b) \vee (a \wedge c) && \text{(because } c = a \wedge c) \\
&= (a \wedge b) \vee (a \wedge (b \wedge c)) && \text{(because } c = b \wedge c) \\
&= (a \wedge b) \vee ((a \wedge b) \wedge c) && \text{(by associativity)} \\
&= a \wedge b && \text{(by absorption)},
\end{aligned}
$$

that is, $c \leq a \wedge b$, completing the proof of $a \wedge b = \inf\{a, b\}$.

(iii) It is enough to observe that the orderings of $\mathfrak{L}$ and $(\mathfrak{L}^{\mathrm{alg}})^{\mathrm{ord}}$ are identical to get (iii).

(iv) The proof of (iv) is similar to the proof of (iii).    □

The proof of Theorem 3, and even the statement of Theorem 3, are subject to criticisms:

- In the definition of a lattice as an algebra, we require eight identities; idempotency is redundant.

- The last step of the proof of (ii) can be made neater by first proving that

$$
b = a \vee b \quad \text{iff} \quad a = a \wedge b.
$$

- Theorem 3 should be preceded by a similar theorem for "semilattices" (see Exercise 1.41).

All these objections will be dealt with in the Exercises that follow this section.

Finally, note that for lattices as algebras, the Duality Principle takes on the following very simple form.

**Duality Principle for Lattices.** *Let $\Phi$ be a statement about lattices expressed in terms of $\vee$ and $\wedge$. The* dual *of $\Phi$ is the statement $\Phi^{\delta}$ we get from $\Phi$ by interchanging $\vee$ and $\wedge$. If $\Phi$ is true for all lattices, then $\Phi^{\delta}$ is also true for all lattices.*

To prove this we only have to observe that if $\mathfrak{L} = (L; \vee, \wedge)$, then the dual of $\mathfrak{L}^{\mathrm{ord}}$ is $(L; \wedge, \vee)^{\mathrm{ord}}$.

Most of the time, $\Phi$ involves $\leq$, and maybe 0 and 1, in addition to $\vee$ and $\wedge$. When dualizing such a $\Phi$, we interchange $\vee$ and $\wedge$, replace $\leq$ by $\geq$, and interchange 0 and 1.

For a lattice $L$, we denote its dual by $L^{\delta}$.

Treating lattices as algebras makes it natural to define conditions on lattices that are like the eight properties named above, holding in some lattices but not in others, for instance,

$$
\begin{aligned}
x \wedge (y \vee z) &= (x \wedge y) \vee (x \wedge z), \\
x \vee (y \wedge z) &= (x \vee y) \wedge (x \vee z).
\end{aligned}
$$

Lattices satisfying these are called *distributive* and the class of distributive lattices is denoted by **D**. As another example take

$$(x \wedge y) \vee (x \wedge z) = x \wedge (y \vee (x \wedge z)).$$

Lattices satisfying this are called *modular* and the class of modular lattices is denoted by **M**. These are discussed in more detail in Section 4.3, and in Chapters II and V.

We can also take classes of algebras that are lattices with additional operations, such as the class of *boolean algebras*, $(B; \vee, \wedge,', 0, 1)$, where $(B; \vee, \wedge)$ is a distributive lattice with zero, 0, and unit, 1, in which

$$a \vee a' = 1,$$
$$a \wedge a' = 0.$$

The corresponding lattice, $(B; \vee, \wedge)$, is called a *boolean lattice*. This class will be discussed in detail in Section 6.1 and in Chapter II.

**Exercises**

### Orders

1.1. Define $x < y$ to mean $x \leq y$ and $x \neq y$. Prove that, in an order, $x < x$ for no $x$, and that $x < y$ and $y < z$ imply that $x < z$.

1.2. Let the binary relation $<$ satisfy the conditions of Exercise 1.1. Define $x \leq y$ to mean $x < y$ or $x = y$. Show that $\leq$ is an ordering.

1.3. Prove the following extension of antisymmetry:
If $x_0 \leq x_1 \leq \cdots \leq x_{n-1} \leq x_0$, then $x_0 = x_1 = \cdots = x_{n-1}$.

1.4. Define the binary relation $\Delta = \{ (x, x) \mid x \in A \}$. For a binary relation $\varrho$ show that (Refl) means that $\Delta \subseteq \varrho$.

1.5. For a binary relation $\varrho$, define the binary relation $\varrho^{-1} = \{ (x, y) \mid (y, x) \in \varrho \}$. Show that (ASym) holds for $\varrho$ iff $\varrho \cap \varrho^{-1} \subseteq \Delta$.

1.6. Show that (Lin) holds for $\varrho$ iff $\varrho \cup \varrho^{-1} = A^2$.

1.7. Enumerate all orderings on a five-element set.

1.8. Let $\leq$ be an ordering on $A$ and let $B$ be a subset of $A$. For $a, b \in B$, set $a \leq_B b$ if $a \leq b$. Prove that $\leq_B$ is an ordering on $B$.

1.9. Let $A$ be a set and let $P$ be the set of all orderings on $A$. For $\varrho, \sigma \in P$, set $\varrho \leq \sigma$ if $a \varrho b$ implies that $a \sigma b$ (for all $a, b \in A$). Prove that $(P; \leq)$ is an order.

1.10. Find an example of an order in which $\inf \varnothing$ does not exist.

1.11. Let $P$ be an order and let us assume that $\inf H$ exists for all nonempty subsets $H$. Prove that $\sup \varnothing$ also exists in $P$.

1.12. Prove that the following are examples of orders:

(a) Let $A = \operatorname{Pow} X$, the set of all subsets of a set $X$; let $X_0 \leq X_1$ mean that $X_0 \subseteq X_1$ (set inclusion).

(b) Let $A$ be the set of all real-valued functions defined on a set $X$; for $f, g \in A$, set $f \leq g$ if $f(x) \leq g(x)$ for all $x \in X$.

(c) Let $A$ be the set of all continuous, concave downward, real-valued functions defined on the real interval $[-\infty, +\infty]$; define $f \leq g$ as in example (b).

(d) Let $A$ be the set of all open sets of a topological space; define $\leq$ as in example (a).

(e) Let $A$ be the set of all human beings; let $a < b$ mean that $a$ is a descendant of $b$.

(f) Let $N$ be the set of all natural numbers; let $a < b$ mean that $a$ divides $b$ and $a \neq b$.

1.13. Prove that every order of finite length $n$ is the union of $n$ antichains.

### Basic concepts

1.14. For a nonempty set $A$, show that the map $\varepsilon \mapsto A/\varepsilon$ is a *bijection* between Equ $A$ and Part $A$.

1.15. What is the inverse of the map in Exercise 1.14?

1.16. Let $\varepsilon_1$ and $\varepsilon_2$ be binary relations on the nonempty set $A$. Define a binary relation $\varepsilon$ on $A$ as follows:

$(x, y) \in \varepsilon$ if there is a sequence $x = z_0, z_1, \ldots, z_{n-1} = y$ of elements of $A$ such that for each $i$ with $0 \leq i < n - 1$, either $(z_i, z_{i+1}) \in \varepsilon_1$ or $(z_i, z_{i+1}) \in \varepsilon_2$.

Prove that if $\varepsilon_1$ and $\varepsilon_2$ are equivalence relations, then so is $\varepsilon$ and $\varepsilon = \sup(\{\varepsilon_1, \varepsilon_2\})$ in Equ $A$.

1.17. For a binary relation $\varrho$ on $A$ satisfying only (Refl) and the condition that $\varrho$ has no cycles, we defined on $A$ an ordering, the *transitive closure* tran($\varrho$) of $\varrho$. Extend the definition of tran($\varrho$) to binary relations failing (Refl).

1.18. What is the appropriate definition of tran($\varrho$) if $\varrho$ has cycles?

### Numeric invariants

1.19. Show that a join-semilattice $(L; \vee)$ has breadth at most $n$, for a positive integer $n$, iff for every nonempty finite subset $X$ of $L$, there exists a nonempty $Y \subseteq X$ with at most $n$ elements such that

$$\bigvee X = \bigvee Y;$$

see Section 1.7.

1.20. Using the statement of Exercise 1.19, we can define two concepts of breadth for a lattice $(L; \vee, \wedge)$: the breadth of $(L; \vee)$, $\operatorname{breadth}_\vee(L)$, and the breadth of $(L; \wedge)$, breadth$_\wedge(L)$. Show that

$$\text{breadth}_\vee(L) = \text{breadth}_\wedge(L)$$

is the breadth of $L$.

1.21. Can you find a finite planar order of order-dimension 3? (Hint: use four elements.)

1.22. Prove the inequality breadth$(P) \leq \dim(P)$.

1.23. Find finite lattices of breadth 2 and order-dimension 3.

### Lattices

1.24. Which of the examples in Exercise 1.12 are lattices? For those that are lattices, compute $a \vee b$ and $a \wedge b$.

1.25. Show that every chain is a lattice.

1.26. Let $A$ be the set of all subgroups of a group $G$; for $X, Y \in A$, set $X \leq Y$ to mean $X \subseteq Y$. Prove that $(A; \leq)$ is a lattice; compute $X \vee Y$ and $X \wedge Y$. What about normal subgroups?

1.27. Let $(P; \leq)$ be an order in which inf $H$ exists *for all* $H \subseteq P$. Show that $(P; \leq)$ is a lattice. (Hint: For $a, b \in P$, let $H$ be the set of all upper bounds of $\{a, b\}$. Prove that $\sup\{a, b\} = \inf H$.) Relate this to Exercise 1.26.

### Lattices as algebras

1.28. Prove that the absorption identities imply the idempotency of $\vee$ and $\wedge$. (Hint: simplify $a \vee (a \wedge (a \vee a))$ in two ways to yield $a = a \vee a$.)

1.29. Show that by invoking the Duality Principle, we can eliminate the second part of the proof of Theorem 3(ii).

1.30. Let the algebra $(A; \vee, \wedge)$ be a lattice. Define $a \leq_\vee b$ if $a \vee b = b$ and define $a \leq_\wedge b$ if $a \wedge b = a$. Prove that $a \leq_\vee b$ iff $a \leq_\wedge b$.

1.31. Prove that the algebra $(A; \vee, \wedge)$ is a lattice iff $(A; \vee)$ and $(A; \wedge)$ are semilattices and $b = a \vee b$ is equivalent to $a = a \wedge b$. Verify that if $(A; \vee, \wedge_1)$ and $(A; \vee, \wedge_2)$ are both lattices, then the operations $\wedge_1$ and $\wedge_2$ are the same.

1.32. Are the three identities defining semilattices independent (meaning that none follows from the others)?

1.33. Prove that an algebra $(A; \vee, \wedge)$ is a lattice iff it satisfies the two identities

$$(w \wedge x) \vee x = x,$$
$$(((x \wedge w) \wedge z) \vee u) \vee v = (((w \wedge z) \wedge x) \vee v) \vee ((t \vee u) \wedge u).$$

(See J. A. Kalman [457]. The first definition of lattices by two identities was found by R. Padmanabhan, see the abstract Notices Amer. Math.

Soc. **14** (1967), No. 67T-468, and published in full in R. Padman-abhan [567]. Kalman's two identities are slightly improved versions of those of Padmanabhan. See Section 6.4 for more results and references.)

1.34. Is there a natural class of lattices that could be viewed as an algebra of type $(2, 2, 0, 0)$?

## Semilattices

1.35. An order is a *join-semilattice* (dually, *meet-semilattice*) if sup$\{a, b\}$ (dually, inf$\{a, b\}$) exists for all pairs of elements $a, b$. Prove that the dual of a join-semilattice is a meet-semilattice, and conversely.

1.36. Let $A$ be the set of finitely generated subgroups of a group $G$, ordered by set inclusion (as in Exercise 1.26). Prove that $(A; \subseteq)$ is a join-semilattice, but not necessarily a lattice.

1.37. Let $C$ be the set of all continuous, strictly convex (strictly concave downward), real-valued functions defined on the real interval $[0, 1]$. For $f, g \in C$, set $f \leq g$ if $f(x) \leq g(x)$ for all $x \in [0, 1]$. Prove that $(C; \leq)$ is a meet-semilattice, but not a join-semilattice. Is this true if "strictly" is omitted?

1.38. Show that an order $(P; \leq)$ is a lattice iff it is both a join- and meet-semilattice.

1.39. Let the order $(P; \leq)$ be a join-semilattice. Show that the algebra $(P; \circ)$ is a semilattice, where $a \circ b = a \vee b = \sup\{a, b\}$. State the analogous result for meet-semilattices.

1.40. Let the algebra $(A; \circ)$ be a semilattice. Define the binary relations $\leq_\vee$ and $\leq_\wedge$ on $A$ as follows: $a \leq_\vee b$ if $b = a \circ b$ and $a \leq_\wedge b$ if $a = a \circ b$. Prove that $(A; \leq_\vee)$ is an order, as an order it is a join-semilattice, and $a \vee b = a \circ b$; that $(A; \leq_\wedge)$ is an order, as an order it is a meet-semilattice, and $a \wedge b = a \circ b$. Show that the dual of the order $(A; \leq_\vee)$ is the order $(A; \leq_\wedge)$.

1.41. Prove the following statements:

(a) Let the order $\mathfrak{A} = (A; \leq)$ be a join-semilattice. Set $a \vee b = \sup\{a, b\}$. Then the algebra $\mathfrak{A}^{\mathrm{sml}} = (A; \vee)$ is a semilattice.

(b) Let the algebra $\mathfrak{A} = (A; \circ)$ be a semilattice. Set $a \leq b$ if $a \circ b = b$. Then $\mathfrak{A}^{\mathrm{ord}} = (A; \leq)$ is an order, and the order $\mathfrak{A}^{\mathrm{ord}}$ is a join-semilattice.

(c) Let the order $\mathfrak{A} = (A; \leq)$ be a join-semilattice. Then

$$(\mathfrak{A}^{\mathrm{sml}})^{\mathrm{ord}} = \mathfrak{A}.$$

(d) Let the algebra $\mathfrak{A} = (A; \circ)$ be a semilattice. Then

$$(\mathfrak{A}^{\mathrm{ord}})^{\mathrm{sml}} = \mathfrak{A}.$$

1.42. Formulate and prove the analog of Theorem 3 for meet-semilattices. (The book I. Chajda, R. Halaš, and J. Kühr [86] presents the folklore of semilattice theory and how it relates to lattice theory.)

## Miscellany

1.43. Let $\pi$ be a set of sets. For $A, B \in \pi$, define $A \varrho B$ to mean that there is a one-to-one map from $A$ to $B$. Is $\varrho$ a preorder? What is the order associated with $\varrho$?

1.44. Let $\circ$ be an associative binary operation on the set $A$. Give a rigorous proof of the statement that any meaningful bracketing of $a_0 \circ \cdots \circ a_{n-1}$ will yield the same element.

1.45. Suppose that in the order $P$ the joins $b \vee c$, $a \vee (b \vee c)$, and $a \vee b$ exist. Prove that the join $(a \vee b) \vee c$ exists and that $a \vee (b \vee c) = (a \vee b) \vee c$.

1.46. Prove that if $a \vee b$ exists in the order $P$, so does $a \wedge (a \vee b)$.

1.47. Let $A$ and $B$ be subsets of the order $P$. Let $C = A \cup B$. Let us assume that $a = \sup A$, $b = \sup B$, and $c = \sup C$ exist. Verify that $a \vee b$ exists in $P$ and equals $c$.

1.48. In an order $P$, define the *comparability relation* $\gamma$: For $a, b \in P$, the relation $a \gamma b$ holds if $a \leq b$ or $b \leq a$. In this and the next exercise, we establish conditions on an arbitrary binary relation $\varrho$ on a set $P$ for it to be the comparability relation $\gamma$ arising from some order on $P$. Take a sequence $a_1, \ldots, a_k$ of elements of $P$ satisfying the following conditions (the indices are taken modulo $k$):

(a) $a_i \neq a_{i+1}$, $a_i \varrho a_{i+1}$, for all $i = 1, \ldots, k-1$.
(b) for no $i < j < k$ does $a_i = a_j$, $a_{i+1} = a_{j+1}$ hold;
(c) for no $1 \leq i \leq k-2$ does $a_i \varrho a_{i+2}$ hold.

Prove that $k$ is even.

*1.49. Prove that a reflexive binary relation $\varrho$ on a set $A$ is the comparability relation $\gamma$ of some order $(A; \leq)$ iff $\varrho$ satisfies the condition of Exercise 1.48 (A. Ghouila-Houri [224]; see also P. C. Gilmore and A. J. Hofman [231] and M. Aigner [27]).

## Dilworth's Chain Decomposition Theorem

**Theorem.** *Every finite order $P$ is the union of* width($P$) *chains.*

The following exercises outline this result of R. P. Dilworth [157] as presented by G. M. Bergman, based on H. Tverberg [686]. See also F. Galvin [216], K. P. Bogart, C. Greene, and J. P. S. Kung [78], K. P. Bogart, R. Freese and J. P. S. Kung [4], and the monograph N. Caspard, B. Leclerc, and B. Monjardet [81].

1.50. Let us assume that $P$ is an order of width $n$ and that the order $P$ can be written as a union of chains $C_1, \ldots, C_n$. Let $A$ be an $n$-element antichain in $P$.

   (i) Show that $C_i \cap A$ is a singleton for $i = 1, \ldots, n$.

   (ii) Deduce that each element of $A$ lies in one and only one member of $\{C_1, \ldots, C_n\}$.

1.51. Let us assume that

   (a) $|P| > 2$;

   (b) the theorem has been established for all orders of cardinality smaller than $|P|$;

   (c) the order $P$ has an $n$-element antichain $A$, where $n = \text{width}(P)$;

   (d) the antichain $A$ does not consist entirely of maximal elements or entirely of minimal elements of $P$.

   Let

   $$P_1 = \{\, y \in P \mid y \geq x, \text{ for some } x \in A \,\},$$
   $$P_2 = \{\, y \in P \mid y \leq x, \text{ for some } x \in A \,\}.$$

   Verify the following statements:

   (i) The orders $P_1$ and $P_2$ are proper subsets of $P$, both containing the antichain $A$. Deduce that each is a union of $n$ chains.

   (ii) For each $a \in A$, piece together the unique chain in $P_1$ containing $a$ (see Exercise 1.50) and the unique chain in $P_2$ containing $a$.

   (iii) Statements (i) and (ii) provide $n$ chains with union $P$.

1.52. Let us assume conditions (a)–(c) of Exercise 1.51. Let us further assume that each $n$-element antichain in $P$ consists either entirely of maximal elements or entirely of minimal elements.

   (i) Let $x$ be any maximal element of $P$; show that there is a minimal element $y \leq x$.

   (ii) Show that $P - \{x, y\}$ has width $< n$, and deduce that it is a union of $n - 1$ chains.

   (iii) Conclude that these $n - 1$ chains and the chain $\{x, y\}$ are $n$ chains with union $P$.

1.53. Prove Dilworth's Chain Decomposition Theorem.

1.54. Show that every finite order has an antichain that meets every maximal chain. But not every finite order has a chain that meets every maximal antichain (G. M. Bergman).

1.55. Let $\omega_1$ denote the first uncountable ordinal and let $D = \omega_1 \times \omega_1$. Prove that every antichain in $D$ is finite, but $D$ is not the union of countably many chains (M. A. Perles [578]).

1.56. Prove that every order (not necessarily finite) of width $n < \omega$ is the union of $n$ chains. (Hint: You need the Compactness Theorem of logic—see Exercise VI.1.28—or Zorn's Lemma, see Section II.1.4.)

## 2. How to Describe Lattices

To illustrate results and refute conjectures, we have to describe a large number of examples of lattices. In this section, we list a few ways of doing this.

### 2.1 Lattice diagrams

A finite lattice $L$ is a finite order, so it has a diagram, see Figure 4 for a small example. Since a finite lattice $L$ has a zero and a unit, a lattice diagram always has a unique "lowest" element and a unique "highest" element—contrast this with the diagrams in Figure 1.

A diagram is *optimal* if the number of pairs of intersecting lines is minimal; if the number is zero, the diagram is *planar* (as defined in Section 1.5).

Figures 4, 5, and 6 show planar diagrams. Observe that diagram (a) of Figure 7 is optimal, but not planar; diagram (c) of Figure 7 is not optimal. As a rule, optimal diagrams are the easiest to visualize.

For more on planar lattices, see Exercises 6.39–6.44, Exercises II.1.49–1.52, Exercises V.1.9–1.11.

### 2.2 Join- and meet-tables

A finite lattice can always be described by a *join-table* and a *meet-table*. For example, the following two tables describe the lattice of Figure 4:

| $\vee$ | 0 | $a$ | $b$ | 1 |
|---|---|---|---|---|
| 0 | 0 | $a$ | $b$ | 1 |
| $a$ | $a$ | $a$ | 1 | 1 |
| $b$ | $b$ | 1 | $b$ | 1 |
| 1 | 1 | 1 | 1 | 1 |

| $\wedge$ | 0 | $a$ | $b$ | 1 |
|---|---|---|---|---|
| 0 | 0 | 0 | 0 | 0 |
| $a$ | 0 | $a$ | 0 | $a$ |
| $b$ | 0 | 0 | $b$ | $b$ |
| 1 | 0 | $a$ | $b$ | 1 |

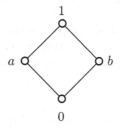

Figure 4. A small lattice diagram

We see that much of the information provided by the tables is redundant. Since both operations are commutative, the tables are symmetric with respect to the diagonal. Furthermore, $x \vee x = x$ and $x \wedge x = x$ for all $x \in L$; thus the diagonals themselves do not provide information. Therefore, the two tables can be condensed into one:

| $\vee^{\wedge}$ | 0 | $a$ | $b$ | 1 |
|---|---|---|---|---|
| 0 |   | 0 | 0 | 0 |
| $a$ | $a$ |   | 0 | $a$ |
| $b$ | $b$ | 1 |   | $b$ |
| 1 | 1 | 1 | 1 |   |

It should be emphasized that the part above the diagonal determines the part below the diagonal since either determines the ordering. To show that this table defines a lattice, we have only to check the associative and absorption identities.

## 2.3  Combinations

We shall describe lattices by combining the methods described above. The lattice $\mathsf{N}_5$ in Figure 5 has five elements: $o, a, b, c, i$, and satisfies the relations $b < a$, $b \vee c = i$, and $a \wedge c = 0$. This description is complete; all the relations follow from the ones given. The lattice $\mathsf{M}_3$ in Figure 5 also has five elements: $o, a, b, c, i$, and

$$a \vee b = a \vee c = b \vee c = i,$$
$$a \wedge b = a \wedge c = b \wedge c = o.$$

We can also start with some elements (say $a, b, c$) with some relations (such as $b \leq a$), and ask for the "most general" (or "least constrained") lattice that can be formed *without* specifying the elements to be used. (The exact meaning of "most general" will be given in Section 5.1.) In this case, we continue to form joins and meets until we get a lattice. We identify a new join (or meet) with an element that we already have if this is forced by the lattice axioms and by the given relations. The lattice we get from $a, b, c$ with the relation $b \leq a$ is shown in Figure 6.

To illustrate these ideas, we give a part of the computation that goes into the construction of the most general lattice $L$ generated by $a, b, c$ with $b \leq a$. We start by constructing the joins and meets $a \vee c$, $b \vee c$, $a \wedge c$, $b \wedge c$; note that

$$a \vee (b \vee c) = (a \vee b) \vee c = a \vee c,$$

since $a \vee b = a$; similarly, $b \wedge (a \wedge c) = b \wedge c$. Next we have to show that the seven elements

$$a, b, c, a \vee c, b \vee c, a \wedge c, b \wedge c$$

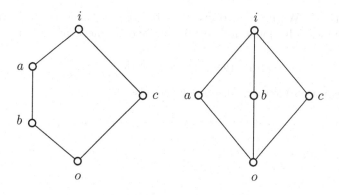

Figure 5. The lattices $N_5$ and $M_3$

that we already have, are all distinct. Remember that two were equal if such equality would follow from the relation ($b \le a$) and the lattice axioms. Therefore, to show a pair of them distinct, it is enough to find a lattice $K$ with $a, b, c \in K$ with $b \le a$, where the two elements are distinct. For instance, to show $a \neq a \vee c$, take the lattice $\{0, 1, 2\}$ with $0 < 1 < 2$ and $b = 0$, $a = 1$, $c = 2$.

The next step is to form one further join and one further meet: $b \vee (a \wedge c)$ and $a \wedge (b \vee c)$ and claim that the nine elements

$$a, b, c, a \vee c, b \vee c, a \wedge c, b \wedge c, b \vee (a \wedge c), a \wedge (b \vee c)$$

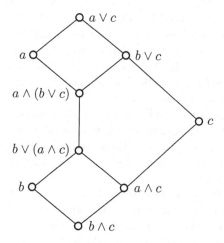

Figure 6. The most general lattice generated by $b \le a$ and $c$

form a lattice. We have to prove that by joins and meets we cannot get anything new. The joins and meets we have to check are all trivial. For instance,

$$b \wedge (a \wedge (b \vee c)) = b \wedge a \wedge (b \vee c) = b,$$

by the absorption identity, since $a \wedge b = b$; also

$$c \wedge (a \wedge (b \vee c)) = (c \wedge a) \wedge (b \vee c) = a \wedge c,$$

since $c \wedge a \leq b \vee c$.

**Exercises**

2.1. Give the meet- and join-table of the lattice in Exercise 1.12(a) for one-, two-, and three-element sets $X$.

2.2. Give the set $\leq$ for the lattices in Exercise 2.1. Which is simpler: the meet- and join-table or the ordering?

2.3. Describe a practical method of checking associativity in a join-table and in a meet-table.

2.4. Let $(P; \leq)$ be an order, $a, b \in P$ with $a < b$, and let $C(a, b)$ denote the set of all chains $H$ in $P$ with smallest element $a$ and largest element $b$. Let $H_0 \leq H_1$ mean $H_0 \subseteq H_1$ for $H_0, H_1 \in C(a, b)$. Show that $(C(a, b); \leq)$ is an order with zero, $\{a, b\}$.

2.5. The order $(Q; \leq)$ satisfies the *Ascending Chain Condition* if all increasing chains terminate; that is, if for all $i = 0, 1, 2, \ldots$, the element $x_i$ is in $Q$, and $x_0 \leq x_1 \leq \cdots \leq x_i \leq \cdots$, then the equalities $x_m = x_{m+1} = \cdots$ hold for some $m$. Show that the Ascending Chain Condition implies the existence of maximal elements (see the definition preceding Exercise 1.50) and that, in fact, every element is majorized by a maximal element. Is the converse statement also true?

2.6. Dualize Exercise 2.5. (The dual of maximal is *minimal* and the dual of ascending is *descending*.)

2.7. If $(Q; \leq)$ is a lattice and $x$ is a maximal element, then $x$ is the unit. Show that this statement is not, in general, true in an order.

2.8. Give examples of orders without maximal elements and of orders with maximal elements in which not every element is majorized by a maximal element.

2.9. Let $(P; \leq)$ be an order and let $a < b \in P$. Assume that all chains in $P$ with smallest element $a$ and largest element $b$ are finite. Does the order $(C(a, b); \leq)$ (defined in Exercise 2.4) formed from $(P; \leq)$ satisfy the Ascending Chain Condition?

2.10. Extend Lemma 1 to orders satisfying the hypothesis of Exercise 2.9 (combine Exercises 2.5 and 2.9).

2.11. Could the result of Exercise 2.10 be proved using the reasoning of Lemma 1?

2.12. Is the result of Exercise 2.10 the best possible?

2.13. Describe a method of finding the meet- and join-table of a lattice given by a diagram.

2.14. Are the orders (a) and (b) of Figure 7 lattices?

2.15. Show that the diagrams (c) and (d) of Figure 7 represent the same lattice.

2.16. Simplify diagram (e) of Figure 7.

2.17. Simplify diagram (f) of Figure 7. What is the number of pairs of intersecting lines in an optimal diagram?

2.18. Draw the diagrams of $\operatorname{Pow} X$ for $|X| = 3, 4$. Does $\operatorname{Pow} X$ have a planar diagram for $|X| = 3$?

2.19. How many elements are there in the lattice of binary relations on $X$ (ordered by set inclusion) for $|X| \leq 3$.

2.20. Describe the most general lattice generated by $a, b, c$ such that $a \geq b$, $a \vee c = b \vee c$, and $a \wedge c = b \wedge c$.

*2.21. Describe the most general lattice generated by $a, b, c, d$ such that $a \geq b \geq c$. (Hint: see Figure 117 on page 468.)

*2.22. Show that the most general lattice generated by $a, b, c, d$ with $a \geq b$ and $c \geq d$ is infinite. (Hint: see Figure 122 on page 491.)

2.23. Let $\mathbb{N}$ be the set of positive integers, $L = \{ (n, i) \mid n \in \mathbb{N}, \ i = 0, 1 \}$. Set $(n, i) \leq (m, j)$ if $n \leq m$ and $i \leq j$. Show that $L$ is a lattice and draw the "generalized diagram" of $L$.

2.24. Draw the diagrams of all lattices with at most six elements.

2.25. Let $A$ be a finite order such that no two incomparable elements of $A$ have a common upper bound. For subsets $X, Y$ of $A$, define $X \leq Y$ to mean that there exists a one-to-one mapping $\varphi \colon X \to Y$ such that $x \leq \varphi(x)$ for all $x \in X$. Prove that all subsets of $A$ form a lattice with respect to this ordering (G. Czédli and Gy. Pollák [115]).

$$* \qquad * \qquad *$$

To dispel the impression that may have been created by Sections 1 and 2, namely, that it is always easy to prove that an order is a lattice, we present Exercises 2.26–2.36 showing that the order $\mathsf{T}_n$ defined in Exercise 2.34 is a lattice. This is a result of D. Tamari [672], first published in H. Friedman and D. Tamari [204]. The present proof is based on S. Huang and D. Tamari [401]. See also D. Huguet [402]. A second such example is presented in Exercises 3.86–3.89.

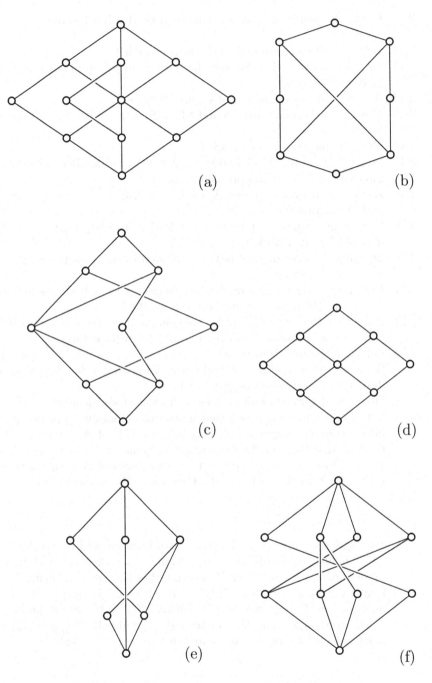

Figure 7. Which of these are lattice diagrams?

2.26. Let $T_n$ denote the set of all possible *binary bracketings* of $x_0 x_1 \cdots x_n$; for instance,

$T_0 = \{x_0\}$,
$T_1 = \{(x_0 x_1)\}$,
$T_2 = \{((x_0 x_1) x_2), (x_0 (x_1 x_2))\}$,
$T_3 = \{(x_0 (x_1 (x_2 x_3))), (x_0 ((x_1 x_2) x_3)), ((x_0 x_1)(x_2 x_3)), ((x_0 (x_1 x_2)) x_3),$
$\quad (((x_0 x_1) x_2) x_3)\}$.

Give a formal (inductive) definition of $T_n$.

2.27. Replacing consistently all occurrences of $(AB)$ by $A(B)$ in a binary bracketing, we get the *right bracketing* of the expression. For instance, the right bracketing of $((x_0 (x_1 x_2)) x_3)$ is $x_0 (x_1 (x_2))(x_3)$ and of $((x_0 x_1)(x_2 x_3))$ is $x_0 (x_1)(x_2 (x_3))$. Give a formal (inductive) definition of right bracketing and prove that there is a one-to-one correspondence between binary and right bracketings.

2.28. Show that in a right bracketing of $x_0 x_1 \cdots x_n$, there is one and only one opening bracket immediately preceding any $x_i$ for any $1 \le i \le n$.

2.29. Associate with a right bracketing of $x_0 x_1 \cdots x_n$ a *bracketing function*

$$\tau \colon \{1, \ldots, n\} \to \{1, \ldots, n\}$$

defined as follows: For every $1 \le i \le n$, there is, by Exercise 2.28, an opening bracket before $x_i$; let this bracket close following $x_j$; set $\tau(i) = j$. Show that $\tau$ has the following properties:

(a) $i \le \tau(i)$ for all $1 \le i \le n$;
(b) $i \le j \le \tau(i)$ imply that $\tau(j) \le \tau(i)$ for all $1 \le i \le j \le n$.

*2.30. Show that (a) and (b) of Exercise 2.29 characterize the bracketing functions.

2.31. Let $E_n$ denote the set of all bracketing functions defined on $\{1, \ldots, n\}$. For the bracketing functions $\alpha, \beta \in E_n$, set $\alpha \le \beta$ if $\alpha(i) \le \beta(i)$ for all $i$ with $1 \le i \le n$. Show that $\le$ is an ordering, the order $E_n$ is a lattice, and $(\alpha \wedge \beta)(i) = \inf\{\alpha(i), \beta(i)\}$ for $\alpha, \beta \in E_n$.

2.32. The *semiassociative identity* applied at the place $i$ is $\sigma_i \colon T_n \to T_n$, a map, defined as follows: If $E = \cdots (A(BC)) \cdots$, where the first variable in $B$ and $C$ is $x_i$ and $x_j$, respectively, then

$$\sigma_i(E) = \cdots ((AB)C) \cdots$$

If $E$ is not of such form, then $\sigma_i(E) = E$. Let $\xi$ and $\psi$ denote the bracketing functions associated with $E$ and $\sigma_i(E)$, respectively. Show that $\xi(k) = \psi(k)$ for all $i < k \le n$, and $\psi(i) = j - 1$; and conclude that $\xi > \psi$.

2.33. Show the converse of Exercise 2.32.

2.34. For $E, F \in T_n$, define $E < F$ to mean the existence of a sequence
$E = X_0, X_1, \ldots, X_k = F$ with $X_i \in T_n$ and $0 \le i \le k$, such that $X_{i+1}$
can be obtained from $X_i$ by some application of the semiassociative
law for any $0 \le i < k$. Let $E \le F$ mean $E = F$ or $E < F$. Show that
$\le$ is an ordering. Verify that Figure 8 displays the diagram of $T_3$ and
$T_4$. Is the diagram of $T_4$ optimal? (No.)

*2.35. Let $X, Y \in T_n$ and $X \succ Y$. Let $\alpha$ and $\beta$ be the bracketing functions
associated with $X$ and $Y$, respectively. Show that $\beta = \sigma_i(\alpha)$ for
some $i$.

2.36. Show that $T_n$ is a lattice for each $n \ge 0$. (Compare $T_n$ and $E_n$.)

The lattices $T_n$ are called *Tamari lattices*. For recent literature, search
*MathSciNet* with `Anywhere` set to `Tamari lattice`. In the abstracts
found, click on `From References`.

## 3.   Some Basic Concepts

### 3.1   The concept of isomorphism

The purpose of any algebraic theory is the investigation of algebras up to
isomorphism. We can introduce two concepts of isomorphism for lattices.

The lattices $\mathfrak{L}_0 = (L_0; \le)$ and $\mathfrak{L}_1 = (L_1; \le)$ *are isomorphic* (in symbols,
$\mathfrak{L}_0 \cong \mathfrak{L}_1$) and the map $\varphi \colon L_0 \to L_1$ is an *isomorphism* if $\varphi$ is a bijection and

$$a \le b \text{ in } \mathfrak{L}_0 \quad \text{iff} \quad \varphi(a) \le \varphi(b) \text{ in } \mathfrak{L}_1;$$

that is, if they are isomorphic as orders as defined in Section 1.3.

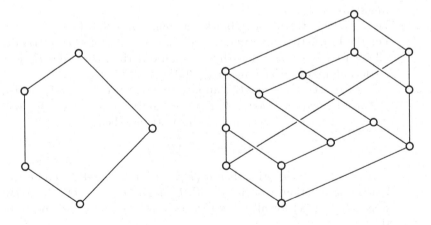

Figure 8. The lattices $T_3$ and $T_4$

The lattices $\mathfrak{L}_0 = (L_0; \vee, \wedge)$ and $\mathfrak{L}_1 = (L_1; \vee, \wedge)$ are *isomorphic* (in symbols, $\mathfrak{L}_0 \cong \mathfrak{L}_1$), and the map $\varphi \colon L_0 \to L_1$ is an *isomorphism* if $\varphi$ is a bijection and

$$\varphi(a \vee b) = \varphi(a) \vee \varphi(b),$$
$$\varphi(a \wedge b) = \varphi(a) \wedge \varphi(b),$$

in agreement with the universal algebraic definition; see Section 1.9.

Having these isomorphism concepts, we can augment Theorem 3 with two obvious statements:

**Lemma 4.**

(v) *Let the orders $\mathfrak{K} = (K; \leq)$ and $\mathfrak{L} = (L; \leq)$ be isomorphic lattices. Then the algebras $\mathfrak{K}^{\mathrm{alg}}$ and $\mathfrak{L}^{\mathrm{alg}}$ are lattices and they are isomorphic.*

(vi) *Let the algebras $\mathfrak{K} = (K; \vee, \wedge)$ and $\mathfrak{L} = (L; \vee, \wedge)$ be isomorphic lattices. Then the orders $\mathfrak{K}^{\mathrm{ord}}$ and $\mathfrak{L}^{\mathrm{ord}}$ are lattices and they are isomorphic.*

This lemma says formally that for lattices we have only one concept of isomorphism, not two. However, when we generalize these to homomorphism concepts, we get various nonequivalent notions. In order to avoid confusion, they will be given different names in Section 3.2.

If the map $\varphi \colon L_0 \to L_1$ is an isomorphism, so is the *inverse map*

$$\varphi^{-1} \colon L_1 \to L_0$$

defined as follows: $\varphi^{-1}(x) = y$ iff $\varphi(y) = x$, for $x \in L_1$.

An isomorphism of a lattice with the dual of another lattice is called a *dual isomorphism*.

An isomorphism of a lattice with itself is called an *automorphism*. The automorphisms of a lattice (or order) $L$ form a group, $\operatorname{Aut} L$, called the *automorphism group*; the product $\varphi\psi$ of two automorphisms $\varphi$ and $\psi$ is defined as the map $x \mapsto \varphi(\psi(x))$ for all $x \in L$. For a characterization of automorphism groups of lattices, see Section II.1.6.

From now on, we shall seldom use the precise notation $(L; \vee, \wedge)$ and $(L; \leq)$ for lattices and orders; we shall denote a lattice by $L$, the symbol for its underlying set.

Note that the first definition of isomorphism can be applied to orders $P_0$ and $P_1$, thus yielding an isomorphism concept for arbitrary orders.

Let $L$ be a lattice and let $P$ be an order. If $L$ as an order is isomorphic to $P$, then $P$ is a lattice; moreover, $L$ as a lattice and $P$ as a lattice are isomorphic. This is a trivial statement, but an important fact to keep in mind.

Recall that (see Section 1.3) $\mathsf{C}_n$ denotes the $n$-element chain, defined on the set $\{0, \dots, n-1\}$ and ordered by $0 < 1 < 2 < \cdots < n-1$. If $C = \{x_0, \dots, x_{n-1}\}$ is any $n$-element chain with $x_0 < x_1 < \cdots < x_{n-1}$, then $\varphi \colon i \mapsto x_i$ is an isomorphism between $\mathsf{C}_n$ and $C$. Therefore, $n$-element chains are unique up to isomorphism.

## 3.2  Homomorphisms

The isomorphism of orders generalizes as follows. The map $\varphi\colon P_0 \to P_1$ is an *isotone map* (also called a *monotone map* or an *order-preserving map*) of the order $P_0$ into the order $P_1$ if $a \leq b$ in $P_0$ implies that $\varphi(a) \leq \varphi(b)$ in $P_1$. An isotone bijection with an isotone inverse is an isomorphism.

The dual of isotone is antitone. The map $\varphi\colon P_0 \to P_1$ is an *antitone map* of the order $P_0$ into the order $P_1$ if $a \leq b$ in $P_0$ implies that $\varphi(a) \geq \varphi(b)$ in $P_1$.

Let us define a *homomorphism* of the semilattice $(S_0; \circ)$ into the semilattice $(S_1; \circ)$ as a map $\varphi\colon S_0 \to S_1$ satisfying $\varphi(a \circ b) = \varphi(a) \circ \varphi(b)$. Since a lattice $\mathfrak{L} = (L; \vee, \wedge)$ is a semilattice both under $\vee$ and under $\wedge$, we get two homomorphism concepts, *join-homomorphism* ($\vee$-homomorphism) and *meet-homomorphism* ($\wedge$-homomorphism). A *homomorphism* is a map that is both a join-homomorphism and a meet-homomorphism. Thus a homomorphism $\varphi$ of the lattice $L_0$ into the lattice $L_1$ is a map of $L_0$ into $L_1$ satisfying both

$$\varphi(a \vee b) = \varphi(a) \vee \varphi(b),$$
$$\varphi(a \wedge b) = \varphi(a) \wedge \varphi(b).$$

A homomorphism of a lattice into itself is called an *endomorphism*. A one-to-one homomorphism will also be called an *embedding*.

Note that join-homomorphisms, meet-homomorphisms, and (lattice) homomorphisms are all isotone. Let us prove this statement for join-homomorphisms. If $\varphi\colon L_0 \to L_1$ and $\varphi(a \vee b) = \varphi(a) \vee \varphi(b)$ for all $a, b \in L_0$, and if $x, y \in L_0$ with $x \leq y$, then $y = x \vee y$; thus $\varphi(y) = \varphi(x \vee y) = \varphi(x) \vee \varphi(y)$, and $\varphi(x) \leq \varphi(y)$ follows in $L_1$. Note that the converse fails, and there is no connection between meet- and join-homomorphisms.

Figure 9 shows three maps of $C_2 \times C_2 = C_2^2$ of Figure 4 into the three-element chain $C_3$. The first map of Figure 9 is isotone but is neither a meet- nor a join-homomorphism. The second map is a join-homomorphism but is not a meet-homomorphism, thus not a homomorphism. The third map of Figure 9 is a homomorphism.

If the (semi) lattices $L_0$ and $L_1$ have zeros, and they are preserved under a homomorphism $\varphi$, we call $\varphi$ a $\{0\}$-homomorphism. Similarly, we shall talk

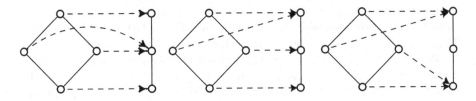

Figure 9. Examples of homomorphisms

about $\{0, 1\}$-homomorphisms, $\{\vee, 0\}$-homomorphisms, and so on. The list of homomorphism concepts will be further extended in Section 6.1.

## 3.3   Sublattices and extensions

Another basic algebraic concept is that of a subalgebra. Here is the lattice version.

The lattice $\mathfrak{K} = (K; \vee, \wedge)$ is a *sublattice* of the lattice $\mathfrak{L} = (L; \vee, \wedge)$ if $K$ is a subset of $L$ with the property that $a, b \in K$ implies that $a \vee b, a \wedge b \in K$ (the operations $\vee$ and $\wedge$ are taken in $\mathfrak{L}$), and the $\vee$ and the $\wedge$ of $\mathfrak{K}$ are restrictions to $K$ of the $\vee$ and the $\wedge$ of $\mathfrak{L}$; in symbols, $\mathfrak{K} \leq \mathfrak{L}$, and when there is no danger of ambiguity, $K \leq L$.

To put this in simpler language, we take a nonempty subset $K$ of a lattice $L$ such that $K$ is closed under $\vee$ and $\wedge$ of $\mathfrak{L}$. Clearly, $K$ is a lattice under the same $\vee$ and $\wedge$.

If $K$ is a sublattice of $L$, then $L$ is an *extension* of $K$; in notation, $L \geq K$. An extension is *proper* if it has at least one more element.

If $K \leq L$ and there is an endomorphism $\varphi \colon L \to K$ satisfying $\varphi(x) = x$ for all $x \in K$, then $\varphi$ is a *retraction* and $K$ is a *retract* of $L$.

If $\varphi$ is a homomorphism of the lattice $K$ into the lattice $L$, then

$$\varphi(K) = \{\, \varphi(x) \mid x \in K \,\}$$

is a sublattice of $L$; if $\varphi$ is one-to-one, then this sublattice is isomorphic to $K$.

The concept of a lattice as an order would suggest the following sublattice concept: Take a nonempty subset $K$ of the lattice $L$; if the suborder $K$ is a lattice, call $K$ a sublattice of $L$. This concept is different from the previous one (see Exercise 3.5) and we shall not use it at all; some use was made of this in D. Kelly and I. Rival [470].

Let $L$ be a lattice and let $A_\lambda \leq L$ for $\lambda \in \Lambda$. Then $\bigcap (A_\lambda \mid \lambda \in \Lambda)$ (the set theoretic intersection of $A_\lambda$ for $\lambda \in \Lambda$) is also closed under $\vee$ and $\wedge$ and hence it is a lattice if it is nonempty; thus for every $H \subseteq L$ with $H \neq \varnothing$, there is a smallest subset of $L$ containing $H$ and closed under $\vee$ and $\wedge$; we denote it by $\mathrm{sub}(H)$. The sublattice $\mathrm{sub}(H)$ is called the *sublattice of $L$ generated by $H$*, and $H$ is called a *generating set* of $\mathrm{sub}(H)$. If $H = \{a, b, c, \ldots\}$, then we write $\mathrm{sub}(a, b, c, \ldots)$ for $\mathrm{sub}(H)$. In many papers, $\mathrm{sub}(H)$ is denoted by $[H]$.

A lattice $L$ is called *locally finite* if $\mathrm{sub}(H)$ is finite for all finite $H \subseteq L$.

The subset $K$ of the lattice $L$ is called *convex* if $a, b \in K$, $c \in L$, and $a \leq c \leq b$ imply that $c \in K$. The concept of a convex sublattice is a typical example of the interplay between the algebraic and order theoretic concepts.

## 3.4   Ideals

The most important example of a convex sublattice is an ideal. A subset $I$ of a lattice $L$ is called an *ideal* if it is a sublattice of $L$ and $x \in I$ and $a \in L$

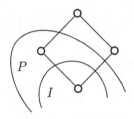

Figure 10. An ideal and a prime ideal

imply that $x \wedge a \in I$. An ideal $I$ of $L$ is *proper* if $I \neq L$. A proper ideal $I$ of $L$ is *prime* if $a, b \in L$ and $a \wedge b \in I$ imply that $a \in I$ or $b \in I$. In Figure 10, $I$ is an ideal and $P$ is a prime ideal; note that $I$ is not prime.

Since the intersection of any number of convex sublattices (ideals) is a convex sublattice (ideal) unless void, we can define the *convex sublattice generated by a subset $H$*, and the *ideal generated by a subset $H$* of the lattice $L$, provided that $H \neq \varnothing$. The ideal generated by a subset $H$ will be denoted by $\mathrm{id}(H)$, and if $H = \{a\}$, we write $\mathrm{id}(a)$ for $\mathrm{id}(\{a\})$; we shall call $\mathrm{id}(a)$ a *principal ideal*. A commonly used notation for $\mathrm{id}(H)$ is $(H]$ and for $\mathrm{id}(a)$ it is $(a]$.

**Lemma 5.** *Let $L$ be a lattice and let $H$ and $I$ be nonempty subsets of $L$.*

(i) *$I$ is an ideal iff the following two conditions hold:*

    (i$_1$) *$a, b \in I$ implies that $a \vee b \in I$,*
    (i$_2$) *$I$ is a down-set.*

(ii) *$I = \mathrm{id}(H)$ iff*

$$I = \{\, x \mid x \leq h_0 \vee \cdots \vee h_{n-1}$$
$$\text{for some } n \geq 1 \text{ and } h_0, \ldots, h_{n-1} \in H \,\}.$$

(iii) *For $a \in L$,*

$$\mathrm{id}(a) = {\downarrow}a = \{\, x \wedge a \mid x \in L \,\}.$$

*Proof.*

(i) Let $I$ be an ideal. Then $a, b \in I$ implies that $a \vee b \in I$, since $I$ is a sublattice, verifying (i$_1$). If $x \leq a \in I$, then $x = x \wedge a \in I$, and (i$_2$) is verified. Conversely, let $I$ satisfy (i$_1$) and (i$_2$). Let $a, b \in I$. Then $a \vee b \in I$ by (i$_1$), and, since $a \wedge b \leq a \in I$, we also have $a \wedge b \in I$ by (i$_2$); thus $I$ is a sublattice. Finally, if $x \in L$ and $a \in I$, then $a \wedge x \leq a \in I$, thus $a \wedge x \in I$ by (i$_2$), proving that $I$ is an ideal.

(ii) Let $I_0$ be the set on the right side of the displayed formula in (ii). Using (i), it is clear that $I_0$ is an ideal, and obviously $H \subseteq I_0$. Finally, if

$H \subseteq J$ and $J$ is an ideal, then $I_0 \subseteq J$, and thus $I_0$ is the smallest ideal containing $H$; that is, $I = I_0$.

(iii) This proof is obvious directly, or by applying (ii). □

Let Id $L$ denote the set of all ideals of $L$ and let $\mathrm{Id}_0\, L = \mathrm{Id}\, L \cup \{\varnothing\}$. We call Id $L$ the *ideal lattice* and $\mathrm{Id}_0\, L$ the *augmented ideal lattice* of $L$.

**Corollary 6.** Id $L$ *and* $\mathrm{Id}_0\, L$ *are orders under set inclusion, and as orders they are lattices.*

In fact, for $I, J \in \mathrm{Id}\, L$,

$$I \vee J = \mathrm{id}(I \cup J).$$

This formula also holds for $I, J \in \mathrm{Id}_0\, L$ if we agree that $\mathrm{id}(\varnothing) = \varnothing$. From Lemma 5(ii), we see that for $I, J \in \mathrm{Id}\, L$, the element $x \in I \vee J$ iff $x \leq i \vee j$ for some $i \in I$ and $j \in J$.

Now observe the formulas

$$\mathrm{id}(a) \vee \mathrm{id}(b) = \mathrm{id}(a \vee b),$$
$$\mathrm{id}(a) \wedge \mathrm{id}(b) = \mathrm{id}(a \wedge b).$$

Since $a \neq b$ implies that $\mathrm{id}(a) \neq \mathrm{id}(b)$, these yield:

**Corollary 7.** *The lattice* $L$ *can be embedded in* Id $L$ *(and also in* $\mathrm{Id}_0\, L$*), and the map* $a \mapsto \mathrm{id}(a)$ *is such an embedding.*

Let us connect homomorphisms and ideals (recall that $\mathsf{C}_2$ denotes the two-element chain with elements 0 and 1).

**Lemma 8.**

(i) $I$ *is a proper ideal of* $L$ *iff there is a join-homomorphism* $\varphi$ *of* $L$ *onto* $\mathsf{C}_2$ *such that* $I = \varphi^{-1}(0)$, *the inverse image of* 0, *that is,*

$$I = \{\, x \mid \varphi(x) = 0 \,\}.$$

(ii) $I$ *is a prime ideal of* $L$ *iff there is a homomorphism* $\varphi$ *of* $L$ *onto* $\mathsf{C}_2$ *with* $I = \varphi^{-1}(0)$.

*Proof.*

(i) Let $I$ be a proper ideal and define $\varphi$ by

$$\varphi(x) = \begin{cases} 0, & \text{if } x \in I, \\ 1, & \text{if } x \notin I; \end{cases}$$

obviously, this $\varphi$ is a join-homomorphism onto $\mathsf{C}_2$.

Conversely, let $\varphi$ be a join-homomorphism of $L$ onto $C_2$ and $I = \varphi^{-1}(0)$. Then $\varphi(a) = \varphi(b) = 0$ holds for all $a, b \in I$; thus

$$\varphi(a \vee b) = \varphi(a) \vee \varphi(b) = 0 \vee 0 = 0,$$

and so $a \vee b \in I$. If $a \in I, x \in L, x \leq a$, then $\varphi(x) \leq \varphi(a) = 0$, that is, $\varphi(x) = 0$; thus $x \in I$. Finally, $\varphi$ is onto, therefore, $I \neq L$.

(ii) If $I$ is prime, take the $\varphi$ constructed in the proof of (i) and note that $\varphi$ can violate the property of being a homomorphism only with $a, b \notin I$. However, since $I$ is prime, $a \wedge b \notin I$; consequently, $\varphi(a \wedge b) = 1 = \varphi(a) \wedge \varphi(b)$, and so $\varphi$ is a homomorphism. Conversely, let $\varphi$ be a homomorphism of $L$ onto $C_2$ and let $I = \varphi^{-1}(0)$. If $a, b \notin I$, then $\varphi(a) = \varphi(b) = 1$, thus $\varphi(a \wedge b) = \varphi(a) \wedge \varphi(b) = 1$, and therefore, $a \wedge b \notin I$, so $I$ is prime. $\qquad\square$

By dualizing, we get the concept of a *filter* (also called a *dual ideal*). A subset $F$ of a lattice $L$ is called an *filter* if it is a sublattice and if $x \in F$ and $a \in L$ then $x \wedge a \in F$. By dualizing, we also get the concepts of a *filter generated by $H$*, fil$(H)$, *principal filter*, fil$(a)$, *proper filter*, *prime filter*, *the lattice of filters*, Fil $L$, ordered by set inclusion, and Fil$_0 L = (\text{Fil } L) \cup \{\varnothing\}$, ordered by set inclusion. (The usual notation for fil$(H)$ is $[H]$.) Note that in Fil $L$ (and in Fil$_0 L$) the largest element is $L$; if $L$ is bounded, then $L = \text{fil}(0)$ is the largest and $\{1\} = \text{fil}(1)$ is the smallest element of Fil $L$. Furthermore,

$$\text{fil}(a) \vee \text{fil}(b) = \text{fil}(a \wedge b),$$
$$\text{fil}(a) \wedge \text{fil}(b) = \text{fil}(a \vee b),$$

for all $a, b \in L$.

**Lemma 9.** *Let $I$ be an ideal and let $D$ be a filter. If $I \cap D \neq \varnothing$, then $I \cap D$ is a convex sublattice. Every convex sublattice can be expressed in this form in one and only one way.*

*Proof.* The first statement is obvious. To prove the second, let $C$ be a convex sublattice and set $I = \text{id}(C)$ and $D = \text{fil}(C)$. Then $C \subseteq I \cap D$. If $t \in I \cap D$, then $t \in I$, and thus by (ii) of Lemma 5, $t \leq c$ for some $c \in C$; also, $t \in D$; therefore, by the dual of (ii) of Lemma 5, $t \geq d$ for some $d \in C$. This implies that $t \in C$ since $C$ is convex, and so $C = I \cap D$.

Suppose now that $C$ has a representation, $C = I_1 \cap D_1$. Since $C \subseteq I_1$, the inclusion $\text{id}(C) \subseteq I_1$ holds. Let $a \in I_1$ and let $c$ be an arbitrary element of $C$. Then $a \vee c \in I_1$ and $a \vee c \geq c \in D_1$, so $a \vee c \in D_1$, thus $a \vee c \in I_1 \cap D_1 = C$. Finally, $a \leq a \vee c \in C$; therefore, $a \in \text{id}(C)$. This shows that $I_1 = \text{id}(C)$. The dual argument shows that $D_1 = \text{fil}(C)$. Hence the uniqueness of such a representation. $\qquad\square$

## 3.5    Intervals

An *interval* of a lattice $L$ is a convex sublattice with a smallest and largest element. Equivalently, an interval is a subset of $L$ of the form

$$[a, b] = \{\, x \mid a \leq x \leq b \,\}$$

for some $a, b \in L$ with $a \leq b$. An interval $[a, b]$ is *trivial* if $a = b$; otherwise, it is *nontrivial*. An interval $\mathfrak{p} = [a, b]$ is *prime* if $a \prec b$ in $L$.

Two intervals $[a, b]$ and $[c, d]$ are *perspective* (in some sense, "similarly positioned"), in notation, $[a, b] \sim [c, d]$ if either

$$d = b \vee c \quad \text{and} \quad a = b \wedge c,$$

or dually

$$b = a \vee d \quad \text{and} \quad c = a \wedge d,$$

see Figure 11. So $\sim$ is a symmetric relation. If we want to emphasize that from $[a, b]$ we go up to $[c, d]$ (the left diagram in the figure), then we say that $[a, b]$ is *up-perspective* to $[c, d]$, in notation, $[a, b] \overset{\mathrm{up}}{\sim} [c, d]$, and the down variant is: $[a, b]$ is *down-perspective* to $[c, d]$, in notation, $[a, b] \overset{\mathrm{dn}}{\sim} [c, d]$.

Perspective intervals need not be isomorphic. In the lattice $N_5$ (see Figure 5), $[o, a] \sim [c, i]$, but the intervals are not isomorphic. There is, however, something to be said about their structure, see the Exercises.

As you will see, perspectivities play a dominant role for some classes of lattices. A slight generalization, called congruence-perspectivity, is the basic concept for congruences.

The transitive extension of perspectivity, $\sim$, is projectivity, $\approx$. If for some natural number $n$ and intervals $[e_i, f_i]$, for $0 \leq i \leq n$,

$$[a, b] = [e_0, f_0] \sim [e_1, f_1] \sim \cdots \sim [e_n, f_n] = [c, d],$$

then we call $[a, b]$ *projective* to $[c, d]$, and we write $[a, b] \approx [c, d]$. If we want to emphasize that the projectivity is in $n$ steps, we write $[a, b] \overset{n}{\approx} [c, d]$.

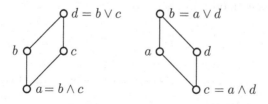

Figure 11. $[a, b] \sim [c, d]$ ($[a, b] \overset{\mathrm{up}}{\sim} [c, d]$ on the left, $[a, b] \overset{\mathrm{dn}}{\sim} [c, d]$ on the right)

## 3.6   Congruences

An equivalence relation $\alpha$ on a lattice $L$ is called a *congruence relation* if the following two *Substitution Properties* hold:

(SP$_\vee$)   $a_0 \equiv b_0 \pmod{\alpha}$   and   $a_1 \equiv b_1 \pmod{\alpha}$

$$\text{imply that} \quad a_0 \vee a_1 \equiv b_0 \vee b_1 \pmod{\alpha}$$

and the dual property

(SP$_\wedge$)   $a_0 \equiv b_0 \pmod{\alpha}$   and   $a_1 \equiv b_1 \pmod{\alpha}$

$$\text{imply that} \quad a_0 \wedge a_1 \equiv b_0 \wedge b_1 \pmod{\alpha}.$$

Trivial examples of congruence relations are **0** and **1** (often denoted by $\omega$ and $\iota$, respectively, in the literature), defined by

$$x \equiv y \pmod{\mathbf{0}} \qquad\qquad \text{iff } x = y;$$
$$x \equiv y \pmod{\mathbf{1}}, \qquad\qquad \text{for all } x, y \in L.$$

If the lattice $L$ has only the two trivial congruences, **0** and **1**, then it is called *simple*.

As introduced in Section 1.2, for an element $a \in L$, the (congruence) block containing $a$ (often called a *congruence class*) is denoted by $a/\alpha$ (often denoted by $[a]\alpha$ in the literature), that is,

$$a/\alpha = \{\, x \mid x \equiv a \pmod{\alpha} \,\}.$$

**Lemma 10.** *Let $\alpha$ be a congruence relation of $L$. Then the block $a/\alpha$ is a convex sublattice for every $a \in L$.*

*Proof.* Let $x, y \in a/\alpha$; then $x \equiv a \pmod{\alpha}$ and $y \equiv a \pmod{\alpha}$. Therefore,

$$x \vee y \equiv a \vee a = a \pmod{\alpha},$$
$$x \wedge y \equiv a \wedge a = a \pmod{\alpha},$$

proving that $a/\alpha$ is a sublattice. If $x \le t \le y$ and $x, y \in a/\alpha$, then $x \equiv a \pmod{\alpha}$ and $y \equiv a \pmod{\alpha}$. Therefore,

$$t = t \wedge y \equiv t \wedge a \pmod{\alpha},$$

and so

$$t = t \vee x \equiv (t \wedge a) \vee x \equiv (t \wedge a) \vee a = a \pmod{\alpha},$$

proving that $a/\alpha$ is convex.                                              □

Sometimes a long computation is required to prove that a given binary relation is a congruence relation (for instance, in the proof of Theorem 12). Such computations are often facilitated by the following lemma (G. Grätzer and E. T. Schmidt [334] and F. Maeda [520]):

**Lemma 11.** *A reflexive binary relation* $\alpha$ *on a lattice* $L$ *is a congruence relation iff the following three properties are satisfied for any* $x, y, z, t \in L$:

(i) $x \equiv y \pmod{\alpha}$   *iff*   $x \wedge y \equiv x \vee y \pmod{\alpha}$.

(ii) *Let* $x \leq y \leq z$; *then* $x \equiv y \pmod{\alpha}$ *and* $y \equiv z \pmod{\alpha}$ *imply that* $x \equiv z \pmod{\alpha}$.

(iii) $x \leq y$ *and* $x \equiv y \pmod{\alpha}$ *imply that* $x \vee t \equiv y \vee t \pmod{\alpha}$ *and* $x \wedge t \equiv y \wedge t \pmod{\alpha}$.

*Remark.* Condition (ii) is transitivity of $\alpha$, but only for three-element chains. Condition (iii) states the Substitution Properties for $\alpha$, but only for comparable elements.

*Proof.* The "only if" part being trivial, assume now that a reflexive binary relation $\alpha$ satisfies conditions (i)–(iii).

We start with an easy claim. Let $b, c \in [a, d]$ and $a \equiv d \pmod{\alpha}$; then $b \equiv c \pmod{\alpha}$.

Indeed, $a \equiv d \pmod{\alpha}$ and $a \leq d$ imply by (iii) that

$$b \wedge c = a \vee (b \wedge c) \equiv d \vee (b \wedge c) = d \pmod{\alpha}.$$

Now $b \wedge c \leq d$ and (iii) imply that

$$b \wedge c = (b \wedge c) \wedge (b \vee c) \equiv d \wedge (b \vee c) = b \vee c \pmod{\alpha};$$

thus by (i), $b \equiv c \pmod{\alpha}$, verifying the claim.

To prove that $\alpha$ is transitive, let $x \equiv y \pmod{\alpha}$ and $y \equiv z \pmod{\alpha}$. Then by (i), $x \wedge y \equiv x \vee y \pmod{\alpha}$, and by (iii),

$$y \vee z = (y \vee z) \vee (x \wedge y) \equiv (y \vee z) \vee (x \vee y) = x \vee y \vee z \pmod{\alpha},$$

and similarly, $x \wedge y \wedge z \equiv y \wedge z \pmod{\alpha}$. Therefore,

$$x \wedge y \wedge z \equiv y \wedge z \equiv y \vee z \equiv x \vee y \vee z \pmod{\alpha}$$

and

$$x \wedge y \wedge z \leq y \wedge z \leq y \vee z \leq x \vee y \vee z.$$

Thus applying (ii) twice, we get $x \wedge y \wedge z \equiv x \vee y \vee z \pmod{\alpha}$. Now we apply the statement of the previous paragraph with $a = x \wedge y \wedge z, b = x, c = z$, $d = x \vee y \vee z$, to conclude that $x \equiv z \pmod{\alpha}$.

Let $x \equiv y \pmod{\alpha}$; we will prove that $x \vee t \equiv y \vee t \pmod{\alpha}$. Indeed, $x \wedge y \equiv x \vee y \pmod{\alpha}$ by (i); thus by (iii), $(x \wedge y) \vee t \equiv x \vee y \vee t \pmod{\alpha}$. Since

$$x \vee t, y \vee t \in [(x \wedge y) \vee t, x \vee y \vee t],$$

we conclude, by the claim above, that $x \vee t \equiv y \vee t \pmod{\alpha}$.

To prove the Substitution Property for $\vee$, let $x_0 \equiv y_0 \pmod{\alpha}$ and $x_1 \equiv y_1 \pmod{\alpha}$. Then

$$x_0 \vee x_1 \equiv x_0 \vee y_1 \equiv y_0 \vee y_1 \pmod{\alpha},$$

implying that $x_0 \vee x_1 \equiv y_0 \vee y_1 \pmod{\alpha}$, since $\alpha$ is transitive. The Substitution Property for $\wedge$ is similarly proved.    $\square$

Let $\operatorname{Con} L$ denote the set of all congruence relations on $L$ ordered by set inclusion; recall that every $\alpha \in \operatorname{Con} L$ is a subset of $L^2$. As a first application of Lemma 11, we prove

**Theorem 12.** *The order* $\operatorname{Con} L$ *is a lattice. For* $\alpha, \beta \in \operatorname{Con} L$,

$$\alpha \wedge \beta = \alpha \cap \beta.$$

*In fact, for* $A \subseteq \operatorname{Con} L$, *the relation* $\bigcap A$ *is a congruence.*

*The join,* $\alpha \vee \beta$, *can be described as follows:*

$x \equiv y \pmod{\alpha \vee \beta}$ *iff there is a sequence*

$$x \wedge y = z_0 \leq z_1 \leq \cdots \leq z_{n-1} = x \vee y$$

*of elements of* $L$ *such that, for each* $i$ *with* $0 \leq i < n-1$, *either* $z_i \equiv z_{i+1} \pmod{\alpha}$ *or* $z_i \equiv z_{i+1} \pmod{\beta}$.

*Remark.* $\operatorname{Con} L$ is called the *congruence lattice* of $L$; it was denoted by $\Theta(L)$ and $C(L)$ in many papers. Observe that $\operatorname{Con} L$ is a sublattice of $\operatorname{Equ} L$ (see Exercise 1.16 and Exercises 3.59–3.60); that is, the join and meet of congruence relations as congruence relations and as equivalence relations (partitions) coincide.

*Proof.* Clearly, $\alpha \cap \beta$ is the infimum of $\alpha$ and $\beta$, therefore, $\alpha \wedge \beta = \alpha \cap \beta$ is obvious. So is the second statement.

To verify the description for the join, let $\gamma$ be the binary relation described in Theorem 12. Then $\alpha \subseteq \gamma$ and $\beta \subseteq \gamma$ are obvious. If $\delta$ is a congruence relation with $\alpha \subseteq \delta$, $\beta \subseteq \delta$, and $x \equiv y \pmod{\gamma}$, then for each $i$, either $z_i \equiv z_{i+1} \pmod{\alpha}$ or $z_i \equiv z_{i+1} \pmod{\beta}$; thus $z_i \equiv z_{i+1} \pmod{\delta}$ for all $i$. By the transitivity of $\delta$, the congruence $x \wedge y \equiv x \vee y \pmod{\delta}$ holds; thus $x \equiv y \pmod{\delta}$. Therefore, $\gamma \subseteq \delta$. This shows that if $\gamma$ is a congruence relation, then $\gamma = \alpha \vee \beta$.

The relation $\gamma$ is obviously reflexive and satisfies Lemma 11(i). Let us assume that $x \equiv y \pmod{\gamma}$, $y \equiv z \pmod{\gamma}$, and $x \leq y \leq z$. Then $x \equiv z$ $\pmod{\gamma}$ is established by joining the sequences showing that $x \equiv y \pmod{\gamma}$ and $y \equiv z \pmod{\gamma}$; this verifies Lemma 11(ii). To show Lemma 11(iii), let $x \equiv y \pmod{\gamma}$ with $x \leq y$; let the sequence $z_0, \ldots, z_{n-1}$ establish this, and let $t \in L$. Then $x \vee t \equiv y \vee t \pmod{\gamma}$ and $x \wedge t \equiv y \wedge t \pmod{\gamma}$ is demonstrated by the sequences $z_0 \vee t, \ldots, z_{n-1} \vee t$ and $z_0 \wedge t, \ldots, z_{n-1} \wedge t$, respectively. Thus the hypotheses of Lemma 11 hold for $\gamma$, and we conclude that $\gamma$ is a congruence relation. $\qquad\square$

Although the first statement of Theorem 12 is trivial, it allows us to introduce a crucial concept.

**Lemma 13.** *Let $L$ be a lattice and let $\varnothing \neq H \subseteq L^2$. Then there exists a smallest congruence relation $\alpha$ such that $a \equiv b \pmod{\alpha}$ for all $(a, b) \in H$ (equivalently, $H \subseteq a/\alpha$ for all $a \in H$).*

*Remark.* We denote this congruence by $\mathrm{con}(H)$ and call it the *congruence relation generated by $H$*.

*Proof.* Let

$$\beta = \bigcap (\, \alpha \in \mathrm{Con}\, L \mid a \equiv b \pmod{\alpha}, \quad \text{for all } (a, b) \in H \,).$$

By Theorem 12, $\beta \in \mathrm{Con}\, L$. It is obvious that the congruence $a \equiv b \pmod{\beta}$ holds for all $(a, b) \in H$. Thus $\beta = \mathrm{con}(H)$. $\qquad\square$

We shall use special notation in two cases: If $H = \{(a, b)\}$, we write $\mathrm{con}(a, b)$ for $\mathrm{con}(H)$. If $H = I^2$, where $I$ is an ideal, we write $\mathrm{con}(I)$ for $\mathrm{con}(H)$. The congruence relation $\mathrm{con}(a, b)$ is called *principal*; its importance is revealed by the following formula.

**Lemma 14.** $\mathrm{con}(H) = \bigvee (\, \mathrm{con}(a, b) \mid (a, b) \in H \,)$.

*Proof.* The proof is obvious. $\qquad\square$

Note that $\mathrm{con}(a, b)$ is the smallest congruence relation under which $a \equiv b$, whereas $\mathrm{con}(I)$ is the smallest congruence relation under which $I$ is contained in a single block.

We shall investigate principal congruences for distributive lattices in Section II.3.1, for general lattices in Section III.1.2, and for modular lattices in Section V.1.4.

Congruence blocks can be considered in arbitrary algebras, see the book I. Chajda, G. Eigenthaler, and H. Länger [85].

## 3.7   Congruences and homomorphisms

Homomorphisms and congruence relations express two sides of the same phenomenon. To establish this fact, we first define quotient lattices. Let $L$ be a lattice and let $\alpha$ be a congruence relation on $L$. Let $L/\alpha$ denote the set of blocks of $\alpha$, that is,

$$L/\alpha = \{\, a/\alpha \mid a \in L \,\}.$$

Set

$$a/\alpha \vee b/\alpha = (a \vee b)/\alpha,$$
$$a/\alpha \wedge b/\alpha = (a \wedge b)/\alpha.$$

This defines $\vee$ and $\wedge$ on $L/\alpha$. Indeed, if $a/\alpha = a_1/\alpha$ and $b/\alpha = b_1/\alpha$, then $a \equiv a_1 \pmod{\alpha}$ and $b \equiv b_1 \pmod{\alpha}$; therefore, $a \vee b \equiv a_1 \vee b_1 \pmod{\alpha}$, that is, $(a \vee b)/\alpha = (a_1 \vee b_1)/\alpha$. Thus $\vee$, and dually $\wedge$, are well defined on $L/\alpha$. The lattice axioms are easily verified. The lattice $L/\alpha$ is the *quotient lattice* of $L$ modulo $\alpha$.

**Lemma 15.** *The map*

$$\alpha\colon x \mapsto x/\alpha, \qquad \text{for } x \in L,$$

*is a homomorphism of $L$ onto $L/\alpha$.*

*Remark.* The lattice $K$ is a *homomorphic image* of the lattice $L$ if there is a homomorphism of $L$ *onto* $K$. Lemma 15 states that any quotient lattice is a homomorphic image.

*Proof.* The proof is trivial.     □

**Theorem 16 (The Homomorphism Theorem).** *Let $L$ be a lattice. Any homomorphic image of $L$ is isomorphic to a suitable quotient lattice of $L$. In fact, if $\varphi\colon L \to L_1$ is a homomorphism of $L$ onto $L_1$ and if $\alpha$ is the congruence relation of $L$ defined by $x \equiv y \pmod{\alpha}$ if $\varphi(x) = \varphi(y)$, then*

$$L/\alpha \cong L_1;$$

*an isomorphism (see Figure 12) is given by*

$$\psi\colon x/\alpha \mapsto \varphi(x), \qquad x \in L.$$

*Proof.* It is easy to check that $\alpha$ is a congruence relation. To prove that $\psi$ is an isomorphism, we have to check that the map $\psi$ (i) is well defined, (ii) is one-to-one, (iii) is onto, and (iv) preserves the operations.

(i) Let $x/\alpha = y/\alpha$. Then $x \equiv y \pmod{\alpha}$; thus $\varphi(x) = \varphi(y)$, that is, $\psi(x/\alpha) = \psi(y/\alpha)$.

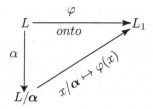

Figure 12. The Homomorphism Theorem illustrated

(ii) Let $\psi(x/\alpha) = \psi(y/\alpha)$, that is $\varphi(x) = \varphi(y)$. Then $x \equiv y \pmod{\alpha}$; and so $x/\alpha = y/\alpha$.

(iii) Let $a \in L_1$. Since $\varphi$ is onto, there is an $x \in L$ with $\varphi(x) = a$. Thus $\psi(x/\alpha) = a$.

(iv)
$$\psi(x/\alpha \vee y/\alpha) = \psi((x \vee y)/\alpha) = \varphi(x \vee y)$$
$$= \varphi(x) \vee \varphi(y) = \psi(x/\alpha) \vee \psi(y/\alpha).$$

The computation for $\wedge$ is dual.    □

For a homomorphism $\varphi \colon L \to L_1$ (not necessarily onto), the relation $\alpha$ described in Theorem 16 is called the *congruence kernel* of the homomorphism $\varphi$; it will be denoted by $\mathrm{Ker}(\varphi)$. If $L_1$ has a zero, 0, then $\varphi^{-1}(0)$ is an ideal of $L$, called the *ideal kernel* of the homomorphism $\varphi$.

Let $L$ be a lattice and let $\alpha$ be a congruence relation of $L$. If $L/\alpha$ has a zero, $a/\alpha$, then $a/\alpha$ as a subset of $L$ is an ideal, called the *ideal kernel* of the congruence relation $\alpha$.

In contrast with group and ring theory, all three kernel concepts are useful in lattice theory. Note the obvious connections; for instance, if $I$ is the ideal kernel of $\varphi$, then it is the ideal kernel of the congruence $\mathrm{Ker}(\varphi)$.

## 3.8  Congruences and extensions

Let $L$ be an extension of the lattice $K$. A congruence $\alpha$ of $L$ naturally defines a congruence $\alpha\rceil K$ of $K$, called the *restriction* of $\alpha$ to $K$, as follows: for $a, b \in K$, let $a \equiv b \pmod{\alpha\rceil K}$ if $a \equiv b \pmod{\alpha}$ holds in $L$. Restriction obviously preserves meets:

$$(\alpha \wedge \beta)\rceil K = \alpha\rceil K \wedge \beta\rceil K,$$

but not the joins (see Exercise 3.19). So the map $\varphi \colon \alpha \mapsto \alpha\rceil K$ is a meet-homomorphism of $\mathrm{Con}\,L$ into $\mathrm{Con}\,K$. If $\varphi$ is an isomorphism, we call $L$ a *congruence-preserving extension* of $K$. For instance, if $L$ is the lattice of Figure 4 and $K = \{0, a, 1\}$, then $L$ is a congruence-preserving extension of $K$. On the other hand, if we take the lattice $L = \mathsf{N}_5$ (see Figure 5) and the sublattice $K = \{o, a, b, i\}$, then $L$ is not a congruence-preserving extension of $K$.

In many instances, if the lattice $K$ has a natural embedding $\varepsilon$ into the lattice $L$ (for instance, if $L$ is a sum of $K$ and another lattice; more generally, a gluing of $K$ and another lattice, see Section IV.2.1), then we call $L$ a congruence-preserving extension of $K$ if it is a congruence-preserving extension of $\varepsilon(K)$.

**Lemma 17.** *Let $L \geq K$. Let $\beta$ be a congruence of $K$. Then there is a smallest congruence $\alpha$ of $L$ such that*

$$\beta \subseteq \alpha \rceil K.$$

*Proof.* Indeed, $\beta \subseteq K^2 \subseteq L^2$, so—as in Lemma 13—we can form the congruence $\alpha = \mathrm{con}_L(\beta)$ in $L$. Clearly, $\alpha$ satisfies $\beta \subseteq \alpha \rceil K$ and it is the smallest such congruence of $L$.                                            $\square$

A lattice $L$ has the *Congruence Extension Property* (often abbreviated as CEP) if for any sublattice $K$ of $L$, any congruence of $K$ extends to $L$; of course, a congruence may have many extensions (G. Grätzer and H. Lakser [298]). A class $\mathbf{K}$ of lattices has CEP if every lattice in $K$ has CEP. For example, the class $\mathbf{D}$ of distributive lattices has CEP, see Theorem 144.

### 3.9    Congruences and quotients

Let $K$ and $L$ be lattices and let $\varphi$ be a lattice homomorphism of $K$ onto $L$ with $\mathrm{Ker}(\varphi) = \alpha$. Then for every congruence $\beta \geq \alpha$ of $K$, we can define a congruence relation $\beta/\alpha$ of $L$: for $x, y \in L$, let $x \equiv y \pmod{\beta/\alpha}$ in $L$ if $a \equiv b \pmod{\beta}$ for some $a, b \in K$ with $\varphi(a) = x$ and $\varphi(b) = y$. Utilizing that $\beta \geq \alpha$, it is trivial that $\beta/\alpha$ is well-defined and it is a congruence.

**Lemma 18.** *The relation $\beta/\alpha$ is a congruence on $L$ and every congruence of $L$ has a unique representation in the form $\beta/\alpha$ with $\beta \geq \alpha$.*

**Corollary 19.** *The congruence lattice of $L$ is isomorphic to the interval $[\alpha, 1]$ of the congruence lattice of $K$.*

Con applied to a lattice $L$ produces a distributive lattice, the congruence lattice of $L$, see Theorem 149.

We can also apply Con naturally to a homomorphism. Let $K$ and $L$ be lattices and let $\varphi \colon K \to L$ be a lattice homomorphism. Then the map

$$\mathrm{Con}(\varphi) \colon \mathrm{Con}\, K \to \mathrm{Con}\, L$$

is defined by setting

$$(\mathrm{Con}(\varphi))(\alpha) = \mathrm{con}_L(\{\, (\varphi(x), \varphi(y)) \mid x, y \in K,\ x \equiv y\,(\alpha) \,\}),$$

for each $\alpha \in \mathrm{Con}\, K$.

**Lemma 20.** *The map* $\mathrm{Con}(\varphi)$ *is a* 0-*preserving join-homomorphism of* $\mathrm{Con}\,K$ *to* $\mathrm{Con}\,L$.

*Proof.* This is trivial.    □

### 3.10   ◇ Tolerances

We define a congruence as an equivalence relation $\alpha$ with the Substitution Properties. If instead, we start with a reflexive and symmetric (but not necessarily transitive) binary relation with the Substitution Properties, we get the concept of a tolerance.

More formally, let $L$ be a lattice and let $\alpha$ be a reflexive and symmetric binary relation on $L$. Then $\alpha$ is a *tolerance relation* on $L$ if the following two *Substitution Properties* hold:

$$a_0 \equiv b_0 \pmod{\alpha} \quad \text{and} \quad a_1 \equiv b_1 \pmod{\alpha}$$

imply that

$(\mathrm{SP}_\vee)$ $\qquad\qquad\qquad a_0 \vee a_1 \equiv b_0 \vee b_1 \pmod{\alpha},$
$(\mathrm{SP}_\wedge)$ $\qquad\qquad\qquad a_0 \wedge a_1 \equiv b_0 \wedge b_1 \pmod{\alpha}.$

While for congruences, we only need the unary versions of $(\mathrm{SP}_\vee)$ and $(\mathrm{SP}_\wedge)$, as witnessed by Lemma 11, for tolerances we need the binary version.

By definition, all congruences are tolerances. Consider the lattice $\mathsf{M}_{3,3}$ of Figure 13 and define the binary relation $\alpha$ on the lattice as follows:

$$x \equiv y \pmod{\alpha} \quad \text{iff} \quad x, y \leq b \text{ or } x, y \geq a.$$

It is easy to see that $\alpha$ is a tolerance relation but it is not transitive, since $o \equiv b$ and $b \equiv i$ but $o \equiv i$ fails, so it is not a congruence relation.

We can construct many tolerances from congruences. Let $\varphi$ be a homomorphism from the lattice $K$ onto the lattice $L$, with congruence kernel $\alpha$, so $L \cong K/\alpha$. Let $\beta$ be a congruence on $K$. We define a binary relation $\varrho$ on $L$ as follows: for $x, y \in L$, $x \equiv y \pmod{\varrho}$ if there are $u, v \in K$ such that $u \equiv v \pmod{\beta}$ and $\varphi(u) = x$, $\varphi(v) = y$. We use the notation $\varrho = \beta/\alpha$ just as for congruences, except that we do not have to assume that $\beta \geq \alpha$.

The following statement is trivial.

**Lemma 21.** *The relation* $\beta/\alpha$ *is a tolerance on* $L$.

Let $K$ be a lattice and let $\varrho$ be a tolerance on $K$. A *block* of $\varrho$ in $K$ is a subset $X \subseteq K$ *maximal* with respect to the property that $x \equiv y \pmod{\varrho}$ for all $x, y \in X$. We denote by $K/\varrho$ the blocks of $K$.

Following G. Czédli [109]], we introduce an ordering on $K/\varrho$: for the blocks $A$ and $B$, let $A \leq B$ if $\mathrm{id}(A) \subseteq \mathrm{id}(B)$ (modeled after G. Grätzer and G. H. Wenzel [365]):

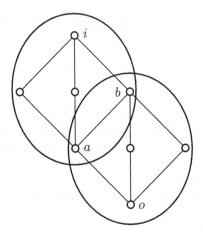

Figure 13. A tolerance example

$\diamond$ **Theorem 22.** *Let $K$ be a lattice and let $\varrho$ be a tolerance on $K$. Then $K/\varrho$ is a lattice. Moreover, in $K/\varrho$, the block $A \vee B$ is the* unique *block of $\varrho$ that includes $\{\, a \vee b \mid a \in A,\ b \in B \,\}$, and dually.*

As an application of this theorem, we can prove the converse of Lemma 21 (G. Czédli and G. Grätzer [113]):

**Theorem 23.** *Let $K$ and $L$ be lattices, let $\alpha$ be a congruence of $K$, and let $L = K/\alpha$. Then for every congruence relation $\beta$ of $K$, the relation $\beta/\alpha$ is a tolerance on $L$.*

*Conversely, for a lattice $L$ and a tolerance $\varrho$ of $L$, there is a lattice $K$, congruences $\alpha$ and $\beta$ on $K$, and a lattice isomorphism $\varphi\colon K/\alpha \to L$ such that $\varrho = \varphi(\beta/\alpha)$.*

*Proof.* Define a lattice

$$K = \{\, (A, x) \mid A \in L/\varrho,\ x \in A \,\},$$

with the operations

$$(A, x) \vee (B, y) = (A \vee B, x \vee y),$$

and dually. Then

$$\alpha = \{\, ((A, x), (B, y)) \mid A = B \,\}$$

is a congruence on $K$. Clearly, $(A, x) \mapsto x$ defines a homomorphism $\varphi$ of $K$ onto $L$. From Zorn's Lemma (see Section II.1.4), we infer that $(x, y) \in \varrho$ iff $\{x, y\} \subseteq A$ for some $A \in L/\varrho$. Hence $\varrho = \alpha/\gamma$. $\qquad\square$

For a survey of this field, see the book I. Chajda [84]. *MathSciNet* gives 14 more references (not including Chajda's book) to tolerance lattices and 78 references to tolerances in lattice theory. The closely related field of tolerances in universal algebras has 105 references.

## 3.11   Direct products

The final algebraic concept introduced in this section is that of direct product. Let $L$, $K$ be lattices and define $\vee$ and $\wedge$ on $L \times K$ "componentwise":

$$(a_0, b_0) \vee (a_1, b_1) = (a_0 \vee a_1, b_0 \vee b_1),$$
$$(a_0, b_0) \wedge (a_1, b_1) = (a_0 \wedge a_1, b_0 \wedge b_1).$$

This makes $L \times K$ into a lattice, called the *direct product* of $L$ and $K$; for an example, see Figure 14.

**Lemma 24.** *Let $L, L_1, K, K_1$ be lattices, and let $L \cong L_1$ and $K \cong K_1$. Then*

$$L \times K \cong L_1 \times K_1 \cong K_1 \times L_1.$$

*Remark.* This means that $L \times K$ is determined up to isomorphism if we know $L$ and $K$ up to isomorphism, and the direct product is determined up to isomorphism by the factors; the order in which they are given is irrelevant.

*Proof.* Let $\varphi \colon L \to L_1$ and $\psi \colon K \to K_1$ be isomorphisms; for $a \in L$ and $b \in K$, define $\chi(a, b) = (\varphi(a), \psi(b))$. Then $\chi \colon L \times K \to L_1 \times K_1$ is an isomorphism. Of course, $L_1 \times K_1 \cong K_1 \times L_1$ is proved by showing that $(a, b) \mapsto (b, a)$, for $a \in L_1$ and $b \in K_1$, is an isomorphism.    $\square$

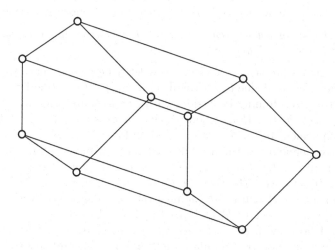

Figure 14. A direct product, $C_2 \times N_5$

More generally, if $L_i$, for $i \in I$, is a family of lattices, we first form the *cartesian product* $\prod( L_i \mid i \in I )$ of the sets, which is defined as the set of all functions

$$f \colon I \to \bigcup( L_i \mid i \in I )$$

such that $f(i) \in L_i$ for all $i \in I$. We then define $\vee$ and $\wedge$ "componentwise", that is, $f \vee g = h$ and $f \wedge g = k$ mean that

$$f(i) \vee g(i) = h(i),$$
$$f(i) \wedge g(i) = k(i)$$

for all $i \in I$. The resulting lattice is the *direct product* $\prod( L_i \mid i \in I )$. If $L_i = L$, for all $i \in I$, we get the *direct power* $L^I$. For a natural number $n$, we define $L^n$ as

$$\underbrace{(\cdots (L \times L) \times \cdots) \times L}_{n\text{-times}}.$$

We shall use the convention that a one-element lattice is the direct product of the empty family of lattices.

For instance, $\operatorname{Pow} A$ is isomorphic to the direct power $(\operatorname{Pow} X)^{|A|}$.

The map

$$\pi_i \colon \prod( L_i \mid i \in I ) \to L_i$$

defined by $\pi_i(f) \mapsto f(i)$, for $f \in \prod( L_i \mid i \in I )$ and $i \in I$, is called the *i*-th *projection map*.

A very important property of direct products is:

**Theorem 25.** *Let $L$ and $K$ be lattices, let $\alpha$ be a congruence relation of $L$, and let $\beta$ be a congruence relation of $K$. Define the relation $\alpha \times \beta$ on $L \times K$ by*

$$(a, b) \equiv (c, d) \pmod{\alpha \times \beta} \quad \text{if} \quad a \equiv c \pmod{\alpha} \text{ and } b \equiv d \pmod{\beta}.$$

*Then $\alpha \times \beta$ is a congruence relation on $L \times K$. Conversely, every congruence relation of $L \times K$ is of this form.*

*Proof.* The first statement is obvious. Now let $\gamma$ be a congruence relation on $L \times K$. For $a, b \in L$, define $a \equiv b \pmod{\alpha}$ if $(a, c) \equiv (b, c) \pmod{\gamma}$ for some $c \in K$. Let $d \in K$. Joining both sides with $(a \wedge b, d)$ and then meeting with $(a \vee b, d)$, we get $(a, d) \equiv (b, d) \pmod{\gamma}$; thus $(a, c) \equiv (b, c) \pmod{\gamma}$, for *some* $c \in K$, is equivalent to $(a, c) \equiv (b, c) \pmod{\gamma}$ for *all* $c \in K$.

Similarly, define $a \equiv b \pmod{\beta}$ if $(c, a) \equiv (c, b) \pmod{\gamma}$ for $a, b \in K$ and for any/for all $c \in L$. It is easily seen that $\alpha$ and $\beta$ are congruences. Let $(a, b) \equiv (c, d) \pmod{\alpha \times \beta}$; then $(a, x) \equiv (c, x) \pmod{\gamma}$ and $(y, b) \equiv (y, d) \pmod{\gamma}$ for all $x \in K$ and $y \in L$. Joining the two congruences with $y = a \wedge c$ and $x = b \wedge d$, we get $(a, b) \equiv (c, d) \pmod{\gamma}$. Finally, let $(a, b) \equiv (c, d) \pmod{\gamma}$. Meeting with $(a \vee c, b \wedge d)$, we get $(a, b \wedge d) \equiv (c, b \wedge d) \pmod{\gamma}$; therefore, $a \equiv c \pmod{\alpha}$. Similarly, $b \equiv d \pmod{\beta}$, and so $(a, b) \equiv (c, d) \pmod{\alpha \times \beta}$, proving that $\gamma = \alpha \times \beta$. $\square$

## 3.12  Closure systems

Closure operators, Galois connections, complete lattices, and algebraic lattices are examples of important concepts that are not algebraic in nature, that is, they do not deal with algebraic properties of a lattice. Nevertheless, they are critical tools of lattice theory. We introduce these concepts in the next four sections.

We have a number of examples of "closure operators" in the preceding sections: $\mathrm{sub}(H)$, $\mathrm{con}(H)$, $\mathrm{id}(H)$, assigning to a set a minimal larger one with specific properties. There are two natural ways of introducing this for general lattices. The first looks at the properties of closure, the second at the properties of the closed sets.

**Definition 26.** Let $L$ be a lattice and let $^-$ be a unary operation on $L$, that is, $\bar{x}$ is an element of $L$ for all $x \in L$. We call

$$L^{\mathrm{op}} = (L, {}^-)$$

("op" for "operational definition") a *closure system* if the following three conditions are satisfied for all $x, y \in L$:

(i)  $x \le \bar{x}$          (extensive);
(ii) If $x \le y$, then $\bar{x} \le \bar{y}$   (isotone);
(iii) $\bar{x} = \bar{\bar{x}}$          (idempotent).

In a closure system $(L, {}^-)$, the elements of the form $\bar{x}$ are called *closed* and the set of closed elements is denoted by $\mathrm{Cld}\, L^{\mathrm{op}}$ or simply $\mathrm{Cld}$.

**Definition 27.** Let $L$ be a lattice and let $\mathrm{Cld}$ be a nonempty subset of $L$. We call

$$L^{\mathrm{set}} = (L, \mathrm{Cld})$$

("set" for "'set definition'") a *closure system* if the following two conditions are satisfied:

(a) if $x, y \in \mathrm{Cld}$, then $x \wedge y \in \mathrm{Cld}$;

(b) for every $x \in L$, there is a smallest $y \in \mathrm{Cld}$ majorizing $x$, that is, such that if $x \le z$, for some $z \in \mathrm{Cld}$, then $y \le z$.

In a closure system $(L, \mathrm{Cld})$, the elements of $\mathrm{Cld}$ are called *closed*. For every $x \in L$, (b) asserts the existence of a unique element $y \in L$; we shall denote this unique $y$ by $\bar{x}$.

It is easy to verify that the two definitions describe the same situations. Even though this is pretty trivial, we do a bit of the computation.

**Lemma 28.** *Let L be a lattice.*

(i) *If* $(L, {}^-)$ *is a closure system with the closure operator* $^-$, *then*

$$(L, {}^-)^{\text{set}} = (L, \text{Cld}(L, {}^-))$$

*is a closure system with closed sets* $\text{Cld} = \text{Cld}(L, {}^-)$.

(ii) *If* $(L, \text{Cld})$ *is a closure system with closed sets* $\text{Cld}$, *then*

$$(L, \text{Cld})^{\text{op}} = (L, {}^-),$$

*where the operation* $^-$ *is introduced right after Definition 27, is a closure system.*

(iii) *If* $(L, {}^-)$ *is a closure system with the closure operator* $^-$, *then*

$$(L, {}^-) = ((L, {}^-)^{\text{set}})^{\text{op}}.$$

(iv) *If* $(L, \text{Cld})$ *is a closure system with closed sets* $\text{Cld}$, *then*

$$(L, \text{Cld}) = ((L, \text{Cld})^{\text{op}})^{\text{set}}.$$

*Proof.*
(i) So let $(L, {}^-)$ be a closure system as in Definition 26. We only have to see that if $x, y \in \text{Cld}(L, {}^-)$, then $x \wedge y \in \text{Cld}(L, {}^-)$. Indeed, since $x \wedge y \leq x$, by Definition 26(ii) and (iii), it follows that

$$\overline{x \wedge y} \leq \overline{x} = x.$$

Similarly, $\overline{x \wedge y} \leq y$. Hence $\overline{x \wedge y} \leq x \wedge y$, so combining with Definition 26(i), we see that $\overline{x \wedge y} = x \wedge y$; therefore, $x \wedge y \in \text{Cld}(L, {}^-)$.

(ii) Let $(L, \text{Cld})$ be a closure system as in Definition 27. We only have to see that the $\overline{x}$ defined in Definition 27(b) satisfies Definition 26(i)–(iii). But this is evident.

(iii) and (iv) . Easy computations. $\qquad\qquad\qquad\qquad\qquad\qquad\square$

**Corollary 29.** *The closed elements in a closure system form a lattice.*

*Proof.* Let $(L, \text{Cld})$ be a closure system. Then $\text{Cld}$ is a meet-semilattice by Definition 27(b). For $x, y \in \text{Cld}$, the join obviously exists by the formula

$$x \vee_{\text{Cld}} y = \overline{x \vee y}. \qquad\qquad\qquad\qquad\qquad\square$$

In most definitions and results of this section, the lattice $L$ is the power set of a set $A$, for instance, for the closure operators $\text{sub}(H), \text{con}(H), \text{id}(H)$; so we restate Definition 26 in this important special case.

**Definition 30.** Let $A$ be a set and let $^-$ be a unary operation on the subsets of $A$, that is, $\overline{X}$ is a subset of $A$ for all $X \subseteq A$. We call the unary operation $^-$ a *closure operator on the set* $A$ if $^-$ is a closure operator on the lattice Pow $A$ as defined in Definition 26.

Equivalently, we require that the following three conditions be satisfied for all $X, Y \subseteq A$:

(i)   $X \subseteq \overline{X}$                    (extensive);

(ii)  If $X \subseteq Y$, then $\overline{X} \subseteq \overline{Y}$    (isotone);

(iii) $\overline{X} = \overline{\overline{X}}$                    (idempotent).

The subsets of $A$ of the form $\overline{X}$ are called *closed*.

For more on closure systems, see the next two subsections and the Exercises.

### 3.13   Galois connections

Closure systems pop up very often in lattices. Galois connections provide one source.

**Definition 31.** Let $K$ and $L$ be lattices and let $\alpha \colon K \to L$ and $\beta \colon L \to K$ satisfy the following conditions.

(i)   For $x, y \in K$, if $x \le y$, then $\alpha(x) \ge \alpha(y)$.
(ii)  For $x, y \in L$, if $x \le y$, then $\beta(x) \ge \beta(y)$.
(iii) For $x \in K$, $\beta\alpha(x) \ge x$.
(iv)  For $x \in L$, $\alpha\beta(x) \ge x$.

Then $(\alpha, \beta)$ is a *Galois connection* between $K$ and $L$.

The conditions in words: the maps $\alpha$ and $\beta$ are antitone, the maps $\beta\alpha$ and $\alpha\beta$ are extensive.

This concept was introduced in G. Birkhoff [65]; see also O. Ore [564]. It is modeled after the correspondence between subgroups of the Galois group and subfields of a separable field extension, which is the subject of the classical Fundamental Theorem of Galois Theory.

The following lemma sets up a deep relationship between closure systems and Galois connections.

**Lemma 32.** *Let $(\alpha, \beta)$ be a Galois connection between the lattices $K$ and $L$. Then the map $\beta\alpha$ is a closure operator on $K$ and the map $\alpha\beta$ is a closure operator on $L$.*

*Proof.* The maps $\alpha$ and $\beta$ are antitone, so the map $\beta\alpha$ is an isotone map of $K$ into $K$. By assumption, the map $\beta\alpha$ is extensive. So Definition 26(i) and (ii) hold for the map $\beta\alpha$. Similarly, for the map $\alpha\beta$.

Since $\beta\alpha$ is extensive, $\beta(b) \leq \beta\alpha\beta(b)$ for all $b \in L$. Using that $\alpha$ is antitone, we obtain that $\alpha\beta(b) \geq \alpha\beta\alpha\beta(b)$. Finally, since $\alpha\beta$ is extensive, $\alpha\beta(b) \leq \alpha\beta\alpha\beta(b)$ also holds, so $\alpha\beta(b) = \alpha\beta\alpha\beta(b)$, verifying Definition 26(iii) for $\alpha\beta$. The proof for $\beta\alpha$ is similar.                         $\square$

A particularly useful application is the following result:

**Theorem 33.** *Let $A$ and $B$ be nonempty sets and let $\varrho \subseteq A \times B$. Define the map $\alpha\colon \operatorname{Pow} A \to \operatorname{Pow} B$ by*

$$\alpha(X) = \{\, y \in B \mid (x, y) \in \varrho, \text{ for all } x \in X \,\},$$

*and symmetrically, define $\beta\colon \operatorname{Pow} B \to \operatorname{Pow} A$ by*

$$\beta(Y) = \{\, x \in A \mid (x, y) \in \varrho, \text{ for all } y \in Y \,\}.$$

*Then $(\alpha, \beta)$ is a Galois connection between $\operatorname{Pow} A$ and $\operatorname{Pow} B$.*

*Let $\operatorname{Cld}_A$ denote the closed sets in $\operatorname{Pow} A$ for the closure map $\beta\alpha$ and let $\operatorname{Cld}_B$ denote the closed sets in $\operatorname{Pow} B$ for the closure map $\alpha\beta$. Then the lattices $\operatorname{Cld}_A$ and $\operatorname{Cld}_B$ are dually isomorphic.*

*Proof.* Clearly, $(\alpha, \beta)$ is a Galois connection between $\operatorname{Pow} A$ and $\operatorname{Pow} B$.

Let $X \in \operatorname{Cld}_A$, that is, let $X = \beta\alpha(X)$. Then

$$\alpha\beta(\alpha(X)) = \alpha(\beta\alpha(X)) = \alpha(X),$$

so $\alpha(X) \in \operatorname{Cld}_A$. Therefore, $\alpha\!\restriction\!\operatorname{Cld}_A$ maps $\operatorname{Cld}_A$ into $\operatorname{Cld}_B$ and, similarly, $\beta\!\restriction\!\operatorname{Cld}_B$ maps $\operatorname{Cld}_B$ into $\operatorname{Cld}_A$. Clearly, $X = \beta\alpha(X)$, for any $X \in \operatorname{Cld}_A$, and $Y = \alpha\beta(Y)$, for any $Y \in \operatorname{Cld}_B$. So the maps $\alpha\!\restriction\!\operatorname{Cld}_A$ and $\beta\!\restriction\!\operatorname{Cld}_B$ are inverses of each other, concluding the proof.                         $\square$

For a development of the subject of Galois connections, where the underlying orders are $\operatorname{Pow} X$ and $\operatorname{Pow} Y$, for sets $X$ and $Y$, with basic results stated in detail, and several interesting examples, the reader should consult Section 5.5 of G. M. Bergman [58].

### 3.14  Complete lattices

A lattice $L$ is called *complete* if $\bigvee H$ and $\bigwedge H$ exist for *any* subset $H \subseteq L$. The concept is selfdual, and half of the hypothesis is redundant.

**Lemma 34.** *Let $P$ be an order in which $\bigwedge H$ exists for all $H \subseteq P$. Then $P$ is a complete lattice.*

*Proof.* For $H \subseteq P$, let $K$ be the set of all upper bounds of $H$. By hypothesis, $\bigwedge K$ exists; set $a = \bigwedge K$. If $h \in H$, then $h \leq k$ for all $k \in K$; therefore, $h \leq a$ and $a \in K$. Thus $a$ is the smallest member of $K$, that is, $a = \bigvee H$.                         $\square$

We can apply this lemma to obtain a large, and very important, class of complete lattices.

**Corollary 35.** *For every lattice $L$, the lattice $\mathrm{Id}_0\, L$ is complete.*

So $\mathrm{Id}_0\, L$ is a complete lattice; but we know much more: we can describe the complete joins and meets.

For $\lambda \in \Lambda$, let $I_\lambda$ be ideals of $L$. Then these have a supremum

$$I = \bigvee(I_\lambda \mid \lambda \in \Lambda),$$

and an infimum

$$J = \bigwedge(I_\lambda \mid \lambda \in \Lambda),$$

in fact,

$$\bigvee(I_\lambda \mid \lambda \in \Lambda) = \mathrm{id}\Big(\bigcup(I_\lambda \mid \lambda \in \Lambda)\Big),$$
$$\bigwedge(I_\lambda \mid \lambda \in \Lambda) = \bigcap(I_\lambda \mid \lambda \in \Lambda).$$

Combining the first formula with Lemma 5(ii), we obtain the following result:

**Corollary 36.** *Let $I_\lambda$, for all $\lambda \in \Lambda$, be ideals and let $I = \bigvee(I_\lambda \mid \lambda \in \Lambda)$. Then $x \in I$ iff $x \le j_{\lambda_0} \vee \cdots \vee j_{\lambda_{n-1}}$ for some integer $n \ge 1$ and for some $\lambda_0, \ldots, \lambda_{n-1} \in \Lambda$ with $j_{\lambda_i} \in I_{\lambda_i}$.*

Here is another important class of complete lattices:

**Theorem 37.** *The congruence lattice $\mathrm{Con}\, L$ of a lattice $L$ is a complete lattice. For $A \subseteq \mathrm{Con}\, L$,*

$$\bigwedge A = \bigcap A.$$

*The join, $\bigvee A$, can be described as follows:*

$x \equiv y \pmod{\bigvee A}$ *iff there is a sequence*

$$x \wedge y = z_0 \le z_1 \le \cdots \le z_{n-1} = x \vee y$$

*of elements of $L$ such that for each $i$ with $0 \le i < n - 1$, there exists a congruence $\alpha_i \in A$ such that $z_i \equiv z_{i+1} \pmod{\alpha_i}$.*

For a further application of Lemma 34, let $\mathrm{Sub}\, L$ denote the set of all subsets $A$ of $L$ that are closed under $\vee$ and $\wedge$, ordered under set inclusion. So $\varnothing \in \mathrm{Sub}\, L$, and if $A \in \mathrm{Sub}\, L$ and $A \ne \varnothing$, then $A$ is a sublattice of $L$. Obviously, $\mathrm{Sub}\, L$ is closed under arbitrary intersections.

**Corollary 38.** *Let $L$ be a lattice. The lattices $\mathrm{Id}_0\, L$ and $\mathrm{Con}\, L$ are complete. If $L$ has a zero, the lattice $\mathrm{Id}\, L$ is complete. The lattice $\mathrm{Sub}\, L$ is complete.*

Note the following generalization.

**Lemma 39.** *The closed elements in a closure system of a complete lattice form a complete lattice.*

Every lattice $L$ can be embedded into a complete lattice, for instance, into the lattice $\mathrm{Id}_0 \, L$. There are more economical ways of doing it; the standard method is the MacNeille completion (see Exercises 3.70–3.74, which are from H. M. MacNeille [517]). Unfortunately, this embedding does not even preserve distributivity, see M. Cotlar [100] and N. Funayama [211].

If we define a *completion* $\hat{L}$ of a lattice $L$ as any complete lattice $\hat{L}$ that contains $L$ as a sublattice such that all infinite meets and joins that exist in $L$ are preserved in $\hat{L}$ and $\hat{L}$ is generated as a complete lattice by $L$, then we can ask whether there is *any* distributive completion of a distributive lattice. The answer is, in general, in the negative, see P. Crawley [105].

For a recent paper on the subject, see M. Gehrke and H. A. Priestley [223]. This is a very active field; *MathSciNet* has 489 references on lattice completions, in general, and 118 references for MacNeille completions of lattices, in particular.

A very important property of complete lattices is the following (for a proof, see Exercise 3.76):

$\diamond$ **Theorem 40 (Fixed Point Theorem).** *Any isotone map $f$ of a complete lattice $L$ into itself has a fixed point (that is, $f(a) = a$ for some $a \in L$).*

See A. Tarski [675] and B. Knaster [475]. *MathSciNet* gives 67 additional references.

### 3.15   Algebraic lattices

Many complete lattices in algebra are algebraic in the following sense:

**Definition 41.**

(i) Let $L$ be a complete lattice and let $a$ be an element of $L$. Then $a$ is called *compact* if $a \leq \bigvee X$, for any $X \subseteq L$, implies that $a \leq \bigvee X_1$ for some finite $X_1 \subseteq X$.

(ii) A complete lattice is called *algebraic* if every element is the join of a (possibly infinite) set of compact elements.

To state the next result, we need an ideal concept for join-semilattices. Just as for lattices, a nonempty subset $I$ of a join-semilattice $F$ is an *ideal* if $a \vee b \in I$ exactly if $a$ and $b \in I$ for all $a, b \in F$. Again, $\mathrm{Id} \, F$ is the join-semilattice (not necessarily a lattice) of all ideals of $F$ ordered under set inclusion. If $F$ has a zero, then $\mathrm{Id} \, F$ is a lattice.

Using $\mathrm{Id} \, F$, we give a useful characterization of algebraic lattices:

**Theorem 42.** *A lattice L is algebraic iff it is isomorphic to the lattice of all ideals of a join-semilattice F with zero.*

*Proof.* Let $F$ be a join-semilattice with zero; we prove that $\operatorname{Id} F$ is algebraic. We know that $\operatorname{Id} F$ is complete. We claim that $\operatorname{id}(a)$ $(= {\downarrow} a)$ is a compact element of $\operatorname{Id} F$ for any $a \in F$. Let $X \subseteq \operatorname{Id} F$ and let

$$\operatorname{id}(a) \subseteq \bigvee X.$$

Just as in the proof of Corollary 6,

$$\bigvee X = \{\, x \mid x \le t_0 \vee \cdots \vee t_{n-1},\ t_i \in I_i,\ I_i \in X \,\}.$$

Therefore, $a \le t_0 \vee \cdots \vee t_{n-1}$ with $t_i \in I_i$ and $I_i \in X$. Thus

$$\operatorname{id}(a) \subseteq \bigvee X_1,$$

with $X_1 = \{I_0, \ldots, I_{n-1}\}$, proving the claim.

For every $I \in \operatorname{Id} F$, the equality

$$I = \bigvee (\operatorname{id}(a) \mid a \in I)$$

holds, so we see that $\operatorname{Id} F$ is algebraic.

Now let $L$ be an algebraic lattice and let $F$ be the set of compact elements of $L$. Obviously, $0 \in F$. Let $a, b \in F$ and $a \vee b \le \bigvee X$ for some $X \subseteq L$. Then $a \le a \vee b \le \bigvee X$, and so $a \le \bigvee X_0$ for some finite $X_0 \subseteq X$. Similarly, $b \le \bigvee X_1$ for some finite $X_1 \subseteq X$. Thus $a \vee b \le \bigvee(X_0 \cup X_1)$, and $X_0 \cup X_1$ is a finite subset of $X$. So $a \vee b \in F$.

Therefore, $(F; \vee)$ is a join-semilattice with zero. Consider the map

$$\varphi \colon a \mapsto \{\, x \in F \mid x \le a \,\} \quad \text{for } a \in L.$$

Obviously, $\varphi$ maps $L$ into $\operatorname{Id} F$. By the definition of an algebraic lattice, $a = \bigvee \varphi(a)$, and thus $\varphi$ is one-to-one. To prove that $\varphi$ is onto, let $I \in \operatorname{Id} F$ and $a = \bigvee I$ in $L$. Therefore, $\varphi(a) \supseteq I$. Let $x \in \varphi(a)$. Then $x \le \bigvee I$, so that $x \le \bigvee I_1$ by the compactness of $x$ for some finite $I_1 \subseteq I$. So $x \in I$, proving that $\varphi(a) \subseteq I$. Consequently, $\varphi(a) = I$, and so $\varphi$ is onto. Thus $\varphi$ is an isomorphism. $\square$

Algebraic lattices originated in A. Komatu [479], G. Birkhoff and O. Frink [75], L. Nachbin [538], and J. R. Büchi [79]. Birkhoff and Frink's original definition is as follows:

1. $L$ is complete;
2. every element in $L$ is the join of join-inaccessible elements;
3. $L$ is meet-continuous.

In this definition, an order $H$ is *directed* if for $x, y \in H$, there exists an upper bound $z \in H$; an element $a$ of $L$ is *join-accessible* if there is a nonempty directed subset $H$ of $L$ such that $\bigvee H = a$ and $a \notin H$; otherwise, $a$ is *join-inaccessible*; $L$ is *meet-continuous* if

$$a \wedge \bigvee H = \bigvee ( a \wedge h \mid h \in H ),$$

for any $a \in L$ and directed $H \subseteq L$.

Interestingly, it is sufficient to formulate conditions 1 and 3 of this definition for chains only: a lattice $L$ is complete if $\bigvee C$ and $\bigwedge C$ exist for any chain $C$ of $L$; and a (complete) lattice $L$ is meet-continuous if

$$a \wedge \bigvee C = \bigvee ( a \wedge c \mid \in C ),$$

for any $a \in L$ and for any chain $C$ of $L$. These statements are immediate consequences of the following result of T. Iwamura [422]:

Let $H$ be an infinite directed set. Then, for some ordinal $\alpha$, $H$ has a decomposition

$$H = \bigcup ( H_\gamma \mid \gamma < \alpha )$$

satisfying the following three conditions:

(i)  each $H_\gamma$ is directed;
(ii)  $H_\gamma \subseteq H_\delta$ for all $\gamma < \delta < \alpha$;
(iii)  $|H_\gamma| < |H|$ for all $\gamma < \alpha$.

There are many lattices that are like algebraic lattices but not necessarily complete: finitely generated free lattices, convex bodies in a euclidean space, noncomplete atomic boolean lattices. These have been investigated in K. V. Adaricheva, V. A. Gorbunov, and M. V. Semenova [20].

### 3.16   ◇ Continuous lattices
### by Jimmie D. Lawson

It is a possible, but uncustomary, initial approach to the study of continuous lattices to view them as generalizations of algebraic lattices. In this approach, one first generalizes the notion of a compact element in a complete lattice $L$. Let $a, b \in L$; then call $a \in L$ *relatively compact in* $b$ if $b \leq \bigvee X$, for any $X \subseteq L$, implies that $a \leq \bigvee X_1$ for some finite $X_1 \subseteq X$. A *continuous lattice* is then a complete lattice in which every element is the join of the elements that are relatively compact in it.

As an example, consider the lattice $\operatorname{Open} X$ of open sets of a Hausdorff topological space. If $X$ is locally compact and $U$ is an open subset, then for every point $x \in U$, there exists an open set $V_x$ and a compact set $K_x$ such that $x \in V_x \subseteq K_x \subseteq U$. In every open cover of $U$, there are finitely many members that cover $K_x$ and hence $V_x$, so $V_x$ is relatively compact in $U$. Therefore, $U$

is the join of elements that are relatively compact in it, and so Open $X$ is a continuous lattice. For this lattice to be algebraic, one needs a basis of sets at each point that are simultaneously open and compact, which would force the space to be totally disconnected, equivalently, 0-dimensional. Hence the lattice of open sets of a *continuum*, a connected locally compact Hausdorff space, has a *continuous* lattice of open sets that is not algebraic.

However, its roots in theoretical computer science and the desire to generalize the theory to orders has influenced the theory to take an alternative albeit equivalent approach, both in formulation and in terminology.

**Definition 43.**

(i) Let $L$ be a complete lattice and let $a, b$ be elements of $L$. Then $a$ *approximates* $b$, in symbols, $a \ll b$, if $b \leq \bigvee H$, for any directed $H \subseteq L$, implies that $a \leq d$ for some $d \in H$. The ordering $\ll$ is called the *order of approximation*, or more suggestively, the *way-below relation*.

(ii) A complete lattice is called a *continuous lattice* if every element is the join of the (typically infinite) set of elements approximating it.

We note that $0 \ll c$ for any $c$ and that if $a \ll c$ and $b \ll c$, then $a \vee b \ll c$, so that the set of elements approximating $c$ is always a nonempty directed set. Thus continuous lattices are precisely those complete lattices in which every element is the *directed join* of the elements approximating it, and it is this notion that generalizes to orders.

The study of continuous lattices was initiated by Dana Scott in the late 1960s in order to build models of a domain of computation [638]. The goal was to interpret the elements of such a domain as pieces of information or (partial) results of a computation and to place on them the *information order*, where elements higher in the order extended the information of the elements below them in a consistent fashion.

Suppose, for example, that one programmed a computer to evaluate some computable function on the natural numbers and the computer was constantly spewing out the functional values for larger and larger numbers. At any stage one would have only partial information about the function, a partial function that represented the function for only finite many values. We may think of this partial function as approximating the computable function in the sense that it gives correct partial information about the function and that any computational scheme that computes, over time, all values of the function must at some finite stage yield the information in the partial function. Hence in the information order the original function is the supremum of the directed set of its finite approximations.

The prominent role of the order of approximation (way-below relation) is a distinctive feature that the theory of continuous lattices (and its generalization, *domain theory*) brings to lattice theory. A second is a focus on a general type of morphism, namely, those functions, called *Scott-continuous* functions, that preserve joins of directed sets: $f(\bigvee H) = \bigvee f(H)$ for all directed sets $H$. By considering two-element chains, one observes that such functions must be order-preserving, hence carry directed sets to directed sets. From a computational point of view, we may view directed sets as the stages of a computation and are thus requiring that morphisms preserve outcomes, joins in the information order, of computations.

*Retractions*, self-maps $r: L \to L$ such that $r$ restricted to its image is the identity, play an important role in continuous lattice theory; the image $r(L)$ is called a *retract*. The next result appears in G. Gierz, K. H. Hofmann, K. Keimel, J. D. Lawson, M. Mislove, and D. S. Scott [225, Section II-3], but had its origins in Scott's early work [639].

$\Diamond$ **Theorem 44.**

  (i) *If $r: L \to L$ is a Scott-continuous retraction on a continuous lattice $L$, then $r(L)$ is a continuous lattice as a suborder of $L$.*

  (ii) *Continuous lattices can be characterized (up to isomorphism) as Scott-continuous retracts of some power of the two-element lattice.*

An important motivation for considering Scott continuous maps, both mathematical and for modeling purposes in theoretical computer science, is the following theorem, see [225, Theorem II-2.12] and the precedent [639].

$\Diamond$ **Theorem 45.** *For continuous lattices $L, M$, the function space $[L \to M]$ of Scott-continuous functions from $L$ to $M$ is again a continuous lattice with the pointwise ordering. Furthermore, the category of continuous lattices and maps preserving directed joins is a cartesian-closed category.*

A third distinct feature of continuous lattice theory is the introduction of important "intrinsic topologies", topologies defined directly from the ordering. The first of these is the *Scott topology*. A set $U$ is open in the Scott topology if the following two conditions are satisfied:

  (i) $x \in U$ and $x \leq y$ implies that $y \in U$;
  (ii) if $\bigvee H \in U$ for a directed set $H$, then $d \in U$ for some $d \in H$.

Such sets form, in general, only a $T_0$-topology.

The Scott topology aptly captures topologically many order-theoretic aspects of continuous lattice theory, for instance, a function $f: L \to M$ between complete lattices is Scott-continuous iff it is topologically continuous when the lattices are equipped with their respective Scott topologies. The second important intrinsic topology is the *Lawson topology*, which has as a subbasis of open sets all Scott open sets and all complements of principal filters $\mathrm{fil}(a)$. It arises out of roots of the theory in topological algebra, particularly the theory of topological lattices and semilattices. The next results may be found in [225, Sections III-1, III-2,VI-3], which builds on the work of J. D. Lawson [499] and K. H. Hofmann and A. L. Stralka [398].

◇ **Theorem 46.**

(i) *A complete lattice is compact in the Lawson topology. It is a continuous lattice iff it is a meet-continuous lattice and the Lawson topology is Hausdorff.*

(ii) *With respect to the Lawson topology, a continuous lattice is a topological semilattice with respect to the meet operation, that is, the operation $(x, y) \mapsto x \wedge y$ is continuous, and each point has a basis of neighborhoods that are meet subsemilattices. Furthermore, every compact Hausdorff topological semilattice that has a basis of neighborhoods at each point which are subsemilattices and its ordering is a lattice ordering arises (up to isomorphism) in this way.*

A continuous lattice admits enough Lawson-continuous meet-semilattice homomorphisms into the unit interval $[0, 1]$ with its usual order to separate points, see J. D. Lawson [499] and [225, Theorem IV-3.20]. This fact provides another characterization of continuous lattices.

◇ **Theorem 47.** *Every continuous lattice is isomorphic to a meet subsemilattice of a power of $[0, 1]$ (equipped with the coordinatewise ordering) that is closed under arbitrary meets and directed joins.*

The most comprehensive treatment of the theory of continuous lattices and their generalization, continuous directed complete orders, is the book *Continuous Lattices and Domains* [225] by G. Gierz, K. H. Hofmann, K. Keimel, J. D. Lawson, M. Mislove, and D. S. Scott. For briefer introductions, see P. Johnstone [430] and B. Davey and H. Priestley [131].

### 3.17   ◇ Algebraic lattices in universal algebra

It was observed in G. Birkhoff and O. Frink [75] (see also the earlier lecture G. Birkhoff [69]) that for a universal algebra $\mathfrak{A}$, the subalgebra lattice, Sub $\mathfrak{A}$ (see Exercise 3.82), and the congruence lattice, Con $\mathfrak{A}$ (see Exercise 3.78), are algebraic.

For subalgebra lattices, they proved the converse:

**Theorem 48 (Birkhoff-Frink Theorem).** *Every algebraic lattice $L$ is isomorphic to the subalgebra lattice of a universal algebra $\mathfrak{A}$.*

*Proof.* Let $L$ be an algebraic lattice and let $F$ be the join-semilattice with zero provided by Theorem 42. Define the algebra $\mathfrak{F} = (F; \vee, \{ f_{a,b} \mid a, b \in F \})$, where $f_{a,b}$ is a unary operation on $F$ defined as follows:

$$f_{a,b}(x) = \begin{cases} a, & \text{for } x = a \vee b; \\ 0, & \text{otherwise.} \end{cases}$$

It is easy to see that the subalgebras of $\mathfrak{F}$ are exactly the ideals of $F$, so by Theorem 42, we are done.    $\square$

Dozens of papers have been published on subalgebra lattices. A. A. Iskander [420], [421] proved that the algebra $\mathfrak{A}$ in the Birkhoff-Frink Theorem can always be constructed in the form $\mathfrak{B}^2$, thus representing an algebraic lattice with the binary relations satisfying the Substitution Property on an algebra. For an alternative proof, see G. Grätzer and W. A. Lampe [322]. I. Chajda and G. Czédli [87] pointed out that the Grätzer-Lampe proof is easy to modify to yield the following result:

$\Diamond$ **Theorem 49.** *The lattice of all tolerance relations of an algebra can be characterized as an algebraic lattice.*

If the algebra is unary, the subalgebra lattice is distributive; there have been more than 20 publications on this topic alone. See *MathSciNet* for a complete listing.

In 1948, G. Birkhoff and O. Frink [75] posed the problem whether the converse is also true for congruence lattices of algebras. The same problem was raised again in 1961 in G. Birkhoff [70]. Interestingly, neither publication references the 1945 lecture, G. Birkhoff [69], where the problem was first raised for algebras finitary or infinitary.

In 1963, this problem was resolved in G. Grätzer and E. T. Schmidt [338]:

$\Diamond$ **Theorem 50.** *Every algebraic lattice $L$ is isomorphic to the congruence lattice of a universal algebra $\mathfrak{A}$.*

The proof of this result is long and tedious. I recall, about 50 years ago, I was seriously concerned that somebody would publish a two-page proof modeled after the proof of the Birkhoff-Frink Theorem as presented here: start with the set $\overline{F}$, derived from the compact elements $F$ of $L$, define a set of operations $O$, and the congruences of the algebra $(\overline{F}; O)$ are in a natural correspondence with the ideals of $F$. About half a dozen proofs have since been published, none this brief (see the references in G. Grätzer [270]).

Many stronger forms of this result have been published, see my universal algebra book [263] for some references. In [270], I give an elementary exposition of these problems with a lot of references.

If we allow infinitary operations in the definition of an algebra $\mathfrak{A}$, then Con $\mathfrak{A}$ is a complete lattice but not necessarily algebraic. G. Birkhoff [69] raised the following question in 1945: *Is every complete lattice isomorphic to the congruence lattice of an infinitary algebra?* An affirmative answer was published by W. A. Lampe and the author as Appendix 7 in G. Grätzer [263] (we state a very special case of the result):

$\Diamond$ **Theorem 51.** *Let $L_a$ and $L_c$ be complete lattices with more than one element and let $G$ be a group. Then there exists an infinitary universal algebra $\mathfrak{A}$ such* *that the subalgebra lattice of $\mathfrak{A}$ is isomorphic to the lattice $L_a$, the congruence lattice of $\mathfrak{A}$ is isomorphic to the lattice $L_c$, and the automorphism group of $\mathfrak{A}$ is isomorphic to the group $G$.*

## Exercises

3.1. Prove Lemma 4.

3.2. Let $L_0$ and $L_1$ be lattices. Let $\varphi \colon L_0 \to L_1$ and $\psi \colon L_1 \to L_0$ be homomorphisms. Show that if $\psi\varphi$ is the identity map on $L_0$ and $\varphi\psi$ is the identity map on $L_1$, then $\varphi$ is an isomorphism and $\psi = \varphi^{-1}$ (the *inverse map*, $\varphi^{-1} \colon L_1 \to L_0$). Furthermore, prove that if $\varphi \colon L_0 \to L_1$ and $\psi \colon L_1 \to L_2$ are isomorphisms, then so is $\varphi^{-1}$ and $\psi\varphi$ is an automorphism.

3.3. A bijective lattice homomorphism is an isomorphism. Is it true that a bijective isotone map between two orders is an isomorphism?

3.4. Find a general construction of meet- (join-) homomorphisms that are not homomorphisms.

3.5. Find a subset $H$ of a lattice $L$ such that $H$ is not a sublattice of $L$ but $H$ is a lattice under the ordering of $L$ restricted to $H$.

3.6. Show that a lattice $L$ is a chain iff every nonempty subset of $L$ is a sublattice.

3.7. Prove that a sublattice generated by two distinct elements has two or four elements.

*3.8. Find an infinite lattice generated by three elements.

3.9. Verify that a nonempty subset $I$ of a lattice $L$ is an ideal iff $a \vee b \in I$ is equivalent to $a, b \in I$ for all $a, b \in L$.

3.10. Prove that if $L$ is finite, then $L$ and $\operatorname{Id} L$ are isomorphic. How about $\operatorname{Id}_0 L$?

3.11. Let $L$ be a lattice and $H \subseteq L$. Under what conditions is $\downarrow H = \operatorname{id}(H)$ true?

*3.12. Is there an infinite lattice $L$ such that $L \cong \operatorname{Id} L$, but not every ideal is principal? (There is no such lattice, see D. Higgs [394].)

3.13. Prove the completeness of $\operatorname{Id}_0 L$ without any reference to Lemma 34.

3.14. Let $L$ and $K$ be lattices and let $\varphi \colon L \to K$ be an onto homomorphism. Let $I$ be an ideal of $L$, and let $J$ be an ideal of $K$. Show that $\varphi(I)$ is an ideal of $K$, and $\varphi^{-1}(J) = \{\, a \in L \mid \varphi(a) \in J \,\}$ is an ideal of $L$.

3.15. Is the image $\varphi(P)$ of a prime ideal under a homomorphism $\varphi$ again prime?

3.16. Show that the inverse image $\varphi^{-1}(P)$ of a prime ideal $P$ under an onto lattice homomorphism $\varphi$ is prime again.

3.17. Show that an ideal $P$ is a prime ideal of a lattice $L$ iff $L - P$ is a filter, in fact, a prime filter.

3.18. Let $L$ be a lattice with zero, and $I$ an ideal of $L$. Show that if $I$ contains a prime ideal, then for any two elements $x, y \in L$ with $x \wedge y = 0$, either $x$ or $y$ belongs to $I$. Is the converse true in general? Is it true if $L$ is distributive?

3.19. Take the lattice $\mathsf{N}_5$ and its sublattice $\mathsf{C}_3$. Does the congruence restriction to $\mathsf{C}_3$ preserve joins? Same for $\mathsf{M}_3$ and a sublattice $\mathsf{C}_2 \times \mathsf{C}_2$.

3.20. Let $H$ be a join-subsemilattice of the lattice $L$. Then

$$\mathrm{id}(H) = {\downarrow} H.$$

3.21. Find all tolerances of $\mathsf{M}_{3,3}$, see Figure 13.

3.22. For a lattice $L$ and congruence $\boldsymbol{\alpha}$, we introduced $L/\boldsymbol{\alpha}$ and $a/\boldsymbol{\alpha}$, where $a \in L$. For a lattice $L$ and tolerance $\varrho$, we introduced $L/\varrho$. Why did we not introduce $a/\varrho$ for $a \in L$?

3.23. Verify by example, that the unary versions of $(\mathrm{SP}_\vee)$ and $(\mathrm{SP}_\wedge)$, would not be sufficient in the definition of a tolerance.

3.24. Prove that in a lattice a block of a tolerance relation is a convex sublattice.

3.25. Let $A$ and $B$ be blocks of a congruence $\boldsymbol{\alpha}$ of a lattice $L$. Let $C = A \vee B$ in $L/\boldsymbol{\alpha}$. Prove that $\mathrm{fil}(C) = \mathrm{fil}(A) \cap \mathrm{fil}(B)$.

3.26. Prove that a tolerance of a lattice $L$ is determined by the set of its blocks.

3.27. In a lattice $L$, for the blocks $A$ and $B$ of a tolerance relation, prove that $\mathrm{id}(A) \subseteq \mathrm{id}(B)$ iff $\mathrm{fil}(A) \supseteq \mathrm{fil}(B)$.

3.28. Let $S$ and $L$ be lattices, let $f \colon S \to L$ be a join-homomorphism, and let $g \colon S \to L$ be a meet-homomorphism. Assume that $f(x) \leq g(x)$, for all $x \in S$, and set

$$L = \bigcup (\, [f(x), g(x)] \mid x \in S\,).$$

Define the binary relation $\varrho$ on $L$ as follows:
$u \equiv v \pmod{\varrho}$ if $u, v \in [f(x), g(x)]$ for some $x \in S$.
Prove that $\varrho$ is a tolerance of $L$.

3.29. Assume that both $f$ and $g$ are one-to-one in Exercise 3.28. Show that $[f(x), g(x)]$ is a block of $\varrho$ for every $x \in S$. (Exercise 3.28 and this exercise are from A. Day and C. Herrmann [140].)

3.30. Let $K$ and $L$ be lattices. Show that if $K \leq L$, then $K$ is isomorphic to a sublattice of $\mathrm{Id}\, L$.

*3.31. Verify that the converse of Exercise 3.30 is false even for some finite lattice $K$.

3.32. Let $L$ be a lattice. Prove that if $\mathsf{N}_5$ (see Figure 5) is isomorphic to a sublattice of $\mathrm{Id}\, L$, then $\mathsf{N}_5$ is isomorphic to a sublattice of $L$.

3.33. Find a lattice $L$ and a convex sublattice $C$ of $L$ that cannot be represented as $a/\boldsymbol{\alpha}$ for any congruence relation $\boldsymbol{\alpha}$ of $L$.

3.34. State and prove an analogue of Lemma 11 for *join-congruence relations*, that is, for equivalence relations on a lattice satisfying the Substitution Property for joins.

3.35. Describe $\alpha \vee \beta$ for join-congruences.

3.36. Find a lattice $L$ such that $L \cong L/\alpha$, for all congruences $\alpha \neq 1$, and there are infinitely many such congruences.

3.37. Describe the congruence lattice of $\mathsf{N}_5$.

3.38. Describe the congruence lattice of an $n$-element chain.

3.39. Describe the congruence lattice of the lattice of Figure 6, and list all quotient lattices.

3.40. Construct a lattice that has exactly three congruence relations.

3.41. Construct infinitely many lattices $L$ such that each lattice is isomorphic to its congruence lattice.

*3.42. Can a lattice $L$ as in Exercise 3.41 be infinite?

3.43. Generalize Lemma 24 to the direct product of more than two lattices.

3.44. Let $L$ and $K$ be lattices. Show that $\mathsf{N}_5 \cong L \times K$ implies that $L$ or $K$ has only one element.

3.45. Let $L$ be a lattice. Show that $\operatorname{Id} L$ is *conditionally complete*: every nonempty set $H$ with an upper bound has a supremum, and dually. Verify that $\operatorname{Id} L$ is complete iff $L$ has a zero.

3.46. Let $L$ be a lattice. Show that the ideal kernel of a homomorphism is an ideal of $L$.

3.47. Find an ideal that is the ideal kernel of no homomorphism.

3.48. Find an ideal that is the kernel of several (infinitely many) homomorphisms (congruence relations).

3.49. Prove that every ideal of a lattice $L$ is prime iff $L$ is a chain.

3.50. For the lattices $L$ and $K$, under what conditions is $L \times K$ planar? (Start with Exercise 2.18.)

3.51. Let $L$ and $K$ be lattices and let $\varphi \colon L \to K$ be a bijection satisfying

$$\{\varphi(a \vee b), \varphi(a \wedge b)\} = \{\varphi(a) \vee \varphi(b), \varphi(a) \wedge \varphi(b)\}$$

for all $a, b \in L$. Let $A \in \operatorname{Sub} L$. Show that $\varphi(A) \in \operatorname{Sub} K$; in fact, $A \mapsto \varphi(A)$ is a (lattice) isomorphism between $\operatorname{Sub} L$ and $\operatorname{Sub} K$.

3.52. Prove the converse of Exercise 3.51.

3.53. Generalize Theorem 25 to finitely many lattices.

3.54. Show that the first part of Theorem 25 holds for any family of lattices (finite or infinite) but the second part does not.

3.55. Show that the second statement of Theorem 25 fails for (abelian) groups.

3.56. For the congruence relations $\alpha$ and $\beta$ of the lattice $L$, form their join $\alpha \vee \beta$ in $\operatorname{Equ} L$ as in Exercise 1.16. Compare the formula with the stronger one in Theorem 25. Can you get to the stronger formula from Exercise 1.16?

3.57. If in the previous exercise, $\alpha$ and $\beta$ are congruences of the join-semilattice $L$, what can you say about $\alpha \vee \beta$?

3.58. In Exercise 3.57, what if $\alpha$ and $\beta$ are congruences of the meet-semilattice $L$? Can you generalize these observations to universal algebra?

3.59. For a nonempty set $A$, show that $\operatorname{Equ} A$ is a complete lattice.

3.60. Let $L$ be a lattice. Show that $\operatorname{Con} L$ is a sublattice of $\operatorname{Equ} L$.

3.61. **The First Isomorphism Theorem.** *Let $L$ be a lattice, let $\alpha$ be a congruence relation on $L$, and let $L_1 \leq L$. If, for every $a \in L$, there exists exactly one $b \in L_1$ satisfying $a \equiv b \pmod{\alpha}$, then*

$$L/\alpha \cong L_1.$$

3.62. Let $L_0$ and $L_1$ be lattices and let $\pi_i \colon (x_0, x_1) \mapsto x_i$ be the projection map of $L_0 \times L_1$ onto $L_i$ for $i = 0, 1$. Prove that if $A$ is a lattice and $\varphi_i$ is a homomorphism of $A$ into $L_i$, $i = 0, 1$, then there is a unique homomorphism $\psi \colon A \to L_0 \times L_1$ satisfying $\pi_i \psi = \varphi_i$ for $i = 0, 1$.

3.63. In what way does Exercise 3.62 characterize $L_0 \times L_1$?

3.64. Characterize $\prod(L_i \mid i \in i)$ by projection maps and homomorphisms.

3.65. Define closure operators for orders. Can you prove Lemma 28 for orders?

3.66. Define a Galois connection as a pair of maps between two orders. Can you prove Lemma 32 for orders?

3.67. Let $S$ be a set. Call pairs of subsets $(A_i, B_i)$, for $i \in I$, *implications* and write $A_i \to B_i$. Define $X \subseteq S$ to be *implication-closed* if $A_i \subseteq X$ implies that $B_i \subseteq X$ for all $i \in I$.

Show that the set of all implication-closed subsets of $S$ is a closure system, and that every closure system arises from a such a family of implications. See M. Wild [731] for a theory of closure systems and implications for a finite set $S$. See also M. Wild [733].

3.68. Let $(\alpha, \beta)$ be a *Galois connection* between the lattices $K$ and $L$. Prove that $\alpha$ (respectively, $\beta$) maps onto the lattice of closed elements, and there is a dual isomorphism between these lattices.

$$*\qquad*\qquad*$$

3.69. Let the lattices $K$ and $L$, maps $\alpha \colon K \to L$ and $\beta \colon L \to K$ be given as in Definition 31. Show that this is a Galois connection iff the following condition holds for all $x \in K, y \in L$:

$$\alpha(x) \geq y \quad \text{iff} \quad \beta(y) \geq x.$$

3.70. For a subset $A$ of a lattice $L$, set

$$A^{\mathrm{ub}} = \{\, x \in L \mid x \text{ is an upper bound of } A \,\},$$
$$A^{\mathrm{lb}} = \{\, x \in L \mid x \text{ is a lower bound of } A \,\}.$$

Prove that this sets up a Galois connection and therefore, $(A^{\mathrm{ub}})^{\mathrm{lb}} \supseteq A$, $(A^{\mathrm{lb}})^{\mathrm{ub}} \supseteq A$, $A^{\mathrm{ub}} = ((A^{\mathrm{ub}})^{\mathrm{lb}})^{\mathrm{ub}}$, and $A^{\mathrm{lb}} = ((A^{\mathrm{lb}})^{\mathrm{ub}})^{\mathrm{lb}}$.

3.71. Call an ideal $I$ of lattice $L$ *normal* if $I = (I^{\mathrm{ub}})^{\mathrm{lb}}$. Show that every principal ideal is normal.

3.72. Is every normal ideal an intersection of principal ideals?

3.73. Let $\mathrm{Id}_{\mathrm{N}}\, L$ denote the set of all normal ideals of $L$. Show that $\mathrm{Id}_{\mathrm{N}}\, L$ is a complete lattice but that it is not necessarily a sublattice of $\mathrm{Id}_0\, L$.

3.74. Let $L$ be a lattice. Show that the map: $x \mapsto \mathrm{id}(x)$ is an embedding of $L$ into $\mathrm{Id}_{\mathrm{N}}\, L$, preserving all meets and joins that exist in $L$. (The lattice $\mathrm{Id}_{\mathrm{N}}\, L$ is called the *MacNeille completion* of $L$; see H. M. MacNeille [517].)

3.75. Generalize Theorem 33 to any Galois connection between two complete lattices.

3.76. Let $L$ be a complete lattice and let $f$ be an isotone map $f$ of $L$ into itself. Define
$$a = \bigvee (\, b \mid b \in L, \ b \le f(b)\,).$$
Prove that $f(a) = a$, verifying the Fixed Point Theorem (A. Tarski [675] and B. Knaster [475]).

3.77. Prove that if the Fixed Point Theorem holds in a lattice $L$, then $L$ is complete (A. C. Davis [134]).

*    *    *

3.78. Introduce and examine isomorphisms and homomorphisms for algebras of arbitrary types. Introduce the congruence lattice $\mathrm{Con}\,\mathfrak{A}$. Show that it is a lattice.

3.79. Describe the formula for the join of two congruences for algebras and compare it with the lattice case.

3.80. Introduce and examine $\mathrm{con}(H)$ for algebras.

3.81. Introduce and examine subalgebras for algebras. Can you describe $\mathrm{sub}(H)$?

3.82. Introduce the subalgebra lattice $\mathrm{Sub}\,\mathfrak{A}$ for algebras. Note the special treatment of $\varnothing$ to make $\mathrm{Sub}\,\mathfrak{A}$ a lattice.

3.83. Prove the Homomorphism Theorem for algebras.

3.84. Introduce direct products for algebras. Find the analogues of Exercises 3.62–3.64 for algebras.

3.85. Verify the First Isomorphism Theorem (Exercise 3.61) for algebras.

*    *    *

The reader will recall that in Exercises 2.26–2.36 we showed that the order $\mathsf{T}_n$ is a lattice for which the lattice axioms are difficult to verify. The following exercises, based on G. Grätzer, R. W. Quackenbush, and E. T. Schmidt [328], provide another such example. A more complicated variant of this construction is in G. Grätzer and H. Lakser [310] and G. Grätzer, H. Lakser, and R. W. Quackenbush [313].

Let $A$ be a finite lattice with zero and unit. Let us call $A$ *separable* if it has an element $v$ which is a *separator*, that is, $0 \prec v \prec 1$.

Let $P$ be a finite order with a family $S_p$, for $p \in P$, of separable lattices, and let $S = \prod( S_p \mid p \in P )$ with ordering $\leq_S$. Denote the zero, the unit, and a fixed separator of $S_p$ by $0_p, 1_p, v_p$, respectively for any $p \in P$. An element $\mathbf{s} \in S$ is written in the form $(\mathbf{s}_p)_{p \in P}$. For $q \in P$, let $\mathbf{u}^q \in S$ be defined by $(\mathbf{u}^q)_q = 1_q$ and $(\mathbf{u}^q)_p = 0_q$, otherwise. For $q \in P$, let $\mathbf{v}^q \in S$ be defined by $(\mathbf{v}^q)_q = v_q$ and otherwise, $(\mathbf{v}^q)_p = 0_p$. Let $B$ be the sublattice of $S$ generated by the set $\{ \mathbf{u}^p \mid p \in P \}$; it is a boolean sublattice of $S$. For a subset $Q$ of $P$, set

$$\mathbf{u}^Q = \bigvee_S \{\mathbf{u}^p \mid p \in Q\}$$

with complement$(\mathbf{u}^Q)' = \mathbf{u}^{P-Q}$ in $B$ (and in $S$).

On the set $S$, we define a binary relation $\leq$. For the elements

$$\mathbf{a} = (a_p)_{p \in P},$$
$$\mathbf{b} = (b_p)_{p \in P}$$

of $S$, let $\mathbf{a} \leq \mathbf{b}$ in $S$ (recall that $\leq_S$ is the direct product ordering in $S$) if the following two conditions hold:

(i) $\mathbf{a} \leq_S \mathbf{b}$;

(ii) if $p < p'$ in $P$ and $a_p = v_p = b_p$, then $a_{p'} = b_{p'}$.

3.86. Show that the binary relation $\leq$ is an ordering on $S$. (In Figure 15, we show the representation of the order $P = \{p, q, r\}$ with $p < q$ and $r < q$ with the lattices $S_p = S_q = S_r = \mathsf{C}_3 = \{0, v, 1\}$. Some edges of $\mathsf{C}_3^3$ are missing in $L$; on the diagram these are marked with dashed lines.)

*3.87. In the order $L = (S; \leq)$, let $\mathbf{a} = (a_p)_{p \in P}$, $\mathbf{b} = (b_p)_{p \in P} \in L$, and let $q \in P$; we shall call $q$ an $\{\mathbf{a}, \mathbf{b}\}$-*fork*, if $a_q = b_q = v$ and $a_{q'} \neq b_{q'}$ for some $q' > q$. Let $\mathbf{a} = (a_p)_{p \in P}$, $\mathbf{b} = (b_p)_{p \in P} \in L$. Define $\mathbf{a} + \mathbf{b} \in S$ by

$$(\mathbf{a} + \mathbf{b})_p = \begin{cases} 1_p, & \text{if } a_p \vee b_p = v \text{ and, for some } p' \geq p, \\ & \quad (1) \ p' \text{ is an } \{\mathbf{a}, \mathbf{b}\}\text{-fork, or} \\ & \quad (2) \ b_p \leq a_p \text{ and } b_{p'} \not\leq a_{p'}, \text{ or} \\ & \quad (3) \ a_p \leq b_p \text{ and } a_{p'} \not\leq b_{p'}; \\ a_p \vee b_p, & \text{otherwise.} \end{cases}$$

Prove that $\mathbf{a} + \mathbf{b} = \sup\{a, b\}$ in $L$, and therefore, $L$ is a join-semilattice.

3.88. Prove that $L = (S; \leq)$ is a lattice.

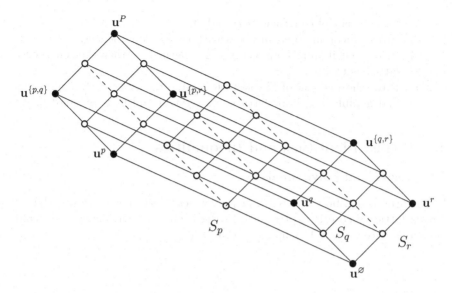

Figure 15. A small example of the construction for Exercises 3.86–3.89

3.89. What is the formula for $\mathbf{a} \wedge \mathbf{b}$?

*        *        *

The *characteristic* $\mathfrak{m}$ of the infinitary universal algebra $(A; F)$ of type $(\delta_0, \delta_1, \ldots, \delta_\gamma, \ldots)$, for $\gamma < \nu$, is the smallest regular cardinal $\mathfrak{m}$ such that $\delta_\gamma < \mathfrak{m}$, for every $\gamma < \nu$. (A cardinal $\mathfrak{m}$ is called regular if any family of fewer than $\mathfrak{m}$ cardinals, each of which is less than $\mathfrak{m}$, has a sum less than $\mathfrak{m}$; for instance $\aleph_0$ is regular).

3.90. Define subalgebras and the lattice of subalgebras for (infinitary) algebras. Define $\mathfrak{m}$-*algebraic lattices* so that the subalgebra lattice of an algebra of characteristic $\mathfrak{m}$ is $\mathfrak{m}$-algebraic.

3.91. Generalize the Birkhoff-Frink Theorem to algebras of characteristic $\mathfrak{m}$ (G. Grätzer [252]).

*3.92. Generalize Theorem 50 to algebras of characteristic $\mathfrak{m}$. (See Appendix 7 by the author and W. A. Lampe in G. Grätzer [263].)

*        *        *

The algebra $(A; F)$ has *type 3 congruences* (see also Section V.4.2) if

$$\alpha \vee \beta = \alpha \circ \beta \circ \alpha \circ \beta$$

whenever $\alpha, \beta \in \mathrm{Con}(A; F)$. Type 2 means that

$$\alpha \vee \beta = \alpha \circ \beta \circ \alpha$$

for any pair of congruences $\alpha$ and $\beta$.

3.93. Prove Theorem 50 for algebras with type 3 congruences.

3.94. Prove that if an algebra has type 2 congruences, then the congruence lattice is modular.

*3.95. Prove the analogue of Theorem 50 for algebras with type 2 congruences and modular algebraic lattices (G. Grätzer and E. T. Schmidt [338]).

# 4.  Terms, Identities, and Inequalities

## 4.1  Terms and polynomials

From variable symbols $x_0, x_1, \ldots, x_{n-1}$, we can form lattice terms, that is, formal lattice-theoretic expressions, in the usual manner, using $\vee$, $\wedge$, and, of course, parentheses. Examples of terms are:

$$x_0$$
$$x_3$$
$$x_0 \vee x_0$$
$$(x_0 \wedge x_2) \vee (x_3 \wedge x_0)$$
$$(x_0 \wedge x_1) \vee ((x_0 \vee x_2) \wedge (x_1 \vee x_2))$$

A formal definition is:

**Definition 52.** The set $\mathrm{Term}(n)$ of *n-ary lattice terms* is the smallest set satisfying (i) and (ii):

(i) $x_i \in \mathrm{Term}(n)$ for $0 \leq i < n$;
(ii) if $p, q \in \mathrm{Term}(n)$, then $(p \vee q), (p \wedge q) \in \mathrm{Term}(n)$.

A *term* is an *n*-ary term for some *n*; this *n* refers to the number of variable symbols used, as specified in (i).

*Remark.* We shall omit the outside parentheses and also the internal parentheses in iterated meets and iterated joins; so we write $p_1 \vee p_2 \vee \cdots \vee p_n$ for $(\cdots (p_1 \vee p_2) \vee \cdots \vee p_n)$, and the same for $\wedge$. Thus we write $x_0 \vee x_1$ for $(x_0 \vee x_1)$ and $x_0 \vee x_1 \vee x_2$ for $((x_0 \vee x_1) \vee x_2)$. Note that if $n \leq m$, then $\mathrm{Term}(n) \subseteq \mathrm{Term}(m)$.

In $p_1 \vee p_2 \vee \cdots \vee p_n$, the $p_i$-s are the *joinands*; we define *meetands* dually.

It is easy to define formally what it means to substitute $x_j$ for all occurrences of $x_i$ in a term $p$, for instance, substituting $x_4$ for $x_1$, the term $p = (x_1 \vee x_2) \wedge (x_1 \wedge x_3)$ becomes $q = (x_4 \vee x_2) \wedge (x_4 \wedge x_3)$; observe that $\mathbf{p} \in \mathrm{Term}(4)$ but $\mathbf{q} \notin \mathrm{Term}(4)$. In general, for $p = p(x_0, \ldots, x_{n-1}) \in \mathrm{Term}(n)$, the substituted term $p(x_{i_0}, \ldots, x_{i_{n-1}})$ is also in $\mathrm{Term}(m)$ for every $m > i_0, \ldots, i_{n-1}$.

In many publications, terms are called *words* or *polynomials* We use "polynomial" for a slightly different concept.

By Definition 52, a term is just a sequence of symbols. However, using such a sequence of symbols, we can define a function on *any* lattice:

**Definition 53.** An $n$-ary term $p$ defines a function $p$ in $n$ variables, called a *term function*, on a lattice $L$ by the following rules ($a_0, \ldots, a_{n-1} \in L$):

(i) If $p = x_i$, then $p(a_0, \ldots, a_{n-1}) = a_i$ for any $0 \le i < n$.

(ii) If $p(a_0, \ldots, a_{n-1}) = a$, $q(a_0, \ldots, a_{n-1}) = b$, and $p \vee q = r$, $p \wedge q = t$, then $r(a_0, \ldots, a_{n-1}) = a \vee b$ and $t(a_0, \ldots, a_{n-1}) = a \wedge b$.

Thus if $p = (x_0 \wedge x_1) \vee (x_2 \vee x_1)$, then $p(a_0, a_1, a_2) = (a_0 \wedge a_1) \vee (a_2 \vee a_1) = a_1 \vee a_2$.

To simplify our notation, we shall not use bold symbols for terms and variables. This blurs the distinction between terms and term functions, but there is little danger of confusion. We shall also use the variables $x, y, z, \ldots$.

We also need a larger class of functions over a lattice $L$, which we obtain by substituting elements of $L$ for variables into term functions. We call these functions *polynomials* over $L$.

More formally, let $p$ be an $n$-ary term and let $p$ be the associated term function. For a lattice $L$ for $0 \le j < n$, and for an element $b \in L$, we denote by $q$ the expression

$$q = p(x_0, \ldots, x_{j-1}, b, x_{j+1}, \ldots, x_{n-1})$$

we obtain from $p$ by substituting $b$ for all occurrences of $x_j$. Now if

$$a_0, \ldots, a_{j-1}, a_{j+1}, \ldots, a_{n-1} \in L,$$

then by Definition 53,

$$p(a_0, \ldots, a_{j-1}, b, a_{j+1}, \ldots, a_{n-1})$$

is well defined, so $q$ is a function on $L$ in the variables

$$x_0, \ldots, x_{j-1}, x_{j+1}, \ldots, x_{n-1}.$$

We call $q$ a *polynomial* over $L$. Proceeding thus, substituting more elements of $L$ for variables, we obtain all the *polynomials* over $L$.

Examples of polynomials over a lattice $L$ with element $a$ and $b$ are:

$$x$$

$$a$$

$$x \vee a$$

$$(x \wedge a) \vee (y \wedge b)$$

$$(a \wedge b) \vee ((x \vee y) \wedge (a \vee z))$$

The analogous concept of a polynomial over a ring $R$ is just the usual concept of a ring polynomial with coefficients from $R$.

These definitions are quite formal but their meaning is very straightforward.

We shall prove statements on terms by induction on $\mathrm{rank}(p)$, the *rank of a term p*. The rank of $x_i$ is 1; the rank of $p \vee q$ and of $p \wedge q$ is the sum of the ranks of $p$ and $q$.

Now we are in a position to describe when an element belongs to $\mathrm{sub}(H)$, the sublattice generated by $H \neq \varnothing$:

**Lemma 54.** $a \in \mathrm{sub}(H)$ *iff* $a = p(h_0, \ldots, h_{n-1})$, *for some integer* $n \geq 1$, *some n-ary term p, and some* $h_0, \ldots, h_{n-1} \in H$.

*Proof.* First we must show that if $a = p(h_0, \ldots, h_{n-1})$, for $h_0, \ldots, h_{n-1} \in H$, then $a \in \mathrm{sub}(H)$, which can be easily accomplished by induction on the rank of $p$. Then we form the set

$$\{\, a \mid a = p(h_0, \ldots, h_{n-1}),\ n \geq 1,\ h_0, \ldots, h_{n-1} \in H,\ p \in \mathrm{Term}(n) \,\}.$$

Observe that this set contains $H$ and that it is closed under $\vee$ and $\wedge$. Since it is contained in $\mathrm{sub}(H)$, it has to equal $\mathrm{sub}(H)$. $\square$

**Corollary 55.** $|\mathrm{sub}(H)| \leq |H| + \aleph_0$.

*Proof.* By Lemma 54, every element of $\mathrm{sub}(H)$ can be associated with a finite sequence of elements of $H \cup \{(, ), \vee, \wedge\}$, and there are no more than $|H| + \aleph_0$ such sequences. $\square$

### 4.2    Identities and inequalities

We start with the basic definition.

**Definition 56.** A *lattice identity* (resp., *inequality*) is an expression of the form $p = q$ (resp., $p \leq q$), where $p$ and $q$ are lattice terms.

An *identity*, $p = q$ (resp., *inequality*, $p \leq q$), *holds in the lattice* $L$ if $p(a_0, \ldots, a_{n-1}) = q(a_0, \ldots, a_{n-1})$ (resp., $p(a_0, \ldots, a_{n-1}) \leq q(a_0, \ldots, a_{n-1})$) holds for every $a_0, \ldots, a_{n-1} \in L$.

Section 1.10 provides a number of examples of identities: the eight identities defining lattices, the two distributive identities, and the modular identity.

*Remark.* For the concept of identity, equation and equality are also used in the literature. We shall use identity and for us *equality* simply means that two things, mostly elements, are equal. So in $B_2 = \{0, a, b, 1\}$, see Figure 4, $a \vee b = 1$ is an equality, while distributivity is defined by two identities:

$$x \wedge (y \vee z) = (x \wedge y) \vee (x \wedge z),$$
$$x \vee (y \wedge z) = (x \vee y) \wedge (x \vee z).$$

Unfortunately, we do not have such naming variants for inequality. So we may say that the inequality $a \vee b \leq 1$ holds in $\mathsf{B}_2$, comparing two elements as in an equality, or we may say that the inequality

$$x \wedge (y \vee z) \leq (x \wedge y) \vee (x \wedge z)$$

holds in $\mathsf{B}_2$, meaning that it holds for all $x, y, z \in \mathsf{B}_2$, just as for an identity. It will always be clear from the context which one we mean.

An identity $p = q$ is equivalent to the two inequalities: $p \leq q$ and $q \leq p$, and the inequality $p \leq q$ is equivalent to the identity $p \vee q = q$ (and to $p \wedge q = p$). Frequently, the validity of an identity is shown by verifying that these two inequalities hold.

One of the most basic properties of term functions is:

**Lemma 57.** *A term function $p$ is* isotone *in each variable. Furthermore,*

$$x_0 \wedge \cdots \wedge x_{n-1} \leq p(x_0, \ldots, x_{n-1}) \leq x_0 \vee \cdots \vee x_{n-1}.$$

*Proof.* We prove the first statement by induction on the rank of $p$. It is certainly true for $p = x_i$. Suppose that it is true for $q$ and $r$ and that

$$p(x_0, \ldots, x_{n-1}) = q(x_0, \ldots, x_{n-1}) \vee r(x_0, \ldots, x_{n-1}).$$

Then for elements $a_0 \leq b_0, \ldots, a_{n-1} \leq b_{n-1}$ in a lattice $L$, compute:

$$
\begin{aligned}
&p(a_0, \ldots, a_{n-1}) \vee p(b_0, \ldots, b_{n-1}) \\
&= (q(a_0, \ldots, a_{n-1}) \vee r(a_0, \ldots, a_{n-1})) \vee (q(b_0, \ldots, b_{n-1}) \vee r(b_0, \ldots, b_{n-1})) \\
&= (q(a_0, \ldots, a_{n-1}) \vee q(b_0, \ldots, b_{n-1})) \vee (r(a_0, \ldots, a_{n-1}) \vee r(b_0, \ldots, b_{n-1})) \\
&= q(b_0, \ldots, b_{n-1}) \vee r(b_0, \ldots, b_{n-1}) = p(b_0, \ldots, b_{n-1});
\end{aligned}
$$

thus $p(a_0, \ldots, a_{n-1}) \leq p(b_0, \ldots, b_{n-1})$. The proof is dual for $p = q \wedge r$.

Since

$$x_0 \wedge \cdots \wedge x_{n-1} \leq x_i \leq x_0 \vee \cdots \vee x_{n-1},$$

for $0 \leq i \leq n - 1$, using the idempotency of $\vee$ and $\wedge$, we obtain:

$$
\begin{aligned}
x_0 \wedge \cdots \wedge x_{n-1} = p(x_0 \wedge \cdots \wedge x_{n-1}, \ldots, x_0 \wedge \cdots \wedge x_{n-1}) &\leq p(x_0, \ldots, x_{n-1}) \\
\leq p(x_0 \vee \cdots \vee x_{n-1}, \ldots, x_0 \vee \cdots \vee x_{n-1}) &= x_0 \vee \cdots \vee x_{n-1}
\end{aligned}
$$

proving the second statement.     $\square$

A simple application is:

**Lemma 58.** *Let $p_i = q_i$, for all $0 \leq i < k$, be lattice identities. Then there is a single identity $p = q$ such that all $p_i = q_i$, for all $0 \leq i < k$, hold in a lattice $L$ iff $p = q$ holds in $L$.*

*Proof.* Let us take two identities, $p_0 = q_0$ and $p_1 = q_1$. Suppose that all terms are $n$-ary and consider the *merged identity* (we use now $2n$-ary terms formed by substitution, see Remark to Definition 52):

$$p_0(x_0, \ldots, x_{n-1}) \wedge p_1(x_n, \ldots, x_{2n-1})$$
$$= q_0(x_0, \ldots, x_{n-1}) \wedge q_1(x_n, \ldots, x_{2n-1}).$$

It is obvious that if $p_0 = q_0$ and $p_1 = q_1$ hold in $L$, then the merged identity holds in $L$. Now let the merged identity hold in $L$ and let $a_0, \ldots, a_{n-1} \in L$. Substitute $x_0 = a_0, \ldots, x_{n-1} = a_{n-1}, x_n = \cdots = x_{2n-1} = a_0 \vee \cdots \vee a_{n-1} = a$ in the merged identity. By Lemma 57,

$$p_0(a_0, \ldots, a_{n-1}) \leq p_0(a, \ldots, a) = a = p_1(a, \ldots, a),$$

whence $p_0(a_0, \ldots, a_{n-1}) = p_0(a_0, \ldots, a_{n-1}) \wedge p_1(a, \ldots, a)$, and similarly for $q_0$ and $q_1$; thus the merged identity yields $p_0(a_0, \ldots, a_{n-1}) = q_0(a_0, \ldots, a_{n-1})$. The second identity is derived similarly. The proof for $k$ identities is similar.    $\square$

The most important (and, in fact, characteristic) properties of identities are given by

**Lemma 59.** *Identities are preserved under the formation of sublattices, homomorphic images, direct products, and ideal lattices.*

*Remark.* More formally, if $L$ is a lattice, then every identity $p = q$ satisfied by $L$ is also satisfied by every sublattice of $L$, by every homomorphic image of $L$, and by the ideal lattice $\operatorname{Id} L$; and if $(L_i \mid i \in I)$ is a (finite or infinite) family of lattices all satisfying an identity $p = q$, then $\prod(L_i \mid i \in I)$ also satisfies the identity $p = q$.

*Proof.* Let the terms $p$ and $q$ both be $n$-ary and let $p = q$ hold in $L$. If $L_1 \leq L$, then $p = q$ obviously holds in $L_1$. Let $\varphi \colon L \to K$ be an onto homomorphism. A simple induction shows that

$$\varphi(p(a_0, \ldots, a_{n-1})) = p(\varphi(a_0), \ldots, \varphi(a_{n-1})),$$

and the similar formula for $q$. Therefore,

$$p(\varphi(a_0), \ldots, \varphi(a_{n-1})) = \varphi(p(a_0, \ldots, a_{n-1}))$$
$$= \varphi(q(a_0, \ldots, a_{n-1})) = q(\varphi(a_0), \ldots, \varphi(a_{n-1})),$$

and so $p = q$ holds in $K$. The statement for direct products is also obvious.

The last statement is an easy corollary of the following formula. Let $p$ be an $n$-ary term and let $I_0, \ldots, I_{n-1}$ be ideals of $L$. Then $I_0, \ldots, I_{n-1} \in \operatorname{Id} L$;

thus we can substitute the $I_j$ into $p$: $p(I_0, \ldots, I_{n-1})$ is also in Id $L$, that is, an ideal of $L$. This ideal can be described by a simple formula:

$$p(I_0, \ldots, I_{n-1})$$
$$= \{ x \in L \mid x \le p(i_0, \ldots, i_{n-1}), \text{ for some } i_0 \in I_0, \ldots, i_{n-1} \in I_{n-1} \}.$$

This follows easily by induction on the rank of $p$ from Lemma 57 and the formula in Section 3.4 describing $I \vee J$. This formula shows that if $p = q$ holds in $L$, then it holds in Id $L$. □

See another result on preserving identities at the end of the next section.

## 4.3 Distributivity and modularity

Now we list a few important inequalities:

**Lemma 60.** *The following inequalities hold in any lattice:*

(i) $$(x \wedge y) \vee (x \wedge z) \le x \wedge (y \vee z),$$

(ii) $$x \vee (y \wedge z) \le (x \vee y) \wedge (x \vee z),$$

(iii) $$(x \wedge y) \vee (y \wedge z) \vee (z \wedge x) \le (x \vee y) \wedge (y \vee z) \wedge (z \vee x),$$

(iv) $$(x \wedge y) \vee (x \wedge z) \le x \wedge (y \vee (x \wedge z)).$$

*Remark.* (i)–(iii) are called *distributive inequalities*, and (iv) is the *modular inequality*.

*Proof.* Note that in each of (i)-(iv), the left-hand side is a join and the right-hand side is a meet. Now for elements $a, b, c$ of a lattice, $a \vee b \le c$ holds iff $a \le c$ and $b \le c$ (by the characterization of $a \vee b$ as the least upper bound of $a$ and $b$); and similarly $a \le b \wedge c$ iff $a \le b$ and $a \le c$. Hence to verify each of (i)-(iv), we need only verify that each joinand on the left is $\le$ each meetand on the right.

These inequalities are all easy to verify. For instance, in (i), the first joinand is $\le$ the last meetand because $x \wedge y \le y \le y \vee z$.

We now prove (iv) and leave the rest to the reader. Since $x \wedge y \le x$ and $x \wedge z \le x$, we get

$$(x \wedge y) \vee (x \wedge z) \le x.$$

Moreover, $x \wedge y \le y \le y \vee (x \wedge z)$ and $x \wedge z \le y \vee (x \wedge z)$, therefore,

$$(x \wedge y) \vee (x \wedge z) \le y \vee (x \wedge z).$$

Meeting the two displayed inequalities, we obtain (iv). □

**Lemma 61.** *Consider the following two identities and the inequality:*

(i) $$x \wedge (y \vee z) = (x \wedge y) \vee (x \wedge z),$$

(ii) $$x \vee (y \wedge z) = (x \vee y) \wedge (x \vee z),$$

(iii) $$(x \vee y) \wedge z \leq x \vee (y \wedge z).$$

(i), (ii), *and* (iii) *are equivalent in any lattice* $L$.

*Remark.* The identities (i) and (ii) are called the *distributive identities*. A lattice satisfying a distributive identity is called *distributive*. As we noted before, the class of distributive lattices is denoted by $\mathbf{D}$.

Note that (i) and (ii) are *not* equivalent for fixed elements; that is, (i) can hold for three elements $a, b, c$ of a lattice $L$, while (ii) does not.

*Proof.* Let (i) hold in $L$, and let $a, b, c \in L$; then, using (i) with $x = a \vee b$, $y = a, z = c$, we obtain that

$$(a \vee b) \wedge (a \vee c) = ((a \vee b) \wedge a) \vee ((a \vee b) \wedge c).$$

Now the first joinand on the right-hand side simplifies to $a$ by absorption, while we can apply (i) (with the order of meetands reversed) to the second; thus the above expression equals

$$a \vee (a \wedge c) \vee (b \wedge c) = a \vee (b \wedge c),$$

verifying (ii).

That (ii) implies (i) follows by duality.

Let (i) hold in $L$; then

$$x \vee (y \wedge z) = (x \vee y) \wedge (x \vee z) \geq (x \vee y) \wedge z,$$

since $x \vee z \geq z$, verifying (iii).

Let (iii) hold in $L$. Then with $x = a, y = b, z = a \vee c$ in (iii), we obtain that

$$(a \vee b) \wedge (a \vee c) \leq a \vee (b \wedge (a \vee c)) = a \vee ((a \vee c) \wedge b).$$

We now apply (iii) to the last joinand in this formula, and conclude that this expression is

$$\leq a \vee (a \vee (c \wedge b)) = a \vee (c \wedge b).$$

This, combined with Lemma 60(ii), gives (ii).    □

**Corollary 62.** *The dual of a distributive lattice is distributive.*

**Lemma 63.** *The identity*

$$(x \wedge y) \vee (x \wedge z) = x \wedge (y \vee (x \wedge z))$$

*is equivalent to the condition*

$$x \geq z \text{ implies that } (x \wedge y) \vee z = x \wedge (y \vee z).$$

*Remark.* The identity of this lemma is called the *modular identity*. A lattice satisfying the modular identity is called *modular.* As we noted before, the class of modular lattices is denoted by **M**.

*Proof.* If $x \geq z$, then $z = x \wedge z$; thus the implication follows from the identity. Conversely, if the implication holds, then since $x \geq x \wedge z$, we have

$$(x \wedge y) \vee (x \wedge z) = x \wedge (y \vee (x \wedge z)).$$ □

Since the condition in Lemma 63 is selfdual, we conclude that the dual of the identity in Lemma 63, namely,

$$(x \vee y) \wedge (x \vee z) = x \vee (y \wedge (x \vee z))$$

also defines modularity.

**Corollary 64.** *The dual of a modular lattice is modular.*

The classes **D** and **M** of lattices are examples of *varieties*, classes of lattices defined by identities. We introduce varieties in Section 5.1 and discuss them in depth in Chapter VI.

M. Wild [728] proves a surprising preservation of identities. Let $\mathfrak{A}$ and $\mathfrak{B}$ be algebras of the same type, and let $\varphi$ be a homomorphism of $\mathfrak{A}$ onto $\mathfrak{B}$. Then if Sub $\mathfrak{A}$ satisfies a so called meet-week identity (for instance, distributivity, modularity), then Sub $\mathfrak{B}$ satisfies the same identity.

**Exercises**

4.1. Give a formal definition of substituting $x_j$ for all occurrences of $x_i$ in a term **p**. Prove formally the statement of the last sentence of the Remark to Definition 52 and the statement of the first sentence of the proof of Lemma 54.

4.2. Let $H = \{h_0, \ldots, h_{n-1}\}$. Prove that $a \in \text{sub}(H)$ iff there exists an $n$-ary term $p$ with $a = p(h_0, \ldots, h_{n-1})$.

4.3. Show that the upper bound in Corollary 55 is best possible if $|H| \geq 3$. Give the best estimates for $|H| \leq 2$. (Recall Exercise 3.8.)

4.4. Give a more detailed proof of Lemma 59.

4.5. Prove (without reference to Lemma 59) that if $L$ is distributive (modular), then so is Id $L$.

4.6. Show that the dual of a modular lattice is modular.

4.7. Prove that $L$ is distributive iff the identity

$$(x \wedge y) \vee (y \wedge z) \vee (z \wedge x) = (x \vee y) \wedge (y \vee z) \wedge (z \vee x)$$

holds in $L$.

4.8. Prove that every distributive lattice is modular, but not conversely. Find the smallest modular but nondistributive lattice.

4.9. Find an identity $p = q$ characterizing distributive lattices such that neither $p \le q$ nor $q \le p$ holds in a general lattice.

4.10. Show that in any lattice

$$\bigvee(\bigwedge(x_{ij} \mid i < m) \mid j < n) \le \bigwedge(\bigvee(x_{ij} \mid j < n) \mid i < m).$$

4.11. Prove that the following identity holds in a distributive lattice:

$$\bigvee(\bigwedge(x_{ij} \mid j < n) \mid i < m) = \bigwedge(\bigvee(x_{i\,f(i)} \mid i < m) \mid f \in F),$$

where $F$ is the set of all functions

$$f: \{0, 1, \ldots, m-1\} \to \{0, 1, \ldots, n-1\}.$$

4.12. Derive (i)–(iii) of Lemma 60 from Exercise 4.10.

4.13. Verify that any chain is a distributive lattice.

4.14. Let $L$ be a lattice with more than one element, and let $L' = \mathsf{C}_1 + L$. Show that $L'$ is then a lattice and that an identity $p = q$ holds in $L$ iff it holds in $L'$.

4.15. Generalize Exercise 4.14 to the sum and the glued sum of two lattices.

4.16. Show that if the identity $x_0 = p(x_0, \ldots, x_{n-1})$ holds in $\mathsf{C}_2$, then it holds in every lattice.

4.17. Prove that $\operatorname{Pow} X$ is a distributive lattice.

4.18. Show that $\operatorname{Part} X$ is distributive iff $|X| \le 2$ and modular iff $|X| \le 3$. (See Exercise 3.59.)

4.19. Find an identity that holds in $\mathsf{C}_2$ but not in $\mathsf{N}_5$.

4.20. Find an identity that holds in $\mathsf{M}_3$ but not in $\mathsf{N}_5$.

4.21. Examine the statements of Lemma 59 for properties of the form, "If $p_0 = q_0$, then $p_1 = q_1$."

4.22. Show that $(L; \vee, \wedge)$ is a lattice satisfying the identity $p = q$ iff it satisfies the identities

$$(w \wedge x) \vee x = x,$$

$$(((x \wedge p) \wedge z) \vee u) \vee v = (((q \wedge z) \wedge x) \vee v) \vee ((t \vee u) \wedge u),$$

where $x, z, u, v, w$ are variables that do not occur in $p$ or $q$ (R. Padmanabhan [566]). (Hint: use Exercise 1.33.)

*4.23. Show that the result of Exercise 4.22 is the best possible: If $p = q$ is an identity such that it is *not* satisfied by some lattice, then the two identities of Exercise 4.22 cannot be replaced by one (R. N. McKenzie [511]).

4.24. Show that the lattice $L$ is modular iff the inequality

$$x \wedge (y \vee z) \le y \vee ((x \vee y) \wedge z)$$

holds in $L$.

$$* \quad * \quad *$$

4.25. For algebras of a fixed type, introduce terms, term functions, and polynomials.

4.26. Prove Lemma 54 for algebras.

4.27. Rephrase and prove Corollary 55 for algebras of a fixed type.

# 5.  Free Lattices

## 5.1   The formal definition

Though it is quite easy to develop a feeling for the most general lattice (generated by a set of elements and satisfying some relations) of Section 2.3 by way of some examples, a general definition seems hard to formulate. So we ask the reader to withhold judgment on whether Definition 66 expresses any intuitive feelings until the theory is developed and we present further examples.

The most general lattice will be called *free*. (In group theory, the terminology "groups presented by generators and relations" is used.) We shall be interested, for example, not only in the most general *lattice* generated by $a, b, c$ and satisfying $b \le a$, but also in the most general *distributive lattice* generated by $a, b, c$ and satisfying $b \le a$, and the most general such lattice in other classes of lattices defined by identities. Let us name such classes.

**Definition 65.** Let $p_i = q_i$ be identities for $i \in I$. The class **K** of all lattices satisfying the identities $p_i = q_i$, for $i \in I$, is called a *variety of lattices*. A variety is *trivial* if it contains one-element lattices only.

*Remark.* Varieties are also called *equational classes*. By Lemma 59, varieties are closed under the formation of sublattices, homomorphic images, and direct products. For the converse statement, see Theorem 469.

The class **L** of all lattices, the class **D** of all distributive lattices, and the class **M** of all modular lattices are examples of varieties of lattices.

Next we have to agree on what kinds of relations to allow in a generating set. Can we prescribe only relations of the form $b \le a$, or do we allow relations of the form $a \wedge b = c$ or $d \vee e = f$? Lemma 73 (below) can be rephrased to furnish an example in which the four generators $a, b, c, d$ are required to satisfy $a \vee b = a \vee c = b \vee c = d$, showing the usefulness of relations of the form $a \vee b = d$. Let us, therefore, agree that for a generating set we take an order $P$, and for relations we take all $a \le b$ that hold in $P$, all $a \wedge b = c$, where $\inf\{a, b\} = c$ in $P$, and all $a \vee b = c$, where $\sup\{a, b\} = c$ in $P$. (A more liberal approach will be presented later in this section.)

Let $P$ be an order and let $N$ be a lattice. In this section, a map $\varphi \colon P \to N$ will be called a *homomorphism* if it is an isotone map with the following two properties for all $a, b, c \in P$:

(1) if $\sup\{a, b\} = c$ in $P$, then $\varphi(a) \vee \varphi(b) = \varphi(c)$ in $L$;

(2) if $\inf\{a, b\} = c$ in $P$, then $\varphi(a) \wedge \varphi(b) = \varphi(c)$ in $L$.

If $P \subseteq N$, then the identity map $\varepsilon \colon x \mapsto x$ is called the *inclusion map*.
    Now we are ready to formulate our basic concept.

**Definition 66.** Let $P$ be an order and let $\mathbf{K}$ be a variety of lattices. A lattice $F$ is called a *free lattice* over $\mathbf{K}$ *generated by* $P$ if the following conditions are satisfied:

(i) $F \in \mathbf{K}$.

(ii) $P \subseteq F$ and the inclusion map is a homomorphism.

(iii) $P$ generates $F$.

(iv) Let $L \in \mathbf{K}$ and let $\varphi \colon P \to L$ be a homomorphism. Then there exists a (lattice) homomorphism $\psi \colon F \to L$ extending $\varphi$ (that is, satisfying $\varphi(a) = \psi(a)$ for all $a \in P$).

If a free lattice $F$ over $\mathbf{K}$ generated by $P$ exists, we denote it by $\mathrm{Free}_{\mathbf{K}}\, P$.

    Let $\varepsilon$ denote the inclusion map on $P$; then the crucial condition (iv) can be expressed by Figure 16. In that and in all similar *commutative diagrams*, the capital letters represent lattices or orders, and the arrows indicate homomorphisms (solid arrows denote maps that are given in the hypothesis and dashed arrows denote maps whose existence is asserted in the conclusion), or maps with certain properties, so that the maps compose as indicated; in this case, $\psi\varepsilon = \varphi$, which is condition (iv).
    If $P$ is an unordered set and $|P| = \mathfrak{m}$, we shall write $\mathrm{Free}_{\mathbf{K}}(\mathfrak{m})$ for $\mathrm{Free}_{\mathbf{K}}\, P$ and call it a *free lattice on* $\mathfrak{m}$ *generators over* $\mathbf{K}$. In case $\mathbf{K} = \mathbf{L}$, we may omit "over $\mathbf{L}$"; thus "*free lattice generated by* $P$" shall mean "free lattice over $\mathbf{L}$ generated by $P$", or, in notation, $\mathrm{Free}\, P$; similarly, we write $\mathrm{Free}(\mathfrak{m})$.

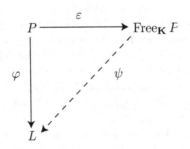

Figure 16. Property (iv) illustrated

It should be noted that if $b \in \text{Free}_\mathbf{K} P$, then by (iii) and by Lemma 54, $b = p(a_0, \ldots, a_{n-1})$, where $p$ is a term and $a_0, \ldots, a_{n-1} \in P$. Given a map $\psi$ as in (iv) (in fact, if $\varphi$ is any map of $P$ into $L$ and if $\psi \colon \text{Free}_\mathbf{K} P \to L$ is any homomorphism extending $\varphi$), then we must have

$$\psi(b) = \psi(p(a_0, \ldots, a_{n-1})) \qquad \text{(and since } \psi \text{ is a homomorphism)}$$
$$= p(\psi(a_0), \ldots, \psi(a_{n-1})) = p(\varphi(a_0), \ldots, \varphi(a_{n-1})),$$

since $\psi(a_i) = (\psi\varepsilon)(a_i) = \varphi(a_i)$. From this we conclude that there is at most one homomorphism $\psi \colon \text{Free}_\mathbf{K} P \to L$ extending $\varphi$, whence:

**Corollary 67.** *The homomorphism $\psi$ in condition* (iv) *is unique.*

This corollary is used to prove:

**Corollary 68.** *Let both* $\text{Free}_\mathbf{K} P$ *and* $\text{Free}_\mathbf{K}^* P$ *satisfy the conditions of Definition 66. Then there exists an isomorphism* $\chi \colon \text{Free}_\mathbf{K} P \to \text{Free}_\mathbf{K}^* P$, *and* $\chi$ *can be chosen so that* $\chi(a) = a$ *for all* $a \in P$. *In other words, free lattices* (*over* $\mathbf{K}$ *generated by* $P$) *are unique up to isomorphism if they exist.*

*Proof.* Let us use Figure 16 with $L = \text{Free}_\mathbf{K}^* P$ and $\varphi = \varepsilon$. Then there exist $\psi_1 \colon \text{Free}_\mathbf{K} P \to \text{Free}_\mathbf{K}^* P$ and $\psi_2 \colon \text{Free}_\mathbf{K}^* P \to \text{Free}_\mathbf{K} P$ such that $\psi_1\varepsilon = \varepsilon$ and $\psi_2\varepsilon = \varepsilon$. Thus $\psi_1\psi_2 \colon \text{Free}_\mathbf{K} P \to \text{Free}_\mathbf{K} P$ is the identity map $\varepsilon$ on $P$. By the statement preceding Corollary 67, $\varepsilon$ has a unique extension to a homomorphism $\text{Free}_\mathbf{K} P \to \text{Free}_\mathbf{K} P$; the identity map on $\text{Free}_\mathbf{K} P$ is one such extension. Therefore, $\psi_2\psi_1$ is the identity map on $\text{Free}_\mathbf{K} P$. Similarly, $\psi_1\psi_2$ is the identity map on $\text{Free}_\mathbf{K}^* P$, and so (see Exercise 3.2) the map $\psi_1$ is the required isomorphism. □

## 5.2 Existence

The previous result settles the uniqueness. How about existence? Naturally, in general, $\text{Free}_\mathbf{K} P$ need not exist. For instance, $\text{Free}_\mathbf{D} N_5$ should be $N_5$, since $\text{Free}_\mathbf{D} P = P$ if $P$ is a lattice, by (ii) and (iii) of Definition 66, but $N_5 \notin \mathbf{D}$, so (i) is violated.

**Theorem 69.** *Let $P$ be an order and let $\mathbf{K}$ be a variety of lattices. Then* $\text{Free}_\mathbf{K} P$ *exists iff the following condition is satisfied:*

(Ex) *There exists a lattice $L$ in $\mathbf{K}$ such that $P \subseteq L$ and the inclusion map is a homomorphism.*

*Proof.* Condition (Ex) is obviously necessary for the existence of $\text{Free}_\mathbf{K} P$; indeed, if $\text{Free}_\mathbf{K} P$ exists, (Ex) can always be satisfied with $L = \text{Free}_\mathbf{K} P$ by (i) and (ii) of Definition 66.

Now assume that (Ex) is satisfied. Obviously, Definition 66(iv) holds iff it holds under the additional assumption $L = \text{sub}(\varphi(P))$.

Let $(L, \varphi)$ denote this situation, that is, $L \in \mathbf{K}$, the map $\varphi \colon P \to L$ is a homomorphism, and $L = \mathrm{sub}(\varphi(P))$. Then $\mathrm{Free}_\mathbf{K}\, P$ or, more precisely $(\mathrm{Free}_\mathbf{K}\, P, \varepsilon)$, has the property that, for every $(L, \varphi)$, there exists a $\psi \colon \mathrm{Free}_\mathbf{K}\, P \to L$ with $\varphi = \psi \varepsilon$. To construct $\mathrm{Free}_\mathbf{K}\, P$, we have to construct a lattice having this property for *all* $(L, \varphi)$.

How would we construct such a lattice for two pairs $(L, \varphi)$-s?

Let $(L_1, \varphi_1)$ and $(L_2, \varphi_2)$ be given. Form $L_1 \times L_2$ and define a map

$$\varphi \colon P \to L_1 \times L_2$$

by $\varphi(p) = (\varphi_1(p), \varphi_2(p))$; set $L = \mathrm{sub}(\varphi(P))$. The fact that $\varphi$ is a homomorphism is easy to check. A simple example is illustrated in Figure 17. Now we define $\psi_i \colon (x_1, x_2) \mapsto x_i$. Obviously, $\psi_i \varphi(p) = \varphi_i(p)$ for any $p \in P$ and $\psi_i \colon L \to L_i$.

If we are given any number of $(L_i, \varphi_i)$ for $i \in I$, we can proceed as before and get $(L, \varphi)$; if one of the $(L_i, \varphi_i)$ is the $(L, \varepsilon)$ given by (Ex), then (ii) of Definition 66 will also be satisfied. There is only one problem: All the pairs $(L_i, \varphi_i)$ do not form a set, so their direct product cannot be formed. The $(L_i, \varphi_i)$ do not form a set because a lattice and all its isomorphic copies do not form a set; therefore, if we can somehow avoid taking too many isomorphic copies, the previous procedure can be followed. Observe that, by Corollary 55, for every pair $(L_i, \varphi_i)$, the inequality

$$|L| \leq |\varphi(P)| + \aleph_0 \leq |P| + \aleph_0$$

holds. Thus by choosing a large enough set $S$ and taking only those $(L_i, \varphi_i)$ that satisfy $L_i \subseteq S$, we can solve our problem.

Now we are ready to proceed with the formal proof. Choose a set $S$ satisfying $|P| + \aleph_0 = |S|$. Let $Q$ be the set of all pairs $(M, \psi)$, where $M \subseteq S$. Form

$$A = \prod (\, M \mid (M, \psi) \in Q \,),$$

and, for each $p \in P$, let $f_p \in A$ be defined by

$$f_p((M, \psi)) = \psi(p).$$

Finally, set

$$N = \mathrm{sub}(\{\, f_p \mid p \in P \,\}).$$

We claim that if, for all $p \in P$, we identify $p$ with $f_p$, then $N$ satisfies (i)–(iv) of Definition 66, and thus $N = \mathrm{Free}_\mathbf{K}\, P$.

(i) $N$ is constructed from members of $\mathbf{K}$ by forming a direct product and by taking a sublattice. By Lemma 59, the lattice $N$ is in $\mathbf{K}$, since $\mathbf{K}$ is a variety.

(ii) Let $\sup\{a, b\} = c$ in $P$. Then $\psi(a) \vee \psi(b) = \psi(c)$, for every $(M, \psi) \in Q$, so

$$f_a((M, \psi)) \vee f_b((M, \psi)) = f_c((M, \psi)),$$

that is, $f_a \vee f_b = f_c$. Since $p$ is identified with $f_p$, we conclude that $a \vee b = c$ in $N$.

Conversely, let $a \vee b = c$ in $N$, that is, $f_a \vee f_b = f_c$. Let $L$ be a lattice given by (Ex) and let $\varepsilon$ be the inclusion map on $P$. We can assume that $L = \mathrm{sub}(P)$; thus we can form $(L, \varepsilon)$. By Corollary 55, $|L| \leq |S|$, so there is a one-to-one map $\alpha \colon L \to S$.

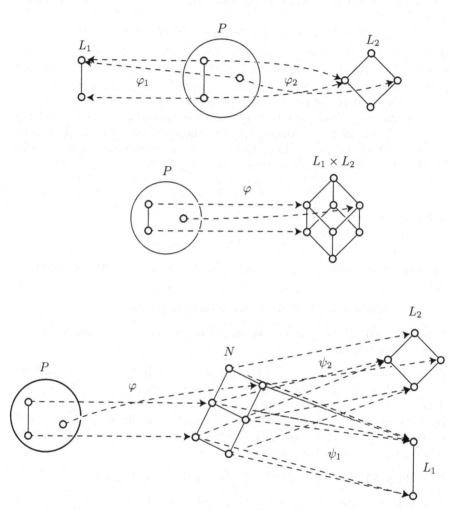

Figure 17. Constructing Free$_{\mathbf{K}}$ $P$, an example

Let $L_1 = L_\alpha$ and make $L_1$ into a lattice by defining

$$\alpha(a) \vee \alpha(b) = \alpha(a \vee b),$$
$$\alpha(a) \wedge \alpha(b) = \alpha(a \wedge b).$$

Then $L \cong L_1$ and we can form the pair $(L_1, \alpha_1)$, where $\alpha_1$ is the restriction of $\alpha$ to $P$ ($\subseteq L$). Clearly, $(L_1, \alpha_1) \in Q$, since $L_1 \subseteq S$. Now $f_a \vee f_b = f_c$ yields that

$$f_a((L_1, \alpha_1)) \vee f_b((L_1, \alpha_1)) = f_c((L_1, \alpha_1)),$$

that is, $\alpha(a) \vee \alpha(b) = \alpha(c)$, which in turn gives that $a \vee b = c$, since $\alpha$ is an isomorphism. By (Ex), $a \vee b = c$ in $L$ implies that $\sup\{a, b\} = c$. The second part of (ii) follows by duality.

(iii) This part of the proof is obvious by the definition of $N$.

(iv) Take $(L, \varphi)$; we have to find a homomorphism $\psi \colon N \to L$ satisfying $\varphi(a) = \psi(a)$ for all $a \in P$. Using $|L| \leq |S|$, the argument given in (ii) can be repeated to find $(L_1, \varphi_1)$, an isomorphism $\alpha \colon L \to L_1$ such that $\alpha\varphi(a) = \varphi_1(a)$ for all $a \in P$ and $L_1 \subseteq S$. Therefore, $(L_1, \varphi_1) \in Q$. Set

$$\psi_1 \colon f \mapsto f((L_1, \varphi_1)), \quad f \in N.$$

Then, for all $a \in P$,

$$\psi_1(a) = \psi_1(f_a) = f_a((L_1, \varphi_1)) = \varphi_1(a) = \alpha\varphi(a).$$

Thus the homomorphism $\psi = \psi_1 \alpha^{-1} \colon N \to L$ will satisfy the requirement of (i).     $\square$

Two consequences of this theorem are very important:

**Corollary 70.** *For any nontrivial variety* **K** *and for any cardinal* $\mathfrak{m}$, *the free lattice* $\mathrm{Free}_{\mathbf{K}}(\mathfrak{m})$ *exists.*

*Proof.* It suffices to find an $L \in \mathbf{K}$ with $X \subseteq L$, such that $|X| = \mathfrak{m}$, and, for all $x, y \in X$ with $x \neq y$, the elements $x$ and $y$ are incomparable. This is easily done. Since $\mathbf{K}$ is nontrivial, there exists an $N \in \mathbf{K}$ with $|N| > 1$; thus $\mathsf{C}_2 \leq N$. By Lemma 59, $\mathsf{C}_2^I \in \mathbf{K}$ for every set $I$. Let $|I| = \mathfrak{m}$, let $L = \mathsf{C}_2^I$; for $i \in I$, define $f_i \in L$ by $f_i(i) = 1$ and $f_i(j) = 0$ for all $i \neq j$, and set $X = \{\, f_i \mid i \in I \,\}$. Obviously, $X$ satisfies the condition.     $\square$

**Corollary 71.** *For any order* $P$, *the free lattice* $\mathrm{Free}\, P$ *exists.*

*Proof.* Take an order $P$ and define $\mathrm{Id}_0 P$ to be the set of all subsets $I$ of $P$ satisfying the condition: let $a, b \in P$ such that $\sup\{a, b\}$ exists; then $\sup\{a, b\} \in I$ iff $a, b \in I$. Ordering $\mathrm{Id}_0 P$ by set inclusion makes $\mathrm{Id}_0 P$ a lattice. Identifying $a \in P$ with $\{\, x \in P \mid x \leq a \,\}$, we see that $\mathrm{Id}_0 P$ contains $P$ and so $P$ satisfies (Ex) of Theorem 69.

The detailed computation is almost the same as that for Theorem 84 below, so it will be omitted.     $\square$

The argument proving Corollary 67 shows that whenever $L$ is generated by $P$, any homomorphism $\varphi$ of $P$ has at most one extension to $L$, and if there is one, it is given by

$$\psi\colon p(a_0,\ldots,a_{n-1}) \mapsto p(\varphi(a_0),\ldots,\varphi(a_{n-1})).$$

This formula gives a homomorphism iff $\psi$ is well defined; in other words, iff

$$p(a_0,\ldots,a_{n-1}) = q(b_0,\ldots,b_{m-1})$$

implies that

$$p(\varphi(a_0),\ldots,\varphi(a_{n-1})) = q(\varphi(b_0),\ldots,\varphi(b_{m-1})),$$

for any $a_0,\ldots,a_{n-1},b_0,\ldots,b_{m-1} \in P$ and $\varphi\colon P \to N \in \mathbf{K}$.

This yields a very practical method of finding free lattices and verifying their freeness.

**Theorem 72.** *In the definition of* $\mathrm{Free}_{\mathbf{K}}\, P$, *condition* (iv) *can be replaced by the following condition:*

(iv′) *If $b \in \mathrm{Free}_{\mathbf{K}}\, P$ has two representations,*

$$b = p(a_0,\ldots,a_{n-1}),$$
$$b = q(b_0,\ldots,b_{m-1}),$$

*where $a_0,\ldots,a_{n-1},b_0,\ldots,b_{m-1} \in P$, then*

$$p(a_0,\ldots,a_{n-1}) = q(b_0,\ldots,b_{m-1})$$

*can be derived from the identities defining* $\mathbf{K}$ *and the relations of $P$ of the form $a \vee b = c$ and $a \wedge b = c$.*

*Remark.* The last phrase means that in proving $p = q$, we can use only the join- and meet-table of $P$, but we cannot use $a \neq b$ or $a \neq b \vee c$, and so on.

*Proof.* If $F$ satisfies (iv′) and we are given $L$ and $\varphi$ as in (iv), then every relation holding in $F$ among the elements of $P$, being derivable as indicated, will also hold among the images under $\varphi$ of the same elements of $P$. By the discussion preceding the proof of the theorem, we get a well-defined homomorphism $F \to L$ as in (iv).

For the converse, verify that the set of terms in the elements of $P$, modulo the set of equalities derivable in the indicated way, can be made into a lattice $L$ with a homomorphism of $P$ into it. Assuming that $F$ satisfies (iv), we get a homomorphism $F \to L$ making a commuting triangle; hence no relations that are not so derivable can hold in $F$.     $\square$

## 5.3   Examples

We illustrate Theorem 72 first by determining $\mathrm{Free_D}(3)$. The following simple observation will be useful:

**Lemma 73.** *Let $x, y, z$ be elements of a lattice $L$ and let the elements $x \vee y$, $y \vee z, z \vee x$ be pairwise incomparable. Then $\{x \vee y, y \vee z, z \vee x\}$ generates a sublattice of $L$ isomorphic to $\mathrm{C}_2^3$ (see Figure 18).*

*Proof.* Almost all the meets and joins are obvious; by symmetry, the non-obvious ones are typified by the following two:

$$((x \vee y) \wedge (y \vee z)) \vee ((x \vee y) \wedge (z \vee x)) = x \vee y,$$
$$((x \vee y) \wedge (y \vee z)) \vee (z \vee x) = x \vee y \vee z.$$

Since $y \leq (x \vee y) \wedge (y \vee z)$ and $x \leq (x \vee y) \wedge (z \vee x)$, we get

$$x \vee y \leq ((x \vee y) \wedge (y \vee z)) \vee ((x \vee y) \wedge (z \vee x)),$$

and $\geq$ is trivial. The second equality follows from $y \leq (x \vee y) \wedge (y \vee z)$.

As an example of how we prove the eight elements distinct, assume that

$$(x \vee y) \wedge (y \vee z) = (x \vee y) \wedge (y \vee z) \wedge (z \vee x).$$

By joining both sides with $z \vee x$, this would imply that $x \vee y \vee z = z \vee x$; thus $x \vee y \leq z \vee x$, contradicting the assumption that these elements are incomparable.                                            □

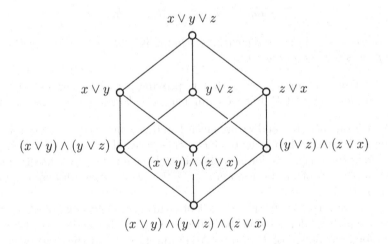

Figure 18. Illustrating Lemma 73

**Theorem 74.** *A free distributive lattice on three generators,* $\text{Free}_D(3)$, *has eighteen elements (see Figure 19).*

*Proof.* Let $x, y, z$ be the free generators. The top eight and the bottom eight elements form sublattices by Lemma 73 and its dual; note that

$$(x \wedge y) \vee (y \wedge z) \vee (z \wedge x) = (x \vee y) \wedge (y \vee z) \wedge (z \vee x)$$

by Exercise 4.7.

According to Theorem 72, we have only to verify that the lattice $L$ of Figure 19 is a distributive lattice, and that if $p, q, r$ are terms representing elements of $L$ and $p \vee q = r$ in $L$, then $p \vee q = r$ in every distributive lattice and similarly for $\wedge$. The first statement is easily proved by representing $L$ by sets (see Exercise 5.12). The second statement requires a complete listing of all triples $p, q, r$ with $p \vee q = r$. If $p, q, r$ belong to the top eight or bottom eight elements, the statement follows from Lemma 73. The remaining cases are all trivial except when $p$ or $q$ is one of $x, y, z$. By symmetry, only $p = x$, $q = y \wedge z$, $r = (x \vee y) \wedge (x \vee z)$ is left to discuss, but then $p \vee q = r$ is the distributive law.   □

The following result was obtained in R. Dedekind [149].

**Theorem 75.** *A free modular lattice on three generators,* $\text{Free}_M(3)$, *has twenty-eight elements (see Figure 20).*

*Proof.* Let $x, y, z$ be the free generators. Again, the modularity of the lattice of Figure 20 can easily proved by a representation (see Exercise 5.15). Theorem 74 takes care of most meets and joins not involving $x_1, y_1, z_1$. Of the rest, only one relation (and the symmetric and the dual cases) is nontrivial to prove: $x_1 \wedge y_1 = v$. This we do now, leaving the rest to the reader.

Compute:

$$
\begin{aligned}
x_1 \wedge y_1 &= ((x \wedge v) \vee u) \wedge ((y \wedge v) \vee u) && \text{(since } u \leq (y \wedge v) \vee u) \\
&= ((x \wedge v) \wedge ((y \wedge v) \vee u)) \vee u && \text{(by modularity)} \\
&= ((x \wedge v) \wedge (y \vee u) \wedge v) \vee u && \text{(substitute } u \text{ and } v) \\
&= (x \wedge (y \vee z) \wedge (y \vee (x \wedge z))) \vee u \\
&= (x \wedge y) \vee (x \wedge z) \vee u \\
&= u. && \square
\end{aligned}
$$

## 5.4   Partial lattices

Consider the lattice represented by Figure 21. We would like to say that it is freely generated by $\{0, a, b, 1\} = P$, but this is clearly not the case according to Definition 66, since $\sup\{a, b\} = 1$ in $P$, whereas in the lattice, $a \vee b < 1$. So to

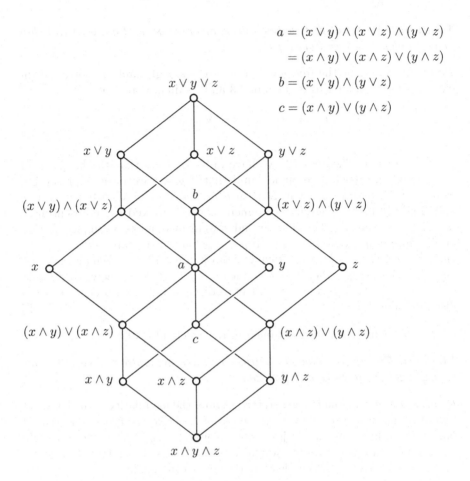

$$a = (x \vee y) \wedge (x \vee z) \wedge (y \vee z)$$
$$= (x \wedge y) \vee (x \wedge z) \vee (y \wedge z)$$
$$b = (x \vee y) \wedge (y \vee z)$$
$$c = (x \wedge y) \vee (y \wedge z)$$

Figure 19. The lattice $\mathrm{Free_D}(3)$

get the most general lattice of Section 2.3, we have to enlarge the framework of our discussion by introducing partial lattices. Of course, the study of partial lattices is also important for other purposes.

**Definition 76.** Let $L$ be a lattice, $H \subseteq L$, and restrict $\vee$ and $\wedge$ to $H$ as follows:

> For all $a, b, c \in H$, if $a \vee b = c$ (respectively, $a \wedge b = c$), then we say that $a \vee b$ (respectively, $a \wedge b$) is defined in $H$ and it equals $c$; and for all $a, b \in H$, if $a \vee b \notin H$ (respectively, $a \wedge b \notin H$), then we say that $a \vee b$ (respectively, $a \wedge b$) is not defined in $H$.

Thus $(H; \vee, \wedge)$ is a set with two binary partial operations, it is called a *partial sublattice* (also a *relative sublattice*) of $L$.

Thus every subset of a lattice determines a partial lattice. The second part of this section is devoted to an internal characterization of partial lattices, based on N. Funayama [212].

We now analyze the way the eight identities (Idem), (Comm), (Assoc), (Absorp) of Section 1.8 that were used to define lattices hold in partial lattices:

**Lemma 77.** *Let $(H; \vee, \wedge)$ be a partial lattice, $a, b, c \in H$.*

(i) *$a \vee a$ exists and $a \vee a = a$.*

(ii) *If $a \vee b$ exists, then $b \vee a$ exists, and $a \vee b = b \vee a$.*

$$u = (x \wedge y) \vee (y \wedge z) \vee (x \wedge z)$$
$$v = (x \vee y) \wedge (y \vee z) \wedge (x \vee z)$$
$$x_1 = (x \wedge v) \vee u$$
$$y_1 = (y \wedge v) \vee u$$
$$z_1 = (z \wedge v) \vee u$$

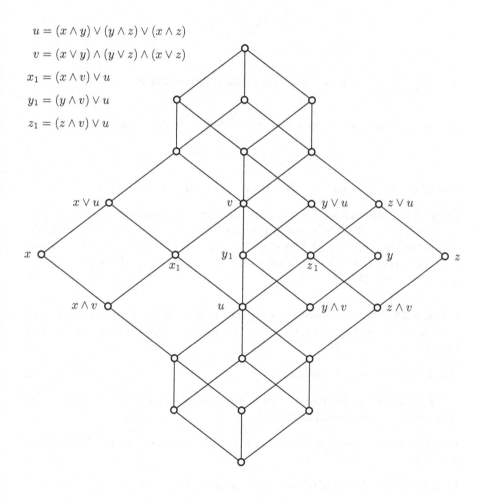

Figure 20. The lattice $\mathrm{Free_M}(3)$

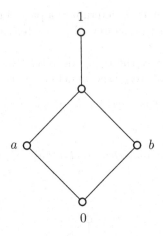

Figure 21. A small distributive lattice

(iii) *If $a \vee b$, $(a \vee b) \vee c$, and $b \vee c$ exist, then $a \vee (b \vee c)$ exists, and*

$$(a \vee b) \vee c = a \vee (b \vee c).$$

(iv) *If $a \wedge b$ exists, then $a \vee (a \wedge b)$ exists, and $a = a \vee (a \wedge b)$.*

*Proof.* As an illustration let us prove the first statement of (iii). Let $H \subseteq L$ as in Definition 76. The assumption that $a \vee b$, $(a \vee b) \vee c$, $b \vee c$ exist in $H$ means that $a, b, c, a \vee b, (a \vee b) \vee c, b \vee c \in H$. But $(a \vee b) \vee c = a \vee (b \vee c)$ in $L$, and so $a \vee (b \vee c) \in H$; that is, $a \vee (b \vee c)$ exists in $H$, and, of course, $(a \vee b) \vee c = a \vee (b \vee c)$.    □

**Lemma 72′.** *Let us denote by (i′)–(iv′) the statements we get from (i)–(iv) of Lemma 77 by interchanging $\vee$ and $\wedge$. Then (i′)–(iv′) hold in any partial lattice.*

*Proof.* This is trivial by duality.    □

Statements (i)–(iv) and (i′)–(iv′) give the required interpretation of the eight identities for partial lattices.

**Definition 78.** A *weak partial lattice* is a set with two binary partial operations satisfying (i)–(iv) and (i′)–(iv′).

**Corollary 79.** *Every partial lattice is a weak partial lattice.*

Based on Figure 22, we give the following example of a weak partial lattice that is not a partial lattice.

Consider the lattice $Y = \{0, a, b, c, d, e, f, g, h, 1\}$ of Figure 22. We define a weak partial lattice $\mathfrak{Y} = (Y; \vee, \wedge)$. Let the $\wedge$ operation of $\mathfrak{Y}$ be the same as the $\wedge$ operation of $Y$.

Let the $\vee$ partial operation of $\mathfrak{Y}$ be defined as follows.

Let $x \vee y = z$ in $\mathfrak{H}$ if $x, y$ is on the following list and $z$ is as stated:

$$x \leq y \text{ in } Y \qquad\qquad \text{and } z = y,$$
$$y \leq x \text{ in } Y \qquad\qquad \text{and } z = x,$$
$$\{x, y\} = \{a, c\} \qquad\qquad \text{and } z = f,$$
$$\{x, y\} = \{b, d\} \qquad\qquad \text{and } z = g,$$
$$\{x, y\} = \{f, g\} \qquad\qquad \text{and } z = 1.$$

Then $(H; \vee, \wedge)$ is a weak partial lattice (check the axioms). Now suppose that there exists a lattice $L$ and $H \subseteq L$, such that $(H; \vee, \wedge)$ is a partial sublattice of $L$. Then $1 = (a \vee c) \vee (b \vee d)$ in $L$, and thus $1 = \sup\{a, b, c, d\}$. Since $e \geq a, b$ and $h \geq c, d$ in $L$, and $1 \geq e, h$, we get $1 = \sup\{e, h\}$ in $L$. The fact that $e, h, 1 \in H$ implies that $e \vee h$ is defined in $H$ (and equals 1), contrary to the definition of $(H; \vee, \wedge)$. (Compare this with Lemma 83.)

To avoid such anomalies we shall introduce two further conditions. To prepare for them, we point out that although the concept of a weak partial lattice is not strong enough to lead to a lattice containing $P$, it is strong enough for a partial ordering on $P$.

**Lemma 80.** *Let $(H; \vee, \wedge)$ be a weak partial lattice. We define an ordering $\leq$ on $H$ as follows:*

*Let $a \leq b$ if $a \vee b$ exists and $a \vee b = b$.*

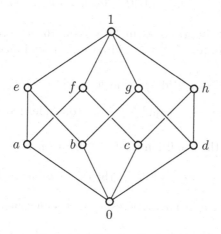

Figure 22. The lattice $Y$

*Under this ordering, if $a \vee b$ exists in $(H; \vee, \wedge)$, then $a \vee b = \sup\{a, b\}$. If $a \wedge b$ exists, then $a \wedge b = \inf\{a, b\}$. Also, $a \leq b$ iff $a \wedge b = a$.*

*Proof.* This proof is the same as the proof of the corresponding parts of Theorem 3, except that the arguments are a bit longer.    $\square$

Note that, in a partial lattice, $\sup\{a, b\}$ may exist but $a \vee b$ does not. For instance, let $L$ be the lattice of Figure 21, $H = \{0, a, b, 1\}$. Then $\sup\{a, b\} = 1$ in $H$ but $a \vee b$ is not defined in $H$ because $a \vee b \notin H$.

**Definition 81.** An *ideal* of a weak partial lattice $H$ is a nonempty subset $I$ of $H$ satisfying the following conditions:

(i) if $a, b \in I$ and $a \vee b$ exists, then $a \vee b \in I$;
(ii) $I$ is a down-set.

Filters are defined dually. Let $\mathrm{Id}_0 \, H$ be the lattice consisting of $\varnothing$ and all ideals of $H$ (ordered by $\subseteq$), let $\mathrm{Fil}_0 \, H$ be the lattice consisting of $\varnothing$ and all filters of $H$ (ordered by $\subseteq$). For $K \subseteq H$, the ideal generated by $K$ is denoted by $\mathrm{id}(K)$, and $\mathrm{fil}(K)$ is the filter generated by $K$. Again, we set

$$\mathrm{id}(a) = \{\, x \mid x \leq a \,\} = {\downarrow} a.$$

**Corollary 82.** *Let $H$ and $L$ be given as in Definition 76. Let $I$ be an ideal of $L$. Then $I \cap H$ is an ideal of $H$ provided that $I \cap H \neq \varnothing$.*

**Lemma 83.** *Any partial lattice $H$ satisfies the following condition:*

(Idl) *If $\mathrm{id}(a) \vee \mathrm{id}(b) = \mathrm{id}(c)$ in $\mathrm{Id}_0 \, H$, then $a \vee b$ exists in $H$ and equals $c$.*

*Proof.* Let $H$ and $L$ be given as in Definition 76, let $a, b, c \in H$, and let $\mathrm{id}(a) \vee \mathrm{id}(b) = \mathrm{id}(c)$ in $\mathrm{Id}_0 \, H$. Set $I = \mathrm{id}(a \vee b)_L$. Then $\mathrm{id}(a)_H, \mathrm{id}(b)_H \subseteq I \cap H$, thus

$$\mathrm{id}(c)_H = \mathrm{id}(a)_H \vee \mathrm{id}(b)_H \subseteq \mathrm{id}(a \vee b)_L;$$

that is, $c \leq a \vee b$. Since $a \leq c$ and $b \leq c$, we conclude that $a \vee b = c$.    $\square$

Let (Fil) denote the condition dual to (Idl), namely:

(Fil) *If $\mathrm{fil}(a) \vee \mathrm{fil}(b) = \mathrm{fil}(c)$ in $\mathrm{Fil}_0 \, H$, then $a \wedge b$ exists in $H$ and equals $c$.*

Now we characterize partial lattices as in N. Funayama [212].

**Theorem 84.** *A partial lattice can be characterized as a weak partial lattice satisfying conditions* (Idl) *and* (Fil).

*Proof.* Corollary 79, Lemma 83, and its dual prove that a partial lattice is a weak partial lattice satisfying (Idl) and (Fil). Conversely, let $(H; \vee, \wedge)$ be a weak partial lattice satisfying (Idl) and (Fil). Consider the map

$$\varphi \colon x \mapsto (\mathrm{id}(x), \mathrm{fil}(x)),$$

sending $H$ into $\mathrm{Id}_0 H \times (\mathrm{Fil}_0 H)^\delta$, where $(\mathrm{Fil}_0 H)^\delta$ is the dual of $\mathrm{Fil}_0 H$. This map $\varphi$ is one-to-one. If $x \vee y = z$, then

$$\mathrm{id}(x) \vee \mathrm{id}(y) = \mathrm{id}(z)$$

in $\mathrm{Id}_0 H$ and

$$\mathrm{fil}(x) \vee \mathrm{fil}(y) = \mathrm{fil}(z)$$

in $(\mathrm{Fil}_0 H)^\delta$, thus

$$\varphi(x) \vee \varphi(y) = \varphi(x \vee y).$$

Conversely, if $\varphi(x) \vee \varphi(y) = \varphi(z)$, then $\mathrm{id}(x) \vee \mathrm{id}(y) = \mathrm{id}(z)$ in $\mathrm{Id}_0 H$. Therefore, by Lemma 83, $x \vee y$ exists and equals $z$, so $\varphi(x) \vee \varphi(y) = \varphi(z)$ implies that $x \vee y = z$. A similar argument shows that $x \wedge y = z$ iff $\varphi(x) \wedge \varphi(y) = \varphi(z)$. Thus we can identify $x$ with $\varphi(x)$, getting $H \subseteq L = \mathrm{Id}_0 H \times (\mathrm{Fil}_0 H)^\delta$. We have just proved that $(H; \vee, \wedge)$ is a partial sublattice of $L$. □

We shall need some further definitions.

**Definition 85.** Let $(A; \vee, \wedge)$ and $(B; \vee, \wedge)$ be weak partial lattices, and let $\varphi \colon A \to B$ be a map. We call $\varphi$ a *homomorphism* if whenever $a \vee b$ exists, for any $a, b \in A$, then $\varphi(a) \vee \varphi(b)$ exists and $\varphi(a \vee b) = \varphi(a) \vee \varphi(b)$, and the dual condition holds for $\wedge$. A one-to-one homomorphism $\varphi$ is an *embedding* provided that $a \vee b$ exists iff $\varphi(a) \vee \varphi(b)$ exists, and the dual condition holds for $\wedge$. If $\varphi$ is onto and it is an embedding, then $\varphi$ is an *isomorphism*.

## 5.5   Free lattices over partial lattices

Now we are ready again to define the most general lattices of Section 2.3.

**Definition 86.** Let $\mathfrak{A} = (A; \vee, \wedge)$ be a partial lattice and let $\mathbf{K}$ be a variety of lattices. The lattice $\mathrm{Free}_\mathbf{K}\, \mathfrak{A}$ (or simply, $\mathrm{Free}_\mathbf{K}\, A$) is a *free lattice over* $\mathbf{K}$ *generated by* $\mathfrak{A}$ if the following conditions are satisfied:

(i)  $\mathrm{Free}_\mathbf{K}\, A$ is a lattice in $\mathbf{K}$.

(ii)  $A \subseteq \mathrm{Free}_\mathbf{K}\, A$, and $A$ is a partial sublattice of $\mathrm{Free}_\mathbf{K}\, A$.

(iii)  $A$ generates $\mathrm{Free}_\mathbf{K}\, A$.

(iv)  If $L \in \mathbf{K}$ and $\varphi \colon A \to L$ is a homomorphism, then there exists a homomorphism $\psi \colon \mathrm{Free}_\mathbf{K}\, A \to L$ extending $\varphi$ (that is, $\varphi(a) = \psi(a)$ for all $a \in A$).

If $\mathbf{K} = \mathbf{L}$, we write Free $\mathfrak{A}$ for Free$_\mathbf{L}$ $\mathfrak{A}$.

To relate the general Definition 86 to the very special Definition 66, we describe a natural way to define a "maximal" partial lattice on an order. At the same time, we also describe the "minimal" one.

**Definition 87.** Let $P$ be an order.

(i) We define the partial operation $\vee^{\max}$ on $P$ as follows: $a \vee^{\max} b = c$ if $\sup\{a, b\} = c$, and we define $\wedge^{\max}$ dually. Then the partial lattice $P^{\max}$ is defined as
$$P^{\max} = (P; \vee^{\max}, \wedge^{\max}).$$

(ii) We define the partial operation $\vee^{\min}$ on $P$ as follows: $a \vee^{\min} b = c$ if $a$ and $b$ *are comparable* and $\sup\{a, b\} = c$, and we define $\wedge^{\min}$ dually. Then the partial lattice $P^{\min}$ is defined as
$$P^{\min} = (P; \vee^{\min}, \wedge^{\min}).$$

**Lemma 88.** $P^{\min}$ *and* $P^{\max}$ *are both partial lattices.*

*Proof.* It is easy to verify that (Idl), (Fil), (i)–(iv) of Lemma 77, and (i′)–(iv′) of Lemma 77′ hold for $P^{\min}$ and $P^{\max}$. $\square$

Using the partial lattice $P^{\max}$, Definition 66 becomes a special case of Definition 86.

The general theory can be developed exactly as it was in the first part of this section. The final result is:

**Theorem 89.** *Let* $\mathfrak{A} = (A; \vee, \wedge)$ *be a partial lattice and let* $\mathbf{K}$ *be a variety. Then* Free$_\mathbf{K}$ $\mathfrak{A}$ *exists iff there exists a lattice* $L$ *in* $\mathbf{K}$ *such that* $\mathfrak{A}$ *is a partial sublattice of* $L$.

See Exercise 5.39 for a somewhat stronger form of this theorem.

**Corollary 90.** *Let* $\mathfrak{A} = (A; \vee, \wedge)$ *be a partial lattice. Then* Free $\mathfrak{A}$ *exists.*

As in Section 1.8, with a partial lattice $\mathfrak{A}$, we can associate an order $\mathfrak{A}^{\mathrm{ord}}$. However, there is no equivalence between partial lattices and "orders as partial lattices". This breaks down in two ways: first, many different partial lattices are associated with the same order; second, all orders are associated with some partial lattices. Both of these points are clarified with the following definition.

**Definition 91.** Let $P$ be an order. The lattice CFree $P =$ Free $P^{\min}$ is called a *completely free lattice*, or a lattice *completely freely generated by* $P$.

CF($P$) is the usual notation for CFree $P$ in the literature.

**Theorem 92.** *For any order* $P$, *the lattices* CFree $P$ *and* CFree$_\mathbf{D}$ $P$ *exist.*

*Proof.* By the definition of CFree $P$ and the definition of $P^{\min}$, we have to find for the order $P$ a distributive lattice $L$ such that $P$ is a suborder of $L$ and for any two incomparable elements $a, b \in P$, the elements $a \vee b$ and $a \wedge b$ (the join and meet formed in $L$) are not in $P$.

Let $L_1 = \text{Down}\, P$ (see Section 1.6) and embed $P$ into $\text{Down}\, P$ by $p \mapsto \downarrow p$ for $p \in P$. Given any two incomparable elements $a, b \in P$, the join in $\text{Down}\, P$ is

$$\downarrow a \vee \downarrow b = \downarrow a \cup \downarrow b,$$

and clearly it is not in $P$.

Proceeding dually starting with $L_1$, we form $L$, which is the distributive lattice we need. $\qquad\qquad\square$

In Section II.1.5, we prove that if a variety $\mathbf{V}$ contains a lattice with more than one element (a *nontrivial variety*), then $\mathbf{D} \subseteq \mathbf{V}$ (see also the discussion in Section VI.2.1). Therefore, we can introduce the concept of a *lattice completely freely generated in* $\mathbf{V}$ *by* $P$, in notation, CFree$_{\mathbf{V}}$ $P$, and conclude that CFree$_{\mathbf{V}}$ $P$ exists for every nontrivial variety $\mathbf{V}$ (see Theorem 121).

It is hard to overemphasize the importance of free lattices; they provide one of the most important research tools of lattice theory. Two typical applications to modular lattices can be found in G. Grätzer and E. T. Schmidt [336] and G. Grätzer [253]. We discuss free lattices in great detail in Section VII.2 and we use the method of constructing special lattices Free$_{\mathbf{V}}$ $P$ throughout the book.

## 5.6 ◇ Finitely presented lattices

Forming Free $P$ for a finite partial lattice $P$, we obtain a *finitely presented lattice*. Finitely presented lattices are the closest to finite lattices and they have very rich structure.

Every element of Free $P$ can be represented in infinitely many ways in the form $p(a_0, \ldots, a_{n-1})$, where $p$ is a term and $a_0, \ldots, a_{n-1} \in P$. There is a "normal form" (G. Grätzer, A. P. Huhn, and H. Lakser [282]; this 1981 paper relates earlier works going back to 1970)—just as for free products of lattices, see Section VII.1.9—a "canonical form" (R. Freese [184] in 1989). R. Freese and J. B. Nation [190] applied "canonical forms" to solve affirmatively Problem 12 of [257] from 1970 (see also as Problem VI.35 of [262] and [269]):

◇ **Theorem 93.** *The automorphism group of a finitely presented lattice is finite.*

Using "normal forms", an alternative proof of this result is presented in G. Grätzer and A. P. Huhn [278].

The following result of A. P. Huhn [405] illustrates how differently the variety $\mathbf{M}$ behaves:

◇ **Theorem 94.** *There is a finite partial lattice $A$ for which $\mathrm{Free}_M\, A$ exists and has an infinite automorphism group.*

Normal forms also yield a structure theorem, see G. Grätzer, A. P. Huhn, and H. Lakser [282], showing how close finitely presented lattices are to free lattices:

◇ **Theorem 95.** *Let $L$ be a finitely presented lattice. Then there is a congruence $\alpha$ of $L$ such that $L/\alpha$ is finite and every block of this congruence is embeddable in a free lattice.*

K. V. Adaricheva, V. A. Gorbunov, and M. V. Semenova [20] prove that finitely generated lattices are noncomplete algebraic lattices, but finitely presented lattices may not be algebraic; in fact, R. Freese [183] presents an example of a finitely presented lattice with no covers. On the other hand, every finitely presented lattice is both meet- and join-continuous, see [20].

The *MathSciNet* listing for `finitely presented lattices` yields 20 references to a variety of topics not mentioned here, for instance, when is a finitely presented lattice finite?

**Exercises**

5.1. Show that a variety is closed under the formation of sublattices, homomorphic images, and direct products. (Utilize Lemma 59.)

5.2. Show that the class $\mathbf{T}$ of all one-element lattices is a variety. For any variety $\mathbf{K}$, prove that $\mathbf{K} \supseteq \mathbf{T}$.

5.3. Let $\mathbf{K}_i$, for all $i \in I$, be varieties. Show that $\bigcap(\mathbf{K}_i \mid i \in I)$ is again a variety.

5.4. Let $\mathbf{A}$ be the variety defined by the identity

$$(x \vee y) \wedge (x \vee z) \wedge (x \vee u)$$
$$= x \vee ((x \vee y) \wedge z \wedge u) \vee ((x \vee z) \wedge y \wedge u) \vee ((x \vee u) \wedge y \wedge z).$$

Prove that $\mathbf{D} \subset \mathbf{A} \subset \mathbf{M}$.

5.5. Let $\mathbf{K}$ be a nontrivial variety. Show that $\mathbf{K}$ contains arbitrarily large lattices.

5.6. Let $\mathbf{P} = \{0, a, b\}$ with $\inf\{a, b\} = 0$, and let $\mathbf{K}$ be a nontrivial variety. Show that $\mathrm{Free}_K\, P \cong \mathrm{Free}(2)$.

5.7. Find a variety $\mathbf{K}$, an order $P$, and an automorphism $\varphi$ on $\mathrm{Free}_K\, P$ that (i) is *not* the identity map on $P$ or (ii) does not map $P$ into itself.

5.8. Let $P$ be an order, let $\mathbf{K}$ and $\mathbf{N}$ be varieties, and assume that the lattices $\mathrm{Free}_K\, P$ and $\mathrm{Free}_N\, P$ exist. Prove that if $\mathbf{K} \supseteq \mathbf{N}$, then there exists a homomorphism $\varphi$ from $\mathrm{Free}_K\, P$ onto $\mathrm{Free}_N\, P$ such that $\varphi$ is the identity map on $P$.

5.9. Work out a proof of the existence of Free(3) without any reference to Theorem 69.

5.10. Formulate and prove the form of Theorem 72 that is used in the proofs of Theorems 74 and 75.

5.11. Prove that $\mathrm{Free_M}(4)$ is infinite (G. Birkhoff [61]). (Hint: Let $\mathbb{R}$ be the set of real numbers and let $L$ be the lattice of vector subspaces of $\mathbb{R}^3$. Set

$$a = \{\,(x, 0, x) \mid x \in \mathbb{R}\,\}, \qquad b = \{\,(0, x, x) \mid x \in \mathbb{R}\,\},$$
$$c = \{\,(0, 0, x) \mid x \in \mathbb{R}\,\}, \qquad d = \{\,(x, x, x) \mid x \in \mathbb{R}\,\}.$$

Then $\mathrm{sub}(\{a, b, c, d\})$ is infinite.)

5.12. Let $T$ be the set of points of a nondegenarate triangle in the plane, with sides $x, y, z$ (all three sides are infinite sets of points). Show that the sublattice of $\mathrm{Pow}\,T$ generated by $\{x, y, z\}$ is isomorphic to the lattice of Figure 19. Using Exercise 4.17, deduce that the lattice of Figure 19 is distributive.

5.13. Show that in Exercise 5.12, the set $T$ can be replaced by a 6-element subset. Conclude that the lattice of Figure 19 is isomorphic to a sublattice of $\mathsf{C}_2^6$.

5.14. Show that the lattice of Figure 19 cannot be embedded in the power-set lattice of a set with fewer than six elements.

5.15. Represent the lattice of Figure 20 as a sublattice of $L \times \mathsf{M}_3$, where $L$ is the lattice of Exercise 5.12 and $\mathsf{M}_3$ is given in Figure 5.

5.16. Show that the condition (Ex) of Theorem 69 is equivalent to the following:

For any $a, b, c \in P$ not satisfying $\inf\{a, b\} = c$, there exist a lattice $L$ in $\mathbf{K}$ and a homomorphism $\varphi\colon P \to L$ with $\varphi(a) \wedge \varphi(b) \neq \varphi(c)$, and dually.

5.17. The statement "$\mathfrak{A} = (A; \vee, \wedge)$ is a *partial algebra*" means that $A$ is a nonempty set and that $\vee$ and $\wedge$ are partial binary operations on $A$. For an $n$-ary term $p$ and elements $a_0, \dots, a_{n-1} \in A$, interpret $p(a_0, \dots, a_{n-1})$. (When is it defined and what is its value?)

5.18. Let $p$ be the term $x \wedge ((x \vee y) \vee (x \vee z))$ and $q$ the term $x$. In the weak partial lattice $\mathfrak{Y} = (Y; \vee, \wedge)$ (derived from the lattice $Y$ of Figure 22), the term $p$ is not defined for $x = 0$, $y = e$, $z = h$. Verify that in every lattice the identity $p = q$ holds, and $q$ is defined in every partial lattice.

5.19. An identity $p = q$ *holds in the partial algebra* $\mathfrak{A} = (A; \vee, \wedge)$ if the following three conditions are satisfied:

(i) If $p(a_0, \dots, a_{n-1})$ and $q(a_0, \dots, a_{n-1})$ are defined, then

$$p(a_0, \dots, a_{n-1}) = q(a_0, \dots, a_{n-1})$$

for $a_0, \ldots, a_{n-1} \in A$.

(ii) If $p(a_0, \ldots, a_{n-1})$ is defined, $q = q_0 * q_1$, where $*$ is $\wedge$ or $\vee$, and both $q_0(a_0, \ldots, a_{n-1})$ and $q_1(a_0, \ldots, a_{n-1})$ are defined, then

$$q_0(a_0, \ldots, a_{n-1}) * q_1(a_0, \ldots, a_{n-1})$$

is defined.

(iii) This condition is the same as (ii) with $p$ and $q$ interchanged.

Check that Lemmas 77 and 77 give this interpretation to the lattice axioms.

5.20. Let

$$p = (((x \vee z) \vee (y \vee u)) \vee v) \vee w,$$
$$q = ((v \vee x) \vee (v \vee y)) \vee ((w \vee z) \vee (w \vee u)).$$

Show that $p = q$ in any lattice. Show that $p = q$ does not hold in the weak partial lattice defined in connection with Figure 22.

5.21. Let $I_0$ and $I_1$ be ideals of a weak partial lattice. Set

$$J_0 = I_0 \cup I_1,$$
$$J_n = \{\, x \mid x \le y \vee z, \text{ for some } y, z \in J_{n-1} \,\} \quad \text{for } n = 1, 2, \ldots$$
$$J = \bigcup(\, J_i \mid i = 0, 1, 2, \ldots).$$

Show that $J = I_0 \vee I_1$.

5.22. Let the weak partial lattice $L$ violate (Idl) (of Lemma 83); that is, $\mathrm{id}(a) \vee \mathrm{id}(b) = \mathrm{id}(c)$, but $a \vee b$ is undefined. Let $I_0 = \mathrm{id}(a)$, $I_1 = \mathrm{id}(b)$, and $c \in J_n$ (see Exercise 5.21). Generalizing Exercise 5.20, find an identity $p = q$ that holds in any lattice but not in $L$. (Exercise 5.20 is the special case $n = 2$.)

5.23. Prove that a partial algebra $\mathfrak{A} = (A; \vee, \wedge)$ is a partial lattice iff every identity $p = q$ holding in any lattice also holds in $\mathfrak{A}$ (in the sense of Exercise 5.19).

5.24. A *homomorphism* of a partial algebra $\mathfrak{A} = (A; \vee, \wedge)$ into a lattice $L$ is a map $\varphi \colon A \to L$ such that $\varphi(a \vee b) = \varphi(a) \vee \varphi(b)$, whenever $a \vee b$ exists, and the same for $\wedge$. Prove that there exists a one-to-one homomorphism of $\mathfrak{A}$ into some lattice $L$ iff there exists an ordering $\le$ on $A$ satisfying $a \wedge b = \inf\{a, b\}$, whenever $a \wedge b$ is defined in $\mathfrak{A}$, and $a \vee b = \sup\{a, b\}$, whenever $a \vee b$ is defined in $\mathfrak{A}$.

5.25. Show that every weak partial lattice satisfies the condition of Exercise 5.24, but not conversely.

5.26. Let $A = \{0, a, b, c, 1\}$ with $0 \le a, b, c \le 1$. For $x \le y$, define

$$x \vee y = y \vee x = y,$$
$$x \wedge y = y \wedge x = x,$$

and define

$$a \vee b = b \vee a = 1,$$
$$a \wedge b = b \wedge a = 0.$$

Show that $(A; \vee, \wedge)$ is a partial lattice.

5.27. Let $A = \{0, a, b, c, d, 1\}$ with $0 \leq a, b, c, d \leq 1$. For $x \leq y$, define

$$x \vee y = y \vee x = y,$$
$$x \wedge y = y \wedge x = x,$$

and define

$$a \vee b = b \vee a = c \vee d = d \vee c = 1,$$
$$a \wedge b = b \wedge a = c \wedge d = d \wedge c = 0.$$

Show that $(A; \vee, \wedge)$ is a partial lattice.

5.28. Let $L$ and $K$ be lattices and let $L \cap K$ be a sublattice of $L$ and of $K$. For $x, y \in L \cup K$ define $x \wedge y = z$ if $x, y, z \in L$ and $x \wedge y = z$ in $L$, or $x, y, z \in K$ and $x \wedge y = z$ in $K$; define $x \vee y$ similarly. Is $(L \cup K; \vee, \wedge)$ a partial lattice?

5.29. Are the eight axioms of a weak partial lattice independent?

5.30. Are the ten axioms of a partial lattice independent?

5.31. Define weak partial semilattice and partial semilattice; prove the analogue of Theorem 84 for partial semilattices.

5.32. Let $A$ be a weak partial lattice in which $a \wedge b$ exists for all $a, b \in A$. Then $A$ is a partial lattice iff $a \mapsto \mathrm{id}(a)$ is an embedding of $A$ into the lattice $\mathrm{Id}_0 A$.

5.33. Let $\mathbf{T}$ be the variety of all one-element lattices. Show that $\mathrm{Free}_{\mathbf{T}} A$ exists iff $|A| = 1$.

5.34. Let $P = \mathsf{B}_2$. Show that for any nontrivial variety $\mathbf{K}$, the free lattices $\mathrm{Free}_{\mathbf{K}} P$ and $\mathrm{CFree}_{\mathbf{K}} P$ exist and that always $\mathrm{Free}_{\mathbf{K}} P \ncong \mathrm{CFree}_{\mathbf{K}} P$.

5.35. Determine $\mathrm{Free}\, P$, $\mathrm{CFree}\, P$, and $\mathrm{Free}_{\mathbf{D}}\, P$, where $P = \{a, b, c\}$, $a < b$, and $c$ is incomparable to $a, b$.

5.36. Discuss the set of all weak partial lattices on $A$ inducing a given ordering on $A$.

5.37. Repeat Exercise 5.36 for partial lattices.

5.38. Show that a one-to-one homomorphism of weak partial lattices need not be an embedding.

5.39. Show that in Theorem 89, the condition "there exists a lattice $L$ in $\mathbf{K}$ such that $\mathfrak{A}$ is a partial sublattice of $L$" can be replaced by the following condition:

For all $a, b, c \in A$ for which $a \vee b = c$ does not hold, there exists a lattice $L$ in $\mathbf{K}$, and a homomorphism $\varphi$ of $A$ into $L$ such that $\varphi(a) \vee \varphi(b) \neq \varphi(c)$, and the same condition for $\wedge$.

5.40. Find the "most economical" proof of Theorem 92 by (repetitive) adding a single element. (Hint: see Section IV.1.2 for doubling an element.)

5.41. Let $(A; \vee, \wedge)$ be a partial algebra, let $\mathbf{K}$ be a variety of lattices, let $L \in \mathbf{K}$, and let $M$ be a partial sublattice of $L$. Then $M$ is called a *maximal homomorphic image of $A$ in $\mathbf{K}$* if there is a homomorphism $\varphi$ of $A$ onto $M$ such that whenever $\psi$ is a homomorphism of $A$ into $N \in \mathbf{K}$, then there is a homomorphism $\alpha \colon M \to N$ such that $\alpha\varphi = \psi$. Prove that a maximal homomorphic image is unique up to isomorphism, provided it exists.

5.42. Starting with an arbitrary partial algebra $(A; \vee, \wedge)$, carry out the construction of Theorem 69 (Theorem 89).

5.43. Are there orders $P_0 \subset P$ such that CFree $P_0 = $ CFree $P$?

5.44. After M. M. Gluhov [234], a finite partial lattice $\mathfrak{A}$ is a *basis* of a lattice $L$ if $L = $ Free $\mathfrak{A}$, but there is no $A_0 \subset A$ such that $L = $ Free $\mathfrak{A}_0$. Show that the lattice $L = $ Free $\mathfrak{A}$ has more than one basis, where $\mathfrak{A}$ is defined as follows: Let $(A; \leq)$ be the order given by Figure 23; for $x \leq y$, let $x \wedge y = y \wedge x = x$; furthermore, the join of any elements is defined in $\{0, a, b, c, d, e, f, 1\}$ as supremum, and $1 \vee g = g \vee 1 = 1$. (This example, which is due to C. Herrmann [389], contradicts M. M. Gluhov's statement.)

5.45. Let $L$ and $L_1$ be lattices, let $L = \mathrm{sub}(A)$, and let $\varphi$ be a map of $A$ into $L_1$. Then $\varphi$ can be extended to a homomorphism $\psi$ of $L$ into $L_1$ iff

$$p(a_0, \ldots, a_{n-1}) = q(a_0, \ldots, a_{n-1})$$

implies that

$$p(\varphi(a_0), \ldots, \varphi(a_{n-1})) = q(\varphi(a_0), \ldots, \varphi(a_{n-1}))$$

holds for any $a_0, \ldots, a_{n-1} \in A$, for every integer $n$, and for any pair $p, q$ of $n$-ary terms.

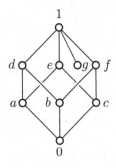

Figure 23. A diagram for Exercise 5.44

5.46. Under the hypotheses of Exercise 5.45, describe $\psi$.

5.47. Characterize **K**-free lattices using Exercise 5.45.

5.48. Under the hypotheses of Exercise 5.45, show that if $A$ is finite and $\varphi$ is isotone, then $\varphi$ can always be extended to an isotone map of $L$ into $L_1$.

5.49. Let $P$ be an order. Find a partial lattice $P'$ so that Free $P$ is isomorphic to Free $P'$.

$$* \qquad * \qquad *$$

5.50. Define varieties for a fixed type of algebras.

5.51. Define $\text{Free}_{\mathbf{K}}(\mathfrak{m})$ for a variety **K** of algebras.

5.52. Prove the existence of $\text{Free}_{\mathbf{K}}(\mathfrak{m})$ for a nontrivial variety **K** of algebras.

5.53. Can one develop the universal algebraic version of "Free lattices over partial lattices"?

# 6.    Special Elements

## 6.1    Complements

In a bounded lattice $L$, the element $a$ is a *complement* of the element $b$ if

$$a \vee b = 1,$$
$$a \wedge b = 0.$$

For instance, in $\mathsf{M}_3$ (see Figure 5), all elements other than the zero and the unit have two complements each.

**Lemma 96.** *In a bounded distributive lattice, an element can have only one complement.*

*Proof.* If $b_0$ and $b_1$ are both complements of $a$, then

$$b_0 = b_0 \wedge 1 = b_0 \wedge (a \vee b_1) = (b_0 \wedge a) \vee (b_0 \wedge b_1) = 0 \vee (b_0 \wedge b_1) = b_0 \wedge b_1;$$

similarly, $b_1 = b_0 \wedge b_1$, thus $b_0 = b_1$.                                    □

Let $a \in [b, c]$; $x$ is a *relative complement* of $a$ in $[b, c]$ if

$$a \vee x = c,$$
$$a \wedge x = b.$$

**Lemma 97 (De Morgan's Identities).** *In a bounded distributive lattice, if the elements $a$ and $b$ have complements, $a'$ and $b'$, respectively, then $a \vee b$ and $a \wedge b$ have complements, $(a \vee b)'$ and $(a \wedge b)'$, respectively, and*

$$(a \vee b)' = a' \wedge b',$$
$$(a \wedge b)' = a' \vee b'.$$

*Proof.* By Lemma 96, it suffices to prove that

$$(a \wedge b) \wedge (a' \vee b') = 0,$$
$$(a \wedge b) \vee (a' \vee b') = 1$$

to verify the second identity; the first is dual. Compute:

$$(a \wedge b) \wedge (a' \vee b') = (a \wedge b \wedge a') \vee (a \wedge b \wedge b') = 0 \vee 0 = 0$$

and

$$(a \wedge b) \vee (a' \vee b') = (a \vee a' \vee b') \wedge (b \vee a' \vee b') = 1 \wedge 1 = 1. \qquad \square$$

The next two lemmas were also originally proved for distributive lattices but they also hold for modular lattices.

**Lemma 98.** *In a bounded modular lattice, if the element a has a complement, then it also has a relative complement in any interval containing it.*

*Proof.* Let $d$ be a complement of $a$. Then

$$x = (d \vee b) \wedge c = b \vee (d \wedge c)$$

(these two expressions are equal by modularity) is a relative complement of $a$ in the interval $[b, c]$, provided that $b \le a \le c$. Indeed,

$$a \vee x = a \vee ((d \vee b) \wedge c) \qquad \text{(use modularity with } a \le c\text{)}$$
$$= (a \vee d \vee b) \wedge c = (1 \vee b) \wedge c = c,$$

and $a \wedge x = b$, by duality. $\qquad \square$

A *complemented lattice* is a bounded lattice in which every element has a complement. A *relatively complemented lattice* is a lattice in which every element has a relative complement in any interval containing it.

Often, especially if we are interested in congruences, it suffices to assume that the lattice is sectionally complemented. We call the lattice $L$ *sectionally complemented* if it has a zero and for all $a \le b \in L$, there exists an element $c \in L$ satisfying $a \vee c = b$ and $a \wedge c = 0$.

The following statements follow immediately from Lemma 98.

**Lemma 99.**

(i) *A modular lattice with zero is sectionally complemented iff it is relatively complemented*

(ii) *A bounded modular lattice is complemented iff it is sectionally complemented iff it is relatively complemented.*

A *boolean lattice* is a complemented distributive lattice. Thus in a boolean lattice $B$, every element $a$ has a unique complement, and $B$ is relatively complemented. We denote by $\mathsf{B}_1$ the two-element boolean lattice and by $\mathsf{B}_n$ the boolean lattice $(\mathsf{B}_1)^n$.

A *boolean algebra* is a boolean lattice in which 0, 1, and $'$ (complementation) are also considered to be operations. Thus a boolean algebra is a system: $(B; \vee, \wedge, ', 0, 1)$, where $\vee$ and $\wedge$ are binary operations, $'$ is a unary operation, and 0, 1 are nullary operations. (A nullary operation on $B$ picks out an element of $B$.) A *homomorphism* $\varphi$ of a boolean algebra is a lattice homomorphism preserving 0, 1 and $'$; that is, it is a $\{0, 1\}$-homomorphism satisfying

$$\varphi(x)' = \varphi(x').$$

A *subalgebra* of a boolean algebra is a $\{0, 1\}$-sublattice closed under complementation.

## 6.2   Pseudocomplements

Note that in a bounded distributive lattice $L$, if $b$ is a complement of $a$, then $b$ is the largest element $x$ of $L$ with $a \wedge x = 0$. More generally, let $L$ be a lattice with zero; an element $a^*$ is a *pseudocomplement* of $a$ $(\in L)$ if $a \wedge a^* = 0$ and $a \wedge x = 0$ implies that $x \leq a^*$. An element can have at most one pseudocomplement.

A *pseudocomplemented lattice* is one in which every element has a pseudocomplement. Every finite distributive lattice is pseudocomplemented. The lattice $\mathsf{M}_3$ is not pseudocomplemented.

A *homomorphism* $\varphi$ of a pseudocomplemented lattice into another pseudocomplemented lattice is a lattice homomorphism additionally preserving $0, 1, *$; that is, it is a $\{0, 1\}$-homomorphism satisfying

$$\varphi(x)^* = \varphi(x^*).$$

The concept of pseudocomplement involves only the meet operation. Thus we can also define *pseudocomplemented semilattices*, with the obvious homomorphism concept.

A more general concept is that of a relative pseudocomplement. Let $L$ be a meet-semilattice and let $a, b \in L$. The *pseudocomplement of $a$ relative to $b$* is an element $a * b$ of $L$ satisfying $a \wedge x \leq b$ iff $x \leq a * b$. In addition, $L$ is called a *relatively pseudocomplemented meet-semilattice*, if $a * b$ exists for all $a, b \in L$.

**Theorem 100.** *Let $L$ be a pseudocomplemented meet-semilattice. We define the skeleton of $L$:*
$$\operatorname{Skel} L = \{\, a^* \mid a \in L \,\}.$$

*Then the ordering of $L$ orders $\operatorname{Skel} L$ and makes $\operatorname{Skel} L$ into a boolean lattice. For $a, b \in \operatorname{Skel} L$, the meet, $a \wedge b$, is in $\operatorname{Skel} L$; the join in $\operatorname{Skel} L$ is described* *as follows:*

$$a \vee_{\mathrm{Skel}} b = (a^* \wedge b^*)^*.$$

*The complement of $a$ in $\mathrm{Skel}\,L$ is $a^*$.*

*Remark.* V. Glivenko [233] proved this result for complete distributive lattices and O. Frink [206] published it in its full generality. Both proofs used special axiomatizations of boolean algebras to get around the difficulty of proving distributivity. The present proof is direct and was first published in [257]; the last paragraph is an improvement from T. Katriňák [464].

Note that even if $L$ is a lattice, the join in $L$ need not be the same as the join in $\mathrm{Skel}\,L$.

*Proof.* We start with the following observations:

(1)
$$a \le a^{**}.$$

(2)
$$a \le b \quad \text{implies that} \quad a^* \ge b^*.$$

(3)
$$a^* = a^{***}.$$

(4)
$$a \in \mathrm{Skel}\,L \quad \text{iff} \quad a = a^{**}.$$

(5)
$$a, b \in \mathrm{Skel}\,L \quad \text{implies that} \quad a \wedge b \in \mathrm{Skel}\,L.$$

(6)
$$\text{For } a, b \in \mathrm{Skel}\,L, \quad \sup_{\mathrm{Skel}\,L}\{a, b\} = (a^* \wedge b^*)^*.$$

Formulas (1) and (2) follow from the definitions.

Formulas (1) and (2) yield $a^* \ge a^{***}$, and by (1) $a^* \le a^{***}$, thus (3). If $a \in \mathrm{Skel}\,L$, then $a = b^*$ for some element $b$; therefore,

$$a^{**} = b^{***} = b^* = a$$

by (3). Conversely, if $a = a^{**}$, then $a = b^*$ with $b = a^*$; thus $a \in \mathrm{Skel}\,L$, proving (4).

If $a, b \in \mathrm{Skel}\,L$, then $a = a^{**}$ and $b = b^{**}$, and so $a \ge (a \wedge b)^{**}$ and $b \ge (a \wedge b)^{**}$, thus $a \wedge b \ge (a \wedge b)^{**}$; by (1), $a \wedge b = (a \wedge b)^{**}$, thus $a \wedge b \in \mathrm{Skel}\,L$. If $x \in \mathrm{Skel}\,L$ with $x \le a$ and $x \le b$, then $x \le a \wedge b$; therefore,

$$a \wedge b = \inf_{\mathrm{Skel}\,L}\{a, b\},$$

proving (5).

Clearly, $a^* \ge a^* \wedge b^*$, thus by (2) and (4), we conclude that $a \le (a^* \wedge b^*)^*$. Similarly, we obtain that $b \le (a^* \wedge b^*)^*$. If $a \le x$ and $b \le x$ (for $x \in \mathrm{Skel}\,L$), then $a^* \ge x^*$ and $b^* \ge x^*$ by (2). By (2) and (4), we get that $(a^* \wedge b^*)^* \le x$, proving (6).

For $a, b \in \mathrm{Skel}\,L$, define

$$a \vee_{\mathrm{Skel}} b = (a^* \wedge b^*)^*.$$

By (5) and (6), $(\mathrm{Skel}\,L; \vee_{\mathrm{Skel}}, \wedge)$ is a lattice with bounds $0$ and $0^*$. The lattice $\mathrm{Skel}\,L$ is complemented since

$$a \vee_{\mathrm{Skel}} a^* = (a^* \wedge a^{**})^* = 0^* = 1,$$
$$a \wedge a^* = 0,$$

for $a \in \mathrm{Skel}\,L$.

The lattice $\mathrm{Skel}\,L$ is relatively pseudocomplemented, in fact, $a*b = (a \wedge b^*)^*$, for all $a, b \in \mathrm{Skel}\,L$. Indeed, for any $x \in L$, $a \wedge x \leq b$ is equivalent to $0 = (a \wedge x) \wedge b^*$, because $b = b^{**}$, and so by associativity, $a \wedge x \leq b$ is equivalent to $0 = x \wedge (a \wedge b^*)$. So the greatest $x$ with $a \wedge x \leq b$ equals the greatest $x$ with $0 = x \wedge (a \wedge b^*)$; that is, $a * b = (a \wedge b^*)^*$, as desired.

Thus $\mathrm{Skel}\,L$ is relatively pseudocomplemented, and by Exercise 6.30, it is distributive.                                                                    □

Observe that for a pseudocomplemented meet-semilattice $L$, we can define a map $\varphi \colon L \to \mathrm{Skel}\,L$:

$$\varphi \colon x \mapsto x^{**}.$$

It is easy to see that $\varphi$ maps $L$ onto $\mathrm{Skel}\,L$, and more precisely, $\varphi$ is a $\{0,1\}$-meet-homomorphism satisfying $\varphi(a^*) = \varphi(a)^*$. Moreover, if $L$ is a pseudocomplemented lattice, then $\varphi$ is a $\{0,1\}$-homomorphism satisfying $\varphi(a^*) = \varphi(a)^*$ for every $a \in L$. Now, the congruence kernel $\gamma$ of $\varphi$ satisfies

$$x \equiv y \pmod{\gamma} \quad \text{iff} \quad x^* = y^* \quad \text{iff} \quad x^{**} = y^{**}.$$

If $L$ is a pseudocomplemented lattice, then $\gamma$ is compatible with $\vee$, $\wedge$, and $^*$. This congruence $\gamma$ is called the *Glivenko congruence* on $L$.

Another important subset of a pseudocomplemented meet-semilattice $L$ is the *dense set*

$$\mathrm{Dns}\,L = 1/\gamma = \{\, a \mid a^* = 0 \,\}.$$

The elements of $\mathrm{Dns}\,L$ are called *dense*. Of course, you can define $\mathrm{Dns}\,L$ for any lattice $L$ with zero:

$$\mathrm{Dns}\,L = \{\, x \mid x \wedge y > 0, \ \text{for all } y > 0 \,\}.$$

## 6.3  Other types of special elements

An element $a$ of a lattice $L$ is an *atom* if $a \succ 0$; we call it a *dual atom*, if $a \prec 1$. A lattice $L$ is called *atomic* if every nonzero element majorizes an atom. A lattice is called *atomistic* if every element is a join of atoms. We denote by $\mathrm{Atom}(L)$ the set of all atoms of $L$, and by $\mathrm{Atom}(x)$ the set of all atoms $p$ in $L$ with $p \leq x$. The chain $\mathsf{C}_3$ is atomic but it is not atomistic.

A lattice $L$ is called *relatively atomic* (often called *weakly atomic*) if every proper interval $[a, b]$ (that is, $a < b \in L$) contains a covering pair $u \prec v$.

An element $a$ of a lattice $L$ is *join-irreducible* if $a \neq 0$ and $a = b \vee c$ implies that $a = b$ or $a = c$; it is *meet-irreducible*, if $a \neq 1$ and $a = b \wedge c$ implies that $a = b$ or $a = c$. An element which is both join- and meet-irreducible is called *doubly irreducible*. In a finite lattice $L$, a join-irreducible element $x$ covers a unique element, denoted by $x_*$.

For complete lattices, we need infinitary variants of these concepts. An element $a$ of a complete lattice $L$ is *completely join-irreducible* if $a \neq 0$ and $a = \bigvee( b_i \mid i \in I )$ implies that $a = b_i$ for some $i \in I$. The element $a$ is *completely meet-irreducible*, if $a \neq 1$ and $a = \bigwedge( b_i \mid i \in I )$ implies that $a = b_i$ for some $i \in I$.

For a finite lattice $L$, let $\mathrm{Ji}\,L$ denote the set of all join-irreducible elements, regarded as an order under the ordering of $L$. Clearly,

$$a = \bigvee( x \in \mathrm{Ji}\,L \mid x \leq a ) = \mathrm{id}(a) \cap \mathrm{Ji}\,L.$$

Dually, we can form $\mathrm{Mi}\,L$, the order of meet-irreducible elements of $L$. Examples are given in the Exercises.

### 6.4    ◇ Axiomatic games

**Lattices** can be defined by identities in innumerable ways. Of the eight identities—(Idem), (Comm), (Assoc), (Absorp)—we used to define lattices in Section 1.10, two (the identities in (Idem)) can be dropped. Ju. I. Sorkin [657] reduces six to four, R. Padmanabhan [567] to two, and finally, in R. N. McKenzie [511], a single identity is found characterizing lattices. Ju. I. Sorkin's identities use only three variables; the others use more. More recently, R. Padmanabhan [568] has found two identities in three variables characterizing lattices. It is easy to see that two variables would not suffice: take the lattice of Figure 21 and redefine the join of the two atoms to be 1; otherwise keep all the joins and meets. The resulting algebra is not a lattice, but every subalgebra generated by two elements is a lattice. Therefore, lattices cannot be defined by identities in two variables.

A result of A. Tarski [676] states that, given any integer $n$, there exists a set of $n$ identities defining lattices such that no identity can be dropped from the set (an *irredundant* set).

Problem I.17 of G. Grätzer [257] proposed to find the shortest single identity characterizing lattices.

R. N. McKenzie's result [511] was already known, but his one identity was very, very long; according to W. Taylor (see his Appendix 4 in [269]), it was of length about $300,000$.

R. Padmanabhan and his collaborators have been working over the last 40 years to improve on McKenzie's result, to produce a single identity characterizing lattices with fewer variables and shorter length:

|  | variables | length |
|---|---|---|
| R. N. McKenzie [511] | 34 | 300, 000 |
| R. Padmanabhan [570] | 7 | 241 |
| R. Padmanabhan and W. McCune [572] | 7 | 77 |
| R. Padmanabhan, W. McCune, and R. Veroff [573] | 8 | 27 |

And here is the lattice identity of length 27:

$$(((y \vee x) \wedge x) \vee (((z \wedge (x \vee x)) \vee (u \wedge x)) \wedge v)) \wedge (w \vee ((s \vee x) \wedge (x \vee t))) = x.$$

According to an unpublished result of R. Padmanabhan and W. McCune, length 27 *cannot be improved* to 24.

The book W. McCune and R. Padmanabhan [527] explains how McCune's Otter and MACE computer programs aid in finding such identities.

Many of the examples of lattice identities are selfdual. The identity

$$a = (a \wedge b) \vee (a \wedge c),$$

where

$$a = x \wedge ((x \wedge y) \vee (y \wedge z) \vee (z \wedge x)),$$
$$b = (x \wedge y) \vee (y \wedge z),$$
$$c = (x \wedge z) \vee (y \wedge z),$$

is an example of a non-selfdual identity. This identity holds in a lattice $L$ iff $L$ does not have a sublattice isomorphic to the dual of the fifth (center) lattice of Figure 110; see H. F. Löwig [505].

**Modular and distributive lattices** can be defined (as algebras $(L; \vee, \wedge)$) by two identities, see Exercise 4.22. Nicer sets of identities for these cases can be found in M. Kolibiar [478]; for instance, the two identities

$$(a \vee (b \wedge b)) \wedge b = b,$$
$$((a \wedge b) \wedge c) \vee (a \wedge d) = ((d \wedge a) \vee (c \wedge b)) \wedge a$$

characterize modular lattices. By M. Sholander [644],

$$a \wedge (a \vee b) = a,$$
$$a \wedge (b \vee c) = (c \wedge a) \vee (b \wedge a)$$

characterize distributive lattices. See also B. Riečan [610].

**Pseudocomplementation** can be described by identities (involving *) as first pointed out by P. Ribenboim [609]. A. Monteiro [534] accomplished the same for relative pseudocomplementation; see also R. Balbes and A. Horn [47] and R. Balbes and P. Dwinger [45]. This fact is applied in G. Grätzer [255].

**Boolean algebras** are usually axiomatized using identities in three variables. It is proved in A. H. Diamond and J. C. C. McKinsey [151] that two variables would not suffice. Finite algebras in which every two-generated subalgebra is boolean were investigated in R. W. Quackenbush [602].

E. V. Huntington [409] provides one of the most useful axiomatizations of boolean algebras: *A boolean algebra is a complemented lattice in which the complementation is pseudocomplementation.* Contrast this with Corollary 502. Observe that a proof of Huntington's result is implicit in the proof of Theorem 100.

One of the briefest axiom systems for boolean algebras in terms of the operations $\wedge$ and $'$ is due to L. Byrne [80]:

$$a \wedge b = b \wedge a,$$
$$a \wedge (b \wedge c) = (a \wedge b) \wedge c,$$
$$a \wedge b' = c \wedge c' \quad \text{iff} \quad a \wedge b = a.$$

Characterization of boolean algebras by identities is usually longer. It was observed by R. N. McKenzie, A. Tarski, and the author that boolean algebras can be defined by a single identity (see A. Tarski [676], and G. Grätzer and R. N. McKenzie [323]). A thorough survey of the axiom systems of boolean algebras is given in S. Rudeanu [620]; see also F. M. Sioson [649] and S. Rudeanu [621]. The only known irredundant selfdual axiom system can be found in R. Padmanabhan [571].

For semilattices, lattices, modular lattices, distributive lattices, and boolean algebras, the book R. Padmanabhan and S. Rudeanu [574] discusses all known axiom systems.

## Exercises

6.1. Find a homomorphism of bounded lattices that is not a $\{0, 1\}$-homomorphism, and a sublattice that is not a $\{0, 1\}$-sublattice.

6.2. Find a modular lattice in which every element $x \neq 0, 1$ has exactly $\mathfrak{m}$ complements.

6.3. Let $L$ be a distributive lattice, $a, b \in L$. Prove that if $a \wedge b$ and $a \vee b$ have complements, so do $a$ and $b$.

6.4. In a bounded lattice $L$, let $x$ be a relative complement of $a$ in $[b, c]$; let $y$ be a relative complement of $c$ in $[x, 1]$; let $z$ be a relative complement of $b$ in $[0, x]$; and let $t$ be a relative complement of $x$ in $[z, y]$. Verify that $t$ is a complement of $a$.

6.5. Let $B_0$ and $B_1$ be boolean algebras and let $\varphi$ be a $\{0, 1\}$-(lattice) homomorphism of $B_0$ into $B_1$. Show that $\varphi$ is a homomorphism of the boolean algebras.

6.6. Let $L$ be a lattice with zero. Show that if $L$ is distributive then Id $L$ is pseudocomplemented.

6.7. Is the converse of Exercise 6.6 true?

6.8. Show that the following lattices are pseudocomplemented but not complemented: any bounded chain of more than two elements; the lattice $N_5$; the lattice of Figure 21; the lattice of open subsets of the real line. In each case, give an element $x$ for which $x^{**} \neq x$.

6.9. In each of the pseudocomplemented lattices of Exercise 6.7, describe the set of dense elements.

6.10. Show that every finite distributive lattice is pseudocomplemented.

6.11. Give an example of a bounded distributive lattice that is not pseudo-complemented.

6.12. Let $L$ be a pseudocomplemented lattice. Show that

$$a^{**} \vee_{\mathrm{Skel}} b^{**} = (a \vee b)^{**}.$$

6.13. Find arbitrarily large pseudocomplemented lattices in which

$$\mathrm{Skel}\, L = \{0, 1\}.$$

6.14. Prove that in a boolean lattice $B$, $x \neq 0$ is join-irreducible iff $x$ is an atom, so $\mathrm{Ji}\, B = \mathrm{Atom}(B)$.

6.15. Show that in a finite lattice every element is the join of join-irreducible elements.

6.16. Verify that "finite lattice" in Exercise 6.15 can be replaced by "lattice satisfying the Descending Chain Condition". (A lattice $L$ or in general, an order, satisfies the *Descending Chain Condition* if $x_0, x_1, x_2, \ldots \in L$ and $x_0 \geq x_1 \geq x_2 \geq \cdots$ imply that $x_n = x_{n+1} = \cdots$ for some $n$.)

*6.17. Show that some form of the Axiom of Choice must be used to verify Exercise 6.16.

6.18. Prove that if a lattice $L$ satisfies the Descending Chain Condition, then $L$ is atomic.

6.19. The dual of the Descending Chain Condition is the Ascending Chain Condition; see Exercise 2.5. Dualize Exercises 6.15–6.18.

6.20. Show that an order satisfies the Ascending Chain Condition and the Descending Chain Condition iff all chains are finite.

6.21. Find a lattice in which all chains are finite but the lattice contains a chain of $n$ elements for every natural number $n$.

6.22. Find a lattice in which there are no join- or meet-irreducible elements.

6.23. Let $L$ be a finite lattice. Prove the inequalities

$$\mathrm{breadth}(L) \leq \dim(L) \leq \mathrm{width}(\mathrm{Ji}\, L)$$

(see Exercise 1.22) and that both these inequalities become equalities in case $L$ is distributive. (Use Dilworth's Chain Decomposition Theorem, page 49.)

6.24. Prove that the Ascending Chain Condition (Descending Chain Condition) holds in a lattice $L$ iff every ideal (filter) of $L$ is principal.

6.25. Let $L$ be a lattice and let $C$ be a chain with $|C| \leq \aleph_0$. Prove that if there is a homomorphism of $L$ onto $C$, then $L$ contains an isomorphic copy of $C$. How does this result extend other lattices $C$?

6.26. Does the result of Exercise 6.25 hold for the chain $C$ of real numbers? (No).

6.27. Let $\wedge$ be a binary operation on $L$, let $*$ be a unary operation on $L$ (that is, $a^* \in L$, for every $a \in L$), and let $0$ be a nullary operation (that is, $0 \in L$). Show that $(L, \wedge, 0, ^*)$ is a meet-semilattice with least element $0$ and pseudocomplement operation $*$ iff the following identities hold (R. Balbes and A. Horn [46]):

$$a \wedge b = b \wedge a,$$
$$(a \wedge b) \wedge c = a \wedge (b \wedge c),$$
$$a \wedge a = a,$$
$$0 \wedge a = 0,$$
$$a \wedge (a \wedge b)^* = a \wedge b^*,$$
$$a \wedge 0^* = a,$$
$$(0^*)^* = 0.$$

6.28. Let $L$ be a pseudocomplemented meet-semilattice and let $a, b \in L$. Verify the formula

$$(a \wedge b)^* = (a^{**} \wedge b)^* = (a^{**} \wedge b^{**})^*.$$

6.29. Let $L$ be a meet-semilattice and let $a, b \in L$. Show that $a * b$, the pseudocomplement of $a$ relative to $b$, is unique if it exists; show that $a * a$ exists iff $L$ has a unit.

Furthermore, let $L$ be a relatively pseudocomplemented meet-semilattice with zero and let $a \in L$. Show that $L$ is pseudocomplemented with $a^* = a * 0$.

6.30. Prove that if $L$ is a relatively pseudocomplemented lattice, then the following implication holds:

If $a \wedge p \leq b$ and $a \wedge q \leq b$, then $a \wedge (p \vee q) \leq b$.

Deduce that $L$ is distributive.

6.31. Let $L$ be a relatively pseudocomplemented meet-semilattice. Show that $L$ has a unit and that

$$a * (b * c) = (a \wedge b) * c,$$
$$a * (b * c) = (a * b) * (a * c)$$

for all $a, b, c \in L$. Furthermore, if $L$ is a lattice, then $L$ is distributive.

6.32. Let $L$ be a pseudocomplemented distributive lattice. Prove that id($a$) is a pseudocomplemented distributive lattice for each $a \in L$; in fact, the pseudocomplement of $x \in$ id($a$) in id($a$) is $x^* \wedge a$.

*6.33. Let $L$ be a pseudocomplemented distributive lattice. Let Skel($a$) denote the set of elements of the form $x^* \wedge a$ with $x \leq a$. Then Skel($a$) is a boolean algebra by Theorem 100. Let $\vee_a$ denote the join in Skel($a$). Show that if $x, y \in$ Skel($a$) and $x, y \in$ Skel($b$), for $a, b \in$ Skel $L$, then

$$x \vee_a y = x \vee_b y.$$

*6.34. Let $b \in$ Skel($a$). Prove that Skel($b$) $\subseteq$ Skel($a$). (The results of Exercises 6.33 and 6.34 first appeared in G. Grätzer [257].)

6.35. Show that $\mathsf{T}_n$ (see Exercise 2.36) is complemented.

6.36. Show that every interval is pseudocomplemented in $\mathsf{T}_n$. (The last two exercises are due to H. Lakser.)

6.37. For a finite lattice $L$, let Irr $L$ denote the set of all doubly irreducible elements. Prove that $|L| \geq 2(\text{len}(L)+1) - |\text{Irr }L|$. (Exercises 6.37–6.40 are based on I. Rival [614].)

6.38. Show that for a finite lattice $L$,

$$\text{len}(\text{Sub } L) = |\text{Irr } L| + \text{len}(\text{Sub }(L - \text{Irr } L)).$$

6.39. A finite lattice $L$ of $n$ elements is *dismantlable* if there is a chain $L_1 \subset L_2 \subset \cdots \subset L_n = L$ of sublattices satisfying $|L_i| = i$. Show that every lattice with at most seven elements is dismantlable. (Hint: use Exercise 6.37.)

6.40. Show that, for every integer $n \geq 8$, there is a lattice of $n$ elements which is not dismantlable.

6.41. A finite lattice is *planar* if it has a planar diagram (see Section 2.1). Show that every finite planar lattice is dismantlable.

6.42. For every integer $n \geq 9$, construct an $n$-element dismantlable lattice which is not planar. (For Exercises 6.41–6.42, see K. A. Baker, P. C. Fishburn, and F. S. Roberts [42].)

6.43. If a dismantlable lattice $L$ is not a chain, then it contains two incomparable doubly irreducible elements.

6.44. Prove that every sublattice and homomorphic image of a dismantlable lattice is dismantlable. Is this also true for planar lattices? (Exercises 6.43 and 6.44 are from D. Kelly and I. Rival [469].)

6.45. Let $L$ be a pseudocomplemented meet-semilattice. Show that $x \mapsto x^{**}$ is a closure map on $L$.

6.46. Let $L$ be a pseudocomplemented meet-semilattice. Let $\gamma$ denote the Glivenko congruence on $L$. Show that the following conditions are equivalent:

(i) $L$ is a boolean lattice;

(ii) $\gamma = \mathbf{0}$;

(iii) $L$ satisfies identity $x = x^{**}$.

6.47. Let $L$ be a pseudocomplemented meet-semilattice. Is Dns $L$ a filter of $L$?

6.48. Let $L$ be a bounded lattice. Then $L$ is relatively pseudocomplemented if the following two conditions hold:

(i) $L$ is distributive and pseudocomplemented;

(ii) Dns $L$ is relatively pseudocomplemented.

6.49. For a finite lattice $L$, form the triple $(\text{Ji}\,L, \text{Mi}\,L, \leq)$ (following the pattern established in Theorem 33). Show that $L$ is uniquely determined by the triple $(\text{Ji}\,L, \text{Mi}\,L, \leq)$ (R. Wille [739] and B. Ganter and R. Wille [219]).

6.50. Prove that an algebra $(L; \vee, \wedge, {}')$ is Boolean iff $(L; \vee, \wedge)$ is a lattice and it satisfies the self-dual identity

$$(x \vee y) \wedge (x \vee y') = (x \wedge y) \vee (x \wedge y').$$

6.51. Prove that the following identity and its dual form an independent basis for lattices

$$(((x \vee y) \wedge y) \vee (z \wedge y)) \wedge (u \vee ((v \vee y) \wedge (y \vee w))) = y$$

(Exercises 6.50 and 6.51 are due to W. McCune and R. Padmanabhan; see R. Padmanabhan and S. Rudeanu [574]).

6.52. Show that every uniquely complemented lattice in the variety generated by the least non-modular lattice $N_5$ is distributive (R. Padmanabhan [569]).

# Chapter

# II

# Distributive Lattices

## 1. Characterization and Representation Theorems

### 1.1 Characterization theorems

The two typical examples of nondistributive lattices are $N_5$ and $M_3$, whose diagrams are given in Figure 24. Our next result characterizes distributivity by the absence of these lattices as sublattices.

We introduce special names and notation for these lattices. A sublattice $A$ of a lattice $L$ is called a *pentagon*, respectively a *diamond*, if $A$ is isomorphic to $N_5$, respectively to $M_3$. If we say that $e_0, e_1, e_2, e_3, e_4$ is a pentagon (respectively, a diamond), we also assume that $e_0 \mapsto o$, $e_1 \mapsto a$, $e_2 \mapsto b$, $e_3 \mapsto c$, $e_4 \mapsto i$ is an isomorphism of $A$ with $N_5$ (respectively, with $M_3$).

The characterization theorem will be stated in two forms. Theorem 101 is a striking and useful characterization of distributive lattices; Theorem 102 is a more detailed version of Theorem 101 with some additional information.

**Theorem 101.** *A lattice $L$ is distributive iff $L$ does not contain a pentagon or a diamond.*

**Theorem 102.**

(i) *A lattice $L$ is modular iff it does not contain a pentagon.*

(ii) *A modular lattice $L$ is distributive iff it does not contain a diamond.*

*Proof.*

(i) If $L$ is modular, then every sublattice of $L$ is also modular; $N_5$ is not modular, thus it cannot be isomorphic to a sublattice of $L$.

G. Grätzer, *Lattice Theory: Foundation*, DOI 10.1007/978-3-0348-0018-1_2,
© Springer Basel AG 2011

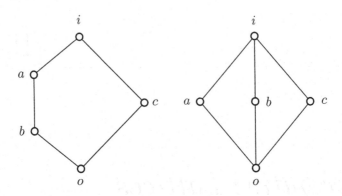

Figure 24. The lattices $N_5$ and $M_3$

Conversely, let $L$ be nonmodular, let $a, b, c \in L$ with $a \geq b$ and let

$$(a \wedge c) \vee b \neq a \wedge (c \vee b).$$

The free lattice generated by $a, b, c$ with $a \geq b$ is shown in Figure 6. Therefore, the sublattice of $L$ generated by $a, b, c$ must be a homomorphic image of the lattice of Figure 6. Observe that if two of the five elements

$$a \wedge c, \ (a \wedge c) \vee b, \ a \wedge (b \vee c), \ b \vee c, c$$

are identified under a homomorphism, then so are $(a \wedge c) \vee b$ and $a \wedge (b \vee c)$. Consequently, these five elements are distinct in $L$, and they form a pentagon.

(ii) Let $L$ be modular, but nondistributive, and choose $x, y, z \in L$ such that

$$x \wedge (y \vee z) \neq (x \wedge y) \vee (x \wedge z).$$

The free modular lattice generated by $x, y, z$ is shown in Figure 20. By inspecting the diagram we see that the elements $u, x_1, y_1, z_1, v$ form a diamond. Thus in *any* modular lattice, they form a sublattice isomorphic to a quotient lattice of $M_3$. But $M_3$ has only two quotient lattices: $M_3$ and the one-element lattice. In the former case, we have finished the proof. In the latter case, note that if $u$ and $v$ collapse, then so do $x \wedge (y \vee z)$ and $(x \wedge y) \vee (x \wedge z)$, contrary to our assumption. $\qquad \square$

Naturally, Theorems 101 and 102 could be proved without any reference to free lattices. A routine proof of (ii) runs as follows: Take $x, y, z$ in a modular lattice $L$ such that $x \wedge (y \vee z) \neq (x \wedge y) \vee (x \wedge z)$ and define the elements $u, x_1, y_1, z_1, v$ as the corresponding terms of Figure 20. Then a direct computation shows that $u, x_1, y_1, z_1, v$ form a diamond. There are some very natural objections to such a proof. How are the appropriate terms found? How is it possible to guess the result? And there is only one answer: by working it out in the free lattice.

For some special classes of lattices, Theorems 101 and 102 have various stronger forms that claim the existence of very large or very small pentagons and diamonds. For instance, a bounded relatively complemented nonmodular lattice always contains a pentagon as a $\{0, 1\}$-sublattice. The same is true of the diamond in certain complemented modular lattices; such results are implicit in J. von Neumann [552], [553]. If the lattice is finite, modular, and nondistributive, then it contains a *cover-preserving* diamond, that is, a diamond in which $a, b, c$ cover $o$, and $i$ covers $a, b, c$. (See E. Fried, G. Grätzer, and H. Lakser [201] for related results.) If $L$ is finite and nonmodular, then the pentagon it contains can be required to satisfy $a \succ b$.

**Corollary 103.** *A lattice $L$ is distributive iff every element has at most one relative complement in any interval.*

*Proof.* The "only if" part was proved in Section I.6.1. If $L$ is nondistributive, then, by Theorem 102, it contains a pentagon or a diamond, and each has an element with two relative complements in some interval.  □

**Corollary 104.** *A lattice $L$ is distributive iff, for any two ideals $I, J \in L$:*

$$I \vee J = \{\, i \vee j \mid i \in I,\ j \in J \,\}.$$

*Proof.* Let $L$ be distributive. By Lemma 5(ii), if $t \in I \vee J$, then $t \leq i \vee j$ for some $i \in I$ and $j \in J$. Therefore,

$$t = t \wedge (i \vee j) = (t \wedge i) \vee (t \wedge j), \qquad t \wedge i \in I, t \wedge j \in J.$$

Conversely, if $L$ is nondistributive, then $L$ contains elements $a, b, c$ as in Figure 24. Let $I = \mathrm{id}(b)$ and $J = \mathrm{id}(c)$; observe that $a \in I \vee J$, since $a \leq b \vee c$. However, $a$ has no representation as required in this corollary, because if $a = b_1 \vee c_1$ with $b_1 \in \mathrm{id}(b)$ and $c_1 \in \mathrm{id}(c)$, then $c_1 \leq a \wedge c = o$ would give that $a = b_1 \vee c_1 \leq b_1 \vee o = b_1 \leq b$, that is, $a \leq b$, a contradiction.  □

Another important property of ideals of a distributive lattice is the following statement.

**Lemma 105.** *Let $I$ and $J$ be ideals of a distributive lattice $L$. If $I \wedge J$ and $I \vee J$ are principal, then so are $I$ and $J$.*

*Proof.* Let $I \wedge J = \mathrm{id}(x)$ and $I \vee J = \mathrm{id}(y)$. Then $y = i \vee j$ for some $i \in I$ and $j \in J$ by Corollary 104. Set $c = x \vee i$ and $b = x \vee j$; note that $c \in I$ and $b \in J$ (since $x \in I \wedge J = I \cap J$). We claim that $I = \mathrm{id}(c)$ and $J = \mathrm{id}(b)$. Indeed, if for instance, $J \neq \mathrm{id}(b)$, then there is an $a > b$ with $a \in J$. It is easy to see that the elements $x, a, b, c, y$ form a pentagon.  □

**Theorem 106.** *Let $L$ be a distributive lattice and let $a \in L$. Then the map*

$$\varphi\colon x \mapsto (x \wedge a, x \vee a), \quad x \in L,$$

*is an embedding of $L$ into $\mathrm{id}(a) \times \mathrm{fil}(a)$; it is an isomorphism if $a$ has a complement.*

*Proof.* The map $\varphi$ is one-to-one, since if $\varphi(x) = \varphi(y)$, then $x$ and $y$ are both relative complements of $a$ in the same interval; thus $x = y$ by Corollary 103. Distributivity implies that $\varphi$ is a homomorphism.

   If $a$ has a complement $b$ and $(u, v) \in \mathrm{id}(a) \times \mathrm{fil}(a)$, then $\varphi(x) = (u, v)$ for $x = (u \vee b) \wedge v$; therefore, $\varphi$ is an isomorphism.  $\square$

### 1.2   Structure theorems, finite case

We start the detailed investigation of the structure of distributive lattices with the finite case. Our basic tool is the concept of down-sets introduced in Section I.1.6. Note that $\mathrm{Down}\, P$ is a lattice in which join and meet are union and intersection, respectively, and thus $\mathrm{Down}\, P$ is distributive.

   In Section I.6.3, we introduced the order $\mathrm{Ji}\, L$ of nonzero join-irreducible elements of a lattice $L$. Set

$$\mathrm{spec}(a) = \{\, x \in \mathrm{Ji}\, L \mid x \leq a \,\} = \mathrm{id}(a) \cap \mathrm{Ji}\, L = {\downarrow} a \cap \mathrm{Ji}\, L,$$

the *spectrum* of $a$. (We give a variant of this definition in the proof of Theorem 119.)

   The structure of finite distributive lattices is revealed by the following result:

**Theorem 107.** *Let $L$ be a finite distributive lattice. Then the map*

$$\varphi\colon a \mapsto \mathrm{spec}(a)$$

*is an isomorphism between $L$ and $\mathrm{Down}\, \mathrm{Ji}\, L$.*

*Proof.* Since $L$ is finite, every element is the join of nonzero join-irreducible elements; thus

$$a = \bigvee \mathrm{spec}(a),$$

showing that the map $\varphi$ is one-to-one. Obviously,

$$\mathrm{spec}(a) \cap \mathrm{spec}(b) = \mathrm{spec}(a \wedge b),$$

and so $\varphi(a \wedge b) = \varphi(a) \wedge \varphi(b)$. The formula $\varphi(a \vee b) = \varphi(a) \vee \varphi(b)$ is equivalent to

$$\mathrm{spec}(a \vee b) = \mathrm{spec}(a) \cup \mathrm{spec}(b).$$

To verify this formula, note that $\mathrm{spec}(a) \cup \mathrm{spec}(b) \subseteq \mathrm{spec}(a \vee b)$ is trivial. Now let $x \in \mathrm{spec}(a \vee b)$. Then

$$x = x \wedge (a \vee b) = (x \wedge a) \vee (x \wedge b);$$

therefore, $x = x \wedge a$ or $x = x \wedge b$, since $x$ is join-irreducible. Thus $x \in \mathrm{spec}(a)$ or $x \in \mathrm{spec}(b)$, that is, $x \in \mathrm{spec}(a) \cup \mathrm{spec}(b)$.

Finally, we have to show that if $A \in \mathrm{Down}\,\mathrm{Ji}\,L$, then $\varphi(a) = A$ for some $a \in L$. Set $a = \bigvee A$. Then $\mathrm{spec}(a) \supseteq A$ is obvious. Let $x \in \mathrm{spec}(a)$; then

$$x = x \wedge a = x \wedge \bigvee A = \bigvee (x \wedge y \mid y \in A).$$

Since $x$ is join-irreducible, it follows that $x = x \wedge y$, for some $y \in A$, implying that $x \in A$, since $A$ is a down-set. $\qquad\square$

**Corollary 108.** *The correspondence $L \mapsto \mathrm{Ji}\,L$ makes the class of all finite distributive lattices with more than one element correspond to the class of all finite orders; isomorphic lattices correspond to isomorphic orders, and vice versa.*

*Proof.* This is obvious from $\mathrm{Ji}\,\mathrm{Down}\,P \cong P$ and $\mathrm{Down}\,\mathrm{Ji}\,L \cong L$. $\qquad\square$

A sublattice $S$ of $\mathrm{Pow}\,A$ is called a *ring of sets*. Since $\mathrm{Down}\,\mathrm{Ji}\,L$ is a ring of sets, we obtain:

**Corollary 109.** *A finite lattice is distributive iff it is isomorphic to a ring of sets.*

If $Q$ is unordered, then $\mathrm{Down}\,Q = \mathrm{Pow}\,Q$; if $B$ is finite and boolean, then $\mathrm{Ji}\,B = \mathrm{Atom}(B)$ and therefore, $\mathrm{Ji}\,B$ is unordered. Thus we get:

**Corollary 110.** *A finite lattice is boolean iff it is isomorphic to the boolean lattice of all subsets of a finite set.*

For an element $a$ of a lattice $L$, the representation

$$a = x_0 \vee \cdots \vee x_{n-1}$$

is *redundant* if

$$a = x_0 \vee \cdots \vee x_{i-1} \vee x_{i+1} \vee \cdots \vee x_{n-1},$$

for some $0 \leq i < n$; otherwise it is *irredundant*.

**Corollary 111.** *Every element of a finite distributive lattice has a unique irredundant representation as a join of join-irreducible elements.*

*Proof.* The existence of such a representation is obvious. If

$$a = x_0 \vee \cdots \vee x_{n-1}$$

is an irredundant representation, then

$$\operatorname{spec}(a) = \bigcup(\,\operatorname{spec}(x_i) \mid 0 \le i < n\,).$$

Thus $x$ occurs in such a representation iff $x$ is a maximal element of $\operatorname{spec}(a)$; hence the uniqueness. $\qquad\square$

**Corollary 112.** *Every maximal chain $C$ of a finite distributive lattice $L$ is of length $|\operatorname{Ji} L|$.*

*Proof.* For $a \in \operatorname{Ji} L$, let $m(a)$ be the smallest member of $C$ majorizing $a$. Then

$$\varphi \colon a \mapsto m(a)$$

is a one-to-one map of $\operatorname{Ji} L$ onto the nonzero elements of $C$.

   To prove that $\varphi$ is one-to-one, let $a \ne b \in \operatorname{Ji} L$ and $m(a) = m(b)$. If $m(a) = m(b) = 0$, then $a = b = 0$, contradicting that $a \ne b$. So let $m(a) = m(b) > 0$. Then $m(a) \succ x$ for an element $x \in C$. Therefore, $x \vee a = m(a) = m(b) = x \vee b$; and so $a = a \wedge (x \vee b) = (a \wedge x) \vee (a \wedge b)$, implying that $a \le x$ or $a \le b$, because $a$ is join-irreducible. But $a \le x$ implies that $m(a) \le x < m(a)$, a contradiction. Consequently, $a \le b$; similarly, $b \le a$; thus $a = b$.

   To prove that $\varphi$ is onto, let $y \succ z$ in $C$. Then $\operatorname{spec}(y) \supset \operatorname{spec}(z)$, by Theorem 107, and so $y = m(a)$ for every $a \in \operatorname{spec}(y) - \operatorname{spec}(z)$. $\qquad\square$

   Corollary 112 and its dual yield

$$|\operatorname{Ji} L| = |\operatorname{Mi} L|.$$

This also holds in the modular case, see Section V.5.13.

   For a finite distributive lattice $L$, what is the smallest $k$ such that $L$ is embeddable in a direct product of $k$ chains? For $a \in L$, let $n_a$ be the number of elements of $L$ covering $a$. Then $k = \max\{\, n_a \mid a \in L \,\}$. This is an easy application of the result of R. P. Dilworth [157], discussed for $k \le 2$ in Exercises 1.50–1.53 and in its full generality in Section 5.13. Note also that $k$ is the same as the width of $\operatorname{Ji} L$.

   It seems hard to generalize the uniqueness of an irredundant join-representation of an element of a finite distributive lattice. The most useful generalization is in R. P. Dilworth [153] (utilized, for instance, in the theory of finite convex geometries). In my opinion, the best generalization is that of R. P. Dilworth and P. Crawley [161] to relatively atomic, distributive, algebraic lattices. See the survey article by R. P. Dilworth [160] and S. Kinugawa and J. Hashimoto [472]. Some results on, and references to, the modular and semimodular cases can be found in Chapter V.

## 1.3  ◇ Structure theorems, finite case, categorical variant

The following version of Corollary 108 gives a wealth of additional information on the correspondence between finite orders and finite distributive lattices.

◇ **Theorem 113.** *Let $P$ and $Q$ be finite orders. Let*

$$L = \text{Down}\, P$$
$$K = \text{Down}\, Q$$

*Then*

(i) *With every $\{0,1\}$-homomorphism $f\colon L \to K$ we can associate an isotone map $\text{Ji}(f)\colon Q \to P$ defined by*

$$\text{Ji}(f)(y) = \inf\{\, x \in P \mid y \in f(\downarrow x)\,\},$$

*for $y \in Q$.*

(ii) *With every isotone map $\psi\colon Q \to P$ we can associate a $\{0,1\}$-homomorphism $\text{Down}(\psi)\colon L \to K$ defined by*

$$\text{Down}(\psi)(a) = \psi^{-1}(a),$$

*for $a \in L$.*

(iii) *The constructions of (i) and (ii) are inverse to one another, and so yield together a bijection between $\{0,1\}$-homomorphisms $L \to K$ and isotone maps $Q \to P$.*

(iv) *$f$ is one-to-one iff $\text{Ji}(f)$ is onto.*

(v) *$f$ is onto iff $\text{Ji}(f)$ is an order-embedding.*

This result tells us that there is a close relationship, like an isomorphism, between finite orders and finite distributive lattices. Category theory provides the language to formulate this mathematically. We give here an informal description of how this is done.

The finite orders form a *category* $\mathbf{Ord}_{\text{fin}}$; the *objects* are the finite orders, and for the finite orders $P$ and $Q$, the category contains the set of *morphisms* $\text{Hom}(P, Q)$, the set of isotone maps from $P$ to $Q$. If $\alpha \in \text{Hom}(P, Q)$ and $\beta \in \text{Hom}(Q, R)$, we can form the *composition* $\beta \circ \alpha$. We write $\beta\alpha$ for $\beta \circ \alpha$. Composition is associative.

Similarly, finite distributive lattices form a category $\mathbf{D}_{\text{fin}}$, where the morphisms are $\{0,1\}$-homomorphisms.

We have a *contravariant functor* $\text{Down}\colon \mathbf{Ord}_{\text{fin}} \to \mathbf{D}_{\text{fin}}$, that is, Down maps the objects of $\mathbf{Ord}_{\text{fin}}$ to objects of $\mathbf{D}_{\text{fin}}$ and if it maps $P$ to $L$ and $Q$ to $K$, then it maps $\text{Hom}(P, Q)$ to $\text{Hom}(K, L)$. (This reversal of direction is what is meant by calling this functor "contravariant". Functors that preserve the order of morphisms are called "covariant".) Similarly, we have a contravariant functor Ji: $\mathbf{D}_{\mathrm{fin}} \to \mathbf{Ord}_{\mathrm{fin}}$, that is, Ji maps the objects of $\mathbf{D}_{\mathrm{fin}}$ to objects of $\mathbf{Ord}_{\mathrm{fin}}$ and if it maps $L$ to $P$ and $K$ to $Q$, then it maps $\mathrm{Hom}(L, K)$ to $\mathrm{Hom}(Q, P)$.

Clearly, the composition Down $\circ$ Ji is a *covariant functor* from the category $\mathbf{D}_{\mathrm{fin}}$ into itself; it is covariant, because if it maps $L$ to $L'$ and $K$ to $K'$, then it maps $\mathrm{Hom}(L, K)$ to $\mathrm{Hom}(L', K')$.

Note that in Theorem 107, we get an isomorphism $\varphi_L$ between $L$ and $(\mathrm{Down} \circ \mathrm{Ji})(L)$.

Let $\mathrm{Id}_{\mathbf{D}_{\mathrm{fin}}}$ be the identity functor on $\mathbf{D}_{\mathrm{fin}}$.

**Theorem 114.** *The family of isomorphisms $(\varphi_L \mid L \in \mathbf{D}_{fin})$ is a* natural isomorphism *between the functors* $\mathrm{Id}_{\mathbf{D}_{fin}}$ *and* Down $\circ$ Ji, *meaning that if $\varphi \in \mathrm{Hom}(L, K)$, then the diagram*

$$
\begin{array}{ccc}
L & \xrightarrow{\ \varphi\ } & K \\[4pt]
\cong \downarrow \varphi_L & & \cong \downarrow \varphi_K \\[4pt]
(\mathrm{Down} \circ \mathrm{Ji})(L) & \xrightarrow{(\mathrm{Down} \circ \mathrm{Ji})(\varphi)} & (\mathrm{Down} \circ \mathrm{Ji})(K)
\end{array}
$$

*is commutative.*

And there is an analogous statement for a natural isomorphism between the functors $\mathrm{Id}_{\mathbf{Ord}_{\mathrm{fin}}}$ and Ji $\circ$ Down.

### 1.4   Structure theorems, infinite case

The crucial Theorem 107 and its most important consequence, Corollary 109, depend on the existence of sufficiently many join-irreducible elements in a finite distributive lattice. In an infinite distributive lattice, there may be no join-irreducible element. Note that in a distributive lattice $L$, a nonzero element $a$ is join-irreducible iff $L - \mathrm{fil}(a)$ is a prime ideal. In the infinite case, the role of join-irreducible elements is taken by prime ideals. The crucial result is the existence of sufficiently many prime ideals (as illustrated in Figure 25).

For a distributive lattice $L$ with more than one element, let $\mathrm{Spec}\, L$ (the "spectrum" of $L$) denote the set of all prime ideals of $L$, regarded as an order under $\subseteq$. The importance of $\mathrm{Spec}\, L$ should be clear from the following results. Topologies on $\mathrm{Spec}\, L$ will be discussed in Section 5.

We start with the fundamental result of M. H. Stone [668]]:

**Theorem 115.** *Let $L$ be a distributive lattice, let $I$ be an ideal, let $D$ be a filter of $L$, and let $I \cap D = \varnothing$. Then there exists a prime ideal $P$ of $L$ such that $P \supseteq I$ and $P \cap D = \varnothing$.*

*Proof.* Some form of the Axiom of Choice is needed to prove this statement. The most convenient form for this proof is:

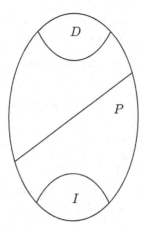

Figure 25. Illustrating Theorem 115

**Zorn's Lemma.** *Let $A$ be a set and let $\mathcal{X}$ be a nonempty subset of* Pow $A$. *Let us assume that $\mathcal{X}$ has the following property: If $\mathcal{C}$ is a chain in $(\mathcal{X}; \subseteq)$, then $\bigcup \mathcal{C} \in \mathcal{X}$. Then $\mathcal{X}$ has a maximal member.*

We define

$$\mathcal{X} = \bigcap ( P \in \operatorname{Spec} L \mid P \supseteq I, P \cap D = \varnothing )$$

and verify that $\mathcal{X}$ satisfies the hypothesis of Zorn's Lemma. The set $\mathcal{X}$ is nonempty, since $I \in \mathcal{X}$. Let $\mathcal{C}$ be a chain in $\mathcal{X}$ and let $M = \bigcup \mathcal{C}$. If $a, b \in M$, then $a \in X$ and $b \in Y$ for some $X, Y \in \mathcal{C}$. Since $\mathcal{C}$ is a chain, either $X \subseteq Y$ or $Y \subseteq X$ hold. If say, $X \subseteq Y$, then $a, b \in Y$, and so $a \vee b \in Y \subseteq M$, since $Y$ is an ideal. Also, if $b \leq a \in M$, then $a \in X \in \mathcal{C}$; since $X$ is an ideal, $b \in X \subseteq M$. Thus $M$ is an ideal. It is obvious that $M \supseteq I$ and $M \cap D = \varnothing$, verifying that $M \in \mathcal{X}$. Therefore, by Zorn's Lemma, $\mathcal{X}$ has a maximal element $P$.

We claim that $P$ is a prime ideal. Indeed, if $P$ is not prime, then there exist $a, b \in L$ such that $a, b \notin P$ but $a \wedge b \in P$. The maximality of $P$ yields that $(P \vee \operatorname{id}(a)) \cap D \neq \varnothing$ and $(P \vee \operatorname{id}(b)) \cap D \neq \varnothing$. Thus there are $p, q \in P$ such that $p \vee a \in D$ and $q \vee b \in D$. Then $x = (p \vee a) \wedge (q \vee b) \in D$, since $D$ is a filter. Expanding by distributivity,

$$x = (p \wedge q) \vee (p \wedge b) \vee (a \wedge q) \vee (a \wedge b) \in P;$$

thus $P \cap D \neq \varnothing$, a contradiction. $\qquad\square$

**Corollary 116.** *Let $L$ be a distributive lattice, let $I$ be an ideal of $L$, and let $a \in L$ and $a \notin I$. Then there is a prime ideal $P$ such that $P \supseteq I$ and $a \notin P$.*

*Proof.* Apply Theorem 115 to $I$ and $D = \operatorname{fil}(a)$. $\qquad\square$

**Corollary 117.** *Let $L$ be a distributive lattice, $a, b \in L$ and $a \neq b$. Then there is a prime ideal containing exactly one of $a$ and $b$.*

*Proof.* Either $\mathrm{id}(a) \cap \mathrm{fil}(b) = \varnothing$ or $\mathrm{fil}(a) \cap \mathrm{id}(b) = \varnothing$, so we can apply Corollary 116. $\qquad\square$

**Corollary 118.** *Every ideal $I$ of a distributive lattice is the intersection of all prime ideals containing it.*

*Proof.* Let

$$I_1 = \bigcap (\, P \in \mathrm{Spec}\, L \mid P \supseteq I \,).$$

Clearly, every $a \in I$ belongs to $I_1$. Conversely, if $a \notin I$, then by Corollary 117, there is a $P$ in the family $\{\, P \in \mathrm{Spec}\, L \mid P \supseteq I \,\}$ not containing $a$, so $a \notin I_1$. $\quad\square$

As a final application, we get the celebrated result of G. Birkhoff [61] and M. H. Stone [668]:

**Theorem 119.** *A lattice is distributive iff it is isomorphic to a ring of sets.*

*Proof.* Let $L$ be a distributive lattice. For $a \in L$, set

$$\mathrm{spec}(a) = \{\, P \in \mathrm{Spec}\, L \mid a \notin P \,\}.$$

the *spectrum* of $a$. Then the family of sets $\{\, \mathrm{spec}(a) \mid a \in L \,\}$ is a ring of sets, and the map $a \mapsto \mathrm{spec}(a)$ is an isomorphism. The details are similar to the proof of Theorem 107, except for the first step, which now uses Corollary 117. $\qquad\square$

### 1.5  Some applications

**Corollary 120.** *Let $L$ be a distributive lattice with more than one element. An identity holds in $L$ iff it holds in the two-element chain, $\mathsf{C}_2$.*

*Proof.* Let $p = q$ hold in $L$. Since $|L| > 1$, clearly $\mathsf{C}_2 \leq L$, and so $p = q$ holds in $\mathsf{C}_2$. Conversely, let $p = q$ hold in $\mathsf{C}_2$. Note that $\mathsf{C}_2 = \mathrm{Pow}\, X$ with $|X| = 1$, and that $\mathrm{Pow}\, A$ is isomorphic to the direct power $(\mathrm{Pow}\, X)^{|A|}$. Therefore, $p = q$ holds in any $\mathrm{Pow}\, A$. By Theorem 119, $L$ is a sublattice of some $\mathrm{Pow}\, A$; thus $p = q$ holds in $L$. $\qquad\square$

So now we have the result we claimed in Section I.5.5:

**Theorem 121.** *For any order $P$, a lattice completely freely generated by $P$, $\mathrm{CFree}_{\mathbf{V}}\, P$, exists for any variety $\mathbf{V}$ containing a two-element lattice.*

We can adapt Theorem 119 to boolean lattices, see M. H. Stone [668], using the concept of a *field of sets*: a ring of sets closed under set complementation.

**Corollary 122.** *A lattice is boolean iff it is isomorphic to a field of sets.*

*Proof.* Use the representation of Theorem 119. Obviously,

$$\mathrm{spec}(a') = \mathrm{Spec}\, L - \mathrm{spec}(a),$$

and thus complements are also preserved.     □

Some interesting properties of $L$ are reflected in $\mathrm{Spec}\, L$. An important result of this type is the following theorem of L. Nachbin [537] (see also L. Rieger [611]):

**Theorem 123.** *Let $L$ be a bounded distributive lattice with $0 \neq 1$. Then $L$ is a boolean lattice iff $\mathrm{Spec}\, L$ is unordered.*

*Proof.* Let $L$ be boolean, $P, Q \in \mathrm{Spec}\, L$, and $P \subset Q$. Choose $a \in Q - P$. Since $a \in Q$, clearly $a' \notin Q$, and thus $a' \notin P$. Therefore, $a, a' \notin P$, but $a \wedge a' = 0 \in P$, a contradiction, showing that $\mathrm{Spec}\, L$ is unordered. This proof, in fact, verifies that in a boolean algebra every prime ideal is maximal.

Now let $\mathrm{Spec}\, L$ be unordered and $a \in L$, and let us assume that $a$ has no complement. Set

$$D = \{\, x \mid a \vee x = 1 \,\}.$$

By distributivity, $D$ is a filter. Take

$$D_1 = D \vee \mathrm{fil}(a) = \{\, x \mid x \geq d \wedge a, \text{ for some } d \in D \,\}.$$

The filter $D_1$ does not contain 0, since $0 = d \wedge a$ and $a \vee d = 1$ would mean that $d$ is a complement of $a$. Thus there exists a prime ideal $P$ disjoint from $D_1$.

Note that $1 \notin \mathrm{id}(a) \vee P$, otherwise $1 = a \vee p$, for some $p \in P$, contradicting that $P \cap D = \varnothing$. Thus some prime ideal $Q$ contains $\mathrm{id}(a) \vee P$; and so $P \subset Q$, which is impossible since $\mathrm{Spec}\, L$ is unordered.     □

According to Corollary 118, every ideal is an intersection of prime ideals. When is this representation unique? This question was answered in J. Hashimoto [375].

**Theorem 124.** *Let $L$ be a bounded distributive lattice with $0 \neq 1$. Every ideal has a unique representation as an intersection of prime ideals iff $L$ is a finite boolean lattice.*

*Proof.* If $L$ is a finite boolean lattice, then $P$ is a prime ideal iff $P = \mathrm{id}(a)$, where $a$ is a dual atom; the uniqueness follows from Corollary 111 (or it is obvious by direct computation).

Now let every ideal of $L$ have a unique representation as a meet of prime ideals. We claim that $\mathrm{Id}\, L$ is boolean. Let $I \in \mathrm{Id}\, L$; define

$$J = \bigcap (\, P \in \mathrm{Spec}\, L \mid P \not\supseteq I \,).$$

Then

$$I \wedge J = \bigcap (P \mid P \in \operatorname{Spec} L) = \operatorname{id}(0).$$

If $L \neq I \vee J$, then there is a prime ideal $P_0 \supseteq I \vee J$, and consequently $J$ has two representations:

$$\bigcap (P \mid P \not\supseteq I) = P_0 \cap \bigcap (P \mid P \not\supseteq I).$$

Thus $L = I \vee J$ and $J$ is a complement of $I$ in $\operatorname{Id} L$.

So $I \vee J = L = \operatorname{id}(1)$ and $I \wedge J = \operatorname{id}(0)$, both principal. Thus by Lemma 105, every ideal of $L$ is principal. We conclude that $L \cong \operatorname{Id} L$, and so $L$ is boolean. By Exercise I.6.24, $L$ satisfies the Ascending Chain Condition; thus every element of $L$ other than the unit is majorized by a dual atom. Since the complement of a dual atom is an atom, by taking complements, we find that every nonzero element of $L$ majorizes an atom.

If $p_0, p_1, \dots, p_n, \dots \in \operatorname{Atom}(L)$, then the ascending chain

$$p_0, p_0 \vee p_1, \dots, p_0 \vee p_1 \vee \cdots \vee p_n, \dots$$

does not terminate, contradicting that $L$ satisfies the Ascending Chain Condition. Thus $\operatorname{Atom}(L)$ is finite, $\operatorname{Atom}(L) = \{p_0, \dots, p_{n-1}\}$. Define the element $a = p_0 \vee \cdots \vee p_{n-1}$. If $a' \neq 0$, then $a'$ has to majorize an atom, which is impossible. Therefore, $a' = 0$, $a = 1$, and $L \cong \operatorname{Pow} X$ with $|X| = n$. $\qquad\square$

## 1.6  Automorphism groups

Let $L$ be a lattice and let $\operatorname{Aut} L$ be the automorphism group of $L$ (see Section I.3.1). In this section, we prove the characterization theorem of automorphism groups, in fact, as in G. Birkhoff [68], we prove here more. (This proof is from G. Grätzer, E. T. Schmidt, and D. Wang [351].)

**Theorem 125.** *Every group $G$ can be represented as the automorphism group of a distributive lattice $D$. If $G$ is finite, $D$ can be chosen to be finite.*

*Proof.* Let $G = \{ g_\gamma \mid \gamma < \alpha \}$ with $g_0 = 1$, the unit element of the group; we assume that $|G| > 1$. We view ordinals as well-ordered chains. In particular, $\gamma \cong \delta$ iff $\gamma = \delta$ for any ordinals $\gamma$ and $\delta$.

For every $x, y \in G$ with $y \neq 1$ (equivalently, with $x \neq yx$), we construct the order $P(x, y)$ of Figure 26, defined on the set $\{x, yx\} \cup \{ (x, y, a_\gamma) \mid \gamma < \beta \}$, where $y = g_\beta$. Note that $G \cap P(x, y) = \{x, yx\}$, where $yx$ is the product of $y$ and $x$ in $G$. We order this set by

$$x < (x, y, 1) < (x, y, 2) < \cdots < (x, y, \gamma) < \cdots, \quad \text{for } \gamma < \beta,$$
$$yx < (x, y, 0) < (x, y, 1).$$

The two minimal elements of $P(x, y)$ are $x$ and $yx$, both in $G$.

Let $P = \bigcup(P(x,y) \mid x,y \in G,\ y \neq 1)$ be ordered by $u < v$ in $P$ iff $u < v$ in some $P(x,y)$. It is sufficient to prove that $\operatorname{Aut} P \cong G$. Indeed, let $L$ be the distributive lattice completely freely generated by $P$; this lattice $L$ exists by Theorem 92. Then $\operatorname{Aut} P \cong \operatorname{Aut} L$; moreover, if $G$ is finite, then both $P$ and $L$ are finite.

To prove that $\operatorname{Aut} P \cong G$, let $\sigma$ be an automorphism of $P$. The set of minimal elements of $P$ is $G$; it follows that the map $\sigma$ permutes $G$. Let $a = \sigma(1)$ and let $b \in G$. We want to show that $\sigma(b) = ba$.

If $b = 1$, this holds by the definition of $a$. So let us assume that $b \neq 1$. Let $b = g_\beta$ with $\beta < \alpha$. Then the order $P(1, b)$, with minimal elements $1$ and $b$, is defined (since $1 \neq b$). Also, $\sigma(b) \neq a$ and so $\sigma(b) = ua$ for some $u \in G$ with $u \neq 1$. Therefore, $P(a, u)$ with minimal elements $a = \sigma(1)$ and $ua = \sigma(b)$, is defined.

Thus $\sigma$ takes the minimal elements of $P(1, b)$ into the minimal elements of $P(a, u)$, hence it must take all of $P(1, b)$ to $P(a, u)$, so $P(1, b) \cong P(a, u)$. Thus the top chain of $P(a, u)$ is the same as the top chain of $P(1, b)$, that is, $\beta$, and so $u = b$, proving that $\sigma(b) = ba$.

For every $u \in G$, define the permutation $\sigma_u$ of $G$ by $\sigma_u(v) = vu$. Then we have just proved that every automorphism of $P$ restricted to $G$ is of this form; the converse is trivial. This completes the proof of the theorem.    □

For an alternative short proof, producing a surprisingly nice distributive lattice, see G. Grätzer, H. Lakser, and E. T. Schmidt [315], Exercises 1.55–1.57.

Small lattices with given automorphism groups are considered in R. Frucht [207] and [208]. R. N. McKenzie and J. Sichler have some related results for lattices of finite length. Two sample results: every group is the automorphism

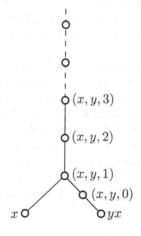

Figure 26. The order $P(x, y)$

group of a lattice of finite length; for every lattice $L$, there exists a bounded lattice $K$ such that $\operatorname{End} L \cong \operatorname{End}_{\{0,1\}} K$ and if $L$ is finite or finite length, then so is $K$, where $\operatorname{End} L$ is the endomorphism monoid (that is, semigroup with identity) of $L$ and $\operatorname{End}_{\{0,1\}} K$ is the monoid of those endomorphism of $K$ that fix 0 and 1. See also J. Sichler [645] and Section VII.3.4.

### 1.7  ◇ Distributive lattices and general algebra

R. Dedekind found the distributive identity by investigating ideals of number fields. Rings with a distributive lattice of ideals have been investigated by E. Noether [554], L. Fuchs [210] (who named such rings *arithmetical rings*— *MathSciNet* lists 61 papers on arithmetical rings alone), I. S. Cohen [93], and C. U. Jensen [426]. Varieties of rings with distributive ideal lattices were considered in G. Michler and R. Wille [530] and in H. Werner and R. Wille [718]. E. A. Behrens [52] and [53] considered rings in which one-sided ideals form a distributive lattice. Rings with a distributive lattice of subrings were classified in P. A. Freĭdman [192]. In this context, G. M. Bergman's work on the distributive-divisor-lattice of free algebras should be mentioned, see Chapter 4 of P. M. Cohn [94] and P. M. Cohn [95].

For an overview of distributive modules and rings, see A. A. Tuganbaev [682].

H. L. Silcock [648] proved that every finite distributive lattice is isomorphic to the lattice of normal subgroups of a group $G$. P. P. Pálfy [575] improved this result: $G$ may be taken to be finite solvable. P. Růžička, J. Tůma, and F. Wehrung [623] proved that every distributive algebraic lattice with at most $\aleph_1$ compact elements is isomorphic to the normal subgroup lattice of some locally finite group and to the submodule lattice of some right module (over a non-commutative ring). Furthermore, they proved that the $\aleph_1$ bound is optimal: for example, the congruence lattice of the free lattice on $\aleph_2$ generators is not isomorphic to the congruence lattice of any congruence-permutable algebra.

The subgroup lattice of a group $G$ is distributive iff $G$ is locally cyclic, see O. Ore [559] and [560].

The distributivity of congruence lattices of lattices has a number of important consequences, for instance, Jónsson's Lemma (Theorem 475). B. Jónsson [444] discovered that many of these results hold for arbitrary universal algebras with distributive congruence lattices. His result has found applications that go far beyond lattice theory—it has been applied to lattice-ordered algebras, closure algebras, nonassociative lattices, cylindric algebras, monadic algebras, lattices with pseudocomplementation, primal algebras, and multi-valued logics. (Jónsson's Lemma is referenced in 55 papers according to *MathSciNet*.)

The foregoing examples show the central role played by distributive lattices in applications of the lattice concept.

**Exercises**

1.1. Consider the three lattices whose diagrams are shown in Figure 7. Which are distributive? Show that the nondistributive ones contain a pentagon.

1.2. Work out a direct proof of Theorem 102(i).

1.3. Work out a direct proof of Theorem 102(ii).

1.4. Let $K$ be a five-element distributive lattice. Is there an identity $p = q$ such that $p = q$ holds in a lattice $L$ iff $L$ has no sublattice isomorphic to $K$?

1.5. Does the property stated in Lemma 105 characterize distributive lattices?

1.6. Let $L$ be a distributive lattice with zero and unit. Prove that the direct decompositions $L_0 \times L_1$ of $L$ are in one-to-one correspondence with the complemented elements of $L$.

1.7. Prove that the complemented elements of a distributive lattice form a sublattice.

1.8. Let $L$ be a distributive lattice with zero and unit. Let

$$L \cong L_0 \times L_1 \cong K_0 \times K_1.$$

Show that there is a direct decomposition

$$L \cong A_0 \times A_1 \times A_2 \times A_3$$

such that

$$A_0 \times A_1 \cong L_0,$$
$$A_2 \times A_3 \cong L_1,$$
$$A_0 \times A_2 \cong K_0,$$
$$A_1 \times A_3 \cong K_1.$$

1.9. Let $L = \mathsf{B}_3$. Describe the orders $\operatorname{Ji} L$ and $\operatorname{Down} \operatorname{Ji} L$.

1.10. Let $L = \operatorname{Free}_{\mathsf{D}}(3)$, see Figure 19. Describe $\operatorname{Ji} L$ and $\operatorname{Down} \operatorname{Ji} L$. Compare $|\operatorname{Free}_{\mathsf{D}}(3)|$ with $|\operatorname{Ji} \operatorname{Free}_{\mathsf{D}}(3)|$.

1.11. Verify Theorem 107 for the distributive lattices of Exercises 1.9 and 1.10.

1.12. Does Theorem 107 hold for countable chains?

1.13. Consider the modular lattice $L = \operatorname{Free}_{\mathsf{M}}(3)$. How many diamonds are in $L$?

1.14. Extend Theorem 107 to distributive lattices satisfying the Descending Chain Condition (see Exercise I.1.16).

1.15. Extend Corollary 108 to distributive lattices satisfying the Descending Chain Condition.

1.16. Can Exercises 1.14 and 1.15 be further sharpened?

1.17. Let $L$ be a distributive lattice with zero and unit. Let $C_0$ and $C_1$ be finite chains in $L$. Show that there exist chains $D_0 \supseteq C_0$ and $D_1 \supseteq C_1$ such that $|D_0| = |D_1|$.

1.18. Derive from Exercise 1.17 the result that all maximal chains of a finite distributive lattice have the same length.

1.19. Find examples showing that Exercise 1.17 is not valid if "finite" is omitted.

1.20. For a finite distributive lattice $L$ and $a \in L$, let $n_a$ be the number of elements of $L$ covering $a$. Prove that

$$\max\{\, n_a \mid a \in L \,\} = \operatorname{width}(\operatorname{Ji} L).$$

1.21. For a finite distributive lattice $L$, what is the smallest $k$ such that $L$ is embeddable in a direct product of $k$ chains? (Hint: the number in Exercise 1.20.)

1.22. Prove the theorem "$L$ is modular iff $\operatorname{Id} L$ is modular" by showing that "$L$ contains a pentagon iff $\operatorname{Id} L$ contains a pentagon".

*1.23. Is the second statement of Exercise 1.22 true for the diamond rather than for the pentagon?

1.24. Let $L$ be a distributive lattice, $a, b, c \in L$, and $a \leq b$. Is it true that $[a, b]$ is boolean iff $[a \wedge c, b \wedge c]$ and $[a \vee c, b \vee c]$ are boolean?

1.25. For an order $P$, let $\operatorname{Down}_{\mathrm{fin}} P$ denote the lattice of all subsets of $P$ of the form $\downarrow H$, where $H \subseteq P$ is finite. Does Theorem 107 hold for $\operatorname{Down}_{\mathrm{fin}} P$?

1.26. Show that the Ascending Chain Condition is equivalent to the Descending Chain Condition for boolean lattices.

1.27. Show that Exercise 1.26 fails to hold for *generalized boolean lattices* (that is, relatively complemented distributive lattices with zero).

1.28. Let $L$ be a lattice, let $P$ be a prime ideal of $L$, and let $a, b, c \in L$. Prove that if $a \vee (b \wedge c) \in P$, then $(a \vee b) \wedge (a \vee c) \in P$.

1.29. Using Exercise 1.28, show that the lattice $L$ is distributive iff, for all $x, y \in L$ with $x < y$, there exists a prime ideal $P$ satisfying $x \in P$ and $y \notin P$.

1.30. Verify the statement of Exercise 1.29 using Theorem 101.

1.31. Let $L$ be a distributive lattice. Then $L$ is relatively complemented iff $\operatorname{Spec} L$ is unordered.

1.32. Prove Theorem 115 by well-ordering the lattice, $L = \{\, a_\gamma \mid \gamma < \alpha \,\}$, and deciding one by one for each $a_\gamma$ whether $a_\gamma \in P$ or $a_\gamma \notin P$ (M. H. Stone [668]).

1.33. Let $L$ be a distributive lattice with unit. Show that every prime ideal $P$ is contained in a *maximal prime ideal* $Q$ (a prime ideal $R$ is maximal if $R \subseteq S \in \operatorname{Spec} L$ implies that $R = S$).

1.34. Let $L$ be a distributive lattice with zero. Verify that every prime ideal $P$ contains a *minimal prime ideal* $Q$ (a prime ideal $Q$ is minimal if

$Q \supseteq S \in \operatorname{Spec} L$ implies that $Q = S$).

1.35. Find a distributive lattice $L$ with no minimal and no maximal prime ideals.

1.36. Investigate the connections among the Ascending Chain Condition (and Descending Chain Condition) for a distributive lattice $L$, for the ideal lattice $\operatorname{Id} L$, and for the order $\operatorname{Spec} L$.

1.37. Let $L$ be a distributive lattice with zero and let $I \in \operatorname{Id} L$. Show that

$$\{\, x \mid \operatorname{id}(x) \wedge I = \operatorname{id}(0) \,\}$$

is the pseudocomplement of the ideal $I$ in $\operatorname{Id} L$. Conclude that $\operatorname{Id} L$ is pseudocomplemented.

1.38. Let $L$ be a distributive lattice with zero and let $I \in \operatorname{Id} L$. Prove that $I = I^{**}$, for every $I \in \operatorname{Id} L$, iff $L$ is a generalized boolean lattice satisfying the Descending Chain Condition.

1.39. The congruence relations $\alpha$ and $\beta$ *permute* if $\alpha \circ \beta = \beta \circ \alpha$. Show that the congruences of a relatively complemented lattice permute.

1.40. Prove the converse of Exercise 1.39 for distributive lattices.

1.41. Generalize Theorem 124 to distributive lattices without 0 and 1.

*1.42. Let $L$ be a distributive lattice, let $a \in L$, let $S \leq L$, and let $a \notin S$. Show that there exists a prime ideal $P$ and a prime filter $Q$ such that $a \notin P \cup Q \supseteq S$, provided that $a$ is not the 0 or 1 of $L$ (J. Hashimoto [375]).

1.43. Let $L$ be a relatively complemented distributive lattice. A sublattice $K$ of $L$ is *proper* if $K \neq L$. Show that every proper sublattice of $L$ can be extended to a maximal proper sublattice of $L$ (K. Takeuchi [671]; see also J. Hashimoto [375] and G. Grätzer and E. T. Schmidt [333]).

1.44. Show that the statement of Exercise 1.43 is not valid in general if $L$ is not relatively complemented (K. Takeuchi [671]; see also M. E. Adams [2]).

1.45. Generalize Corollary 111 to infinite distributive lattices, claiming the unique irredundant representation of certain ideals as a meet of prime ideals.

1.46. If $P$ is a prime ideal of $L$, then $\operatorname{id}(P)$ is a principal prime ideal of $\operatorname{Id} L$. Is the converse true?

1.47. Show that Corollary 117 characterizes distributivity.

1.48. Let $C$ be a chain in an order $P$. If $C \subseteq D$ implies that $C = D$, for every chain $D$ in $P$, then $C$ is called *maximal*. Using Zorn's Lemma, show that every chain is contained in a maximal chain.

1.49. Prove that a finite distributive lattice is planar iff no element is covered by three elements.

1.50. Show that a finite distributive lattice is planar iff it is dismantlable (see Exercise I.6.39).

1.51. Show that we can obtain every planar distributive lattice $D$ in the following way. We start with a direct product of two finite chains, $L_0 = C_1 \times C_2$. We obtain $L_1$ by removing a doubly irreducible element from the *boundary* of $L_0$. For $i > 1$, we obtain $L_i$ by removing a doubly irreducible element from the *boundary* of $L_{i-1}$. In finitely many steps, we obtain $D$.

1.52. Show that $D$ is a *cover-preserving sublattice* of $C_1 \times C_2$ in Exercise 1.51, that is, if $a \prec b$ in $D$, then $a \prec b$ in $C_1 \times C_2$.

1.53. Let $S$ be a sublattice of the finite lattice $L$. Then $S$ can be represented in the form

$$L - \bigcup([a_i, b_i] \mid i \in I),$$

where $a_i$ is join-irreducible and $b_i$ is meet-irreducible for all $i \in I$.

1.54. Prove the converse of Exercise 1.53 for distributive lattices. (Exercises 1.53 and 1.54 are from I. Rival [612]; see also I. Rival [613].)

1.55. Derive from Exercises VII.3.1–VII.3.16, that for every finite group $G$, there exists a finite graph $(V; E)$ (that is, $V$ is a nonempty set and $E \subseteq V^2$, as in Section I.1.5) such that $G$ is isomorphic to the automorphism group of $(V; E)$.

1.56. Let $G$ and $V$ be as in Exercise 1.55. Let $F$ be the free distributive lattice generated by $V$ with zero and unit. Define in $F$:

$$o = \bigvee(x \wedge y \mid \{x, y\} \in E).$$

Define the finite distributive lattice

$$D = [o, 1].$$

Prove that $\operatorname{Aut} D \cong G$ (G. Grätzer, H. Lakser, and E. T. Schmidt [315]).

1.57. Extend the construction of Exercise 1.56 to arbitrary groups, re-proving Birkhoff's result (Theorem 125).

## 2.  Terms and Freeness

### 2.1  Terms for distributive lattices

We can introduce an equivalence relation $\equiv_{\mathbf{D}}$ for lattice terms: for $p, q \in \operatorname{Term}(n)$, let $p \equiv_{\mathbf{D}} q$ iff $p$ and $q$ define the same functions in the class $\mathbf{D}$ of distributive lattices. More formally, if $p$ and $q$ are $n$-ary terms (see Section I.4.1), then $p \equiv_{\mathbf{D}} q$ if, for every *distributive lattice* $L$ and $a_1, \dots, a_n \in L$, the equality $p(a_1, \dots, a_n) = q(a_1, \dots, a_n)$ holds (see Definitions 52 and 53).

For an $n$-ary lattice term $p$, let $p/\mathbf{D}$ denote the set of all $n$-ary lattice terms $q$ satisfying $p \equiv_{\mathbf{D}} q$ and let $\operatorname{Term}_{\mathbf{D}}(n)$ denote the set of all these blocks, that is,

$$\operatorname{Term}_{\mathbf{D}}(n) = \{\, p/\mathbf{D} \mid p \in \operatorname{Term}(n) \,\}.$$

Observe that, for any $p, p_1, q, q_1 \in \text{Term}(n)$, if $p \equiv_{\mathbf{D}} p_1$ and $q \equiv_{\mathbf{D}} q_1$, then $p \vee q \equiv_{\mathbf{D}} p_1 \vee q_1$ and $p \wedge q \equiv_{\mathbf{D}} p_1 \wedge q_1$. Thus

$$p/\mathbf{D} \vee q/\mathbf{D} = (p \vee q)/\mathbf{D},$$
$$p/\mathbf{D} \wedge q/\mathbf{D} = (p \wedge q)/\mathbf{D}$$

define the operations $\vee$ and $\wedge$ on $\text{Term}_{\mathbf{D}}(n)$. It is easily seen that $\text{Term}_{\mathbf{D}}(n)$ is a distributive lattice and $p/\mathbf{D} \leq q/\mathbf{D}$ iff the inequality $p \leq q$ holds in the class $\mathbf{D}$.

To describe the structure of $\text{Term}_{\mathbf{D}}(n)$, for $n > 0$, let $Q(n)$ denote the dual of the order of all proper nonempty subsets of $\{0, 1, \ldots, n-1\}$.

**Theorem 126.** *Let $n > 0$. Then*

(i) $\text{Term}_{\mathbf{D}}(n)$ *is a free distributive lattice on $n$ generators.*

(ii) $\text{Term}_{\mathbf{D}}(n)$ *is isomorphic with $\text{Down}\, Q(n)$.*

(iii) $2^n - 2 \leq |\text{Term}_{\mathbf{D}}(n)| \leq 2^{2^n - 2}$.

(iv) *A finitely generated distributive lattice is finite.*

*Proof.*

(i) Let $L$ be a distributive lattice, $a_0, \ldots, a_{n-1} \in L$. Then the map $x_i \mapsto a_i$ can be extended to the homomorphism

$$p/\mathbf{D} \mapsto p(a_0, \ldots, a_{n-1}),$$

proving (i).

(ii) A lattice term $p$ is called a *meet-term* if it is of the form $x_{i_0} \wedge \cdots \wedge x_{i_{k-1}}$. (Recall from Section I.4 that we omit the outside parentheses and also the internal parentheses in iterated meets and iterated joins.)

For $\varnothing \neq J \subseteq \{0, \ldots, n-1\}$, set

$$p_J = \bigwedge (x_i \mid i \in J).$$

We claim that, for any nonempty $J, K \subseteq \{0, \ldots, n-1\}$, the inequality

$$p_J/\mathbf{D} \leq p_K/\mathbf{D}$$

holds iff the containment $J \supseteq K$ holds. The "if" part is obvious. Now assume that $J \not\supseteq K$; then there exists an $i \in K$ such that $i \notin J$. Consider the two-element chain $\mathbf{C}_2$ and substitute $x_i = 0$ and $x_j = 1$ for all $j \neq i$. Obviously, $p_J = 1$ and $p_K = 0$; thus the inequality $p_J \leq p_K$ fails in $\mathbf{C}_2$, and therefore, in $\mathbf{D}$.

We claim that every lattice term is equivalent under $\equiv_{\mathbf{D}}$ to one of the form $\bigvee p_J$ for some family of nonempty sets $J \subseteq \{0, \ldots, n-1\}$.

Indeed, every $x_i$ is of this form (for a single $J$, which is a singleton), so it suffices to show that the set of terms equivalent to terms of this form is closed under $\vee$ and $\wedge$. Closure under $\vee$ is clear. To see closure under $\wedge$, we note that by distributivity,

$$\bigvee p_{J_i} \wedge \bigvee p_{K_j} \equiv_{\mathbf{D}} \bigvee (p_{J_i} \wedge p_{K_j})$$

and

$$p_{J_i} \wedge p_{K_j} \equiv_{\mathbf{D}} p_{J_i \cup K_j}.$$

Next we claim that $p/\mathbf{D}$ is join-irreducible in $\mathrm{Term}_{\mathbf{D}}(n)$ iff it is a $p_J/\mathbf{D}$. Since every $p/\mathbf{D} \in \mathrm{Term}_{\mathbf{D}}(n)$ is a join of terms $p_J/\mathbf{D}$, it suffices to prove that each $p_J/\mathbf{D}$ is join-irreducible. Let

$$p_J \equiv_{\mathbf{D}} \bigvee (p_{J_k} \mid k \in K),$$

where each $J_k$ satisfies that $\varnothing \subseteq J_k \subset \{0, \ldots, n-1\}$. Then $J \subseteq J_k$ follows from $p_J/\mathbf{D} \geq p_{J_k}/\mathbf{D}$. If $p_J/\mathbf{D} > p_{J_k}/\mathbf{D}$, holds for some $k \in K$, then $J \subset J_k$ holds.

In $C_2$, put $x_i = 1$, for all $i \in J$, and $x_i = 0$, otherwise. Then $p_J = 1$, and $\bigvee (p_{J_i} \mid i \in K) = 0$, which is a contradiction.

A reference to Theorem 107 completes the proof of (ii).

(iii) This proof is obvious from (ii).

(iv) This proof is obvious from (iii).    □

Figure 19 is a diagram of $\mathrm{Term}_{\mathbf{D}}(3)$.

The problem of determining $|\mathrm{Free}_{\mathbf{D}}(n)|$ goes back to R. Dedekind [149]. For a modern survey of the field, see A. D. Korshunov [480]; the article has 356 references.

Free distributive lattices (on a finite or infinite generating set) have many interesting properties. All chains are finite or countable (the proof of this is similar to that of Theorem 550). If $a$ and $H$ are such that $x \wedge y = a$, for all $x, y \in H$ with $x \neq y$, call $H$ $a$-disjoint. In a free distributive lattice, all $a$-disjoint sets are finite, see R. Balbes [44].

## 2.2    Boolean terms

*Boolean terms* are defined exactly like lattice terms except that all five operations $\vee$, $\wedge$, $'$, $0$, $1$ are used in the formation of the terms. A formal definition is the same as Definition 52 with two clauses added: If $p$ is a boolean term, so is $p'$; $0$ and $1$ are boolean terms. An $n$-ary boolean term $p$ defines a function in $n$ variables on any boolean algebra $B$; we define $p(a_0, \ldots, a_{n-1})$ imitating Definition 53.

For the boolean terms $p$ and $q$, set $p \equiv_{\mathbf{B}} q$ if, for every boolean algebra $B$ and $a_0, \ldots, a_{n-1} \in B$, the equality $p(a_0, \ldots, a_{n-1}) = q(a_0, \ldots, a_{n-1})$ holds. Let $p/\mathbf{B}$ denote the block containing $p$. Observe that $p \equiv_{\mathbf{B}} q$ is equivalent to the identity $p = q$ holding in the class $\mathbf{B}$ of all boolean algebras.

Let $\mathrm{Term}_{\mathbf{B}}(n)$ denote the set of all $p/\mathbf{B}$, where $p$ is an $n$-ary boolean term. It is easily seen that we can define the boolean operations on $\mathrm{Term}_{\mathbf{B}}(n)$:

$$p/\mathbf{B} \vee q/\mathbf{B} = (p \vee q)/\mathbf{B},$$
$$p/\mathbf{B} \wedge q/\mathbf{B} = (p \wedge q)/\mathbf{B},$$
$$(p/\mathbf{B})' = p'/\mathbf{B},$$
$$0 = 0/\mathbf{B},$$
$$1 = 1/\mathbf{B};$$

thus $\mathrm{Term}_{\mathbf{B}}(n)$ is a boolean algebra.

**Theorem 127.**

  (i) $\mathrm{Term}_{\mathbf{B}}(n)$ *is a free boolean algebra on $n$ generators.*
  (ii) $\mathrm{Term}_{\mathbf{B}}(n)$ *is isomorphic to* $(\mathsf{B}_1)^{2^n}$.
  (iii) $|\mathrm{Term}_{\mathbf{B}}(n)| = 2^{2^n}$.
  (iv) *A finitely generated boolean algebra is finite.*

*Proof.* The proof of (i) is routine (same proof as in Theorem 126).

A boolean term in $x_0, \ldots, x_{n-1}$ is called *atomic* if it is of the form

$$x_0^{i_0} \wedge \cdots \wedge x_{n-1}^{i_{n-1}},$$

where $i_j = 0$ or $1$, $x^0$ denotes $x$, and $x^1$ denotes $x'$. There is an atomic term $p_J$, for every $J \subseteq \{0, \ldots, n-1\}$, for which $i_j = 0$ iff $j \in J$. The crucial statement is:

$$p_{J_0}/\mathbf{B} \leq p_{J_1}/\mathbf{B} \quad \text{iff} \quad J_0 = J_1.$$

Indeed, let $J_0 \neq J_1$. We make the following substitutions in $\mathsf{B}_1$:
$x_i = 1$ if $i \in J_0$ and $x_i = 0$ if $i \notin J_0$.
This makes $p_{J_0} = 1$ and $p_{J_1} = 0$, contradicting that $p_{J_0}/\mathbf{B} \leq p_{J_1}/\mathbf{B}$.

Let $B(n)$ be the set of all boolean terms that are equivalent to one of the form $\bigvee(p_{J_i} \mid i \in K)$. Then $B(n)$ is closed under $\vee$ and $\wedge$, since

$$\bigvee p_{J_i} \wedge \bigvee p_{I_k} \equiv_{\mathbf{B}} \bigvee(p_{J_i} \wedge p_{I_k})$$

and $p_{J_i} \wedge p_{I_k} \equiv_{\mathbf{B}} p_{J_i}$, if $J_i = I_k$, and $p_{J_i} \wedge p_{I_k} \equiv_{\mathbf{B}} 0$, otherwise.

Now we prove by induction on $n$ that $x_i, x_i' \in B(n)$ for all $i < n$. If $n = 1$, then $x_0$ and $x_0'$ are atomic terms, so $x_0, x_0' \in B(1)$. By induction,

$$x_0 \equiv_{\mathbf{B}} \bigvee(p_{J_i} \mid i \in K),$$

where the $p_{J_i}$ are atomic $(n-1)$-ary terms; then

$$x_0 \equiv_{\mathbf{B}} x_0 \wedge (x_{n-1} \vee x_{n-1}') \equiv_{\mathbf{B}} (x_0 \wedge x_{n-1}) \vee (x_0 \wedge x_{n-1}')$$
$$\equiv_{\mathbf{B}} \bigvee(p_{J_i} \wedge x_{n-1} \mid i \in K) \vee \bigvee(p_{J_i} \wedge x_{n-1}' \mid i \in K),$$

and similarly for $x_0'$. Thus $x_0, x_0' \in B(n)$, and, by symmetry, $x_i, x_i' \in B(n)$ for all $i < n$. Since

$$(p_J)' \equiv_\mathbf{B} \bigvee(x_i' \mid i \in J) \vee \bigvee(x_i \mid i \notin J),$$

we conclude that $p_J' \in B(n)$; therefore, $B(n)$ is closed under $'$. Thus $B(n)$ is closed under $\vee, \wedge, '$ and clearly, under $0, 1$. Since $B(n)$ includes $x_i$, for all $i < n$, it is the set of all $n$-ary boolean terms.

Consequently, every $p/\mathbf{B}$ is a join of atomic terms, the $p/\mathbf{B}$ for $p$ atomic terms are unordered and $2^n$ in number, implying (ii) and (iii). Finally, (iv) follows trivially from (iii). $\qquad\square$

I. Reznikoff [608] and A. Horn [400] prove that all chains of a free boolean algebra are finite or countable.

Infinitary boolean terms are considered in H. Gaifman [215] and A. W. Hales [369]; they prove that free complete boolean algebras on infinitely many generators do not exist.

### 2.3    Free constructs

We can use Theorems 126 and 127 to characterize free distributive lattices and free boolean algebras, respectively.

**Theorem 128.** *Let $L$ be a distributive lattice generated by $I$. The lattice $L$ is distributive freely generated by $I$ iff the validity in $L$ of a relation of the form*

$$\bigwedge I_0 \leq \bigvee I_1$$

*implies that $I_0 \cap I_1 \neq \varnothing$ for finite nonempty subsets $I_0$ and $I_1$ of $I$.*

*Proof.* The "only if" part can be easily verified by using substitutions in $\mathbf{C}_2$. For the converse, let $F$ be the distributive lattice freely generated by $I$, and let $\varphi$ be the homomorphism of $F$ into (in fact, onto) $L$ satisfying $\varphi(i) = i$ for all $i \in I$. It suffices to prove that for the lattice terms $p$ and $q$, the inequality $\varphi(p) \leq \varphi(q)$ implies that $p/\mathbf{D} \leq q/\mathbf{D}$. (We think of the elements of $F$ as blocks of terms in $I$.)

Let

$$p \equiv_\mathbf{D} \bigvee(\bigwedge I_j \mid j \in J),$$
$$q \equiv_\mathbf{D} \bigwedge(\bigvee K_t \mid t \in T).$$

Then $\varphi(p) \leq \varphi(q)$ takes the form

$$\bigvee(\bigwedge I_j \mid j \in J) \leq \bigwedge(\bigvee K_t \mid t \in T)$$

in $L$, which is equivalent to

$$\bigwedge I_j \leq \bigvee K_t$$

for all $j \in J$ and $t \in T$. By assumption, this implies that $I_j \cap K_t \neq \varnothing$ for all $j \in J$ and $t \in T$; thus

$$\bigwedge J_j \leq \bigvee K_t$$

as polynomials for all $j \in J$ and $t \in T$. This implies that $p/\mathbf{D} \leq q/\mathbf{D}$. $\square$

**Theorem 129.** *Let $B$ be a boolean algebra generated by $I$. Then $B$ is freely generated by $I$ iff, whenever $I_0, I_1, J_0, J_1$ are finite subsets of $I$ with $I_0 \cup I_1 = J_0 \cup J_1$ and $I_0 \cap I_1 = \varnothing$, then*

$$\bigwedge I_0 \wedge \bigwedge I_1 \leq \bigwedge J_0 \wedge \bigwedge J_1$$

*implies that $I_0 = J_0$ and $I_1 = J_1$.*

*Proof.* Again, the "only if" part is by substitution into $\mathsf{B}_1$. On the other hand, clearly $B$ is freely generated by $I$ iff, for every finite subset $K$ of $I$, the subalgebra $\mathrm{sub}(K)$ is freely generated by $K$. By Theorem 127, the latter holds iff the substitution map

$$\mathrm{Free}_B(|K|) \to \mathrm{sub}(K)$$

is one-to-one, equivalently, iff $\mathrm{sub}(K)$ has $2^{2^{|K|}}$ elements, which, in turn by Corollary 111, is equivalent to $\mathrm{sub}(K)$ having $2^{|K|}$ atoms. Using the proof of Theorem 127 and the present hypothesis for $I_0 \cup I_1 = K$, we can see that the elements of the form

$$\bigwedge I_0 \wedge \bigwedge I_1,$$

where $I_0 \cup I_1 = K$ and $I_0 \cap I_1 = \varnothing$, are distinct atoms in $\mathrm{sub}(K)$, thus completing the proof. $\square$

## 2.4 Boolean homomorphisms

Now we turn our attention to an important application of terms: finding homomorphisms of boolean algebras.

**Theorem 130.** *Let the boolean algebra $B$ be generated by the subalgebra $D_1$ and the element $a$. Let $D_2$ be a boolean algebra and let $\varphi$ be a homomorphism of $D_1$ into $D_2$. The extensions of $\varphi$ to homomorphisms of $B$ into $D_2$ are in one-to-one correspondence with the elements $p$ of $D_2$ satisfying the following conditions:*

  (i) *If $x \in D_1$ and $x \leq a$, then $\varphi(x) \leq p$.*
  (ii) *If $x \in D_1$ and $x \geq a$, then $\varphi(x) \geq p$.*

To prepare for the proof of this theorem we verify a simple lemma, in which + denotes the *symmetric difference*; that is,

$$x + y = (x' \wedge y) \vee (x \wedge y').$$

**Lemma 131.** *Let the boolean algebra $B$ be generated by the subalgebra $D_1$ and the element $a$. Then every element $x$ of $B$ can be represented in the form*

$$x = (a \wedge x_0) \vee (a' \wedge x_1), \quad x_0, x_1 \in D_1.$$

*This representation is not unique. Rather,*

$$(a \wedge x_0) \vee (a' \wedge x_1) = (a \wedge y_0) \vee (a' \wedge y_1), \quad x_0, x_1, y_0, y_1 \in D_1,$$

*iff*

$$a \leq (x_0 + y_0)' \quad and \quad x_1 + y_1 \leq a.$$

*Proof.* Let $D_0$ denote the set of all elements of $B$ having such a representation. If $x \in D_1$, then $x = (a \wedge x) \vee (a' \wedge x)$; thus $D_1 \subseteq D_0$. Also $a = (a \wedge 1) \vee (a' \wedge 0)$, and so $a \in D_0$. Therefore, to show that $D_0 = B$, it suffices to verify that $D_0$ is a subalgebra, which is left as an exercise. Now note that for all $p, q \in B$, the equality $p = q$ holds iff $p \wedge a = q \wedge a$ and $p \wedge a' = q \wedge a'$; thus

$$(a \wedge x_0) \vee (a' \wedge x_1) = (a \wedge y_0) \vee (a' \wedge y_1)$$

iff

$$a \wedge x_0 = a \wedge y_0 \quad and \quad a' \wedge x_1 = a' \wedge y_1.$$

However, $a \wedge x_0 = a \wedge y_0$ is equivalent to $(a \wedge x_0) + (a \wedge y_0) = 0$; that is, to $a \wedge (x_0 + y_0) = 0$ (see Exercise 2.13), which is the same as $a \leq (x_0 + y_0)'$. Similarly, $a' \wedge x_1 = a' \wedge y_1$ iff $x_1 + y_1 \leq a$. $\square$

*Proof of Theorem 130.* Let $p$ be an element as specified and define the map $\psi \colon B \to D_2$ as follows:

$$(a \wedge x_0) \vee (a' \wedge x_1) \mapsto (p \wedge \varphi(x_0)) \vee (p' \wedge \varphi(x_1)).$$

By Lemma 131, the set of values at which $\psi$ is defined is all of $B$. The map $\psi$ is well defined, because if

$$(a \wedge x_0) \vee (a' \wedge x_1) = (a \wedge y_0) \vee (a' \wedge y_1),$$

then

$$x_1 + y_1 \leq a \leq (x_0 + y_0)';$$

thus

$$\varphi(x_1 + y_1) \leq p \leq (\varphi(x_0 + y_0))',$$

and therefore
$$\varphi(x_1) + \varphi(y_1) \le p \le (\varphi(x_0) + \varphi(y_0))',$$
implying that

$$(p \wedge \varphi(x_0)) \vee (p' \wedge \varphi(x_1)) = (p \wedge \varphi(y_0)) \vee (p' \wedge \varphi(y_1)).$$

It is routine to check that $\psi$ is a homomorphism. Conversely, if $\psi$ is an extension of $\varphi$ to $B$, then $\psi$ is uniquely determined by $p = \psi(a)$, and $p$ satisfies (i) and (ii).     $\square$

**Corollary 132.** *Let us assume the conditions of Theorem 130. In addition, let $D_2$ be complete. Set*

$$x_0 = \bigvee(\,\varphi(x) \mid x \in D_1, \ x \le a\,),$$
$$x_1 = \bigwedge(\,\varphi(x) \mid x \in D_1, \ x \ge a\,).$$

*Then the extensions of $\varphi$ to $B$ are in one-to-one correspondence with the elements of the interval $[x_0, x_1]$. In particular, there is always at least one such extension.*

A more general form of Theorem 130 can be found in R. Sikorski [647]; for the universal algebraic background, see Theorem 12.2 in G. Grätzer [254].

## 2.5   $\Diamond$ Polynomial completeness of lattices
## by Kalle Kaarli

Let $D$ be a bounded distributive lattice. Let us say that an $n$-ary function $f$ on $L$ is *congruence compatible* if for any congruence $\theta$ of $L$ and

$$a_i \equiv b_i \pmod{\boldsymbol{\theta}}, \text{ for } i = 1, \dots, n,$$

the congruence

$$f(a_1, \dots, a_n) \equiv f(b_1, \dots, b_n) \pmod{\boldsymbol{\theta}}$$

holds.

In Section I.4.1, we introduced polynomials. Clearly, polynomials are congruence compatible. The converse often fails. If $D$ is boolean, the unary function $f(x) = x'$ is congruence compatible but is not a polynomial, in fact, it is not even isotone.

Let us call the lattice $D$ *affine complete* if every congruence compatible function on $D$ is a polynomial.

$\Diamond$ **Theorem 133.** *A bounded distributive lattice $D$ is affine complete iff it does not have any nontrivial boolean interval.*

This result of G. Grätzer [251], see also J. D. Farley [177], started an interesting chapter in lattice theory and universal algebra, covered in great depth in the book K. Kaarli and A. F. Pixley [455]. To provide some of the highlights, we start with some definitions.

An $n$-ary function on a lattice $L$ is a *local polynomial* if its restriction to any finite subset $H$ of $L^n$ equals a polynomial restricted to $H$. We denote by $\mathcal{P}(L)$ and $\mathcal{LP}(L)$ the set of all polynomial and local polynomial functions on $L$, respectively. Obviously, $\mathcal{P}(L) \subseteq \mathcal{LP}(L)$ for every lattice $L$.

Let us assume that we assign to every lattice $L$, a set $\mathcal{F}(L)$ of finitary functions on $L$ containing $\mathcal{LP}(L)$. The lattice $L$ is $\mathcal{F}$-*polynomially complete* if $\mathcal{F}(L) = \mathcal{P}(L)$, that is, every function in $\mathcal{F}(L)$ is a polynomial; similarly, the lattice $L$ is *locally $\mathcal{F}$-polynomially complete* if $\mathcal{F}(L) = \mathcal{LP}(L)$, that is, every function in $\mathcal{F}(L)$ is a local polynomial.

In our example results, $\mathcal{F} \in \{\mathcal{O}, \mathcal{I}, \mathcal{C}, \mathcal{I} \cap \mathcal{C}\}$ where:

$\mathcal{O}(L)$     is the set of all functions on $L$;

$\mathcal{C}(L)$     is the set of all congruence compatible functions on $L$;

$\mathcal{I}(L)$     is the set of all isotone functions on $L$.

The polynomial completeness properties corresponding to these sets of functions are named as follows:

$\mathcal{O}$             (local) polynomial completeness;

$\mathcal{C}$             (local) affine completeness;

$\mathcal{I}$             (local) order polynomial completeness;

$\mathcal{I} \cap \mathcal{C}$         (local) order affine completeness.

Obviously no nontrivial lattice can be (locally) polynomially complete because all (local) polynomial functions on lattices are isotone. As Theorem 133 states, the class of affine complete lattices does contain nontrivial lattices. This result was generalized in different directions by D. Dorninger, G. Eigenthaler, and M. Ploščica. The first two of them proved in [166] that $\mathcal{LP}(L) = \mathcal{C}(L) \cap \mathcal{I}(L)$ for any distributive lattice $L$. Since we can construct from nontrivial boolean intervals congruence compatible functions that are not isotone (see G. Grätzer [251]), D. Dorninger and G. Eigenthaler [166] obtained the following result.

$\Diamond$ **Theorem 134.** *A distributive lattice is locally affine complete iff it has no nontrivial boolean intervals.*

An ideal $I$ of a lattice $L$ is *almost principal* if its intersection with any principal ideal of $L$ is principal. If $L$ has a unit, then every almost principal ideal of $L$ is principal. An *almost principal filter* of $L$ is defined dually. It is easy to observe that, from almost principal but not principal ideals and filters, we can construct locally polynomial functions that are not polynomials. The converse also holds by M. Ploščica [585].

◇ **Theorem 135.** *A distributive lattice is affine complete iff it has no nontrivial boolean interval and all of its almost principal ideals and almost principal filters are principal.*

The necessity part of this stronger form of Theorem 133 holds for arbitrary lattices, thus every affine complete lattice must be infinite. Indeed, such a lattice cannot have prime intervals; in particular, it has no atoms or dual atoms. An obvious example of an affine complete distributive lattice is the chain $\mathbb{R}$. It is not known whether there exist nondistributive affine complete lattices.

We know considerably more about order polynomially complete lattices, see M. Kindermann [471].

◇ **Theorem 136.** *A finite lattice is order polynomially complete iff it has no nontrivial tolerances.*

In the modular case, R. Wille [737] provides the following nice description.

◇ **Theorem 137.** *A finite, simple, modular lattice is order polynomially complete iff it is complemented.*

Thus all finite irreducible projective geometries viewed as lattices are order polynomially complete. It also follows from Theorem 136 that every finite order polynomially complete lattice is simple. The question whether there exist infinite order polynomially complete lattices remained open until M. Goldstern and S. Shelah [236, 237] answered it in the negative.

It was proved by K. Kaarli and A. Pixley [455] that the "local" versions of Theorems 136 and 137 remain valid for lattices of finite height.

Next we consider (locally) order affine complete lattices. In view of Theorems 2.5 and 135, and the observations preceding them, we have the following result.

◇ **Theorem 138.** *Every distributive lattice is locally order affine complete. Every bounded distributive lattice is order affine complete.*

The theory of non-distributive (locally) order affine complete lattices is based on the following two observations:

(1) An isotone function on a lattice $L$ is a local polynomial iff it preserves all tolerances of $L$.

(2) A sublattice $L$ of a direct product $L_1 \times \cdots \times L_n$ is locally order affine complete iff so are all 2-fold coordinate projections $L_{ij}$ of $L$.

Observation (1) first appeared for finite lattices in M. Kindermann [471]; it was used for proving Theorem 136. R. Wille's characterization of finite order affine complete lattices in [738] is based on the same observation; it says, in essence, that a finite lattice is order affine complete iff all of its tolerances are obtainable in a certain way from congruences. A more general version of this result is presented in K. Kaarli and A. F. Pixley [455, Theorem 5.3.28].

Observation (2) for finite lattices appeared in R. Wille [738]; the general form is due to K. Kaarli and V. Kuchmei [454]. It allows one to reduce the study of locally order affine complete lattices of finite height to the case of subdirect products of two subdirectly irreducible lattices. (For this concept, see Section 6.5.) By K. Kaarli and V. Kuchmei [454], relatively few subdirect products of two simple lattices are locally order affine complete.

$\Diamond$ **Theorem 139.** *Let $L$ be a subdirect product of simple lattices $L_1$ and $L_2$ of finite height. The lattice $L$ is locally order affine complete iff $L_1$ and $L_2$ have no nontrivial tolerances and one of the following cases occurs:*

(1) *$L = L_1 \times L_2$;*
(2) *$L$ is a maximal sublattice of $L_1 \times L_2$;*
(3) *$L$ is the intersection of two maximal sublattices of $L_1 \times L_2$, one containing $(0, 1)$ and the other $(1, 0)$.*

This result is especially useful for modular lattices because subdirectly irreducible modular lattices of finite height are simple.

In conclusion, we consider another version of local order affine completeness that is defined using partial functions and has several good properties. We call a lattice $L$ *strictly locally order affine complete* if any isotone congruence compatible function $f \colon X \to L$, where $X$ is a finite meet (or join) subsemilattice of some power $L^n$, is the restriction of some polynomial of $L$. The following results were obtained by K. Kaarli and K. Täht [456].

$\Diamond$ **Theorem 140.**

(1) *A lattice is strictly locally order affine complete iff all of its tolerances are congruences.*
(2) *Every relatively complemented lattice is strictly locally order affine complete.*
(3) *Every strictly locally order affine complete lattice is congruence permutable.*
(4) *A modular lattice of finite height is strictly locally order affine complete iff it is relatively complemented.*

## Exercises

2.1. Regard Term($n$) as an algebra with the binary operations $\vee$ and $\wedge$. Define the concept of a congruence relation $\alpha$ on Term($n$) and the corresponding *quotient algebra* Term($n$)/$\alpha$.

Prove that $\equiv_{\mathbf{D}}$ is a congruence relation on $\mathrm{Term}(n)$ and the corresponding quotient algebra is isomorphic to $\mathrm{Term}_{\mathbf{D}}(n)$.

2.2. Work out Exercise 2.1 for Boolean terms.

2.3. Get lower and upper bounds for $|\mathrm{Term}_{\mathbf{B}}(n)|$ that are sharper than those given by Theorem 126(iii).

2.4. Work out the details of the last steps in the proof of Theorem 126.

2.5. Let $p_J$ be an atomic boolean term. Show that under the substitution $x_i = 1$, for all $i \in J$, and $x_i = 0$, for all $i \notin J$, we get $p_J = 1$ and $p_{J_0} = 0$ for all $J_0 \neq J$.

2.6. Let $f\colon \{0,1\}^n \to \{0,1\}$. Prove that there is an $n$-ary boolean term $p$ that defines the function $f$ on $\mathsf{B}_1$, as in Definition 53.

2.7. Let $B$ be a boolean algebra. Prove that there is a one-to-one correspondence between $n$-ary boolean terms over $B$ (up to equivalence) and maps $\{0,1\}^n \to \{0,1\}$. In other words, all $\{0,1\}$ substitutions take 0 and 1 as values and determine the term $p$.

2.8. A *polynomial* over a boolean algebra $B$ is built up inductively from the variables and the elements of $B$ using $\vee$, $\wedge$, and $'$. A polynomial on $B$ defines a function on $B$, called a *boolean polynomial*. Show that the $n$-ary polynomials are in one-to-one correspondence with maps $\{0,1\}^n \to B$.

2.9. Let $p$ be an $n$-ary term over the boolean algebra $B$ and $\boldsymbol{\alpha}$ a congruence relation on $B$. Show that $a_i \equiv b_i \pmod{\boldsymbol{\alpha}}$, for all $i < n$, implies that $p(a_0, \ldots, a_{n-1}) \equiv p(b_0, \ldots, b_{n-1}) \pmod{\boldsymbol{\alpha}}$. ($p$ has the Substitution Property.)

2.10. Show that the property described in Exercise 2.9 characterizes boolean polynomials (G. Grätzer [248]).

*2.11. Use the property described in Exercise 2.9 to define *boolean polynomials* over a distributive lattice. Show that, for bounded distributive lattices, Exercise 2.8 holds without any change (G. Grätzer [251]).

2.12. Show that a free boolean algebra on countably many generators has no atoms.

2.13. Show that $a \wedge (b + c) = (a \wedge b) + (a \wedge c)$ holds in any boolean algebra.

2.14. Let $B$ be the boolean algebra freely generated by $I$. Let $L$ be the sublattice generated by $I$. Prove that $L$ is the free distributive lattice freely generated by $I$.

2.15. Let $L$ and $L_1$ be distributive lattices, let $L = \mathrm{sub}(A)$, and let $\varphi$ be a map of $A$ into $L_1$. Show that there is a homomorphism of $L$ into $L_1$ extending $\varphi$ iff, for every pair of finite nonempty subsets $A_1$ and $A_2$ of $A$,

$$\bigwedge A_1 \leq \bigvee A_2 \quad \text{implies that} \quad \bigwedge \varphi(A_1) \leq \bigvee \varphi(A_2).$$

(Compare this with Exercise I.5.45.)

2.16. State and prove Exercise 2.15 for boolean algebras.

2.17. Interpret Lemma 131 using Exercise 2.16.

2.18. Extend the last statement of Corollary 132 to the case in which $D_1$ is generated by $B$ and $a_0, \dots, a_{n-1} \in D_1$ for some $n > 1$.

2.19. Let $p$ and $q$ be lattice terms. Since $p$ and $q$ can also be regarded as boolean terms, $p \equiv q$ was defined in two ways: as $p \equiv_{\mathbf{D}} q$ and as $p \equiv_{\mathbf{B}} q$. Show that the two definitions are equivalent for lattice terms.

2.20. Define $\equiv_{\mathbf{K}}$ for lattice terms with respect to a class $\mathbf{K}$ of lattices closed under isomorphisms. Show that $\mathrm{Term}_{\mathbf{K}}(n) \in \mathbf{K}$ iff the free lattice over $\mathbf{K}$ with $n$ generators exists, in which case $\mathrm{Term}_{\mathbf{K}}(n)$ is a free lattice with $n$ generators.

## 3.   Congruence Relations

### 3.1   Principal congruences

In distributive lattices, the following description of the principal congruence $\mathrm{con}(a, b)$ (the notation introduced in Section I.3.6) is important (G. Grätzer and E. T. Schmidt [334]):

**Theorem 141.** *Let $L$ be a distributive lattice, $a, b, x, y \in L$, and let $a \le b$. Then*

$$x \equiv y \pmod{\mathrm{con}(a, b)} \quad iff \quad x \vee b = y \vee b \ and \ x \wedge a = y \wedge a.$$

*Remark.* This result is illustrated in Figure 27.

*Proof.* Let $\beta$ denote the binary relation under which $x \equiv y \pmod{\beta}$ iff $x \vee b = y \vee b$ and $x \wedge a = y \wedge a$. The binary relation $\beta$ is obviously an equivalence relation. If $x \equiv y \pmod{\beta}$ and $z \in L$, then

$$(x \vee z) \wedge a = (x \wedge a) \vee (z \wedge a) = (y \wedge a) \vee (z \wedge a) = (y \vee z) \wedge a,$$

and

$$(x \vee z) \vee b = z \vee (x \vee b) = z \vee (y \vee b) = (y \vee z) \vee b;$$

thus $x \vee z \equiv y \vee z \pmod{\beta}$. Similarly, $x \wedge z \equiv y \wedge z \pmod{\beta}$. We conclude that $\beta$ is a congruence relation. The congruence $a \equiv b \pmod{\beta}$ is obvious. Finally, let $\alpha$ be any congruence relation such that $a \equiv b \pmod{\alpha}$ and let $x \equiv y \pmod{\beta}$. Then

$$x \vee a = y \vee a,$$
$$x \wedge b = y \wedge b,$$
$$x \vee a \equiv x \vee b \pmod{\alpha},$$
$$x \wedge b \equiv x \wedge a \pmod{\alpha}.$$

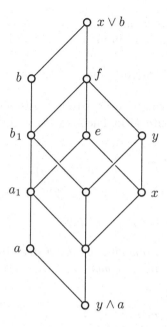

Figure 27. $x \equiv y \pmod{\operatorname{con}(a, b)}$ in a distributive lattice for $a \leq b$ and $x \leq y$

Computing modulo $\boldsymbol{\alpha}$, we obtain

$$x = x \vee (x \wedge a) = x \vee (y \wedge a) = (x \vee y) \wedge (x \vee a) \equiv (x \vee y) \wedge (x \vee b)$$
$$= (x \vee y) \wedge (y \vee b) = y \vee (x \wedge b) \equiv y \vee (x \wedge a) = y \vee (y \wedge a) = y,$$

that is, $x \equiv y \pmod{\boldsymbol{\alpha}}$, proving that $\boldsymbol{\beta} \leq \boldsymbol{\alpha}$.                                □

*Explanation.* Since $a \equiv b$ implies that $(a \vee p) \wedge q \equiv (b \vee p) \wedge q$, we must have $x \equiv y \pmod{\operatorname{con}(a, b)}$ if

$$x \vee y = (b \vee p) \wedge q,$$
$$x \wedge y = (a \vee p) \wedge q.$$

It is easy to check (see Exercise 3.1) that the $x$ and $y$ satisfying the conditions of Theorem 141 are exactly the same as those for which such $p$ and $q$ exist. Thus Theorem 141 can be interpreted as follows: We get all pairs $x \leq y$ with $x \equiv y \pmod{\operatorname{con}(a, b)}$, by applying the Substitution Property "twice" to a subinterval of $[a, b]$. No further application of the Substitution Property is required nor is transitivity needed.

    From the point of view of perspectivity and projectivity of intervals, see Section I.3.5, we get that in a distributive lattice $L$ with elements $a \leq b$ and

$x \le y$, the congruence $x \equiv y \pmod{\mathrm{con}(a, b)}$ holds iff the interval $[a, b]$ has a subinterval $[a_1, b_1]$ and there is an interval $[e, f]$ such that

$$[a_1, b_1] \stackrel{\mathrm{up}}{\sim} [e, f] \stackrel{\mathrm{dn}}{\sim} [x, y];$$

in particular, projectivity is equivalent to a two-step projectivity.

The description of $\mathrm{con}(a, b)$ in Theorem 141 is equivalent to the following two conditions:

$$x \vee y \le b \vee (x \wedge y),$$
$$(a \vee (x \wedge y)) \wedge (x \vee y) = x \wedge y.$$

Some applications of Theorem 141 follow.

**Corollary 142.** *Let $I$ be an ideal of the distributive lattice $L$. Then the following two conditions are equivalent:*

(i) $x \equiv y \pmod{\mathrm{con}(I)}$;
(ii) $x \vee y = (x \wedge y) \vee i$ *for some $i \in I$.*

*Therefore, $I$ is a block modulo $\mathrm{con}(I)$.*

*Remark.* This situation is illustrated in Figure 28, in which the dotted line indicates congruence modulo $\mathrm{con}(I)$.

Some properties of $\mathrm{con}(I)$ can be generalized to certain ideals of a general lattice, see Chapter III.

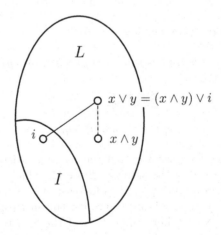

Figure 28. $x$ and $y$ are congruent modulo $\mathrm{con}(I)$ in a distributive lattice

*Proof.* If $x \vee y = (x \wedge y) \vee i$, then

$$x \equiv y \pmod{\mathrm{con}(x \wedge y \wedge i, i)}$$

with $x \wedge y \wedge i, i \in I$, and so $x \equiv y \pmod{\mathrm{con}(I)}$. Conversely,

$$\mathrm{con}(I) = \bigvee (\mathrm{con}(u, v) \mid u, v \in I)$$

by Lemma 14. However,

$$\mathrm{con}(u, v) \vee \mathrm{con}(u_1, v_1) \le \mathrm{con}(u \wedge v \wedge u_1 \wedge v_1, u \vee v \vee u_1 \vee v_1);$$

therefore,

$$\mathrm{con}(I) = \bigcup (\mathrm{con}(u, v) \mid u, v \in I).$$

If $x \equiv y \pmod{\mathrm{con}(u, v)}$, for $u, v \in I$ with $u \le v$, then $x \vee v = y \vee v$, and so $(x \wedge y) \vee (v \wedge (x \vee y)) = x \vee y$; thus Corollary 142(ii) is satisfied with $i = v \wedge (x \vee y) \in I$. Finally, if $a \in I$ and $a \equiv b \pmod{\mathrm{con}(I)}$, then $a \vee b = (a \wedge b) \vee i$ for some $i \in I$; so $a \vee b \in I$ and $b \in I$, showing that $I$ is a full block.    $\square$

**Corollary 143.** *Let $L$ be a distributive lattice, $x, y, a, b \in L$, and let*

$$x \le y \le a \le b$$

*or*

$$a \le b \le x \le y.$$

*Then $x \equiv y \pmod{\mathrm{con}(a, b)}$ implies that $x = y$.*

## 3.2    Prime ideals

A very important congruence relation has already been used in the proof of Lemma 8(ii): Given a prime ideal $P$ of the lattice $L$, we can construct a congruence relation that has exactly two blocks, $P$ and $L - P$. This statement can be generalized as follows: Let $\mathcal{A}$ be a set of prime ideals of a lattice $L$ and let us call two elements $x$ and $y$ congruent modulo $\mathcal{A}$ if either $x, y \in P$ or $x, y \in L - P$ for every $P \in \mathcal{A}$; this describes a congruence relation on $L$. For instance, if $\mathcal{A} = \{P, Q, R\}$ with $Q \subset P$ and $R \subset P$, then we get five blocks as shown in Figure 29; the quotient lattice is shown in Figure 21.

This principle will be used often. An interesting application to the Congruence Extension Property (see Section I.3.8) is the following statement:

**Theorem 144.** *A distributive lattice $L$ has the Congruence Extension Property. Therefore, the class $\mathbf{D}$ of distributive lattices has the CEP.*

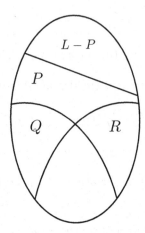

Figure 29. An important congruence

*Proof.* Let $\alpha$ be a congruence of $K$ and let $\alpha\colon x \mapsto x/\alpha$ be the natural homomorphism of $K$ onto $K/\alpha$; then $\alpha^{-1}(P)$ is a prime ideal of $K$ for every prime ideal $P$ of $K/\alpha$. Therefore, $\mathrm{id}(\alpha^{-1}(P))$ is an ideal of $L$, $\mathrm{fil}(K - \alpha^{-1}(P))$ is a filter of $L$, and they are disjoint. Thus by Theorem 115, we can choose a prime ideal $P_1$ of $L$ such that $P_1 \supseteq \alpha^{-1}(P)$ and $P_1 \cap (K - \alpha^{-1}(P)) = \varnothing$.

For every prime ideal $P$ of $K/\alpha$, we choose such a prime ideal $P_1$ of $L$. Let $\mathcal{A}$ denote the collection of all such prime ideals. Let $\beta$ be the congruence relation associated with $\mathcal{A}$, as previously described.

Now for $x, y \in K$, the congruence $x \equiv y \pmod{\alpha}$ is equivalent to the condition $\alpha(x) = \alpha(y)$, and so, for every $P_1 \in \mathcal{A}$, either $x, y \in P_1$ or $x, y \notin P_1$; thus $x \equiv y \pmod{\beta}$. Conversely, if $x \equiv y \pmod{\beta}$, then, for every $P_1 \in \mathcal{A}$, either $x, y \in P_1$ or $x, y \notin P_1$, and so either $\alpha(x), \alpha(y) \in P$ or $\alpha(x), \alpha(y) \notin P$. Since every pair of distinct elements of $K/\alpha$ is separated by a prime ideal (Corollary 117), we conclude that $\alpha(x) = \alpha(y)$ and thus $x \equiv y \pmod{\alpha}$.    $\square$

An alternative proof would proceed as in Lemma 17. Form in $L$ the congruence

$$\beta = \mathrm{con}(\alpha) = \bigvee(\,\mathrm{con}(x, y) \mid x, y \in \alpha\,).$$

Now if $\beta\rceil_K = \alpha$ fails, then some $\mathrm{con}_L(x, y)\rceil_K = \mathrm{con}_K(x, y)$ would fail, an easy contradiction with Theorem 141.

## 3.3    Boolean lattices

It is well known that in rings, ideals are in a one-to-one correspondence with congruence relations. In one class of lattices the situation is exactly the same.

**Theorem 145.** *Let $L$ be a boolean lattice. Then $\alpha \mapsto 0/\alpha$ is a one-to-one correspondence between congruence relations and ideals of $L$.*

*Proof.* By Corollary 142, the map is onto; therefore, we have only to prove that it is one-to-one, that is, that $I = 0/\alpha$ determines $\alpha$. This fact, however, is obvious, since $a \equiv b \pmod{\alpha}$ iff $a \wedge b \equiv a \vee b \pmod{\alpha}$, which, in turn, is equivalent to $c \equiv 0 \pmod{\alpha}$, where $c$ is the relative complement of $a \wedge b$ in $[0, a \vee b]$ (see Figure 30). Thus $a \equiv b \pmod{\alpha}$ iff $c \in 0/\alpha$. $\qquad\square$

This proof does not make full use of the hypothesis that $L$ is a complemented distributive lattice. In fact, all we need to make the proof work is that $L$ has a zero and is relatively complemented. Such a distributive lattice is called a *generalized boolean lattice*. The following result of J. Hashimoto [375] demonstrates the importance of the class of generalized boolean lattices. For the proof we present, see G. Grätzer and E. T. Schmidt [334].

**Theorem 146.** *Let $L$ be a lattice. There is a one-to-one correspondence between ideals and congruence relations of $L$ under which the ideal $I$ corresponding to a congruence relation $\alpha$ is a whole block under $\alpha$ iff $L$ is a generalized boolean lattice.*

*Proof.* The "if" part is in the proof of Theorem 145. We proceed with the "only if" part. The ideal corresponding to **0** has to be $\{0\}$, and thus $L$ has a 0. If $L$ contains a diamond, $\{o, a, b, c, i\}$, then $\mathrm{id}(a)$ cannot be a block, because $a \equiv o$ implies that

$$i = a \vee c \equiv o \vee c = c,$$
$$b = b \wedge i \equiv b \wedge c = o.$$

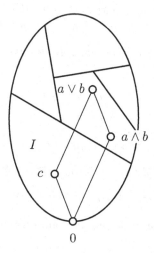

Figure 30. Illustrating the proof of Theorem 145

But $o \in \mathrm{id}(a)$, and thus any block containing $\mathrm{id}(a)$ contains $b \notin \mathrm{id}(a)$. Similarly, if $L$ contains a pentagon, $\{o, a, b, c, i\}$, and a block contains $\mathrm{id}(b)$, then $b \equiv o$; thus

$$i = b \vee c \equiv o \vee c = c,$$

and so

$$a = a \wedge i \equiv a \wedge c = o.$$

Therefore, this block has to contain $a$, and $a \notin \mathrm{id}(b)$. Thus by Theorem 101, $L$ is distributive. Let $a < b$ and $I = 0/\mathrm{con}(a, b)$. By Corollary 142, $\mathrm{con}(I)$ is also a congruence relation of $L$ having $I$ as a whole block; consequently, we obtain that $\mathrm{con}(I) = \mathrm{con}(a, b)$, and so $a \equiv b \pmod{\mathrm{con}(I)}$. Thus again by Corollary 142, $b = a \vee i$ and $i \equiv 0 \pmod{\mathrm{con}(a, b)}$ for some $i \in I$. The latter is equivalent to $i \vee b = 0 \vee b$ and $i \wedge a = 0 \wedge a$. We conclude that $a \vee i = b$ and $a \wedge i = 0$, and so $i$ is a relative complement of $a$ in $[0, b]$.    □

It is no coincidence that, in the class of generalized boolean lattices, congruences and ideals behave as they do in rings. Indeed, generalized boolean lattices are rings in disguise as demonstrated in M. H. Stone [668]:

**Theorem 147.**

(i) *Let $\mathfrak{B} = (B; \vee, \wedge)$ be a generalized boolean lattice. Define the binary operations $\cdot$ and $+$ on $B$ by setting*

$$x \cdot y = x \wedge y$$

*and by defining $x + y$ as a relative complement of $x \wedge y$ in $[0, x \vee y]$ (see Figure 31). Then $\mathfrak{B}^{\mathrm{ring}} = (B; +, \cdot)$ is a boolean ring—that is, an (associative) ring satisfying $x^2 = x$, for all $x \in B$ (and, consequently, satisfying $xy = yx$ and $x + x = 0$ for all $x, y \in B$).*

(ii) *Let $\mathfrak{B} = (B; +, \cdot)$ be a boolean ring. Define the binary operations $\vee$ and $\wedge$ in $B$ by*

$$x \vee y = x + y + x \cdot y,$$
$$x \wedge y = x \cdot y.$$

*Then $\mathfrak{B}^{\mathrm{lat}} = (B; \vee, \wedge)$ is a generalized boolean lattice.*

(iii) *Let $\mathfrak{B}$ be a generalized boolean lattice. Then $(\mathfrak{B}^{\mathrm{ring}})^{\mathrm{lat}} = \mathfrak{B}$.*

(iv) *Let $\mathfrak{B}$ be a boolean ring. Then $(\mathfrak{B}^{\mathrm{lat}})^{\mathrm{ring}} = \mathfrak{B}$.*

The proof of this theorem is purely computational. Some steps will be given in the Exercises.

The method given in Theorem 147 is not the only one used to introduce ring operations in a generalized boolean lattice. G. Grätzer and E. T. Schmidt

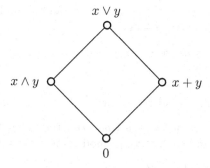

$$x \vee y$$

$$x \wedge y \qquad\qquad x + y$$

$$0$$

Figure 31. Defining $x + y$

[334] prove that ring operations $+$ and $\cdot$ can be introduced on a distributive lattice $L$ such that $+$ and $\cdot$ satisfy the Substitution Property iff $L$ is relatively complemented. Furthermore, $+$ and $\cdot$ are uniquely determined by the zero of the ring, which can be an arbitrary element of $L$.

The correspondence between boolean rings and generalized boolean lattices preserves many algebraic properties.

**Theorem 148.** *Let $\mathfrak{B}_0$ and $\mathfrak{B}_1$ be generalized boolean lattices.*

(i) *Let $I \subseteq B_0$. Then $I$ is an ideal of $\mathfrak{B}_0$ iff $I$ is an ideal of $\mathfrak{B}_0^{\mathrm{ring}}$.*

(ii) *Let $\varphi\colon B_0 \to B_1$. Then $\varphi$ is a $\{0\}$-homomorphism of $\mathfrak{B}_0$ into $\mathfrak{B}_1$ iff $\varphi$ is a homomorphism of $\mathfrak{B}_0^{\mathrm{ring}}$ into $\mathfrak{B}_1^{\mathrm{ring}}$.*

(iii) *$\mathfrak{B}_0$ is a $\{0\}$-sublattice of $\mathfrak{B}_1$ iff $\mathfrak{B}_0^{\mathrm{ring}}$ is a subring of $\mathfrak{B}_1^{\mathrm{ring}}$.*

The proof is again left to the reader.

## 3.4  Congruence lattices

N. Funayama and T. Nakayama [214] proves that congruence relations on an arbitrary lattice have an interesting connection with distributive lattices:

**Theorem 149.** *Let $L$ be an arbitrary lattice. Then $\operatorname{Con} L$, the lattice of all congruence relations of $L$, is distributive.*

*Proof.* Let $\alpha, \beta, \gamma \in \operatorname{Con} L$. Since

$$\alpha \wedge (\beta \vee \gamma) \ge (\alpha \wedge \beta) \vee (\alpha \wedge \gamma),$$

it suffices to prove that

$$a \equiv b \pmod{\alpha \wedge (\beta \vee \gamma)} \quad \text{implies that} \quad a \equiv b \pmod{(\alpha \wedge \beta) \vee (\alpha \wedge \gamma)}.$$

So let $a \equiv b \pmod{\alpha \wedge (\beta \vee \gamma)}$; that is, $a \equiv b \pmod{\alpha}$ and $a \equiv b \pmod{\beta \vee \gamma}$. By Theorem 12, there exists a sequence

$$a \wedge b = z_0 \leq \cdots \leq z_n = a \vee b$$

such that

$$z_i \equiv z_{i+1} \pmod{\beta} \quad \text{or} \quad z_i \equiv z_{i+1} \pmod{\gamma}$$

for every $0 \leq i < n$. Since $a \equiv b \pmod{\alpha}$, the congruence $a \wedge b \equiv a \vee b \pmod{\alpha}$ also holds, and so $z_i \equiv z_{i+1} \pmod{\alpha}$ for every $0 \leq i < n$. Thus

$$z_i \equiv z_{i+1} \pmod{\alpha \wedge \beta} \quad \text{or} \quad z_i \equiv z_{i+1} \pmod{\alpha \wedge \gamma},$$

for every $0 \leq i < n$, implying that

$$a \equiv b \pmod{(\alpha \wedge \beta) \vee (\alpha \wedge \gamma)}. \qquad \square$$

Now we connect the foregoing with algebraic lattices, see Definition 41.

**Lemma 150.** *Every principal congruence relation is compact.*

*Proof.* Let $L$ be a lattice, let $a, b \in L$. Let $\Lambda \subseteq \operatorname{Con} L$ and

$$\operatorname{con}(a, b) \leq \bigvee \Lambda.$$

Then $a \equiv b \pmod{\bigvee \Lambda}$, and thus (just as in Theorem 37) there exists a sequence

$$a = x_0, \ x_1, \ \ldots, \ x_n = b$$

with

$$x_i \equiv x_{i+1} \pmod{\alpha_i},$$

for some $\alpha_i \in \Lambda$, and for all $i$ with $0 \leq i < n$. Therefore, $a \equiv b \pmod{\bigvee \Lambda_0}$, where

$$\Lambda_0 = \{\alpha_0, \ldots, \alpha_{n-1}\},$$

and so $\operatorname{con}(a, b) \leq \bigvee \Lambda_0$, where $\Lambda_0$ is a finite subset of $\Lambda$. $\qquad \square$

**Theorem 151.** *Let $L$ be an arbitrary lattice. Then $\operatorname{Con} L$ is an algebraic lattice.*

*Proof.* For every $\alpha \in \operatorname{Con} L$,

$$\alpha = \bigvee (\operatorname{con}(a, b) \mid a \equiv b \pmod{\alpha}).$$

Consequently, this theorem follows from Lemma 150. $\qquad \square$

Combining Theorems 149 and 151 we get:

**Corollary 152.** *Let $L$ be an arbitrary lattice. Then $\operatorname{Con} L$ is a distributive algebraic lattice.*

The converse of Corollary 152 for the finite case is proved in Section IV.4.1.

## Exercises

3.1. Let $L$ be a distributive lattice and let $u, v, a, b \in L$. Prove that if $a \leq b$ and $x \leq v$, then

$$u \vee b = v \vee b \quad \text{and} \quad u \wedge a = v \wedge a$$

is equivalent to

$$(a \vee p) \wedge q = u \quad \text{and} \quad (b \vee p) \wedge q = v$$

for some $p, q$ in $L$.

3.2. Use Theorem 141 to prove that the class **D** of distributive lattices has the CEP (Theorem 144).

3.3. Verify Corollary 142 directly.

3.4. Let $K$ be a sublattice of the distributive lattice $L$ and let $P$ be a prime ideal of $K$. Prove that there exists a prime ideal $Q$ of $L$ with $Q \cap K = P$.

3.5. Prove that if Corollary 143 holds for a lattice $L$, then $L$ is distributive.

3.6. Show that if Theorem 144 holds for a lattice $L$, then $L$ is distributive.

3.7. Let $L$ be a *sectionally complemented lattice* (see Section I.6.1). Prove that $\alpha \mapsto 0/\alpha$ is a one-to-one correspondence between congruences and *certain* ideals of $L$.

*3.8. Show that the "certain ideals" that appear in Exercise 3.7 form a sublattice of $\operatorname{Id} L$. (See Section III.3.)

3.9. Prove that every (principal) ideal of $L$ is of the form $0/\alpha$ for a suitable congruence $\alpha$ of $L$ iff $L$ is distributive.

3.10. Let $L$ be a distributive lattice and let $I$ be an ideal of $L$. Define a binary relation $\beta(I)$ on $L$:

$$x \equiv y \pmod{\beta(I)} \quad \text{iff}$$

there is no $a \in L$ with $a \leq x \vee y$, $x \wedge y \wedge a \in I$, $a \notin I$.

Prove that $\beta(I)$ is the largest congruence relation of $L$ under which the ideal $I$ is a block.

3.11. Let $L$ be a distributive lattice with zero. Prove that there is a one-to-one correspondence between ideals and congruence relations (in the sense of Theorem 146) iff $\operatorname{con}(I) = \beta(I)$ for all $I \in \operatorname{Id} L$.

3.12. Prove Theorem 146 using Exercises 3.10 and 3.11 (G. Ya. Areškin [31]).

*3.13. Let $L$ be a lattice and let $a$ be an element of $L$. Show that every convex sublattice of $L$ containing $a$ is a block under exactly one congruence relation iff $L$ is distributive and all the intervals $[b, a]$ ($b \in L$ with $b \leq a$) and $[a, c]$ ($c \in L$ with $a \leq c$) are complemented (G. Grätzer and E. T. Schmidt [334]).

3.14. Derive Theorem 146 (and also, a variant of Theorem 146) by taking $a = 0$ (and arbitrary $a \in L$) in Exercise 3.13.

3.15. Let $L$ be a relatively complemented lattice, let $I, J \in \operatorname{Id} L$, and let $I \subseteq J$. Prove that if $I$ is an intersection of prime ideals, then so is $J$ (J. Hashimoto [375]).

3.16. Use Exercises 3.14 and 3.15 to get the following theorem: Let $L$ be a relatively complemented lattice. Then $L$ is distributive iff, for some element $a$ of $L$, the ideal $\operatorname{id}(a)$ is an intersection of prime ideals and the filter $\operatorname{fil}(a)$ is an intersection of prime filters (J. Hashimoto [375]).

3.17. Prove that the verification of Theorem 147(i) can be reduced to the boolean lattice case and that in this case

$$x + y = (x \wedge y') \vee (x' \wedge y).$$

3.18. Let $B$ be a boolean lattice. Verify that

$$x + y = (x \vee y) \wedge (x' \vee y').$$

3.19. Let $B$ be a boolean lattice. Verify that

$$(x + y) + z = (x \wedge y' \wedge z') \vee (x' \wedge y \wedge z') \vee (x' \wedge y' \wedge z)$$

and conclude that $+$ is associative.

3.20. Prove that $x(y + z) = xy + xz$ in a boolean lattice.

3.21. Prove Theorem 147(i).

3.22. Prove Theorem 147(ii).

3.23. Let $\mathfrak{B}$ be a generalized boolean lattice. For any $x, y \in B$, observe that the meet $x \wedge y$ is the same in $\mathfrak{B}$ as in $(\mathfrak{B}^{\mathrm{ring}})^{\mathrm{lat}}$ (namely, $x \cdot y$); conclude that $\mathfrak{B} = (\mathfrak{B}^{\mathrm{ring}})^{\mathrm{lat}}$.

3.24. Verify Theorem 147(iv).

3.25. Verify Theorem 148.

3.26. Show that, using the concept of a distributive semilattice (see Section 5.1), Corollary 152 can be reformulated as follows: Let $L$ be an arbitrary lattice. Then there exists a distributive join-semilattice $F$ with zero such that $\operatorname{Con} L$ is isomorphic to $\operatorname{Id} F$.

3.27. Characterize the lattice of all ideals of a lattice using the concept of an algebraic lattice.

3.28. Characterize the lattice of all ideals of a boolean lattice as a special type of algebraic lattices.

*3.29. Show that a chain $C$ is the congruence lattice of a lattice iff $C$ is algebraic. (This exercise and the next are from G. Grätzer and E. T. Schmidt [332].)

3.30. Prove that a boolean lattice $B$ is the congruence lattice of a lattice iff $B$ is algebraic.

3.31. Let $L$ be a distributive lattice. Show that $a \mapsto \operatorname{con}(\operatorname{id}(a))$ embeds $L$ into $\operatorname{Con} L$.

3.32. Let $L$ be a bounded distributive lattice. For $a, b \in L$ with $a \leq b$, show that the congruence $\mathrm{con}(a, b)$ has a complement in $\mathrm{Con}\, L$, namely, the congruence $\mathrm{con}(0, a) \vee \mathrm{con}(b, 1)$.

3.33. Generalize Exercise 3.32 to arbitrary distributive lattices.

3.34. Let $L$ be a bounded distributive lattice. Show that the compact elements of $\mathrm{Con}\, L$ form a boolean lattice (J. Hashimoto [375], G. Grätzer and E. T. Schmidt [335]).

3.35. Let $B$ be a boolean algebra freely generated by $X$ and let $L \leq B$ be the sublattice generated by $X$. Is $L$ freely generated by $X$ in **D** (see Exercise 2.14)?

# 4. Boolean Algebras R-generated by Distributive Lattices

## 4.1   Embedding results

The following result is the fundamental embedding theorem for distributive lattices.

**Theorem 153.** *Every distributive lattice can be embedded in a boolean lattice.*

*Proof.* By Theorem 119, every distributive lattice $L$ is isomorphic to a ring of subsets of some set $X$. Obviously, $L$ can be embedded into $\mathrm{Pow}\, X$.   $\square$

**Definition 154.** Let $L$ be a $\{0\}$-sublattice of the generalized boolean lattice $B$. Then $B$ is *R-generated* by $L$ if $L$ generates $B$ as a ring.

Note that if $L$ has a unit element, then the same element is the unit element of $B$; equivalently, if $\bigvee L$ exists, then $\bigvee B$ exists and $\bigvee L = \bigvee B$.

Our goal is to show the uniqueness of the generalized boolean lattice R-generated by $L$. The first result is essentially due to H. M. MacNeille [518]:

**Lemma 155.** *Let $B$ be R-generated by $L$. Then every $a \in B$ can be expressed in the form*

$$a_0 + a_1 + \cdots + a_{n-1}, \qquad a_0 \leq a_1 \leq \cdots \leq a_{n-1}, \qquad a_0, a_1, \ldots, a_{n-1} \in L.$$

*Remark.* Let $B$ be the boolean lattice shown in Figure 32 with the sublattice $L = \{0, a_0, a_1, a_2\}$. Then $L$ R-generates $B$.

*Proof.* Let $B_1$ denote the set of all elements that can be represented in the form $a_0 + \cdots + a_{n-1}$, where $a_0, \ldots, a_{n-1} \in L$. Then $L \subseteq B_1$, and $B_1$ is closed under $+$ and $-$ (since $x - y = x + y$). Furthermore,

$$(a_0 + \cdots + a_{n-1})(b_0 + \cdots + b_{m-1}) = \sum a_i b_j,$$

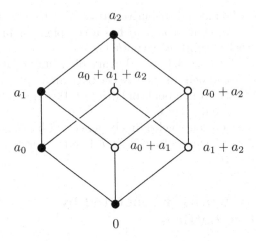

Figure 32. Illustrating "R-generated"

and each term $a_i b_j = a_i \wedge b_j \in L$, so $B_1$ is closed under multiplication. We conclude that $B_1 = B$.

Note that $L$ is a sublattice of $B$; therefore, for the elements $a, b \in L$, the join $a \vee b$ in $L$ is the same as the join in $B$. Thus $a \vee b = a + b + ab$, and so

$$a + b = ab + (a \vee b) = (a \wedge b) + (a \vee b).$$

Take $a_0 + \cdots + a_{n-1} \in B$. We prove by induction on $n$ that the summands can be made to form an increasing sequence. For $n = 1$, this is obvious. Let us assume that $a_1 \leq \cdots \leq a_{n-1}$. Then

$$
\begin{aligned}
& a_0 + a_1 + \cdots + a_{n-1} \\
&= (a_0 \wedge a_1) + (a_0 \vee a_1) + a_2 + \cdots + a_{n-1} \\
&= (a_0 \wedge a_1) + ((a_0 \vee a_1) \wedge a_2) + (a_0 \vee a_2) + a_3 + \cdots + a_{n-1} \\
&= (a_0 \wedge a_1) + ((a_0 \vee a_1) \wedge a_2) + ((a_0 \vee a_2) \wedge a_3) + (a_0 \vee a_3) + \cdots + a_{n-1} \\
& \quad \cdots \\
&= (a_0 \wedge a_1) + ((a_0 \vee a_1) \wedge a_2) + \cdots + ((a_0 \vee a_{n-2}) \wedge a_{n-1}) + (a_0 \vee a_{n-1}),
\end{aligned}
$$

and

$$a_0 \wedge a_1 \leq (a_0 \vee a_1) \wedge a_2 \leq \cdots \leq (a_0 \vee a_{n-2}) \wedge a_{n-1} \leq a_0 \vee a_{n-1}. \qquad \square$$

**Lemma 156.** *Let $L$ be a distributive lattice with zero. Then there exists a generalized boolean lattice $B$ freely R-generatedby $L$, that is, a generalized boolean lattice $B$ with the following properties:*

(i) *$B$ is R-generated by $L$.*

(ii) *If $B_1$ is R-generated by $L$, then there is a homomorphism $\varphi$ of $B$ onto $B_1$ that is the identity map on $L$.*

*Proof.* The existence of $B$ can be proved by copying the proof of Theorem 69 (or Theorem 89), *mutatis mutandis.*    □

An interesting property of generalized boolean lattices R-generated by distributive lattices is proved in J. Hashimoto [375].

**Lemma 157.** *Let $B$ be a generalized boolean lattice R-generated by the distributive lattice $L$ with zero. Then $B$ is a congruence-preserving extension of $L$.*

*Proof.* Let $\alpha$ be a congruence of $L$. The existence of an extension of $\alpha$ to $B$ was proved in Theorem 144. By Theorems 145 and 148(i), the following statement implies the uniqueness of the extension:

*If $I$ and $J$ are (ring) ideals of $B$ with $I \subset J$, then there are elements $a, b \in L$ with $a \neq b$, such that $a \equiv b \pmod{J}$ and $a \not\equiv b \pmod{I}$.*

Indeed, let $x \in J - I$. By Lemma 155, $x$ can be represented in the form

$$x = x_0 + \cdots + x_{n-1}, \qquad x_0 \leq \cdots \leq x_{n-1}, \qquad x_0, \ldots, x_{n-1} \in L.$$

If $n$ is odd, then $x_0 = x \cdot x_0 \leq x \in J$, and thus $x_0 \in J$; also,

$$x_0 + x_1 + x_2 = x \cdot x_2 \in J,$$

therefore

$$x_1 + x_2 = x_0 + (x_0 + x_1 + x_2) \in J.$$

Similarly,

$$x_3 + x_4, x_5 + x_6, \ldots \in J.$$

Since

$$x_0 + (x_1 + x_2) + (x_3 + x_4) + \cdots \in J - I,$$

we conclude that either $x_0 \in J - I$, or $x_{2i-1} + x_{2i} \in J - I$ for some $2i < n$. If $n$ is even, then we obtain $x_0 + x_1, x_2 + x_3, \ldots \in J$ (by multiplying $x$ by $x_1$, $x_3$, $\ldots$), and we conclude that $x_{2i-1} + x_{2i} \in J - I$ for some $2i < n$.

Now if $x_{2i-1} + x_{2i} \in J - I$, then $x_{2i-1} \equiv x_{2i} \pmod{J}$, but $x_{2i-1} \not\equiv x_{2i}$ (mod $I$) with $x_{2i-1}, x_{2i} \in L$. Finally, if $x_0 \in J - I$, then $x_0 \equiv 0 \pmod{J}$ and $x_0 \not\equiv 0 \pmod{I}$.    □

**Theorem 158.** *If $D_1$ and $D_2$ are generalized boolean lattices R-generated by a distributive lattice $L$ with zero, then $D_1$ and $D_2$ are isomorphic.*

*Proof.* Let $B$ be a free generalized boolean lattice R-generated by $L$ (as defined in Lemma 156). Let $\varphi$ be a homomorphism of $B$ onto $D_1$ such that $\varphi$ is the identity on $L$, see Lemma 156(ii). We want to show that $\varphi$ is an isomorphism. Indeed, if $\varphi$ is not an isomorphism, then the ideal kernel $I$ of $\varphi$ is not 0. Thus by Lemma 157, $a \equiv b \pmod{I}$ for some $a, b \in L$ with $a \neq b$. This means that $\varphi(a) = \varphi(b)$, contrary to our assumptions. Similarly, there is an isomorphism $\psi$ between $B$ and $D_2$. Obviously, $\psi\varphi^{-1}$ is an isomorphism between $D_1$ and $D_2$.    $\square$

*Remark.* For a distributive lattice $L$ with zero, we shall denote by $\mathrm{BR}\, L$ a generalized boolean lattice R-generated by $L$.

**Corollary 159.** *Let $L_0$ and $L_1$ be distributive lattices with zero and let $\varphi$ be a $\{0\}$-homomorphism of $L_0$ onto $L_1$. Then $\varphi$ can be extended to a homomorphism of $\mathrm{BR}\, L_0$ onto $\mathrm{BR}\, L_1$.*

*Proof.* Let $\alpha$ be the congruence kernel of $\varphi$, and let $\overline{\alpha}$ be the extension of $\alpha$ to $\mathrm{BR}\, L_0$ (by Lemma 157). Then $(\mathrm{BR}\, L_0)/\overline{\alpha}$ is a generalized boolean lattice R-generated by $L_0/\alpha \cong L_1$. Thus

$$(\mathrm{BR}\, L_0)/\overline{\alpha} \cong \mathrm{BR}\, L_1$$

by Theorem 158. Now it is trivial to prove this corollary.    $\square$

**Corollary 160.** *Let $L_0$ be a $\{0\}$-sublattice of the distributive lattice $L_1$ with zero. Let $B$ denote the subalgebra of $\mathrm{BR}\, L_1$ R-generated by $L_0$. Then $\mathrm{BR}\, L_0 \cong B$.*

*Proof.* The proof is trivial.    $\square$

Let $L_0$ and $L_1$ be given as in Corollary 160. It is natural to ask: Under what conditions does $L_0$ R-generate $\mathrm{BR}\, L_1$? Let $\overline{L}_0$ denote the generalized boolean sublattice of $\mathrm{BR}\, L_1$ R-generated by $L_0$. We can answer our query by determining $L_1 \cap \overline{L}_0$.

**Lemma 161.** *Let $L_0$ and $L_1$ be given as in Corollary 160. Then $L_1 \cap \overline{L}_0$ is the smallest sublattice of $L_1$ containing $L_0$ that is closed under taking relative complements in $L_1$. Therefore, $L_0$ R-generates $\mathrm{BR}\, L_1$ iff the smallest sublattice of $L_1$ containing $L_0$ and closed under relative complementation in $L_1$ is $L_1$ itself.*

*Proof.* It is obvious that $L_0 \subseteq L_1 \cap \overline{L}_0$. If $a, b, c \in L_1 \cap \overline{L}_0$, $d \in L_1$, and $d$ is a relative complement of $b$ in $[a, c]$, then $d = a + b + c \in L_1 \cap \overline{L}_0$, since (see Figure 33) $d$ is a relative complement of $a + b$ in the interval $[0, c]$. Thus $d \in L_1 \cap \overline{L}_0$. Now let us assume that $L$ is a sublattice of $L_1$ containing $L_0$

and closed under relative complementation in $L_1$. If $x \in L_1 \cap \overline{L}_0$, then by Lemma 155, we can represent $x$ as

$$x = a_0 + \cdots + a_{n-1}, \quad a_0, \ldots, a_{n-1} \in L_0, \quad a_0 \leq \cdots \leq a_{n-1}.$$

We prove that $x \in L$ by induction on $n$.

If $n = 1$, then $x = a_0 \in L_0 \subseteq L$.

If $n = 2$, then $x$ is a relative complement of $a_0$ in $[0, a_1]$ with $0, a_0, a_1 \in L_0$, thus $x \in L$.

If $n = 3$, then (see Figure 32) $x = a_0 + a_1 + a_2$ is a relative complement of $a_1$ in $[a_0, a_2]$, and so $x \in L$.

Now let $n > 3$, and let $y \in L$ be proved for all $y = b_0 + \cdots + b_{k-1}$ for the elements $b_0, \ldots, b_{k-1} \in L_0$ with $b_0 \leq \cdots \leq b_{k-1}$ and $k < n$. Note that $x \in L_1$ and $a_{n-3} \in L_0$ imply that

$$x a_{n-3} = a_0 + \cdots + a_{n-3} + a_{n-3} + a_{n-3} = a_0 + \cdots + a_{n-3} \in L_1$$

and

$$x \vee a_{n-3} = x + a_{n-3} + x a_{n-3}$$
$$= a_0 + \cdots + a_{n-1} + a_{n-3} + a_0 + \cdots + a_{n-3}$$
$$= a_{n-3} + a_{n-2} + a_{n-1} \in L_1.$$

By the induction hypothesis,

$$a_0 + \cdots + a_{n-3} \in L \text{ and } a_{n-3} + a_{n-2} + a_{n-1} \in L;$$

therefore, $x$ is a relative complement in $L_1$ of an element (namely, of $a_{n-3}$) of $L$ in an interval in $L$, namely, in

$$[a_0 + \cdots + a_{n-3}, a_{n-3} + a_{n-2} + a_{n-1}],$$

and so, by assumption, $x \in L$. Thus $L_1 \cap \overline{L}_0 \subseteq L$.  □

Some of the results presented above were first published in G. Grätzer [257].

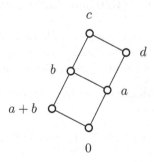

Figure 33. Illustrating the position of $d$

## 4.2    The complete case

In Theorem 153, we embedded $L$ into $\operatorname{Pow} X$, which is a complete boolean lattice. The question arises whether we can require this embedding to be complete, that is, to preserve arbitrary joins and meets, if they exist in $L$.

It is easy to see that not every complete distributive lattice has a complete embedding into a complete boolean lattice (J. von Neumann [552], [553]).

**Lemma 162.** *Let $B$ be a complete boolean lattice. Then $B$ satisfies the Join Infinite Distributive Identity*

$$\text{(JID)} \qquad\qquad x \wedge \bigvee Y = \bigvee (\, x \wedge y \mid y \in Y \,),$$

*for $x \in B$ and $Y \subseteq B$, and its dual, the Meet Infinite Distributive Identity,* (MID).

Of course, (JID) is not an identity in the sense of Section I.4.2, only an infinitary analogue. The proof of Theorem 149 easily yields that (JID) holds for $\operatorname{Con} L$, the lattice of all congruence relations of the lattice $L$.

*Proof.* Obviously, $\bigvee (\, x \wedge y \mid y \in Y \,) \leq x \wedge \bigvee Y$. Now let $u$ be any upper bound of $\{\, x \wedge y \mid y \in Y \,\}$, that is, $x \wedge y \leq u$ for all $y \in Y$. Then

$$y = y \wedge (x \vee x') = (y \wedge x) \vee (y \wedge x') \leq u \vee x',$$

and so $\bigvee Y \leq u \vee x'$. Thus

$$x \wedge \bigvee Y \leq x \wedge (u \vee x') = (x \wedge u) \vee (x \wedge x') = x \wedge u \leq u,$$

showing that $x \wedge \bigvee Y$ is the least upper bound for $\{\, x \wedge y \mid y \in Y \,\}$. By duality, condition (MID) follows. $\qquad\square$

**Corollary 163.** *Any complete distributive lattice that has a complete embedding into a complete boolean lattice satisfies both* (JID) *and* (MID).

Easy examples show that (JID) and (MID) need not hold in a complete distributive lattice.

Our task now is to show the converse of Corollary 163 (N. Funayama [213]). The construction of V. Glivenko [233] depends on a property of $\operatorname{BR} L$, on Theorem 100, and on the concept of the skeleton introduced in Section I.6.2.

**Lemma 164.** *Let $L$ be a distributive lattice with zero. Then $\operatorname{Id} L$ is a pseudo-complemented lattice in which*

$$I^* = \{\, x \mid x \wedge i = 0, \text{ for all } i \in I \,\}.$$

*Let*

$$\operatorname{Skel}(\operatorname{Id} L) = \{\, I^* \mid I \in \operatorname{Id} L \,\}.$$

*If $L$ is a boolean lattice, then* Skel(Id $L$) *is a complete boolean lattice and the map $a \mapsto \mathrm{id}(a)$ embeds $L$ into* Skel(Id $L$); *this embedding preserves all existing meets and joins.*

*Proof.* The first statement is trivial. Now let $L$ be boolean. It follows from Theorem 100 that Skel(Id $L$) is a boolean lattice. Furthermore, it is easily seen that for any $X \subseteq \mathrm{Id}\, L$, the sup and inf of $X$ in Skel(Id $L$) are $(\bigvee X)^{**}$ and $\bigwedge X$, respectively, where $\bigvee$ and $\bigwedge$ are the join and meet of $X$ in Id $L$, respectively. For $x, a \in L$, observe that $x \wedge a' = 0$ iff $x \leq a$, and so

$$\mathrm{id}(a) = \mathrm{id}(a')^* \in \mathrm{Skel}(\mathrm{Id}\, L).$$

Since

$$\bigwedge (\,\mathrm{id}(x) \mid x \in X\,) = \mathrm{id}(\inf X),$$

whenever $\inf X$ exists in $L$, the map $a \mapsto \mathrm{id}(a)$ of $L$ into Skel(Id $L$) preserves all existing meets in $L$. Now let $a = \sup X$ in $L$ and set

$$I = \mathrm{id}(X) \quad (= \bigvee(\,\mathrm{id}(x) \mid x \in X\,)).$$

To show that $x \mapsto \mathrm{id}(x)$ is join-preserving, we have to verify that $I^{**} = \mathrm{id}(a)$, or equivalently, that $I^* = \mathrm{id}(a')$.

Indeed, if $b \in I^*$, then $b \wedge x = 0$, for all $x \in I$, and thus $x \leq b'$. Therefore,

$$a = \sup X \leq b',$$

proving $a' \geq b$, that is, $b \in \mathrm{id}(a')$. Conversely, let $b \in \mathrm{id}(a')$. Then $b' \geq a$; therefore,

$$b' \geq a = \sup X \geq x,$$

for all $x \in X$, and so $b \wedge x = 0$ for all $x \in X$. This shows that $b \in I^*$, proving that $I^* = \mathrm{id}(a')$.    □

**Lemma 165.** *Let $L$ be a complete lattice satisfying* (JID) *and* (MID). *Then the identity map is a complete embedding of $L$ into* BR $L$.

*Proof.* Let us write $a \in \mathrm{BR}\, L$ in the form

$$a = a_0 + \cdots + a_{n-1}, \quad a_0 \leq \cdots \leq a_{n-1}, \quad a_0, \ldots, a_{n-1} \in L.$$

If $n$ is even, let us replace $a_0$ by $0 + a_0$; thus we can assume that $n$ is odd.

We claim that, for any $x \in L$ and $a \in \mathrm{BR}\, L$, the inequality $x \leq a$ holds iff $x \wedge a_0 = x \wedge a_1$ and $x \leq a_2 + \cdots + a_{n-1}$.

Indeed, let $x \leq a$. Then

$$xa_1 = xa_1(a_0 + \cdots + a_{n-1}) = x(a_0 + a_1 + a_1 + \cdots + a_1) = xa_0;$$

therefore, $x \wedge a_0 = x \wedge a_1$. Thus

$$x(a_2 + \cdots + a_{n-1}) = (xa_0 + xa_1) + x(a_2 + \cdots + a_{n-1}) = xa = x,$$

and so $x \leq a_2 + \cdots + a_{n-1}$. Conversely, if $x \wedge a_0 = x \wedge a_1$ and $x \leq a_2 + \cdots + a_{n-1}$, then

$$xa = xa_0 + xa_1 + x(a_2 + \cdots + a_{n-1}) = x,$$

proving that $x \leq a$.

Now a simple induction proves that $x \leq a$ holds iff

$$x \wedge a_0 = x \wedge a_1,$$
$$x \wedge a_2 = x \wedge a_3,$$
$$\cdots$$
$$x \wedge a_{n-3} = x \wedge a_{n-2},$$
$$x \leq a_{n-1}.$$

Let $X \subseteq L$, let $y = \sup X$ in $L$, and let $a \in \mathrm{BR}\, L$. If $x \leq a$, for all $x \in X$, then the formulas last displayed hold for all $x$, hence by (JID), for the element $y$, proving that $y \leq a$. Thus $y = \sup X$ in $\mathrm{BR}\, L$. The dual argument, using (MID), completes the proof. □

So finally we obtained the following result of N. Funayama [213].

**Theorem 166.** *A complete lattice $L$ has a complete embedding into a complete boolean lattice iff $L$ satisfies* (JID) *and* (MID).

*Proof.* Combine Lemma 162, Corollary 163, and Lemmas 164, 165. □

### 4.3   Boolean lattices generated by chains

The representation for $a \in \mathrm{BR}\, L$ given in Lemma 155 is not unique in general; the only exception is when $L$ is a chain. Since this case is of special interest, we shall investigate it in detail.

Repeating the definition, a boolean lattice $B$ is R-*generated by a chain* $C$ with zero if $B = \mathrm{BR}\, C$. This concept is due to A. Mostowski and A. Tarski [535] and can be extended to distributive lattices as follows.

A distributive lattice $L$ with zero is R-*generated by a chain* $C$ ($\subseteq L$) with zero if $C$ R-generates $\mathrm{BR}\, L$.

**Lemma 167.** *Let $L$ be a distributive lattice with zero and let $C$ be a chain in $L$ with $0 \in C$. Then $C$ R-generates $L$ iff $L$ is the smallest sublattice of itself containing $C$ and closed under formation of relative complements.*

*Proof.* Apply Lemma 161 to $C$. □

An explicit representation of $\operatorname{BR} C$ is given as follows: for a chain $C$ with zero, let $B[C]$ be the set of all subsets of $C$ of the form

$$\operatorname{id}(a_0) + \operatorname{id}(a_1) + \cdots + \operatorname{id}(a_{n-1}),\ 0 < a_0 < a_1 < \cdots < a_{n-1},\ a_0,\ldots,a_{n-1} \in C,$$

where $+$ is the symmetric difference in $\operatorname{Pow} X$. We consider $B[C]$ as an order with the ordering $\subseteq$. We identify $a \in C$ with $\operatorname{id}(a) \cap (C - \{0\})$. Thus $C \subseteq B[C]$.

**Lemma 168.** $B[C]$ *is the generalized boolean lattice R-generated by* $C$.

*Proof.* The proof is obvious, by construction and by Theorem 158.     □

Note that every nonempty element $a$ of $B[C]$ can be represented in the form

$$a = (b_0, a_0] \cup (b_1, a_1] \cup \cdots \cup (b_{n-1}, a_{n-1}],$$
$$0 \le b_0 < a_0 < b_1 < a_1 < \cdots < b_{n-1} < a_{n-1},$$

where the union is disjoint union and $(x, y]$ is a *half-open interval*:

$$(x, y] = \{\, t \mid x < t \le y \,\}$$

for the elements $x, y \in C$. Note that $(x, y]$ can be written $\operatorname{id}(x) + \operatorname{id}(y)$, which, under our identification of an element $x \in C$ with the ideal $\operatorname{id}(x) \in B[C]$, becomes $x + y$. Thus

$$a = a_0 + b_1 + a_1 + \cdots + b_{n-1} + a_{n-1},$$

and so we conclude:

**Corollary 169.** *In* $\operatorname{BR} C$, *every nonzero element* $a$ *has a unique representation in the form*

$$a = a_0 + a_1 + \cdots + a_{n-1}, \quad 0 < a_0 < a_1 < \cdots < a_{n-1}, \quad a_0, a_1, \ldots, a_{n-1} \in C.$$

The following results show that many distributive lattices can be R-generated by chains.

**Lemma 170.** *Every finite boolean lattice* $B$ *can be R-generated by a chain;* *in fact,* $B = \operatorname{BR} C$ *for every maximal chain* $C$ *of* $B$.

*Proof.* Let $B_1$ be the subalgebra of $B$ R-generated by $C$. Using the notation of Corollary 112, the length of $C$ equals $|\operatorname{Ji} B|$; also, the length of $C$ equals $|\operatorname{Ji} B_1|$; thus $|\operatorname{Ji} B| = |\operatorname{Ji} B_1| = n$. We conclude that both $B$ and $B_1$ have $2^n$ elements. Since $B_1 \subseteq B$, we conclude that $B = B_1$.     □

**Corollary 171.** *Every finite distributive lattice* $L$ *can be R-generated by a chain, in fact, by any maximal chain of* $L$.

*Proof.* Let $C$ be a maximal chain in $L$ and let $B = \mathrm{BR}\,L$. Then $|\,\mathrm{Ji}\,L\,| = |\,\mathrm{Ji}\,B\,|$. By Corollary 112, $C$ is maximal in $B$. Thus $B = \mathrm{BR}\,C \supseteq L$.    □

**Theorem 172.** *Let $L$ be a countable distributive lattice with zero. Then $L$ can be R-generated by a chain.*

*Proof.* Let
$$L = \{a_0 = 0, a_1, a_2, \ldots, a_n, \ldots\},$$
and let $L_n$ be the sublattice of $L$ generated by $a_0, \ldots, a_n$. Let $A_0$ be a maximal chain of $L_0$, and, inductively, let $A_n$ be a maximal chain of $L_n$ containing $A_{n-1}$. Set
$$A = \bigcup(\, A_i \mid i < \omega\,).$$
Obviously, $0 \in A$. We claim that $A$ R-generates $L$. Take $a \in \mathrm{BR}\,L$;
$$a = x_0 + \cdots + x_{m-1}, \quad x_0, \ldots, x_{m-1} \in L.$$
Clearly,
$$L = \bigcup(\, L_i \mid i < \omega\,);$$
thus $x_0, \ldots, x_{m-1} \in L_n$, for some $n$, and so $a \in \mathrm{BR}\,L_n$. Since $L_n$ is finite, we get $\mathrm{BR}\,L_n = \mathrm{BR}\,A_n$; therefore, $a \in \mathrm{BR}\,A_n \subseteq \mathrm{BR}\,A$, which proves that $L \subseteq \mathrm{BR}\,A$.    □

**Corollary 173.** *The correspondence $C \mapsto \mathrm{BR}\,C$ maps the class of countable chains with zero onto the class of countable generalized boolean lattices. Under this correspondence, $\{0\}$-subchains and $\{0\}$-homomorphic images correspond to $\{0\}$-subalgebras and $\{0\}$-homomorphic images.*

Note, however, that $C \cong C'$ is *not* implied by $\mathrm{BR}\,C \cong \mathrm{BR}\,C'$ (see Exercise 4.25).

W. Hanf [372] proves that there is no algorithmic way to find a generating chain in all countable boolean algebras. However, by R. S. Pierce [580], there always are generating chains of a rather special order type.

Much is known about countable chains. Utilizing the previous results, such information can be used to prove results on countable generalized boolean lattices.

**Lemma 174.** *Every countable chain $C$ can be embedded in the chain $\mathbb{Q}$ of rational numbers. Every countable chain not containing any prime interval is isomorphic to one of the intervals $(0,1)$, $[0,1)$, $(0,1]$, and $[0,1]$ of $\mathbb{Q}$.*

*Proof.* Let $C = \{x_0, x_1, \ldots, x_{n-1}, \ldots\}$. We define the map $\varphi$ inductively as follows: Pick an arbitrary $r_0 \in \mathbb{Q}$ and set $\varphi(x_0) = r_0$. If $\varphi(x_0), \ldots, \varphi(x_{n-1})$ have already been defined, we define $\varphi(x_n)$ as follows: Let
$$L_n = \bigcup(\,\mathrm{id}(\varphi(x_i)) \mid x_i < x_n,\ i < n\,),$$
$$U_n = \bigcup(\,\mathrm{fil}(\varphi(x_i)) \mid x_i > x_n,\ i < n\,);$$

observe that $L_n = \varnothing$ or $U_n = \varnothing$ is possible. Note that if $L_n \neq \varnothing$, then it has a greatest element $l_n$, and if $U_n \neq \varnothing$, then it has a smallest element $u_n$. If both are nonempty, then $l_n < u_n$. In any case, we can choose an $r_n \in \mathbb{Q}$ satisfying $r_n \notin L_n \cup U_n$. We set $\varphi(x_n) = r_n$. Obviously, $\varphi$ is an embedding.

To prove the second assertion, we may adjoin a zero and/or a unit if these are absent; it is then easy to see that the desired result is equivalent to the statement that any two *bounded* countable chains $C$ and $D$ with no prime intervals satisfy $C \cong D$.

To prove this, let $C = \{c_0, c_1, \ldots\}$ and $D = \{d_0, d_1, \ldots\}$. We define two maps: $\varphi \colon C \to D$ and $\psi \colon D \to C$.

Let us assume that $c_0 = 0$, $c_1 = 1$ and $d_0 = 0$, $d_1 = 1$. For each $n < \omega$, we shall define inductively two finite chains $C^{(n)} \subseteq C$ and $D^{(n)} \subseteq D$, and an isomorphism $\varphi_n \colon C^{(n)} \to D^{(n)}$ with inverse $\psi_n \colon D^{(n)} \to C^{(n)}$.

Set $C^{(0)} = \{c_0, c_1\} = \{0, 1\}$ and $D^{(0)} = \{d_0, d_1\} = \{0, 1\}$; set $\varphi_0(i) = i$ and $\psi_0(i) = i$ for $i = 0, 1$.

Given $C^{(n)}$, $D^{(n)}$, $\varphi_n$, $\psi_n$, and $n$ even, let $k$ be the smallest integer with $c_k \notin C^{(n)}$. Define

$$u_k = \bigwedge (\mathrm{fil}(c_k) \cap C^{(n)}),$$
$$l_k = \bigvee (\mathrm{id}(c_k) \cap C^{(n)}).$$

Then $l_n < c_k < u_k$, and so $\varphi_n(l_k) < \varphi_n(u_k)$. Since $D$ contains no prime intervals, we can choose a $d \in D$ satisfying the inequalities

$$\varphi_n(l_k) < d < \varphi_n(u_k).$$

Since $\psi_n$ is isotone, $d \notin D^{(n)}$. Define

$$C^{(n+1)} = C^{(n)} \cup \{c_k\},$$
$$D^{(n+1)} = D^{(n)} \cup \{d\}.$$

Let $\varphi_{n+1}$ restricted to $C^{(n)}$ be $\varphi_n$, and let $\varphi_{n+1}(c_k) = d$. Let $\psi_{n+1}$ restricted to $D^{(n)}$ be $\psi_n$ and let $\psi_{n+1}(d) = c_k$. If $n$ is odd, then we proceed in a similar way, but we interchange the role of $C$ and $D$, $C^{(n)}$ and $D^{(n)}$, $\varphi_n$ and $\psi_n$, respectively.

Finally, put $\varphi = \bigcup(\varphi_n \mid n < \omega)$. Clearly,

$$C = \bigcup (C^{(n)} \mid n < \omega),$$
$$D = \bigcup (D^{(n)} \mid n < \omega),$$

and $\varphi$ is the required isomorphism. $\qquad\square$

**Corollary 175.** *Up to isomorphism, there is exactly one countable boolean lattice with no atoms and exactly one countable generalized boolean lattice with no atoms and no unit element,* BR $[0, 1)_{\mathbb{Q}}$.

*Proof.* Take the rational intervals $[0, 1]$ and $[0, 1)$. The generalized boolean lattices in question are $\mathrm{BR}\,[0, 1]$ and $\mathrm{BR}\,[0, 1)$. This follows from the observation that $[a, b]$ is a prime interval in $C$ iff $a + b$ is an atom in $\mathrm{BR}\,C$. The results follow from Lemmas 168 and 174 and Theorem 172.                      □

**Theorem 176.** *Let $B$ be a countable boolean algebra. Then $B$ has either $\aleph_0$ or $2^{\aleph_0}$ prime ideals.*

*Remark.* This is obvious if we assume the Continuum Hypothesis. Interestingly, we can give a proof without it.

*Proof.* For a boolean algebra $B$ and an ideal $I$ of $B$, we shall write $B/I$ for $B/\mathrm{con}(I)$. If $J$ is an ideal of $B$ with $J \supseteq I$, then

$$J/I = \{\, x/\mathrm{con}(I) \mid x \in J \,\}$$

is an ideal of $B/I$. ($J/I$ is the usual notation in ring theory.)

Let $B$ be a boolean algebra. We define the ideals $I_\gamma$ by transfinite induction. Let $I_0 = \mathrm{id}(0)$, and let $I_1$ be the ideal generated by the set $\mathrm{Atom}(B)$. Given $I_\gamma$, let $I$ be the ideal of $B/I_\gamma$ generated by $\mathrm{Atom}(B/I_\gamma)$. Let

$$\varphi\colon x \mapsto x + I_\gamma$$

be the homomorphism of $B$ onto $B/I_\gamma$; we set

$$I_{\gamma+1} = \varphi^{-1}(I).$$

Finally, if $\gamma$ is a limit ordinal, set

$$I_\gamma = \bigcup (\, I_\delta \mid \delta < \gamma \,).$$

The *rank* of $B$ is defined to be the smallest ordinal $\alpha$ such that $I_\alpha = I_{\alpha+1}$.

No element of $B$ can have an image which is an atom in more than one of the quotient lattices $B/I_\gamma$, hence, the cardinality of $\alpha$ is at most $|B|$.

**Claim 177.** *Let $B$ be countable. If $I_\alpha \neq B$, then $|\mathrm{Spec}\,B| = 2^{\aleph_0}$.*

Indeed, if $I_\alpha \neq B$, then $\mathrm{Atom}(B/I_\alpha) = \varnothing$, hence by the proof of Corollary 175, the isomorphism $B/I_\alpha \cong \mathrm{BR}\,C$ holds, where $C$ is the rational interval $[0, 1]$. By Lemma 157 and Exercise 4.32,

$$|B/I_\alpha| = |\mathrm{Spec}\,C| = |\mathrm{Id}\,C| = 2^{\aleph_0}.$$

**Claim 178.** *Let $B$ be countable. If $I_\alpha = B$, then $|\mathrm{Spec}\,B| = \aleph_0$.*

Indeed, for an ordinal $\gamma < \alpha$, let $\operatorname{Spec}_\gamma B$ be the set of prime ideals $P$ of $B$ for which $I_\gamma \subseteq P$ and $I_{\gamma+1} \not\subseteq P$. It is easy to see that every prime ideal of $B$ lies in one of the sets $\operatorname{Spec}_\gamma B$. Since $\alpha$ is finite or countable, it suffices to show that $|\operatorname{Spec}_\gamma B| = \aleph_0$. If $P \in \operatorname{Spec}_\gamma B$, then, by Corollary 116 and Theorem 123, the equality $P \vee I_{\gamma+1} = B$ holds. It follows that

$$P \cap I_{\gamma+1} \neq Q \cap I_{\gamma+1}$$

for $P, Q \in \operatorname{Spec}_\gamma B$ with $P \neq Q$. Thus

$$P \mapsto (P \cap [I_{\gamma+1}]_R)/I_\gamma$$

is a one-to-one correspondence of $\operatorname{Spec}_\gamma B$ into (in fact, onto) $\operatorname{Spec}([I_{\gamma+1}]/I_\gamma)$; but $[I_{\gamma+1}]_R/I_\gamma$ is just the generalized boolean lattice of all finite subsets of a countable set. Therefore, $|\operatorname{Spec}_\gamma B| = \aleph_0$. $\qquad\square$

In order to avoid giving the impression that most boolean algebras can be R-generated by chains, we state:

**Lemma 179.** *Let $B$ be a complete boolean algebra R-generated by a chain $C$ with zero. Then $B$ is finite.*

*Proof.* Let $B = \operatorname{BR} C$ and let the chain $C$ be infinite. Then by Exercise I.6.20, $C$ cannot satisfy both the Ascending and the Descending Chain Conditions. Assume that the former fails. (If the latter fails, replace $C$ by the chain of complements of its elements.) Thus $C$ contains a subchain

$$0 < x_0 < x_1 < \cdots < x_n < \cdots .$$

Then we define

$$a_n = x_0 + x_1 + \cdots + x_{2n} + x_{2n+1}$$

for all $n < \omega$. We claim that $\bigvee(a_n \mid n < \omega)$ does not exist. Indeed, let $a$ majorize $\{a_n \mid n < \omega\}$. By the remarks immediately following Lemma 168, we can represent each $a_n$ by a set

$$(x_0, x_1] \cup (x_2, x_3] \cup \cdots \cup (x_{2n}, x_{2n+1}],$$

and we can represent $a$ in the form

$$a = (b_0, a_0] \cup (b_1, a_1] \cup \cdots \cup (b_{m-1}, a_{m-1}],$$

where $m < \omega$ and

$$0 \leq b_0 < a_0 < b_1 < a_1 < \cdots < b_{m-1} < a_{m-1}$$

with $a_i, b_i \in C$ for all $i < m$. Since $a$ contains each $a_n$, there must exist an $n$ and a $j < m$ such that both $(x_{2n}, x_{2n+1}]$ and $(x_{2n+2}, x_{2n+3}]$ are contained in $(b_{j-1}, b_j]$ or in $(0, b_0]$. Therefore, the interval $(x_{2n+1}, x_{2n+2}]$ can be deleted from $a$, and it will still contain all the $a_n$, that is, $a + x_{2n+1} + x_{2n+2}$ majorizes both $\{a_n \mid n < \omega\}$ and $a + x_{2n+1} + x_{2n+2} < a$. We conclude that $\{a_n \mid n < \omega\}$ does not have a least upper bound. $\qquad\square$

Next we consider which chains with zero can be R-generating chains of a given distributive lattice.

**Lemma 180.** *Let $L$ be a distributive lattice with zero and let $C$ be a chain in $L$ with $0 \in C$. If $L$ is R-generated by $C$, then $C$ is maximal in $L$.*

*Proof.* If $C$ is not maximal in $L$, then we can find an element $a$ in $L$ not in $C$, such that $C \cup \{a\}$ is a chain. Write

$$a = a_0 + a_1 + \cdots + a_{n-1},$$

with $0 < a_0 < a_1 < \cdots < a_{n-1}$ and $a_i \in C$ for all $i < n$. Since $a \notin C$, it follows that $n > 1$. Now

$$a \wedge a_0 = a_0 + a_0 + \cdots + a_0,$$

which is $a_0$ if $n$ is odd and $0$ if $n$ is even. But $a_0 \neq a$ and $0 \neq a$, therefore, since $a$ and $a_0$ are comparable, $a \wedge a_0 = a_0$ and $n$ is odd. Then

$$a \wedge a_1 = a_0 + a_1 + \cdots + a_1 = a_0,$$

contradicting the comparability of $a$ and $a_1$. $\qquad\square$

The converse of Lemma 180 is false by Lemma 179. To settle the matter, we need a new concept.

**Definition 181.** Let $L$ be a distributive lattice with zero and let $C$ be a chain in $L$ with $0 \in C$. The chain $C$ is called *strongly maximal in $L$* if, for every homomorphism $\varphi$ of $L$ onto a distributive lattice $L_1$, the chain $\varphi(C)$ is maximal in $L_1$.

Now the following theorem resolves our problem.

**Theorem 182.** *Let $L$ be a distributive lattice with zero and let $C$ be a chain in $L$ with $0 \in C$. Then $C$ R-generates $L$ iff $C$ is strongly maximal in $L$.*

*Proof.* If $C$ R-generates $L$, then $\varphi(C)$ R-generates $\varphi(L)$ for every onto homomorphism $\varphi$. By Lemma 180, $\varphi(C)$ is maximal in $\varphi(L)$, so $C$ is strongly maximal in $L$.

Next assume that $C$ is strongly maximal in $L$ but does not R-generate $L$. Without any loss of generality, we can assume that $L$ and $C$ have a greatest element. (Otherwise, add one. Then $C \cup \{1\}$ is strongly maximal in $L \cup \{1\}$ but does not R-generate $L \cup \{1\}$.) Let $B_1 = \mathrm{BR}\, L$ and let $B_0 = \mathrm{BR}\, C$. By hypothesis, $B_0 \neq B_1$, so there exists an $a \in B_1 - B_0$.

We claim that there exist prime ideals $P_1 \neq P_2$ of $B_1$ with

$$B_0 \cap P_1 = B_0 \cap P_2.$$

With $I = \mathrm{id}(\mathrm{id}(a) \cap B_0)$ and $D = \mathrm{fil}(a)$ (formed in $B_1$), the equality $I \cap D = \varnothing$ holds, so by Theorem 115, there is a prime ideal $P_1$ such that $I \subseteq P_1$ and $P_1 \cap D = \varnothing$. Then let $I_1 = \mathrm{id}(a)$ and $D_1 = \mathrm{fil}(B_0 - P_1)$. Since $\mathrm{id}(a) \cap B_0 \subseteq P_1$, it follows that $I_1 \cap D_1 = \varnothing$. Let $P_2$ be a prime ideal with $I_1 \subseteq P_2$ and $P_2 \cap D_1 = \varnothing$. Then $a \in P_2 - P_1$, so $P_1 \neq P_2$. Because $P_2 \cap (B_0 - P_1) = \varnothing$, it follows that $P_2 \cap B_0 \subseteq P_1 \cap B_0$. Since prime ideals of a boolean lattice are unordered (Theorem 123), it follows that $P_1 \cap B_0 = P_2 \cap B_0$, proving our claim.

Now we can map $B_1$ onto $\mathsf{B}_2$ by a homomorphism $\psi$:

$$
\psi(x) = \begin{cases}
(0,0), & \text{for } x \in P_1 \cap P_2; \\
(0,1), & \text{for } x \in P_2 - P_1; \\
(1,0), & \text{for } x \in P_1 - P_2; \\
(1,1), & \text{for } x \notin P_1 \cup P_2.
\end{cases}
$$

Since $\psi(C) \subseteq \psi(B_0) = \{(0,0),(1,1)\}$ is not maximal, we conclude that $C$ is not strongly maximal in $L$. $\qquad\square$

Thus a distributive lattice $L$ is R-generated by a chain iff it has a strongly maximal chain. Theorem 172 shows that such chains exist if $L$ is countable, while Lemma 179 shows that they do not always exist.

**Corollary 183.** *Let $C$ and $D$ be strongly maximal chains of the distributive lattice $L$ with zero. Then $|C| = |D|$ and $|\operatorname{Id} C| = |\operatorname{Id} D|$.*

*Proof.* If $L$ is finite, these conclusions follow from Corollary 112. If $|L|$ is infinite, then $C$ and $D$ generate $\mathrm{BR}\,L$ as a generalized boolean lattice, and so $|C| = |D| = |L|$. By Lemma 157,

$$
|\operatorname{Spec} C| = |\operatorname{Spec}(\mathrm{BR}\,L)| = |\operatorname{Spec} D|;
$$

also $\operatorname{Spec} C = \operatorname{Id} C$ and $\operatorname{Spec} D = \operatorname{Id} D$, hence the second statement. $\qquad\square$

Corollary 183 is the strongest known extension of Corollary 112 to the infinite case. The second statement of Corollary 183 is from G. Grätzer and E. T. Schmidt [330].

A. Mostowski and A. Tarski [535] were the first to investigate boolean algebras generated by chains. Theorem 172 for boolean lattices and Theorem 176 were communicated to the author by J. R. Büchi. These results have been known for some time in topology (via the Stone topological representation theorem, see Section 5.2). Some of the other results appeared first in G. Grätzer [257].

## Exercises

4.1. In a boolean lattice $B$, prove that
$$b_0 + \cdots + b_{n-1} = \sum_{1 \le m \le n} \bigvee (b_{i_0} \wedge \cdots \wedge b_{i_{m-1}} \mid i_0 < \cdots < i_{m-1})$$

for $b_0, \ldots, b_{n-1} \in B$. Observe that the terms being summed on the right side of the formula form a chain (G. M. Bergman).

4.2. Use Exercise 4.1 to prove Lemma 155.

4.3. Give a detailed proof of Lemma 156.

4.4. Try to describe the most general situation to which the idea of the proof of Theorem 69 (Theorem 89) could be applied.

4.5. Can you redefine "the boolean lattice $B$ generated by a distributive lattice $L$", so that Lemma 157 remains valid?

4.6. Find necessary and sufficient conditions on a distributive lattice $L$ in order that $L$ have a boolean congruence-preserving extension $B$.

4.7. Work out Corollaries 159 and 160 for the boolean lattice R-generated by a distributive lattice $L$.

4.8. Let $B$ be a generalized boolean lattice and let $L$ be a sublattice of $B$. Let $x \in L$ be written in $B$ as the sum of a chain of elements of $L$:
$x = x_0 + \cdots + x_{n-1}$ with $x_0 \le \cdots \le x_{n-1}$ and $x_0, \ldots, x_{n-1} \in L$.
Then
$$x_0 + \cdots + x_{m-1} \in L$$
holds for each $m \le n$ with $m \equiv n \pmod 2$.

4.9. Use Exercise 4.8 to prove Lemma 161.

4.10. Let $L$ be a finite lattice. Under what conditions on $L$ is the map
$$x \mapsto \{ u \in \mathrm{Mi}\, L \mid x \not\le u \}$$
a meet-embedding into the boolean lattice $\mathrm{Pow}(\mathrm{Ji}\, L)$.

4.11. How do you modify Exercise 4.10 to get a join-embedding? (M. Wild [729] characterizes finite lattices that have a cover-preserving embedding into a boolean lattice; but the embedding usually is neither join- nor meet-embedding.)

4.12. Let $L$ be the lattice of closed subsets of the real unit interval $[0, 1]$. Does (JID) or (MID) hold in $L$?

4.13. Show that in any complete distributive lattice, (JID) holds whenever $x$ is a complemented element.

4.14. Can you generalize Exercise 4.13 to "dual semi complements", that is, to elements $\bar{x}$ with $x \vee \bar{x} = 1$?

*4.15. The *Complete Infinite Distributive Identity* is (for $I, J \ne \varnothing$):

(CID)
$$\bigwedge (\bigvee (a_{ij} \mid j \in J) \mid i \in I)$$
$$= \bigvee (\bigwedge (a_{i\,\varphi(i)} \mid i \in I) \mid \varphi \colon I \to J).$$

Show that (CID) holds in a complete boolean lattice $B$ iff it is atomic (A. Tarski [673]). (Hint: apply (CID) to $\bigwedge( a \vee a' \mid a \in B ) = 1$.)

4.16. Prove that (CID) is selfdual for Boolean lattices.

4.17. Let $B$ be a boolean lattice and let $I$ be an ideal of $B$. Show that $I$ is normal iff $I = I^{**}$ (for these concepts, see Exercise I.3.71–I.3.74 and Lemma 164).

4.18. Prove that the boolean lattice $\mathrm{Skel}(\mathrm{Id}\, L)$ of Lemma 164 is the Mac-Neille completion of the boolean lattice $L$.

*4.19. Show that the MacNeille completion of a distributive lattice need not even be modular.

4.20. Let $L$ be a distributive algebraic lattice. Show that $L$ satisfies (JID). (Thus $\mathrm{Con}\, K$ satisfies (JID) for every lattice $K$.)

4.21. Let $L$ be a distributive lattice, $a_i, b_i \in L$, for all $i < \omega$, and

$$[a_0, b_0] \supset [a_1, b_1] \supset \cdots .$$

Define

$$\alpha = \bigvee( \mathrm{con}(a_0, a_i) \vee \mathrm{con}(b_0, b_i) \mid i < \omega ).$$

Show that

$$\alpha \vee \bigwedge( \mathrm{con}(a_i, b_i) \mid i < \omega ) \neq \bigwedge( \alpha \vee \mathrm{con}(a_i, b_i) \mid i < \omega ).$$

4.22. Let $L$ be a distributive lattice. Use Exercise 4.21 to show that (MID) holds in $\mathrm{Con}\, L$ iff every interval in $L$ is finite (G. Grätzer and E. T. Schmidt [332]).

*4.23. Prove the converse of Lemma 170: If every maximal chain R-generates the boolean lattice $B$, then $B$ is finite.

4.24. Why is it not possible to use transfinite induction to extend Theorem 172 to the uncountable case?

4.25. Let $C$ be a bounded chain and let $a \in C - \{0, 1\}$. Define

$$C' = [a, 1] \dotplus [0, a].$$

Then $C'$ is a chain, and $\mathrm{BR}\, C \cong \mathrm{BR}\, C'$, but, in general, $C \cong C'$ does not hold.

4.26. Describe a countable family of pairwise nonisomorphic countable boolean algebras.

4.27. Prove that a distributive lattice $L_1$ is R-generated by a sublattice $L_0$ iff distinct prime ideals of $L_1$ restrict to distinct prime ideals of $L_0$.

4.28. Relate Exercise 4.27 to Theorem 182.

4.29. Give an example of a bounded distributive lattice $L$ with a maximal chain $C$ such that $C$ is not maximal in $\mathrm{BR}\, L$ (G. W. Day [142]).

4.30. Let $L_0$ be the $[0, 1]$ rational interval and let $L_1$ be the $[0, 1]$ real interval. Let

$$C = \{ (x, x) \mid 0 \leq x \leq 1,\ x \text{ rational} \}.$$

Then $C$ is a maximal chain in $L_0 \times L_1$. Show that $C$ is not strongly maximal (G. Grätzer and E. T. Schmidt [330]).

4.31. In $L_0 \times L_1$ of Exercise 4.30, find a maximal chain of cardinality $\aleph_0$; find another of cardinality $2^{\aleph_0}$. Show that $L_0 \times L_1$ has strongly maximal chains. What are their cardinalities?

4.32. Let $L$ be a distributive lattice with zero and let $B = \mathrm{BR}\, L$. Show that

$$P \mapsto P \cap L, \quad \text{for } P \in \mathrm{Spec}\, B$$

is a one-to-one correspondence between the prime ideals of $L$ and $B$.

4.33. Let $A$ be a countably infinite set, and let $B = \mathrm{Pow}\, A$. Prove that $B$ has maximal chains of cardinality $\aleph_0$ and $2^{\aleph_0}$.

*4.34. Using the Generalized Continuum Hypothesis, generalize Exercise 4.33 to arbitrary infinite sets.

4.35. Construct an example in which the sequence of ideals $I_\gamma$ of Theorem 176 does not terminate in finitely many steps.

4.36. Let $C$ be the $[0,1]$ interval of the rational numbers. Show that $\mathrm{BR}\, C$ is $\mathrm{Free}_{\mathbf{B}}(\aleph_0)$.

4.37. Let $L$ be a distributive lattice with zero and let $B$ be a generalized boolean lattice. Then every $\{0\}$-homomorphism $\varphi \colon L \to B$ can be extended to a unique $\{0\}$-homomorphism $\overline{\varphi} \colon \mathrm{BR}\, L \to B$.

4.38. Let $L_0$ and $L_1$ be distributive lattices with zero. Then every $\{0\}$-homomorphism $\varphi \colon L_0 \to L_1$ can be extended to a unique $\{0\}$-homomorphism

$$\mathrm{BR}(\varphi) \colon \mathrm{BR}\, L_0 \to \mathrm{BR}\, L_1.$$

4.39. The assignment $L \mapsto \mathrm{BR}\, L$, $\varphi \mapsto \mathrm{BR}(\varphi)$ described in Exercise 4.38 is a functor from the category of distributive lattices with zero with $\{0\}$-homomorphisms to the category of generalized boolean lattices with ring homomorphisms. Show that this functor preserves direct limits (F. Wehrung).

## 5.  Topological Representation

The order $\mathrm{Spec}\, L$ of prime ideals does give a great deal of information about the distributive lattice $L$, but obviously it does not characterize $L$. For instance, for a countably infinite boolean algebra $L$, the order $\mathrm{Spec}\, L$ is an unordered set of cardinality $\aleph_0$ or $2^{\aleph_0}$, whereas there are surely more than two such boolean algebras up to isomorphism.

Therefore, it is necessary to endow $\mathrm{Spec}\, L$ with more structure if we want it to characterize $L$. M. H. Stone [669] endowed $\mathrm{Spec}\, L$ with a topology; see also L. Rieger [611]. In most of this section, we shall discuss this approach in a slightly more general but, in our opinion, more natural framework. Then we follow H. A. Priestley and also endow $\mathrm{Spec}\, L$ with an ordering: $\subseteq$. (See Section VI.2.8 for a related topic.)

These two approaches use topology to better understand distributive lattices. We conclude this section with a brief section on frames; how distributive lattices can be used to better understand topology.

## 5.1  Distributive join-semilattices

Let us call a join-semilattice $L$ *distributive* if

$$a \leq b_0 \vee b_1, \qquad \text{for } a, b_0, b_1 \in L,$$

implies that

$$a = a_0 \vee a_1 \qquad \text{for some } a_0, a_1 \in L \text{ with } a_0 \leq b_0 \text{ and } a_1 \leq b_1;$$

see Figure 34. Note that $a_0$ and $a_1$ need not be unique.

Some elementary properties of a distributive join-semilattice are as follows (see the basic concepts following Definition 41):

**Lemma 184.**

(i) *If $(L; \vee, \wedge)$ is a lattice, then the join-semilattice $(L; \vee)$ is distributive iff the lattice $(L; \vee, \wedge)$ is distributive.*

(ii) *If a join-semilattice $L$ is distributive, then for every $a, b \in L$, there is an element $d \in L$ with $d \leq a$ and $d \leq b$. Consequently, $\operatorname{Id} L$ is a lattice.*

(iii) *A join-semilattice $L$ is distributive iff $\operatorname{Id} L$, as a lattice, is distributive.*

*Proof.*

(i) If $(L; \vee, \wedge)$ is distributive, and $a \leq b_0 \vee b_1$, then with $a_0 = a \wedge b_0$ and $a_1 = a \wedge b_1$, we obtain that $a = a_0 \vee a_1$. Conversely, if $(L; \vee)$ is distributive, and the lattice $L$ contains a diamond or a pentagon $\{o, a, b, c, i\}$, then $a \leq b \vee c$,

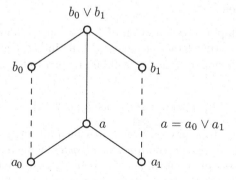

Figure 34. The distributivity of a semilattice

but $a$ cannot be represented as $a = a_0 \vee a_1$ with $a_0 \leq b$ and $a_1 \leq c$, a contradiction.

(ii) $a \leq a \vee b$, thus $a = a_0 \vee b_0$, where $a_0 \leq a$ and $b_0 \leq b$. Since, in addition, the inequality $b_0 \leq a$ holds, it follows that $b_0$ is a lower bound for $a$ and $b$.

(iii) First we observe that, for $I, J \in \operatorname{Id} L$,

$$I \vee J = \{\, i \vee j \mid i \in I, \; j \in J \,\}$$

follows from the assumption that the join-semilattice $L$ is distributive. Therefore, the distributivity of $\operatorname{Id} L$ can be easily proved. Conversely, if $\operatorname{Id} L$ is distributive and $a \leq b_0 \vee b_1$, then

$$\operatorname{id}(a) = \operatorname{id}(a) \wedge (\operatorname{id}(b_0) \vee \operatorname{id}(b_1)) = (\operatorname{id}(a) \wedge \operatorname{id}(b_0)) \vee (\operatorname{id}(a) \wedge \operatorname{id}(b_1)),$$

and so $a = a_0 \vee a_1$ with $a_0 \in \operatorname{id}(b_0)$ and $a_1 \in \operatorname{id}(b_1)$, which is distributivity for the join-semilattice $L$.   $\square$

A nonempty subset $D$ of a join-semilattice $L$ is called a *filter* if the following two conditions hold:

(i) $a, b \in D$ implies that there exists a lower bound $d \in D$ of $a$ and $b$;
(ii) $a \in D$, $x \in L$, and $x \geq a$ imply that $x \in D$.

An ideal $I$ of $L$ is *prime* if $I \neq L$ and $L - I$ is a filter. Again, let $\operatorname{Spec} L$ denote the set of all prime ideals of $L$.

**Lemma 185.** *Let $I$ be an ideal and let $D$ be a filter of a distributive join-semilattice $L$. If $I \cap D = \varnothing$, then there exists a prime ideal $P$ of $L$ with $P \supseteq I$ and $P \cap D = \varnothing$.*

*Proof.* The proof is a routine modification of the proof of Theorem 115.   $\square$

In the rest of this section, unless stated otherwise, let $L$ stand for a distributive join-semilattice with zero.

## 5.2   Stone spaces

We shall now develop a representation of $L$ as a join-semilattice of subsets of the set $\operatorname{Spec} L$. Note that larger elements of $L$ lie in fewer prime ideals; hence under our representation, each $a \in L$ will be mapped to the set of prime ideals *not* containing it:

$$\operatorname{spec}(a) = \{\, P \in \operatorname{Spec} L \mid a \notin P \,\}.$$

We shall introduce a topology on $\operatorname{Spec} L$ in which all sets $\operatorname{spec}(a)$ are open.

We also denote by $\operatorname{Spec} L$ the *topological space* defined on $\operatorname{Spec} L$ by postulating that the sets of the form $\operatorname{spec}(a)$ be a *subbase* for the open sets; we shall call $\operatorname{Spec} L$ the *Stone space* of $L$. (Exercises 5.1–5.23 review the basic topological concepts used in this section.)

**Lemma 186.** *For an ideal $I$ of $L$, define*

$$\operatorname{spec}(I) = \{\, P \in \operatorname{Spec} L \mid P \not\supseteq I \,\}.$$

*Then $\operatorname{spec}(I)$ is open in $\operatorname{Spec} L$. Conversely, every open set $U$ of $\operatorname{Spec} L$ can be uniquely represented as $\operatorname{spec}(I)$ for some ideal $I$ of $L$.*

*Proof.* We simply observe that

$$\operatorname{spec}(I) \cap \operatorname{spec}(J) = \operatorname{spec}(I \wedge J),$$
$$\operatorname{spec}\Big(\bigvee (I_j \mid j \in K)\Big) = \bigcup (\operatorname{spec}(I_j) \mid j \in K),$$

and $\operatorname{spec}(\operatorname{id}(a)) = \operatorname{spec}(a)$, from which it follows that the $\operatorname{spec}(I)$ form the smallest collection of sets closed under finite intersection and arbitrary union containing all the sets $\operatorname{spec}(a)$ for $a \in L$. Observe that $a \in I$ iff $\operatorname{spec}(a) \subseteq \operatorname{spec}(I)$. Thus $\operatorname{spec}(I) = \operatorname{spec}(J)$ holds iff $a \in I$ is equivalent to $a \in J$, which in turn is equivalent to $I = J$. $\qquad\square$

**Lemma 187.** *The subsets of $\operatorname{Spec} L$ of the form $\operatorname{spec}(a)$ can be characterized as compact open sets.*

*Proof.* Indeed, if a family of open sets $\{\, \operatorname{spec}(I_k) \mid k \in K \,\}$ is a cover for $\operatorname{spec}(a)$, that is,

$$\operatorname{spec}(a) \subseteq \bigcup (\operatorname{spec}(I_k) \mid k \in K) = \operatorname{spec}\Big(\bigvee (I_k \mid k \in K)\Big),$$

then $a \in \bigvee (I_k \mid k \in K)$. This implies that $a \in \bigvee (I_k \mid k \in K_0)$, for some finite $K_0 \subseteq K$, proving that $\operatorname{spec}(a) \subseteq \bigcup (\operatorname{spec}(I_k) \mid k \in K_0)$. Thus $\operatorname{spec}(a)$ is compact. Conversely, if $I$ is not principal, then

$$\operatorname{spec}(I) \subseteq \bigcup (\operatorname{spec}(a) \mid a \in I),$$

but being nonprincipal, $I$ will not be finitely generated, hence

$$\operatorname{spec}(I) \not\subseteq \bigcup (\operatorname{spec}(a) \mid a \in I_0)$$

for any finite $I_0 \subseteq I$. $\qquad\square$

From Lemma 187, we immediately conclude:

**Theorem 188.** *The Stone space $\operatorname{Spec} L$ determines $L$ up to isomorphism.*

## 5.3   The characterization of Stone spaces

Stone spaces are characterized in Theorem 191. To prepare for the proof of Theorem 191, we first prove Lemma 189.

Let $P$ be a prime ideal of $L$. Then $P$ is represented as an element of $\operatorname{Spec} L$ and also by $\operatorname{spec}(P)$, an open set not containing that element. The connection between $P$ and $\operatorname{spec}(P)$ is given in Lemma 189 and is illustrated by Figure 35.

**Lemma 189.** *For every prime ideal $P$ of $L$,*

$$\overline{\{P\}} = \operatorname{Spec} L - \operatorname{spec}(P),$$

*where $\overline{\{P\}}$ is the topological closure of the set $\{P\}$.*

*Proof.* By the definition of closure,

$$
\begin{aligned}
\overline{\{P\}} &= \{\, Q \in \operatorname{Spec} L \mid Q \in \operatorname{spec}(a) \text{ implies that } P \in \operatorname{spec}(a) \,\} \\
&= \{\, Q \in \operatorname{Spec} L \mid Q \supseteq P \,\} = \operatorname{Spec} L - \{\, Q \mid Q \not\supseteq P \,\} \\
&= \operatorname{Spec} L - \operatorname{spec}(P). \qquad\qquad\qquad\qquad\qquad\qquad\qquad \square
\end{aligned}
$$

**Corollary 190.** *If $P \neq Q$, then $\overline{\{P\}} \neq \overline{\{Q\}}$.*

*Proof.* Combine Lemmas 186 and 189. $\qquad\qquad\qquad\qquad\qquad\qquad\qquad\qquad\qquad \square$

So $\operatorname{Spec} L$ is a $T_0$-space.

Lemma 189 also shows that if $P$ is a prime ideal, then $\operatorname{Spec} L - \operatorname{spec}(P)$ must be the closure of a singleton. In other words:

(GC)   Let $U$ be a proper open set. Let us assume that for any pair of compact open sets $U_0$ and $U_1$ satisfying $U_0 \cap U_1 \subseteq U$, it follows that $U_0 \subseteq U$ or $U_1 \subseteq U$. Then $\operatorname{Spec} L - U = \overline{\{P\}}$ for some element $P$.

Now we can state the characterization theorem.

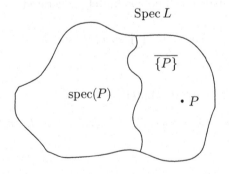

Figure 35. The connection between $P$ and $\operatorname{spec}(P)$

**Theorem 191.** *Let $S$ be a topological space. Then there exists a distributive join-semilattice $L$ such that up to homeomorphism, $\operatorname{Spec} L = S$ iff the following two conditions hold.*

(Stone1) $S$ *is a $T_0$-space in which the compact open sets form a base for the open sets.*

(Stone2) *Let $F$ be a closed set in $S$, let $\{\, U_k \mid k \in K \,\}$ be a dually directed family (that is, $K \neq \varnothing$ and, for every $k, l \in K$, there exists a $t \in K$ such that $U_t \subseteq U_k \cap U_l$) of compact open sets of $S$, and let $U_k \cap F \neq \varnothing$ for all $k \in K$; then $\bigcap(\, U_k \mid k \in K\,) \cap F \neq \varnothing$.*

*Remark.* The meaning of condition (Stone1) is clear. Condition (Stone2) is a complicated way of ensuring that condition (GC) holds and that Lemma 185 holds for the join-semilattice of compact open sets of $\operatorname{Spec} L$.

*Proof.* To show that condition (Stone1) holds for $\operatorname{Spec} L$, we have to verify that the $\operatorname{spec}(a)$, for $a \in L$, form a base (not only a subbase) for the open sets of $\operatorname{Spec} L$. In other words, for $a, b \in L$ and $P \in \operatorname{spec}(a) \cap \operatorname{spec}(b)$, we have to find an element $c \in L$ with $P \in \operatorname{spec}(c)$ and $\operatorname{spec}(c) \subseteq \operatorname{spec}(a) \cap \operatorname{spec}(b)$. By assumption, $a \notin P$ and $b \notin P$. Since $P$ is prime, there exists an element $c \in L$ with $c \notin P$ and with $c \leq a$, $c \leq b$. Then

$$P \in \operatorname{spec}(c), \ \operatorname{spec}(c) \subseteq \operatorname{spec}(a), \ \operatorname{spec}(c) \subseteq \operatorname{spec}(b),$$

as required.

To verify condition (Stone2) for $\operatorname{Spec} L$, let $F = \operatorname{Spec} L - \operatorname{spec}(I)$ and $U_k = \operatorname{spec}(a_k)$. Thus

$$F = \{\, P \mid P \supseteq I \,\},$$
$$U_k = \{\, P \mid a_k \notin P \,\}.$$

The assumption that the $\{\, U_k \mid k \in K \,\}$ is a dually directed family implies that

$$D = \{\, x \mid x \geq a_k \text{ for some } k \in K \,\}$$

is a filter; since $U_k \cap F \neq \varnothing$, we have $\operatorname{spec}(a_k) \nsubseteq \operatorname{spec}(I)$; that is, $a_k \notin I$, showing that $D \cap I = \varnothing$. Therefore, by Lemma 185, there exists a prime ideal $P$ with $P \supseteq I$ and $P \cap D = \varnothing$. Then $a_k \notin P$, and so $P \in \operatorname{spec}(a_k)$ for all $k \in K$. Also $P \supseteq I$, thus $P \notin \operatorname{spec}(I)$, and so $P \in F$, proving that

$$P \in F \cap \bigcap(\, U_k \mid k \in K\,),$$

verifying (Stone2).

Conversely, let $S$ be a topological space satisfying conditions (Stone1) and (Stone2). Let $L$ be the set of compact open sets of $S$. Obviously, $\varnothing \in L$;

moreover, if $A, B \in L$, then $A \cup B \in L$, and thus $L$ is a join-semilattice with zero. Let
$$A \subseteq B_0 \cup B_1, \text{ with } A, B_0, B_1 \in L.$$

Then $A \cap B_i$ is open, and therefore
$$A \cap B_i = \bigcup( A_j^i \mid j \in J_i ), \quad i = 0, 1,$$

where the $A_j^i$ are compact open sets. Since
$$A = (A \cap B_0) \cup (A \cap B_1) \subseteq \bigcup( A_j^i \mid j \in J_0 \cup J_1, \; i = 0, 1 ),$$

by the compactness of $A$, we get
$$A \subseteq \bigcup( A_j^i \mid j \in J_0^* \text{ or } j \in J_1^* ),$$

where $J_i^*$ is a finite subset of $J_i$ for $i = 0, 1$. Set
$$A_i = \bigcup( A_j^i \mid j \in J_i^* ), \quad \text{for } i = 0, 1.$$

Then $A_0, A_1 \in L$, $A = A_0 \cup A_1$, and $A_0 \subseteq B_0$, $A_1 \subseteq B_1$, showing that $L$ is distributive.

It follows from (Stone1) that the open sets of $\mathcal{S}$ are uniquely associated with ideals of $L$: for an ideal $I$ of $L$, let
$$U(I) = \bigcup( a \mid a \in I )$$

(keep in mind that an $a \in L$ is a subset of $\mathcal{S}$, as illustrated in Figure 36). Note that $a \in I$ iff $a \subseteq U(I)$ for any $a \in L$.

Now let $P$ be a prime ideal of $L$, let $F = \mathcal{S} - U(P)$, and let $\{ U_k \mid k \in K \}$ be the set of all compact open sets of $\mathcal{S}$ that have nonempty intersections with $F$. Thus the $U_k$ are exactly those elements of $L$ that are not in $P$. Therefore, by the definition of a prime ideal, given $k, l \in K$, there exists $t \in K$ with $U_t \subseteq U_k$ and $U_t \subseteq U_l$, proving that $F$ and $\{ U_k \mid k \in K \}$ satisfy the hypothesis of (Stone2). By (Stone2), we conclude that there exists an element
$$p \in F \cap \bigcap( U_k \mid k \in K ).$$

If $q \in F$, then $U \cap F \neq \varnothing$ for every compact open set $U$ with $q \in U$; thus $p \in U$, proving that $\overline{\{p\}} = F$. Note that $\mathcal{S}$ is a $T_0$-space; therefore, $p$ is unique. We shall write $p = \varphi(P)$.

Conversely, if $p \in \mathcal{S}$, let
$$I = \{ a \in L \mid a \subseteq \mathcal{S} - \overline{\{p\}} \}.$$

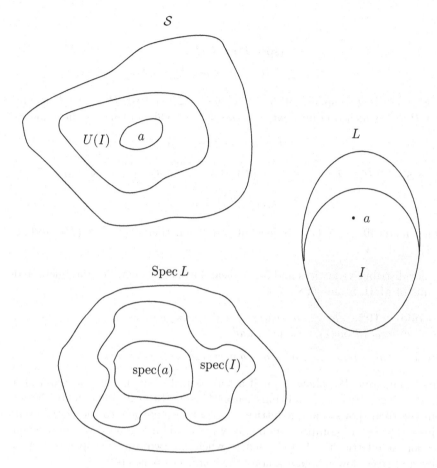

Figure 36. $a$, $\mathrm{spec}(a)$; $I$, $\mathrm{spec}(I)$; $L$, $\mathcal{S}$, and $\mathrm{Spec}\,L$

Then $I$ is an ideal of $L$, and $\mathcal{S} - \overline{\{p\}} = U(I)$. We claim that $I$ is prime. Indeed, if $U, V \in L$, $U \notin I$, $V \notin I$, then $U \cap \overline{\{p\}} \neq \varnothing$, $V \cap \overline{\{p\}} \neq \varnothing$, and therefore, $p \in U$ and $p \in V$. Thus $p \in U \cap V$ and so $U \cap V \nsubseteq U(I)$. By (Stone1), there exists a $W \in L$ with $W \subseteq U \cap V$ and $W \nsubseteq U(I)$. Therefore, $W \notin I$, and so $I$ is prime.

Summing up, the map $\varphi \colon P \mapsto p$ is a bijection between $\mathrm{Spec}\,L$ and $\mathcal{S}$. To show that $\varphi$ is a homeomorphism, it suffices to show that $U$ is open in $\mathrm{Spec}\,L$ iff $\varphi(U)$ is open in $\mathcal{S}$.

A typical open set in $\mathrm{Spec}\,L$ is of the form $\mathrm{spec}(I)$, for $I \in \mathrm{Id}\,L$, and an open set of $\mathcal{S}$ is of the form $U(I)$, therefore, we need only prove that

$$\varphi(\operatorname{spec}(I)) = U(I),$$
$$\varphi^{-1}(U(I)) = \operatorname{spec}(I);$$

in other words, $P \in \operatorname{spec}(I)$ iff $(\varphi(P) =) p \in U(I)$. Indeed, $P \in \operatorname{spec}(I)$ means that $P \not\supseteq I$, which is equivalent to $U(P) \not\supseteq U(I)$; this, in turn, is the same as

$$U(I) \cap (\mathcal{S} - U(P)) \neq \varnothing.$$

Since $\mathcal{S} - U(P) = \overline{\{p\}}$ with $p = \varphi(P)$, the last condition means that

$$U(I) \cap \overline{\{p\}} \neq \varnothing,$$

which holds iff $p \in U(I)$. Indeed, if $p \notin U(I)$, then $U(I) \subseteq U(P)$, and so $U(I) \cap \overline{\{p\}} = \varnothing$.                                     $\square$

For distributive lattices and for boolean lattices, we now get the celebrated results of M. H. Stone [668], [669].

**Corollary 192.** *The Stone spaces of distributive lattices can be characterized by conditions* (Stone1), (Stone2), *and*

(Stone3) *The intersection of two compact open sets is compact.*

*Proof.* Theorem 191 shows that if a topological space is the spectrum of a distributive lattice $L$, conditions (Stone1) and (Stone2) must hold. Theorem 188 then shows that the lattice $L$ must be isomorphic to the distributive join-semilattice of compact open subsets of $\mathcal{S}$. Thus we need to know when this join-semilattice is a lattice; that is, when any two compact open sets $A, B$ have a greatest lower bound among the compact open sets.

Now by condition (Stone1), $A \cap B$ is a union of compact open subsets $U_i$, hence any greatest lower bound of $A$ and $B$ among compact open subsets must contain all these $U_i$, hence equal $A \cap B$. So such a greatest lower bound will exist iff $A \cap B$ is itself a compact open subset, which is guaranteed by condition (Stone3).                                     $\square$

**Corollary 193.** *The Stone spaces of boolean lattices* (*called* boolean spaces) *can be characterized as the compact Hausdorff spaces in which the closed open* (clopen) *sets form a base for the open sets.* (*In other words, they are* totally disconnected *compact Hausdorff spaces.*)

*Proof.* Let $\mathcal{S} = \operatorname{Spec} B$, where $B$ is a boolean lattice. Then $\mathcal{S} = \operatorname{spec}(1)$, and thus $\mathcal{S}$ is compact. Let $P, Q \in \mathcal{S}$ and $P \neq Q$; by symmetry, we can take an element $a \in P - Q$. Then $Q \in \operatorname{spec}(a)$ and $P \in \operatorname{spec}(a')$; therefore, every pair of elements of $\mathcal{S}$ can be separated by clopen sets, verifying that $\mathcal{S}$ is Hausdorff. This also shows that $\mathcal{S}$ is totally disconnected.

Conversely, let $S$ be compact, Hausdorff, and totally disconnected. Then condition (Stone1) is obvious. Condition (Stone2) follows from the observation that $F$ and the $U_i$, for all $i \in I$, are now closed sets having the finite intersection property; therefore, by compactness, they have an element in common. Thus applying Theorem 191, $S$ has the form $\operatorname{Spec} L$ for some distributive join-semilattice $L$; and by Lemma 187, $L$ is the semilattice of compact open subsets of $S$. Now in a compact Hausdorff space, the compact open subsets are the clopen subsets, and form a boolean lattice; so $S$ is indeed homeomorphic to the Stone space of a boolean lattice. $\qquad\square$

## 5.4 Applications

As an interesting application, we prove:

**Theorem 194.** *Let $B$ be an infinite boolean lattice. Then $|\operatorname{Spec} B| \geq |B|$.*

*Proof.* Let $S$ be a totally disconnected compact Hausdorff space. For $a, b \in S$ with $a \neq b$, fix a pair of clopen sets $U_{a,b}$ and $U_{b,a}$ such that $a \in U_{a,b}$, $b \in U_{b,a}$, and $U_{a,b} \cap U_{b,a} = \varnothing$. Now let $U$ be clopen and $a \in U$. Then

$$S - U \subseteq \bigcup (U_{b,a} \mid b \in S - U),$$

and so, by the compactness of $S - U$,

$$S - U \subseteq \bigcup (U_{b,a} \mid b \in X),$$

for some finite $X \subseteq S - U$. Then $V_a = \bigcap (V_a \mid b \in X)$ is open and $a \in V_a \subseteq U$. Thus $U = \bigcup (V_a \mid a \in U)$, so by the compactness of $U$, for some finite $A \subseteq U$, we obtain that $U = \bigcup (V_a \mid a \in A)$.

Thus every clopen set is a finite union of finite intersections of $U_{a,b}$, and so there are no more clopen sets than there are finite sequences of elements of $S$; this cardinality is $|S|$, provided that $|S|$ is infinite. $\qquad\square$

It might be illuminating to compare this to an algebraic proof, see Exercise 5.37.

Theorem 191 and its corollaries provide topological representations for distributive join-semilattices, distributive lattices, and boolean lattices, respectively. It is also possible to give a topological representation for homomorphisms. We do it here only for $\{0, 1\}$-homomorphisms of bounded distributive lattices.

**Lemma 195.** *Let $L_0$ and $L_1$ be bounded distributive lattices and let $\varphi$ be a $\{0, 1\}$-homomorphism of $L_0$ into $L_1$. Then*

$$\operatorname{Spec}(\varphi) \colon P \mapsto \varphi^{-1}(P)$$

*maps $\operatorname{Spec} L_1$ into $\operatorname{Spec} L_0$; the map $\operatorname{Spec}(\varphi)$ is a continuous function with the property that if $U$ is compact open in $\operatorname{Spec} L_0$, then $\operatorname{Spec}(\varphi)^{-1}(U)$ is compact in $\operatorname{Spec} L_1$. Conversely, if $\psi \colon \operatorname{Spec} L_1 \to \operatorname{Spec} L_0$ has these properties, then $\psi = \operatorname{Spec}(\varphi)$ for exactly one $\varphi \colon L_0 \to L_1$.*

*Proof.* If $U = \mathrm{spec}(a)$, for some $a \in L_0$, then

$$\mathrm{Spec}(\varphi)^{-1}(U) = \{\, P \in \mathrm{Spec}\, L_1 \mid \varphi^{-1}(P) \in \mathrm{spec}(a) \,\}$$
$$= \{\, P \in \mathrm{Spec}\, L_1 \mid a \notin \varphi^{-1}(P) \,\}$$
$$= \{\, P \in \mathrm{Spec}\, L_1 \mid \varphi(a) \notin P \,\}$$
$$= \mathrm{spec}(\varphi(a)),$$

and so $\mathrm{Spec}(\varphi)$ is continuous, and has the desired property.

Conversely, if such a map $\psi$ is given and $U = \mathrm{spec}(a)$, for some $a \in L_0$, then $\psi^{-1}(U)$ is compact open, and so $\psi^{-1}(U) = \mathrm{spec}(b)$ for a unique $b \in L_1$. The map $\varphi \colon a \mapsto b$ is a $\{0,1\}$-homomorphism, and $\psi = \mathrm{Spec}(\varphi)$. □

The following interpretation of conditions (Stone1), (Stone2), and (Stone3) will be useful. Let $\mathcal{S}$ be a topological space. The *booleanization of $\mathcal{S}$ is a* topological space $\mathcal{S}^{\mathrm{B}}$ on $\mathcal{S}$ that has the compact open sets of $\mathcal{S}$ *and their complements* as a subbase for open sets. (For a similar construction on the prime spectrum of a commutative ring, see M. Hochster [396].)

**Lemma 196.** *A compact topological space $\mathcal{S}$ satisfies conditions* (Stone1), (Stone2), *and* (Stone3) *iff $\mathcal{S}^{\mathrm{B}}$ is a boolean space.*

*Proof.* Let $\mathcal{S}$ satisfy (Stone1), (Stone2), and (Stone3). Then $\mathcal{S}^{\mathrm{B}}$ is obviously Hausdorff and totally disconnected.

To verify the compactness of $\mathcal{S}^{\mathrm{B}}$, let $\mathcal{F}_0$ be a collection of compact open sets of $\mathcal{S}$, and let $\mathcal{F}_1$ be a collection of complements of compact open sets of $\mathcal{S}$ such that in $\mathcal{F} = \mathcal{F}_0 \cup \mathcal{F}_1$ no finite intersection is void. Because of (Stone3), we can assume that $\mathcal{F}_0$ is closed under finite intersection. Since members of $\mathcal{F}_1$ are closed in $\mathcal{S}$ and $\mathcal{S}$ is compact, the set

$$\bigcap(\, X \mid X \in \mathcal{F}_1 \,) = F$$

is nonempty. Also, $U \cap X$ is closed in $U$, for every $U \in \mathcal{F}_0$ and $X \in \mathcal{F}_1$, and thus by compactness of $U$,

$$U \cap F = \bigcap(U \cap X \mid X \in \mathcal{F}_1) \neq \varnothing.$$

Applying (Stone2) to $F$ and $\mathcal{F}_0$, we conclude that $\bigcap \mathcal{F} \neq \varnothing$, which, by Alexander's Theorem (see Exercise 5.15), proves compactness.

Conversely, if $\mathcal{S}^{\mathrm{B}}$ is boolean, then the compact open sets of $\mathcal{S}^{\mathrm{B}}$ form a boolean lattice $L$. Moreover, every compact open subset of $\mathcal{S}$ is closed in $\mathcal{S}^{\mathrm{B}}$, hence so is the intersection of any two such subsets, hence such an intersection, as a closed subset of the compact space $\mathcal{S}^{\mathrm{B}}$, will be compact in the topology of $\mathcal{S}^{\mathrm{B}}$. Hence it must also be compact in the weaker topology of $\mathcal{S}$, showing that $\mathcal{S}$ satisfies (Stone3). Hence the compact open sets of $\mathcal{S}$ form a sublattice $L_1$ of $L$. Thus $L_1$ is a distributive lattice, and by Theorem 191, the homeomorphism $\mathcal{S} \cong \mathrm{Spec}\, L_1$ holds, and $\mathcal{S}$ also satisfies conditions (Stone1) and (Stone2). □

## 5.5  Free distributive products

Let $L_i$, for $i \in I$, be pairwise disjoint distributive lattices. Then

$$Q = \bigcup(L_i \mid i \in I)$$

is a partial lattice. A free lattice generated by $Q$ over the class $\mathbf{D}$ of all distributive lattices is called a *free distributive product* of the $L_i$ for $i \in I$. To prove the existence of free distributive products, it suffices by Theorem 89 to show that there exists a distributive lattice $L$ containing $Q$ as a partial sublattice. This is easily done: Let $L$ be the direct product of the $L_i \cup \{0\}$, for $i \in I$, where 0 is a new zero element of $L_i$. Identify $x \in L_i$ with $f \in L$ defined by $f(i) = x$ and $f(j) = 0$ for all $j \neq i$. Then $Q$ becomes a partial sublattice of $L$.

An equivalent definition is:

**Definition 197.** Let $\mathbf{K}$ be a class of lattices and let $L_i$, for $i \in I$, be lattices in $\mathbf{K}$. A lattice $L$ in $\mathbf{K}$ is called a *free $\mathbf{K}$-product* of the $L_i$, for $i \in I$, if every $L_i$ has an embedding $\varepsilon_i$ into $L$ such that:

(i) $L$ is generated by $\bigcup(\varepsilon_i(L_i) \mid i \in I)$.

(ii) If $K$ is any lattice in $\mathbf{K}$ and $\varphi_i$ is a homomorphism of $L_i$ into $K$, for all $i \in I$, then there exists a homomorphism $\varphi$ of $L$ into $K$ satisfying $\varphi_i = \varphi \varepsilon_i$ for all $i \in I$ (see Figure 37).

For distributive lattices, this is equivalent to the first definition. In most cases, we will assume that each $L_i \leq L$ and that $\varepsilon_i$ is the inclusion map; then (ii) will simply state that the $\varphi_i$ have a common extension. Note that in all cases we shall consider, (i) can be replaced by the requirement that the $\varphi$ in (ii) be unique.

If, in Definition 197, $\mathbf{K}$ is a class of bounded lattices and all homomorphisms are assumed to be $\{0, 1\}$-homomorphisms, we get the concept of a *free $\mathbf{K}$*

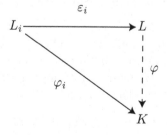

Figure 37. Illustrating condition (ii) in Definition 197

$\{0, 1\}$-*product*. In particular, if $\mathbf{K} = \mathbf{L}$, we get the concept of a *free* $\{0, 1\}$-*product*, see Section VII.1.12, and if $\mathbf{K} = \mathbf{D}$, we obtain the concept of a *free* $\{0, 1\}$-*distributive product*.

Our final result is the existence and description of a free $\{0, 1\}$-distributive product of a family of bounded distributive lattices, see A. Nerode [548].

A *Stone space* is a topological space satisfying the conditions (Stone1), (Stone2), and (Stone3).

**Theorem 198.** *Let $L_i$, for $i \in I$, be distributive lattices with zero and unit. Let $\mathcal{S} = \prod(\operatorname{Spec} L_i \mid i \in I)$ (see Exercise 5.16). Then $\mathcal{S}$ is a Stone space, and thus $\mathcal{S} \cong \operatorname{Spec} L$ for some distributive lattice $L$. Such a lattice $L$ is a free $\{0, 1\}$-distributive product of the $L_i$ for $i \in I$.*

The proof of Theorem 198 will be preceded by two lemmas.

**Lemma 199.** *Let $\mathcal{S}_i$, for $i \in I$, be compact Stone spaces. Then*

$$\prod(\mathcal{S}_i^{\mathrm{B}} \mid i \in I) = (\prod(\mathcal{S}_i \mid i \in I))^{\mathrm{B}}.$$

*Proof.* For $U \subseteq \mathcal{S}_j$, let

$$E(U) = \{\, f \in \prod \mathcal{S}_i \mid f(j) \in U \,\}$$

(see Exercise 5.16). The compact open sets form a base for open sets in $\mathcal{S}_j$; therefore,

$$\{\, E(U) \mid U \text{ compact open in some } \mathcal{S}_j \,\}$$

is a subbase for open sets in $\prod(\mathcal{S}_i \mid i \in I)$. Note that all the sets $E(U)$ in the above family are compact open in $\prod \mathcal{S}_i$; therefore, $V \subseteq \prod \mathcal{S}_i$ is compact open iff it is a finite union of finite intersections of some of the $E(U)$. Consequently, declaring also the complements of compact open sets to be open (when forming $(\prod \mathcal{S}_i)^{\mathrm{B}}$) is equivalent to making the complements of the sets $E(U)$ open. But the complement of $E(U)$ is $E(\mathcal{S}_i - U)$, and $\mathcal{S}_i - U$ is an open set of $\mathcal{S}_i^{\mathrm{B}}$. Thus $\prod \mathcal{S}_i^{\mathrm{B}}$ and $(\prod \mathcal{S}_i)^{\mathrm{B}}$ have the same topology. $\qquad\square$

**Lemma 200.** *A product of compact Stone spaces is again a compact Stone space.*

*Proof.* Let $\mathcal{S}_i$, for $i \in I$, be Stone spaces. Then $\mathcal{S} = \prod \mathcal{S}_i$ is $T_0$ and compact (use Exercises 5.17 and 5.22). Since $\mathcal{S}_j^{\mathrm{B}}$ is boolean (see Lemma 196), so is $\prod \mathcal{S}_i^{\mathrm{B}}$ (by Exercises 5.21–5.23). By Lemma 199, the homeomorphism $\mathcal{S}^{\mathrm{B}} = \prod \mathcal{S}_i^{\mathrm{B}}$ holds, and thus $\mathcal{S}^{\mathrm{B}}$ is boolean. Therefore, $\mathcal{S}$ is a Stone space by Lemma 196. $\quad\square$

*Proof of Theorem 198.* Let $e_i$ be the $i$th projection ($e_i \colon \operatorname{Spec} L \to \operatorname{Spec} L_i$ is given by $e_i(f) = f(i)$). By Lemma 195, there is a unique $\{0, 1\}$-homomorphism $\varepsilon_i \colon L_i \to L$ satisfying $\operatorname{Spec}(\varepsilon_i) = e_i$. It is easy to visualize $\varepsilon_i$; think of the

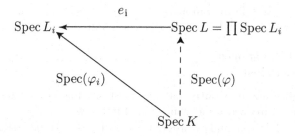

Figure 38. Proving Theorem 198

elements of $L_i$ as compact open sets of $\mathcal{S}_i$; then $\varepsilon_i(U) = E(U) = e_i^{-1}(U)$. It is obvious from this that the map $\varepsilon_i$ is an embedding.

Now let $K$ be a bounded distributive lattice and let $\varphi_i : L_i \to K$ be $\{0,1\}$-homomorphisms as in Figure 37. By applying Spec, we obtain Figure 38, where the dashed arrow is a continuous map; we have yet to show that it satisfies the conditions of the last sentence of Lemma 195, and so arises from a lattice homomorphism $\varphi$.

Thus the method of defining $\mathrm{Spec}(\varphi)$ is clear. For $x \in \mathrm{Spec}\,K$, the element $\mathrm{Spec}(\varphi)(x)$ is a member of $\prod \mathrm{Spec}\,L_i$, and

$$\mathrm{Spec}(\varphi)(x)(i) = \mathrm{Spec}(\varphi_i)(x)$$

for $i \in I$.

To show that this correspondence is indeed of the form $\mathrm{Spec}(\varphi)$, for some homomorphism $\varphi \colon L \to K$, we have to verify the following:

(a) the map we labeled $\mathrm{Spec}(\varphi)$ is continuous (this statement follows from Exercise 5.19),

(b) if $V$ is compact open in $\mathrm{Spec}\,L$, then $\mathrm{Spec}(\varphi)^{-1}(V)$ is compact open in $\mathrm{Spec}\,K$.

Let us first verify (b) for $V = E(U)$, where $U$ is compact open in some $\mathrm{Spec}\,L_i$. In this case

$$\mathrm{Spec}(\varphi)^{-1}(V) = \mathrm{Spec}(\varphi)^{-1}(E(U)) = \mathrm{Spec}(\varphi)^{-1}(e_i^{-1}(U))$$
$$= (e_i\,\mathrm{Spec}(\varphi))^{-1}(U) = \mathrm{Spec}(\varphi_i)^{-1}(U),$$

and therefore, $\mathrm{Spec}(\varphi)^{-1}(V)$ is compact open since $\mathrm{Spec}(\varphi_i)$ satisfies the condition of Lemma 195.

Next consider the case where $V$ is a finite intersection of sets $E(U)$. From the facts that inverse images respect intersections, and that $\mathrm{Spec}\,K$ satisfies condition (Stone3) by Corollary 192, we see that $\mathrm{Spec}(\varphi)^{-1}(V)$ will again be compact. Finally, let $V$ be an arbitrary compact open subset of $\mathrm{Spec}\,L$. Then because it is open, it is a union of such finite intersections, hence by compactness, it is a union of finitely many of them; and since inverse images respect unions, and finite unions of compact sets are compact, we again conclude that $\mathrm{Spec}(\varphi)^{-1}(V)$ is compact, as required.    $\square$

## 5.6   ◇ Priestley spaces
## by Hilary A. Priestley

Priestley duality for distributive lattices is Stone duality in different clothes. But in terms of outward appearance the difference is significant. In outline, Priestley's formulation makes order overt and has a Hausdorff topology in place of a $T_0$-topology. This better reveals how the finite and boolean cases fit into the overall picture: for the former we need only order (the topology is discrete and can be suppressed) and for the latter we need only topology (the order is discrete and can be suppressed). At the time of Stone's pioneering work [669], non-Hausdorff topologies were rather an alien concept and Stone's representation for distributive lattices was relatively little exploited. The $T_0$-spaces it uses came into vogue only much later, through the development of domain theory (see Section I.3.16).

We focus on the class $\mathbf{D}$ of bounded distributive lattices with $\{0,1\}$-preserving homomorphisms, leaving aside the adaptations to encompass lattices lacking one or both bounds. So consider a member $L$ of $\mathbf{D}$ and its spectrum $\mathrm{Spec}\, L$ of prime ideals. To obtain Priestley's representation, we order $\mathrm{Spec}\, L$ by inclusion and take the topology $\mathfrak{T}$ having as a subbase for the open sets the sets of the forms

$$\mathrm{spec}(a) \quad \text{and} \quad (\mathrm{Spec}\, L - \mathrm{spec}(b)) \quad\quad (a, b \in L).$$

We now form the ordered space $X_L = \langle \mathrm{Spec}\, L; \leq, \mathfrak{T} \rangle$, where $\leq$ is the inclusion ordering on prime ideals. Then $X_L$ is a *Priestley space*: $\mathfrak{T}$ is compact and the space $X_L$ is *totally order-disconnected* in the sense that given $x \not\leq y$ in $X_L$ there is a $\mathfrak{T}$-clopen down-set $U$ with $x \in U$ and $y \notin U$. Here the latter property is immediate since we can just take $U = \mathrm{spec}(a)$ with $a \in x - y$; compactness is proved via Alexander's Subbase Lemma. Priestley's representation theorem for $\mathbf{D}$ then asserts that each $L$ in $\mathbf{D}$ is isomorphic to the lattice of all clopen down-sets of its Priestley dual space $X_L$. Furthermore, every Priestley space is isomorphic, topologically and order-theoretically, to the dual space of its lattice of clopen down-sets. We arrive at a dual equivalence between $\mathbf{D}$ and the category $\mathbf{P}$ of Priestley spaces (in which the morphisms are the continuous order-preserving maps). With this equivalence to hand many other results tumble out. Some of the most useful are collected together in Theorem 201. An account, with proofs, of Priestley duality in simple dress can be seen in the textbook by B. A. Davey and H. A. Priestley [131].

**Theorem 201.** *Let $L$ be a bounded distributive lattice and*

$$X_L = \langle \mathrm{Spec}\, L; \leq, \mathfrak{T} \rangle,$$

*as defined above, be its Priestley dual space. Then, up to isomorphism,*

(i) *the order dual $L^\delta$ of $L$ is the lattice of clopen up-sets;*

(ii) *the minimal boolean extension of $L$ is the lattice of all clopen sets;*

(iii) *the ideal lattice* $\mathrm{Id}\, L$ *is the lattice of open down-sets, with principal and prime ideals corresponding to open down-sets which are, respectively, closed and of the form $X_L - {\uparrow}x$ $(x \in X_L)$;*

(iv) *the filter lattice* $\mathrm{Fil}\, L$ *is the order dual of the lattice of closed down-sets;*

(v) *the congruence lattice* $\mathrm{Con}\, L$ *is the lattice of open sets, that is,* $\mathcal{T}$.

To clarify the relationship between Stone duality and Priestley duality we indicate how to pass to and fro between Priestley spaces and spectral spaces, using a bijection under which the Priestley dual space of $L$ in **D** corresponds to its Stone space $\mathrm{Spec}\, L$. By a *spectral space* we mean a compact space satisfying conditions (Stone1), (Stone2) and (Stone3) from Section II.5.3 (see Corollary 192). Given a Priestley space $\langle X; \leq, \mathcal{T}\rangle$, the space $\langle X; \tau\rangle$ is a spectral space, where $\tau$ is the topology consisting of the $\mathcal{T}$-open down-sets.

In the other direction, let $\langle X; \tau\rangle$ be a spectral space. Let $\leq_\tau$ be the associated specialization order: $x \leq_\tau y$ iff $x \in cl_\tau\{y\}$. Consider the *dual topology* $\tau^*$. This has as a subbase for its closed sets the $\tau$-compact sets which are saturated with respect to $\geq_\tau$, that is, which are intersections of $\tau$-open sets. Then $\langle X; \tau^*\rangle$ is also a spectral space. Let $\mathcal{T}$ be the *patch topology* formed by taking the join of $\tau$ and $\tau^*$. Then, finally, $\langle X; \geq_\tau, \mathcal{T}\rangle$ is a Priestley space whose family of open down-sets is exactly $\tau$.

An interesting example of the above correspondence comes from domain theory. An algebraic lattice, or more generally a Scott domain, is a spectral space in its Scott topology and the associated Priestley space topology is the Lawson topology; see Section I.3.16 and [225]. More formally, the category of Stone spaces (the morphisms being the continuous maps under which inverse images of compact open subsets are compact) is isomorphic (and not merely equivalent) to the category **P** of Priestley spaces; see W. H. Cornish [98].

It is a moot point in duality theory whether it is preferable to order $\mathrm{Spec}\, L$ by $\subseteq$ or its opposite, and whether to use down-sets or up-sets. Indeed, the Priestley representation for $L$ can equally well be set up so as to identify $L$ with the clopen up-sets of a Priestley space, rather than the clopen down-sets. The research literature concerning Priestley duality and its applications is divided roughly equally between these alternatives—a source of minor irritation.

The down-sets version of the duality fits well with Birkhoff's representation for the finite case; see Section II.1.2. The up-sets version naturally arises if one bases the representation of a lattice $L$ not on the prime ideals but on the prime filters or equivalently on the $\{0, 1\}$-homomorphisms into the lattice $2 = \{0, 1\}$ with $0 < 1$. This last approach was the one initially adopted by H. A. Priestley [591] and it is functorially by far the smoothest. An account is given by D. M. Clark and B. A. Davey in [91, Chapter 1], where, adapted to the non-bounded case, it is used as a prototype example for the theory of natural dualities (see also Section VI.2.8 by B. A. Davey and M. Haviar).

The passage from $T_0$-spaces to ordered spaces with a compact Hausdorff topology is critical here. Natural duality theory (in its vanilla form) applies to quasivarieties $\mathbf{ISP}(M)$, where $M$ is a finite algebra. The dual structures belong to a category $\mathbf{IS_cP^+}(\underset{\sim}{M})$ of structured boolean spaces built from an *alter ego* $\underset{\sim}{M}$, which is a relational structure on the underlying set of $M$ carrying the discrete topology. Here $\mathbf{IS_cP^+}(\underset{\sim}{M})$ is the class of isomorphic copies of closed substructures of nonzero powers of $\underset{\sim}{M}$.

Stone duality for boolean algebras is an example of a natural duality, but because no relational structure on the alter ego is needed the way to generalize this duality was not clearly apparent. And in [669] Stone had, in a different way, also concealed the role of relational structures by working with topology alone.

It is well known that $\mathbf{D} = \mathbf{ISP}(2)$. It can also be shown (see for example [91]) that $\mathbf{P} = \mathbf{IS_cP^+}(\underset{\sim}{2})$, where $\underset{\sim}{2} = \langle\{0,1\}; \leqslant, \mathcal{T}\rangle$, with $\leqslant$ the underlying ordering and $\mathcal{T}$ the discrete topology. We can now present Priestley duality for $\mathbf{D}$ in full regalia.

**Theorem 202.** *There are natural hom-functors* $D\colon \mathbf{D} \to \mathbf{P}$ *and* $E\colon \mathbf{P} \to \mathbf{D}$:

*on objects:*     $D\colon L \mapsto \mathbf{D}(L,2) \leqslant \underset{\sim}{2}^L$   *and*   $E\colon X \mapsto \mathbf{P}(X,\underset{\sim}{2}) \leqslant 2^X$;

*on morphisms:*   $D\colon f \mapsto -\circ f$   *and*   $E\colon \phi \mapsto -\circ\phi$.

*Then $D$ and $E$ set up a dual equivalence between $\mathbf{D}$ and $\mathbf{P}$ with the unit and co-unit the evaluation maps* $e_L\colon L \to ED(L)$ *and* $\varepsilon_X\colon X \to DE(X)$, *where $e_L(a)(x) = x(a)$ (for all $a \in L$, $x \in X$) and $\varepsilon_X(x)(\alpha) = \alpha(x)$ (for all $\alpha \in \mathbf{P}(X,\underset{\sim}{2})$ and $x \in X$).*

*In addition, the free lattice in $\mathbf{D}$ on $\kappa$ generators is (isomorphic to) $E(\underset{\sim}{2}^\kappa)$ and, more generally, coproducts in $\mathbf{D}$ correspond to direct (concrete) products in $\mathbf{P}$.*

We conclude this section with a few remarks on the application of Priestley duality to lattice-based algebras. Given a variety of algebras having $\mathbf{D}$-reducts, one may seek to enrich the Priestley dual spaces with operations or relations capturing the non-lattice operations and to find a first-order description of the resulting dual structures. No comprehensive account exists of the myriad of dualities developed for $\mathbf{D}$-based algebras. A full bibliography up to 1985 was compiled by M. E. Adams and W. Dziobiak [6]. Among later work we draw attention to W. H. Cornish's systematic treatment of dualities for classes of algebras whose non-lattice operations are dual endomorphisms; his monograph [99] encompasses *inter alia* De Morgan algebras, Kleene algebras, Stone algebras and more generally Ockham algebras. We also note R. Goldblatt's paper [235] which investigates $n$-ary operations which are coordinatewise $\vee$- or $\wedge$-preserving. This paper includes too a Priestley-type duality for Heyting algebras, also obtained independently by M. E. Adams [3]. A Priestley space $\langle X; \leq, \mathcal{T} \rangle$ is a *Heyting space*, that is, the dual of a Heyting algebra, iff $\uparrow U$ is $\mathcal{T}$-open whenever $U$ is $\mathcal{T}$-open; for $a, b$ clopen down-sets, $a \to b$ is given by $X - \uparrow(a - b)$. Here the topological condition is exactly what is needed to ensure that the formula for the relative pseudocomplement valid in the (topology-free) finite case also works in general.

Building on Boole's original ideas on classical propositional calculus, lattice-based algebras are extensively used in logic as models for propositional logics. Join and meet model disjunction and conjunction and additional operations are used to model a non-classical negation or implication, or, traditionally on boolean algebras, a modal operator. In particular Heyting algebras model IPC (intuitionistic propositional calculus) and unary operations preserving join or meet represent modalities.

Fifty years ago S. Kripke famously introduced relational semantics for modal logic and for IPC. Kripke's ideas were hugely influential in modal logic, leading to the development of powerful semantic techniques in a subject which had hitherto been studied syntactically; see the textbook by P. Blackburn, M. de Rijke, and Y. Venema [76] and the earlier monograph by A. Chagrov and M. Zakharyaschev [83].

For both modal logic and IPC, Kripke semantics used relational frames of 'possible worlds', which were in each case, sets carrying an 'accessibility relation' $R$. The underlying sets of the frames are the prime filters (ultrafilters in the boolean setting) of the lattice reducts of the algebras they serve to represent. But mathematically the role of $R$ is quite different in the two cases. For Heyting algebras this relation is an order, as in Heyting spaces; we note that these algebras are special in that the Heyting implication is determined by the underlying lattice ordering. For modal logic $R$ is a binary relation used to capture via the frames the modal operator; no ordering is needed since the underlying lattices are boolean.

Our remarks hint at a bigger picture, of which Goldblatt gave a first glimpse in his 1989 paper. Blackburn, de Rijke and Venema [76, pp. 41 and 328] comment on the parallel but separate developments of Kripke semantics on the one hand and Jónsson and Tarski's theory of canonical extensions of boolean algebras with operators on the other. The latter theory has now been vastly extended by B. Jónsson, M. Gehrke, and many others, so that it now encompasses very many classes of lattice-based algebras. It supplies, in a systematized way, purely relational models. Loosely, adding topology gives Priestley-type topologies for these classes. These connections are developed in a monograph in preparation *Lattices in Logic* by M. Gehrke and H. A. Priestley.

## 5.7   ◇ Frames
## by Aleš Pultr

A *frame* is a complete lattice $L$ satisfying (JID), introduced in Section 4.2. A *frame homomorphism* $h: L \to M$ preserves *all* joins and *all finite* meets.

The most important example is given by the lattices Open $X$ of open sets of a topological space $X$, and the maps

$$\text{Open}(f): \text{Open } Y \to \text{Open } X$$

we get from continuous maps $f: X \to Y$ by the formula

$$\text{Open}(f)(U) = f^{-1}(U).$$

Viewing spaces as systems of "places" or "spots" with their interrelations—rather than as a structured set of points—was one of the strongest motivations for developing this theory. Instead of frames we often speak of *locales*—which terminology inverts the direction of homomorphisms to bring them into agreement with the continuous maps they represent.

How much information is lost? How well are spaces and continuous maps represented as frames? The answer is pleasing:

◇ **Lemma 203.** *Let $Y$ be a Hausdorff space and let $X$ be an arbitrary topological space. Then the homomorphisms $h$: Open $Y \to$ Open $X$ are precisely the maps* Open$(f)$, *where $f: X \to Y$ is a continuous map. The map $f$ is uniquely determined by $h$.*

This theorem also holds for *sober spaces*, which are more general than the Hausdorff spaces. (In the lattice Open $X$, each filter $U(x)$ of all open neighborhoods of a point $x$ is, trivially, completely prime; the space $X$ is sober if each completely prime filter in Open $X$ is of the form $U(x)$.)

This allows us to reconstruct a space $Y$ from the lattice Open $Y$.

Not every frame is isomorphic to an Open $X$. When studying frames, we deal with a larger class of generalized ("point-free") spaces. Is this good or bad? It has proved to be useful; nevertheless, it is always good to know whether a frame is *spatial*, that is, isomorphic to an Open $X$ (K. H. Hofmann and J. D. Lawson [397]).

◇ **Theorem 204 (Hofmann-Lawson duality).** *The formulas*

$$X, \ f \mapsto \text{Open } X, \ \text{Open}(f)$$

*provide a one-one correspondence between the class of all locally compact sober spaces and their continuous maps, and the locally compact frames and their frame homomorphisms.*

It should be noted that locally compact frames coincide with distributive continuous lattices in the sense of Scott (G. Gierz, K. H. Hofmann, K. Keimel, J. D. Lawson, M. Mislove, and D. S. Scott [225]).

Topological concepts and phenomena (like regularity, complete regularity, normality, or compactness, paracompactness or local compactness) are, as a rule, easily translated into the "point-free" language.

Sometimes the extended classes have better properties than the original topological concepts. For instance: in classical topology, Tychonoff's Theorem (products of compact spaces are compact) is equivalent to the Axiom of Choice. Here is the point-free counterpart:

◇ **Theorem 205.** *Products of compact locales are compact.*

This is fully constructive ("choice-free"), see P. T. Johnstone [428]. Furthermore, the counterpart of the Čech-Stone compactification can be described by a simple formula (B. Banaschewski and C. J. Mulvey [48]); prime filters are not involved.

This result combined with the Hofmann-Lawson duality gives Tychonoff's theorem. The duality is, of course, heavily choice dependent; so the choice aspect of the product of spaces is not in the compactness but rather whether it has enough points—another fact revealed by point-free reasoning.

In the point-free context, we can also work with the richer structures. Thus for instance, a *uniformity* on a frame $L$ can be viewed as a system of covers (a *cover* of $L$ is a subset $A \subseteq L$ such that $\bigvee A = 1$, the top) with specific natural properties. One has a concept of *completeness*, parallel with the classical one, and of *completion*; like the compactification, this completion is constructive.

Here is an interesting fact that holds in the point-free context but not in the classical one (J. R. Isbell [419]):

◇ **Theorem 206.** *A frame is paracompact iff it admits a complete uniformity.*

While in the classical context, paracompact spaces often misbehave in constructions (even a product of a paracompact space with a metric space is not necessarily paracompact), for locales we have the following nice result.

◇ **Theorem 207.** Paracompact locales are reflective in the category of all locales.

Hence, in particular, paracompactness is preserved by all products (and similar constructions). This is one of the instances where we see that it is useful to have more "spaces" than before; the situation is strongly reminiscent of the extension of reals to complex numbers, allowing solutions of problems unsolvable in the real case.

For the basic ideas, and for the early history of the area, see the excellent surveys P. T. Johnstone [429] and [431].

For more about frames and for further references, see P. T. Johnstone [430], A. Pultr [600], and S. Vickers [691].

**Exercises**

The first 22 exercises review the basics of topology that is utilized in this section.

5.1. A *topological space* is a set $A$ and a collection $\mathcal{T}$ of subsets of $A$, satisfying the properties:

(i) $A \in \mathcal{T}$;
(ii) $\mathcal{T}$ is closed under finite intersections;
(iii) $\mathcal{T}$ is closed under unions (empty, nonempty, finite, infinite).

A member of $\mathcal{T}$ is called an *open set*. Call a set *closed* if its complement is open. Characterize those subsets of Pow $A$ that are the families of all closed subsets under topologies on $A$.

5.2. A family of nonempty sets $\mathcal{B}$ in $\mathcal{T}$ is a *base* for open sets iff every open set is a union of members of $\mathcal{B}$. Show that for a set $A$, a collection $\mathcal{B}$ of subsets of $A$ is a base of open sets of some topological space defined on $A$ iff $\bigcup \mathcal{B} = A$, and for $X, Y \in \mathcal{B}$ and $p \in X \cap Y$, there exists a $Z \in \mathcal{B}$ with $p \in Z$, such that $Z \subseteq X$ and $Z \subseteq Y$.

5.3. A family of nonempty sets $\mathcal{C} \subseteq$ Pow $A$ is a *subbase* for open sets if the finite intersections of members of $\mathcal{C}$ form a base for open sets. Show that $\mathcal{C} \subseteq$ Pow $A$ is a subbase of some topology defined on $A$ iff $\bigcup \mathcal{C} = A$.

5.4. Let $A$ be a topological space and let $X \subseteq A$. Then there exists a smallest closed set $\overline{X}$ containing $X$, called the *closure* of $X$. Show that $\overline{\varnothing} = \varnothing$ and that, for all $X, Y \subseteq A$,

(a) $X \subseteq Y$ implies that $\overline{X} \subseteq \overline{Y}$,
(b) $X \subseteq \overline{X}$,
(c) $\overline{X \cup Y} = \overline{X} \cup \overline{Y}$,
(d) $\overline{\overline{X}} = \overline{X}$.

5.5. In Section I.3.12, we introduced closure operators (Definition 30). Relate Exercise 5.4 to closure operators.

5.6. Prove that the conditions of Exercise 5.4 characterize an operation $^{-}\colon$ Pow $A \to$ Pow $A$ that is the topological closure with respect to a topology on $A$.

5.7. Show that $a \in \overline{X}$ iff every open set (in a given subbase) containing $a$ has a nonempty intersection with $X$.

5.8. A space $A$ is a $T_0$-space if $\overline{\{x\}} = \overline{\{y\}}$ implies that $x = y$ for $x, y \in A$. Show that $A$ is a $T_0$-space iff, for every $x, y \in A$ with $x \neq y$, there exists an open set (in a given base) containing exactly one of $x$ and $y$.

5.9. A space $A$ is a $T_1$-space if $\overline{\{x\}} = \{x\}$ for all $x \in A$. A $T_1$-space is a $T_0$-space. Show that $A$ is a $T_1$-space iff, for $x, y \in A$ with $x \neq y$, there exists an open set (in a given subbase) containing $x$ but not $y$.

5.10. Let $A$ and $B$ be topological spaces and $f\colon A \to B$. Then $f$ is called *continuous* if $f^{-1}(U)$ is open in $A$ for every open set $U$ of $B$. The map $f$ is a *homeomorphism* if $f$ is a bijection and if both $f$ and $f^{-1}$ are continuous. Show that continuity can be checked by considering only those $f^{-1}(U)$, where $U$ belongs to a given subbase.

5.11. Show that $f\colon A \to B$ is continuous iff $f(\overline{X}) \subseteq \overline{f(X)}$ for all $X \subseteq A$.

5.12. A subset $X$ of a topological space $A$ is *compact* if whenever

$$X \subseteq \bigcup(U_i \mid i \in I),$$

where the $U_i$, for $i \in I$, are open sets, implies that

$$X \subseteq \bigcup(U_i \mid i \in I')$$

for some finite $I' \subseteq I$. The space $A$ is *compact* if $X = A$ is compact. Show that $A$ is compact, iff, for every family $\mathcal{F}$ of closed sets, if $\bigcap \mathcal{F}_1 \neq \varnothing$, for all finite $\mathcal{F}_1 \subseteq \mathcal{F}$, then $\bigcap \mathcal{F} \neq \varnothing$.

5.13. Let $A$ be a compact topological space and let $X$ be a closed set in $A$. Show that $X$ is compact.

5.14. Prove that a space $A$ is compact iff, in the lattice of closed sets of $A$, every maximal filter is principal.

*5.15. Show that a space $A$ is compact iff it has a subbase $\mathcal{C}$ of closed sets (that is, $\{A - X \mid X \in \mathcal{C}\}$ is a subbase for open sets) with the property: If $\bigcap \mathcal{D} = \varnothing$ for some $\mathcal{D} \subseteq \mathcal{C}$, then $\bigcap \mathcal{D}_1 = \varnothing$ for some finite $\mathcal{D}_1 \subseteq \mathcal{D}$ (J. W. Alexander [28]).

5.16. Let $A_i$, for $i \in I$, be topological spaces and set $A = \prod(A_i \mid i \in I)$. For $U \subseteq A_i$, set $E(U) = \{f \in A \mid f(i) \in U\}$. The *product topology* on $A$ is the topology determined by taking all the sets $E(U)$ as a subbase for open sets, where $U$ ranges over all open sets of $A_i$ for all $i \in I$. Show that the projection map $e_i\colon f \mapsto f(i)$ is a continuous map of $A$ onto $A_i$. (As a rule, a product of topological spaces will be understood to have the product topology.)

5.17. Show that if $A_i$, for $i \in I$, are $T_0$-spaces ($T_1$-spaces), so is

$$A = \prod(A_i \mid i \in I).$$

5.18. A map $f\colon A \to B$ is *open* if $f(U)$ is open in $B$ for every open $U \subseteq A$. Show that the projection maps (see Exercise 5.16) are open.

5.19. Prove that a function $f\colon B \to \prod A_i$ is continuous iff $e_i f\colon B \to A_i$ is continuous for every $i \in I$.

5.20. A space $A$ is a *Hausdorff space* ($T_2$-space) if, for all $x, y \in A$ with $x \neq y$, there exist open sets $U$ and $V$ such that $x \in U$, $y \in V$, $U \cap V = \varnothing$. Show that:

(a) $A$ is Hausdorff iff $\Delta = \{(x, x) \mid x \in A\}$ is closed in $A \times A$.

(b) A compact subset of a $T_2$-space is closed.

5.21. Prove that a product of Hausdorff spaces is a Hausdorff space.

5.22. Show that

**Theorem 208 (Tychonoff's Theorem).** *A product of compact spaces is compact.*

(Hint: use Exercise 5.15.)

5.23. A space $A$ is *totally disconnected* if, for all $x, y \in A$ with $x \neq y$, there exists a clopen set $U$ with $x \in U$ and $y \notin U$. Show that the product of any family of totally disconnected sets is totally disconnected.

$$* \qquad * \qquad *$$

5.24. Let $I$ and $J$ be ideals of a join-semilattice. Verify that

$$I \vee J = \{\, t \mid t \leq i \vee j,\ i \in I,\ j \in J \,\}.$$

5.25. Let $L$ be a join-semilattice. Show that $\operatorname{Id} L$ is a lattice iff any two elements of $L$ have a common lower bound.

5.26. Give a detailed proof of Lemma 185.

5.27. Prove that every join-semilattice can be embedded in a boolean lattice (considered as a join-semilattice).

5.28. Show that a finite distributive join-semilattice is a distributive lattice.

5.29. Let $L$ be a join-semilattice and let $\alpha$ be a *join-congruence*, that is, an equivalence relation on $L$ having the Substitution Property for join. Then $L/\alpha$ is also a join-semilattice. Show that the distributivity of $L$ does not imply the distributivity of $L/\alpha$.

5.30. Let $F$ be a free join-semilattice on a set $S$; let $F_0$ be $F$ with a new zero added. Show that $F_0$ is a distributive join-semilattice.

5.31. Let $\varphi$ be a join-homomorphism of the join-semilattice $F_0$ onto the join-semilattice $F_1$. For distributive join-semilattices $F_0$ and $F_1$, is the proper homomorphism concept the one requiring that if $P$ is a prime ideal of $F_1$, then $\varphi^{-1}(P)$ is a prime ideal of $F_0$?

5.32. Show that there is no "free distributive join-semilattice" with the homomorphism concept of Exercise 5.31.

5.33. Does Theorem 123 generalize to bounded distributive join-semilattices?

5.34. Characterize the Stone spaces of finite boolean lattices and of finite chains.

5.35. Let $\mathcal{S}_0$ and $\mathcal{S}_1$ be disjoint topological spaces; let $\mathcal{S} = \mathcal{S}_0 \cup \mathcal{S}_1$ and call $U \subseteq \mathcal{S}$ open if $U \cap \mathcal{S}_0$ and $U \cap \mathcal{S}_1$ are open. Show that if $\mathcal{S}_0$ and $\mathcal{S}_1$ are Stone spaces, then so is $\mathcal{S}$.

5.36. If in Exercise 5.35, $\mathcal{S}_i = \operatorname{Spec} L_i$, for $i = 0, 1$, then

$$\mathcal{S} = \operatorname{Spec}(L_0 \times L_1).$$

5.37. As an alternative proof of Theorem 194, pick an element

$$a(P,Q) \in P - Q$$

for all $P, Q \in \operatorname{Spec} B$ with $P \neq Q$. Show that the elements $a(P,Q)$ R-generate all of $B$.

5.38. Give necessary and sufficient conditions for a map $\varphi\colon L_0 \to L_1$ to be one-to-one, respectively, onto, in terms of the induced map

$$\operatorname{Spec}(\varphi)\colon \operatorname{Spec} L_1 \to \operatorname{Spec} L_0.$$

5.39. Determine the connection between the Stone space of a lattice and the Stone space of a sublattice.

5.40. Call the Stone space of a generalized boolean lattice a *generalized boolean space*; characterize such spaces. (Compactness of $S$ should be replaced by *local compactness*: For every $p \in S$, there exists an open set $U$ with $p \in U$ and a set $V$ with $U \subseteq V$ such that $V$ is compact.)

5.41. Show that the product of (generalized) boolean spaces is (generalized) boolean.

5.42. Call the join-semilattice $L$ *modular* if, for all elements $a, b, c \in L$ satisfying $a \leq b$ and $b \leq a \vee c$, there exists an element $c_1 \in L$ with $c_1 \leq c$ and $b = a \vee c_1$. Show that a distributive join-semilattice is modular.

5.43. Show that Lemma 184 remains valid if all occurrences of the word "distributive" are replaced by the word "modular".

5.44. Show that the set of all finitely generated normal subgroups of a group (and also the finitely generated ideals of a ring) form a modular join-semilattice.

5.45. The lattice of congruence relations of a join-semilattice $L$ is distributive iff any pair of elements of $L$ with a lower bound is comparable (D. Papert [577], R. A. Dean and R. H. Oehmke [148]).

$$*\qquad *\qquad *$$

5.46. Define the concepts of subalgebra, term, identity, and variety for algebras of a given type $\tau$. Show that if $\mathbf{K}$ is a variety, $\mathfrak{A}$ is an algebra in $\mathbf{K}$, and $\mathfrak{B}$ is a subalgebra of $\mathfrak{A}$, then $\mathfrak{B}$ is in $\mathbf{K}$.

5.47. Define the concepts of homomorphism, homomorphic image, and direct product for algebras of a given type. Show that a variety is closed under the formation of homomorphic images and direct products.

5.48. Let $\mathfrak{A} = (A; F)$ be an algebra, let $H \subseteq A$, and let $H \neq \varnothing$. Show that there exists a smallest subset $\operatorname{sub}(H)$ of $A$ with $\operatorname{sub}(H) \supseteq H$ such that $(\operatorname{sub}(H); F)$ is a subalgebra of $\mathfrak{A}$. (This subalgebra is said to be *generated by $H$*.)

5.49. Show that $|\mathrm{sub}(H)| \leq |H| + |F| + \aleph_0$.

5.50. Modify Definition 197 for algebras. Show that the $\varphi$ in (ii) is unique.

5.51. Let $\mathfrak{B}$ and $\mathfrak{C}$ be free $\mathbf{K}$-products of $\mathfrak{A}_i$, for $i \in I$, with embeddings $\varepsilon_i$ and $\chi_i$, for $i \in I$, respectively. Show that there exists an isomorphism $\alpha\colon B \to C$ such that $\alpha\varepsilon_i = \chi_i$ for all $i \in I$.

5.52. Let $\mathbf{K}$ be a variety of algebras and let $\mathfrak{A}_i \in \mathbf{K}$ for all $i \in I$. Choose a set $S$ satisfying

$$|S| \geq \sum (\,|A_i| \mid i \in I\,) + |F| + \aleph_0.$$

Let $Q$ be the set of all pairs $(\mathfrak{B}, (\,\varphi_i \mid i \in I\,))$ such that $B \subseteq S$, the map $\varphi_i$ is a homomorphism of $\mathfrak{A}_i$ into $\mathfrak{B}$, and

$$B = \mathrm{sub}(\bigcup(\,\varphi(A_i) \mid i \in I\,)).$$

Form

$$\mathfrak{A} = \prod(\,\mathfrak{B} \mid (\mathfrak{B}, (\,\varphi_i \mid i \in I\,)) \in Q\,)$$

(direct product), and, for all $a \in A_i$, define $f_a \in A$ by

$$f_a((\mathfrak{B}, (\,\varphi_i \mid i \in I\,))) = \varphi_i(a).$$

Finally, let $\mathfrak{N}$ be the subalgebra generated by the $f_a$ for all $a \in A_i$ and $i \in I$. Show that $\mathfrak{N} \in \mathbf{K}$, the map $a \mapsto f_a$ is a homomorphism $\varepsilon_i$ of $\mathfrak{A}_i$ into $\mathfrak{N}$, for every $i \in I$, and that $\mathfrak{N}$ is generated by $\bigcup(\,\varepsilon_i(A_i) \mid i \in I\,)$.

5.53. Show that $\varepsilon_i$ is one-to-one iff, for all $i \in I$ and for all $a, b \in A_i$ with $a \neq b$, there exists an algebra $\mathfrak{C} \in \mathbf{K}$ and homomorphisms $\psi_j\colon \mathfrak{A}_j \to \mathfrak{C}$, for all $j \in I$, such that $\psi_i(a) \neq \psi_i(b)$.

5.54. Combine the previous exercises to prove the following result.

**Theorem 209 (Existence Theorem for Free Products).** *Let $\mathbf{K}$ be a variety of algebras, let $\mathfrak{A}_i$, be algebras in $\mathbf{K}$, for $i \in I$. A free $\mathbf{K}$-product of the algebras $\mathfrak{A}_i$, for $i \in I$, exists iff, for all $i \in I$ and for all $a, b \in A_i$ with $a \neq b$, there exists an algebra $\mathfrak{C} \in \mathbf{K}$, and there exist homomorphisms $\psi_j\colon \mathfrak{A}_j \to \mathfrak{C}$, for all $j \in I$, such that $\psi_i(a) \neq \psi_i(b)$.*

5.55. Show that in proving the existence of free distributive products and free $\{0, 1\}$-distributive products, we can always choose $\mathfrak{C} = \mathsf{C}_2$, the two-element chain, in applying Exercise 5.54.

5.56. Show that the free boolean algebra on $\mathfrak{m}$ generators is a free $\{0, 1\}$-distributive product of $\mathfrak{m}$ copies of the free boolean algebra on one generator.

5.57. Prove that the free boolean algebra on $\mathfrak{m}$ generators can be represented by the clopen subsets of $\{0, 1\}^{\mathfrak{m}}$, where $\{0, 1\}$ is the two-element discrete topological space.

5.58. Find a topological representation for a free distributive lattice on $\mathfrak{m}$ generators (G. Ya. Areškin [32]).

5.59. For an order $P$, let $\mathrm{Down_{fin}}\, P$ denote the set of all subsets of $P$ of the form $\downarrow H$ for a finite set $H \subseteq P$ and order this set by inclusion. Show that $\mathrm{Down_{fin}}\, P$ is a join-semilattice.

5.60. Find examples of orders $P$ for which $\mathrm{Down_{fin}}\, P$ is not a distributive lattice. Is there a "smallest" such example?

5.61. Show that if we define $\mathrm{Ji}\, L$ in the obvious way for a join-semilattice $L$, then for $L = \mathrm{Down_{fin}}\, P$, the isomorphism $\mathrm{Ji}\, L \cong P$ holds.

5.62. Deduce that for any join-semilattice $L$ of the form $\mathrm{Down_{fin}}\, P$, the analog of Corollary 108 holds.

5.63. Let $L$ be a distributive algebraic lattice and let $F$ be the set of compact elements of $L$. Show that the join-semilattice $F$ is distributive.

5.64. What is the converse of Exercise 5.63?

# 6. Distributive Lattices with Pseudocomplementation

## 6.1 Definitions and examples

In this section, we shall deal exclusively with pseudocomplemented distributive lattices. There are two distinct concepts: a lattice, $(L; \vee, \wedge)$, in which every element has a pseudocomplement; and an algebra $(L; \vee, \wedge, {}^*, 0, 1)$, where $(L; \vee, \wedge, 0, 1)$ is a bounded lattice and where, for every $a \in L$, the element $a^*$ is the pseudocomplement of $a$. We shall call the former a *pseudocomplemented lattice* and the latter a *lattice with pseudocomplementation* (as an operation)— the same kind of distinction we make between boolean lattices and boolean algebras.

As defined in the Exercises of Section 5, a pseudocomplemented lattice is an algebra of type $(2, 2)$, whereas a lattice with pseudocomplementation is an algebra of type $(2, 2, 1, 0, 0)$. To see the difference in viewpoint, consider the lattice of Figure 39. As a distributive lattice, it has twenty-five sublattices and eight congruences; as a lattice with pseudocomplementation, it has three subalgebras and five congruences.

Thus for a lattice with pseudocomplementation $L$, a *subalgebra* $L_1$ is a $\{0, 1\}$-sublattice of $L$ closed under $^*$ (that is, $a \in L_1$ implies that $a^* \in L_1$). A *homomorphism* $\varphi$ is a $\{0, 1\}$-homomorphism that also satisfies

$$(\varphi(x))^* = \varphi(x^*).$$

Similarly, a *congruence relation* $\boldsymbol{\alpha}$ shall have the Substitution Property also for $^*$, that is, $a \equiv b \pmod{\boldsymbol{\alpha}}$ implies that $a^* \equiv b^* \pmod{\boldsymbol{\alpha}}$.

A wide class of examples is provided by

**Theorem 210.** *Any complete lattice that satisfies the Join Infinite Distributive Identity* (JID) *is a pseudocomplemented distributive lattice.*

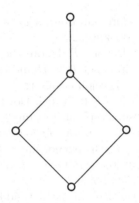

Figure 39. A small example

*Proof.* Let $L$ be such a lattice. For $a \in L$, set

$$a^* = \bigvee(x \in L \mid a \wedge x = 0).$$

Then, by (JID),

$$a \wedge a^* = a \wedge \bigvee(x \mid a \wedge x = 0) = \bigvee(a \wedge x \mid a \wedge x = 0) = \bigvee 0 = 0.$$

Furthermore, if $a \wedge x = 0$, then $x \leq a^*$ by the definition of $a^*$; thus $a^*$ is indeed the pseudocomplement of $a$.                                             $\square$

**Corollary 211.** *Every distributive algebraic lattice is pseudocomplemented.*

*Proof.* Let $L$ be a distributive algebraic lattice. By Theorem 42 and Lemma 184, represent $L$ as $\operatorname{Id} S$, where $S$ is a distributive join-semilattice with zero. Let $I$ and $I_j$, for $j \in J$, be ideals of $S$. Then

$$\bigvee(I \wedge I_j \mid j \in J) \subseteq I \wedge \bigvee(I_j \mid j \in J)$$

is obvious. To prove the reverse inclusion, let

$$a \in I \wedge \bigvee(I_j \mid j \in J),$$

that is, $a \in I$ and $a \in \bigvee(I_j \mid j \in J)$. The latter implies that

$$a \leq t_1 \vee \cdots \vee t_n, \text{ where } t_1 \in I_{j_1}, \ldots, t_n \in I_{j_n}, j_1, \ldots, j_n \in J.$$

Thus $a \in I_{j_1} \vee \cdots \vee I_{j_n}$ and so, using the distributivity of $\operatorname{Id} L$, we obtain that

$$a \in I \wedge (I_{j_1} \vee \cdots \vee I_{j_n}) = (I \wedge I_{j_1}) \vee \cdots \vee (I \wedge I_{j_n}) \subseteq \bigvee(I \wedge I_j \mid j \in J),$$

completing the proof of (JID). The statement now follows from Theorem 210.                                                     $\square$

Note that we have remarked this much already in Exercise 4.20.

Thus the lattice of all congruence relations of an arbitrary lattice and the lattice of all ideals of a distributive lattice (or semilattice) with zero are examples of pseudocomplemented distributive lattices. Note that

$$I^* = \{\, x \in K \mid x \wedge i = 0 \text{ for all } i \in I \,\}$$

for any ideal $I$ of a distributive lattice $K$. Also, any finite distributive lattice is pseudocomplemented. Therefore, our investigations include all finite distributive lattices.

## 6.2  Stone algebras

The class of Stone algebras (named in G. Grätzer and E. T. Schmidt [331]) was the first class of distributive lattices with pseudocomplementation, other than the class of boolean algebras, to be examined in detail. A distributive lattice with pseudocomplementation $L$ is called a *Stone algebra* if it satisfies the *Stone identity*:

$$a^* \vee a^{**} = 1.$$

The corresponding pseudocomplemented lattice is called a *Stone lattice*.

For a Stone algebra $L$, the skeleton $\mathrm{Skel}\, L$ is a subalgebra of $L$:

**Lemma 212.** *For a distributive lattice with pseudocomplementation $L$, the following conditions are equivalent:*

(i) *$L$ is a Stone algebra.*
(ii) *$(a \wedge b)^* = a^* \vee b^*$ for all $a, b \in L$.*
(iii) *$a, b \in \mathrm{Skel}\, L$ implies that $a \vee b \in \mathrm{Skel}\, L$.*
(iv) *$\mathrm{Skel}\, L$ is a subalgebra of $L$.*

*Proof.* The proofs that (ii) implies (iii), that (iii) implies (iv), and that (iv) implies (i) are trivial. To prove that (i) implies (ii), let $L$ be a Stone algebra. We show that $a^* \vee b^*$ is the pseudocomplement of $a \wedge b$, verifying (ii). First,

$$(a \wedge b) \wedge (a^* \vee b^*) = (a \wedge b \wedge a^*) \vee (a \wedge b \wedge b^*) = 0 \vee 0 = 0.$$

Second, if $(a \wedge b) \wedge x = 0$, then $(b \wedge x) \wedge a = 0$, and so $b \wedge x \le a^*$. Meeting both sides by $a^{**}$ yields

$$b \wedge x \wedge a^{**} \le a^* \wedge a^{**} = 0;$$

that is, $x \wedge a^{**} \wedge b = 0$, implying that $x \wedge a^{**} \le b^*$. Then $a^* \vee a^{**} = 1$, by the Stone identity, and thus

$$x = x \wedge 1 = x \wedge (a^* \vee a^{**}) = (x \wedge a^*) \vee (x \wedge a^{**}) \le a^* \vee b^*. \qquad \square$$

This is already enough to yield the structure theorem for finite Stone algebras (G. Grätzer and E. T. Schmidt [331]):

**Corollary 213.** *A finite distributive lattice $L$ is a Stone lattice iff it is the direct product of finite distributive dense lattices, that is, finite distributive lattices with only one atom.*

*Proof.* By Lemma 212, a Stone lattice $L$ has a complemented element $a$ different from 0 and 1 iff $\operatorname{Skel} L \neq \{0, 1\}$; thus the decomposition of Theorem 106 can be repeated until each factor $L_i$ satisfies $\operatorname{Skel} L_i = \{0, 1\}$. In a direct product, the pseudocomplementation $^*$ is formed componentwise; therefore, all the lattices $L_i$ are Stone lattices.

For a finite distributive lattice $K$ with $\operatorname{Skel} K = \{0, 1\}$, the condition that the lattice $K$ has exactly one atom is equivalent to $K$ being a Stone lattice. $\square$

### 6.3    Triple construction

In addition to the skeleton, the *dense set*,

$$\operatorname{Dns} L = \{ \, a \mid a^* = 0 \, \},$$

is another significant subset of a Stone algebra. The elements of $\operatorname{Dns} L$ are called *dense*.

We can easily check that $\operatorname{Dns} L$ is a filter of $L$ and $1 \in \operatorname{Dns} L$; thus $\operatorname{Dns} L$ is a distributive lattice with unit. Since $a \vee a^* \in \operatorname{Dns} L$, for every $a \in L$, we can interpret the identity

$$a = a^{**} \wedge (a \vee a^*)$$

to mean that every $a \in L$ can be represented in the form

$$a = b \wedge c, \quad b \in \operatorname{Skel} L, \ c \in \operatorname{Dns} L.$$

Such an interpretation correctly suggests that if we know $\operatorname{Skel} L$ and $\operatorname{Dns} L$ and the relationships between elements of $\operatorname{Skel} L$ and $\operatorname{Dns} L$, then we can describe $L$. The relationship is expressed by the homomorphism

$$\varphi_L \colon \operatorname{Skel} L \to \operatorname{Fil}(\operatorname{Dns} L)$$

defined by

$$\varphi_L \colon a \mapsto \{ \, x \in \operatorname{Dns} L \mid x \geq a^* \, \}.$$

**Theorem 214.** *Let $L$ be a Stone algebra. Then $\operatorname{Skel} L$ is a boolean algebra, $\operatorname{Dns} L$ is a distributive lattice with unit, and $\varphi_L$ is a $\{0, 1\}$-homomorphism of $\operatorname{Skel} L$ into $\operatorname{Fil}(\operatorname{Dns} L)$. The triple*

$$(\operatorname{Skel} L, \operatorname{Dns} L, \varphi_L)$$

*characterizes $L$ up to isomorphism.*

*Proof.* The first statement is easily verified. To get the characterization result, for $a \in \operatorname{Skel} L$, set

$$F_a = \{\, x \mid x^{**} = a \,\}.$$

The sets $\{\, F_a \mid a \in \operatorname{Skel} L \,\}$ form a partition of $L$; for a small example, see Figure 40. Obviously, $F_0 = \{0\}$ and $F_1 = \operatorname{Dns} L$. The map $x \mapsto x \vee a^*$ sends $F_a$ into $F_1 = \operatorname{Dns} L$; in fact, the map is an isomorphism between the filters $F_a$ and $\varphi_L(a) \subseteq \operatorname{Dns} L$. Thus $x \in F_a$ is completely determined by $a$ and $x \vee a^* \in \varphi_L(a)$, that is, by a pair $(a, z)$, where $a \in \operatorname{Skel} L$ and $z \in \varphi_L(a)$, and every such pair determines one and only one element of $L$. To complete our proof, we have to show how the ordering on $L$ can be determined by such pairs.

Let $x \in F_a$ and $y \in F_b$. Then $x \leq y$ implies that $x^{**} \leq y^{**}$, that is, $a \leq b$. Since $x \leq y$ iff

$$a \vee x \leq a \vee y \quad \text{and} \quad x \vee a^* \leq y \vee a^*,$$

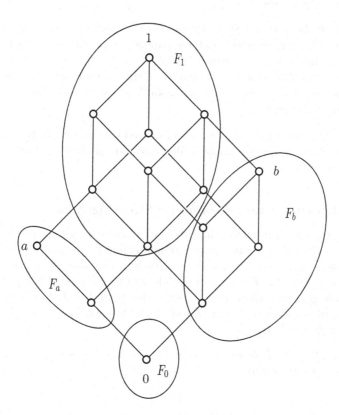

Figure 40. Decomposing a Stone algebra

and since the first of these two conditions is trivial, we obtain that

$$x \leq y \quad \text{iff} \quad a \leq b \text{ and } x \vee a^* \leq y \vee a^*.$$

Identifying $x$ with $(x \vee a^*, a)$ and $y$ with $(y \vee b^*, b)$, we see that the preceding conditions are stated in terms of the components of the ordered pairs, except that $y \vee a^*$ will have to be expressed by the triple.

The map $\varphi_L$ is a $\{0, 1\}$-homomorphism and $a$ is the complement of $a^*$, so we conclude that $\varphi_L(a)$ and $\varphi_L(a^*)$ are complementary filters of $\operatorname{Dns} L$. Thus every $z \in \operatorname{Dns} L$ can be written in a unique fashion in the form $z = \varrho_a(z) \wedge z_1$, where $\varrho_a(z) \in \varphi_L(a)$ and $z_1 \in \varphi_L(a^*)$. Observe that the map $\varrho_a$ is expressed in terms of the triple. Finally,

$$y \vee a^* = y \vee b^* \vee a^* = \varrho_a(y \vee b^*).$$

Thus

$$(u, a) \leq (v, b) \quad \text{iff} \quad a \leq b \text{ and } u \leq \varrho_a(v)$$

holds for $u \in \varphi_L(a)$ and $v \in \varphi_L(b)$.    □

This result shows that a Stone algebra is characterized by its triple, see C. C. Chen and G. Grätzer [89] and [90]; these papers also provides a characterization theorem for triples, see Exercises 6.17–6.31.

Theorem 214 shows that the behavior of the skeleton and the dense set is decisive for Stone algebras. This conclusion leads us to formulate the goal of research for Stone algebras:

> A problem for Stone algebras is considered solved if it can be reduced to two problems: one for boolean algebras and one for distributive lattices with unit.

## 6.4   A characterization theorem for Stone algebras

By applying Zorn's Lemma to prime filters of a lattice with zero, we obtain that every prime filter is contained in a maximal prime filter, or, equivalently, we get that every prime ideal contains a *minimal prime ideal* $P$, that is, a prime ideal $P$ such that $Q \subset P$ for no prime ideal $Q$ (see Exercise 1.34). Minimal prime ideals play an important role in the theory of distributive lattices with pseudocomplementation, as illustrated by the following result in G. Grätzer and E. T. Schmidt [331]:

**Theorem 215.** *Let $L$ be a distributive lattice with pseudocomplementation. Then $L$ is a Stone algebra iff*

$$P \vee Q = L,$$

*for all distinct minimal prime ideals $P$ and $Q$.*

*Proof.* Let $L$ be a Stone algebra and let $P$ and $Q$ be distinct minimal prime ideals. Note that $P \not\subset Q$, since $Q$ is minimal; also, $Q \neq P$, hence $P - Q \neq \varnothing$. So we can choose $a \in P - Q$. Since $a \wedge a^* = 0 \in Q$, utilizing that $a \notin Q$ and $Q$ is prime, we obtain that $a^* \in Q$.

$L - P$ is a maximal dual prime ideal, hence by the dual of Corollary 118, it is a maximal filter of $L$. Thus $(L - P) \vee \text{fil}(a) = L$ and so $0 = a \wedge x$ for some $x \in L - P$. Therefore, $a^* \geq x \in L - P$ and so $a^* \notin P$. Hence $a^* \in Q - P$. Similarly, $a^{**} \in P - Q$, which implies that

$$1 = a^* \vee a^{**} \in P \vee Q,$$

yielding that $P \vee Q = L$.

To prove the converse (for this proof, see J. C. Varlet [689]), let us assume that $L$ is not a Stone algebra and let $a \in L$ such that $a^* \vee a^{**} \neq 1$. Let $R$ be a prime ideal (see Corollary 117) such that $a^* \vee a^{**} \in R$.

We claim that $(L - R) \vee \text{fil}(a^*) \neq L$. Indeed, if $(L - R) \vee \text{fil}(a^*) = L$, then there exists an $x \in L - R$ such that $x \wedge a^* = 0$. Then $a^{**} \geq x \in L - R$, hence $a^{**} \in L - R$, a contradiction. Let $F$ be a maximal dual prime ideal containing $(L - R) \vee \text{fil}(a^*)$ and similarly, let $G$ be a maximal dual prime ideal containing $(L - R) \vee \text{fil}(a^{**})$. We set $P = L - F$ and $Q = L - G$. Then $P$ and $Q$ are minimal prime ideals. Moreover, $P \neq Q$, because $a^* \in F = L - P$ and hence $a^* \notin P$; thus $a^{**} \in P$, while $a^{**} \notin Q$. Finally, $P, Q \subseteq R$, hence $P \vee Q \neq L$. $\square$

## 6.5   Two representation theorems for Stone algebras

We prove two representation theorems for Stone algebras that correspond to the two representation theorems for distributive lattices given in Section 1. The proofs we present use the Subdirect Product Representation Theorem of G. Birkhoff [67]. Direct proofs are possible but we shall present a proof that can be generalized to other varieties of distributive lattices with pseudocomplementation.

In the remainder of this section "algebra" means universal algebra, as defined in Section I.1.9. For the purpose of this book, the reader can substitute "lattice" or "lattice with pseudocomplementation" for "algebra". Just as for orders and lattices, we write $A$ for the algebra $\mathfrak{A} = (A; F)$ if there is no danger of confusion.

**Definition 216.** An algebra $A$ is called *subdirectly irreducible* if there exist elements $u, v \in A$ such that $u \neq v$ and $u \equiv v \pmod{\boldsymbol{\alpha}}$ for all congruences $\boldsymbol{\alpha} > 0$.

In other words, $A$ has at least two elements and

$$\text{Con } A = \{\mathbf{0}\} \cup \text{fil}(\text{con}(u, v)),$$

as illustrated in Figure 41, where the unique atom is the congruence $\mathrm{con}(u, v)$. Intuitively, this means that if we collapse any two distinct elements of $A$, then this congruence spreads to collapse $u$ and $v$.

The congruence $\mathrm{con}(u, v)$ (unique!) is called the *base congruence* of $A$; it is often called the *monolith* in the literature. An equivalent form of this definition is the following (see Section I.6.3 for the concept we are using).

**Corollary 217.** *The algebra $A$ is subdirectly irreducible iff $\mathbf{0}$ is completely meet-irreducible in* $\mathrm{Con}\, A$.

**Example 218.** A distributive lattice $L$ is subdirectly irreducible iff $|L| = 2$.

*Proof.* If $|L| = 1$, then $L$ is not subdirectly irreducible by definition. If $|L| = 2$, then obviously $L$ is subdirectly irreducible.

Let $|L| > 2$. Then there exist $a, b, c \in L$ with $a < b < c$. We claim that $\mathrm{con}(a, b) \wedge \mathrm{con}(b, c) = \mathbf{0}$, which by Corollary 217 shows that $L$ is not subdirectly irreducible. Let

$$x \equiv y \quad (\mathrm{mod}\ \mathrm{con}(a, b) \wedge \mathrm{con}(b, c)).$$

By Theorem 141, this implies that $x \vee b = y \vee b$ and $x \wedge b = y \wedge b$; thus $x = y$ by Corollary 103. □

**Example 219.** $\mathsf{B}_1$ is the only subdirectly irreducible boolean algebra.

*Proof.* Let $B$ be boolean. The statement is obvious for $|B| \leq 2$. If $|B| > 2$, then $B$ has a direct product representation, $B = A_1 \times A_2$ with $|A_1|, |A_2| \geq 2$ (use Exercise 5.6); thus $B$ cannot be subdirectly irreducible. □

We shall need a simple universal algebraic lemma.

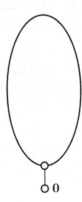

Figure 41. The congruence lattice of a subdirectly irreducible lattice; the unique atom is the base congruence

**Lemma 220 (The Second Isomorphism Theorem).** *Let $A$ be an algebra and let $\alpha$ be a congruence relation of $A$. For any congruence $\beta \geq \alpha$ of $A$, define the relation $\beta/\alpha$ on $A/\alpha$ by*

$$x/\alpha \equiv y/\alpha \pmod{\beta/\alpha} \qquad iff \qquad x \equiv y \pmod{\beta}.$$

*Then $\beta/\alpha$ is a congruence of $A/\alpha$. Conversely, every congruence $\gamma$ of $A/\alpha$ can be (uniquely) represented in the form $\gamma = \beta/\alpha$, for some congruence $\beta \geq \alpha$ of $A$. In particular, the congruence lattice of $A/\alpha$ is isomorphic with the filter $\mathrm{fil}(\alpha)$ of the congruence lattice of $A$.*

*Proof.* We have to prove that $\beta/\alpha$ is well-defined, it is an equivalence relation, and it has the Substitution Property. To represent $\gamma$, define a congruence $\beta$ of $A$ by

$$x \equiv y \pmod{\beta} \qquad iff \qquad x/\alpha \equiv y/\alpha \pmod{\gamma}.$$

Again, we have to verify that $\beta$ is a congruence. Then $\beta/\alpha = \gamma$ follows from the definition of $\beta$. The details are trivial and left to the reader. $\square$

Varieties of universal algebras can be introduced by defining terms and identities, just as in the case of lattices. However, in the next theorem (see G. Birkhoff [67]), the reader can avoid the use of this terminology by substituting for "variety" the phrase "class closed under the formation of subalgebras, homomorphic images, and direct products". (This does not make the result more general, see Theorem 469.)

**Theorem 221 (Birkhoff's Subdirect Representation Theorem).** *Let $\mathbf{K}$ be a variety of algebras. Every algebra $A$ in $\mathbf{K}$ can be embedded in a direct product of subdirectly irreducible algebras in $\mathbf{K}$.*

*Proof.* For $a, b \in A$ with $a \neq b$, let $\mathcal{X}$ denote the set of all congruences $\alpha$ of $A$ satisfying $a \not\equiv b \pmod{\alpha}$. Then $\mathcal{X}$ is not empty since $\mathbf{0} \in \mathcal{X}$. Let $\mathcal{C}$ be a chain in $\mathcal{X}$. Since $\alpha = \bigcup \mathcal{C}$ is a congruence and $a \not\equiv b \pmod{\alpha}$, it follows that every chain in $\mathcal{X}$ has an upper bound. By Zorn's Lemma, there is a maximal element $\gamma(a, b)$ of $\mathcal{X}$.

We claim that $A/\gamma(a, b)$ is subdirectly irreducible; in fact, the elements $u = a/\gamma(a, b)$ and $v = b/\gamma(a, b)$ satisfy the condition of Definition 216. Indeed, if $\alpha$ is a congruence of $A/\gamma(a, b)$ with $\alpha \neq \mathbf{0}$, then by Lemma 220, represent it as $\alpha = \beta/\gamma(a, b)$, where $\beta$ is a congruence of $A$. Since $\alpha \neq \mathbf{0}$, we obtain that $\beta > \gamma(a, b)$, and so $a \equiv b \pmod{\beta}$. Thus $u \equiv v \pmod{\alpha}$, as claimed.

Let

$$B = \prod (\, A/\gamma(a, b) \mid a, b \in A, \ a \neq b\,).$$

Then $B$ is a direct product of subdirectly irreducible algebras. We embed the algebra $A$ into $B$ by the map $\varphi: x \mapsto f_x$, where $f_x$ takes on the value $x/\gamma(a, b)$ in the algebra $A/\gamma(a, b)$. Clearly, $\varphi$ is a homomorphism. To show that $\varphi$ is one-to-one, assume that $f_x = f_y$. Then $x \equiv y \pmod{\gamma(a,b)}$ for all $a, b \in A$ with $a \neq b$. Therefore,

$$x \equiv y \pmod{\bigwedge(\gamma(a,b) \mid a, b \in A, \ a \neq b)},$$

and so $x = y$.    □

We got a little bit more than claimed. If we pick $w \in A/\gamma(a,b)$, then $w = x/\gamma(a,b)$ for some $x \in A$. Thus there is an element in the representation of $A$ whose component in $A/\gamma(a,b)$ is $w$; such a representation is called a *subdirect product*. This concept is so important that we give a formal definition.

**Definition 222.** Let the algebra $B$ be a direct product of the algebras $B_i$, for $i \in I$, with the projection maps $\pi_i \colon B \to B_i$ for $i \in I$. A subalgebra $A$ of $B$ is called a *subdirect product* of the algebras $B_i$, for $i \in I$, if the projection map $\pi_i$ maps $A$ onto $B_i$ for all $i \in I$.

Equivalently, an algebra $A \subseteq \prod(B_i \mid i \in I)$ is a subdirect product of the algebras $B_i$, for $i \in I$ if, for any $i \in I$ and for every $b \in A_i$, there is an element $a \in A$ such that $\pi_i(a) = b$.

**Corollary 223.** *In a variety* **K**, *every algebra can be represented as a subdirect product of subdirectly irreducible algebras in* **K**.

Observe how strong Theorem 221 is. If combined with Example 218, it yields Theorem 119; when combined with Example 219, we obtain Corollary 122.

It is interesting to observe the subtle of the Axiom of Choice in the proof of Birkhoff's Subdirect Representation Theorem. For every $a, b \in A$ satisfying $a \neq b$, we prove that there are maximal congruences under which $a$ and $b$ are not congruent. We pick one such congruence, $\gamma(a,b)$. Since we pick one for every $a \neq b$, we need the Axiom of Choice for this step. In fact, G. Grätzer [264] proves that Birkhoff's Subdirect Representation Theorem is equivalent to the Axiom of Choice.

The readers should note that subdirect representations of an algebra $A$ are in one-to-one correspondence with families $(\alpha_i \mid i \in I)$ of congruence relations of $A$ satisfying

$$\bigwedge(\alpha_i \mid i \in I) = \mathbf{0}.$$

A subdirect representation by subdirectly irreducible algebras corresponds to families

$$(\alpha_i \mid i \in I)$$

of completely meet-irreducible congruences (see Section I.6.3 for this concept).

Thus Lemma 220 and Theorem 221 combine to yield the following statement.

**Corollary 224.** *Every congruence relation of an algebra is a meet of completely meet-irreducible congruences.*

Let $S_1$ denote the three-element chain $\{0, e, 1\}$ $(0 < e < 1)$ as a distributive lattice with pseudocomplementation.

**Theorem 225.** *Up to isomorphism, $B_1$ and $S_1$ are the only subdirectly irreducible Stone algebras.*

*Proof.* $B_1$ and $S_1$ are obviously subdirectly irreducible (the congruence lattice of $S_1$ is a three-element chain).

Now let $L$ be a subdirectly irreducible Stone algebra. By Lemma 212, $\mathrm{Skel}\, L$ is a subalgebra of $L$. By definition, $|L| > 1$. If $|\mathrm{Skel}\, L| > 2$, then $\mathrm{Skel}\, L$ is directly decomposable and therefore, so is $L$. Thus $|\mathrm{Skel}\, L| = 2$, that is,

$$\mathrm{Skel}\, L = \{0, 1\}.$$

If $|\mathrm{Dns}\, L| > 2$, then there exist congruences $\alpha$ and $\beta$ on $\mathrm{Dns}\, L$ such that $\alpha \wedge \beta = \mathbf{0}$ on $\mathrm{Dns}\, L$ (by Example 218). Extend $\alpha$ and $\beta$ to $L$ by defining $\{0\}$ as the only additional block. We conclude that $L$ is subdirectly reducible.

Thus $\mathrm{Skel}\, L = \{0, 1\}$ and so $L = \mathrm{Dns}\, L \cup \{0\}$ and $|\mathrm{Dns}\, L| \leq 2$, yielding that $L \cong B_1$ or $L \cong S_1$.     □

**Corollary 226.** *Every Stone algebra can be embedded in a direct product of two- and three-element chains (regarded as Stone algebras).*

*Proof.* Combine Corollary 223 and Theorem 225.     □

See G. Grätzer [255]; a weaker form of this corollary can be found in T. P. Speed [660].

Every distributive lattice can be embedded in some $\mathrm{Pow}\, X$. O. Frink [206] asked whether every Stone algebra can be embedded in some $\mathrm{Id}(\mathrm{Pow}\, X)$. This problem was solved in G. Grätzer [249].

**Theorem 227.** *A distributive lattice with pseudocomplementation $L$ is a Stone algebra iff it can be embedded into some $\mathrm{Id}(\mathrm{Pow}\, X)$.*

*Proof.* The algebra $\mathrm{Id}(\mathrm{Pow}\, X)$ is a Stone algebra, and therefore, any of its subalgebras is a Stone algebra by Corollary 211.

It is obvious that the class of Stone algebras that can be embedded into some $\mathrm{Id}(\mathrm{Pow}\, X)$ is closed under the formation of direct products and subalgebras. Hence by Corollary 226, it is sufficient to prove that $B_1$ and $S_1$ can be so embedded. For $B_1$ this is obvious. To embed $S_1$, take an infinite set $X$ and embed $S_1$ into $\mathrm{Id}(\mathrm{Pow}\, X)$ as follows:

$$0 \mapsto \{\varnothing\},$$
$$e \mapsto \{\, A \subseteq X \mid |A| < \aleph_0 \,\},$$
$$1 \mapsto \mathrm{Pow}\, X.$$

It is obvious that this is an embedding.     □

## 6.6   ◇ Generalizing Stone algebras

Let $\mathbf{B}_n$ denote the variety of distributive lattices with pseudocomplementation satisfying the identity

$$(\mathrm{L}_n) \qquad (x_1 \wedge \cdots \wedge x_n)^* \vee (x_1^* \wedge \cdots \wedge x_n)^* \vee \cdots \vee (x_1 \wedge \cdots \wedge x_n^*)^* = 1$$

for $n \geq 1$. Then $\mathbf{B}_1$ is the class of Stone algebras. K. B. Lee [500] has proved that $\mathbf{B}_n$, for $-1 \leq n \leq \omega$, is a complete list of varieties of distributive lattices with pseudocomplementation, where $\mathbf{B}_{-1} = \mathbf{T}$ is the trivial class, $\mathbf{B}_0 = \mathbf{B}$ is the class of boolean algebras, and $\mathbf{B}_\omega$ is the class of all distributive lattices with pseudocomplementation. Moreover,

$$\mathbf{B}_{-1} \subset \mathbf{B}_0 \subset \mathbf{B}_1 \subset \cdots \subset \mathbf{B}_n \subset \cdots \subset \mathbf{B}_\omega.$$

In H. Lakser [492] and G. Grätzer and H. Lakser [298] and [300], most of the structure theorems known for Stone algebras have been generalized to the classes $\mathbf{B}_n$. In these papers the amalgamation class of $\mathbf{B}_n$ (in the sense of Section VI.4.3) is also described.

These results, and a lot more, are written up in my book G. Grätzer [257] (which was reprinted in 2008).

## 6.7   ◇ Background

Except for V. Glivenko's early work [233], the study of pseudocomplemented distributive lattices started only in 1956 with a solution of Problem 70 of G. Birkhoff [70] in G. Grätzer and E. T. Schmidt [331], characterizing Stone lattices by minimal prime ideals (for a simplified proof, see J. C. Varlet [689]), see Theorem 215.

The idea of a triple was conceived by the author (in 1961, while visiting O. Frink at Penn State) as a tool to prove Frink's conjecture (see O. Frink [206]). This attempt failed and as a result triples were not utilized until 1969, see C. C. Chen and G. Grätzer [89] and [90] (also Section V.1.8). Frink's conjecture was solved using the Compactness Theorem in G. Grätzer [249] (see Theorem 227). An interesting generalization can be found in H. Lakser [492].

## Exercises

6.1. Show that every bounded chain is a pseudocomplemented distributive lattice.

6.2. Let $L$ be a lattice with unit. Adjoin a new zero to $L$: $L_1 = \mathsf{C}_1 + L$. Show that $L_1$ is a pseudocomplemented lattice.

6.3. Call a lattice with zero *dense* if the element 0 is meet-irreducible. Show that every bounded dense lattice $K$ is pseudocomplemented and that every such lattice can be constructed by the method of Exercise 6.2 with $L = \operatorname{Dns} K$.

6.4. Find an example of a complete distributive lattice $L$ that is not pseudocomplemented.

6.5. Prove that if $L$ is a complete Stone lattice, then so is $\operatorname{Id} L$. (Hint: $I^* = \operatorname{id}(a)$, where $a = \bigwedge(x^* \mid x \in I)$.)

6.6. Show that a distributive pseudocomplemented lattice is a Stone lattice iff

$$(a \vee b)^{**} = a^{**} \vee b^{**}$$

for all $a, b \in L$.

6.7. Find a small set of identities characterizing Stone algebras.

6.8. Let $L$ be a Stone algebra. Show that $\operatorname{Skel} L$ is a *retract* of $L$, that is, there is a homomorphism $\varphi \colon L \to \operatorname{Skel} L$ such that $\varphi(x) = x$ for all $x \in \operatorname{Skel} L$.

6.9. Let $L$ be a Stone algebra, $a, b \in \operatorname{Skel} L$, and $a \leq b$. Prove that

$$x \mapsto (x \vee a^*) \wedge b$$

embeds $F_a$ into $F_b$.

6.10. Let $B$ be a boolean algebra. Define $B^{[2]} \subseteq B^2$ by $(a, b) \in B^{[2]}$ if $a \leq b$. Verify that $B^{[2]} \leq B^2$ but it is not a subalgebra of $B^2$. Show that $B^{[2]}$ is a Stone lattice.

6.11. Let $L$ be a pseudocomplemented distributive lattice. Show that, for all $a, b \in L$,

$$(a \vee b)^* = a^* \wedge b^*,$$
$$(a \wedge b)^{**} = a^{**} \wedge b^{**}.$$

6.12. Prove that a prime ideal $P$ of a Stone algebra $L$ is minimal iff $P$ as an ideal of $L$ is generated by $P \cap \operatorname{Skel} L$.

6.13. Show that a distributive lattice with pseudocomplementation is a Stone algebra iff every prime ideal contains exactly one minimal prime ideal (G. Grätzer and E. T. Schmidt [331]).

*6.14. Prove that an order $Q$ is isomorphic to the order of all prime ideals of a Stone algebra iff

(a) every element of $Q$ contains exactly one minimal element;
(b) for every minimal element $m$ of $Q$, the order $\uparrow m - \{m\}$ is isomorphic to the order of all prime ideals of some distributive lattice with unit.

(See C. C. Chen and G. Grätzer [90].)

6.15. Give a detailed proof of the Second Isomorphism Theorem.

6.16. Prove Corollary 226 directly.

$$* \qquad * \qquad *$$

Exercises 6.17–6.31 are from C. C. Chen and G. Grätzer [89] and [90]. Let $B$ be a boolean algebra, let $D$ be a distributive lattice with unit, and let $\varphi$ be a $\{0, 1\}$-homomorphism of $B$ into Fil $D$. Set

$$L = \{ (x, a) \mid a \in B, \ x \in \varphi(a) \},$$

and define $(x, a) \le (y, b)$ if $a \le b$ and $x \le \varrho_a(y)$, where $\mathrm{fil}(\varrho_a(y)) = \varphi(a) \wedge \mathrm{fil}(y)$.

6.17. Verify the following formulas:

(a) If $a \in B$ and $d \in D$, then $\varrho_a(d) = d$ iff $d \in \varphi(a)$.

(b) $\varrho_a(d) \ge d$ for $a \in B$ and $d \in D$.

(c) $\varrho_a(d) \wedge \varrho_{a'}(d) = d$ for $a \in B$ and $d \in D$ (where $a'$ is the complement of $a$ in $B$).

(d) $\varrho_a \varrho_b = \varrho_{a \wedge b}$ for all $a, b \in C$.

6.18. Prove that:

(a) $\varrho_a(d) \wedge \varrho_b(d) = \varrho_{a \vee b}(d)$ for all $a, b \in B$ and $d \in D$.

(b) $\varrho_{a \wedge b}(d) = \varrho_a(d) \vee \varrho_b(d)$ for all $a, b \in B$ and $d \in D$.

6.19. Show that $L$ is an order under the given ordering.

6.20. For $(x, a), (y, b) \in L$, verify that

$$(x, a) \wedge (y, b) = (\varrho_b(x) \wedge \varrho_a(y), a \wedge b).$$

*6.21. Show that

$$(x, a) \vee (y, b) = ((\varrho_{b'}(x) \wedge y) \vee (x \wedge \varrho_{a'}(y)), a \vee b).$$

6.22. For $(x, a), (y, b), (z, c) \in L$, let

$$U = ((x, a) \wedge (y, b)) \vee (z, c),$$
$$V = ((x, a) \vee (z, c)) \wedge ((y, b) \vee (z, c)).$$

Compute $U$; show that

$$V = (d, (a \vee c) \wedge (b \vee c)),$$

where

$$d = d_0 \vee d_1 \vee d_2 \vee d_3,$$
$$d_0 = \varrho_{b \wedge c'}(x) \wedge \varrho_{a \wedge c'}(y) \wedge z,$$
$$d_1 = \varrho_{b \wedge c'}(x) \wedge \varrho_{a \wedge c}(y) \wedge z,$$
$$d_2 = \varrho_{b \vee c}(x) \wedge \varrho_{a \wedge c'}(y) \wedge z,$$
$$d_3 = \varrho_{b \vee c}(x) \wedge \varrho_{b \vee c}(y) \wedge \varrho_{a' \vee b'}(z).$$

6.23. Show that $d_0 \geq d_1$ and $d_0 \geq d_2$; therefore, $d = d_0 \vee d_3$.

6.24. Show that $L$ is distributive.

6.25. Show that $L$ is a Stone lattice.

6.26. Identify $b \in B$ with $(1, b)$ and $d \in D$ with $(d, 1)$. Verify that

$$\mathrm{Skel}\, L = B,$$
$$\mathrm{Dns}\, L = D,$$
$$\varphi_L = \varphi.$$

In other words, we have proved the following theorem of C. C. Chen and G. Grätzer [89]:

**Theorem 228 (Construction Theorem of Stone Algebras).**
*Given a boolean algebra $B$, a distributive lattice $D$ with unit, and a $\{0,1\}$-homomorphism $\varphi\colon B \to \mathrm{Fil}\, D$, there exists a Stone algebra $L$ whose triple is $(B, D, \varphi)$.*

6.27. Describe isomorphisms and homomorphisms of Stone algebras in terms of triples.

6.28. Describe subalgebras of Stone algebras in terms of triples.

6.29. For a given boolean algebra $B$ with more than one element and distributive lattice $D$ with unit, construct a Stone algebra $L$ with $\mathrm{Skel}\, L \cong B$ and $\mathrm{Dns}\, L \cong D$. (That is, prove that $\mathrm{Skel}\, L$ and $\mathrm{Dns}\, L$ are independent.)

6.30. Show that a Stone algebra $L$ is complete if $\mathrm{Skel}\, L$ and $\mathrm{Dns}\, L$ are complete.

*6.31. Characterize the completeness of Stone algebras in terms of triples.

*6.32. Show that a distributive lattice with pseudocomplementation $L$ has CEP—defined in Section I.3.8 (G. Grätzer and H. Lakser [298]).

6.33. Let $L$ be a lattice or a lattice with additional operations. If $a, b \in L$ and $[b, a]$ is simple, then $\gamma(a, b)$ (defined in the proof of Theorem 221) is unique.

6.34. Use the Subdirect Product Representation Theorem of G. Birkhoff to prove that every distributive lattice is a subdirect product of copies of $\mathsf{C}_2$. Relate this to Theorem 119.

6.35. Find conditions under which a distributive lattice is a subdirect product of copies of $\mathsf{C}_3$.

6.36. Find conditions under which a distributive lattice is a subdirect product of copies of $\mathsf{C}_n$ (F. W. Anderson and R. L. Blair [29]).

# Chapter

# III

---

# *Congruences*

## 1. Congruence Spreading

### 1.1 Congruence-perspectivity

Let $a, b, c, d$ be elements of a lattice $L$. If

$$a \equiv b \pmod{\boldsymbol{\alpha}} \quad \text{implies that} \quad c \equiv d \pmod{\boldsymbol{\alpha}},$$

for any $\boldsymbol{\alpha} \in \operatorname{Con} L$, then we can say that $a \equiv b$ *spreads to* $c \equiv d$.

In Section I.3.6, we saw that $x \equiv y \pmod{\boldsymbol{\alpha}}$ iff $x \wedge y \equiv x \vee y \pmod{\boldsymbol{\alpha}}$; therefore, to investigate how congruences spread, it is enough to deal with comparable pairs, $a \leq b$ and $c \leq d$. By Lemma 10, the congruence blocks are convex sublattices, so instead of comparable pairs, we shall deal with intervals $[a, b]$ and $[c, d]$.

Congruences spread in two ways:

(i) by applying the Substitution Properties $(\mathrm{SP}_\vee)$ and $(\mathrm{SP}_\wedge)$;
(ii) by transitivity.

For instance, in either of the two lattices of Figure 42, if $a \equiv b \pmod{\boldsymbol{\alpha}}$, then by the Substitution Property, it follows that $c \equiv d \pmod{\boldsymbol{\alpha}}$. The second way of spreading simply means that if $a \equiv b \pmod{\boldsymbol{\alpha}}$ spreads to $x \equiv y \pmod{\boldsymbol{\alpha}}$ and $y \equiv z \pmod{\boldsymbol{\alpha}}$, then it spreads to $x \equiv z \pmod{\boldsymbol{\alpha}}$.

To treat these ideas with precision, we introduce congruence-perspective intervals, a slight generalization of perspective intervals of Section I.3.5.

G. Grätzer, *Lattice Theory: Foundation*, DOI 10.1007/978-3-0348-0018-1_3,
© Springer Basel AG 2011

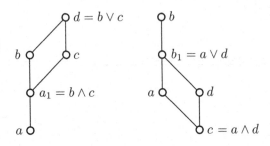

Figure 42. $[a, b] \overset{up}{\twoheadrightarrow} [c, d]$ and $[a, b] \overset{dn}{\twoheadrightarrow} [c, d]$

As illustrated in Figure 42, we say that $[a, b]$ is *up congruence-perspective* to $[c, d]$ and write $[a, b] \overset{up}{\twoheadrightarrow} [c, d]$ if $a \leq c$ and $d = b \vee c$; similarly, $[a, b]$ is *down congruence-perspective* to $[c, d]$ and write $[a, b] \overset{dn}{\twoheadrightarrow} [c, d]$ if $d \leq b$ and $c = a \wedge d$.

In the notation $[a, b] \overset{up}{\twoheadrightarrow} [c, d]$, the arrow points in the direction the congruence spreads. The same important comment applies to the following notation.

If $[a, b] \overset{up}{\twoheadrightarrow} [c, d]$ *or* $[a, b] \overset{dn}{\twoheadrightarrow} [c, d]$, then $[a, b]$ is *congruence-perspective* (or *c-perspective*) to $[c, d]$ and we write $[a, b] \twoheadrightarrow [c, d]$. While perspectivity is symmetric, congruence-perspectivity is not.

Clearly, if $[a, b] \sim [c, d]$, then $[a, b] \twoheadrightarrow [c, d]$, and the same for $\overset{up}{\sim}$ and $\overset{up}{\twoheadrightarrow}$, and for $\overset{dn}{\sim}$ and $\overset{dn}{\twoheadrightarrow}$.

The transitive extension of $\twoheadrightarrow$ is congruence-projectivity, $\Rightarrow$ (or *c-projectivity*). If, for some natural number $n$ and intervals $[e_i, f_i]$, for $0 \leq i \leq n$,

$$[a, b] = [e_0, f_0] \twoheadrightarrow [e_1, f_1] \twoheadrightarrow \cdots \twoheadrightarrow [e_n, f_n] = [c, d],$$

then we call $[a, b]$ *congruence-projective* (or *c-projective*) to $[c, d]$, and we write $[a, b] \Rightarrow [c, d]$. Note that if $a \leq c \leq d \leq b$, then $[a, b] \Rightarrow [c, d]$. Finally, if $[a, b] \Rightarrow [c, d]$ and $[c, d] \Rightarrow [a, b]$ both hold, we write $[a, b] \Leftrightarrow [c, d]$. Note that the intervals used to establish $[a, b] \Rightarrow [c, d]$ may be very different from the ones used to establish $[c, d] \Rightarrow [a, b]$.

**Lemma 229.** *Let $L$ be a lattice, $a, b, c, d \in L$ with $a \leq b$ and $c \leq d$. Then the following conditions are equivalent:*

(i) *$[a, b]$ is c-projective to $[c, d]$.*

(ii) *There is an integer $m$ and there are elements $e_0, \ldots, e_{m-1} \in L$ such that*

$$p_m(a, e_0, \ldots, e_{m-1}) = c,$$
$$p_m(b, e_0, \ldots, e_{m-1}) = d,$$

*where the term $p_m$ is defined by*

$$p_m(x, y_0, \ldots, y_{m-1}) = \cdots (((x \vee y_0) \wedge y_1) \vee y_2) \wedge \cdots .$$

(iii) *There is an integer $n$, there are intervals*

$$[a, b] = [e_0, f_0], [e_1, f_1], \ldots, [e_n, f_n] = [c, d],$$

*and there are elements $e_i'$ with $e_i \leq e_i' \leq f_i$, for all $0 \leq i < n$, such that*
$[e_{i-1}', f_{i-1}] \overset{up}{\sim} [e_i, f_i]$ *if* $[e_{i-1}, f_{i-1}] \overset{up}{\twoheadrightarrow} [e_i, f_i]$ *and* $[e_{i-1}, e_{i-1}'] \overset{dn}{\sim} [e_i, f_i]$ *if*
$[e_{i-1}, f_{i-1}] \overset{dn}{\twoheadrightarrow} [e_i, f_i]$.

*Proof.* By the definitions.                                          □

G. Grätzer and E. T. Schmidt [334] discusses the properties of congruence projectivity in greater detail. For universal algebras, A. I. Mal'cev introduced similar concepts; the difference is that while for lattices it is sufficient to consider unary polynomials of a special form, namely,

$$\cdots (((x \wedge a_0) \vee a_1) \wedge a_2) \cdots ),$$

for universal algebras, in general, we have to consider arbitrary unary polynomials (see, for instance, G. Grätzer [254]). The term $p_2 = (x \vee y_0) \wedge y_1$ is also of special interest; the identities of $p_2$ that hold for lattices imply that the congruence lattices of lattices are distributive. A general condition for the distributivity of congruence lattices of algebras in a given class (variety) of algebras can be found in B. Jónsson [444].

## 1.2   Principal congruences

Intuitively, "$a \equiv b$ congruence spreads to $c \equiv d$" if the interval $[c, d]$ is put together from pieces $[c', d']$ each satisfying $[a, b] \Rightarrow [c', d']$. To state this more precisely, we describe $\mathrm{con}(a, b)$, the smallest congruence relation under which $a \equiv b$, introduced in Section I.3.6; see R. P. Dilworth [157].

**Theorem 230.** *Let $L$ be a lattice and let $a \leq b$ and $c \leq d$ in $L$. Then*

$$c \equiv d \pmod{\mathrm{con}(a, b)}$$

*iff, for some ascending sequence*

$$c = e_0 \leq e_1 \leq \cdots \leq e_m = d,$$

*the $c$-projectivities*

$$[a, b] \Rightarrow [e_j, e_{j+1}]$$

*hold for all $j = 0, \ldots, m - 1$.*

*Proof.* Let $\beta$ denote the following relation on $L$: $x \equiv y \pmod{\beta}$ if $c \leq d$ satisfies the condition of the theorem with $x \wedge y = c$ and $x \vee y = d$.

We first prove that $\beta$ is a congruence relation by verifying the conditions of Lemma 11. The relation $\beta$ is reflexive since, for every $c \in L$, we get

$$[a, b] \overset{\text{up}}{\twoheadrightarrow} [a \vee b, b \vee c] \overset{\text{dn}}{\twoheadrightarrow} [c, c].$$

It is also obvious that if $a_1 \leq a_2 \leq a_3$ and

$$a_1 \equiv a_2 \pmod{\beta},$$
$$a_2 \equiv a_3 \pmod{\beta},$$

then $a_1 \equiv a_3 \pmod{\beta}$. Indeed, take the ascending sequences establishing the two congruences; putting the two sequences together, we get an ascending sequence establishing that $a_1 \equiv a_3 \pmod{\beta}$.

Now let $c \equiv d \pmod{\beta}$ with $c \leq d$ and let $f \in L$. Let

$$c = e_0 \leq e_1 \leq \cdots \leq e_m = d$$

be the ascending sequence establishing $c \equiv d \pmod{\beta}$, that is,

$$[a, b] \Rightarrow [e_i, e_{i+1}]$$

for $i = 0, \ldots, m - 1$. Then

$$c \vee f = e_0 \vee f \leq e_1 \vee f \leq \cdots \leq e_m \vee f = d \vee f,$$
$$[a, b] \Rightarrow [e_i, e_{i+1}] \overset{\text{up}}{\twoheadrightarrow} [e_i \vee f, e_{i+1} \vee f],$$

hence

$$[a, b] \Rightarrow [e_i \vee f, e_{i+1} \vee f]$$

for $i = 0, \ldots, n - 1$; this proves that

$$c \vee f \equiv d \vee f \pmod{\beta}.$$

Similarly,

$$c \wedge f \equiv d \wedge f \pmod{\beta}.$$

Thus by Lemma 11, the relation $\beta$ is a congruence relation.

The congruence $a \equiv b \pmod{\beta}$ obviously holds, so $\beta$ is a congruence relation under which $a \equiv b$.

Now let $\alpha \in \operatorname{Con} L$ satisfy $a \equiv b \pmod{\alpha}$. It is easy to see that for the intervals $[x, y]$ and $[u, v]$, the congruence $x \equiv y \pmod{\alpha}$ and the c-perspectivity $[x, y] \twoheadrightarrow [u, v]$ imply that $u \equiv v \pmod{\alpha}$. By a trivial induction, $x \equiv y \pmod{\alpha}$ and $[x, y] \Rightarrow [u, v]$ imply that $u \equiv v \pmod{\alpha}$. So finally, let $c \equiv d \pmod{\beta}$, established by

$$c \wedge d = e_0 \leq e_1 \leq \cdots \leq e_m = c \vee d.$$

Since $[a, b] \Rightarrow [e_i, e_{i+1}]$, we conclude that the congruences $e_i \equiv e_{i+1} \pmod{\boldsymbol{\alpha}}$ hold for all $i = 0, \ldots, m - 1$. Therefore, by the transitivity of $\boldsymbol{\alpha}$, we obtain that $c \equiv d \pmod{\boldsymbol{\alpha}}$. This proves that $\boldsymbol{\beta}$ is the smallest congruence relation under which $a \equiv b$, and so $\boldsymbol{\beta} = \mathrm{con}(a, b)$. $\qquad\square$

The following is a typical use of congruence-projectivity.

**Lemma 231.** *Let $L$ be a lattice and let $a_i, b_i \in L$ with $a_i < b_i$ for all $i = 1, \ldots, n$. Then*

$$\bigwedge (\, \mathrm{con}(a_i, b_i) \mid i = 1, 2, \ldots, n \,) \neq \mathbf{0}$$

*iff there exist elements $a < b$ in $L$ such that*

$$[a_i, b_i] \Rightarrow [a, b], \quad \text{for all } i = 1, 2, \ldots, n.$$

*Proof.* If such elements $a < b$ exist in $L$, then $a \equiv b \pmod{\mathrm{con}(a_i, b_i)}$ for all $i = 1, 2, \ldots, n$. Hence

$$a \equiv b \pmod{\bigwedge (\, \mathrm{con}(a_i, b_i) \mid i = 1, 2, \ldots, n \,)}$$

and so

$$\bigwedge (\, \mathrm{con}(a_i, b_i) \mid i = 1, 2, \ldots, n \,) \neq \mathbf{0}.$$

We prove the converse, by induction on $n$, in a somewhat stronger form: if $u < v$ and

$$u \equiv v \pmod{\bigwedge (\, \mathrm{con}(a_i, b_i) \mid i = 1, 2, \ldots, n \,)},$$

then there exists a subinterval $[a, b]$ of $[u, v]$ such that $[a_i, b_i] \Rightarrow [a, b]$ for all $i = 1, 2, \ldots, n$.

For $n = 1$, apply Theorem 230. Assuming the statement proved for $n - 1$, we get a proper subinterval $[a', b']$ of $[u, v]$ satisfying $[a_i, b_i] \Rightarrow [a', b']$ for all $i = 1, \ldots, n - 1$. Since

$$u \equiv v \pmod{\mathrm{con}(a_n, b_n)},$$

we get that

$$a' \equiv b' \pmod{\mathrm{con}(a_n, b_n)}.$$

Hence, by Theorem 230, we obtain a proper subinterval $[a, b]$ of $[a', b']$ with $[a_n, b_n] \Rightarrow [a, b]$. Since $[a', b'] \Rightarrow [a, b]$, we also have $[a_n, b_n] \Rightarrow [a, b]$, and so $[a_i, b_i] \Rightarrow [a, b]$ for all $i = 1, 2, \ldots, n$. $\qquad\square$

The congruence $\mathbf{0}$ is meet-irreducible in the congruence lattice of a subdirectly irreducible lattice, therefore, we conclude:

**Corollary 232.** *Let $L$ be a subdirectly irreducible lattice and let $a_i < b_i$ in $L$ for all $i = 1, \ldots, n$. Then there exists $a < b$ of $L$ satisfying $[a_i, b_i] \Rightarrow [a, b]$ for all $i = 1, \ldots, n$.*

In fact, Corollary 232 holds for any lattice in which $\mathbf{0}$ is meet-irreducible.

## 1.3   The join formula

Let $L$ be a lattice and $H \subseteq L^2$. To compute $\mathrm{con}(H)$, the smallest congruence relation $\alpha$ under which $a \equiv b \pmod{\alpha}$, for all $(a,b) \in H$, we use the formula stated in Lemma 14:

$$\mathrm{con}(H) = \bigvee(\,\mathrm{con}(a,b) \mid (a,b) \in H\,),$$

and we need a formula for joins:

**Lemma 233.** *Let $L$ be a lattice and let $\alpha_i \in \mathrm{Con}\,L$ for $i \in I$. Then*

$$a \equiv b \quad (\mathrm{mod}\ \bigvee(\alpha_i \mid i \in I))$$

*iff there is a sequence*

$$z_o = a \wedge b \le z_1 \le \cdots \le z_n = a \vee b$$

*such that, for each $j$ with $0 \le j < n$, there is an $i_j \in I$ with the property: the congruence $z_j \equiv z_{j+1} \pmod{\alpha_{i_j}}$ holds.*

The proof of Lemma 233 is the same as that of Theorems 12 and 37, namely, a direct application of Lemma 11.

By combining Theorem 230 and Lemma 233, we get:

**Corollary 234.** *Let $L$ be a lattice, let $H \subseteq L^2$, and let $a \le b$ in $L$. Then $a \equiv b \pmod{\mathrm{con}(H)}$ iff there exists a sequence*

$$a = c_0 \le c_1 \le \cdots \le c_n = b,$$

*for some integer $n$, and there exist $(d_i, e_i) \in H$ satisfying*

$$[d_i \wedge e_i, d_i \vee e_i] \Rightarrow [c_i, c_{i+1}]$$

*for each $i$ with $0 \le i < n$.*

**Corollary 235.** *Let $L$ be a lattice, let $I$ be an ideal of $L$, and let $a \le b$ in $L$. Then $a \equiv b \pmod{\mathrm{con}(I)}$ iff there exists a sequence*

$$a = c_0 \le c_1 \le \cdots \le c_n = b$$

*and there exist elements $d \le e$ in $I$ such that $[d, e] \Rightarrow [c_i, c_{i+1}]$ for all $0 \le i < n$.*

Recall from Section I.3.7 that an ideal $I$ is called the ideal kernel of a congruence relation $\alpha$ iff $I$ is a block modulo $\alpha$.

**Corollary 236.** *Let $L$ be a lattice and let $I$ be an ideal of $L$. The ideal $I$ is the ideal kernel of a congruence relation iff*

$$[a, b] \Rightarrow [c, d] \text{ and } c \in I \text{ imply that } d \in I$$

*for all $a \le b$ in $I$ and $c \le d$ in $L$.*

*Proof.* Combine Corollary 235 with the observation that $I$ is an ideal kernel of some congruence relation iff it is the kernel of $\mathrm{con}(I)$. □

In distributive lattices, every ideal is the kernel of some congruence relation; in fact, this property characterizes distributivity. In general lattices, we shall introduce various classes of ideals that are kernels and for which the congruence $\mathrm{con}(I)$ can be nicely described. This will be done and applied in Sections 2–4.

## 1.4   Finite lattices

Prime intervals (see Section I.3.5) play a dominant role in the study of the congruences of a finite lattice $L$. Let $L$ be a finite lattice and let $\mathrm{PrInt}(L)$ denote the set of prime intervals of $L$. Let $\mathfrak{p} = [a, b] \in \mathrm{PrInt}(L)$ and let $\boldsymbol{\alpha}$ be a congruence relation of $L$; then we write $\mathfrak{p} \in \boldsymbol{\alpha}$ for $a \equiv b \pmod{\boldsymbol{\alpha}}$ and $\mathrm{con}(\mathfrak{p})$ for $\mathrm{con}(a, b)$.

In a finite lattice $L$, the formula

$$\boldsymbol{\alpha} = \bigvee(\,\mathrm{con}(\mathfrak{p}) \mid \mathfrak{p} \in \boldsymbol{\alpha}\,)$$

immediately yields that the join-irreducible congruences of $L$, which we will denote by $\mathrm{Con}_{\mathrm{Ji}} L$, are the congruences of the form $\mathrm{con}(\mathfrak{p})$ for some $\mathfrak{p} \in \mathrm{PrInt}(L)$. Of course, a join-irreducible congruence $\boldsymbol{\alpha}$ can be expressed, as a rule, in many ways in the form $\mathrm{con}(\mathfrak{p})$. However, Corollary 234 yields the following two crucial statements:

**Lemma 237.** *Let $L$ be a finite lattice, and let $\mathfrak{p} \in \mathrm{PrInt}(L)$. If $\mathfrak{p} \in \mathrm{con}(a, b)$, then there is a prime interval $\mathfrak{q}$ in $[a, b]$ satisfying $\mathfrak{q} \Rightarrow \mathfrak{p}$.*

**Lemma 238.** *Let $L$ be a finite lattice, and let $\mathfrak{p}, \mathfrak{q} \in \mathrm{PrInt}(L)$. Then $\mathrm{con}(\mathfrak{p}) \leq \mathrm{con}(\mathfrak{q})$ iff $\mathfrak{q} \Rightarrow \mathfrak{p}$ and $\mathrm{con}(\mathfrak{p}) = \mathrm{con}(\mathfrak{q})$ iff $\mathfrak{p} \Leftrightarrow \mathfrak{q}$.*

In view of Theorem 107, we get the following:

**Theorem 239.** *Let $L$ be a finite lattice. The relation $\Rightarrow$ is a preordering on $\mathrm{PrInt}(L)$. The blocks under $\Leftrightarrow$ form an order that is dually isomorphic to $\mathrm{Con}_{\mathrm{Ji}} L$.*

We use Figure 43 to illustrate how we compute the congruence lattice of the lattice $\mathsf{N}_5$ using Theorem 239. The lattice $\mathsf{N}_5$ has five prime intervals: $[o, a], [a, b], [b, i], [o, c], [c, i]$. The blocks are

$$\boldsymbol{\alpha} = \{[a, b]\}, \ \ \boldsymbol{\beta} = \{[o, c], [b, i]\}, \ \ \boldsymbol{\gamma} = \{[o, a], [c, i]\}.$$

The ordering is $\boldsymbol{\alpha} < \boldsymbol{\gamma}$ because $[c, i] \overset{\mathrm{dn}}{\twoheadrightarrow} [o, b] \overset{\mathrm{up}}{\twoheadrightarrow} [a, b]$. Similarly, $\boldsymbol{\alpha} < \boldsymbol{\beta}$.

It is important to note that the computation of $\mathfrak{p} \Rightarrow \mathfrak{q}$ may involve non-prime intervals, as in the previous paragraph, computing $[c, i] \Rightarrow [a, b]$.

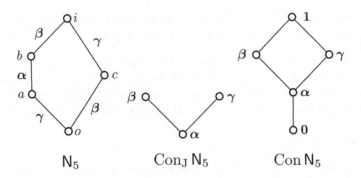

Figure 43. Computing the congruence lattice of $N_5$

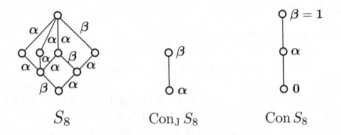

Figure 44. Computing the congruence lattice of $S_8$

As another example, we compute the congruence lattice of $S_8$; see Figure 44.

If the finite lattice $L$ is atomistic, then the join-irreducible congruences are even simpler to find. Indeed if $[a, b]$ is a prime interval, and $p$ is an atom with $p \leq b$ and $p \nleq a$, then $[a, b] \overset{\mathrm{dn}}{\sim} [0, p]$, so $\mathrm{con}(a, b) = \mathrm{con}(0, p)$.

**Corollary 240.** *Let $L$ be a finite atomistic lattice. Then every join-irreducible congruence can be represented in the form* $\mathrm{con}(0, p)$*, where $p$ is an atom. The relation $p \Rightarrow q$, defined as* $[0, p] \Rightarrow [0, q]$*, introduces a preordering on* $\mathrm{Atom}(L)$*. The blocks under the preordering form an order dually isomorphic to* $\mathrm{Con}_{\mathrm{Ji}} L$*.*

### 1.5    Congruences and extensions

Let $L$ be a lattice, and let $K \leq L$. How do the congruences of $L$ relate to the congruences of $K$?

Every congruence $\alpha$ *restricts* to $K$: the relation $\alpha \cap K^2 = \alpha \rceil K$ on $K$ is a congruence of $K$. So we get the *restriction map*:

$$\mathrm{re} \colon \mathrm{Con}\, L \to \mathrm{Con}\, K,$$

that maps a congruence $\alpha$ of $L$ to a congruence $\alpha \rceil K$ of $K$.

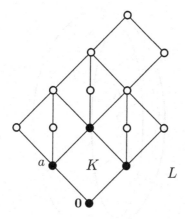

Figure 45. Illustrating the map re

**Lemma 241.** *Let $K \leq L$ be lattices. Then the map*

$$\mathrm{re}\colon \mathrm{Con}\, L \to \mathrm{Con}\, K$$

*is a $\{\wedge, 0, 1\}$-homomorphism.*

For instance, if $K = \{o, a, i\}$ and $L = \mathsf{M}_3$, then $\mathrm{Con}\, K$ is isomorphic to $\mathsf{B}_2$, but only the congruences $\mathbf{0}$ and $\mathbf{1}$ are restrictions of congruences in $L$. As another example, take the lattice $L$ of Figure 45 and its sublattice $K$, the black-filled elements; in this case, $\mathrm{Con}\, L \cong \mathrm{Con}\, K \cong \mathsf{B}_2$, but again only $\mathbf{0}$ and $\mathbf{1}$ are restrictions. There is no natural relationship between the congruences of $K$ and $L$.

If $K$ is an ideal in $L$ (or any convex sublattice), we can say a lot more.

**Lemma 242.** *Let $K \leq L$ be lattices. If $K$ is an ideal of $L$, then the map $\mathrm{re}\colon \mathrm{Con}\, L \to \mathrm{Con}\, K$ is a $\{0, 1\}$-homomorphism.*

*Proof.* By Lemma 241, the map re is a $\{\wedge, 0, 1\}$-homomorphism. Let $\alpha$ and $\beta$ be congruences of $L$; we have to prove that

$$\alpha\rceil K \vee \beta\rceil K = (\alpha \vee \beta)\rceil K.$$

Since the relation $\leq$ is trivial, we prove $\geq$. So let $a, b \in K$ and

$$a \equiv b \pmod{(\alpha \vee \beta)\rceil K};$$

we want to prove that

$$a \equiv b \pmod{\alpha\rceil K \vee \beta\rceil K}.$$

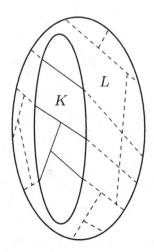

Figure 46. A congruence extends

By Lemma 233, there is a sequence

$$z_0 = a \wedge b \leq z_1 \leq \cdots \leq z_n = a \vee b$$

such that either $z_j \equiv z_{j+1} \pmod{\alpha}$ or $z_j \equiv z_{j+1} \pmod{\beta}$ holds in $L$ for each $j$ with $0 \leq j < n$. Since $a, b \in K$ and $K$ is an ideal, it follows that $z_0, z_1, \ldots, z_n \in K$, so either $z_j \equiv z_{j+1} \pmod{\alpha \rceil K}$ or $z_j \equiv z_{j+1} \pmod{\beta \rceil K}$ holds for each $j$ with $0 \leq j < n$, proving that $a \equiv b \pmod{\alpha \rceil K \vee \beta \rceil K}$.    □

Let $K \leq L$ be lattices, and let $\alpha \in \operatorname{Con} K$. The congruence $\operatorname{con}_L(\alpha)$ (the congruence $\operatorname{con}(\alpha)$ formed in the sublattice $L$) is the smallest congruence $\beta$ of $L$ such that $\alpha \leq \beta \rceil K$. Unfortunately, $\operatorname{con}_L(\alpha) \rceil K$ may be different from $\alpha$, as in the example of Figure 45. We say that the congruence $\alpha$ of $K$ *extends* to $L$, if $\alpha$ is the restriction of $\operatorname{con}_L(\alpha)$. Figure 46 illustrates this in part. If a congruence $\alpha$ extends, then the blocks of $\alpha$ in $K$ extend to blocks in $L$, but there may be blocks in $L$ that are not such extensions.

The *extension map*

$$\operatorname{ext} \colon \operatorname{Con} K \to \operatorname{Con} L$$

maps a congruence $\alpha$ of $K$ to the congruence $\operatorname{con}_L(\alpha)$ of $L$. The map ext is a $\{\vee, 0\}$-homomorphism of $\operatorname{Con} K$ into $\operatorname{Con} L$. In addition, ext maps only the 0 to 0, that is, ext is a $\{0\}$-*separating* $\{\vee, 0\}$-homomorphism. To summarize:

**Lemma 243.** *Let $K \leq L$ be lattices. Then the map* $\operatorname{ext} \colon \operatorname{Con} K \to \operatorname{Con} L$ *is a $\{0\}$-separating $\{\vee, 0\}$-homomorphism.*

## 1.6   Congruence-preserving extensions

We introduced in Section I.3.9 a very strong concept: Let $K$ be a lattice and let $L$ be an extension of $K$. Then $L$ is a *congruence-preserving extension* of $K$ if every congruence $\alpha$ of $K$ has *exactly one* extension $\overline{\alpha}$ to $L$ satisfying $\overline{\alpha}\rceil_K = \alpha$; we also say that $K$ is a *congruence-preserving sublattice*. Of course, $\overline{\alpha} = \mathrm{con}_L(\alpha)$. It follows that $\mathrm{ext}\colon \alpha \mapsto \mathrm{con}_L(\alpha)$ is an isomorphism between $\mathrm{Con}\,K$ and $\mathrm{Con}\,L$.

Two congruence-preserving extensions are shown in Figure 47, while Figure 48 shows two other extensions that are not congruence-preserving.

We can obtain congruence-preserving extensions using gluing, see Section IV.2.

**Lemma 244.** *Let the lattice $L$ be an extension of the lattice $K$. Then $L$ is a congruence-preserving extension of $K$ iff the following two conditions hold:*

(i) $\mathrm{re}(\mathrm{ext}\,\alpha) = \alpha$ *for every congruence $\alpha$ of $K$.*
(ii) $\mathrm{ext}(\mathrm{re}\,\alpha) = \alpha$ *for every congruence $\alpha$ of $L$.*

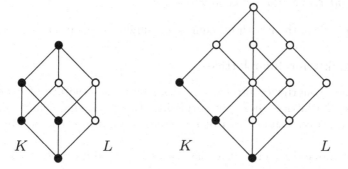

Figure 47. Examples of congruence-preserving extensions

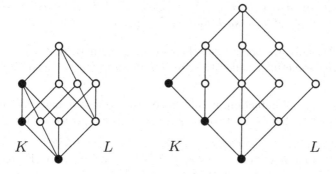

Figure 48. Examples of not congruence-preserving extensions

We can say a lot more for finite lattices:

**Lemma 245.** *Let $L$ be a finite lattice, and let $K \leq L$. Then $L$ is a congruence-preserving extension of $K$ iff the following two conditions hold:*

(a) *Let $\mathfrak{p}$ and $\mathfrak{q}$ be prime intervals in $K$; if $\mathfrak{p} \Rightarrow \mathfrak{q}$ in $L$, then $\mathfrak{p} \Rightarrow \mathfrak{q}$ in $K$.*

(b) *Let $\mathfrak{p}$ be a prime interval of $L$. Then there exists a prime interval $\mathfrak{q}$ in $K$ such that $\mathfrak{p} \Leftrightarrow \mathfrak{q}$ in $L$.*

Condition (ii) of Lemma 244 is very interesting by itself. It says that every congruence $\boldsymbol{\alpha}$ of $L$ is determined by its restriction to $K$. In other words, $\boldsymbol{\alpha} = \operatorname{con}_L(\boldsymbol{\alpha}\rceil K)$. We shall call such a sublattice $K$ a *congruence-determining sublattice*.

We can easily modify Lemma 245 to characterize congruence-determining sublattices in finite lattices:

**Lemma 246.** *Let $L$ be a finite lattice $L$, and $K \leq L$. Then $K$ is a congruence-determining sublattice of $L$ iff for any prime interval $\mathfrak{p}$ in $L$, there is a prime interval $\mathfrak{q}$ in $K$ satisfying $\mathfrak{p} \Leftrightarrow \mathfrak{q}$ in $L$.*

The proofs of the last three lemmas are similar to the proofs in Section 1.4.

## 1.7   Weakly modular lattices

In a sense, congruence projectivity describes the structure of congruence relations of a lattice. It is not surprising, therefore, that many important classes of lattices can be described by congruence projectivities. We give an example.

To introduce this class, the class of weakly modular lattices, we need a lemma for motivation.

**Lemma 247.** *Let $L$ be a lattice, $a < b$ and $c < d$ in $L$, and $[a, b] \Rightarrow [c, d]$. If $L$ is modular or if $L$ is relatively complemented, then there exists a proper subinterval $[a', b']$ of $[a, b]$ such that $[c, d] \Rightarrow [a', b']$.*

*Remark.* Actually, we prove more than we state, namely, that $[a', b']$ is projective to $[a, b]$, sometimes called the *projectivity property*; see also Exercise 1.3.

*Proof.* By a trivial induction, it suffices to prove this statement for the binary relation $\twoheadrightarrow \circ \twoheadrightarrow$, that is, for two steps (in an $n$-step c-projectivity); by duality, we prove it for $\overset{\text{up}}{\twoheadrightarrow} \circ \overset{\text{dn}}{\twoheadrightarrow}$. So let $[a, b] \overset{\text{up}}{\twoheadrightarrow} [e, f] \overset{\text{dn}}{\twoheadrightarrow} [c, d]$, see Figure 49.

Let $L$ be modular. Set $x = e \vee d$, $a' = b \wedge e$, and $b' = b \wedge x$. By modularity, $b' < a'$. Thus $[a', b']$ is a proper subinterval of $[a, b]$ and the projectivity $[c, d] \Rightarrow [a', b']$ is obvious.

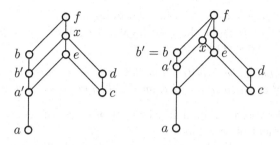

Figure 49. Weak modularity

Let $L$ be relatively complemented. Let $x$ be a relative complement of $e \vee d$ in $[e, f]$ (see the second diagram in Figure 49). Then

$$[c, d] \overset{\text{up}}{\twoheadrightarrow} [e, e \vee d] \overset{\text{up}}{\twoheadrightarrow} [x, f] \overset{\text{dn}}{\twoheadrightarrow} [b \wedge x, b],$$

so $[c, d] \Rightarrow [b \wedge x, b]$. Since $c < d$ by assumption, it follows that $x < f$ and $b \wedge x < b$. Define $a' = b \wedge x$ and $b' = b$ and we are done.  □

Lemma 247 contains two distinct statements. They can be unified using the following concept introduced in G. Grätzer and E. T. Schmidt [334].

**Definition 248.** Let us call a lattice $L$ *weakly modular* if for any pair $[a, b]$ and $[c, d]$ of proper intervals of $L$, the c-projectivity $[a, b] \Rightarrow [c, d]$ implies the existence of a proper subinterval $[a', b']$ of $[a, b]$ with the c-projectivity $[c, d] \Rightarrow [a', b']$.

**Corollary 249.** *Every modular and every relatively complemented lattice is weakly modular.*

Weak modularity is a rather complicated condition. It can be somewhat simplified for finite lattices; a finite lattice is weakly modular iff $[a, b] \Rightarrow [c, d]$ and $a > b \geq c > d$ imply the existence of a proper subinterval $[a', b']$ of $[a, b]$ satisfying $[c, d] \Rightarrow [a', b']$ (G. Grätzer [250]).

The importance of the class of weakly modular lattices will be illustrated in Sections 2–4. *MathSciNet* gives more than 20 additional references on this topic.

## 1.8  Representable congruences

We study c-projectivities to enable us to obtain descriptions of quotient lattices such as $L/\text{con}(a, b)$. Sometimes, however, a quotient lattice $L/\alpha$ can be identified very simply within $L$: if there is a sublattice $L_1$ of $L$ having one and only one element in every block, then $L/\alpha \cong L_1$. Such congruence relations are called *representable*. It happens much more often that we can get a meet-subsemilattice or a join-subsemilattice $L_1$ of $L$ having one and only one element in every block; in such cases we call $\alpha$ *meet-representable* or *join-representable*, respectively.

**Lemma 250.** *Let $L$ be a lattice and let $\alpha$ be a congruence relation of $L$. If every block of $\alpha$ has a minimal element, then $\alpha$ is join-representable.*

*Proof.* Let $L_1$ be the set of minimal elements of blocks of $\alpha$; then $L_1$ contains exactly one element of each block.

Now let $a, b \in L_1$. We show that $a \vee b$ is the smallest element of the block

$$(a \vee b)/\alpha = a/\alpha \vee b/\alpha.$$

Indeed, let us assume that $c < a \vee b$ and $c \equiv a \vee b \pmod{\alpha}$. It follows then that $a \wedge c \equiv a \pmod{\alpha}$ and $b \wedge c \equiv b \pmod{\alpha}$. Since $c < a \vee b$, we obtain that $a \wedge c < a$ or $b \wedge c < b$, say $a \wedge c < a$. Then $a \wedge c < a$ and $a \wedge c \in a/\alpha$ contradict that $a \in L_1$. $\square$

Thus if $L$ satisfies the Descending Chain Condition, then every congruence relation is join-representable, and dually.

**Exercises**

1.1. For all pairs of intervals $[x, y], [u, v]$ of $\mathsf{N}_5$, investigate when $x \equiv y$ spreads to $u \equiv v$. Repeat the investigation for $\mathsf{M}_3$.

1.2. Show that if $L$ is sectionally complemented, then in order to learn the congruence structure of $L$ it is sufficient to consider "$a \equiv b$ spreads to $c \equiv d$" in the special case $b = d = 0$.

1.3. Prove that if a lattice is modular or relatively complemented, then $[a, b] \Rightarrow [c, d]$ iff $[a', b'] \approx [c, d]$ for some subinterval $[a', b']$ of $[a, b]$.

1.4. Show, by an example, that the conclusion of Exercise 1.3 is false for lattices in general.

1.5. Let $L$ be a distributive lattice and let $[a, b] \Rightarrow [c, d]$. Prove that there exists an interval $[e, f]$ and a subinterval $[a', b']$ of $[a, b]$ such that

$$[a', b'] \overset{\mathrm{up}}{\sim} [e, f] \overset{\mathrm{dn}}{\sim} [c, d].$$

1.6. Let $L$ be a lattice, and let $q(x)$ be a unary polynomial. Under what conditions is it true that for all intervals $[a, b]$ of $L$,

$$[a, b] \Rightarrow [q(a), q(b)].$$

1.7. Referring to the proof of Lemma 229, axiomatize those properties of $p_2$ which make it possible to define $p_m$ from $p_2$ to obtain generalizations of Lemma 229 to some varieties of algebras.

1.8. Show that $p_m(x, y_0, \ldots, y_i, y_i, y_{i+2}, \ldots, y_{m-1})$ does not depend on $x, y_0, \ldots, y_{i-1}$ (that is, the value of $p_m(a, e_0, \ldots, e_i, e_i, e_{i+2}, \ldots, e_{m-1})$ is independent of $a, e_0, \ldots, e_{i-1}$).

1.9. Prove that $p_m(x, c, d, c, d, \ldots) = x$ for every $d \leq x \leq c$.

1.10. Rephrase and prove Lemma 229 using $q_m = \cdots((x \wedge y_0) \vee y_1) \wedge \cdots$.

1.11. Show that in any sequence of congruence-perspectivities, any number of steps $\overset{\text{up}}{\twoheadrightarrow}$ and $\overset{\text{dn}}{\twoheadrightarrow}$ can be added. Observe that in any nonredundant sequence of congruence-perspectivities, $\overset{\text{up}}{\twoheadrightarrow}$ and $\overset{\text{dn}}{\twoheadrightarrow}$ have to alternate.

1.12. Find examples to show that $m$ cannot be bounded in Theorem 230. (Make all your examples planar modular lattices.)

1.13. Find examples to show that in Theorem 230, $\Rightarrow$ cannot be replaced by an $n$-step version for any natural number $n$. (Make all your examples planar modular lattices.)

1.14. Prove that an ideal $I$ is a kernel iff $I$ is the kernel of $\mathrm{con}(I)$.

1.15. Let $L$ be a lattice, and let $I$ and $J$ be ideals of $L$ with $J \subseteq I$. Assume that $J$ is a kernel in the lattice $I$ and that $I$ is a kernel in the lattice $L$. Is $J$ a kernel in $L$?

1.16. Show that the ideal kernels of the lattice $L$ form a sublattice of $\mathrm{Id}\, L$, the lattice of all ideals of $L$. In fact, this sublattice is closed under arbitrary joins and under all meets existing in $\mathrm{Id}\, L$.

1.17. Let $L$ be a distributive lattice, let $I$ be an ideal of $L$, and let $[a, b]$ be an interval of $L$. Then $a \equiv b \pmod{\mathrm{con}(I)}$ iff $[a, b] \sim [a', b']$ for some interval $[a', b']$ of $I$.

1.18. Show that Exercise 1.5 characterizes distributivity.

1.19. Show that Exercise 1.17 characterizes distributivity.

1.20. Let $L$ be a lattice and let $\alpha \in \mathrm{Con}\, L$. Verify the formula:

$$\mathrm{con}(a/\alpha, b/\alpha) = (\mathrm{con}(a, b) \vee \alpha)/\alpha.$$

1.21. Let $a_i, b_i \in L$ for all $1 \leq i \leq n$. Show that

$$\bigwedge_{1 \leq i \leq n} \mathrm{con}(a_i/\alpha, b_i/\alpha) = \Big( \bigwedge_{1 \leq i \leq n} (\mathrm{con}(a_i, b_i) \vee \alpha) \Big)/\alpha.$$

1.22. Show that, for ideals $I$ and $J$ of a lattice $L$,

$$\mathrm{con}(I) \vee \mathrm{con}(J) = \mathrm{con}(I \vee J).$$

1.23. Let $L$ be a lattice and let $J_i$, for $i \in I$, be ideals of $L$. Prove that

$$\bigvee(\mathrm{con}(J_i) \mid i \in I) = \mathrm{con}\Big(\bigvee(J_i \mid i \in I)\Big).$$

1.24. Show that $\mathrm{con}(I) \wedge \mathrm{con}(J) = \mathrm{con}(I \wedge J)$ does not hold in general.

1.25. Verify

$$\mathrm{con}(I) \wedge \mathrm{con}(J) = \mathrm{con}(I \wedge J),$$

for ideals $I$ and $J$ of a distributive lattice.

1.26. Show that the formula of Exercise 1.25 holds also under the condition that every ideal is the kernel of at most one congruence relation.

1.27. Show that Corollary 232 does not hold for infinitely many $a_i < b_i$.

1.28. Show that in a finite lattice (or in a lattice in which all chains are finite) every congruence relation is join- and meet-representable.

1.29. Find a congruence relation $\alpha$ of the lattice of Figure 50 which is not representable. (Observe that $\alpha$ is both meet- and join-representable.)

1.30. Let $L$ be a finite lattice and let $L/\alpha \cong M_3$. Show that $\alpha$ is representable.

1.31. Give a formal proof of the statement that if $L_1$ represents the congruence relation $\alpha$ of $L$, then $L/\alpha \cong L_1$.

1.32. A congruence relation $\alpha$ of a lattice $L$ is *order-representable* if there exists a subset $H \subseteq L$ such that $a/\alpha \cap H$ is a singleton, for all $a \in L$, and

$$a \leq b \text{ in } L \quad \text{iff} \quad a/\alpha \leq b/\alpha \text{ in } L/\alpha \quad \text{for } a, b \in H.$$

Show that $H$ as a suborder of $L$ is a lattice and $H \cong L/\alpha$.

1.33. Prove that meet-representability implies order-representability for congruence relations.

1.34. Let $L$ be a lattice and let $\alpha$ be a congruence relation of $L$. Verify that if $L/\alpha$ is finite or countable, then $\alpha$ is order-representable.

1.35. Find a chain $C$, a lattice $L$, and $\alpha \in \operatorname{Con} L$ such that $L/\alpha \cong C$ but $\alpha$ is not order-representable.

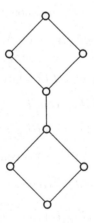

Figure 50. Diagram for Exercise 1.29

## 2. Distributive, Standard, and Neutral Elements

### 2.1 The three element types

The three types of elements of a lattice mentioned in the title of this section were discovered by O. Ore [557], G. Grätzer [246], and G. Birkhoff [66], respectively. It turned out later that all three can be defined using distributive equations only.

**Definition 251.** Let L be a lattice and let $a$ be an element of $L$.

(i) The element $a$ is called *distributive* if

$$a \vee (x \wedge y) = (a \vee x) \wedge (a \vee y)$$

for all $x, y \in L$.

(ii) The element $a$ is called *standard* if

$$x \wedge (a \vee y) = (x \wedge a) \vee (x \wedge y)$$

for all $x, y \in L$.

(iii) The element $a$ is called *neutral* if

$$(a \wedge x) \vee (x \wedge y) \vee (y \wedge a) = (a \vee x) \wedge (x \vee y) \wedge (y \vee a)$$

for all $x, y \in L$.

For instance, the elements $o, i, a, c$ are distributive in $\mathsf{N}_5$; the elements $o, i, a$ are standard but $c$ is not standard; only $o$ and $i$ are neutral. In $\mathsf{M}_3$, only $o$ and $i$ are distributive; they are also standard and neutral. Of course, every element of a distributive lattice is distributive, standard, and neutral.

We can also dualize these definitions and define *dually distributive elements* and *dually standard elements*. Observe that the concept of neutrality is selfdual.

Various useful equivalent forms of these definitions are given in the three theorems that follow.

### 2.2 Distributive elements

Let Distr $L$ denote the set of all distributive elements of the lattice $L$; following O. Ore [557], we give various characterizations of this subset of $L$.

**Theorem 252.** *Let $L$ be a lattice and let $a \in L$. The following conditions on the element $a$ are equivalent:*

(i) $a \in$ Distr $L$.

(ii) *The map*
$$\varphi\colon x \mapsto a \vee x,$$
*for $x \in L$, is a homomorphism of $L$ onto* fil$(a)$.

(iii) *Let $\boldsymbol{\alpha}_a$ be the binary relation on $L$ defined as follows: $x \equiv y \pmod{\boldsymbol{\alpha}_a}$ if $a \vee x = a \vee y$. Then $\boldsymbol{\alpha}_a$ is a congruence relation.*

*Remark.* (i) and (ii) are equivalent since $\varphi$ is a homomorphism iff it preserves meets, while (ii) and (iii) are equivalent because $\boldsymbol{\alpha}_a = \mathrm{Ker}(\varphi)$. A more formal proof follows.

*Proof.*
(i) *implies* (ii). Indeed, $\varphi$ maps $L$ into fil$(a)$ for every element $a$ of $L$; in fact, $\varphi$ is always onto since $\varphi(b) = b$ for all $b \geq a$. The map $\varphi$ is always a join-homomorphism since
$$\varphi(x) \vee \varphi(y) = (a \vee x) \vee (a \vee y) = a \vee (x \vee y) = \varphi(x \vee y).$$

In view of Definition 251(i), if $a$ is distributive, we also have
$$\varphi(x) \wedge \varphi(y) = (a \vee x) \wedge (a \vee y) = a \vee (x \wedge y) = \varphi(x \wedge y),$$

and so $\varphi$ is a homomorphism.

(ii) *implies* (iii). The congruence $\boldsymbol{\alpha}_a$ is the kernel of the homomorphism $\varphi$ and therefore, $\boldsymbol{\alpha}_a$ is a congruence relation.

(iii) *implies* (i). Since $x \vee a = (a \vee x) \vee a$, it follows that $x \equiv a \vee x \pmod{\boldsymbol{\alpha}_a}$. Similarly, $y \equiv a \vee y \pmod{\boldsymbol{\alpha}_a}$. Therefore,
$$x \wedge y \equiv (a \vee x) \wedge (a \vee y) \pmod{\boldsymbol{\alpha}_a}.$$

By the definition of $\boldsymbol{\alpha}_a$, we get that
$$a \vee (x \wedge y) = a \vee ((a \vee x) \wedge (a \vee y)) = (a \vee x) \wedge (a \vee y),$$

proving (i). $\qquad\qquad\qquad\qquad\qquad\qquad\qquad\qquad\qquad\qquad\qquad\qquad\square$

An element $a$ is *dually distributive* in the lattice $L$, if $a$ is a distributive element of $L^\delta$, the dual of $L$. The set of dually distributive elements of the lattice $L$ will be denoted by Distr$^\delta L$. Of course, Distr$^\delta L = $ Distr $L^\delta$.

## 2.3    Standard elements

Let Stand $L$ denote the set of all standard elements of the lattice $L$; following G. Grätzer and E. T. Schmidt [336], we give various characterizations of this subset of $L$.

**Theorem 253.** *Let $L$ be a lattice and let $a \in L$. The following conditions are equivalent:*

(i) $a \in \operatorname{Stand} L$.

(ii) *Let $\alpha_a$ be the binary relation on $L$ defined as follows: $x \equiv y \pmod{\alpha_a}$ if $(x \wedge y) \vee a_1 = x \vee y$ for some $a_1 \leq a$. Then $\alpha_a$ is a congruence relation.*

(iii) $a \in \operatorname{Distr} L$ *and*

$$a \vee x = a \vee y \text{ and } a \wedge x = a \wedge y \quad \text{imply that} \quad x = y$$

*for all $x, y \in L$.*

*Proof.*
   (i) *implies* (ii). Let $a \in \operatorname{Stand} L$ and let $\alpha_a$ be defined as in (ii). We use Lemma 11 to verify that $\alpha_a$ is a congruence relation. By definition, $\alpha_a$ is reflexive and $x \equiv y \pmod{\alpha_a}$ iff $x \wedge y \equiv x \vee y \pmod{\alpha_a}$. If

$$x \leq y \leq z,$$
$$x \equiv y \pmod{\alpha_a},$$
$$y \equiv z \pmod{\alpha_a},$$

then

$$x \vee a_1 = y,$$
$$y \vee a_2 = z$$

for some $a_1, a_2 \leq a$. Hence $x \vee (a_1 \vee a_2) = z$, showing that $x \equiv z \pmod{\alpha_a}$ since $a_1 \vee a_2 \leq a$. Finally, let $x \leq y$ and $x \equiv y \pmod{\alpha_a}$, that is, $x \vee a_1 = y$ with $a_1 \leq a$. The equality $(x \vee t) \vee a_1 = y \vee t$ holds, for any $t \in L$, hence

$$x \vee t \equiv y \vee t \pmod{\alpha_a}.$$

To show the Substitution Property for meet, observe that

$$y \wedge t \leq y = x \vee a_1 \leq x \vee a,$$

and so, using the fact that $a$ is standard we get

$$y \wedge t = (y \wedge t) \wedge (x \vee a) = ((y \wedge t) \wedge x) \vee ((y \wedge t) \wedge a) = (x \wedge t) \vee a_2,$$

where $a_2 = y \wedge t \wedge a \leq a$. Thus the conditions of Lemma 11 are verified, and $\alpha_a$ is a congruence relation.
   (ii) *implies* (iii). Let us assume that the $\alpha_a$ defined in (ii) is a congruence relation. We can show that $a \in \operatorname{Distr} L$ just as in Theorem 252. Now let

$$a \vee x = a \vee y,$$
$$a \wedge x = a \wedge y.$$

Since $y \equiv a \vee y \pmod{\alpha_a}$, meeting both sides with $x$ and using $a \vee y = a \vee x$, we obtain that

$$x \wedge y \equiv x \wedge (a \vee y) = x \wedge (a \vee x) = x \pmod{\alpha_a}.$$

Thus $x = (x \wedge y) \vee a_1$ for some $a_1 \leq a$. Also $a_1 \leq x$, hence $a_1 \leq a \wedge x = a \wedge y$, and so $a_1 \leq x \wedge y$. We conclude that $x = x \wedge y$. Similarly, $y = x \wedge y$, and so $x = y$.

(iii) *implies* (i). Let us assume (iii), let $x, y \in L$, and define

$$b = x \wedge (a \vee y),$$
$$c = (x \wedge a) \vee (x \wedge y).$$

In order to show $b = c$, by (iii), it will be sufficient to prove that

$$a \vee b = a \vee c,$$
$$a \wedge b = a \wedge c.$$

To prove the first, we compute, using the fact that $a \in \mathrm{Distr}\, L$:

$$a \vee b = a \vee ((x \wedge (a \vee y))) = (a \vee x) \wedge (a \vee y)$$
$$= a \vee (x \wedge y) = a \vee (x \wedge a) \vee (x \wedge y) = a \vee c.$$

To prove the second,

$$a \wedge x \leq a \wedge c \qquad\qquad\qquad \text{(using } c \leq b)$$
$$\leq a \wedge b = a \wedge x \wedge (a \vee y) = a \wedge x,$$

and hence $a \wedge c = a \wedge b$. $\qquad\qquad\qquad\qquad\qquad\qquad\qquad\qquad$ $\square$

## 2.4    Neutral elements

Let $\mathrm{Neutr}\, L$ denote the set of all neutral elements of the lattice $L$; we give various characterizations of this subset of $L$.

**Theorem 254.** *Let $L$ be a lattice and let $a \in L$. The following conditions are equivalent:*

(i) $a \in \mathrm{Neutr}\, L$.

(ii) $a \in \mathrm{Distr}\, L$ *and* $a \in \mathrm{Distr}^\delta L$, *and*

$$a \vee x = a \vee y,$$
$$a \wedge x = a \wedge y$$

*imply that* $x = y$ *for every* $x, y \in L$.

(iii) *There is an embedding $\varphi$ of $L$ into a direct product $A \times B$, where $A$ has a unit and $B$ has a zero and $\varphi(a) = (1, 0)$.*

(iv) *For any $x, y \in L$, the sublattice generated by $a, x, y$ is distributive.*

*Remark.* The equivalence of (ii)–(iv) is due to G. Birkhoff [66], who used (iv) as a definition of neutrality. The equivalence of (i) to (ii)–(iv) was conjectured in G. Grätzer and E. T. Schmidt [336]. This was proved in G. Grätzer [247], J. Hashimoto and S. Kinugawa [376], and Iqbalunnisa [415].

*Proof.*

(i) *implies* (ii). Let $a \in \operatorname{Neutr} L$. Then

(1) $$a \vee (x \wedge y) = x \wedge (a \vee y) \quad \text{for } x \geq a.$$

Indeed,

$$a \vee (x \wedge y) = (a \wedge x) \vee (x \wedge y) \vee (y \wedge a)$$

(by Definition 251(iii))

$$= (a \vee x) \wedge (x \vee y) \wedge (y \vee a) = x \wedge (a \vee y).$$

To show that $a \in \operatorname{Distr} L$, compute

$$a \vee (x \wedge y) = a \vee (a \wedge x) \vee (x \wedge y) \vee (y \wedge a)$$

(by Definition 251(iii))

$$= a \vee ((a \vee x) \wedge (x \vee y) \wedge (y \vee a))$$

(apply (1) to $a$, $a \vee x$, and $(x \vee y) \wedge (y \vee a)$)

$$= (a \vee x) \wedge (a \vee ((x \vee y) \wedge (y \vee a)))$$

(apply (1) to $a$, $y \vee a$, and $x \vee y$)

$$= (a \vee x) \wedge (y \vee a) \wedge (a \vee x \vee y) = (a \vee x) \wedge (a \vee y),$$

as claimed. By duality, we get that $a \in \operatorname{Distr}^\delta L$.

Finally, let $a \vee x = a \vee y$ and $a \wedge x = a \wedge y$. Then

$$\begin{aligned}
x &= x \wedge (a \vee x) \wedge (a \vee y) \wedge (x \vee y) \\
&= x \wedge ((a \wedge x) \vee (x \wedge y) \vee (a \wedge y)) \\
&= x \wedge ((a \wedge x) \vee (x \wedge y)) \\
&= (a \wedge x) \vee (x \wedge y) \\
&= (a \wedge x) \vee (a \wedge y) \vee (x \wedge y).
\end{aligned}$$

Since the right-hand side is symmetric in $x$ and $y$, we conclude that $x = y$.

(ii) *implies* (iii). Let (ii) hold for $a$ and define $A = \mathrm{id}(a)$ and $B = \mathrm{fil}(a)$. Let

$$\varphi \colon x \mapsto (x \wedge a, x \vee a).$$

Since $a \in \mathrm{Distr}\, L$ and $a \in \mathrm{Distr}^\delta L$, by Theorem 252(ii) and its dual, the map $\varphi$ is a homomorphism of $L$ into $A \times B$. The map $\varphi$ is one-to-one, since if $\varphi(x) = \varphi(y)$ for $x, y \in L$, then

$$(x \wedge a, x \vee a) = (y \wedge a, y \vee a),$$

that is,

$$x \wedge a = y \wedge a,$$
$$x \vee a = y \vee a;$$

thus $x = y$ by (ii). So $\varphi$ is an embedding, $\varphi(a) = (a, a)$, and $a$ is the unit element of $A$ and the zero of $B$.

(iii) *implies* (iv). The following three statements are obvious:

(iv) holds for the zero and the unit in any lattice.

(iv) holds for $(a_0, a_1)$ in $A_0 \times A_1$ iff it holds for $a_0$ in $A_0$ and $a_1$ in $A_1$.

Let $a \in L_0 \leq L_1$ and let (iv) hold for $a$ in $L_1$; then (iv) holds for $a$ in $L_0$.

(iii) and these three statements prove (iv).

(iv) *implies* (i). Obvious by Exercise I.4.7.    □

## 2.5    Connections

The results stated in Theorems 252–254 make it possible to verify the most important properties of distributive, standard, and neutral elements.

**Theorem 255.** *Let $L$ be a lattice.*

(i) $\mathrm{Neutr}\, L \subseteq \mathrm{Stand}\, L$.

(ii) $\mathrm{Stand}\, L \subseteq \mathrm{Distr}\, L$.

(iii) *If $a \in \mathrm{Stand}\, L$ and $a \in \mathrm{Distr}^\delta L$, then $a \in \mathrm{Neutr}\, L$.*

(iv) *If $a \in \mathrm{Distr}\, L$ or $a \in \mathrm{Stand}\, L$, then the relation $\alpha_a$ defined in Theorem 252(iii) and Theorem 253(ii), respectively, agrees with $\mathrm{con}(\mathrm{fil}(a))$.*

*Proof.*

(i) and (iii). By Theorem 254(ii) and Theorem 253(iii).

(ii) By Theorem 253(iii).

(iv) Let $a$ be a distributive element and let $\boldsymbol{\alpha}$ be a congruence of $L$. If $u \equiv a \pmod{\boldsymbol{\alpha}}$, for every $u \le a$, then $a \vee x = a \vee y$ implies that

$$x = x \vee (a \wedge x) \equiv x \vee a = y \vee a \equiv y \vee (a \wedge y) = y \pmod{\boldsymbol{\alpha}},$$

hence $\boldsymbol{\alpha}_a \le \boldsymbol{\alpha}$ for the relation $\boldsymbol{\alpha}_a$ given by Theorem 252(iii). Similarly, if $(x \wedge y) \vee a_1 = x \vee y$, for some $a_1 \le a$, then

$$x \wedge y = (x \wedge y) \vee (a \wedge x \wedge y) \equiv (x \wedge y) \vee a_1 = x \vee y \pmod{\boldsymbol{\alpha}},$$

and so $x \equiv y \pmod{\boldsymbol{\alpha}}$; hence, $\boldsymbol{\alpha}_a \le \boldsymbol{\alpha}$ for the relation $\boldsymbol{\alpha}_a$ of Theorem 253(ii). $\square$

For a principal ideal $\mathrm{id}(a) = I$, let $\mathrm{con}(I)$ be denoted by $\boldsymbol{\alpha}_a$. Then by Theorem 255(iv), the relation $\boldsymbol{\alpha}_a$ defined in Theorem 252(iii) and the relation $\boldsymbol{\alpha}_a$ of Theorem 253(ii) are indeed the $\boldsymbol{\alpha}_a$ just defined in case $a$ is distributive or standard. Hence Theorem 252(iii) and Theorem 253(ii) are definitions of distributive and standard elements, respectively, in terms of the properties of $\boldsymbol{\alpha}_a$.

As we have already seen, the converse of Theorem 255(i), as well as that of Theorem 255(ii), is false. Two wide classes of lattices in which the converse holds are the class of modular lattices and the class of relatively complemented lattices. In fact, it also holds in their common generalization introduced in Definition 248, see G. Grätzer and E. T. Schmidt [336].

**Theorem 256.** *Let $L$ be a weakly modular lattice. Then* $\mathrm{Distr}\, L = \mathrm{Neutr}\, L$.

Before proving the theorem, we verify a lemma connecting the distributivity of an element with congruence-projectivity.

**Lemma 257.** *Let $L$ be a lattice and let $a$ be an element of $L$. Then $a \in \mathrm{Distr}\, L$ iff $u \le z \le a \le x \le y$ and $[u, z] \Rightarrow [x, y]$ imply that $x = y$.*

*Proof.* Let $a \in \mathrm{Distr}\, L$, let $u \le z \le a \le x \le y$, and let $[u, z] \Rightarrow [x, y]$. Since $u \le z \le a$, we get that $u \equiv z \pmod{\boldsymbol{\alpha}_a}$, hence $x \equiv y \pmod{\boldsymbol{\alpha}_a}$. By Theorem 252(iii), we conclude that $x = x \vee a = y \vee a = y$.

Let $a \notin \mathrm{Distr}\, L$. Then there are elements $b, c \in L$ such that

$$a \vee (b \wedge c) < (a \vee b) \wedge (a \vee c).$$

Set $\boldsymbol{\beta}_a = \mathrm{con}(\mathrm{id}(a))$. From $b \equiv a \vee b \pmod{\boldsymbol{\beta}_a}$ and $c \equiv a \vee c \pmod{\boldsymbol{\beta}_a}$, it follows that $b \wedge c \equiv (a \vee b) \wedge (a \vee c) \pmod{\boldsymbol{\beta}_a}$, thus

$$a \vee (b \wedge c) \equiv (a \vee b) \wedge (a \vee c) \pmod{\boldsymbol{\beta}_a},$$

so by Corollary 235, there exist $x, y, u \in L$ satisfying

$$u < z \le a \le (a \vee b) \wedge (a \vee c) \le x < y \le a \vee (b \wedge c),$$

with $z = a$, and so $[u, z] \Rightarrow [x, y]$ with $x < y$. $\square$

*Proof of Theorem 256.* Let $L$ be a weakly modular lattice and let $a \in \operatorname{Distr} L$. If $a \notin \operatorname{Distr}^\delta L$, then by Lemma 257, there exist $u < z \le a \le x < y$ satisfying $[u, z] \Rightarrow [x, y]$. By weak modularity, there exists a proper subinterval $[u_1, z_1]$ of $[u, z]$ satisfying $[x, y] \Rightarrow [u_1, z_1]$, contradicting that $a$ is distributive, by Lemma 257.

Now let

$$a \vee x = a \vee y,$$
$$a \wedge x = a \wedge y,$$

and $x \ne y$ for some $x, y \in L$. Proceeding just as in the step "(ii) implies (iii)" in the proof of Theorem 253, we conclude that

$$x \equiv y \pmod{\operatorname{con}(a, a \wedge x)},$$

and dually

$$x \equiv y \pmod{\operatorname{con}(a, a \vee x)}.$$

Applying Theorem 230 to

$$x \equiv y \pmod{\operatorname{con}(a, a \wedge x)},$$

we get a proper subinterval $[x_1, y_1]$ of $[x \wedge y, x \vee y]$ satisfying $[a \wedge x, a] \Rightarrow [x_1, y_1]$. Since

$$x \equiv y \pmod{\operatorname{con}(a, a \vee x)},$$

we also have

$$x_1 \equiv y_1 \pmod{\operatorname{con}(a, a \vee x)},$$

and again, by Theorem 230, we get a proper subinterval $[x_2, y_2]$ of $[x_1, y_1]$ such that $[a, a \vee x] \Rightarrow [x_2, y_2]$. Using the weak modularity of $L$, we obtain a proper subinterval $[x_3, y_3]$ of $[a, a \vee x]$ satisfying $[x_2, y_2] \Rightarrow [x_3, y_3]$. Hence

$$[a \wedge x, a] \Rightarrow [x_1, y_1] \Rightarrow [x_2, y_2] \Rightarrow [x_3, y_3],$$

contradicting Lemma 257. By Theorem 254(ii), $a \in \operatorname{Neutr} L$.    □

**Corollary 258.** *Let $L$ be a weakly modular lattice. Then* $\operatorname{Stand} L = \operatorname{Neutr} L$.

## 2.6    The set of distributive, standard, and neutral elements

**Theorem 259.** *Let $L$ be a lattice. Then*

(i) *$\operatorname{Distr} L$ is a join-subsemilattice of $L$.*
(ii) *$\operatorname{Stand} L$ is a sublattice of $L$.*
(iii) *$\operatorname{Neutr} L$ is a sublattice of $L$.*

*Proof.*

(i) Let $a, b \in \operatorname{Distr} L$ and compute

$$(a \vee b) \vee (x \wedge y) = a \vee b \vee (x \wedge y)$$
$$= a \vee ((b \vee x) \wedge (b \vee y))$$
$$= (a \vee b \vee x) \wedge (a \vee b \vee y),$$

so $a \vee b \in \operatorname{Distr} L$.

(ii) Let $a, b \in \operatorname{Stand} L$. First we do the join:

$$x \wedge (a \vee b \vee y) = (x \wedge a) \vee (x \wedge (b \vee y))$$
$$= (x \wedge a) \vee (x \wedge b) \vee (x \wedge y)$$
$$= (x \wedge (a \vee b)) \vee (x \wedge y),$$

proving that $a \vee b \in \operatorname{Stand} L$. Now we verify the formula

$$\boldsymbol{\alpha}_a \wedge \boldsymbol{\alpha}_b = \boldsymbol{\alpha}_{a \wedge b},$$

where $\boldsymbol{\alpha}_a$, $\boldsymbol{\alpha}_b$, and $\boldsymbol{\alpha}_{a \wedge b}$ are the relations described by Theorem 253(ii). Since $\boldsymbol{\alpha}_a \wedge \boldsymbol{\alpha}_b \geq \boldsymbol{\alpha}_{a \wedge b}$ is trivial, let $x \equiv y \pmod{\boldsymbol{\alpha}_a \wedge \boldsymbol{\alpha}_b}$. Then $x \equiv y \pmod{\boldsymbol{\alpha}_a}$, and so $(x \wedge y) \vee a_1 = x \vee y$ for some $a_1 \leq a$. We also have $x \equiv y \pmod{\boldsymbol{\alpha}_b}$, and therefore

$$a_1 = a_1 \wedge (x \vee y) \equiv a_1 \wedge x \wedge y \pmod{\boldsymbol{\alpha}_b}.$$

Thus $a_1 = (a_1 \wedge x \wedge y) \vee b_1$ for some $b_1 \leq b$. Now

$$(x \wedge y) \vee b_1 = (x \wedge y) \vee (a_1 \wedge x \wedge y) \vee b_1 = (x \wedge y) \vee a_1 = x \vee y;$$

since $b_1 \leq b$ and $b_1 \leq a_1 \leq a$, we obtain that $b_1 \leq a \wedge b$; these verify that $x \equiv y \pmod{\boldsymbol{\alpha}_{a \wedge b}}$.

This formula shows that if $a, b \in \operatorname{Stand} L$, then the relation $\boldsymbol{\alpha}_{a \wedge b}$ of Theorem 253(ii) is a congruence relation, hence $a \wedge b \in \operatorname{Stand} L$ by Theorem 253.

(iii) Let $a, b \in \operatorname{Neutr} L$. By Theorem 255, $a, b \in \operatorname{Stand} L$. Hence by (ii), $a \wedge b \in \operatorname{Stand} L$. By Theorems 253 and 254, in order to show $a \wedge b \in \operatorname{Neutr} L$, we have to prove only that $a \wedge b$ is dually distributive. Since $a$ and $b$ are dually distributive, they are distributive elements of $L^\delta$, hence by (i), $a \vee b$ is distributive in $L^\delta$ and so $a \wedge b$ in $L$ is dually distributive. $\qquad \square$

Figure 51 shows that $a, b \in \operatorname{Distr} L$ does not imply that $a \wedge b \in \operatorname{Distr} L$.

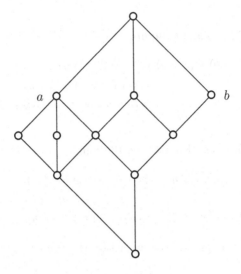

Figure 51. Distr $L$ is not a sublattice

## Exercises

2.1. Let $L$ be a bounded lattice. Show that the elements 0 and 1 are distributive, standard, and neutral.

2.2. Let $L$ be the lattice of Figure 52. Then id$(a)$ is the kernel of some congruence relation but $a$ is not distributive.

2.3. If $a$ is a distributive element of a lattice $L$, then $L/\alpha_a \cong$ fil$(a)$. To what extent does $L/\alpha_a \cong$ fil$(a)$ characterize the distributivity of $a$?

2.4. Find distributive elements that are not dually distributive.

2.5. Investigate the relation $\alpha_a$ of Theorem 252(iii) as a congruence relation of the join-semilattice.

2.6. Let $L$ be a lattice, let $\varphi$ be a homomorphism of $L$ onto $L'$, and let $L_1 \le L'$. Let $a$ be an element of $L$ satisfying $\varphi(a) \in L_1$. Show that if $a$ is a distributive in $L$, then $\varphi(a)$ is distributive in $L_1$.

2.7. Prove the analogues of Exercise 2.6 for standard and neutral elements.

2.8. Let $a$ be a distributive element of $L$. Show that id$(a)$ is a distributive element of Id $L$.

2.9. Prove the analogues of Exercise 2.8 for standard and neutral elements.

2.10. Show that in Theorem 253(iii) (and in Theorem 254(ii)) the condition can be weakened by assuming that $x \le y$.

2.11. Verify that the element $a$ of a lattice $L$ is standard iff $x \le a \vee y$ implies that $x = (x \wedge a) \vee (x \wedge y)$ (G. Grätzer and E. T. Schmidt [336]).

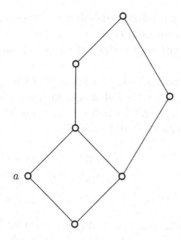

Figure 52. A congruence kernel that is not distributive

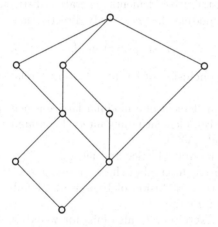

Figure 53. Illustrating Exercise 2.22

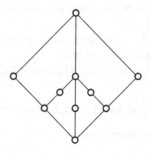

Figure 54. An example for Exercise 2.23

2.12. Prove that the join of two distributive elements is again distributive by verifying the formula $\alpha_a \vee \alpha_b = \alpha_{a \vee b}$.

2.13. Find classes of lattices in which the meet of two distributive elements is distributive.

2.14. Show that the map $a \mapsto \alpha_a$ for standard elements is an embedding of the sublattice of standard elements into the congruence lattice.

2.15. Let $L$ be a lattice, and let $a, b, c \in L$. Show that the elements $a, b, c$ generate a distributive sublattice iff

$$x \wedge (y \vee z) = (x \wedge y) \vee (x \wedge z),$$
$$x \vee (y \wedge z) = (x \vee y) \wedge (x \vee z),$$
$$(x \wedge y) \vee (y \wedge z) \vee (z \wedge x) = (x \vee y) \wedge (y \vee z) \wedge (z \vee x)$$

for every permutation $x, y, z$ of $a, b, c$ (O. Ore [561]).

2.16. Let $L$ be a lattice and let $a, b$ be standard elements of $L$. Show that the elements $a, b, c$ generate a distributive sublattice for every $c \in L$.

2.17. Do three distributive elements generate a distributive sublattice?

2.18. Let $L$ be a modular lattice. Verify directly that

$$\operatorname{Distr} L = \operatorname{Stand} L = \operatorname{Neutr} L.$$

2.19. Verify the conclusion of Exercise 2.18 in a relatively complemented lattice.

2.20. Show that an element of a modular lattice is neutral iff it has at most one relative complement in any interval containing it (G. Grätzer and E. T. Schmidt [336]).

2.21. Let $L$ be a modular lattice and let $a \in L$. Then $a$ is neutral iff $a$ is neutral in every interval of the form $[a \wedge x, a \vee x]$ for all $x \in L$.

2.22. Show that the conclusion of Exercise 2.21 fails in weakly modular lattices. (See Figure 53.)

2.23. Show that Exercise 2.21 also fails for weakly modular lattices and standard elements. (See Figure 54.)

2.24. Let $L$ be a bounded relatively complemented lattice. Then an element $a$ of $L$ is neutral iff it has a unique complement (J. Hashimoto and S. Kinugawa [376]).

2.25. Show that $\operatorname{Neutr} L$ is the intersection of the maximal distributive sublattices of $L$ (G. Birkhoff [66]).

2.26. Prove Theorem 259(iii) using Theorem 254(iii).

# 3.    Distributive, Standard, and Neutral Ideals

### 3.1    Defining the three ideal types

The three types of ideals mentioned in the title of this section derive naturally from the concepts introduced in Section 2.1.

**Definition 260.** An ideal $I$ of a lattice $L$ is called *distributive, standard* or *neutral,* respectively, if $I$ is distributive, standard, or neutral, respectively, as an element of Id $L$, the lattice of all ideals of $L$.

To establish a connection between the type of elements and the type of ideals they generate, we need a lemma:

**Lemma 261.** *Let $L$ be a lattice, let $I$ be an ideal of $L$, and let $p$ and $q$ be $n$-ary terms. Let us assume that, for all $a \in I$, there is an element $b \in I$ such that $a \leq b$ and*

$$p(a, c_1, \ldots, c_{n-1}) = q(b, c_1, \ldots, c_{n-1})$$

*for all $c_1, \ldots, c_{n-1} \in L$. Then*

$$p(I, J_1, \ldots, J_{n-1}) = q(I, J_1, \ldots, J_{n-1})$$

*holds in* Id $L$ *for all $J_1, \ldots, J_{n-1} \in$* Id $L$.

*Proof.* In Section I.4.2 (in the proof of Lemma 59), we proved the formula:

$$p(I_0, \ldots, I_{n-1}) = \{x \mid x \leq p(i_0, \ldots, i_{n-1})$$
$$\text{for some } i_0 \in I_0, \ldots, i_{n-1} \in I_{n-1}\}.$$

Thus we obtain:

$$
\begin{aligned}
p(I, &J_1, \ldots, J_{n-1}) \\
&= \{x \mid x \leq p(a, j_1, \ldots, j_{n-1}) \\
&\qquad \text{for some } a \in I, j_1 \in J_1, \ldots, j_{n-1} \in J_{n-1}\} \\
&= \{x \mid x \leq p(b, j_1, \ldots, j_{n-1}) \\
&\qquad \text{where } b \in I, j_1 \in J_1, \ldots, j_{n-1} \in J_{n-1}, \text{ and } b \text{ satisfies} \\
&\qquad p(a, c_1, \ldots, c_{n-1}) = q(b, c_1, \ldots, c_{n-1}) \\
&\qquad \text{for some } a \in I \text{ and for all } c_1, \ldots, c_{n-1} \in L\} \\
&= \{x \mid x \leq q(b, j_1, \ldots, j_{n-1}) \\
&\qquad \text{for some } b \in I, j_i \in J_1, \ldots, j_{n-1} \in J_{n-1}\} \\
&= q(I, J_1, \ldots, J_{n-1}). \qquad\qquad\qquad \square
\end{aligned}
$$

**Corollary 262.** *Let $L$ be a lattice, and let $a \in L$. Then $a$ is distributive, standard, or neutral, respectively, iff* id$(a)$, *as an ideal, is distributive, standard, or neutral, respectively.*

## 3.2   Characterization theorems

The main characterization theorems can be proved similarly to the proofs of Theorems 252–254.

**Theorem 263.** *Let $L$ be a lattice, and let $I$ be an ideal of $L$. The following conditions on $I$ are equivalent:*

(i) *$I$ is distributive.*

(ii) *The congruence $\mathrm{con}(I)$ on $L$ can be described as follows:*

$$x \equiv y \pmod{\mathrm{con}(I)} \quad \textit{iff} \quad x \vee i = y \vee i \quad \textit{for some } i \in I.$$

*Proof.* If in (ii), we put $\mathrm{id}(x) \vee I = \mathrm{id}(y) \vee I$ in place of $x \vee i = y \vee i$, then the equivalence can be proved as in Section 2 since

$$x \equiv y \pmod{\mathrm{con}(I)}$$

in $L$ iff

$$\mathrm{id}(x) \equiv \mathrm{id}(y) \pmod{\alpha_I}$$

in $\mathrm{Id}\, L$. ($\alpha_I$ is the binary relation defined on $\mathrm{Id}\, L$ as in Theorem 253.) Now, if $x \vee i = y \vee i$, for some $i \in I$, then $\mathrm{id}(x) \vee I = \mathrm{id}(y) \vee I$ is obvious. Conversely, if $\mathrm{id}(x) \vee I = \mathrm{id}(y) \vee I$, then $x \leq y \vee i_0$ and $y \leq x \vee i_1$ for some $i_0, i_1 \in I$. Therefore, $x \vee i = y \vee i$ with $i_0 \vee i_1 = i \in I$. $\qquad\square$

**Theorem 264.** *Let $L$ be a lattice, and let $I$ be an ideal of $L$. The following conditions on $I$ are equivalent:*

(i) *$I$ is standard.*

(ii) *The equality*

$$\mathrm{id}(a) \wedge (I \vee \mathrm{id}(b)) = (\mathrm{id}(a) \wedge I) \vee \mathrm{id}(a \wedge b)$$

*holds for all $a, b \in L$.*

(iii) *For any ideal $J$ of $L$,*

$$I \vee J = \{\, i \vee j \mid i \in I \text{ and } j \in J \,\}.$$

(iv) *$\mathrm{con}(I)$ can be described by*

$$x \equiv y \pmod{\mathrm{con}(I)} \qquad \textit{iff} \qquad (x \wedge y) \vee i = x \vee y \quad \textit{for some } i \in I.$$

(v) *$I$ is a distributive ideal and let*

$$I \vee J = I \vee K \text{ and } I \wedge J = I \wedge K \qquad \textit{imply that} \qquad J = K$$

*for all $J, K \in \mathrm{Id}\, L$*

*Remark.* This result, and all the other unreferenced results in this section, are based on G. Grätzer [246] and G. Grätzer and E. T. Schmidt [336].

*Proof.* The equivalence of the five conditions can be verified as in Section 2. Only (iii) is new. But (iii) follows from (ii): if $I$ satisfies (ii) and $a \in I \vee J$, then $a \leq i \vee j$ for some $i \in I$ and $j \in J$; hence

$$\operatorname{id}(a) = \operatorname{id}(a) \wedge (I \vee \operatorname{id}(j)) = (\operatorname{id}(a) \wedge I) \vee \operatorname{id}(a \wedge j)$$

by (ii). Therefore, $a \leq i_1 \vee j_1$, where $i_1 \in \operatorname{id}(a) \wedge I$ and $j_1 \leq a \wedge j$. Consequently, $a = i_1 \vee j_1$. Finally, observe that when proving the analogue of "(i) implies (iv)", it is sufficient to use (iii). □

**Theorem 265.** *Let $L$ be a lattice, and let $I$ be an ideal of $L$. The following conditions on $I$ are equivalent:*

(i) *$I$ is neutral.*

(ii) *For all $j, k \in L$,*

$$(I \wedge \operatorname{id}(j)) \vee (\operatorname{id}(j) \wedge \operatorname{id}(k)) \vee (\operatorname{id}(k) \wedge I)$$
$$= (I \vee \operatorname{id}(j)) \wedge (\operatorname{id}(j) \vee \operatorname{id}(k)) \wedge (\operatorname{id}(k) \vee I).$$

(iii) *For all $J, K \in \operatorname{Id} L$, the ideals $I, J, K$ generate a distributive sublattice of $\operatorname{Id} L$.*

(iv) *$I$ is distributive and dually distributive, and*

$$I \vee J = I \vee K \text{ and } I \wedge J = I \wedge K \quad \text{imply that} \quad J = K$$

*for all $J, K \in \operatorname{Id} L$.*

*Proof.* We can verify that (i) is equivalent to (ii) by using the argument of Lemma 261. The rest of the proof is the same as in Section 2. □

Observe that every distributive ideal $I$ of a lattice $L$ is the kernel of the congruence $\operatorname{con}(I)$. Indeed, if $i \in I$, $a \in L$, and $i \equiv a \pmod{\operatorname{con}(I)}$, then $i \vee j = a \vee j$, for some $j \in I$, thus $a \leq i \vee j \in I$, and so $a \in I$. This explains why $L/\operatorname{con}(I)$ can always be described. This description is especially effective if $I$ is principal, $I = \operatorname{id}(a)$. Then $\operatorname{con}(I) = \boldsymbol{\alpha}_a$ is a representable congruence relation and $\operatorname{fil}(a)$ represents $\boldsymbol{\alpha}_a$, hence

$$L/\boldsymbol{\alpha}_a \cong \operatorname{fil}(a).$$

If $I$ is not principal, then we describe $\operatorname{Id}(L/\operatorname{con}(I))$ as follows:

**Theorem 266.** *Let $I$ be a distributive ideal of a lattice $L$. Then $\operatorname{Id}(L/\operatorname{con}(I))$ is isomorphic with the lattice of all ideals of $L$ containing $I$, that is, with the interval $[I, L]$ in $\operatorname{Id} L$.*

*Proof.* Let $\alpha = \mathrm{con}(I)$. Let $\varphi$ be the homomorphism $x \mapsto x/\alpha$ of $L$ onto $L/\alpha$. Then the map $\psi \colon K \to \varphi^{-1}(K)$ maps $\mathrm{Id}\,(L/\alpha)$ into $[I, L]$. To show that this map is onto, it is sufficient to see that $J/\alpha = J$ for all $J \supseteq I$. Indeed, if $j \in J$, $a \in L$, and $j \equiv a \pmod{\mathrm{con}(I)}$, then $j \vee i = a \vee i$, for some $i \in I$, and so $a \leq i \vee j \in J$ and $a \in J$, as claimed. Since $\psi$ is obviously isotone and one-to-one, we conclude that it is an isomorphism.     $\square$

**Corollary 267.** $L/\mathrm{con}(I)$ *is determined by the interval* $[I, L]$ *of* $\mathrm{Id}\,L$, *provided that $I$ is a distributive ideal of $L$.*

*Proof.* We know from Section II.2 that $\mathrm{Id}\,(L/\mathrm{con}(I))$ determines $L/\mathrm{con}(I)$ up to isomorphism.     $\square$

### 3.3   The associated congruences

Let us call a congruence relation $\alpha$ *distributive*, *standard*, *neutral*, respectively, if $\alpha = \mathrm{con}(I)$, where $I$ is a distributive, standard, or neutral ideal, respectively. The following result was first established by J. Hashimoto [375] for neutral congruences, and by G. Grätzer and E. T. Schmidt [336] in its present form.

The congruences $\alpha$ and $\beta$ of a lattice $L$ *permute* (or are *permutable*) if $\alpha \vee \beta = \alpha \circ \beta$. An equivalent definition is that $\alpha$ and $\beta$ permute if $\alpha \circ \beta = \beta \circ \alpha$, which, in turn, is equivalent to $\alpha \circ \beta$ being a congruence relation.

**Theorem 268.** *Any two standard congruences of a lattice permute.*

*Proof.* Let $\alpha$ and $\beta$ be standard congruences of the lattice $L$, that is, $\alpha = \mathrm{con}(I)$ and $\beta = \mathrm{con}(J)$, where $I$ and $J$ are standard ideals of $L$. Let

$$a \equiv b \pmod{\alpha} \quad \text{and} \quad b \equiv c \pmod{\beta}, \quad a, b, c \in L.$$

We want to show that there exists a $d \in L$ satisfying

$$a \equiv d \pmod{\beta} \quad \text{and} \quad d \equiv c \pmod{\alpha}.$$

If

$$x \leq y \leq z, \quad x \equiv y \pmod{\alpha}, \quad \text{and } y \equiv z \pmod{\beta},$$

then $y = x \vee i$, for some $i \in I$, and $z = y \vee j$ for some $j \in J$. Then $z = u \vee i$ with $u = x \vee j$, and hence $x \equiv u \pmod{\beta}$ and $u \equiv z \pmod{\alpha}$.

Applying this observation to

$$a \equiv a \vee b \pmod{\alpha},$$
$$a \vee b \equiv a \vee b \vee c \pmod{\beta},$$

and to

$$c \equiv b \vee c \pmod{\beta},$$
$$b \vee c \equiv a \vee b \vee c \pmod{\alpha},$$

we obtain an element $e$ with $a \le e \le a \vee b \vee c$ and an element $f$ with $c \le f \le a \vee b \vee c$ satisfying

$$a \equiv e \pmod{\beta},$$
$$e \equiv a \vee b \vee c \pmod{\alpha},$$
$$c \equiv f \pmod{\alpha},$$
$$f \equiv a \vee b \vee c \pmod{\beta}.$$

Set $d = e \wedge f$. Then

$$a \equiv e = e \wedge (a \vee b \vee c) \equiv e \wedge f = d \pmod{\beta},$$
$$c \equiv f = f \wedge (a \vee b \vee c) \equiv f \wedge e = d \pmod{\alpha}. \qquad \square$$

Theorem 268 is significant because in a wide class of lattices all congruences are standard. This class will be introduced in Theorem 272. To state and prove Theorem 272, we need some definitions and lemmas.

**Definition 269.** Let $x$ and $y$ be elements in a lattice $L$ with zero and let $I$ be an ideal of $L$. Then

(i) The element $x$ is *perspective to the element* $y$, in symbols, $x \sim y$, if there exists an element $z \in L$ such that $x \wedge z = y \wedge z = 0$ and $x \vee z = y \vee z$.

(ii) The ideal $I$ is *perspectivity closed* if $x \sim y$ and $y \in I$ imply that $x \in I$.

(iii) The element $x$ is *subperspective to the element* $y$, in notation, $x \lesssim y$, if there exists an element $z \in L$ such that $x \wedge z = y \wedge z = 0$ and $x \le y \vee z$.

(iv) The ideal $I$ is *subperspectivity closed* if $x \lesssim y$ and $y \in I$ imply that $x \in I$.

See Exercises 3.21–3.23, for alternative definitions of the relations $\sim$ and $\lesssim$.

Perspectivity of elements and perspectivity of intervals are closely related. Indeed, if $x \sim y$, then $[0, x] \overset{\text{up}}{\sim} [z, x \vee y] \overset{\text{dn}}{\sim} [0, y]$. So $x \sim y$ is the easiest way to guarantee that $[0, x] \Rightarrow [0, y]$.

The material from here to the end of this section is due to F. Wehrung.

**Definition 270.**

(a) In a lattice $L$ with zero, define a binary relation $\le^{\oplus}$ as follows:

For $x, y \in L$, let $x \le^{\oplus} y$ if $x \le y$ and $x$ has a relative complement in $[0, y]$.

(b) The lattice $L$ is *sectionally decomposing* if for all $x \le y$ in $L$, there is a natural number $n$ and elements $z_0, z_1, \ldots, z_n \in L$ such that the relations

$$x = z_0 \le^{\oplus} z_1 \le^{\oplus} \cdots \le^{\oplus} z_n = y$$

hold in $L$.

**Lemma 271.** *Let $L$ be a lattice with zero. If one of the following two conditions hold:*

(i) *$L$ is sectionally complemented;*
(ii) *$L$ is atomistic and every closed interval of $L$ has a finite maximal chain,*

*then $L$ is sectionally decomposing.*

*Proof.* The conclusion is trivial if condition (i) holds; take $n = 1$. Now let condition (ii) hold and let $x \leq y$ in $L$. By (ii), there is a natural number $n$ and there is a maximal chain $x = z_0 \prec z_1 \prec \cdots \prec z_n = y$. Since $L$ is atomistic, there exists, for each $i < n$, an atom $p_i \leq z_{i+1}$ such that $p_i \nleq z_i$. Thus $p_i \wedge z_i = 0$ and $p_i \vee z_i = p_{i+1}$. It follows that $z_0 = x, z_n = y$, and $z_i \leq^\oplus z_{i+1}$ for all $i < n$. $\qquad\square$

The following result provides a convenient way to verify that an ideal is standard in a sectionally complemented lattice or in an atomistic lattice of finite length.

**Theorem 272.** *Let $L$ be a sectionally decomposing lattice. Then every congruence of $L$ is standard. An ideal $I$ of $L$ is standard iff it is subperspectivity closed.*

*Proof.* Let $\boldsymbol{\alpha}$ be a congruence of $L$. Then $I = 0/\boldsymbol{\alpha}$ is an ideal of $L$. If $x \lesssim y$ and $y \in I$, then choose $z \in L$ such that $x \wedge z = 0$ and $x \leq y \vee z$. Clearly, $x \leq z$ and $x \equiv z \pmod{\boldsymbol{\alpha}}$. Since $x \wedge z = 0$, it follows that $x \equiv 0 \pmod{\boldsymbol{\alpha}}$, and so $x \in I$, proving that $0/\boldsymbol{\alpha}$ is subperspectivity closed. This applies, in particular, to $\mathrm{con}(I)$ for a standard ideal $I$ of $L$, thus every standard ideal is subperspectivity closed.

Conversely, let the ideal $I$ be subperspectivity closed. To prove that $I$ is standard, we use the characterization provided by Theorem 264(ii).

Let $a, b \in L$ and let $x \in I$; we must prove that $a \wedge (b \vee x)$ belongs to $\mathrm{id}(a \wedge b) \vee (\mathrm{id}(a) \wedge I)$. Replacing $x$ by a sectional complement of $b \wedge x$ in $[0, x]$ affects neither the value of $b \vee x$ nor the validity of the statement $x \in I$, so we may assume that $b \wedge x = 0$. Since $L$ is sectionally decomposing, there is a natural number $n$ and there are elements such that

$$a \wedge b = c_0 \leq^\oplus c_1 \leq^\oplus \cdots \leq^\oplus c_n = a \wedge (b \vee x).$$

For each $i < n$, select a sectional complement $t_i$ of $c_i$ in $[0, c_{i+1}]$. Observe that $t_i \leq a \wedge (b \vee x) \leq a$. Since $c_i \wedge t_i = 0$ and $t_i \leq a$, it follows that

$$b \wedge t_i = a \wedge b \wedge t_i \leq c_i \wedge t_i = 0 = b \wedge x.$$

Also $t_i \leq b \vee x$, so it follows that $t_i \lesssim x$. Since $x \in I$, we obtain that $t_i \in I$. Therefore, the join $t$ of all $t_i$ also belongs to $\mathrm{id}(a) \wedge I$. Finally, $a \wedge (b \vee x) = (a \wedge b) \vee t$ and therefore, $a \wedge (b \vee x) \in \mathrm{id}(a \wedge b) \vee (\mathrm{id}(a) \wedge I)$.

Now let $\alpha$ be a congruence of $L$. We have already proved that $I = 0/\alpha$ is a standard ideal. Let $x \leq y$ in $L$ satisfy that $x \equiv y \pmod{\alpha}$. Since $L$ is sectionally decomposing, there is a natural number $n$ and there is a decomposition

$$x = z_0 \leq^{\oplus} z_1 \leq^{\oplus} \cdots \leq^{\oplus} z_n = y.$$

For each $i < n$, select a sectional complement $t_i$ of $z_i$ in $[0, z_{i+1}]$. Since $z_i \equiv z_{i+1} \pmod{\alpha}$, for all $i < n$, it follows that all $t_i \in I$, thus the join $t$ of all $t_i$-s also belongs to $I$. In view of $y = x \vee t$, we obtain that $x \equiv y \pmod{\alpha}$. Hence $\alpha = \mathrm{con}(I)$ is a standard congruence and $I$ is a standard ideal. $\qquad\square$

With the help of Lemma 271, we now get the following two statements.

**Corollary 273.** *Let $L$ be a lattice that is either sectionally complemented or atomistic with every interval containing a finite maximal chain. Then every congruence of $L$ is standard, and an ideal of $L$ is distributive iff it is standard iff it is subperspectivity closed.*

**Corollary 274.** *Let $L$ be a weakly modular, sectionally complemented lattice. If $L$ satisfies the Ascending Chain Condition, then all congruences of $L$ are neutral, in fact, of the form $\alpha_a$, where $a$ is a neutral element.*

*Proof.* Let $\alpha$ be a congruence relation of the lattice $L$. By Theorem 272, we obtain that $\alpha = \mathrm{con}(I)$, where $I$ is a standard ideal of $L$. By the Ascending Chain Condition, $I = \mathrm{id}(a)$. By Corollary 262, the element $a$ is standard. Since the lattice $L$ is weakly modular, by Theorem 256, the element $a$ is neutral. Hence $\alpha = \alpha_a$ with $a$ neutral. $\qquad\square$

## Exercises

3.1. Prove that an ideal generated by a set of distributive elements is distributive.

3.2. Show that an ideal generated by a set of standard elements is standard.

3.3. Does the analogue of Exercises 3.1 and 3.2 hold for neutral ideals?

3.4. Verify that the converse of Exercise 3.2 does not hold. (Hint: Consider the lattice of Figure 55.)

3.5. Prove Corollary 262 directly.

3.6. Consider the lattice $L$ as a sublattice of $\mathrm{Id}\, L$ under the natural embedding $x \mapsto \mathrm{id}(x)$. Show that every congruence relation of $L$ can be extended to $\mathrm{Id}\, L$.

3.7. Characterize those congruence relations of $\mathrm{Id}\, L$ that are extensions of congruences of $L$.

3.8. For any ideal $I$ of a lattice $L$, relate the congruence relation $\mathrm{con}(I)$ of $L$ with the congruence relation $\alpha_I$ of $\mathrm{Id}\, L$.

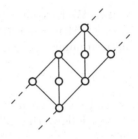

Figure 55. A lattice for Exercise 3.4

3.9. Characterize a standard ideal $I$ of $L$ in terms of the congruence relation $\alpha_I$ on $\operatorname{Id} L$.

3.10. Show that in Theorem 264 it is sufficient to assume that condition (iii) holds for principal ideals.

3.11. Construct a lattice $L$ and an ideal $I$ of $L$ such that Theorem 264(v) holds for all principal ideals $J$ and $K$, yet $I$ is not standard (Iqbalunnisa [416]).

3.12. Show that we can assume that $J \subseteq K$ in Theorem 264(v) and Theorem 265(iv).

3.13. Show that the congruence relations $\alpha$ and $\beta$ of the lattice $L$ permute iff, for all $a, b, c \in L$ with

$$a \le b \le c, \qquad a \equiv b \pmod{\alpha}, \qquad b \equiv c \pmod{\beta},$$

there exists a $d \in L$ satisfying

$$a \le d \le c, \qquad a \equiv d \pmod{\beta}, \qquad d \equiv c \pmod{\alpha}.$$

3.14. Use Exercise 3.13 to reprove Theorem 268.

3.15. Show that $I \mapsto \operatorname{con}(I)$ is an isomorphism between the lattice of standard ideals and the lattice of standard congruence relations.

3.16. Construct a lattice $L$ and standard ideals $I_j$, for $j \in J$, of $L$ such that $I = \bigwedge(I_j \mid j \in J) \ne \varnothing$ and $I$ is not a standard ideal. (Hint: Consider the lattice of Figure 56.)

3.17. Let $L$ be a lattice and let $I$ and $J$ be ideals of $L$. Let $I$ be standard and let us assume that $I \vee J$ and $I \wedge J$ are principal. Prove that $J$ is principal.

3.18. For a lattice $L$ and standard ideal $I$ of $L$, let $L/I$ denote the quotient lattice $L/\operatorname{con}(I)$. Verify the First Isomorphism Theorem for Standard Ideals: For any ideal $J$ of $L$, the ideal $I \wedge J$ is a standard ideal of $I$ and

$$I \vee J/I \cong J/I \wedge J.$$

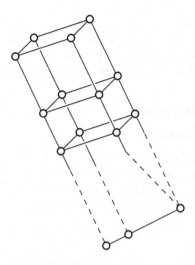

Figure 56. A lattice for Exercise 3.16

3.19. Prove the Second Isomorphism Theorem for Standard Ideals: Let $L$ be a lattice, let $I$ and $J$ be ideals of $L$ with $J \subseteq I$, and let $J$ be standard. Then $I$ is standard iff $I/J$ is standard in $L/J$, and in this case

$$L/I \cong (L/J)/(I/J).$$

3.20. State and verify the Second Isomorphism Theorem for Neutral Ideals (J. Hashimoto [375]).

3.21. Let $x$ and $y$ be elements in a sectionally complemented modular lattice $L$. Prove that the following are equivalent:

(i) $x \sim y$;

(ii) there exists an element $z \in L$ such that

$$x \wedge z = y \wedge z = 0,$$
$$x \vee z = y \vee z = x \vee y.$$

3.22. Let $x$ and $y$ be elements in a sectionally complemented lattice $L$. Prove that

(i) $x \sim y$ iff there exists $z \in L$ such that

$$x \wedge z = y \wedge z,$$
$$x \vee z = y \vee z.$$

(ii) $x \lesssim y$ iff there exists $z \in L$ such that

$$x \wedge z \le y \wedge z,$$
$$x \vee z \le y \vee z.$$

Deduce that if both $L$ and $L^\partial$ are sectionally complemented, then $x \sim y$ in $L$ iff $x \sim y$ in $L^\partial$, while $x \lesssim y$ in $L$ iff $y \lesssim x$ in $L^\partial$.

3.23. Let $L$ be a bounded relatively complemented lattice. Then the elements $x, y \in L$ are perspective iff they have a common complement.

## 4.  Structure Theorems

### 4.1  Direct decompositions

Let $A$ and $B$ be bounded lattices and $L = A \times B$. Then $a = (1, 0)$ and $b = (0, 1)$ are neutral elements, and they are complementary. Conversely, let $L$ be a bounded lattice, $a, b \in L$, let $a$ and $b$ be complementary elements, and let $a$ be neutral; then, by Theorem 254, the map

$$\varphi \colon x \mapsto (x \wedge a, x \vee a)$$

embeds $L$ into $\mathrm{id}(a) \times \mathrm{fil}(a)$. Observe that $\varphi$ is also onto: assuming that $(u, v) \in \mathrm{id}(a) \times \mathrm{fil}(a)$, we conclude that $\varphi(x) = (u, v)$ for $x = u \vee (v \wedge b)$. Indeed, the first component of $\varphi(x)$ is

$$(u \vee (v \wedge b)) \wedge a = (u \wedge a) \vee (v \wedge a \wedge b) = u,$$

since $u \leq a$, $a \wedge b = 0$, and $a$ is dually distributive. Similarly,

$$x \vee a = u \vee (v \wedge b) \vee a = u \vee ((v \vee a) \wedge (a \vee b)) = v,$$

since $a$ is distributive, $u \leq a \leq v$, and $a \vee b = 1$.

**Theorem 275.** *The direct decompositions of a bounded lattice $L$ into two factors are (up to isomorphism) in one-to-one correspondence with the complemented neutral elements of $L$.*

The proof given above is slightly artificial. Its essence will come out better if we consider lattices with zero.

Let $L$ be a lattice with zero and let $L = A \times B$. Then both $A$ and $B$ are lattices with zero, so we can set

$$I = \{\, (a, 0) \mid a \in A \,\},$$
$$J = \{\, (0, b) \mid b \in B \,\}.$$

It is easily seen that $I$ and $J$ are ideals of $L$, the map $a \mapsto (a, 0)$ is an isomorphism between $A$ and $I$, and the map $b \mapsto (0, b)$ is an isomorphism between $B$ and $J$. Moreover,

$$I \wedge J = \{0\},$$
$$I \vee J = L.$$

The last equality holds because $(a, b) = (a, 0) \vee (0, b)$. For the same reason, every ideal $K$ of $L$ has a unique representation of the form

$$K = I_1 \vee J_1,$$

where $I_1 \subseteq I$, $J_1 \subseteq J$, and $I_1, J_1 \in \operatorname{Id} L$ (with $I_1 = I \cap K$ and $J_1 = J \cap K$), and every such pair $(I_1, J_1)$ occurs (indeed, for $K = I_1 \vee J_1$). Therefore, the map

$$K \to (K \cap I, K \cap J)$$

is a direct product representation of $\operatorname{Id} L$ and so $I$ and $J$ are neutral ideals by Theorem 254.

Conversely, let $I$ and $J$ be complementary neutral ideals of $L$. Then, for any $x \in L$,

$$x \in \operatorname{id}(x) = \operatorname{id}(x) \wedge (I \vee J) = (\operatorname{id}(x) \cap I) \vee (\operatorname{id}(x) \cap J),$$

and so

$$x \leq i \vee j, \quad i \in \operatorname{id}(x) \cap I, \quad j \in \operatorname{id}(x) \cap J.$$

But $i \leq x$ and $j \leq x$, hence $x = i \vee j$. So every element $x$ of $L$ has a representation of the form

$$x = i \vee j, \quad i \in I, \, j \in J.$$

This representation is unique since if $x = i \vee j$ with $i \in I$ and $j \in J$, then

$$\operatorname{id}(x) \cap I = \operatorname{id}(i \vee j) \cap I = (\operatorname{id}(i) \vee \operatorname{id}(j)) \cap I = \operatorname{id}(i) \vee (\operatorname{id}(j) \cap I) = \operatorname{id}(i),$$

and similarly,

$$\operatorname{id}(x) \cap J = \operatorname{id}(j).$$

Since every pair $(i, j)$ occurs in some representation, we obtain that

$$x \mapsto (i, j)$$

is an isomorphism between $L$ and $I \times J$. Generalizing slightly we get the following:

**Theorem 276.** *Let $L$ be a lattice with zero. There is (up to isomorphism) a one-to-one correspondence between direct decompositions*

$$L = A_1 \times \cdots \times A_n$$

*and $n$-tuples of neutral ideals $(I_1, \ldots, I_n)$ that satisfy*

$$I_i \wedge I_j = 0, \quad \text{for } i \neq j,$$
$$I_1 \vee \cdots \vee I_n = L.$$

*Let L be a lattice with zero. Let L have two direct decompositions:*

$$L = A_1 \times \cdots \times A_n,$$
$$L = B_1 \times \cdots \times B_m.$$

*Then L has a direct decomposition*

$$L = C_{1,1} \times C_{1,2} \times \cdots \times C_{1,n}$$
$$\times C_{2,1} \times C_{2,2} \times \cdots \times C_{2,n}$$
$$\cdots$$
$$\times C_{m,1} \times C_{m,2} \times \cdots \times C_{m,n}$$

*such that*
$$A_i \cong C_{1,i} \times C_{2,i} \times \cdots \times C_{m,i}, \quad \text{for all } 1 \leq i \leq n,$$

*and*
$$B_j \cong C_{j,1} \times C_{j,2} \times \cdots \times C_{j,n}, \quad \text{for all } 1 \leq j \leq m.$$

*Proof.* Let $(I_1, I_2, \ldots, I_n)$ and $(J_1, J_1, \ldots, J_m)$ be the neutral ideals associated with the two decompositions as in Theorem 276. Then

$$(I_1 \wedge J_1, I_2 \wedge J_1, \ldots, I_n \wedge J_1, \ldots, I_n \wedge J_1, \ldots, I_n \wedge J_m)$$

is again a sequence of neutral ideals satisfying the conditions of Theorem 276. (We only need Theorem 259 to help verify this.) The direct decomposition associated with this sequence will yield the decomposition required.    □

### 4.2    Indecomposable and simple factors

A lattice $L$ is called *directly indecomposable* if $L$ has no representation in the form $L = A \times B$, where both $A$ and $B$ have more than one element.

**Corollary 277.** *Let L be a lattice with zero. If L has a representation*

$$L = A_1 \times \cdots \times A_n,$$

*where $A_i$, for each $1 \leq i \leq n$, is directly indecomposable, then for any other decomposition*

$$L = B_1 \times \cdots \times B_m$$

*of L into directly indecomposable factors, the equality $n = m$ holds and there exists a permutation $\pi$ of $\{1, \ldots, n\}$ such that $A_i \cong B_{\pi(i)}$ for all $1 \leq i \leq n$.*

Results analogous to Corollary 277 also hold for general lattices, see the Exercises.

In trying to sharpen Corollary 277 to a structure theorem, there are two difficulties we have to overcome. A lattice need not have a decomposition into directly indecomposable factors. Directly indecomposable lattices are hard to accept as "building blocks"; one would rather have *simple lattices*, that is, lattices with only the two trivial congruences, $0$ and $1$.

The first difficulty is easy to overcome by chain conditions. Observe that if $A$ and $B$ are lattices with zero, and we are given the chains $C = \{0, a_1, a_2, \ldots, a_n\} \subseteq A$ and $D = \{0, b_1, \ldots, b_m\} \subseteq B$, then

$$E = \{(0,0), (a_1, 0), \ldots, (a_n, 0), (a_n, b_1), \ldots, (a_n, b_m)\}$$

is a chain in $A \times B$ of length $n + m$. This easily implies that $\operatorname{len}(A \times B) = \operatorname{len}(A) + \operatorname{len}(B)$ for lattices $A$ and $B$ of finite length.

**Lemma 278.** *Let $L$ be a bounded lattice. If $L$ is of finite length, then $L$ is isomorphic to a direct product of directly indecomposable lattices.*

The passage from "directly indecomposable" to "simple" requires stronger hypotheses.

**Theorem 279.** *Let $L$ be a sectionally complemented, weakly modular lattice. If $L$ is of finite length, then $L$ can be represented as a direct product of simple lattices.*

*Proof.* First observe that if $L \cong A \times B$ and $L$ satisfies the hypotheses of this theorem, then so do $A$ and $B$; only the statement about weak modularity has to be verified and it follows from the observation that

$$[(a_1, a_2), (b_1, b_2)] \Rightarrow [(c_1, c_2), (d_1, d_2)] \quad \text{implies that} \quad [a_1, b_1] \Rightarrow [c_1, d_1].$$

By Lemma 278, we decompose the lattice: $L \cong L_1 \times \cdots \times L_k$, where all $L_i$ are directly indecomposable. By the observation above, all $L_i$ satisfy the hypotheses of this theorem. So it is sufficient to prove that if $L$ satisfies the hypotheses of the theorem and $L$ is directly indecomposable, then $L$ is simple. Indeed, if $\boldsymbol{\alpha}$ is a congruence relation of $L$, then $\boldsymbol{\alpha} = \operatorname{con}(I)$, where $I$ is a standard ideal by Corollary 273. By the chain condition, $I = \operatorname{id}(a)$, and $a$ is standard by Corollary 262. By Theorem 256, $a$ is neutral. Since $L$ is complemented, by Theorem 275, $\operatorname{id}(a)$ is a direct factor of $L$. But $L$ is directly indecomposable, hence $a = 0$ or $a = 1$, yielding $\boldsymbol{\alpha} = \boldsymbol{0}$ or $\boldsymbol{\alpha} = \boldsymbol{1}$, verifying that $L$ is simple. $\qquad\square$

Since modular lattices and relatively complemented lattices are special cases of weakly modular lattices, we obtain two famous structure theorems as special cases (G. Birkhoff [64] and K. Menger [529]):

**Corollary 280 (The Birkhoff-Menger Theorem).** *Let $L$ be a complemented modular lattice. If $L$ is of finite length, then $L$ is isomorphic to a direct product of simple lattices.*

We shall see in Chapter V that these simple lattices are exactly $C_2$ and the nondegenerate projective geometries of finite dimension.

The Birkhoff-Menger Theorem was generalized in R. P. Dilworth [157]:

**Corollary 281.** *Let $L$ be a relatively complemented lattice. If $L$ is of finite length, then $L$ is isomorphic to a direct product of simple lattices.*

### 4.3    Boolean congruence lattices

If $L$ is a direct product of finitely many simple lattices $L_1, \ldots, L_k$, then by Theorem 25,

$$\operatorname{Con} L \cong \operatorname{Con} L_1 \times \cdots \times \operatorname{Con} L_k \cong C_2^k = B_k,$$

and so $\operatorname{Con} L$ is a boolean lattice. This led G. Birkhoff to suggest (in G. Birkhoff [70]) the study of lattices $L$ for which $\operatorname{Con} L$ is boolean. We are going to present an approach to this problem.

Let $L$ be a lattice and let $\boldsymbol{\alpha}$ be a congruence relation of $L$. We call $\boldsymbol{\alpha}$ a *separable congruence* if, for all $a, b \in L$ with $a < b$, there exists a sequence $a = x_0 < x_1 < \cdots < x_n = b$ such that $x_i \equiv x_{i+1} \pmod{\boldsymbol{\alpha}}$ or $u \equiv v \pmod{\boldsymbol{\alpha}}$ for *no proper subinterval* $[u, v]$ of $[x_i, x_{i+1}]$ for all $0 \le i < n$. We call $L$ a *congruence separable lattice* if all congruences of $L$ are separable.

G. Grätzer and E. T. Schmidt [334] characterize lattices with boolean congruence lattices using these concepts.

**Theorem 282.** *Let $L$ be a lattice. Then $\operatorname{Con} L$ is boolean iff $L$ is weakly modular and congruence separable.*

*Proof.* Let $\operatorname{Con} L$ be boolean.

We first show that $L$ is weakly modular. Let

$$[a, b] \Rightarrow [c, d], \quad a, b, c, d \in L, \quad a < b, \ c < d.$$

We have to find a proper subinterval $[a', b']$ of $[a, b]$ satisfying $[c, d] \Rightarrow [a', b']$. Let $\boldsymbol{\alpha} = \operatorname{con}(c, d)$. Since $\operatorname{Con} L$ is boolean, the congruence $\boldsymbol{\alpha}$ has a complement $\boldsymbol{\alpha}'$ in $\operatorname{Con} L$. Then $\boldsymbol{\alpha} \vee \boldsymbol{\alpha}' = \mathbf{1}$ and so $a \equiv b \pmod{\boldsymbol{\alpha} \vee \boldsymbol{\alpha}'}$. By Theorem 12, there is a sequence $a = x_0 < x_1 < \cdots < x_n = b$ such that

$$x_i \equiv x_{i+1} \pmod{\boldsymbol{\alpha}} \quad \text{or} \quad x_i \equiv x_{i+1} \pmod{\boldsymbol{\alpha}'}$$

holds for all $0 \le i < n$. If $x_i \equiv x_{i+1} \pmod{\boldsymbol{\alpha}'}$, for all $0 \le i < n$, then $a \equiv b \pmod{\boldsymbol{\alpha}'}$. Since $[a, b] \Rightarrow [c, d]$ this would imply that $c \equiv d \pmod{\boldsymbol{\alpha}'}$. By definition, $c \equiv d \pmod{\boldsymbol{\alpha}}$, so $c \equiv d \pmod{\boldsymbol{\alpha} \wedge \boldsymbol{\alpha}'}$, that is, $c \equiv d \pmod{\mathbf{0}}$, and so $c = d$, a contradiction.

Therefore, $x_i \equiv x_{i+1} \pmod{\boldsymbol{\alpha}}$ for some $0 \le i < n$. Applying Theorem 230 to $x_i \equiv x_{i+1} \pmod{\operatorname{con}(c, d)}$, we obtain a proper subinterval $[a', b']$ of $[x_{i+1}, x_i]$ satisfying $[c, d] \Rightarrow [a', b']$, proving that $L$ is weakly modular.

We next show that $L$ is separable. Now let $\alpha$ be a congruence relation of $L$ and let $a, b \in L$ with $a < b$. Since $\operatorname{Con} L$ is boolean, the congruence $\alpha$ has a complement $\alpha'$ in $\operatorname{Con} L$. Then $a \equiv b \pmod{\alpha \vee \alpha'}$. By Theorem 12, there exists an ascending sequence $a = x_0 < x_1 < \cdots < x_n = b$ such that

$$x_i \equiv x_{i+1} \pmod{\alpha} \quad \text{or} \quad x_i \equiv x_{i+1} \pmod{\alpha'}$$

for all $0 \leq i < n$. We claim that the same sequence establishes the separability of $\alpha$. Indeed, if $x_i \not\equiv x_{i+1} \pmod{\alpha}$, for some $i$, and $u, v \in [x_i, x_{i+1}]$ satisfies $u \equiv v \pmod{\alpha}$, then $x_i \equiv x_{i+1} \pmod{\alpha'}$ and so $u \equiv v \pmod{\alpha'}$, implying that $u \equiv v \pmod{\alpha \wedge \alpha'}$, that is $u = v$.

To prove the converse, let us start out with a weakly modular lattice $L$ and a congruence relation $\alpha$ of $L$. We claim:

*The binary relation $\beta$ defined on $L$ by*

$a \equiv b \pmod{\beta}$ *if* $u \equiv v \pmod{\alpha}$ *for no proper subinterval* $[u, v]$ *of* $[a \wedge b, a \vee b]$

*is a congruence relation; in fact, $\beta = \alpha^*$, the pseudocomplement of $\alpha$.*

We prove the first part of this claim by verifying that $\beta$ satisfies conditions (i)–(iii) of Lemma 11. Condition (i) is clear by the definition of $\beta$. To verify condition (ii), let

$$a, b, c \in L, \quad a \leq b \leq c, \quad a \equiv b \pmod{\beta}, \quad b \equiv c \pmod{\beta};$$

we wish to show that $a \equiv c \pmod{\beta}$. Assume to the contrary that $a \not\equiv c \pmod{\beta}$; then $u \equiv v \pmod{\alpha}$ for some proper subinterval $[u, v]$ of $[a, c]$. Since

$$u \equiv v \pmod{\operatorname{con}(a, c)},$$
$$\operatorname{con}(a, c) = \operatorname{con}(a, b) \vee \operatorname{con}(b, c),$$

by Theorem 230 and Lemma 233, there exists a proper subinterval $[x, y]$ of $[u, v]$ satisfying $[a, b] \Rightarrow [x, y]$ or $[c, d] \Rightarrow [x, y]$, say, $[a, b] \Rightarrow [x, y]$. Therefore, $a < b$. By weak modularity, there exists a proper subinterval $[a', b']$ of $[a, b]$, for which the projectivity $[x, y] \Rightarrow [a', b']$ holds. But $x, y \in [u, v]$ and $u \equiv v \pmod{\alpha}$, so $x \equiv y \pmod{\alpha}$. Hence $a' \equiv b' \pmod{\alpha}$, contradicting that $a', b' \in [a, b]$ and $a \equiv b \pmod{\beta}$. This shows that (ii) holds.

To verify condition (iii), assume to the contrary that

$$a, b, c \in L, \quad a \leq b, \quad a \equiv b \pmod{\beta}, \quad \text{but} \quad a \vee c \not\equiv b \vee c \pmod{\beta}.$$

Then there exists a proper subinterval $[u, v]$ of $[a \vee c, b \vee c]$ satisfying $u \equiv v \pmod{\alpha}$. It follows that

$$[a, b] \Rightarrow [a \vee c, b \vee c] \Rightarrow [u, v],$$

hence $[a, b] \Rightarrow [u, v]$; so $a < b$ and by weak modularity, $[u, v] \Rightarrow [a', b']$, where $[a', b']$ is a proper subinterval of $[a, b]$. Since $u \equiv v \pmod{\alpha}$, we obtain that $a' \equiv b' \pmod{\alpha}$, contradicting that $a \equiv b \pmod{\beta}$.

By duality, $a \wedge c \equiv b \wedge c \pmod{\beta}$.

This shows that $\beta$ is a congruence relation. To verify that $\beta$ is a pseudo-complement of $\alpha$ in $\operatorname{Con} L$, observe that $\alpha \wedge \beta = 0$ is trivial. Now if

$$\alpha \wedge \gamma = 0, \quad a \equiv b \pmod{\gamma}, \quad u, v \in [a \wedge b, a \vee b], \quad \text{and } u \equiv v \pmod{\alpha},$$

then also $u \equiv v \pmod{\gamma}$, hence $u = v$, showing that $\gamma \subseteq \beta$. This completes the proof of the claim.

Now to complete the proof of the theorem, let $L$ be weakly modular and let the congruences of $L$ be separable. Let $\alpha$ be a congruence of $L$. Let $\beta = \alpha^*$ be the congruence defined in the claim. For $a, b \in L$ with $a < b$, let

$$a = x_0 < x_1 < \cdots < x_n = b$$

be the chain establishing the separability of $\alpha$. Then, for each $0 \le i < n$, either $x_i \equiv x_{i+1} \pmod{\alpha}$, or by the definitions of separability and of $\beta$, we obtain that $x_i \equiv x_{i+1} \pmod{\beta}$. Therefore, $a \equiv b \pmod{\alpha \vee \beta}$, and so $\alpha \vee \beta = 1$. This shows that $\beta$ is a complement of $\alpha$. By Theorem 149, $\operatorname{Con} L$ is distributive, proving that $\operatorname{Con} L$ is boolean. $\qquad\square$

We get a large number of corollaries from Theorem 282 yielding that $\operatorname{Con} L$ is boolean for special classes of lattices. Some of these will be discussed in the Exercises.

In the proof of Theorem 282, we describe the pseudocomplement $\beta$ of a congruence relation $\alpha$ of a weakly modular lattice. Iqbalunnisa [417] observes that the converse also holds: if in a lattice $L$, the relation $\beta$ given by any congruence relation $\alpha$ (as described in the proof of Theorem 282) is always a congruence relation, then $L$ is weakly modular.

The *center* of a bounded lattice $L$ is the sublattice $\operatorname{Cen} L$ of complemented neutral elements. It is obvious that $0, 1 \in \operatorname{Cen} L$ and $\operatorname{Cen} L$ is a boolean lattice. In some chapters of lattice theory, it is important to know when $\operatorname{Cen} L$ is a complete lattice. (If this is the case, $L$ can sometimes be "coordinatized" over $\operatorname{Cen} L$.) This was established for continuous geometries by J. von Neumann [552], [553].

Let $L$ be a complete lattice and $K \subseteq L$. Then $K$ is a *complete sublattice* of $L$ if $\bigvee X$ and $\bigwedge X \in K$ for all $X \subseteq K$ (note that $\bigvee X$ and $\bigwedge X$ are formed in $L$).

**Theorem 283.** *Let $L$ be a complete, sectionally complemented, dually sectionally complemented, weakly modular lattice. Then the center of $L$ is a complete sublattice of $L$.*

*Proof.* Let $X \subseteq \operatorname{Cen} L$. Set $a = \bigwedge X$ and $\boldsymbol{\alpha}_x = \operatorname{con}(0, x)$ for all $x \in X$. Then $\operatorname{id}(a)$ is the kernel of the congruence relation $\bigwedge(\boldsymbol{\alpha}_x \mid x \in X)$. Hence, by Theorem 272, the ideal $\operatorname{id}(a)$ is standard; by Corollary 262, the element $a$ is standard. Now Corollary 258 tells us that the element $a$ is neutral. Since the element $a$ is complemented, it follows that $a \in \operatorname{Cen} L$.

By duality, we obtain that $\bigvee X \in \operatorname{Cen} L$.    $\square$

There is another variant of this result in M. F. Janowitz [425].

**Corollary 284.** *The center of a complete, relatively complemented lattice is a complete sublattice.*

This result has some interesting implications concerning direct decompositions of complete relatively complemented lattices, see S. Maeda [523].

## 4.4 ◇ Infinite direct decompositions of complete lattices by Friedrich Wehrung

A lattice $L$ is *totally decomposable* if it is isomorphic to a (not necessarily finite) direct product of directly indecomposable factors—see Section 4.2.

By Lemma 278, every lattice of finite length is totally decomposable. Can this lemma be made stronger? Let us call an element $p$ in a lattice $L$ *completely join-irreducible* if $p = \bigvee X$ implies that $p \in X$ for any $X \subseteq L$. Equivalently, $p$ has exactly one lower cover $p_*$ majorizing all elements of $L$ below $p$.

**Definition 285.** A lattice $L$ is

(i) *spatial* if every element of $L$ is a join of completely join-irreducible elements of $L$;

(ii) *finitely spatial* if every element of $L$ is a join of join-irreducible elements of $L$.

For example, $L$ is spatial iff, for any $a, b \in L$ such that $a \not\leq b$, there exists a completely join-irreducible element $p$ such that $p \leq a$ and $p \not\leq b$. As every atom is completely join-irreducible, every atomistic lattice is spatial.

L. Libkin [503] offers the first direct decomposition result for infinite lattices.

◇ **Theorem 286.** *Every spatial algebraic lattice is totally decomposable.*

Neither of the two conditions on the lattice $L$ (algebraic or spatial) can be removed from the statement of Theorem 286, not even in the distributive case, see Example 3.10 in F. Wehrung [706].

Observe that every algebraic lattice is dually spatial (see G. Gierz, K. H. Hofmann, K. Keimel, J. D. Lawson, M. Mislove, and D. S. Scott [225, Theorem I.4.25] or V. A. Gorbunov [243, Lemma 1.3.2]), but not necessarily spatial. Hence, by Theorem 286, *Every algebraic and dually algebraic lattice is totally decomposable.* Theorem 286 trivially implies Theorem 393, namely, that every geometric lattice is totally decomposable.

Theorem 286 does not say anything about the *structure* of directly in-decomposable geometric lattices, in particular, whether they are *subdirectly irreducible*. Indeed, a finite, directly indecomposable, atomistic lattice is not necessarily subdirectly irreducible; as an example take the lattice of all convex subsets of $B_2$, see G. Birkhoff and M. K. Bennett [74].

A. Walendziak [694] provides an interesting extension of Theorem 286 (see also J. Jakubík [424]):

◇ **Theorem 287.** *Let $L$ be an algebraic lattice with unit. If the unit of $L$ is a join of join-irreducible elements, then $L$ is totally decomposable.*

The following results are from F. Wehrung [706].

◇ **Lemma 288.** *Let $L$ be a complete lattice. The following are equivalent:*

(i) *$L$ is totally decomposable;*

(ii) *$\operatorname{Cen} L$ is a complete sublattice of $L$, it is atomistic, and the following condition holds:*

(J)     $$x = \bigvee ( x \wedge u \mid u \in \operatorname{Atom}(\operatorname{Cen} L) ) \quad \text{for each } x \in L.$$

Example 2.9 of the paper cited shows that the condition (J) in Lemma 288 is not redundant, even if $L$ is distributive and selfdual. The same paper contains the following extension of Theorem 286.

◇ **Theorem 289.** *Every complete lattice that is both finitely spatial and dually finitely spatial is totally decomposable.*

Observe that Theorem 289 no longer requires that the lattice be algebraic.

## Exercises

4.1. Show that the representation of a lattice $L$ with zero as a direct product of two lattices are (up to isomorphism) in one-to-one correspondence with pairs of ideals $(I, J)$ satisfying $I \cap J = \{0\}$ such that every element $a$ of $L$ has exactly one representation of the form $a = i \vee j$ with $i \in I$ and $j \in J$.

4.2. Let $L = A_1 \times A_2$. Define the binary relation $\alpha_i$ on $L$ by

$$(a_1, a_2) \equiv (b_1, b_2) \pmod{\alpha_i}$$

if $a_i = b_i$ for $i = 1, 2$. Show that $\alpha_1$ and $\alpha_2$ are congruence relations of $L$, and

$$\alpha_1 \wedge \alpha_2 = \mathbf{0},$$
$$\alpha_1 \vee \alpha_2 = \mathbf{1}.$$

4.3. Show that the congruence relations $\alpha_1$ and $\alpha_2$ of Exercise 4.2 are permutable.

4.4. Prove that the representations of a lattice $L$ as a direct product of two lattices are (up to isomorphism) in a one-to-one correspondence with pairs of congruence relations $(\alpha_1, \alpha_2)$ that are complementary and permutable.

4.5. Show that if, in Exercise 4.4, we pass from two to $n$ factors, then we get an $n$-tuple of congruences $(\alpha_1, \ldots, \alpha_n)$ such that

$$\alpha_1 \wedge \cdots \wedge \alpha_n = \mathbf{0},$$
$$(\alpha_1 \wedge \cdots \wedge \alpha_{i-1}) \vee \alpha_i = \mathbf{1},$$
$$\alpha_1 \wedge \cdots \wedge \alpha_{i-1} \text{ and } \alpha_i \text{ permute}$$

for $i = 2, \ldots, n$.

4.6. Can we replace, in Exercise 4.5, the condition "$\alpha_1 \wedge \cdots \wedge \alpha_{i-1}$ and $\alpha_i$ permute" by "any pair $\alpha_i, \alpha_j$ permute"?

4.7. Use Exercise 4.5 to verify Corollary 277 for arbitrary lattices.

4.8. Verify Corollary 277 for arbitrary lattices not using Exercise 4.5.

4.9. Relate Exercise 4.5 to Theorem 276.

4.10. Find a complete lattice $L$ and a sublattice $K$ of $L$ such that $K$ is a complete lattice but not a complete sublattice of $L$.

4.11. Let $L$ be a complemented modular lattice. Then $\operatorname{Con} L$ is boolean iff all neutral ideals of $L$ are principal (Shih-chiang Wang [695]).

4.12. Let $B$ be the boolean lattice R-generated by the rational interval $[0, 1]$ and let $n$ be a natural number. Show that $B$ has a representation as a direct product of $n$ lattices $L_1, \ldots, L_n$ with $|L_i| > 1$, for all $i = 1, \ldots, n$, but $B$ has no representation as a direct product of directly indecomposable lattices.

4.13. Construct a lattice that is not of finite length but every chain in the lattice is finite.

4.14. Four statements of this section (numbered 278–281) deal with lattices of finite length. Which of these statements remain valid for lattices in which every chain is finite?

$$* \qquad * \qquad *$$

The following exercises are based on G. Grätzer and E. T. Schmidt [334].

4.15. Prove that every congruence relation of a lattice of finite length is separable.

4.16. Verify that if in a lattice $L$, for every $a, b \in L$ with $a < b$, there is a finite maximal chain in $[a, b]$, then all congruence relations of $L$ are separable. This holds, in particular, if all intervals are finite.

4.17. Let $L$ be a distributive lattice, let $a, b, a_1, a_2, \cdots \in L$ satisfy $a = a_1 < a_2 < a_3 < \cdots < b$. Then $\alpha = \bigvee(\operatorname{con}(a_{2i-1}, a_{2i}) \mid i = 1, 2, \ldots)$ is not a separable congruence relation.

4.18. For a distributive lattice $L$, the congruence lattice $\operatorname{Con} L$ is boolean iff $L$ is *locally finite*, that is, every finitely generated sublattice of $L$ is finite. (Use Exercises 4.16 and 4.17, see J. Hashimoto [375].)

4.19. The separable congruences of a lattice $L$ form a sublattice of $\operatorname{Con} L$.

4.20. Let $L$ be a lattice with unit and let $I$ be an ideal of $L$. Show that a neutral congruence $\operatorname{con}(I)$ is separable iff $I$ is principal.

# Chapter

# IV

---

# *Lattice Constructions*

## 1. Adding an Element

What can you do by adding a single element to a lattice? It turns out, quite a lot. Of course, for a lattice $L$, you can form $C_1 + L$ to add a (new) zero and $L + C_1$ to add a (new) unit; and you can do both. In this section, we discuss some less trivial constructions.

### 1.1 One-Point Extension

Let $L$ be a lattice (with the ordering $\leq_L$) and let $\mathbf{I} = [u, v]$ denote a proper interval in $L$. We define the One-Point Extension $L_\mathbf{I}$ of a lattice $L$ by adjoining a new element $m_\mathbf{I} = m$ to $L$ and by requiring that $u \prec m \prec v$. If you start with the chain $C_3$ and the interval $[0, 1]$, you obtain the extension $B_2$. On the other hand, if you start with the boolean lattice $B_2$ and the interval $[0, 1]$, the lattice extension is $M_3$. In the first example, all congruences extend, in the second, only the congruences $\mathbf{0}$ and $\mathbf{1}$ extend.

We associate with $x \in L_\mathbf{I}$ the elements $\underline{x}$ and $\overline{x}$ of $L$: for $x \in L$, set $\underline{x} = \overline{x} = x$, $\underline{m} = u$, and $\overline{m} = v$. We then define the relation $\leq$ on the set $L_\mathbf{I}$ as follows: $x \leq y$ if $x = y$ or $\overline{x} \leq_L \underline{y}$.

The following lemma is straightforward.

**Lemma 290.** $(L_\mathbf{I}; \leq)$ *is a lattice; it is an extension of $L$.*

This construction has many applications. We list here a few.

G. Grätzer, *Lattice Theory: Foundation*, DOI 10.1007/978-3-0348-0018-1_4,
© Springer Basel AG 2011

**Corollary 291.**

(i) *Every lattice L can be embedded into a sectionally complemented lattice.*

(ii) *Every lattice L can be embedded into a relatively complemented lattice.*

(iii) *Every bounded subdirectly irreducible lattice L can be embedded into a simple lattice K with $|K - L| \leq 3$.*

Recall that subdirectly irreducible lattices are defined in Definition 216.

To illustrate the proofs, we verify (iii). Let $\mathrm{con}(u, v)$ be the base congruence of $L$, where $u < v$. Apply the construction to the interval $[0, v]$, to obtain the lattice $L \cup \{m\}$. Now apply the construction twice more to the interval $[0, 1]$ to obtain the elements $x$ and $y$, and set $K = L \cup \{m, x, y\}$. We claim that $K$ is simple. Indeed, let $\boldsymbol{\alpha}$ be a congruence of $K$ with $\boldsymbol{\alpha} \neq \mathbf{0}$. Then there exist elements $a < b \in K$ such that $a \equiv b \pmod{\boldsymbol{\alpha}}$. If $a, b \in L$ fails, then, say, $a = x$ and $b = 1$ (the other cases are handled similarly), and $x \equiv 1 \pmod{\boldsymbol{\alpha}}$ implies that $v \equiv 0 \pmod{\boldsymbol{\alpha}}$, so $u \equiv v \pmod{\boldsymbol{\alpha}}$. Therefore, $m \equiv 0 \pmod{\boldsymbol{\alpha}}$ and so $x \equiv 1 \pmod{\boldsymbol{\alpha}}$ and $y \equiv 1 \pmod{\boldsymbol{\alpha}}$; finally we obtain that

$$x \wedge y = 0 \equiv 1 \pmod{\boldsymbol{\alpha}},$$

that is, $\boldsymbol{\alpha} = \mathbf{1}$. On the other hand, if $a, b \in L$, then $a \equiv b \pmod{\boldsymbol{\alpha}}$ implies that $u \equiv v \pmod{\boldsymbol{\alpha}}$ because $\mathrm{con}(u, v)$ is the base congruence. By the definition of the element $m$, the congruence $0 \equiv m \pmod{\boldsymbol{\alpha}}$ holds, and we proceed as in the previous case.

Next we shall investigate which congruences of $L$ extend to $L_{\mathbf{I}}$, based on G. Grätzer and H. Lakser [306].

**Theorem 292 (One-Point Extension Theorem).** *Let $\mathbf{I} = [u, v]$ be a nontrivial, nonprime interval in the lattice $L$, and let $\boldsymbol{\alpha}$ be a congruence relation on $L$. Then $\boldsymbol{\alpha}$ has an extension $\boldsymbol{\alpha}_{\mathbf{I}}$ to $L_{\mathbf{I}}$ iff $\boldsymbol{\alpha}$ satisfies the following condition and its dual:*

(OP) *For $y < v$ and $u < x$,*

$$y \equiv v \pmod{\boldsymbol{\alpha}} \quad \textit{implies that} \quad v \vee x \equiv x \pmod{\boldsymbol{\alpha}}.$$

*If $\boldsymbol{\alpha}$ has an extension, then it is unique.*

*Proof.* Let us assume that congruence relation $\boldsymbol{\alpha}$ of the lattice $L$ has an extension $\boldsymbol{\alpha}'$ to the lattice $L_{\mathbf{I}}$. Let $x$ and $y$ be given as in condition (OP). Note that $y \not\geq m$ and $x \not\leq m$. Now $y \equiv v \pmod{\boldsymbol{\alpha}'}$, and in view of $y \wedge m = y \wedge u$, by taking the meet with $m$, we conclude that $y \wedge u \equiv m \pmod{\boldsymbol{\alpha}'}$. Therefore, $m \equiv u \pmod{\boldsymbol{\alpha}'}$. Joining with $x$, we obtain that

$$v \vee x = m \vee x \equiv x \pmod{\boldsymbol{\alpha}},$$

since $\alpha'$ is an extension of $\alpha$, establishing condition (OP).

By duality, we obtain the dual of condition (OP).

Now let the congruence relation $\alpha$ on $L$ satisfy condition (OP) and its dual. We extend $\alpha$ to a binary relation $\beta$ on the pairs of comparable elements of $L_I$: set $m \equiv m \pmod{\beta}$; if $a < m$, set $a \equiv m \pmod{\beta}$ if

$(L_a)$

$a \equiv u \pmod{\alpha}$ and there is $y_a \in L$ with $y_a < v$ and $y_a \equiv v \pmod{\alpha}$;

dually, if $m < b$, set $m \equiv b \pmod{\beta}$ if

$(U_b)$

$v \equiv b \pmod{\alpha}$ and there is $x_b \in L$ with $x_b > u$ and $u \equiv x_b \pmod{\alpha}$.

We now prove that conditions (ii) and (iii) of Lemma 11 hold for the relation $\beta$. We first verify Lemma 11(ii).

Let $a, b, c \in L_I$, and let

$$a < b < c, \quad a \equiv b \pmod{\beta}, \quad \text{and} \quad b \equiv c \pmod{\beta}.$$

If $m \notin \{a, b, c\}$, there is nothing to do. So we can assume that $m \in \{a, b, c\}$. If $m = a$, then $b \equiv c \pmod{\alpha}$ and condition $(U_b)$ holds. Then $v \equiv c \pmod{\alpha}$, and, setting $x_c = x_b$, we get condition $(U_c)$, that is, $a \equiv c \pmod{\beta}$. The dual argument holds if $m = c$.

We are then left with the case $m = b$. Then by condition $(L_a)$, the congruence

$$a \equiv u \pmod{\alpha}$$

holds and by condition $(U_c)$, the congruence

$$v \equiv c \pmod{\alpha}$$

holds. We then need only verify that $u \equiv v \pmod{\alpha}$. By conditions $(L_a)$ and $(U_c)$, there are $y_a, x_c \in L$ with $y_a < v$, with $y_a \equiv v \pmod{\alpha}$, with $x_c > u$, and with $u \equiv x_c \pmod{\alpha}$. By condition (OP), $x_c \equiv v \vee x_c \pmod{\alpha}$ and so the congruence $u \equiv v \vee x_c \pmod{\alpha}$ holds. Then certainly, $u \equiv v \pmod{\alpha}$, concluding the verification of condition (ii) for the relation $\beta$.

We now establish Lemma 11(iii) for the relation $\beta$. Let $a, b, c \in L_I$ with $a < b$ and $a \equiv b \pmod{\beta}$. To establish that $a \vee c \equiv b \vee c \pmod{\beta}$, we may assume that $a < c$ and $b \not\leq c$. We then must show that $c \equiv b \vee c \pmod{\beta}$.

If $m \notin \{a, b, c\}$, then we are in the sublattice $L$ with the congruence relation $\alpha$, and nothing needs to be done. We can, therefore, assume that $m \in \{a, b, c\}$.

First, let $m = a$. Then since $a < b, c$, we get that $v < b, c$. Furthermore, since $a \equiv b \pmod{\beta}$, by condition $(U_b)$, we get that $v \equiv b \pmod{\alpha}$. Then $c \equiv b \vee c \pmod{\alpha}$, that is, $c \equiv b \vee c \pmod{\beta}$.

Next, let $m = c$. Since $b \not\leq c$, it follows that $b \vee c = b \vee v$. We must establish that $m \equiv b \vee v \pmod{\beta}$, that is, that condition $(U_{b \vee v})$ holds. Now $a \leq u$, since $a < c = m$. Thus $u \equiv b \vee u \pmod{\boldsymbol{\alpha}}$ and $v \equiv b \vee v \pmod{\boldsymbol{\alpha}}$. Since $b \not\leq c = m$, it follows that $b \vee u > u$. Thus setting $x_{b \vee v} = b \vee u$, establishes condition $(U_{b \vee v})$.

Finally, let $m = b$. There are then two subcases, the subcase $c < b$ and the subcase $c \parallel b$. In either subcase, we have condition $(L_a)$, that is, we have $a \leq u$ with $a \equiv u \pmod{\boldsymbol{\alpha}}$, and $y_a \in L$ with $y_a < v$, and $y_a \equiv v \pmod{\boldsymbol{\alpha}}$.

In the subcase $c < b$, we wish to establish that $c \equiv b \pmod{\boldsymbol{\beta}}$, that is, since $b = m$, condition $(L_c)$. Now $c \leq u$ since $c < b = m$. Also, since $a < c$ and $a \equiv u \pmod{\mathbf{g}}$, it follows that $c \equiv u \pmod{\boldsymbol{\alpha}}$. Thus, setting $y_c = y_b$, establishes condition $(L_c)$.

We finally consider the subcase where $c$ and $b = m$ are incomparable. Then $b \vee c = v \vee c$. Now,

$$c \equiv u \vee c \pmod{\boldsymbol{\alpha}},$$

since $a < c$ and $a \equiv u \pmod{\boldsymbol{\alpha}}$. Furthermore, since $c$ and $m$ are incomparable, it follows that $u < u \vee c$. Then by condition (OP) with $y = y_a$ and $x = u \vee c$, we obtain that

$$v \vee c \equiv u \vee c \pmod{\boldsymbol{\alpha}}.$$

Thus $c \equiv v \vee c \pmod{\boldsymbol{\alpha}}$, that is, $c \equiv b \vee c \pmod{\boldsymbol{\beta}}$ in this final subcase.

Therefore, we have verified Lemma 11(iii). The dual of Lemma 11(iii) holds by duality.

Consequently, setting

$$x \equiv y \pmod{\boldsymbol{\alpha_I}} \quad \text{iff} \quad x \wedge y \equiv x \vee y \pmod{\boldsymbol{\beta}}$$

(so as to satisfy also Lemma 11(i)) yields a congruence $\boldsymbol{\alpha_I}$ on $L_I$ extending the congruence $\boldsymbol{\alpha}$ of $L$.

Now assume that the interval $\mathbf{I}$ is not prime. We show that the extension of $\boldsymbol{\alpha}$ to $L_I$ is unique. Let the congruence relation $\boldsymbol{\alpha'}$ on $L_I$ be an extension of $\boldsymbol{\alpha}$. By duality, it suffices to show that if $a \in L$ with $a < m$ and $a \equiv m \pmod{\boldsymbol{\alpha'}}$, then $a \equiv m \pmod{\boldsymbol{\beta}}$, that is, that condition $(L_a)$ holds, and, conversely, that condition $(L_a)$ implies that $a \equiv m \pmod{\boldsymbol{\alpha'}}$. Since $a \leq u < m$ and $\boldsymbol{\alpha'}$ is an extension of $\boldsymbol{\alpha}$, it follows that

$$a \equiv u \pmod{\boldsymbol{\alpha}}.$$

Since $\mathbf{I}$ is not prime in $L$, there is a $y \in L$ with $u < y < v$. We claim that $y$ is the required $y_a$ in condition $(L_a)$. Indeed $y$ is a relative complement of $m$ in the interval $\mathbf{I}$ of $L_I$ and so

$$y \equiv v \pmod{\boldsymbol{\alpha}}$$

since $u \equiv m \pmod{\boldsymbol{\alpha'}}$. Thus condition $(L_a)$ is established, that is, the congruence $a \equiv m \pmod{\boldsymbol{\beta}}$ holds.

On the other hand, condition $(L_a)$ implies $a \equiv m \pmod{\boldsymbol{\alpha}'}$ whether or not $\mathbf{I}$ is prime; indeed, $y_a \equiv v \pmod{\boldsymbol{\alpha}'}$ implies that $y_a \wedge m \equiv m \pmod{\boldsymbol{\alpha}'}$. But $y_a \wedge m = y_a \wedge u$, whereby $u \equiv m \pmod{\boldsymbol{\alpha}'}$, and so $a \equiv m \pmod{\boldsymbol{\alpha}'}$.

Thus if $\mathbf{I}$ is not prime, the extension $\boldsymbol{\alpha}_{\mathbf{I}}$ is unique. □

**Corollary 293.** *Under the conditions of the One-Point Extension Theorem, the lattice $L_{\mathbf{I}}$ is a congruence-preserving extension of $L$.*

For more complicated applications, see G. Grätzer and H. Lakser [306] and G. Grätzer and R. W. Quackenbush [327].

There is a much more complicated scheme to construct new lattices (called *attachment*), see G. Grätzer and D. Kelly [288] and [289] and the Exercises.

## 1.2 Doubling elements and intervals

In the paper G. Grätzer [260], instead of adding an element to the lattice $L$, an element of $L$ is "doubled". This is how it works: Take the lattice $L$ and an element $a \in L$, $a \neq 0, 1$. (This does not mean that $L$ must have 0 or 1, only that if it has either, then $a$ is different.) Now we add to $L$ the "double" of $a$: the element $a^{\vee}$. We introduce the order $\leq^{\vee}$ on the set $L^{\vee} = L \cup \{a^{\vee}\}$ as follows: For $x, y \in L$, let $x \leq^{\vee} y$ if $x \leq y$; for $x \leq a$, let $x <^{\vee} a^{\vee}$; for $a < x$, let $a^{\vee} \leq^{\vee} x$.

**Lemma 294.** *The order $L^{\vee}$ is a lattice. In $L^{\vee}$, the element $a$ is meet-irreducible and the element $a^{\vee}$ is join-irreducible.*

Here is an illustration of how such a result can be used (see Exercise 1.7).

◇ **Theorem 295.** *Every lattice $L$ can be embedded into the lattice of all ideals of some lattice $K$ with no doubly reducible element.*

Call a finite lattice $L$ *transferable* (G. Grätzer [256]) if whenever $L$ has an embedding $\varphi$ into the ideal lattice $\operatorname{Id} K$ of a lattice $K$, then $L$ has an embedding into the lattice $K$.

**Corollary 296.** *A transferable lattice cannot have a doubly reducible element.*

For a recent application of doubling of an element, see G. Grätzer, D. S. Gunderson, and R. W. Quackenbush [277].

I introduced the doubling of an element when I gave a series of lectures at McMaster University in the Winter of 1964 and 1965 (the lectures were published in [254]) as a means of eliminating X-failures: elements $x, y, u, v$ satisfying $x \vee y = u \wedge v$, see Figure 57. (X refers to the way the diagram looks.) A. Day (who was a graduate student at McMaster at the time) noticed that a very similar construction, the doubling of an interval, could be used to eliminate W-failures, four elements satisfying $x \vee y < u \wedge v$ (W stands for Whitman; see Definition 523); see A. Day [135].

Here is Day's definition: Let $L$ be a lattice and $I = [a, b]$ be an interval of $L$. We form $I \times \mathsf{C}_2$, where $\mathsf{C}_2 = \{0, 1\}$ is the two-element chain and we form the set $L[I] = (L - I) \cup (I \times \mathsf{C}_2)$ with the ordering for $x, y \in L$ and $i, j \in \mathsf{C}_2$:

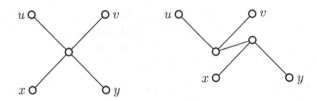

Figure 57. X and W

$x \leq y$ if $x \leq y$ in $L$;
$(x, i) \leq y$ if $x \leq y$ in $L$;
$x \leq (y, j)$ if $x \leq y$ in $L$;
$(x, i) \leq (y, j)$ if $x \leq y$ in $L$ and $i \leq j$ in $\mathsf{C}_2$.

A crucial property of this construction is stated in the following obvious statement.

**Lemma 297.** *There is a congruence $\delta$ on $L[I]$ such that $L[I]/\delta \cong L$ and every block of $\delta$ is a singleton or a two-element chain.*

For more about the doubling construction, see Sections VI.2 and VII.2, and the survey article J. B. Nation [543].


**Exercises**

1.1. Can we get from $\mathsf{C}_3$ to $\mathsf{N}_5$ with two applications of the One-Point Extension?

1.2. Prove that every lattice $L$ can be embedded into a sectionally complemented lattice.

1.3. Prove that every lattice $L$ can be embedded into a sectionally complemented simple lattice.

1.4. Prove that every lattice $L$ can be embedded into a relatively complemented, bounded, simple lattice.

1.5. Let $A$ and $B$ be lattices, let $B$ be bounded. On $A \times B$, define the ordering $(a, b) \leq (a', b')$ if $a <_A a'$, or $a = a'$ and $b \leq_B b'$. Show that $A \oslash B = (A \times B; \leq)$ is a lattice.

1.6. Investigate the map $I \mapsto I \oslash B$ of $\mathrm{Id}\, A$ into $\mathrm{Id}(A \oslash B)$.

1.7. Prove Theorem 295 (G. Grätzer [260]).

$$*\qquad *\qquad *$$

The following exercises are based on G. Grätzer and H. Lakser [306].

1.8. State and prove the One-Point Extension Theorem for a prime interval $\Lambda$.

1.9. State and prove the One-Point Extension Theorem for a collection of nonprime intervals.

1.10. To what extent does the One-Point Extension Theorem preserve the completeness of the lattice?

$$* \qquad * \qquad *$$

The following exercises are based on G. Grätzer and D. Kelly [288].

A family $\mathcal{A} = ((A_i; \leq_i) \mid i \in I)$ of orders is *attachable* to an order $P$ if the following two conditions are satisfied:

(Ord) The relations $\leq_P$ and $\leq_i$ agree on $A_i \cap P$ for all $i \in I$.
(Int) $A_i \cap A_j \subseteq P$ for all $i \neq j$ in $I$.

Given such a family, we define the order

$$Q = (P \cup \bigcup (A_i \mid i \in I); \leq_Q)$$

by defining $x \leq_Q y$ to hold if one of the following five conditions holds:

(S) $x, y \in P$ and $x \leq_P y$.

(A) $x, y \in A_i$ and $x \leq_i y$ for some $i \in I$.

(AS) $x \in A_i$, for some $i \in I$, $y \in P$, and there is an element $t \in A_i \cap P$ such that $x \leq_i t$ and $t \leq_P y$.

(SA) $x \in P$, $y \in A_i$, for some $i \in I$, and there is an element $t \in A_i \cap P$ such that $x \leq_P t$ and $t \leq_i y$.

(AA) $x \in A_i$, $y \in A_j$, for some $i \neq j \in I$, and there are elements $u, v \in P$ such that $x \leq_i u$, $u \leq_P v$, and $v \leq_j y$.

1.11. Prove that given an attachable family $\mathcal{A}$ for an order $P$, the relation $\leq_Q$ is an order relation. Moreover, the restriction of $\leq_Q$ to $P$ is $\leq_P$ and the restriction of $\leq_Q$ to $A_i$ equals $\leq_i$ for each $i \in I$.
The order $(Q; \leq_Q)$ is denoted by $P[\mathcal{A}]$.
A family $\mathcal{A} = ((A_i; \vee_i, \wedge_i, 0_i, 1_i) \mid i \in I)$ of nontrivial bounded lattices is *lattice-attachable* to a lattice $K$ when the following four conditions are satisfied:

(i) $\mathcal{A}$ is attachable to $K$ as orders.

(ii) $K$ contains $0_i$ and $1_i$ for every $i \in I$.

(iii) $A_i \cap K$ is a sublattice of both $K$ and $A_i$, whenever $i \in I$.

(iv) The following two conditions are satisfied:

(C1) For $i \in I$ and $a \in A_i$, there is in $\mathrm{id}(a)_{Ai} \cap K$ a greatest element $\underline{a}$ and there is in $\mathrm{fil}(a)_{Ai} \cap K$ a least element $\overline{a}$.

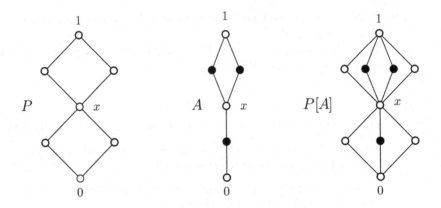

Figure 58. A lattice attachment

> (C2) Let $i \in I$ and $x \in K$. If $0_i \leq x$, then $\mathrm{id}(x)_K \cap A_i$ contains a greatest element $x_{(i)}$. If $x \leq 1_i$, then $\mathrm{fil}(x)_K \cap A_i$ contains a smallest element $x^{(i)}$.

1.12. Verify that Figure 58 pictures a lattice attachment.

1.13. For a lattice attachment, prove that the order $L = K[\mathcal{A}]$ contains $K$, the *skeleton*, and each attachment as a sublattice.

*1.14. Find necessary and sufficient conditions under which a lattice attachment $L = K[\mathcal{A}]$ is a lattice.

1.15. Let $\mathcal{A} = (A_i \mid i \in I)$ be a lattice attachable family for the lattice $K$. The order $L = K[\mathcal{A}]$ is a lattice if the following two conditions are satisfied.

(a) For each $i \in I$, $A_i \cap K$ is a chain or equals $[0_i, 1_i]_K$.
(b) For $i \neq j \in I$, $A_i \cap A_j$ is a chain (possibly empty).

*1.16. Use attachments to describe all subdirectly irreducible modular lattices of order-dimension 2 (G. Grätzer and R. W. Quackenbush [327]).

## 2.  Gluing

In Section I.1.6, we introduced glued sums of orders, $P \overset{\cdot}{+} Q$, applied to an order $P$ with largest element $1_P$ and an order $Q$ with smallest element $0_Q$. This can be applied to a lattice $K$ with unit and a lattice $L$ with zero to obtain a lattice $K \overset{\cdot}{+} L$.

A natural generalization of this construction is gluing, first introduced in R. P. Dilworth [154]; see also R. P. Dilworth and M. Hall [163]. For an overview of gluing (and some of its generalizations) for modular lattices up to 1989, see A. Day and R. Freese [139].

## 2.1 Definition

Let $K$ and $L$ be lattices, let $F$ be a filter of $K$, and let $I$ be an ideal of $L$. If $F$ is isomorphic to $I$ (with $\psi$ the isomorphism), then we can form the lattice $G$, the *gluing* of $K$ and $L$ over $F$ and $I$ (with respect to $\psi$), defined as follows:

We form the disjoint union $K \cup L$, and identify $a \in F$ with $\psi(a) \in I$, for all $a \in F$, to obtain the set $G$. We order $G$ as follows (see Figure 59):

$$a \leq b \quad \text{if} \quad \begin{cases} a \leq_K b, & \text{for } a, b \in K; \\ a \leq_L b, & \text{for } a, b \in L; \\ a \leq_K x \text{ and } \psi(x) \leq_L b, & \text{for } a \in K \text{ and } b \in L, \\ & \text{and for some } x \in F. \end{cases}$$

**Lemma 298.** *$G$ is an order, in fact, $G$ is a lattice. The join in $G$ is described by*

$$a \vee b = \begin{cases} a \vee_K b, & \text{for } a, b \in K; \\ a \vee_L b, & \text{for } a, b \in L; \\ \psi(a \vee_K x) \vee_L b, & \text{for } a \in K \text{ and } b \in L, \text{ for any } b \geq x \in F, \end{cases}$$

*and dually for the meet. If $L$ has a zero, $0_L$, then the last clause for the join may be rephrased:*

$$a \vee b = \psi(a \vee_K 0_L) \vee_L b, \quad \text{for } a \in K \text{ and } b \in L.$$

*The lattice $G$ contains $K$ and $L$ as sublattices; in fact, $K$ is an ideal and $L$ is a filter of $G$.*

*Proof.* Let $\leq_K$, $\leq_L$, $\leq_m$ denote the binary relations defined by the three lines of the formula describing $\leq$ and let $\varrho = \leq_K \cup \leq_L \cup \leq_m$.

Clearly, the transitive closure of $\varrho$ is an order. So to prove the formula, we have to verify that the transitive closure of $\varrho$ is itself.

On $G = K \cup L$, the relation $\leq$ is obviously reflexive.

To show that $\leq$ is antisymmetric, let $a, b \in G$, let $a \leq b$ and $b \leq a$. If $a, b \in K$ or $a, b \in L$, then $a = b$, because $\leq_K$ and $\leq_L$ are antisymmetric. By symmetry, we can assume that $a \in K - L$ and $b \in L - K$, in which case $b \leq a$ must fail.

To show that $\leq$ is transitive, let $a, b, c \in G$, and let $a \leq b$ and $b \leq c$. As in the previous paragraph, there are many cases: (i) $a, b, c \in K$, (ii) $a, b, c \in L$, (iii) $a, b \in K$ and $c \in L - K$, (iv) $a \in K - L$ and $b, c \in L$. All four cases are trivial.

We leave to the reader to verify that $G$ is a lattice, the lattice operations are as described, and $G$ contains $K$ as an ideal and $L$ as a filter. $\qquad \square$

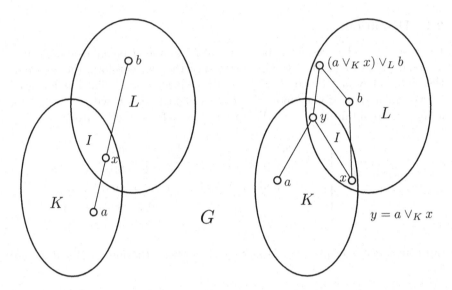

Figure 59. Defining gluing

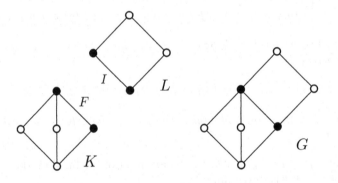

Figure 60. An easy gluing example

A small example of gluing is shown in Figure 60. For a more interesting example see Figure 82.

**Lemma 299.** *Let $K, L, F, I, G$ be given as above. Let $A$ be a lattice containing $K$ and $L$ as sublattices so that $K \cap L = I = F$. Then $K \cup L$ is a sublattice of $A$ and it is isomorphic to $G$.*

*Proof.* We are assuming that $\leq = \leq_A$ when restricted to $K$ is $\leq_K$, when restricted to $L$ is $\leq_L$, and $\psi$ is the identity map. The join formula (and its dual) of Lemma 298 shows that $K \cup L$ is a sublattice of $A$ and it is isomorphic to $G$.                                   $\square$

## 2.2   Congruences

Let $K, L, F, I, G$ be given as in Section 2.1. We can easily describe the congruences of $G$.

If $\alpha_K$ is a reflexive binary relation on $K$ and $\alpha_L$ is a reflexive binary relation on $L$, we then define the *reflexive product* $\alpha_K \overset{r}{\circ} \alpha_L$, a relation on $G$, as $\alpha_K \cup \alpha_L \cup (\alpha_K \circ \alpha_L)$.

**Lemma 300.** *A congruence* $\alpha$ *of* $G$ *can be uniquely written in the form*

$$\alpha = \alpha_K \overset{r}{\circ} \alpha_L,$$

*where*

(i) $\alpha_K$ *is a congruence of* $K$;

(ii) $\alpha_L$ *is a congruence of* $L$;

(iii) $\alpha_K$ *restricted to* $F$ *equals* $\alpha_L$ *restricted to* $I$ (*under the identification of the elements by* $\psi$).

*Conversely, if* $\alpha_K$ *is a congruence of* $K$ *and* $\alpha_L$ *is a congruence of* $L$ *satisfying the condition that* $\alpha_K$ *restricted to* $F$ *equals* $\alpha_L$ *restricted to* $I$ (*under the identification of the elements by* $\psi$), *then* $\alpha = \alpha_K \overset{r}{\circ} \alpha_L$ *is a congruence of* $G$.

*Proof.* Let $\alpha$ be a congruence of $G$. Define $\alpha_K = \alpha\rceil K$ and $\alpha_L = \alpha\rceil L$. Obviously, (i)–(iii) hold. We verify that

$$\alpha = \alpha_K \overset{r}{\circ} \alpha_L.$$

Clearly, $\alpha_K \overset{r}{\circ} \alpha_L \subseteq \alpha$. So let $a \equiv b \pmod{\alpha}$. We can assume that $a \in K$ and $b \in L$ with $a \leq b$. So by Lemma 298, there is an $x \in K \cap L$ such that $a \leq x \leq b$. Therefore, $a \equiv x \pmod{\alpha}$ and $a, x \in K$ implies that $a \equiv x \pmod{\alpha\rceil K}$. Similarly, $x \equiv b \pmod{\alpha\rceil L}$. So $a \equiv b \pmod{\alpha_K \overset{r}{\circ} \alpha_L}$, proving that $\alpha = \alpha_K \overset{r}{\circ} \alpha_L$.     $\square$

We can also obtain congruence-preserving extensions using gluing, based on the following result:

**Lemma 301.** *Let* $K$ *and* $L$ *be lattices, let* $F$ *be a filter of* $K$, *and let* $I$ *be an ideal of* $L$. *Let* $\psi$ *be an isomorphism between* $F$ *and* $I$. *Let* $G$ *be the gluing of* $K$ *and* $L$ *over* $F$ *and* $I$ *with respect to* $\psi$. *If* $L$ *is a congruence-preserving extension of* $I$, *then* $G$ *is a congruence-preserving extension of* $K$.

*Proof.* Let $\gamma$ be a congruence of $K$. We have to show that it extends to $G$ and extends uniquely. Consider $\gamma\rceil F$ as a congruence of $I$. Since $L$ is a congruence-preserving extension of $I$, the congruence $\gamma\rceil F$ extends (uniquely) to a congruence $\delta$ of $L$. By definition, $\gamma\rceil F = \delta\rceil I$. By Lemma 300, there is a congruence $\varphi$ on $G$ that is an extension of $\gamma$. The uniqueness is obvious.     $\square$

If $I$ and $L$ are simple and $I$ has more than one element, then $L$ is a congruence-preserving extension of $I$. So we obtain:

**Corollary 302.** *Let $K, L, F, I$, and $\psi$ be given as above, and let $I$ have more than one element. If $I$ and $L$ are simple lattices, then $G$ is a congruence-preserving extension of $K$.*

### 2.3 ◇ Generalizations

The most remarkable feature of gluing is Lemma 299. In the first edition of this book I suggested the following field of study (Exercise 12 of Section V.4 in [262]): Find further examples of the phenomenon observed in Lemma 299.

This was taken up in V. Slavík [653]–[655]. He introduced a construction, utilized in A. Day and J. Ježek [141], which was named pasting in E. Fried and G. Grätzer [195], [196]:

Let $L$ be a lattice. Let $A, B, S \leq L$, and let $A \cap B = S$ and $A \cup B = L$. Let $\varphi_A$ and $\varphi_B$ be the embeddings of $A$ and $B$, respectively, into $L$. Then $L$ *pastes* $A$ and $B$ together over $S$ if, whenever $\psi_A$ and $\psi_B$ are embeddings of $A$ and $B$ into a lattice $K$ satisfying $\psi_A(x) = \psi_B(x)$, for all $x \in S$, then there is an embedding $\eta$ of L into K satisfying $\eta\varphi_A = \psi_A$ and $\eta\varphi_B = \psi_B$ (see Figure 61).

Every gluing is a pasting. For a very small example of a pasting that is not a gluing, take the lattices $A = \mathsf{C}_3 = \{0, a, 1\}$ and $B = \mathsf{B}_2 = \{0, u, v, a\}$ ($a$ is the unit element), and set $S = \{0, a\}$. The pasted lattice is $L = \mathsf{B}_2 + \mathsf{C}_1$.

Some lattice properties are preserved under gluing; for example, modularity, see E. Fried and G. Grätzer [196]:

◇ **Theorem 303.** *The class of modular lattices is closed under pasting.*

See also L. Beran [55], E. Fried and G. Grätzer [196], G. Grätzer [265], F. Micol and G. Takách [531], E. T. Schmidt [631].

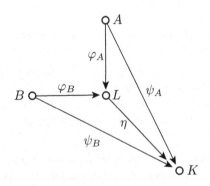

Figure 61. Pasting

C. Herrmann [388] extends the gluing construction to $S$-glued sums, where $S$ is a lattice of finite length, to provide a structural tool for examining modular lattices of finite length. Here are the basic concepts.

Let $\mathcal{S} = (L_s \mid s \in S)$ be a family of lattices of finite length indexed by a lattice $S$ of finite length. The system $\mathcal{S}$ is called an S-*glued system* if the following conditions are satisfied:

(i) For all $s, t \in S$, if $s \leq t$, then either $L_s \cap L_t = \varnothing$ or $L_s \cap L_t$ is a filter in $L_s$ and an ideal in $L_t$.

(ii) For all $s, t \in S$ with $s \leq t$ and for all $a, b \in L_s \cap L_t$, the relation $a \leq b$ holds in $L_s$ iff it holds in $L_t$.

(iii) $L_s \cap L_t \neq \varnothing$ for all $s \prec t \in S$.

(iv) $L_s \cap L_t \subseteq L_{s \vee t} \cap L_{s \wedge t}$ for all $s, t \in S$.

For an S-glued system $\mathcal{S}$, let $L = \bigcup(L_s \mid s \in S)$. Define an ordering $\leq$ in $L$ as follows: for $a, b \in L$, let $a \leq b$ iff there exist a sequence

$$a = x_0, x_1, \ldots, x_n = b$$

of elements of $L$ and an increasing sequence

$$s_0 \leq s_1 \leq \cdots \leq s_{n-1}$$

of elements of $S$ such that $x_{i-1} \leq x_i$ in $L_{s_i}$ for $i = 1, \ldots, n$. Then $L$ is a lattice, the *S-glued sum* of $\mathcal{S}$.

S-glued sums work very well for modular lattices of finite length, see C. Herrmann [388]:

◇ **Theorem 304.** *Every modular lattice $L$ of finite length is the S-glued sum of its maximal complemented intervals.*

There is a further generalization of S-glued sums in A. Day and C. Herrmann [140]. *Multipasting* is a common generalization of pasting and $S$-glued sums, see E. Fried, G. Grätzer, and E. T. Schmidt [202].

**Exercises**

2.1. In the definition of gluing, do we have a choice of how we define $\leq$?

2.2. Can $\mathsf{N}_5$ or $\mathsf{M}_3$ be represented as the gluing of two lattices?

2.3. For a lattice $G$, find conditions under which it can be represented as the gluing of two lattices.

$$* \qquad * \qquad *$$

In the next three exercises, let $K$ and $L$ be lattices, let $F$ be a filter of $K$, and let $I$ be an ideal of $L$ with the isomorphism $\psi\colon F \to J$. Let $G$ be the gluing of $K$ and $L$ over $F$ and $I$.

2.4. What can you say about the length of $G$ in terms of the lengths of $K$ and $L$.

2.5. Show that if $K$ and $L$ are modular, so is $G$.

2.6. Show that if $K$ and $L$ are distributive, so is $G$.

$$* \qquad * \qquad *$$

2.7. Find a lattice identity that is not preserved under gluing.

2.8. Let $L$ be a finite lattice. Let $A, B, S$ be sublattices of $L$, let $L = A \cup B$ and $S = A \cap B$. Prove that $L$ pastes $A$ and $B$ together over $S$ iff the following two conditions hold:

   (i) For $a \in A$ and $b \in B$, if $a < b$, then there exists an $s \in S$ satisfying $a \le s \le b$; and dually.

   (ii) For $s \in S$, all the covers of $s$ in $L$ are in $A$ or all are in $B$; and dually

   (V. Slavík [653], A. Day and J. Ježek [141]).

2.9. Let the finite lattice $L$ paste $A$ and $B$ together over $S$. Let $u, v \in L$ with $u \le v$ and $S \cap [u,v] \ne \varnothing$. Set $L_1 = [u,v]$, $A_1 = A \cap [u,v]$, $B_1 = B \cap [u,v]$, $S_1 = S \cap [u,v]$. Then $L_1$ pastes $A_1$ and $B_1$ over $S_1$ (E. Fried and G. Grätzer [195]).

*2.10. Let $L$ be obtained by pasting together two finite modular lattices. Prove that $L$ is modular (E. Fried and G. Grätzer [195]).

*2.11. Generalize Exercise 2.10 to infinite lattices (E. Fried and G. Grätzer [196]).

2.12. Let the lattice $L$ paste $A$ and $B$ together over $S$. Let $C$ be a convex sublattice of $L$ with $S \cap C \ne \varnothing$. Set $A_1 = A \cap C$, $B_1 = B \cap C$, $S_1 = S \cap C$. Then $C$ pastes $A_1$ and $B_1$ over $S_1$ (E. Fried and G. Grätzer [196]).

$$* \qquad * \qquad *$$

In the next few exercises, let the lattice $L$ be the $S$-glued sum of $S = (L_s \mid s \in S)$.

2.13. Show that if $S = C_2 = \{0,1\}$, then $L$ can be obtained as a gluing of $L_0$ and $L_1$.

2.14. We call the lattices $L_s$, for $s \in S$, the blocks of $L$. Prove that any block is an interval in $L$.

2.15. Show that if $A$ and $B$ are blocks indexed by comparable elements of $S$, then $A \cup B$ is a sublattice $C$ of L. Prove that $C$ is the gluing of A and B, except if $A$ and $B$ are disjoint.

2.16. Let $a \prec b$ in $L$. Verify that $a \prec b$ in some block.

2.17. Let $s, t \in S$. Let $L_s = [a, b]$ and $L_t = [c, d]$. Prove that $L_{s \vee t}$ is of the form $[a \vee c, e]$ for some $e \in L$.

# 3. Chopped Lattices

Finite chopped lattices were introduced by G. Grätzer and H. Lakser, published in the 1978 edition of this book. See also G. Grätzer and E. T. Schmidt [343] and [344], where the infinite case is also discussed.

## 3.1 Basic definitions

An ($n$-ary) *partial operation* on a nonempty set $A$ is a map from a subset of $A^n$ to $A$. For $n = 2$, we call the partial operation *binary*. A *partial algebra* is a nonempty set $A$ with partial operations defined on $A$.

A finite meet-semilattice $(M; \wedge)$ may be regarded as a partial algebra, $(M; \wedge, \vee)$, called a *chopped lattice*, where $\wedge$ is the meet operation and $\vee$ is a partial operation: $a \vee b$ is the least upper bound of $a$ and $b$, provided that $a \vee b$ exists.

In a chopped lattice $M$, in view of the finiteness of $M$, if $a, b \in M$ are majorized by $c \in M$, then $a \vee b$ exists.

We can obtain an example of a chopped lattice by taking a finite lattice with unit, 1, and chopping off the unit element: $M = L - \{1\}$. The converse also holds: by adding a new unit, 1, to a chopped lattice $M$, we obtain a finite lattice $L$, and chopping off the unit element, we get $M$ back.

A more useful example is obtained with merging. Let $C$ and $D$ be lattices such that $J = C \cap D$ is an ideal in both $C$ and $D$. Then, with the natural ordering, $\mathrm{Merge}(C, D) = C \cup D$, called the *merging* of $C$ and $D$, is a chopped lattice. Note that if $a \vee b = c$ in $\mathrm{Merge}(C, D)$, then either $a, b, c \in C$ and $a \vee b = c$ in $C$ or $a, b, c \in D$ and $a \vee b = c$ in $D$.

We define an equivalence relation $\boldsymbol{\alpha}$ to be a *congruence* of a chopped lattice $M$ as we defined it for lattices in Section I.3.6: we require that the Substitution Properties (SP$_\vee$) and (SP$_\wedge$) hold, the former with the proviso: whenever $a_0 \vee a_1$ and $b_0 \vee b_1$ exist. The set $\mathrm{Con}\, M$ of all congruence relations of $M$ ordered by set inclusion is a lattice.

**Lemma 305.** *Let $M$ be a chopped lattice and let $\boldsymbol{\alpha}$ be an equivalence relation on $M$ satisfying the following two conditions for all $x, y, z \in M$:*

(i) *If $x \equiv y \pmod{\boldsymbol{\alpha}}$, then $x \wedge z \equiv y \wedge z \pmod{\boldsymbol{\alpha}}$.*

(ii) *If $x \equiv y \pmod{\boldsymbol{\alpha}}$ and $x \vee z$ and $y \vee z$ exist, then $x \vee z \equiv y \vee z \pmod{\boldsymbol{\alpha}}$.*

*Then $\boldsymbol{\alpha}$ is a congruence relation on $M$.*

*Proof.* Condition (i) states that $\alpha$ preserves $\wedge$.

Now let $x, y, u, v \in S$ with $x \equiv y \pmod{\alpha}$ and $u \equiv v \pmod{\alpha}$; let us assume that $x \vee u$ and $y \vee v$ exist. Then $x \equiv x \wedge y \equiv y \pmod{\alpha}$ and $(x \wedge y) \vee u$ and $(x \wedge y) \vee v$ exist. Thus by condition (ii),

$$x \vee u \equiv (x \wedge y) \vee u \equiv (x \wedge y) \vee v \equiv y \vee v \pmod{\alpha}. \qquad \square$$

A nonempty subset $I$ of the chopped lattice $M$ is an *ideal* if it is a down-set with the property:

(Id) $a, b \in I$ implies that $a \vee b \in I$, provided that $a \vee b$ exists in $M$.

The set $\operatorname{Id} M$ of all ideals of $M$ ordered by set inclusion is a lattice. For $I, J \in \operatorname{Id} M$, the meet is $I \cap J$, but the join is a bit more complicated to describe.

**Lemma 306.** *Let $I$ and $J$ be ideals of the chopped lattice $M$. Define the set $U(I, J)_i \subseteq M$ inductively for all $0 < i < \omega$. Let*

$$U(I, J)_0 = I \cup J.$$

*If $U(I, J)_{i-1}$ is defined, then let $U(I, J)_i$ be the set of all $x \in M$ for which there are $u, v \in U(I, J)_{i-1}$ such that $u \vee v$ is defined in $M$ and $x \le u \vee v$. Then*

$$I \vee J = \bigcup (U(I, J)_i \mid i < \omega).$$

*Proof.* Set $U = \bigcup (U(I, J)_i \mid i < \omega)$. If $K$ is an ideal of $M$, then $I \subseteq K$ and $J \subseteq K$ imply—by induction—that $U(I, J)_i \subseteq K$, therefore, $U(I, J) \subseteq K$. So it is sufficient to prove that $U$ is an ideal of $M$.

Obviously, $U$ is a down-set. Also, $U$ has property (Id), since if $a, b \in U$ and $a \vee b$ exists in $M$, then $a, b \in U(I, J)_i$, for some $0 < i < \omega$, and so $a \vee b \in U(I, J)_{i+1} \subseteq U$. $\qquad \square$

Most lattice concepts and notation will be used for chopped lattices without further explanation.

In the literature, infinite chopped lattices are also defined, see the Exercises.

### 3.2  Compatible vectors of elements

Let $M$ be a chopped lattice, and let $\operatorname{Max}(M)$ (Max if $M$ is understood) be the set of maximal elements of $M$. Then $M = \bigcup (\operatorname{id}(m) \mid m \in \operatorname{Max})$ and each $\operatorname{id}(m)$ is a (finite) lattice. A *vector* (associated with $M$) is of the form $(i_m \mid m \in \operatorname{Max})$, where $i_m \le m$ for all $m \in M$. We order the vectors componentwise.

With every ideal $I$ of $M$, we can associate the vector $(i_m \mid m \in \operatorname{Max})$ defined by $I \cap \operatorname{id}(m) = \operatorname{id}(i_m)$. Clearly, $I = \bigcup (\operatorname{id}(i_m) \mid m \in M)$. Such vectors are easy to characterize. Let us call the vector $(j_m \mid m \in \operatorname{Max})$ *compatible* if $j_m \wedge n = j_n \wedge m$ for all $m, n \in \operatorname{Max}$.

**Lemma 307.** *Let $M$ be a chopped lattice.*

(i) *There is a one-to-one correspondence between ideals and compatible vectors of $M$.*

(ii) *Given any vector $\mathbf{g} = (g_m \mid m \in \mathrm{Max})$, there is a smallest compatible vector $\overline{\mathbf{g}} = (i_m \mid m \in \mathrm{Max})$ majorizing $\mathbf{g}$.*

(iii) *Let $I$ and $J$ be ideals of $M$, with corresponding compatible vectors*

$$(i_m \mid m \in \mathrm{Max}),$$
$$(j_m \mid m \in \mathrm{Max}).$$

*Then*

(a) *$I \le J$ in $\mathrm{Id}\, M$ iff $i_m \le j_m$ for all $m \in \mathrm{Max}$.*

(b) *The compatible vector corresponding to $I \wedge J$ is $(i_m \wedge j_m \mid m \in \mathrm{Max})$.*

(c) *Let $\mathbf{a} = (i_m \vee j_m \mid m \in \mathrm{Max})$. Then the compatible vector corresponding to $I \vee J$ is $\overline{\mathbf{a}}$.*

*Proof.*
(i) Let $I$ be an ideal of $M$. Then $(i_m \mid m \in \mathrm{Max})$ is compatible since $i_m \wedge m \wedge n$ and $i_n \wedge m \wedge n$ both generate the principal ideal $I \cap \mathrm{id}(m) \cap \mathrm{id}(n)$ for all $m, n \in \mathrm{Max}$.

Conversely, let $(j_m \mid m \in \mathrm{Max})$ be compatible, and define

$$I = \bigcup(\,\mathrm{id}(j_m) \mid m \in \mathrm{Max}\,).$$

Observe that

(1) $$I \cap \mathrm{id}(m) = \mathrm{id}(j_m)$$

for all $m \in \mathrm{Max}$. Indeed, if $x \in I \cap \mathrm{id}(m)$ and $x \in \mathrm{id}(j_n)$, for some $n \in \mathrm{Max}$, then $x \le m \wedge j_n = n \wedge j_m$ (since $(j_m \mid m \in \mathrm{Max})$ is compatible), so $x \le j_m$, that is, $x \in \mathrm{id}(j_m)$. The reverse inclusion is obvious.

$I$ is obviously a down-set. To verify property (Id) for $I$, let $a, b \in I$, and let us assume that $a \vee b$ exists in $M$. Then $a \vee b \le m$, for some $m \in \mathrm{Max}$, so $a \le m$ and $b \le m$. By (1), we get $a \le j_m$ and $b \le j_m$, so $a \vee b \le j_m \in I$. Since $I$ is a down-set, it follows that $a \vee b \in I$, verifying property (Id).

(ii) Obviously, the vector $(m \mid m \in \mathrm{Max})$ majorizes all other vectors and it is compatible. Since the componentwise meet of compatible vectors is compatible, the statement follows.

(iii) is obvious since, by (ii), we are dealing with a closure system (see Section I.3.12). $\qquad\square$

### 3.3  Compatible congruence vectors

Let $M$ be a chopped lattice. With any congruence $\alpha$ of $M$, we can associate the *restriction vector* $(\alpha]_m \mid m \in \mathrm{Max})$, where $\alpha]_m$ is the restriction of $\alpha$ to $\mathrm{id}(m)$. The restriction $\alpha]_m$ is a congruence of the lattice $\mathrm{id}(m)$.

Let $\beta_m$ be a congruence of $\mathrm{id}(m)$ for all $m \in \mathrm{Max}$. The *congruence vector* $(\beta_m \mid m \in \mathrm{Max})$ is called *compatible* if $\beta_m$ restricted to $\mathrm{id}(m \wedge n)$ is the same as $\beta_n$ restricted to $\mathrm{id}(m \wedge n)$ for all $m, n \in \mathrm{Max}$. Obviously, a restriction vector is compatible. The converse also holds.

**Lemma 308.** *Let $(\beta_m \mid m \in \mathrm{Max})$ be a compatible congruence vector of a chopped lattice $M$. Then there is a unique congruence $\alpha$ of $M$ such that the restriction vector of $\alpha$ agrees with $(\beta_m \mid m \in \mathrm{Max})$.*

*Proof.* Let $(\beta_m \mid m \in \mathrm{Max})$ be a compatible congruence vector. We define a binary relation $\alpha$ on $M$ as follows:

Let $m, n \in \mathrm{Max}$. For $x \in \mathrm{id}(m)$ and $y \in \mathrm{id}(n)$, let $x \equiv y \pmod{\alpha}$ if $x \equiv x \wedge y \pmod{\beta_m}$ and $y \equiv x \wedge y \pmod{\beta_n}$.

Obviously, $\alpha$ is reflexive and symmetric. To prove transitivity, choose the elements $m, n, k \in \mathrm{Max}$, and $x \in \mathrm{id}(m)$, $y \in \mathrm{id}(n)$, $z \in \mathrm{id}(k)$. Let us assume that $x \equiv y \pmod{\alpha}$ and $y \equiv z \pmod{\alpha}$, that is,

$$(2) \qquad\qquad x \equiv x \wedge y \pmod{\beta_m},$$

$$(3) \qquad\qquad y \equiv x \wedge y \pmod{\beta_n},$$

$$(4) \qquad\qquad y \equiv y \wedge z \pmod{\beta_n},$$

$$(5) \qquad\qquad z \equiv y \wedge z \pmod{\beta_k}.$$

Then meeting the congruence (4) with $x$ (in the lattice $\mathrm{id}(n)$), we get that

$$(6) \qquad\qquad x \wedge y \equiv x \wedge y \wedge z \pmod{\beta_n},$$

and from (3), by meeting with $z$, we obtain that

$$(7) \qquad\qquad y \wedge z \equiv x \wedge y \wedge z \pmod{\beta_n}.$$

Since $x \wedge y$ and $x \wedge y \wedge z \in \mathrm{id}(m)$, by compatibility, (6) implies that

$$(8) \qquad\qquad x \wedge y \equiv x \wedge y \wedge z \pmod{\beta_m}.$$

Now (2) and (8) yield that

$$(9) \qquad\qquad x \equiv x \wedge y \wedge z \pmod{\beta_m}.$$

Similarly,

$$(10) \qquad\qquad z \equiv x \wedge y \wedge z \pmod{\beta_k}.$$

$\beta_m$ is a lattice congruence on $\mathrm{id}(m)$ and $x \wedge y \wedge z \leq x \wedge z \leq x$, so

(11)
$$x \equiv x \wedge z \quad (\mathrm{mod}\ \beta_m).$$

Similarly,

(12)
$$z \equiv x \wedge z \quad (\mathrm{mod}\ \beta_k).$$

Equation (11) and (12) yield that $x \equiv z$ (mod $\alpha$), proving transitivity.

$(\mathrm{SP}_\wedge)$ is easy: choose the elements $x \in \mathrm{id}(m)$, $y \in \mathrm{id}(n)$, $z \in M$; if the congruence $x \equiv y$ (mod $\alpha$) holds, then $x \wedge z \equiv y \wedge z$ (mod $\alpha$) because $x \wedge z \equiv x \wedge y \wedge z$ (mod $\beta_m$) and $y \wedge z \equiv x \wedge y \wedge z$ (mod $\beta_n$).

Finally, we verify $(\mathrm{SP}_\vee)$. Let $x \equiv y$ (mod $\alpha$) and $z \in M$, and let us assume that $x \vee z$ and $y \vee z$ exist. Then there are $p, q \in \mathrm{Max}$ such that $x \vee z \in \mathrm{id}(p)$ and $y \vee z \in \mathrm{id}(q)$. By compatibility, $x \equiv x \wedge y$ (mod $\beta_p$), so $x \vee z \equiv (x \wedge y) \vee z$ (mod $\beta_p$). Since $(x \wedge y) \vee z \leq (x \vee z) \wedge (y \vee z) \leq x \vee z$, we also have

$$x \vee z \equiv (x \vee z) \wedge (y \vee z) \quad (\mathrm{mod}\ \beta_p).$$

Similarly,

$$y \vee z \equiv (x \vee z) \wedge (y \vee z) \quad (\mathrm{mod}\ \beta_q).$$

The last two displayed equations show that $x \vee z \equiv y \vee z$ (mod $\alpha$).    □

## 3.4  From the chopped lattice to the ideal lattice

The map $m \mapsto \mathrm{id}(m)$ embeds the chopped lattice $M$ with zero into its ideal lattice, $\mathrm{Id}\,M$, so we can regard $\mathrm{Id}\,M$ as an extension. It is, in fact, a congruence-preserving extension (G. Grätzer and H. Lakser [296], proof first published in the book [257]):

**Theorem 309.** *Let $M$ be a chopped lattice. Then $\mathrm{Id}\,M$ is a congruence-preserving extension of $M$.*

*Proof.* Let $\alpha$ be a congruence relation of $M$. If $I, J \in \mathrm{Id}\,M$, define

$$I \equiv J \quad (\mathrm{mod}\ \overline{\alpha}) \quad \text{if}\quad I/\alpha = J/\alpha.$$

Obviously, $\overline{\alpha}$ is an equivalence relation. Let $I \equiv J$ (mod $\overline{\alpha}$), and choose $N \in \mathrm{Id}\,M$ and $x \in I \cap N$. Then $x \equiv y$ (mod $\alpha$), for some $y \in J$, and so $x \equiv x \wedge y$ (mod $\alpha$) and $x \wedge y \in J \cap N$. This shows that $(I \cap N)/\alpha \subseteq (J \cap N)/\alpha$. Similarly, $(J \cap N)/\alpha \subseteq (I \cap N)/\alpha$, so $I \cap N \equiv J \cap N$ (mod $\overline{\alpha}$).

To prove that

$$I \vee N \equiv J \vee N \quad (\mathrm{mod}\ \overline{\alpha}),$$

it is sufficient to verify that $I \vee N \subseteq (J \vee N)/\alpha$ by symmetry. By Lemma 306, this is equivalent to proving that

$$U_n(I, N) \subseteq (J \vee N)/\alpha,$$

for all $n < \omega$. This is obvious for $n = 0$.

Now assume that

$$U_{n-1}(I, N) \subseteq (J \vee N)/\alpha$$

and let $x \in U_n(I, N)$. Then $x \leq t_1 \vee t_2$ for some $t_1, t_2 \in U_{n-1}(I, N)$. Thus

$$t_1 \equiv u_1 \pmod{\alpha},$$
$$t_2 \equiv u_2 \pmod{\alpha},$$

for some $u_1, u_2 \in J \vee N$, and so

$$t_1 \equiv t_1 \wedge u_1 \pmod{\alpha},$$
$$t_2 \equiv t_2 \wedge u_2 \pmod{\alpha}.$$

Observe that $t_1 \vee t_2$ majorizes the elements $t_1 \wedge u_1, t_2 \wedge u_2$; consequently, $(t_1 \wedge u_1) \vee (t_2 \wedge u_2)$ exists. Therefore,

$$t_1 \vee t_2 \equiv (t_1 \wedge u_1) \vee (t_2 \wedge u_2) \pmod{\alpha}.$$

Finally,

$$x \equiv x \wedge (t_1 \vee t_2) = x \wedge ((t_1 \wedge u_1) \vee (t_2 \wedge u_2)) \pmod{\alpha},$$

and

$$x \wedge ((t_1 \wedge u_1) \vee (t_2 \wedge u_2)) \in J \vee N.$$

Thus $x \in (J \vee N)/\alpha$, completing the induction, verifying that $\overline{\alpha}$ is a congruence relation of $\operatorname{Id} M$.

If $a \equiv b \pmod{\alpha}$ and $x \in \operatorname{id}(a)$, then $x \equiv x \wedge b \pmod{\alpha}$. Therefore, $\operatorname{id}(a) \subseteq \operatorname{id}(b)/\alpha$. Similarly, $\operatorname{id}(b) \subseteq \operatorname{id}(a)/\alpha$, and so $\operatorname{id}(a) \equiv \operatorname{id}(b) \pmod{\overline{\alpha}}$. Conversely, if $\operatorname{id}(a) \equiv \operatorname{id}(b) \pmod{\overline{\alpha}}$, then $a \equiv b_1 \pmod{\alpha}$ and $a_1 \equiv b \pmod{\alpha}$ for some $a_1 \leq a$ and $b_1 \leq b$. Forming the join of these two congruences, we get the congruence $a \equiv b \pmod{\alpha}$. Thus $\overline{\alpha}$ has all the properties required by Lemma 306.

To show the uniqueness, let $\beta$ be a congruence relation of $\operatorname{Id} M$ satisfying $\operatorname{id}(a) \equiv \operatorname{id}(b) \pmod{\beta}$ iff $a \equiv b \pmod{\alpha}$. Let the congruence $I \equiv J \pmod{\beta}$ hold for $I, J \in \operatorname{Id} M$, and choose $x \in I$. Then

$$\operatorname{id}(x) \cap I \equiv \operatorname{id}(x) \cap J \pmod{\beta},$$
$$\operatorname{id}(x) \cap I = \operatorname{id}(x),$$
$$\operatorname{id}(x) \cap J = \operatorname{id}(y)$$

for some $y \in J$. Thus $\operatorname{id}(x) \equiv \operatorname{id}(y) \pmod{\beta}$, and so $x \equiv y \pmod{\alpha}$, proving that $I \subseteq J/\alpha$. Similarly, $J \subseteq I/\alpha$, and so $I \equiv J \pmod{\overline{\alpha}}$. Conversely, if $I \equiv J \pmod{\overline{\alpha}}$, then take all $x \in I$ and $y \in J$ with $x \equiv y \pmod{\alpha}$. By our assumption regarding $\beta$, we get the congruence $\operatorname{id}(x) \equiv \operatorname{id}(y) \pmod{\beta}$, and by our definition of $\overline{\alpha}$, the join of all these congruences yields $I \equiv J \pmod{\overline{\alpha}}$. Thus $\beta = \overline{\alpha}$. $\qquad\square$

This result is very useful. It means that in order to construct a finite lattice $L$ to represent a given finite distributive lattice $D$ as a congruence lattice, it is sufficient to construct a chopped lattice $M$ with $\operatorname{Con} M \cong D$, since $\operatorname{Con} M \cong \operatorname{Con}(\operatorname{Id} M) = \operatorname{Con} L$, where $L = \operatorname{Id} M$ and $L$ is a finite lattice.

This result also allows us to construct congruence-preserving extensions.

**Corollary 310.** *Let $M = \operatorname{Merge}(A, B)$ be a chopped lattice with $A = \operatorname{id}(a)$ and $B = \operatorname{id}(b)$. If $a \wedge b > 0$ and $B$ is simple, then the lattice $\operatorname{Id} M$ is a congruence-preserving extension of the lattice $A$.*

*Proof.* Let $\alpha$ be a congruence of $A$. Then $(\alpha, \beta)$ is a compatible congruence vector iff

$$\beta = \begin{cases} 0, & \text{if } \alpha \text{ is discrete on } [0, a \wedge b]; \\ 1, & \text{otherwise.} \end{cases}$$

So $\beta$ is determined by $\alpha$ and the statement follows.     $\square$

## Exercises

3.1. Take a finite lattice $L$ and an up-set $A \subset L$, and define $M = L - A$. Is $M$ a chopped lattice? Is this a typical example?

3.2. Let $M_1$ and $M_2$ be chopped lattices. Let $p_1$ be an atom of $M_1$ and let $p_2$ be an atom of $M_2$. Define the set $M$ as the disjoint union $M_1 \cup M_2$ with the zeroes identified, and also $p_1$ identified with $p_2$. Then $M$ is a chopped lattice containing $M_1$ and $M_2$ as chopped sublattices.

*3.3. Let $L$ be an arbitrary finite lattice with more than one element. Then $L$ has a finite congruence-preserving extension $K$ such that $K$ is atomistic and has no proper automorphism (G. Grätzer and E. T. Schmidt [343]).

$$* \qquad * \qquad *$$

The following exercises are based on G. Grätzer and E. T. Schmidt [344].

In the general (not finite) case, define a *chopped lattice* $M$ as a partial lattice we obtain from a bounded lattice by removing the unit element. As in the finite case, we introduce ideals, congruences, and merging: $\operatorname{Merge}(C, D) = C \cup D$, called the *merging* of $C$ and $D$, where $C$ and $D$ are lattices with zero and $J = C \cap D$ is an ideal in both $C$ and $D$. A chopped lattice $M$ satisfies condition (FG) if every finitely generated ideal is a finite union of principal ideals.

3.4. Give an internal characterization of a chopped lattice $M$ as a partial algebra $(A; \vee, \wedge)$.

3.5. Is there any connection between the congruence lattice of a chopped lattice $M$ and the congruence lattice of its ideal lattice Id $M$?

3.6. Let $C$ and $D$ be lattices with zero and let $J = C \cap D$ be an ideal in both $C$ and $D$. Verify that $\mathrm{Merge}(C, D)$ is a chopped lattice.

3.7. Show that every congruence of a chopped lattice $M$ has an extension to Id $M$.

3.8. Find an example of a chopped lattice $M$ with countably many congruences such that Id $M$ has the power of the continuum.

3.9. Let $\mathrm{Id}_{\mathrm{fg}}\, M$ denote the order of finitely generated ideals of the chopped lattice $M$. Show that if $M$ satisfies (FG), then $\mathrm{Id}_{\mathrm{fg}}\, M$ is a sublattice of Id $M$.

3.10. Show that if the chopped lattice $M$ satisfies (FG), then $\mathrm{Id}_{\mathrm{fg}}\, M$ is a congruence-preserving extension of $M$.

3.11. When is $\mathrm{Merge}(C, D)$ a chopped lattice?

3.12. Find conditions under which $\mathrm{Merge}(C, D)$ satisfies (FG).

# 4.  Constructing Lattices with Given Congruence Lattices

## 4.1  The finite case

The Funayama-Nakayama Theorem (see Theorem 149), states that the congruence lattice of a lattice is distributive. As an application of chopped lattices, we prove the converse for finite lattices, first published in G. Grätzer and E. T. Schmidt [337].

**Theorem 311 (The Dilworth Theorem).** *Every finite distributive lattice $D$ can be represented as the congruence lattice of a finite lattice $L$.*

Applying Theorem 309, it is sufficient to verify the following:

**Theorem 312.** *Let $D$ be a finite distributive lattice. Then there exists a chopped lattice $M$ such that $\mathrm{Con}\, M$ is isomorphic to $D$.*

Using the equivalence of nontrivial finite distributive lattices and finite orders (see Corollary 108) and using the notation $\mathrm{Con_{Ji}}\, M$ (see Section III.1.4) for the order of join-irreducible congruences of $M$, we can rephrase Theorem 312 as follows:

**Theorem 313.** *Let $P$ be a finite order. Then there exists a chopped lattice $M$ such that $\mathrm{Con_{Ji}}\, M$ is isomorphic to $P$.*

The basic "gadget" for the construction is the lattice $\mathsf{N_6} = \mathsf{N}(p, q)$ of Figure 62. The lattice $\mathsf{N}(p, q)$ has three congruence relations, namely, $\mathbf{0}, \mathbf{1}, \alpha$, where $\alpha$ is the congruence relation with blocks $\{0, q_1, q_2, q\}$ and $\{p_1, p(q)\}$, indicated by the dashed line in Figure 62. Thus $\mathrm{con}(p_1, 0) = \mathbf{1}$. In other words, $p_1 \equiv 0$ "implies" that $q_1 \equiv 0$, but $q_1 \equiv 0$ "does not imply" that $p_1 \equiv 0$. We will use the "gadget" $N_6 = N(p, q)$ to achieve such congruence-forcing.

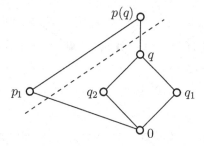

Figure 62. The lattice $N_6 = N(p, q)$ and the congruence $\alpha$

To convey the idea of how to prove Theorem 313, we present three small examples in which we construct the chopped lattice $M$ from copies of $N(p, q)$.

*Example 1: The three-element chain.* Let $P = \{a, b, c\}$ with $c \prec b \prec a$. We take two copies of the gadget, $N(a, b)$ and $N(b, c)$; they share the ideal $I = \{0, b_1\}$, see Figure 63. So we can merge them and form the chopped lattice

$$M = \mathrm{Merge}(N(a, b), N(b, c))$$

as shown in Figure 63.

The congruences of $M$ are easy to find. The isomorphism $P \cong \mathrm{Con_{Ji}}\, M$ is given by the map $x \mapsto \mathrm{con}(0, x_1)$ for $x \in P$.

A congruence of $M$ can be described by a vector $(\alpha_{a,b}, \alpha_{b,c})$ (a compatible congruence vector), where $\alpha_{a,b}$ is a congruence of the lattice $N(a, b)$ and $\alpha_{b,c}$ is a congruence of the lattice $N(b, c)$, subject to the condition that $\alpha_{a,b}$ and $\alpha_{b,c}$ agree on $I$. Looking at Figure 62, we see that if the shared congruence on $I$ is $\mathbf{0}$ $(= \mathbf{0}_I)$, then we must have

$$\alpha_{a,b} = \mathbf{0}\ (= \mathbf{0}_{N(a,b)}),$$
$$\alpha_{b,c} = \mathbf{0}\ (= \mathbf{0}_{N(b,c)})\quad \text{or}\quad \alpha_{b,c} = \alpha\ \text{on}\ N(b, c).$$

If the shared congruence on $I$ is $\mathbf{1}$ $(= \mathbf{1}_I)$, then we must have

$$\alpha_{a,b} = \alpha\quad \text{or}\quad \alpha_{a,b} = \mathbf{1}\ (= \mathbf{1}_{N(a,b)})\ \text{on}\ N(a, b),$$
$$\alpha_{b,c} = \mathbf{1}\ (= \mathbf{1}_{N(b,c)})\ \text{on}\ N(b, c).$$

So there are three congruences distinct from $\mathbf{0}$: $(\mathbf{0}, \alpha), (\alpha, \mathbf{1}), (\mathbf{1}, \mathbf{1})$. Thus the join-irreducible congruences form the three-element chain.

*Example 2: The three-element order $P_V$ of Figure 64.* (We call $P_V$ the "order $V$".) We take two copies of the gadget, $N(b, a)$ and $N(c, a)$; they share

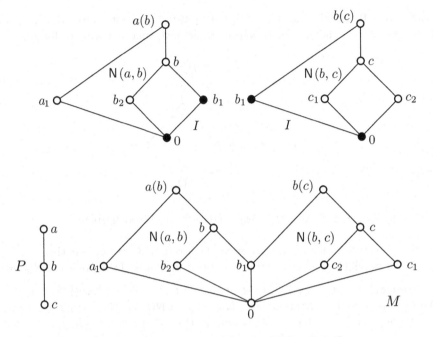

Figure 63. The chopped lattice $M$ for $P = \mathsf{C}_3$

the ideal $J = \{0, a_1, a_2, a\}$; we merge them to form the chopped lattice

$$M_V = \mathrm{Merge}(\mathsf{N}(b, a), \mathsf{N}(c, a)),$$

see Figure 64. Again, the isomorphism $P_V \cong \mathrm{Con}_{\mathrm{Ji}} M_V$ is given by the map $x \mapsto \mathrm{con}(0, x_1)$ for $x \in P_V$.

*Example 3: The three-element order $P_H$ of Figure 65.* (We call $P_H$ the "order hat".) We take two copies of the gadget, $\mathsf{N}(a, b)$ and $\mathsf{N}(a, c)$; they share the ideal $J = \{0, a_1\}$; we merge them to form the chopped lattice

$$M_V = \mathrm{Merge}(\mathsf{N}(a, b), \mathsf{N}(a, c)),$$

see Figure 65. Again, the isomorphism $P_H \cong \mathrm{Con}_{\mathrm{Ji}} M_H$ is given by the map $x \mapsto \mathrm{con}(0, x_1)$ for $x \in P_V$.

The reader should now be able to picture the general proof: instead of the few atoms in these examples, we start with enough atoms to reflect the structure of $P$, see Figure 66. Whenever $b \prec a$ in $P$, we build a copy of $\mathsf{N}(a, b)$, see Figure 67.

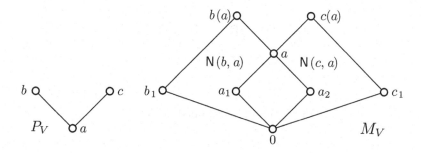

Figure 64. The chopped lattice for the order $V$

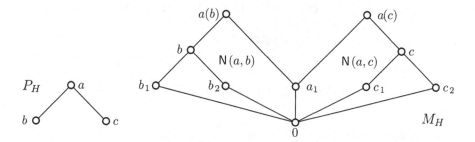

Figure 65. The chopped lattice for the order hat

## 4.2   Construction and proof

For a finite order $P$, let Max be the set of maximal elements in $P$. We form the set

$$M_0 = \{0\} \cup \{\, p_1 \mid p \in \text{Max}\,\} \cup \bigcup(\{a_1, a_2\} \mid a \in P - \text{Max}\,)$$

consisting of 0, the maximal elements of $P$ indexed by 1, and two copies of the nonmaximal elements of $P$, indexed by 1 and 2. We make $M_0$ a meet-semilattice by defining $\inf\{x, y\} = 0$ if $x \neq y$, as illustrated in Figure 66. Note that $x \equiv y \pmod{\alpha}$ and $x \neq y$ imply that $x \equiv 0 \pmod{\alpha}$ and $y \equiv 0 \pmod{\alpha}$ in $M_0$; therefore, the congruence relations of $M_0$ are in one-to-one correspondence with subsets of $P$. Thus $\text{Con}\, M_0$ is a boolean lattice whose atoms are associated with atoms of $M_0$; the congruence $\Phi_x$ associated with the atom $x$ has only one nontrivial block $\{0, x\}$.

We construct an extension $M$ of $M_0$ as follows:

The chopped lattice $M$ consists of four kinds of elements: (i) the zero, 0; (ii) for all maximal elements $p$ of $P$, the element $p_1$; (iii) for any nonmaximal element $p$ of $P$, three elements: $p, p_1, p_2$; (iv) for each pair $p, q \in P$ with $p \succ q$,

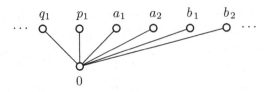

Figure 66. The chopped lattice $M_0$

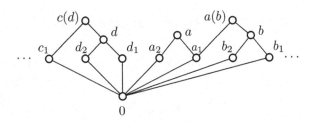

Figure 67. The chopped lattice $M$

a new element, $p(q)$. For $p, q \in P$ with $p \succ q$, we set

$$N_6 = N(p, q) = \{0, p_1, q, q_1, q_2, p(q)\}.$$

For $x, y \in M$, let us define $x \leq y$ to mean that $x, y \in N(p, q)$, for some $p, q \in P$ with $p \succ q$ and $x \leq y$ in the lattice $N(p, q)$. It is easily seen that $x \leq y$ does not depend on the choice of $p$ and $q$, and that $\leq$ is an ordering. Since, under this ordering, all $N(p, q)$ and $N(p, q) \cap N(p', q')$ ($p \succ q$ and $p' \succ q'$ in $P$) are lattices, $x \in N(p, q)$ and $y \leq x$ imply that $y \in N(p, q)$, so we conclude that $M$ is a chopped lattice; in fact, it is a union of the ideals $N(p, q)$ with $p \succ q$ in $P$, and two such distinct ideals intersect in a one-, two-, or four-element ideal.

Since the chopped lattice $M$ is atomistic, Corollary 240 applies. If $p_i \Rightarrow q_j$ in $M$, for $p, q \in P$ and $i, j \in \{1, 2\}$, then $p \geq q$ in $P$, and conversely. So the blocks $\mathrm{Atom}(M)$ under the preordering $\Rightarrow$ form an order isomorphic to $\mathrm{Down}\, P$. This completes the verification that $\mathrm{Con_{Ji}}\, M \cong P$, and therefore, of Theorem 313.

## 4.3  Sectional complementation

We define *sectionally complemented chopped lattices* as for lattices in Section I.6.1: We call the chopped lattice $M$ *sectionally complemented* if it has a zero and for all $a \leq b \in M$, there exist an element $c \in M$ satisfying $a \vee c = b$ and $a \wedge c = 0$.

We illustrate the use of compatible vectors with two results on sectionally complemented chopped lattices. The first result is from G. Grätzer and E. T. Schmidt [345].

**Lemma 314 (Atom Lemma).** *Let $M$ be a chopped lattice with exactly two maximal elements $m_1$ and $m_2$. We assume that $\mathrm{id}(m_1)$ and $\mathrm{id}(m_2)$ are sectionally complemented lattices. If $p = m_1 \wedge m_2$ is an atom, then $\mathrm{Id}\, M$ is sectionally complemented.*

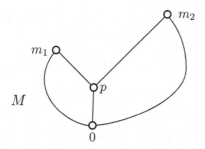

Figure 68. Atom Lemma illustrated

*Proof.* To show that $\mathrm{Id}\, M$ is sectionally complemented, let $I \subseteq J$ be two ideals of $M$, represented by the compatible vectors $(i_1, i_2)$ and $(j_1, j_2)$, respectively. Let $s_1$ be a sectional complement of $i_1$ in $j_1$ and let $s_2$ be a sectional complement of $i_2$ in $j_2$. If $p \wedge s_1 = p \wedge s_2$, then $(s_1, s_2)$ is a compatible vector, representing an ideal $S$ that is a sectional complement of $I$ in $J$. Otherwise, without loss of generality, we can assume that $p \wedge s_1 = 0$ and $p \wedge s_2 = p$. Since the ideal $\mathrm{id}(m_2)$ is sectionally complemented, there is a sectional complement $s_2'$ of $p$ in the interval $[0, s_2]$. Then $(s_1, s_2')$ satisfies $p \wedge s_1 = p \wedge s_2' \, (= 0)$, and so it is compatible; therefore, $(s_1, s_2')$ represents an ideal $S$ of $M$ by Lemma 307. Obviously, $I \wedge S = \{0\}$.

From $p \wedge s_2 = p$, it follows that $p \leq s_2 \leq j_2$. Since $J$ is an ideal and $j_2 \wedge p = p$, it follows that $j_1 \wedge p = p$, that is, $p \leq j_1$. Obviously, $I \vee S \subseteq J$. So to show that $I \vee S = J$, it is sufficient to verify that $j_1, j_2 \in I \vee S$. Evidently, $j_1 = i_1 \vee s_1 \in I \vee S$. Note that $p \leq j_1 = i_1 \vee s_1 \in I \vee S$. Thus $p, s_2', i_2 \in I \vee S$, and therefore

$$p \vee s_2' \vee i_2 = (p \vee s_2') \vee i_2 = s_2 \vee i_2 = j_2 \in I \vee S. \qquad \square$$

The second result (G. Grätzer, H. Lakser, and M. Roddy [314]) shows that the ideal lattice of a sectionally complemented chopped lattice is not always sectionally complemented.

**Theorem 315.** *There is a sectionally complemented chopped lattice $M$ whose ideal lattice $\mathrm{Id}\, M$ is not sectionally complemented.*

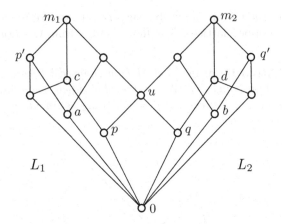

Figure 69. The chopped lattice $M$

*Proof.* Let $M$ be the chopped lattice of Figure 69, where $L_1 = \mathrm{id}(m_1)$ and $L_2 = \mathrm{id}(m_2)$. Note that $p$ is meet-irreducible in $\mathrm{id}(m_2)$ and $q$ is meet-irreducible in $\mathrm{id}(m_1)$.

The unit element of the ideal lattice of $M$ is the compatible vector $(m_1, m_2)$. We show that the compatible vector $(a, b)$ has no complement in the ideal lattice of $M$.

Assume, to the contrary, that the compatible vector $(s, t)$ is a complement of $(a, b)$. Since $(a, b) \le (a \vee u, m_2)$, a compatible vector, $(s, t) \nleq (a \vee u, m_2)$, that is,

$$(13) \qquad\qquad s \nleq a \vee u.$$

Similarly, by considering $(m_1, b \vee u)$, we conclude that

$$(14) \qquad\qquad t \nleq b \vee u.$$

Now $(a, b) \le (p', q')$, which is a compatible vector. Thus $(s, t) \nleq (p', q')$, and so either $s \nleq p'$ or $t \nleq q'$. We may assume that $s \nleq p'$, without loss of generality. It then follows by (13) that $s$ can be only $c$ or $m_1$. Since $s \wedge a = 0$, we conclude that $s = c$. Thus $s \wedge u = p$, and so $t \wedge u = p$. But $p$ is meet-irreducible in $L_2$. Thus $t = p \le b \vee u$, contradicting (14). □

This result illustrates that the Atom Lemma (Lemma 314) cannot be extended to the case where $[0, m_1 \wedge m_2]$ is a four-element boolean lattice.

## 4.4  ◇ Finite lattices
## by J. B. Nation

There are different ways of looking at lattices, each with its own advantages. To study congruences, it is useful to represent a finite lattice as the lattice of closed sets of a closure operator on its set of join-irreducible elements. This is an efficient way to encode the structure.

The techniques described here were developed over a period of time by R. N. McKenzie, B. Jónsson, A. Day, R. Freese and J. B. Nation to deal with various specific questions, see [138], [191], [449], [512], [540], [541].

We need some terminology. Throughout this section, $L$ will denote a finite lattice.

For subsets $X$, $Y \subseteq L$ we say that $X$ *refines* $Y$ (in symbols, $X \ll Y$) if for each element $x \in X$, there exists an element $y \in Y$ with $x \leq y$. It is easy to see that the relation $\ll$ is a preordering, but not, in general, antisymmetric. Note that $X \subseteq Y$ implies that $X \ll Y$.

A *join cover* of an element $p \in L$ is a finite set $A$ such that $p \leq \bigvee A$. A join cover $A$ of $p$ is *minimal* if $\bigvee A$ is an irredundant join and $A$ cannot be properly refined to another join cover of $p$, that is, $p \leq \bigvee B$ and $B \ll A$ implies that $A \subseteq B$.

Define a binary relation $D$ on the set $\operatorname{Ji} L$ of join-irreducibles as follows: $p \mathrel{D} q$ if there exists an element $x \in L$ such that $p \leq q \vee x$ but $p \not\leq q_* \vee x$, where $q_*$ denotes the unique element with $q \succ q_*$ (see Section I.6.3). Equivalently, $p \mathrel{D} q$ if $q \in Q$ for some minimal nontrivial join cover $Q$ of $p$. This relation is central to our analysis of the congruences of a finite lattice.

Every finite lattice can be represented as the lattice of closed sets of a closure operator on the set of its join-irreducible elements. The closed sets are those subsets $A \subseteq \operatorname{Ji} L$ of the form $A = \operatorname{id}(a) \cap \operatorname{Ji} L$ for some $a \in A$. In this representation, a set $A \subseteq \operatorname{Ji} L$ is closed iff the following two conditions hold:

1. $x \leq y \in A$ implies that $x \in A$;
2. if $Y$ is a minimal nontrivial join cover of $x$ and $Y \subseteq A$, then $x \in A$.

The corresponding element is then, of course, $a = \bigvee A$.

This representation is a type of duality for finite lattices. These ideas are developed by L. Santocanale [626], extending [541]. An alternate representation is provided by R. Wille's *formal concept analysis*, which is based on the Galois connection induced by the relation $\leq$ restricted to $\operatorname{Ji} L \times \operatorname{Mi} L$, see [219]. It has also proved to be useful in working with congruence lattices.

Now let us describe the congruence lattice of a finite lattice in terms of the $D$-relation. The reflexive, transitive closure of $D$ is a preorder $\sqsubseteq$ on $\operatorname{Ji} L$. This in turn induces an equivalence $\equiv$, so that $((\operatorname{Ji} L)/{\equiv}; \sqsubseteq)$ is an order $\operatorname{Qu} L$. In this order, if $\overline{p}$ denotes the block $p/{\equiv}$ and $\overline{q}$ denotes $q/{\equiv}$, then $\overline{p} \sqsubseteq \overline{q}$ iff there is a sequence $r_0, \dots, r_k$, for $k \geq 0$, of join-irreducible elements with

$$p = r_0 \mathrel{D} r_1 \mathrel{D} \cdots \mathrel{D} r_{k-1} \mathrel{D} r_k = q.$$

The congruence lattice of $L$ is now easily determined.

◇ **Theorem 316.** *If $L$ is a finite lattice, then the congruence lattice* $\operatorname{Con} L$ *is isomorphic to the lattice* $\operatorname{Down} \operatorname{Qu} L$.

The basic idea here is quite simple. It is easy to see that if $p \, D \, q$, then $\mathrm{con}(p, p_*) \leq \mathrm{con}(q, q_*)$. The congruence $\boldsymbol{\theta}_I$ associated with a down-set $I$ of $\mathrm{Qu}\, L$ should collapse exactly those pairs $(p, p_*)$ with $\overline{p} \in I$, for a join-irreducible element $p$. To extend this to a congruence on $L$, let

$$T = \{\, q \in \mathrm{Ji}\, L \mid \overline{q} \notin I \,\},$$

and define $x \, \boldsymbol{\theta}_I \, y$ iff $\mathrm{id}(x) \cap T = \mathrm{id}(y) \cap T$. It is straightforward to check that $\boldsymbol{\theta}_I$ is a congruence relation with $p \, \boldsymbol{\theta}_I \, p_*$ iff $p \notin T$, that is, $\overline{p} \in I$.

The $D$ relation is easy to determine, so it is not hard to find $\mathrm{Qu}\, L$ for a finite lattice $L$. Hence this result provides a reasonably efficient algorithm for determining the congruence lattice of a finite lattice. Let us turn to some of the consequences of this characterization.

We want to represent a finite distributive lattice $D$ as the congruence lattice of a finite lattice. Now $D \cong \mathrm{Down}\, P$ for an order $P$, so we need only construct a finite lattice $L$ with $\mathrm{Qu}\, L \cong P$. This is an easy exercise (justifying, perhaps, the original appearance of Dilworth's representation theorem as an exercise with an asterisk in Birkhoff's Lattice Theory, [70]).

By a result M. Tischendorf [679], every finite lattice $L$ has a congruence-preserving embedding into a finite atomistic lattice. To verify this, observe that the structure of $L$ is determined by the order on $\mathrm{Ji}\, L$ and its minimal nontrivial join covers, while $\mathrm{Qu}\, L$ and the congruence lattice are determined by the minimal nontrivial join covers alone. So we can form a new lattice $L'$ with the same set of join-irreducible elements, but now ordered as an antichain, and the same minimal nontrivial join covers. Then $\mathrm{Qu}\, L' = \mathrm{Qu}\, L$, clearly $L$ is naturally embedded into $L'$, which is atomistic.

As another application, let $L$ be a finite lattice that is either modular or relatively complemented. For these types of lattices, it is not hard to show that the relation $D$ is symmetric on $\mathrm{Ji}\, L$, and hence $\mathrm{Con}\, L$ is a boolean algebra. Of course, in the relatively complemented case, we can do somewhat better: a relatively complemented finite lattice is a direct sum of simple (relatively complemented) lattices (R. P. Dilworth [158]).

A $D$-cycle in a finite lattice is a sequence $r_0, \ldots, r_{k-1}$, for $k \geq 2$, of join-irreducible elements such that

$$r_0 \, D \, r_1 \, D \cdots D \, r_{k-1} \, D \, r_0.$$

Clearly, the equivalence relation $\equiv$ used in constructing $\mathrm{Qu}\, L$ will be nontrivial iff the lattice $L$ contains a $D$-cycle.

R. N. McKenzie [512] showed that the absence of $D$-cycles has profound consequences. A lattice homomorphism $h \colon K \to L$ is *lower bounded* if, for each element $a \in L$, the preimage $\{\, x \in K \mid h(x) \geq a \,\}$ has a least element. The basic result relating these ideas is the following.

◊ **Theorem 317.** *The following are equivalent for a finite lattice L.*

(i) *L contains no D-cycle.*

(ii) *There exists a lower bounded homomorphism h: F → L from a finitely generated free lattice F onto L.*

(iii) *Every homomorphism g: K → L from a finitely generated lattice K into L is lower bounded.*

Properties (ii) and (iii) of Theorem 317 are equivalent for a finitely generated lattice. A finitely generated lattice with these properties is said to be *lower bounded*. More generally, a lattice L is lower bounded if every finitely generated sublattice of L is lower bounded. The dual property is called *upper bounded*, and lattices that are both upper and lower bounded are called *bounded*. (The terminology conflicts with the other definition of *bounded*, but both are in common use. The context usually makes it clear when these properties are intended, rather than having a greatest and least element.) The theory of finitely generated, lower bounded lattices is developed in the book *Free Lattices* [186], see Section VII.2.7, and for various extensions, K. V. Adaricheva and V. A. Gorbunov [19].

Here, let us just mention two applications of these concepts for finite lattices. To the list in Theorem 557 of Chapter VII of conditions characterizing finite projective lattices, we can add

(vi) *K is bounded and satisfies (W).*

*Splitting lattices*, discussed in Section VI.2.3, are characterized thusly: L is a splitting lattice iff L is a finite, subdirectly irreducible, bounded lattice.

## 4.5   ◊ Finite lattices in special classes

This field is surveyed in the author's book [271]. Most of the results are of the following two types, where **K** is a class of lattices.

**Theorem Scheme A.** *Every finite distributive lattice D can be represented as the congruence lattice of a finite lattice L ∈ **K**.*

**Theorem Scheme B.** *Every finite lattice K has a congruence-preserving extension to a finite lattice L ∈ **K**.*

For **K** = **L**, the class of all lattices, Theorem Scheme A becomes Theorem 311, so it holds. For the class **K** of all sectionally complemented lattices, the validity of Theorem Scheme A is stated in Exercise 4.5.

Let us call a congruence $\alpha$ of a lattice L a *uniform congruence* if all blocks of $\alpha$ have the same cardinality. We call a lattice L a *uniform lattice* if all of its congruences are uniform. In G. Grätzer, E. T. Schmidt, and K. Thomsen [350], Theorem Scheme A is verified for the class **K** of uniform lattices.

There is a lattice property much stronger than uniformity. We call a congruence $\alpha$ of a lattice $L$ an *isoform congruence* if every block of $\alpha$ is isomorphic to every other one. We call lattice $L$ an *isoform lattice* if all of its congruences are isoform. In G. Grätzer and E. T. Schmidt [349], Theorem Scheme A is proved for the class **K** of isoform lattices.

My book [271] presents many more results of this type, for instance, for the class **K** of (planar) semimodular lattices, see G. Grätzer, H. Lakser, and E. T. Schmidt [318].

Theorems of the type Theorem Scheme B are much harder to prove. It was done for the class **K** of all sectionally complemented lattices as stated in Exercise 4.6, for the class **K** of isoform lattices in G. Grätzer, R. W. Quackenbush, and E. T. Schmidt [328], for the class **K** of semimodular lattices in G. Grätzer and E. T. Schmidt [346].

### 4.6  ◇ Two finite lattices

The loosest connection between two lattices $L_1$, $L_2$ is that they are both sublattices of the same lattice $L$. In this case, there is a map

$$\operatorname{Con} L_1 \to \operatorname{Con} L_2$$

obtained by first extending each congruence relation of $L_1$ to $L$ and then restricting the resulting congruence relation to $L_2$. This map is isotone and it preserves 0 (that is, **0**). The converse also holds for finite lattices (G. Grätzer, H. Lakser, and E. T. Schmidt [317]):

◇ **Theorem 318.** *Let $D_1$ and $D_2$ be finite distributive lattices, and let*

$$\psi\colon D_1 \to D_2$$

*be an isotone map that preserves 0. Then there is a finite lattice $L$ with sublattices $L_1$ and $L_2$ and there are isomorphisms*

$$\alpha_1\colon D_1 \to \operatorname{Con} L_1, \qquad \alpha_2\colon D_2 \to \operatorname{Con} L_2$$

*such that the diagram*

$$
\begin{array}{ccc}
D_1 & \xrightarrow{\ \psi\ } & D_2 \\[2pt]
{\scriptstyle\cong}\big\downarrow{\scriptstyle\alpha_1} & & {\scriptstyle\cong}\big\downarrow{\scriptstyle\alpha_2} \\[2pt]
\operatorname{Con} L_1 & \xrightarrow[]{\text{extension}} \operatorname{Con} L \xrightarrow[]{\text{restriction}} & \operatorname{Con} L_2
\end{array}
$$

*is commutative.*

See also G. Grätzer, H. Lakser, and E. T. Schmidt [319].

A tighter connection between two lattices $K$ and $L$ is the sublattice relation: $K \leq L$. How then does $\operatorname{Con} K$ relate to $\operatorname{Con} L$?

As we discussed in Section III.1.5, the relation $K \leq L$ induces a map ext of $\operatorname{Con} K$ into $\operatorname{Con} L$: For a congruence relation $\alpha$ of $K$, the image $\operatorname{ext} \alpha$ is the congruence relation of $L$ generated by $\alpha$, that is, $\operatorname{ext} \alpha = \operatorname{con}_L(\alpha)$. The map ext is a $\{0\}$-separating join-homomorphism.

In 1974, A. P. Huhn in [406] stated the converse:

$\Diamond$ **Theorem 319.** *Let $D$ and $E$ be finite distributive lattices, and let*

$$\psi \colon D \to E$$

*be a $\{0\}$-separating join-homomorphism. Then there are finite lattices $K \leq L$, and isomorphisms $\gamma \colon D \to \operatorname{Con} K$ and $\delta \colon E \to \operatorname{Con} L$ satisfying*

$$\delta\psi = (\operatorname{ext} \operatorname{id}_K)\gamma,$$

*where $\operatorname{id}_K$ is the embedding of $K$ into $L$; that is, such that the diagram*

$$
\begin{array}{ccc}
D & \xrightarrow{\ \psi\ } & E \\[2pt]
{\scriptstyle\cong}\Big\downarrow{\scriptstyle\gamma} & & {\scriptstyle\cong}\Big\downarrow{\scriptstyle\delta} \\[2pt]
\operatorname{Con} K & \xrightarrow{\ \operatorname{ext} \operatorname{id}_K\ } & \operatorname{Con} L
\end{array}
$$

*is commutative.*

A much stronger version of this theorem is in G. Grätzer, H. Lakser, and E. T. Schmidt [316].

If $K$ is an ideal of $L$, then the restriction map $\operatorname{Con} L \to \operatorname{Con} K$ is a $\{0, 1\}$-homomorphism and the converse also holds (G. Grätzer and H. Lakser [304]):

$\Diamond$ **Theorem 320.** *Let $D$ and $E$ be finite distributive lattices; let $D$ have more than one element. Let $\varphi$ be a $\{0, 1\}$-preserving homomorphism of $D$ into $E$. Then there exists a finite lattice $L$ and an ideal $K$ of $L$ such that $D \cong \operatorname{Con} L$, $E \cong \operatorname{Con} K$, and $\varphi$ is represented by re, the restriction map.*

### 4.7   $\Diamond$ More than two finite lattices

The first result involving four lattices is due to J. Tůma [683]. We state it in the following stronger form, see G. Grätzer, H. Lakser, and F. Wehrung [320] (for the notation $\operatorname{Con} \eta$, see Lemma 20; see also Exercise 4.7):

$\Diamond$ **Theorem 321.** *Let $L_0$, $L_1$, $L_2$ be finite lattices and let $\eta_i \colon L_0 \to L_i$, for $i \in \{1, 2\}$, be lattice homomorphisms. Let $D$ be a finite distributive lattice, and, for $i \in \{1, 2\}$, let $\psi_i \colon \operatorname{Con} L_i \to D$ be join-homomorphisms such that*

$$\psi_1 \circ \operatorname{Con} \eta_1 = \psi_2 \circ \operatorname{Con} \eta_2.$$

*There is then a finite atomistic lattice $L$, there are lattice homomorphisms $\varphi_i \colon L_i \to L$, for $i \in \{1, 2\}$, with*

$$\varphi_1 \circ \eta_1 = \varphi_2 \circ \eta_2,$$

*and there is an isomorphism $\gamma \colon \operatorname{Con} L \to D$ such that*

$$\gamma \circ \operatorname{Con} \varphi_i = \psi_i \quad \text{for } i \in \{1, 2\}.$$

*If both $\eta_1$, $\eta_2$ preserve zero, then $\varphi_1$, $\varphi_2$ can be chosen to preserve zero.*

Most of the maps of Theorem 321 are pictured in the following two diagrams:

$$
\begin{array}{ccc}
L_0 & \xrightarrow{\;\eta_1\;} & L_1 \\
{\scriptstyle \eta_2}\big\downarrow & & \big\downarrow{\scriptstyle \varphi_1} \\
L_2 & \xrightarrow{\;\varphi_2\;} & L
\end{array}
\qquad
\begin{array}{ccccc}
\operatorname{Con} L_0 & \xrightarrow{\;\operatorname{Con} \eta_1\;} & \operatorname{Con} L_1 \\
{\scriptstyle \operatorname{Con} \eta_2}\big\downarrow & & \big\downarrow{\scriptstyle \psi_1} \\
\operatorname{Con} L_1 & \xrightarrow{\;\psi_2\;} & D & \xleftarrow{\;\gamma\;} & \operatorname{Con} L
\end{array}
$$

The second diagram is called the *congruence lifting* of the first diagram, see P. Pudlák [597], J. Tůma [683], J. Tůma and F. Wehrung [684], [685], F. Wehrung [702]–[707].

## 4.8 ◇ Independence theorem for finite lattices

In Section II.1.6, we proved Birkhoff's result (Theorem 125), which we now state as follows: *Every (finite) group $G$ can be represented as the automorphism group of a (finite) lattice $L$.*

In this section, we proved Dilworth's result (Theorem 311): *Every finite distributive lattice $D$ can be represented as the congruence lattice of a finite lattice $L$.*

Can we combine these two results? Indeed, we can. This question was raised in 1975 in the first edition of this book (Problem II.18):

*Let $K$ be a lattice with more than one element, and let $G$ be a group. Does there exist a lattice $L$ such that the congruence lattice of $L$ is isomorphic to the congruence lattice of $K$ and the automorphism group of $L$ is isomorphic to $G$? If $K$ and $G$ are finite, can $L$ chosen to be finite?*

For finite lattices, this question was answered in the late 1970s by V. A. Baranskiĭ [50], [51] and A. Urquhart [687]. We now state the Baranskiĭ-Urquhart theorem:

◇ **Theorem 322 (The Independence Theorem).** *Let $D$ be a finite distributive lattice with more than one element, and let $G$ be a finite group. Then there exists a finite lattice $L$ such that the congruence lattice of $L$ is isomorphic to $D$ and the automorphism group of $L$ is isomorphic to $G$.*

Both proofs rely heavily on the representation theorem of finite distributive lattices as congruence lattices of finite lattices and on the representation theorem of finite groups as automorphism groups of finite lattices stated above.

There is a congruence-preserving extension variant, which was published in G. Grätzer and E. T. Schmidt [343]:

◇ **Theorem 323** (**The Strong Independence Theorem**). *Let $K$ be a finite lattice with more than one element and let $G$ be a finite group. Then $K$ has a congruence-preserving extension $L$ whose automorphism group is isomorphic to $G$.*

G. Grätzer and E. T. Schmidt [348] considered the independence problem for modular lattices.

For infinite lattices, the question was answered in the year 2000, see Section 5.4 (Theorem 340).

## 4.9  ◇ General lattices

Anybody familiar with Theorem 149 (N. Funayama and T. Nakayama [214], 1942) and Theorem 48 (G. Birkhoff and O. Frink [75], 1948) would naturally raise the question:

*Can every distributive algebraic lattice $L$ be represented as the congruence lattice of a lattice $K$?*

This became one of the most celebrated problems of lattice theory for about half a century. See G. Grätzer [270] for an elementary survey.

Surprisingly, this problem did not make it into the Birkhoff and Frink paper [75] or Birkhoff's book [71]. When asked, G. Birkhoff and O. Frink in 1961 called this an oversight. Certainly, R. P. Dilworth was aware of this problem. The first time the problem appeared in print was in G. Grätzer and E. T. Schmidt [337] in 1962. Already in 1958, in G. Grätzer and E. T. Schmidt [332], a partial positive solution was given.

We are going to deal with this topic in the companion volume of this book in depth, but here are the highlights.

Let us call a join-semilattice with zero $S$ *representable* if it is isomorphic to the join-semilattice of compact congruences $\mathrm{Con}_c L$ for some lattice $L$.

Now we can state the semilattice formulation of the general problem:

*Is every distributive join-semilattice with zero representable?*

Two typical positive results follow, the first from E. T. Schmidt [630] and the second from A. P. Huhn [407].

◇ **Theorem 324.** *Every distributive lattice with zero is representable.*

◇ **Theorem 325.** *Every distributive join-semilattice with zero of cardinality at most $\aleph_1$ is representable.*

See G. Grätzer, H. Lakser, and F. Wehrung [320], for an elementary proof of Huhn's theorem.

F. Wehrung [711] settled the general problem in the negative:

$\Diamond$ **Theorem 326.** *There exists a distributive join-semilattice with zero of cardinality $\aleph_{\omega+1}$ that is not representable.*

The cardinality $\aleph_{\omega+1}$ in Wehrung's result was improved to $\aleph_2$ by P. Růžička [622], which is, of course, optimal by Huhn's result. M. Ploščica [586] provides two more examples of nonrepresentable distributive join-semilattices with zero of cardinality $\aleph_2$.

For a survey article as of 2002, see J. Tůma and F. Wehrung [685]. For some recent results, see M. Ploščica [587] and F. Wehrung [712].

To relate representation results of large semilattices with results for finitely many finite lattices, many proofs utilize techniques from set theory and infinite combinatorics. For an in depth introduction to the latter, see P. Erdős, A. Hajnal, A. Máté, and R. Rado [172]. A typical such result is Kuratowski's Free Set Theorem, see K. Kuratowski [489] and also Theorem 45.7 in [172].

For a natural number $n$ and a set $X$, we denote by $[X]^n$ and $[X]^{<\omega}$ the set of all $n$-element and the set of all finite subsets of $X$, respectively.

$\Diamond$ **Theorem 327 (Kuratowski's Free Set Theorem).** *Let $n$ be a natural number, let $X$ be a set. Then $|X| \geq \aleph_n$ iff for each $\Phi\colon [X]^n \to [X]^{<\omega}$, there exists an $(n+1)$-element subset $H$ of $X$ such that $x \notin \Phi(H - \{x\})$ for each $x \in H$.*

Another technique is the use of ladders introduced (under a different name) in H. Dobbertin [165]. For a positive integer $n$, we call a lattice $L$ with zero an *n-ladder* if every principal ideal in $L$ is finite and every element of $L$ has at most $n$ lower covers.

Every finite chain is a 1-ladder. The chain $\omega$ of all nonnegative integers is also a 1-ladder.

The following result is due to S. Z. Ditor [164] (we present the proof from G. Grätzer, H. Lakser, and F. Wehrung [320]); it is used in the proof of Theorem 321.

**Lemma 328.** *There exists a 2-ladder of cardinality $\aleph_1$.*

*Proof.* For $\xi < \omega_1$ (the first uncountable ordinal), we construct inductively the lattices $L_\xi$ with no largest element, as follows. Put $L_0 = \omega$. If $\lambda$ is a countable limit ordinal, define $L_\lambda = \bigcup(L_\xi \mid \xi < \lambda)$. So assume that we have constructed $L_\xi$, a countable 2-ladder with no largest element. Then $L_\xi$ has a strictly increasing, countable, cofinal, sequence $(a_n \mid n < \omega)$. Let $(b_n \mid n < \omega)$ be a strictly increasing countable chain with $b_n \notin L_\xi$ for all $n$. Define $L_{\xi+1}$ by

$$L_{\xi+1} = L_\xi \cup \{\, b_n \mid n < \omega \,\},$$

endowed with the least partial ordering containing the ordering of $L_\xi$, the natural ordering of $\{\, b_n \mid n < \omega \,\}$, and all pairs $a_n < b_n$ for all $n < \omega$. It is easy to verify that $L = \bigcup(\, L_\xi \mid \xi < \omega_1 \,)$ is a 2-ladder of cardinality $\aleph_1$.    $\square$

In the paper G. Grätzer, H. Lakser, and F. Wehrung [320]), the existence of a 2-ladder of cardinality $\aleph_1$ together with Theorem 321 is used to prove that every distributive join-semilattice with zero is representable with a locally finite, relatively complemented lattice with zero.

S. Z. Ditor [164] raised the question: Does there exist a 3-ladder of cardinality $\aleph_2$? A positive answer is given in F. Wehrung [713] under additional set-theoretical axioms that are known to be consistent with the usual axiom system ZFC of set theory.

Let $\mathcal{A}$ and $\mathcal{B}$ be varieties of lattices. Define the *critical point* crit$(\mathcal{A}; \mathcal{B})$ as the least cardinality of a join-semilattice with zero isomorphic to $\mathrm{Con}_c L$, for some $L \in \mathcal{A}$, but not for any $L \in \mathcal{B}$. J. Tůma and F. Wehrung [685] asked whether the critical point between two finitely generated varieties of lattices can be $\aleph_1$. P. Gillibert answered this question positively; an example (with two finitely generated modular varieties) can be found in P. Gillibert [228]; see also P. Gillibert [229].

This branch of lattice theory is intractably intertwined with a chapter of universal algebra dealing with congruence lattices in varieties of algebras. For instance, M. Ploščica [587] proves that the congruence lattice of the free majority algebra (a universal algebraic concept) on at least $\aleph_2$ generators is not isomorphic to the congruence lattice of any lattice, thereby providing a simpler example for Wehrung's result.

For this intersection of lattice theory and universal algebra, see M. Ploščica [586] and P. Gillibert [228], [229].

### 4.10    $\Diamond$ Complete lattices

In Section I.3.6, we defined the Substitution Properties. They generalize to complete lattices in a natural way:

$$a_i \equiv b_i \pmod{\boldsymbol{\alpha}}, \quad \text{for } i \in I,$$

imply that

(SP$_\vee$)    $$\bigvee(\, a_i \mid i \in I \,) \equiv \bigvee(\, b_i \mid i \in I \,) \pmod{\boldsymbol{\alpha}},$$

(SP$_\wedge$)    $$\bigwedge(\, a_i \mid i \in I \,) \equiv \bigwedge(\, b_i \mid i \in I \,) \pmod{\boldsymbol{\alpha}}.$$

An equivalence relation $\boldsymbol{\alpha}$ on a complete lattice $L$ is called a *complete congruence relation* of $L$ if these two *Complete Substitution Properties* hold.

For a complete lattice $L$, let $\mathrm{Com}\,L$ denote the lattice of complete congruence relations of $L$. Obviously, $\mathrm{Com}\,L$ is a complete lattice; however, unlike $\mathrm{Con}\,L$, the lattice of congruence relations of a lattice $L$, it is not distributive, in general; an example is presented in K. Reuter and R. Wille [607] (see also G. Grätzer, H. Lakser, and B. Wolk [321]), where the question is raised whether every complete lattice $K$ can be represented in the form $\operatorname{Com} L$ for some complete lattice $L$.

This problem was raised in the mid 1980s. G. Birkhoff [69], in the mid 1940s, raised a related question: does every complete lattice $K$ have a representation as the congruence lattice of some infinitary algebra $(A; F)$.

Birkhoff's problem was solved in the late 1970s by G. Grätzer and W. A. Lampe, published as Appendix 7 in [263]:

◇ **Theorem 329.** *Every complete lattice $K$ is isomorphic to the congruence lattice,* $\operatorname{Con}(A; F)$, *of some (infinitary) algebra* $(A; F)$.

The Reuter-Wille problem was solved for finite lattices in S.-K. Teo [677]. A solution of the general case was announced in G. Grätzer [266] and [267]. The first published proof is in G. Grätzer and H. Lakser [305].

◇ **Theorem 330.** *Every complete lattice $K$ is isomorphic to the complete congruence lattice,* $\operatorname{Com} L$, *of some complete lattice $L$.*

The strongest form of this theorem was published in G. Grätzer and E. T. Schmidt [341] (see also G. Grätzer and E. T. Schmidt [340] and [342]):

◇ **Theorem 331.** *Every complete lattice $K$ is isomorphic to the complete congruence lattice,* $\operatorname{Com} L$, *of some complete* distributive *lattice $L$.*

A number of papers have appeared improving this result, see, for instance, the results in Exercises 4.12–4.14. It was quite unexpected, however, that the techniques developed turned out to be very useful in improving the results for *finite* congruence lattices. For instance, it is a direct consequence of this development that we can prove planarity and minimal size for finite lattices.

**Exercises**

4.1. Let $M$ be the chopped lattice constructed to prove the Dilworth Theorem (Theorem 311). For a set $A \subseteq \operatorname{Atom}(M)$, there is an ideal $U$ with $\operatorname{Atom}(U) = A$ iff $A$ satisfies the condition (the index is computed modulo 2):

(C1)        For $p \succ q$ in $P$, if $p_1$, $q_i \in A$, then $q_{i+1} \in A$.

(Exercises 4.1–4.6 are based on G. Grätzer and E. T. Schmidt [337] and G. Grätzer and H. Lakser [307].)

4.2. The assignment $I \mapsto \mathrm{Atom}(I)$ is a bijection between the ideals of $M$ and closed subsets of $\mathrm{Atom}(M)$, and

$$\mathrm{Atom}(I \wedge J) = \mathrm{Atom}(I) \cap \mathrm{Atom}(J),$$
$$\mathrm{Atom}(I \vee J) = \overline{\mathrm{Atom}(I) \cup \mathrm{Atom}(J)}$$

for $I, J \in \mathrm{Id}\, M$. The inverse map assigns to a closed set $X$ of atoms, the ideal $\mathrm{id}(X)$ of $M$ generated by $X$.

So we can regard $L = \mathrm{Id}\, M$ as the lattice of closed sets in $\mathrm{Atom}(M)$. Let $I \subseteq J \in L$. Let us say, that $q \in P$ *splits over* $(I, J)$ if there exists a covering pair $p \succ q$ in $P$ with $p_1, q_i \in J - I$ and $q_{i+1} \in I$. If there is an element $q \in P$ that splits over $(I, J)$, then $I - J$ is not closed. Let $X = X(I, J)$ be the set of all elements $q_i$ in $J - I$ such that $q$ splits over $(I, J)$. Let

$$S = S(I, J) = (J - I) - X,$$

that is, let $S$ be the set of all elements $q_i$ in $J - I$ such that $q$ does not split over $(I, J)$.

4.3. Prove that $S(I, J) \in L$.

4.4. Prove that $S = S(I, J)$ is a sectional complement of $I$ in $J$.

4.5. Use Exercises 4.1–4.4 to verify the following result of G. Grätzer and E. T. Schmidt [337] (G. Grätzer and M. Roddy [329] and G. Grätzer, G. Klus, and A. Nguyen [290] offer a completely different approach to this result):

**Theorem 332.** *Every finite distributive lattice $D$ can be represented as the congruence lattice of a finite sectionally complemented lattice $L$.*

*4.6. Prove the following much stronger version of the result in Exercise 4.5 (published 37 years later in G. Grätzer and E. T. Schmidt [345]):

**Theorem 333.** *Every finite lattice $K$ has a congruence-preserving extension to a finite sectionally complemented lattice $L$.*

4.7. Prove the following generalization of Theorem 321 (see G. Grätzer, H. Lakser, and F. Wehrung [320]):

Let $L_0, L_1, L_2$ be lattices and let $\eta_i \colon L_0 \to L_i$, for $i \in \{1, 2\}$, be lattice homomorphisms. Let $D$ be a finite distributive lattice, and, for $i \in \{1, 2\}$, let $\psi_i \colon \mathrm{Con}\, L_i \to D$ be $\bigvee$-homomorphisms such that

$$\psi_1 \circ (\mathrm{Con}\, \eta_1) = \psi_2 \circ (\mathrm{Con}\, \eta_2).$$

There is then a lattice $L$, there are lattice homomorphisms $\varphi_i \colon L_i \to L$, for $i \in \{1, 2\}$ with

$$\varphi_1 \circ \eta_1 = \varphi_2 \circ \eta_2,$$

and there is an isomorphism $\gamma\colon \operatorname{Con} L \to D$ such that

$$\gamma \circ (\operatorname{Con} \varphi_i) = \psi_i \quad \text{for } i \in \{1,2\}.$$

If $L_0$, $L_1$, $L_2$ have zero elements and both $\eta_1$, $\eta_2$ preserve the zeros, then $L$ can be chosen to have a zero and $\varphi_1$, $\varphi_2$ can be chosen to preserve the zeros.

4.8. Prove that every $n$-ladder has breadth at most $n$.

4.9. Find a lattice $L$ of breadth 2 in which $\operatorname{id}(a)$ is finite, for all $a \in L$, and which is not an $n$-ladder for any $n$.

*4.10. Let $L$ be a lattice with the property that $\operatorname{id}(a)$ is finite for all $a \in L$. Let $n$ be a positive integer. If $L$ has breadth at most $n$, then $|L| \le \aleph_{n-1}$ (S. Z. Ditor [164]).

*4.11. Prove Theorem 330 using variants of the One-Point Extension Theorem. (G. Grätzer [267], [268], and G. Grätzer and H. Lakser[305]).

*4.12. Let $\mathfrak{m}$ be an infinite regular cardinal. Define the concepts: $\mathfrak{m}$-complete lattice and $\mathfrak{m}$-complete congruence. Prove Theorem 329 for the $\mathfrak{m}$-complete case (G. Grätzer and E. T. Schmidt [339]).

*4.13. Prove Theorem 330 by constructing a complete *modular* lattice $K$ (R. Freese, G. Grätzer, and E. T. Schmidt [185]).

*4.14. Prove Theorem 331.

# 5. Boolean Triples

Let $D$ be a bounded distributive lattice. In 1974, E. T. Schmidt [629] defined the set

$$(15) \qquad \mathsf{M}_3[D] = \{\, (x,y,z) \in D^3 \mid x \wedge y = y \wedge z = z \wedge x \,\},$$

regarded as a suborder of $D^3$. We will prove—at the end of this section—that $\mathsf{M}_3[D]$ is a modular lattice. This construction played a crucial role in a number of results on congruence lattices of modular lattices, see Chapter 10 of [271] and the references therein.

It is easy to construct a proper congruence-preserving extension of a *finite lattice* with more than two elements. In the early 1990s, G. Grätzer and E. T. Schmidt raised the question in [343] whether *every lattice* with more than two elements has a proper congruence-preserving extension.

It took almost a decade for the answer to appear in G. Grätzer and F. Wehrung [357]. For infinite lattices, the affirmative answer was provided by their construction: boolean triples, which is a generalization of the $\mathsf{M}_3[D]$ construction to an arbitrary (not necessarily distributive) lattice $D$.

## 5.1   The general construction

For a lattice $L$, let us call the triple $(x, y, z) \in L^3$ *boolean* if (FP stands for "Fixed Point Definition")

$$
\begin{aligned}
x &= (x \vee y) \wedge (x \vee z), \\
\text{(FP)} \qquad y &= (y \vee x) \wedge (y \vee z), \\
z &= (z \vee x) \wedge (z \vee y).
\end{aligned}
$$

Note that by Lemma 73 if (FP) holds, then $\mathrm{sub}(\{x, y, z\})$ is boolean.

(FP) is a "Fixed point definition" because the triple $(x, y, z)$ satisfies (FP) iff $p(x, y, z) = (x, y, z)$, where

$$
p = ((x \vee y) \wedge (x \vee z), (y \vee x) \wedge (y \vee z), (z \vee x) \wedge (z \vee y)).
$$

We denote by $\mathsf{M}_3[L] \subseteq L^3$ the order of boolean triples of $L$ (ordered as a suborder of $L^3$, that is, componentwise).

Observe that any boolean triple $(x, y, z) \in L^3$ is *balanced*, that is, it satisfies (Bal stands for "Balanced")

$$
\text{(Bal)} \qquad\qquad x \wedge y = y \wedge z = z \wedge x.
$$

Indeed, if $(x, y, z)$ is boolean, then

$$
x \wedge y = y \wedge z = z \wedge x = (x \vee y) \wedge (y \vee z) \wedge (z \vee x).
$$

For a distributive lattice $L$, the reverse also holds. Indeed, if $(x, y, z) \in L^3$ is balanced, that is, it satisfies (Bal), then

$$
\begin{aligned}
(x \vee y) \wedge (x \vee z) &= x \vee (y \wedge z) &\qquad \text{by (Bal)} \\
&= x \vee (x \wedge y) = x,
\end{aligned}
$$

verifying, by symmetry, (FP).

We introduce the notation

$$
\mathsf{M}_3[L]_{\mathrm{bal}} = \{\, (x, y, z) \in L^3 \mid x \wedge y = y \wedge z = z \wedge x \,\}.
$$

If $L$ is a distributive lattice, then $\mathsf{M}_3[L]_{\mathrm{bal}} = \mathsf{M}_3[L]$.

Here are some of the basic properties of boolean triples:

**Lemma 334.** *Let $L$ be a lattice.*

(i) $(x, y, z) \in L^3$ *is boolean iff there is a triple* $(u, v, w) \in L^3$ *satisfying* (Ex stand for "Existential")

$$
\begin{aligned}
x &= u \wedge v, \\
\text{(Ex)} \qquad y &= u \wedge w, \\
z &= v \wedge w.
\end{aligned}
$$

(ii) $M_3[L]$ *is a closure system in* $L^3$. *For* $(x, y, z) \in L^3$, *the closure is*

$$\overline{(x, y, z)} = ((x \vee y) \wedge (x \vee z), (y \vee x) \wedge (y \vee z), (z \vee x) \wedge (z \vee y)).$$

(iii) *If* $L$ *has a zero, then the suborder* $\{ (x, 0, 0) \mid x \in L \}$ *is a sublattice of* $M_3[L]$ *and* $\varphi \colon x \mapsto (x, 0, 0)$ *is an isomorphism between* $L$ *and this sublattice.*

(iv) *If* $L$ *is bounded, then* $M_3[L]$ *has a spanning* $M_3$, *that is, a* $\{0, 1\}$-*sublattice isomorphic to* $M_3$, *namely,*

$$\{(0, 0, 0), (1, 0, 0), (0, 1, 0), (0, 0, 1), (1, 1, 1)\}.$$

*Proof.*
(i) If $(x, y, z)$ is boolean, then $u = x \vee y, v = x \vee z, w = y \vee z$ satisfy (Ex). Conversely, if there is a triple $(u, v, w) \in L^3$ satisfying (Ex), then by Lemma 73, the sublattice generated by $x, y, z$ is isomorphic to a quotient of $B_3$ and $x, y, z$ are the images of the three atoms of $B_3$. Thus $(x \vee y) \wedge (x \vee z) = x$, the first equation in (FP). The other two equations are proved similarly.

(ii) $M_3[L] \neq \varnothing$; for instance, for all $x \in L$, the diagonal element $(x, x, x)$ is in $M_3[L]$.

For $(x, y, z) \in L^3$, define $u = x \vee y, v = x \vee z, w = y \vee z$. Set $x_1 = u \wedge v$, $y_1 = u \wedge w, z_1 = v \wedge w$. Then $(x_1, y_1, z_1)$ is boolean by (i) and $(x, y, z) \le (x_1, y_1, z_1)$ in $L^3$. Now if $(x, y, z) \le (x_2, y_2, z_2)$ in $L^3$ and $(x_2, y_2, z_2)$ is boolean, then

$$
\begin{aligned}
x_2 &= (x_2 \vee y_2) \wedge (x_2 \vee z_2) && \text{(by (FP))} \\
&\ge (x \vee y) \wedge (x \vee z) && \text{(by } (x, y, z) \le (x_2, y_2, z_2)) \\
&= u \wedge v = x_1,
\end{aligned}
$$

and similarly, $y_2 \ge y_1$ and $z_2 \ge z_1$. Thus $(x_2, y_2, z_2) \ge (x_1, y_1, z_1)$, and so $(x_1, y_1, z_1)$ is the smallest boolean triple majorizing $(x, y, z)$.

(iii) and (iv) are obvious.                    $\square$

$M_3[L]$ is difficult to draw in general. Figure 70 shows the diagram of $M_3[C_3]$ with the three-element chain $C_3 = \{0, a, 1\}$. If $C$ is an arbitrary bounded chain, with bounds 0 and 1, it is easy to picture $M_3[C]$, as sketched in Figure 70. The element $(x, y, z) \in C^3$ is boolean iff it is of the form $(x, y, y)$, or $(y, x, y)$, or $(y, y, x)$, where $y \le x$ in $C$. So the diagram is made up of three isomorphic "flaps" overlapping on the diagonal (the elements of the form $(x, x, x)$). Two of the flaps form the "base", a planar lattice: $C^2$, the third one (shaded) comes up out of the plane pointing in the direction of the viewer.
We get some more examples of $M_3[L]$ from the following observation:

**Lemma 335.** *Let* $L = L_1 \times L_2$ *be a direct product decomposition of the lattice* $L$. *Then* $M_3[L] \cong M_3[L_1] \times M_3[L_2]$.

*Proof.* This is obvious, since $((x_1, y_1), (x_2, y_2), (x_3, y_3))$ is a boolean triple iff $(x_1, x_2, x_3)$ and $(y_1, y_2, y_3)$ both are, where $x_i \in L_1$ and $y_i \in L_2$ for $i = 1, 2, 3$.                                             □

## 5.2  Congruence-preserving extension

Let $L$ be a nontrivial lattice with zero and let

$$\varphi \colon x \mapsto (x, 0, 0) \in \mathsf{M}_3[L]$$

be an embedding of $L$ into $\mathsf{M}_3[L]$.

Here is the main result of G. Grätzer and F. Wehrung [357], resolving the problem of G. Grätzer and E. T. Schmidt [343]:

**Theorem 336.** $\mathsf{M}_3[L]$ *is a congruence-preserving extension of* $\varphi(L)$.

The next two lemmas prove this theorem.

For a congruence $\boldsymbol{\alpha}$ of $L$, let $\boldsymbol{\alpha}^3$ denote the congruence of $L^3$ defined componentwise. Let $\mathsf{M}_3[\boldsymbol{\alpha}]$ be the restriction of $\boldsymbol{\alpha}^3$ to $\mathsf{M}_3[L]$.

**Lemma 337.** $\mathsf{M}_3[\boldsymbol{\alpha}]$ *is a congruence relation of* $\mathsf{M}_3[L]$.

*Proof.* $\mathsf{M}_3[\boldsymbol{\alpha}]$ is obviously an equivalence relation on $\mathsf{M}_3[L]$. Since $\mathsf{M}_3[L]$ is a meet-subsemilattice of $L^3$, it is clear that $\mathsf{M}_3[\boldsymbol{\alpha}]$ satisfies $(\mathrm{SP}_\wedge)$. To verify $(\mathrm{SP}_\vee)$ for $\mathsf{M}_3[\boldsymbol{\alpha}]$, let $(x_1, y_1, z_1), (x_2, y_2, z_2) \in \mathsf{M}_3[L]$, let

$$(x_1, y_1, z_1) \equiv (x_2, y_2, z_2) \pmod{\mathsf{M}_3[\boldsymbol{\alpha}]},$$

and let $(u, v, w) \in \mathsf{M}_3[L]$. Set

$$(x_i', y_i', z_i') = (x_i, y_i, z_i) \vee (u, v, w)$$

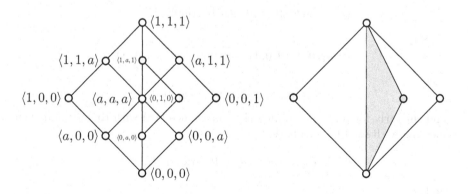

Figure 70. The lattice $\mathsf{M}_3[C_3]$ with a sketch

(the join formed in $M_3[L]$) for $i = 1, 2$.

Then, using Lemma 334(ii) for $x_1 \vee u$, $y_1 \vee v$, and $z_1 \vee w$, we obtain that

$$
\begin{aligned}
x_1' &= (x_1 \vee u \vee y_1 \vee v) \wedge (x_1 \vee u \vee z_1 \vee w) \\
&\equiv (x_2 \vee u \vee y_2 \vee v) \wedge (x_2 \vee u \vee z_2 \vee w) = x_2' \quad (\mathrm{mod}\ M_3[\alpha]),
\end{aligned}
$$

and similarly, $y_1' \equiv y_2'$ (mod $M_3[\alpha]$) and $z_1' \equiv z_2'$ (mod $M_3[\alpha]$), hence

$$
(x_1, y_1, z_1) \vee (u, v, w) \equiv (x_2, y_2, z_2) \vee (u, v, w) \quad (\mathrm{mod}\ M_3[\alpha]). \qquad \square
$$

It is obvious that $M_3[\alpha]$ restricted to $\varphi(L)$ is $\varphi(\alpha)$.

**Lemma 338.** *Every congruence of $M_3[L]$ is of the form $M_3[\alpha]$ for a suitable congruence $\alpha$ of $L$.*

*Proof.* Let $\beta$ be a congruence of $M_3[L]$, and let $\alpha$ denote the congruence of $L$ obtained by restricting $\beta$ to the sublattice $\varphi(L) = \{\, (x, 0, 0) \mid x \in L \,\}$ of $M_3[L]$, that is, $x \equiv y$ (mod $\alpha$) if $(x, 0, 0) \equiv (y, 0, 0)$ (mod $\beta$) for all $x, y \in L$. We prove that $\beta = M_3[\alpha]$.

To show that $\beta \subseteq M_3[\alpha]$, let

(16) $$(x_1, y_1, z_1) \equiv (x_2, y_2, z_2) \quad (\mathrm{mod}\ \beta).$$

Meeting the congruence (16) with $(1, 0, 0)$ yields

$$(x_1, 0, 0) \equiv (x_2, 0, 0) \quad (\mathrm{mod}\ \beta),$$

and so

(17) $$(x_1, 0, 0) \equiv (x_2, 0, 0) \quad (\mathrm{mod}\ M_3[\alpha]).$$

Meeting the congruence (16) with $(0, 1, 0)$ yields that

(18) $$(0, y_1, 0) \equiv (0, y_2, 0) \quad (\mathrm{mod}\ \beta).$$

Since

$$(0, y_1, 0) \vee (0, 0, 1) = \overline{(0, y_1, 1)} = (y_1, y_1, 1),$$

and

$$(y_1, y_1, 1) \wedge (1, 0, 0) = (y_1, 0, 0),$$

and similarly for $(0, y_2, 0)$, joining the congruence (18) with $(0, 1, 0)$ and then meeting with $(1, 0, 0)$, yields that

$$(y_1, 0, 0) \equiv (y_2, 0, 0) \quad (\mathrm{mod}\ \beta),$$

and so

(19) $$(0, y_1, 0) \equiv (0, y_2, 0) \quad (\mathrm{mod}\ M_3[\alpha]).$$

Similarly,

$$(20) \qquad (0,0,z_1) \equiv (0,0,z_2) \pmod{\mathsf{M}_3[\alpha]}.$$

Joining the congruences (17), (19), and (20), we obtain that

$$(21) \qquad (x_1,y_1,z_1) \equiv (x_2,y_2,z_2) \pmod{\mathsf{M}_3[\alpha]},$$

proving that $\beta \subseteq \mathsf{M}_3[\alpha]$.

To prove the converse, $\mathsf{M}_3[\alpha] \subseteq \beta$, take

$$(22) \qquad (x_1,y_1,z_1) \equiv (x_2,y_2,z_2) \pmod{\mathsf{M}_3[\alpha]}$$

in $\mathsf{M}_3[L]$; equivalently,

$$(23) \qquad (x_1,0,0) \equiv (x_2,0,0) \pmod{\beta},$$
$$(24) \qquad (y_1,0,0) \equiv (y_2,0,0) \pmod{\beta},$$
$$(25) \qquad (z_1,0,0) \equiv (z_2,0,0) \pmod{\beta}$$

in $\mathsf{M}_3[L]$.

Joining the congruence (24) with $(0,0,1)$ and then meeting the result with $(0,1,0)$, we get (as in the computation following (18)):

$$(26) \qquad (0,y_1,0) \equiv (0,y_2,0) \pmod{\beta}.$$

Similarly, from (25), we conclude that

$$(27) \qquad (0,0,z_1) \equiv (0,0,z_2) \pmod{\beta}.$$

Finally, joining the congruences (23), (26), and (27), we obtain that

$$(28) \qquad (x_1,y_1,z_1) \equiv (x_2,y_2,z_2) \pmod{\beta},$$

that is, $\mathsf{M}_3[\alpha] \subseteq \beta$. This completes the proof of this lemma. $\qquad \square$

## 5.3   The distributive case

We have already noted that $\mathsf{M}_3[L]_{\mathrm{bal}} = \mathsf{M}_3[L]$ for a distributive lattice $L$.

We now prove the result of E. T. Schmidt [629].

**Theorem 339.** *Let $D$ be a bounded distributive lattice. Then $\mathsf{M}_3[D]$ (the set of balanced triples $(x,y,z) \in D^3$) is a modular lattice. The map*

$$\varphi \colon x \mapsto (x,0,0) \in \mathsf{M}_3[D]$$

*is an embedding of $D$ into $\mathsf{M}_3[D]$, and $\mathsf{M}_3[D]$ is a congruence-preserving extension of $\varphi(D)$.*

*Proof.* This result follows from the results of this section, except for the modularity. A direct computation of this is not so easy—although entertaining. However, we can do it without computation. Observe that it is enough to prove modularity for a finite $D$. Now if $D$ is finite, then by Lemma 335 and Corollary 109, the lattice $M_3[D]$ can be embedded into

$$M_3[B_n] \cong (M_3[C_2])^n \cong M_3^n,$$

a modular  lattice; hence $M_3[D]$ is modular.                                    $\square$

### 5.4  $\Diamond$ Tensor products

The boolean triple construction is closely related to tensor products, introduced in J. Anderson and N. Kimura [30] and G. A. Fraser [180]. Let $A$ and $B$ be $\{\vee, 0\}$-semilattices. We denote by $A \otimes B$ the *tensor product* of $A$ and $B$, defined as the free $\{\vee, 0\}$-semilattice generated by the set $A^- \times B^-$ and subject to the relations

$$(a, b_0) \vee (a, b_1) = (a, b_0 \vee b_1), \qquad \text{for } a \in A^-,\ b_0,\, b_1 \in B^-;$$
$$(a_0, b) \vee (a_1, b) = (a_0 \vee a_1, b), \qquad \text{for } a_0,\, a_1 \in A^-,\ b \in B^-.$$

(For a $\{\vee, 0\}$-semilattice $L$, we denote by $L^-$ the $\vee$-subsemilattice of $L$ defined on $L - \{0\}$.) In general, $A \otimes B$ is not a lattice (it is a $\{\vee, 0\}$-semilattice) except in special cases, for instance, if both $A$ and $B$ are finite. R. W. Quackenbush [603] raised the question whether $A \otimes B$ is always a lattice; G. Grätzer and F. Wehrung [360] provided a negative answer, see Exercise 5.3.

The main result of G. Grätzer, H. Lakser, and R. W. Quackenbush [312] is the statement that

$$\operatorname{Con} A \otimes \operatorname{Con} B \cong \operatorname{Con}(A \otimes B)$$

holds for *finite* lattices $A$ and $B$.

Applying the previous result to the case when $A$ is simple, we obtain that $A \otimes B$ is a congruence-preserving extension of $B$. How to generalize this result to infinite lattices? Since $A \otimes B$ is not a lattice, in general, the box product $A \,\Box\, B$ and lattice tensor product $A \boxtimes B$ constructions were introduced in G. Grätzer and F. Wehrung [358]. For lattice tensor products we get a far-reaching generalization of Theorem 336: if $A$ is a simple bounded lattice, then $A \boxtimes B$ is a congruence-preserving extension of $B$.

For the very technical definitions and proofs, we refer the reader to the survey article G. Grätzer and F. Wehrung [364] and to the original articles: G. Grätzer and F. Wehrung [357]–[363]. See also G. Grätzer and M. Greenberg [272]–[275], G. Grätzer and E. T. Schmidt [347], G. Grätzer, M. Greenberg, and E. T. Schmidt [276].

The best application of tensor products is in G. Grätzer and F. Wehrung [363]:

$\Diamond$ **Theorem 340 (The Strong Independence Theorem for Lattices).**
*Let $L_A$ and $L_C$ be lattices, let $L_C$ have more than one element. Then there exists a lattice $K$ that is an automorphism-preserving extension of $L_A$ and a congruence-preserving extension of $L_C$.*

## 5.5 $\Diamond$ Congruence-permutable, congruence-preserving extensions by Friedrich Wehrung

In this section, we are interested in congruence-preserving extensions of lattices (see Section I.3.8) to *congruence-permutable* lattices.

A convenient characterization of permutability of a given pair of congruences is given in Exercise III.3.13. It implies that every relatively complemented lattice is congruence-permutable, see Exercise II.1.39. Now let $L$ be a lattice which is either sectionally complemented or finite atomistic. By Lemma 271, $L$ is sectionally decomposing. Hence, by Theorem 272, every congruence of $L$ is standard. Therefore, by Theorem 268, $L$ is congruence-permutable.

Let $L$ be a finite lattice. We define a subset $X$ of $\operatorname{Ji} L$ to be *closed* if for every subset $I$ of $X$ and every $p \in \operatorname{Ji} L$, if $I$ is a minimal join-cover of $p$ (see Section 4.4), then $p \in X$. The set $\operatorname{Tisch} L$ of all closed subsets of $\operatorname{Ji} L$, ordered by set inclusion, is a finite atomistic lattice; the atoms of $\operatorname{Tisch} L$ are the singletons $\{p\}$, where $p \in \operatorname{Ji} L$. The assignment

$$x \mapsto \{\, p \in \operatorname{Ji} L \mid p \le x \,\}$$

obviously defines a meet-embedding $\tau$ of $L$ into $\operatorname{Tisch} L$, preserving the bounds. As proved by M. Tischendorf [679] in 1992, much more is true.

$\Diamond$ **Theorem 341.** *The map $\tau$ is a congruence-preserving lattice embedding from $L$ into $\operatorname{Tisch} L$.*

**Corollary 342.** *Every finite lattice has a congruence-permutable, congruence-preserving extension.*

Since 1992, stronger embedding results have been found, for example Theorem 333: *Every finite lattice $K$ has a congruence-preserving extension to a finite sectionally complemented lattice $L$.*

The congruence lattice of a finite *relatively complemented* lattice is Boolean, thus it is not possible to replace "sectionally complemented" by "relatively complemented" in the statement of Theorem 333.

The construction of the finite sectionally complemented extension in Theorem 333 is more complex than the $\operatorname{Tisch} L$ construction. Existence of congruence-preserving extensions of finite lattices to "nice" finite lattices is a whole topic by itself, studied in depth in the monograph G. Grätzer [271]; see also Section 4.5. Thus we shall now shift the focus to *infinite* lattices.

Tischendorf's extension procedure $L \hookrightarrow \operatorname{Tisch} L$ looks "canonical" enough, so one may try to extend it to any (possibly infinite) lattice, or, at least, to any locally finite lattice $L$; express $L$ as a directed union of finite sublattices $L_i$, embed each $L_i$ into Tisch $L_i$, then take the direct limit of all finite, atomistic lattices Tisch $L_i$.

The problem is that the assignment $L \mapsto$ Tisch $L$ *cannot* be extended to a functor. This is implied by the following result, established in P. Růžička, J. Tůma, and F. Wehrung [684], using ideas from M. Ploščica, J. Tůma, and F. Wehrung [623] together with J. Tůma and F. Wehrung [684]. For a variety $\mathbf{V}$ of lattices (resp., of bounded lattices), we denote by $\mathrm{Free}_\mathbf{V}(X)$ the free lattice (resp., bounded free lattice) in $\mathbf{V}$ on a set $X$.

$\Diamond$ **Theorem 343.** *Let $\mathbf{V}$ be a nondistributive variety of lattices (resp., of bounded lattices). For any set $X$ with at least $\aleph_2$ elements, there is no congruence-permutable algebra $A$ such that $\mathrm{Con}\,\mathrm{Free}_\mathbf{V}(X) \cong \mathrm{Con}\,A$.*

In case $\mathbf{V}$ is generated by a single finite lattice (for instance, $\mathsf{M}_3$ or $\mathsf{N}_5$), $\mathrm{Free}_\mathbf{V}(\aleph_2)$ is locally finite, nevertheless there is no congruence-permutable algebra $A$ such that $\mathrm{Con}\,\mathrm{Free}_\mathbf{V}(\aleph_2) \cong \mathrm{Con}\,A$. In particular, $\mathrm{Free}_\mathbf{V}(\aleph_2)$ *has no congruence-permutable, congruence-preserving extension.*

By G. Grätzer, H. Lakser, and F. Wehrung [320], every distributive $\{\vee, 0\}$-semilattice of cardinality at most $\aleph_1$ is isomorphic to $\mathrm{Con}_c\,L$ for some relatively complemented lattice $L$. The problem is quite different for the existence of congruence-preserving extensions. Still for the cardinality $\aleph_1$, the problem was completely solved in P. Gillibert and F. Wehrung [230].

$\Diamond$ **Theorem 344.** *Let $\mathbf{V}$ be a nondistributive variety of lattices (resp., of bounded lattices). Then there is no congruence-permutable, congruence-preserving extension of $\mathrm{Free}_\mathbf{V}(\aleph_1)$.*

Very roughly speaking, the proof of Theorem 344 starts by constructing a *diagram*, indexed by the square $\mathsf{B}_2$, of finite members of $\mathbf{V}$, with no "simultaneous congruence-preserving extension" into congruence-permutable lattices. Then, by using categorical tools called *larders*, the diagram counterexample is turned into an object counterexample. The bound $\aleph_1$ (instead of $\aleph_0$) comes from the fact that the *order-dimension* of the order $\mathsf{B}_2$ is 2. The proof of Theorem 344 is considerably harder than that of Theorem 343.

Although the countable case is not fully solved yet, part of it is already settled. A lattice $L$ is

- *congruence-finite* if $\mathrm{Con}\,L$ is finite;

- *$\omega$-congruence-finite* if $L$ is a countable directed union of congruence-finite sublattices.

In particular, every locally finite, (at most) countable lattice is $\omega$-congruence-finite. The following result is established in G. Grätzer, H. Lakser, and F. Wehrung [320].

$\Diamond$ **Theorem 345.** *Every $\omega$-congruence-finite lattice has an $\omega$-congruence-finite, relatively complemented, congruence-preserving extension.*

The problem whether every *countable* lattice has a relatively complemented (or even congruence-permutable) congruence-preserving extension has been posed in J. Tůma and F. Wehrung [684]. Although Theorem 345 solves it in the locally finite case, the general case is still open.

Completely different methods, involving a Boolean-valued (forcing) approach of congruence lattices of lattices, lead to the following result, established in F. Wehrung [707].

$\Diamond$ **Theorem 346.** *Let $L$ be a lattice such that $\mathrm{Con_c}\, L$ is a lattice. Then $L$ has a congruence-preserving extension to a relatively complemented lattice.*

This shows another discrepancy between the problem of finding congruence-preserving extensions of lattices on the one hand, and representing distributive semilattices on the other hand. Namely,

- Every distributive $\{\vee, 0\}$-semilattice of cardinality at most $\aleph_1$, and every distributive lattice with zero, is isomorphic to $\mathrm{Con_c}\, L$ for some lattice $L$ (A. P. Huhn and E. T. Schmidt, respectively).

- Every lattice $L$ such that either $L$ is (at most) countable and locally finite, or $\mathrm{Con_c}\, L$ is a lattice, has a relatively complemented, congruence-preserving extension.

## Exercises

5.1. Draw $\mathsf{M_3[C_4]}$.
5.2. Can you draw $\mathsf{M_3[B_2]}$?
*5.3. Show that $\mathsf{M_3} \otimes \mathrm{Free}(3)$ is not a lattice (G. Grätzer and F. Wehrung [360]).

$$* \qquad * \qquad *$$

The following exercises are based on G. Grätzer and F. Wehrung [359].

Let us define the terms $p_n, q_n, r_n$, for all $n < \omega$, in the variables $x, y, z$:

$$p_0 = x, \qquad\qquad q_0 = y, \qquad\qquad r_0 = z,$$
$$p_1 = x \vee (y \wedge z), \qquad q_1 = y \vee (x \wedge z), \qquad r_1 = z \vee (x \wedge y),$$
$$\cdots$$
$$p_{n+1} = p_n \vee (q_n \wedge r_n), \quad q_{n+1} = q_n \vee (p_n \wedge r_n), \quad r_{n+1} = r_n \vee (p_n \wedge q_n).$$

Let $(x, y, z) \in L^3$. Define

$$(x, y, z)^{(n)} = (p_n(x, y, z), q_n(x, y, z), r_n(x, y, z))$$

for all $n > 0$. Note that

$$(x, y, z) \leq (x, y, z)^{(1)} \leq \cdots \leq (x, y, z)^{(n)} \leq \cdots.$$

For $n > 0$, define the identity $\mu_n$ as $p_n = p_{n+1}$. Let $\mathbf{M}_n$ be the lattice variety defined by $\mu_n$. The lattices in $\mathbf{M}_n$ are called $n$-modular; lattices in $\mathbf{M}_n - \mathbf{M}_{n-1}$ are called *exactly $n$-modular* or *of modularity rank $n$*. A lattice $L \notin \mathbf{M}_n$, for all $n < \omega$, is *of modularity rank* $\infty$.

5.4. Let $L$ be a lattice. Then $\mathsf{M}_3[L]_{\mathrm{bal}}$ is a lattice iff $\mathsf{M}_3[L]_{\mathrm{bal}}$ is a closure system in $L^3$.

5.5. $L$ is an $n$-modular lattice iff $(a, b, c)^{(n)}$ is the closure of $(a, b, c)$, for all $a, b, c \in L$.

5.6. For every finite lattice $L$, there is an integer $n > 0$ such that $L$ is $n$-modular.

5.7. $\mathbf{M}_1$ is the variety of modular lattices.

5.8. The variety $\mathbf{N}_5$ generated by $\mathsf{N}_5$ is 2-modular.

5.9. Let $n > 0$ and let $L \in \mathbf{M}_n$ be a lattice. Then $\mathsf{M}_3[L]_{\mathrm{bal}}$ is a lattice. Furthermore, if $L$ is bounded, then $\mathsf{M}_3[L]_{\mathrm{bal}}$ has a spanning $\mathsf{M}_3$.

5.10. Let $n > 0$ and let $L$ be a bounded $n$-modular lattice. Then $\mathsf{M}_3[L]_{\mathrm{bal}}$ is a lattice with a spanning $M_3$. The map

$$\varepsilon \colon x \mapsto (x, 0, 0)$$

embeds $L$ into $\mathsf{M}_3[L]_{\mathrm{bal}}$. If we identify $x \in L$ with

$$\varepsilon(x) = (x, 0, 0) \in \mathsf{M}_3[L]_{\mathrm{bal}},$$

then the lattice $\mathsf{M}_3[L]_{\mathrm{bal}}$ is a congruence-preserving extension of $L$.

5.11. Find a lattice $L_n$ in $\mathbf{M}_{n+1}$ but not in $\mathbf{M}_n$.

5.12. Let $L$ be an $n$-modular lattice for some $n > 0$. Then

$$\mathsf{M}_3[\mathrm{Id}\, L]_{\mathrm{bal}} \cong \mathrm{Id}(\mathsf{M}_3[L]_{\mathrm{bal}}).$$

$$*\qquad *\qquad *$$

The following exercises are based on G. Grätzer and M. Greenberg [272]–[274].

Let $A$ be a finite lattice and let $B$ be an arbitrary bounded lattice. We define three unary maps, $m, j, p$, on $B^A$. For $a \in A$ and $\alpha \in B^A$, define in $B$:

$$m_a(\alpha) = \bigwedge \alpha(A - \mathrm{fil}(a)) = \bigwedge_{x \not\geq a} \alpha(x),$$

$$j_a(\alpha) = \bigvee \alpha(A - \mathrm{id}(a)) = \bigvee_{y \not\leq a} \alpha(y),$$

$$p_a(\alpha) = \bigwedge_{x \not\geq a} j_x(\alpha) = \bigwedge_{x \not\geq a} \bigvee \alpha(A - \mathrm{id}(x)) = \bigwedge_{x \not\geq a} \bigvee_{y \not\leq x} \alpha(y).$$

5.13. Verify that for the elements $a, b \in A$ and for the map $\alpha \in B^A$,

$$m_a(\alpha) \wedge m_b(\alpha) = m_{a \vee b}(\alpha),$$
$$j_a(\alpha) \vee j_b(\alpha) = j_{a \wedge b}(\alpha).$$

Moreover, $m_0(\alpha) = 1_B$, $j_1(\alpha) = 0_B$, and $p_0(\alpha) = 1_B$.

Define the map $m \colon B^A \to B^A$ by

$$m(\alpha)(a) = m_a(\alpha),$$

for a map $\alpha \in B^A$. Define the maps $j$ and $p$:

$$j(\alpha)(a) = j_a(\alpha),$$
$$p(\alpha)(a) = p_a(\alpha).$$

Define the new lattice construction:

$$A\langle B \rangle = \{ m(\alpha) \mid \alpha \in B^A \}.$$

Notation: $p(\alpha) = \overline{\alpha}$ for $\alpha \in B^A$, where $\overline{\alpha}$ is the *closure of* $\alpha$ (in $B^A$).

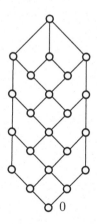

Figure 71. A lattice for Exercise 5.11

5.14. Verify that for $\alpha, \beta \in B^A$,

   (i) $\alpha \leq \overline{\alpha}$.
   (ii) If $\alpha \leq \beta$ and $\beta \in A\langle B \rangle$, then $\overline{\alpha} \leq \beta$.
   (iii) $\alpha \in A\langle B \rangle$ iff $\alpha = \overline{\alpha}$.

5.15. Prove that $A\langle B \rangle$ is a lattice with meets computed pointwise and joins computed as the closures of the pointwise joins, that is, according to the formula

$$\alpha \vee_{A\langle B \rangle} \beta = \overline{\alpha \vee_{B^A} \beta},$$

   for $\alpha, \beta \in A\langle B \rangle$.

5.16. Verify that $\operatorname{Id} A\langle B \rangle \cong A\langle \operatorname{Id} B \rangle$. (For $A = \mathsf{M}_3$ and $B$ $n$-modular, this is in G. Grätzer and F. Wehrung [359].)

*5.17. Let $A$ be a finite simple lattice. Then $A\langle B \rangle$ is a congruence-preserving extension of $B$.

# Chapter

# V

# Modular and Semimodular Lattices

## 1. Modular Lattices

### 1.1 Equivalent forms

**Theorem 347.** *For a lattice $L$, the following conditions are equivalent:*

(i) *$L$ is modular, that is,*

$$x \geq z \quad \text{implies that} \quad x \wedge (y \vee z) = (x \wedge y) \vee z$$

*for all $x, y, x \in L$.*

(ii) *$L$ satisfies the* shearing identity*:*

$$x \wedge (y \vee z) = x \wedge ((y \wedge (x \vee z)) \vee z)$$

*for all $x, y, x \in L$.*

(iii) *$L$ does not contain a pentagon.*

(iv) *Let $a \leq b \in L$ and $c \in L$. Then the elements $a, b, c$ generate a distributive sublattice.*

G. Grätzer, *Lattice Theory: Foundation*, DOI 10.1007/978-3-0348-0018-1_5,
© Springer Basel AG 2011

*Remark.* In Section II.1.1, we have already proved the equivalence of (i) and (iii). The importance, or convenience, of the shearing identity (which was named by I. Halperin) is that it can be applied to any expressions of the form $x \wedge (y \vee z)$ without any assumption. Observe also the dual of the shearing identity:

$$x \vee (y \wedge z) = x \vee ((y \vee (x \wedge z)) \wedge z).$$

*Proof.*
(i) *implies* (ii). Since $x \vee z \geq z$, by modularity,

$$(y \wedge (x \vee z)) \vee z = (y \vee z) \wedge (x \vee z),$$

and so

$$x \wedge ((y \wedge (x \vee z)) \vee z) = x \wedge ((y \vee z) \wedge (x \vee z)) = x \wedge (y \vee z).$$

(ii) *implies* (iii). In $N_5$, (ii) fails; indeed,

$$a \wedge (c \vee b) = a \wedge i = a,$$
$$a \wedge ((c \wedge (a \vee b)) \vee b) = a \wedge ((c \wedge a) \vee b) = a \wedge b = b.$$

Thus (ii) implies (iii).

We prove the remaining implications as in Section II.1.1, utilizing Figure 6. □

A typical elementary computation using modularity is in the proof of Lemma 99: *A modular lattice with zero is sectionally complemented iff it is relatively complemented.*

## 1.2    The Isomorphism Theorem for Modular Lattices

The most important form of modularity is the following:

**Theorem 348 (The Isomorphism Theorem for Modular Lattices).**
*Let $L$ be a modular lattice and let $a, b \in L$. Then*

$$\varphi_b \colon x \mapsto x \wedge b, \qquad x \in [a, a \vee b],$$

*is an isomorphism between the intervals $[a, a \vee b]$ and $[a \wedge b, b]$. The inverse isomorphism is*

$$\psi_a \colon y \mapsto x \vee a, \qquad y \in [a \wedge b, b].$$

(See Figure 72.)

*Remark.* In brief, perspective intervals are isomorphic in a modular lattice; we shall reference this as the Isomorphism Theorem.

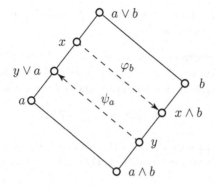

Figure 72. The Isomorphism Theorem for Modular Lattices

*Proof.* It is sufficient to show that $\psi_a\varphi_b(x) = x$ for all $x \in [a, a \vee b]$. Indeed, if this is true, then by duality, $\varphi_b\psi_a(y) = y$, for all $y \in [a \wedge b, b]$, is also true. The isotone maps $\varphi_b$ and $\psi_a$, thus compose into the identity maps, hence they are isomorphisms, as claimed.

So let $x \in [a, a \vee b]$. Then $\psi_a\varphi_b(x) = (x \wedge b) \vee a$. Since $x \in [a, a \vee b]$, the inequality $a \leq x$ holds, and so modularity applies:

$$\psi_a\varphi_b(x) = (x \wedge b) \vee a = x \wedge (b \vee a) = x,$$

because $x \leq a \vee b$. $\qquad\qquad\qquad\qquad\qquad\qquad\qquad\qquad\qquad\Box$

The Isomorphism Theorem is obviously equivalent to modularity. The statement: $[a, a \vee b] \cong [a \wedge b, b]$, for all $a, b \in L$, characterizes the modularity of a lattice $L$ only in special cases: for finite lattices (M. Ward [696]) and, more generally, for algebraic lattices (P. Crawley [103]).

**Corollary 349.** *Let $L$ be a modular lattice and let $a, b \in L$. Then the equality*

$$a \vee (x \wedge y) = (a \vee x) \wedge (a \vee y)$$

*holds for all $x, y \in [a \wedge b, b]$.*

## 1.3   Two applications

Most applications of the Isomorphism Theorem are like Corollary 349: the special form of the isomorphism is used. There are two important applications where the form of the isomorphism plays no role.

### A. Covering Conditions

A lattice $L$ is said to satisfy the *Upper Covering Condition* if $a \preceq b$ implies that $a \vee c \preceq b \vee c$ for all $a, b, c \in L$. The *Lower Covering Condition* is the dual.

**Theorem 350.** *A modular lattice satisfies both the Upper Covering Condition and the Lower Covering Condition.*

*Proof.* Let $L$ be a modular lattice. Let $a, b, c \in L$ and $b \prec a$. If $a \vee c = b \vee c$, we have nothing to prove. If $a \vee c \neq b \vee c$, then $a \nleq b \vee c$, and so $a \wedge (b \vee c) = b$. Applying the Isomorphism Theorem to $a$ and $b \vee c$, we obtain the isomorphism

$$[b, a] \cong [b \vee c, a \vee c].$$

Since $[b, a]$ is a prime interval, so is $[b \vee c, a \vee c]$, that is, $b \vee c \prec a \vee c$, proving the Upper Covering Condition. By duality, we get the Lower Covering Condition. $\square$

## B. The Kuroš-Ore Theorem

The second application deals with representations of elements. In Corollary 111, we proved that in a finite distributive lattice the irredundant representation of an element as a join of join-irreducible elements is unique. This is obviously false in $M_3$. But we have the following results (A. G. Kuroš [490], O. Ore [558]):

**Theorem 351 (The Kuroš-Ore Theorem).** *Let $L$ be a modular lattice and let $a \in L$. If $a = x_0 \vee \cdots \vee x_{n-1}$ and $a = y_0 \vee \cdots \vee y_{m-1}$ are irredundant representations of the element $a$ as joins of join-irreducible elements, then for every $x_i$, there is a $y_j$ such that*

$$a = x_0 \vee \cdots \vee x_{i-1} \vee y_j \vee x_{i+1} \vee \cdots \vee x_{n-1}$$

*and $n = m$.*

*Proof.* Let us prove the first statement, say for $x_0$. Let $x_0^c = x_1 \vee \cdots \vee x_{n-1}$. (See Figure 73.)

Since $y_0 \vee \cdots \vee y_{m-1} = a$, we obtain that

$$(x_0^c \vee y_0) \vee (x_0^c \vee y_1) \vee \cdots \vee (x_0^c \vee y_{m-1}) = a,$$

where $x_0^c \vee y_0, \ldots, x_0^c \vee y_{m-1} \in [x_0^c, a]$. By the Isomorphism Theorem,

$$[x_0^c, a] \cong [x_0 \wedge x_0^c, x_0],$$

and the image of $a$ under any such isomorphism is $x_0$. But $x_0$ is join-irreducible in $L$, and, therefore, in $[x_0 \wedge x_0^c, x_0]$; thus $a$ is join-irreducible in $[x_0^c, a]$. Hence $x_0^c \vee y_j^c = a$, for some $j$, proving the first statement.

Now, let $a = z_0 \vee \cdots \vee z_{k-1}$ be an irredundant representation of $a$ as a join of join-irreducibles with $k$ minimal. Applying the statement we have just proved to $a = z_0 \vee \cdots \vee z_{k-1}$, $z_0$ and $a = x_0 \vee \cdots \vee x_{n-1}$, we obtain that $a = x_{j_0} \vee z_1 \vee \cdots \vee z_{k-1}$. This representation is irredundant, otherwise, the

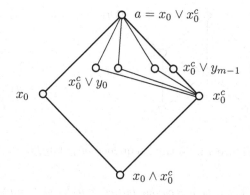

Figure 73. Illustrating the proof of the Kuroš-Ore Theorem

minimality of $k$ would be contradicted. Repeating this, we eventually obtain that $a = x_{j_0} \vee \cdots \vee x_{j_{k-1}}$. However, $a = x_0 \vee \cdots \vee x_{n-1}$ is an irredundant representation and so $\{j_0, \ldots, j_{k-1}\} = \{0, \ldots, n-1\}$. This shows that $n \le k$. Thus $k = n \ (= m)$.    $\square$

It is shown in M. Wild [732] that a sharp lower bound for the number of join-irreducibles in a finite modular lattice $L$ is $2\operatorname{len}(L) - s(L)$, where $s(L)$ is the number of maximal congruences.

The Kuroš-Ore Theorem (Theorem 351) holds for a finite semimodular lattice $L$ iff $L$ is *locally modular*, that is, for every $a \in L$, the interval $[v_a, a]$ is modular, where $v_a = \bigwedge(x \mid a \succ x)$, see R. P. Dilworth [154]. For generalizations to infinite lattices, see the survey article R. P. Dilworth [160] and Chapters 5 and 6 of P. Crawley and R. P. Dilworth [107].

## 1.4  Congruence spreading

Congruence projectivities in modular lattices can be described in terms of projectivities. Let us call a sequence of perspectivities:

$$[x_1, y_1] \sim [x_2, y_2] \sim \cdots \sim [x_n, y_n]$$

*alternating* if, for each $i$ with $1 < i < n$, either

$$[x_{i-1}, y_{i-1}] \overset{\mathrm{up}}{\sim} [x_i, y_i] \overset{\mathrm{dn}}{\sim} [x_{i+1}, y_{i+1}]$$

or

$$[x_{i-1}, y_{i-1}] \overset{\mathrm{dn}}{\sim} [x_i, y_i] \overset{\mathrm{up}}{\sim} [x_{i+1}, y_{i+1}].$$

If, in addition, $y_i = y_{i-1} \vee y_{i+1}$, in the first case, and $x_i = x_{i-1} \wedge x_{i+1}$, in the second, we call the sequence *normal*; see Figure 74.

Now we prove that all sequence of perspectivities can be normalized, see G. Grätzer [253].

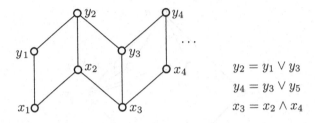

$$y_2 = y_1 \vee y_3$$
$$y_4 = y_3 \vee y_5$$
$$x_3 = x_2 \wedge x_4$$

Figure 74. Normal sequence of perspectivities

**Theorem 352.** *Let $L$ be a modular lattice, and let $[a, b]$ and $[c, d]$ be intervals of $L$. Then $[a, b] \Rightarrow [c, d]$ iff there exists a normal sequence of perspectivities $[x_1, y_1] \sim [x_2, y_2] \sim \cdots \sim [x_n, y_n] = [c, d]$, where $[x_1, y_1]$ is a subinterval of $[a, b]$.*

*Proof.* It follows from the Isomorphism Theorem that if $[x, y] \overset{\text{up}}{\sim} [u, v]$ and $[x_1, y_1]$ is a subinterval of $[x, y]$, then $[x_1, y_1] \overset{\text{up}}{\sim} [u_1, v_1]$, where $u_1 = x_1 \vee u$ and $v_1 = y_1 \vee u$. This, the dual statement, and a simple induction show that if $a/b \Rightarrow c/d$, then there is an alternating sequence of perspectivities

$$[x_1, y_1] \sim [x_2, y_2] \sim \cdots \sim [x_n, y_n] = [c, d],$$

where $[x_1, y_1]$ is a subinterval of $[a, b]$. By duality, we can assume that $[x_{n-1}, y_{n-1}] \overset{\text{dn}}{\sim} [x_n, y_n]$. Starting with this sequence, we show, by induction on $n$, that there is a normal sequence of perspectivities

$$[x_1, y_1] = [u_1, v_1] \sim \cdots \sim [u_n, v_n] = [x_n, y_n] = [c, d]$$

where $[x_1, y_1]$ is a subinterval of $[a, b]$ and with the same up and down perspectivities as the original sequence.

This statement is obvious for $n \leq 1$. By the induction hypothesis, there is a normal sequence

$$[x_1, y_1] = [u_1, v_1] \sim \cdots \sim [u_{n-1}, v_{n-1}] = [x_{n-1}, y_{n-1}]$$

with $[u_{n-2}, v_{n-2}] \overset{\text{up}}{\sim} [u_{n-1}, v_{n-1}]$.

We define

$$u'_{n-1} = (v_{n-2} \vee y_n) \wedge x_{n-1},$$
$$v'_{n-1} = v_{n-2} \vee y_n$$

and form the new sequence as follows:

$$[u_1, v_1] \sim \cdots \overset{\text{dn}}{\sim} [u_{n-2}, v_{n-2}] \overset{\text{up}}{\sim} [u'_{n-1}, v'_{n-1}] \overset{\text{dn}}{\sim} [x_n, y_n].$$

We claim that this is a normal sequence of perspectivities. We have to check two perspectivities:

$$[u_{n-2}, v_{n-2}] \overset{\text{up}}{\sim} [u'_{n-1}, v'_{n-1}],$$
$$[u'_{n-1}, v'_{n-1}] \overset{\text{dn}}{\sim} [x_n, y_n],$$

and that the new sequence is normal at $n-2$ and $n-1$.

We verify the first perspectivity:

$$v_{n-2} \wedge u'_{n-1} = v_{n-2} \wedge ((v_{n-2} \vee y_n) \wedge x_{n-1}) = v_{n-2} \wedge x_{n-1} = u_{n-2},$$

and, since $v_{n-2} \vee y_n \leq y_{n-1}$,

$$v_{n-2} \vee u'_{n-1} = v_{n-2} \vee ((v_{n-2} \vee y_n) \wedge x_{n-1}) = (v_{n-2} \vee y_n) \wedge (v_{n-2} \vee x_{n-1})$$
$$= (v_{n-2} \vee y_n) \wedge y_{n-1} = v_{n-2} \vee y_n.$$

The proof of the second perspectivity is similar.

The new sequence is normal at $n-1$ by the definition of $v'_{n-1}$. To see that the new sequence is normal at $n-2$, we have to verify that $u_{n-3} \wedge u'_{n-1} = u_{n-2}$. Since $u_{n-3} \wedge u_{n-1} = u_{n-2}$ and $u_{n-2} \leq v_{n-2} \leq v_{n-2} \vee y_n$, it follows that

$$u_{n-3} \wedge u'_{n-1} = u_{n-3} \wedge (u_{n-1} \wedge (v_{n-2} \vee y_n)) = u_{n-3} \wedge u_{n-1} \wedge (v_{n-2} \vee y_n)$$
$$= u_{n-2} \wedge (v_{n-2} \vee y_n) = u_{n-2}. \qquad \square$$

We can do even more than this: we can determine the sublattices generated by three consecutive intervals, see G. Grätzer [253].

**Theorem 353.** *Let $L$ be a modular lattice and let*

$$[x_0, y_0] \overset{\text{up}}{\sim} [x_1, y_1] \overset{\text{dn}}{\sim} [x_2, y_2]$$

*be a normal sequence of intervals. Let $A$ be the sublattice of $L$ generated by the elements $x_0, x_1, x_2, y_0, y_1, y_2$. Then either $A$ is distributive, in which case $A$ is the first lattice of Figure 75 or some quotient thereof and we also have*

$$[x_0, y_0] \overset{\text{dn}}{\sim} [(x_0 \wedge x_2), (y_0 \wedge y_2)] \overset{\text{up}}{\sim} [x_2, y_2],$$

*or $A$ is not distributive and $A$ is the second lattice of Figure 75 or some quotient not collapsing the diamond in $A$.*

*Proof.* Since the sequence is normal, the sublattice $A$ is generated by the elements $y_0, x_1, y_2$, hence $A$ is isomorphic to a quotient lattice of $\mathrm{Free}_M(3)$, see Figure 20. The generators satisfy the relation $y_0 \vee x_1 = x_1 \vee y_2 = y_0 \vee y_2$ (the last one because of normality). The corresponding quotient lattice is the second lattice in Figure 75. A quotient of this lattice is distributive iff the diamond is collapsed, yielding the first lattice of Figure 75. $\qquad \square$

**Corollary 354.** *Let $L$ be a modular lattice and let $[a, b] \approx [c, d]$ in $L$. Then there is a shortest normal sequence of intervals*

$$[a, b] = [x_0, y_0] \sim [x_1, y_1] \sim [x_n, y_n] = [c, d]$$

*and either $n \leq 2$ or the sublattice generated by the elements*

$$x_{i-1}, x_i, x_{i+1}, y_{i-1}, y_i, y_{i+1}$$

*is isomorphic to one of the last two lattices of Figure 75 or their duals for each $i$ with $0 < i < n$.*

*Proof.* If $n > 2$, then the sublattice generated by the elements

$$x_{i-1}, x_i, x_{i+1}, y_{i-1}, y_i, y_{i+1}$$

cannot be distributive, otherwise we could interchange the down-perspectivity and the up-perspectivity and get a sequence of length $n - 1$. Thus these elements generate the second or the third lattice of Figure 75, or its dual, or a quotient lattice or its dual. The only nontrivial quotient lattice of the second lattice of Figure 75 is the third lattice of Figure 75.    □

For a stronger form of Theorem 353 and Corollary 354, see D. X. Hong [399].

In Section III.1.1, we considered two very special forms of congruence spreading: perspectivity and subperspectivity of elements.

In the modular case, we find that subperspectivity is indeed "sub perspectivity":

**Lemma 355.** *Let $L$ be a sectionally complemented modular lattice and let $x, y \in L$. Then $x \lesssim y$ iff there exists $y' \leq y$ such that $x \sim y'$.*

*Proof.* If $x \lesssim y$, let $z \in L$ such that $x \wedge z = y \wedge z = 0$ and $x \leq y \vee z$. The element $y' = (x \vee z) \wedge y$ is majorized by $y$, thus $x \wedge z = y' \wedge z = 0$. By modularity,

$$y' \vee z = (x \vee z) \wedge (y \vee z) = x \vee z,$$

so $x \sim y'$.

Conversely let $x \sim y'$ for some $y' \leq y$. Let $z \in L$ with $x \wedge z = y' \wedge z = 0$ and $x \vee z = y' \vee z$. Then $x \wedge z = 0$ and $x \leq y \vee z$. If $z'$ is a sectional complement of $y \wedge z$ in $[0, z]$, then $x \wedge z' = y \wedge z' = 0$ and $x \leq y \vee z = y \vee z'$, and so $x \lesssim y$.    □

**Corollary 356.** *In a sectionally complemented modular lattice $L$, an ideal $I$ is standard iff it is perspectivity closed.*

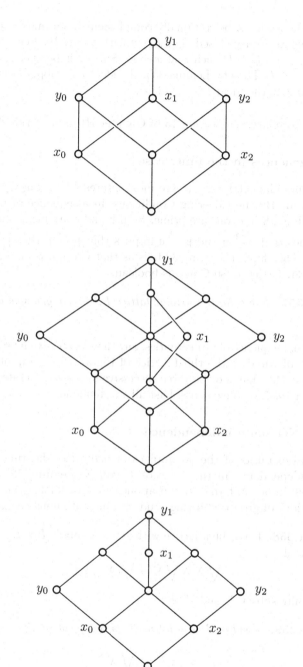

Figure 75. Normal perspectivities

*Proof.* Let the lattice $L$ be sectionally complemented and modular and let the ideal $I$ be perspectivity closed. Let $x \lesssim y$ with $y \in I$. By Lemma 355, there exists $y' \leq y$ (thus $y' \in I$) such that $x \sim y'$. Since $I$ is perspectivity closed, it follows that $x \in I$. Thus by Lemma 355, the ideal $I$ is subperspectivity closed. By Theorem 272, the ideal $I$ is standard. □

For this corollary, see Section 13 of Chapter III in G. Birkhoff [71].

### 1.5   Congruences in the finite case

In a finite modular lattice $L$, we are most interested in $\mathrm{con}(\mathfrak{p})$, where $\mathfrak{p}$ is a prime interval. By the Covering Conditions, in a sequence of projectivities starting with $\mathfrak{p}$, all intervals are prime. So if $\mathfrak{p}$ and $\mathfrak{q}$ are prime intervals, then $\mathfrak{p} \overset{\mathrm{up}}{\to} \mathfrak{q}$ implies that $\mathfrak{p} \overset{\mathrm{up}}{\sim} \mathfrak{q}$ and $\mathfrak{p} \overset{\mathrm{dn}}{\to} \mathfrak{q}$ implies that $\mathfrak{p} \overset{\mathrm{dn}}{\sim} \mathfrak{q}$; thus $\mathfrak{p} \Rightarrow \mathfrak{q}$ implies that $\mathfrak{p} \approx \mathfrak{q}$. Therefore, Theorem 239 tells us that $\mathrm{Con_{Ji}}\, L$ is an antichain for a finite modular lattice $L$, so $\mathrm{Con}\, L$ is boolean.

**Theorem 357.** *For a finite modular lattice $L$, the congruence lattice $\mathrm{Con}\, L$ is boolean.*

In a normal sequence of projectivities, any three successive intervals generate a sublattice pictured as the third lattice of Figure 75, or a quotient thereof, not collapsing the diamond. So three successive intervals generate a very small sublattice of 10, 7, or 5 elements; basically, a diamond.

### 1.6   Von Neumann independence

As a typical example of the advantages of using the shearing identity, we consider independence in the sense of J. von Neumann [552], [553]. This plays a very important role in the applications of lattice theory to direct decompositions of groups and rings and also in continuous geometries.

**Definition 358.** Let $L$ be a lattice with zero. A subset $I$ of $L - \{0\}$ is called *independent* if

$$\bigvee X \wedge \bigvee Y = \bigvee (X \cap Y),$$

for every finite subset $X$ and $Y$ of $I$.

**Corollary 359.** *A subset $I$ of a lattice $L$ is independent iff*

$$\varphi \colon X \mapsto \bigvee X$$

*is an isomorphism between* $\mathrm{sub}(I)$ *and the generalized boolean lattice of all finite subsets of* $I$.

*Proof.* Since

$$\bigvee X \vee \bigvee Y = \bigvee (X \cup Y),$$

it follows that if $I$ is independent, then $\varphi$ is a homomorphism. If $\varphi$ is not one-to-one, then $\bigvee X = \bigvee Y$ for some finite $X, Y \subseteq I$ with $X \neq Y$. Let say, $X \nsubseteq Y$, and let $a \in X - Y$. Then $a \leq \bigvee Y$ and $a \notin Y$. Therefore,

$$a = a \wedge \bigvee Y = \bigvee \{a\} \wedge \bigvee Y = \bigvee (\{a\} \cap Y) = \bigvee \varnothing = 0,$$

a contradiction. Thus $\varphi$ is an isomorphism. The converse is obvious. $\square$

A singleton $\{a\}$, for any $a \in L - \{0\}$, is always independent.

The two-element set $\{a, b\}$ is independent iff $a \wedge b = 0$ for any $a, b \in L - \{0\}$ and $a \neq b$.

Now consider a three-element subset $\{a, b, c\}$ of $L$. We have to require that

$$a \wedge (b \vee c) = 0,$$
$$b \wedge (a \vee c) = 0,$$
$$c \wedge (a \vee b) = 0,$$
$$(a \vee b) \wedge (a \vee c) = a,$$
$$(b \vee c) \wedge (b \vee a) = b,$$
$$(c \vee a) \wedge (c \vee b) = c.$$

For modular lattices, fewer relations will do:

**Theorem 360.** *Let $L$ be a modular lattice with zero. Then an $n$ element set $\{a_1, \dots, a_n\} \subseteq L - \{0\}$ is independent iff*

$$(a_1 \vee \cdots \vee a_i) \wedge a_{i+1} = 0 \quad \text{for } i = 1, 2, \dots, n-1.$$

*Proof.* We obtain the necessity of the condition by setting $X = \{a_1, \dots, a_i\}$ and $Y = \{a_{i+1}\}$. Now assume that $\{a_1, \dots, a_n\}$ satisfies the condition. Let $X, Y \subseteq \{a_1, \dots, a_n\}$ and $X \cap Y = \varnothing$. We claim that

$$\bigvee X \wedge \bigvee Y = 0.$$

Indeed, let $a_k \in X \cup Y$ with $k$ maximal. Let, say, $a_k \in Y$. Apply the shearing identity to $\bigvee X \wedge \bigvee Y$ with $x = \bigvee X$, $y = a_k$, and $z = \bigvee (Y - \{a_k\})$:

$$\bigvee X \wedge \bigvee Y = \bigvee X \wedge \left( a_k \vee \bigvee (Y - \{a_k\}) \right)$$
$$= \left( \bigvee X \wedge \left( a_k \wedge \left( \bigvee X \vee \bigvee (Y - \{a_k\}) \right) \right) \right) \vee \bigvee (Y - \{a_k\})$$
$$= \bigvee X \wedge \bigvee (Y - \{a_k\}),$$

since $a_k \wedge (\bigvee X \vee \bigvee(Y - \{a_k\})) \leq (a_1 \vee \cdots \vee a_{k-1}) \wedge a_k = 0$.

Proceeding thus, we can eliminate all the $a_i$ belonging to $X \cup Y$, getting $\bigvee X \wedge \bigvee Y = \bigvee \varnothing = 0$. Now, in the general case,

$$\bigvee X \wedge \bigvee Y = \bigvee X \wedge (\bigvee(X \cap Y) \vee \bigvee(Y - X))$$

(by modularity)

$$= \bigvee(X \cap Y) \vee (\bigvee X \wedge \bigvee(Y - X))$$

(by $X \cap (Y - X) = \varnothing$)

$$= \bigvee(X \cap Y). \qquad \square$$

The following result is very important; it is also a nice application of the concept of independence.

**Theorem 361.** *Let $L$ be a modular lattice with zero and denote by $F$ the set of joins of all (possibly empty) finite subsets of* $\mathrm{Atom}(L)$. *Then the following statements hold:*

  (i) *Every element of $F$ is the join of an independent set of atoms of $L$.*
  (ii) *$F$ is an ideal of $L$.*

*Proof.* (i) Let $a \in F$ and let $I$ be a finite subset of $\mathrm{Atom}(L)$ satisfying $a = \bigvee I$ and minimal with respect to containment. If $I$ is not independent, then by Theorem 360, there exists $p \in I$ such that $p \wedge \bigvee(I - \{p\}) = 0$. Since $p$ is an atom, it follows that $p \leq \bigvee(I - \{p\})$, and so $a = \bigvee(I - \{p\})$, which contradicts the minimality assumption on $I$.

(ii) For $a \in L$, $b \in F$, and $a \leq b$, we prove, by induction on $\mathrm{len}[a, b]$, that $a$ also belongs to $F$. This is trivial for $\mathrm{len}[a, b] = 0$.

Assume that $\mathrm{len}[a, b] = 1$. By (i), there exists a finite independent subset $I$ of $\mathrm{Atom}(L)$ such that $b = \bigvee I$. Since $a < b$, there exists $p \in I$ such that $p \not\leq a$. Set $c = \bigvee(I - \{p\})$. Then $a \vee p = c \vee p = b$ and $a \wedge p = c \wedge p = 0$, so the elements $a$ and $c$ are perspective. By Theorem 348, the intervals $[0, a]$, $[p, b]$, and $[0, c]$ are pairwise isomorphic. Since $c$ is a finite join of atoms, so is $a$.

In the general case with $a < b$, there exists an element $c$ such that $a \prec c$ and $c \leq b$. By the induction hypothesis, $c$ is a finite join of atoms. Since $\mathrm{len}[a, c] = 1$, $a$ is also a finite join of atoms. $\qquad \square$

As shown by the second lattice of Figure 77, Theorem 361(ii) does not extend to *semimodular* lattices.

## 1.7  Sublattice theorems

We conclude this section with three important "sublattice theorems"; the first is due to J. von Neumann [552], [553].

**Theorem 362.** *Let $L$ be a modular lattice and let $a, b, c \in L$. The sublattice of $L$ generated by the elements $a, b, c$ is distributive iff $a \wedge (b \vee c) = (a \wedge b) \vee (a \wedge c)$.*

*Proof.* By inspection of the diagram of $\mathrm{Free}_M(3)$ (see Figure 20). If $a, b, c$ are the generators and $a \wedge (b \vee c) \equiv (a \wedge b) \vee (a \wedge c) \pmod{\alpha}$, where $\alpha$ is a congruence relation, then $\alpha$ collapses the only diamond and so $\mathrm{Free}_M(3)/\alpha$ is distributive. $\qquad\square$

One can view the definition of modularity as requiring that any sublattice generated by three elements, two of which are comparable, has to be distributive. R. Dedekind [149] and G. Birkhoff [65] proved that this is true in general for every pair of chains.

**Theorem 363.** *Let $L$ be a modular lattice. Let $C$ and $C'$ be chains in $L$. Then $\mathrm{sub}(C \cup C')$, the sublattice of $L$ generated by $C \cup C'$, is distributive.*

*Proof.* Since a lattice is distributive iff every finitely generated sublattice is distributive, it is sufficient to verify this result for finite $C$ and $C'$. Let

$$
\begin{aligned}
C &= \{a_0, \ldots, a_{m-1}\}, \quad a_0 < \cdots < a_{m-1}, \\
C' &= \{b_0, \ldots, b_{n-1}\}, \quad b_0 < \cdots < b_{n-1}.
\end{aligned}
$$

To simplify the notation, we shall write $a_m = b_n = a_{m-1} \vee b_{n-1}$. Let us define

$$
x(r, \alpha, \beta) = (a_{\alpha(1)} \wedge b_{\beta(1)}) \vee \cdots \vee (a_{\alpha(r)} \wedge b_{\beta(r)}),
$$

where

$$
\begin{aligned}
\alpha &= (\alpha(1), \ldots, \alpha(r)), \quad m \geq \alpha(1) > \alpha(2) > \cdots > \alpha(r) \geq 1, \\
\beta &= (\beta(1), \ldots, \beta(r)), \quad 1 \leq \beta(1) < \beta(2) < \cdots < \beta(r) \leq n.
\end{aligned}
$$

Let $A$ denote the set of all elements of $L$ of the form $x(r, \alpha, \beta)$. In view of $a_i = a_i \wedge b_n$ and $b_i = a_m \wedge b_i$, it follows that $C, C' \subseteq A$. It is also easy to see that $A$ is closed under join: if we form $x(r, \alpha, \beta) \vee x(s, \alpha', \beta')$ and we have both $a_i \wedge b_j$ and $a_i \wedge b_k$, then we can eliminate one by absorption; if $a_i < a_j$, $b_k < b_t$ and we have both $a_i \wedge b_k$ and $a_j \wedge b_t$, the former is eliminated by absorption; what is left can be written in the form $x(r', \alpha'', \beta'')$.

Now we prove, by induction, the formula

$$
x(r, \alpha, \beta) = a_{\alpha(1)} \wedge (b_{\beta(1)} \vee a_{\alpha(2)}) \wedge \cdots \wedge (b_{\beta(r-1)} \vee a_{\alpha(r)}) \wedge b_{\beta(r)}
$$

and its dual. The case $r = 1$ is obvious. Let us assume that this formula and its dual have been verified for $r - 1$. Now compute:

$$x(r, \alpha, \beta) = (a_{\alpha(1)} \wedge b_{\beta(1)}) \vee \cdots \vee (a_{\alpha(r)} \wedge b_{\beta(r)})$$

(observe that $a_{\alpha(1)} \geq (a_{\alpha(2)} \wedge b_{\beta(2)}) \vee \cdots \vee (a_{\alpha(r)} \wedge b_{\beta(r)})$ and apply modularity)

$$= a_{\alpha(1)} \wedge (b_{\beta(1)} \vee (a_{\alpha(2)} \wedge b_{\beta(2)}) \vee \cdots \vee (a_{\alpha(r)} \wedge b_{\beta(r)}))$$

(observe that $b_{\beta(r)} \geq b_{\beta(1)} \vee \cdots \vee (a_{\alpha(r-1)} \wedge b_{\beta(r-1)})$ and apply modularity)

$$= a_{\alpha(1)} \wedge (b_{\beta(1)} \vee (a_{\alpha(2)} \wedge b_{\beta(2)}) \vee \cdots \vee (a_{\alpha(r-1)} \wedge b_{\beta(r-1)}) \vee a_{\alpha(r)})$$
$$\wedge b_{\beta(r)}$$

(apply to the expression in parentheses the dual of the formula for $r - 1$)

$$= a_{\alpha(1)} \wedge (b_{\beta(1)} \vee a_{\alpha(2)}) \wedge (b_{\beta(2)} \vee a_{\alpha(3)}) \wedge \cdots \wedge (b_{\beta(r-1)} \vee a_{\alpha(r)})$$
$$\wedge b_{\beta(r)},$$

completing the proof.

Now we easily see that $A$ is a sublattice. Indeed, $x(r, \alpha, \beta) \wedge x(s, \alpha', \beta')$ can be rewritten by the formula as a meet of elements of the form $b_i \vee a_j$; but by the dual of the above formula, the result is again of the form $x(r, \alpha, \beta)$.

So we conclude that $A$ is the sublattice generated by $C \cup C'$ and therefore $|A| \leq$ the number of expressions of the form $x(r, \alpha, \beta)$.

Now consider the set $X = [0, n + 1] \times [0, m + 1]$; let $F$ be the sublattice of $\operatorname{Pow} X$ generated by

$$a_i = \{\, (x, y) \mid y \leq i + 1 \,\}, \quad \text{for } i = 0, 1, \ldots, m - 1,$$
$$b_i = \{\, (x, y) \mid x \leq i + 1 \,\}, \quad \text{for } i = 0, 1, \ldots, n - 1.$$

Let $C = \{a_0, \ldots, a_{m-1}\}$ and $C' = \{b_0, \ldots, b_{n-1}\}$. In the lattice $F$, the equality

$$x(r, \alpha, \beta) = \bigcup (\, \{\, (x, y) \mid x \leq \alpha(i) + 1, \ y \leq \beta(i) + 1) \,\} \mid i = 1, 2, \ldots, r\,)$$

holds, so all the $x(r, \alpha, \beta)$ represent different elements. Consequently, $F$ is the free modular lattice generated by $C \cup C'$. Since $F$ is distributive, any modular lattice generated by $C \cup C'$ is distributive.   □

Finally, we generalize Theorem 106 to modular lattices:

**Theorem 364.** *Let $L$ be a modular lattice and let $a, b \in L$. Then the sublattice of $L$ generated by $[a \wedge b, a] \cup [a \wedge b, b]$ is isomorphic to $[a \wedge b, a] \times [a \wedge b, b]$.*

*Remark.* In the distributive case, the sublattice generated by $[a \wedge b, a] \cup [a \wedge b, b]$ is $[a \wedge b, a \vee b]$; this does not hold for modular lattices, as exemplified by $\mathsf{M}_3$.

*Proof.* The isomorphism we set up is

$$\varphi \colon (x, y) \mapsto x \vee y \quad \text{for } x \in [a \wedge b, a] \text{ and } y \in [a \wedge b, b].$$

Using the formula for the dual of $x(r, \alpha, \beta)$ in the proof of Theorem 363, we obtain

$$x \vee y = (a \vee y) \wedge (b \vee x).$$

Hence

$$a \wedge (x \vee y) = a \wedge (a \vee y) \wedge (b \vee x) = a \wedge (b \vee x) = (a \wedge b) \vee x = x,$$
$$b \wedge (x \vee y) = y,$$

proving that $\varphi$ is one-to-one. The map $\varphi$ is obviously onto and preserves join. Let $x, x_1 \in [a \wedge b, a]$ and $y, y_1 \in [a \wedge b, b]$. Then

$$(x \vee y) \wedge (x_1 \vee y_1) = (a \vee y) \wedge (b \vee x) \wedge (a \vee y_1) \wedge (b \vee x_1)$$
$$= (a \vee y) \wedge (a \vee y_1) \wedge (b \vee x) \wedge (b \vee x_1)$$

(use Corollary 349)

$$= (a \vee (y \wedge y_1)) \wedge (b \vee (x \wedge x_1))$$
$$= (x \wedge x_1) \vee (y \wedge y_1),$$

proving that $\varphi$ is an isomorphism.     □

## 1.8   ◇ Pseudocomplemented modular lattices
### by Tibor Katriňák

Can we generalize the triple construction for Stone algebras to pseudocomplemented modular lattices? Looking at the proofs in Section II.6.5, the heavy use of distributivity and the Stone identity, one would be somewhat pessimistic. So it is surprising that the answer is in the affirmative.

Let us recall that by Theorem 214 (C. C. Chen and G. Grätzer [89]), we can associate with every Stone algebra $L$ a triple (Skel $L$, Dns $L$, $\varphi_L$). Conversely, for every "abstract" triple $(B, D, \varphi)$, where $B$ is a boolean algebra, $D$ is a distributive lattice with unit, and $\varphi \colon B \to \mathrm{Fil}\, D$ is a $\{0, 1\}$-homomorphism, we can construct a Stone algebra $L$, whose associated triple (Skel $L$, Dns $L$, $\varphi_L$) is isomorphic to $(B, D, \varphi)$ (see the "Construction Theorem" in Exercise II.6.26).

This makes it possible to describe properties of Stone algebras as properties of triples.

Let us define a *p-algebra* as a pseudocomplemented lattice, with the pseudo-complementation regarded as a unary operation. T. Katriňák [459] extended the triple representation of Stone algebras to distributive p-algebras.

How about modular p-algebras? This was accomplished in T. Katriňák and P. Mederly [466].

There are various other generalizations. Let us define the algebra PCS as a pseudocomplemented semilattice, with the pseudocomplementation regarded as a unary operation.

T. Katriňák [458] obtained a triple representation for PCS-s. (See also W. C. Nemitz [547] for another triple representation of relatively pseudocomplemented meet-semilattices.) In addition, W. H. Cornish [97] and P. Mederly [528] extended the representation from T. Katriňák [458] to modular pseudocomplemented semilattices.

To what extent is modularity necessary in order to obtain a triple representation? The following definition is crucial.

**Definition 365.** A PCS $S$ is *decomposable* if for every element $x \in S$, there exists an element $d \in \mathrm{Dns}\, S$ such that

$$x = x^{**} \wedge d.$$

In any decomposable PCS $S$, we can define the congruences $a\mathbf{\Phi}(S)$, for $a \in \mathrm{Skel}\, S$, on the filter of dense elements, $\mathrm{Dns}\, S$, as follows:

$$d \equiv e \pmod{a\mathbf{\Phi}(S)} \text{ iff } a^* \wedge d = a^* \wedge e.$$

The map
$$\mathbf{\Phi}(S)\colon \mathrm{Skel}\, S \to \mathrm{Con}(\mathrm{Dns}\, S)$$

is the *structure map*. Clearly, $\mathbf{\Phi}(S)$ is a $\{0, 1\}$-isotone map.

With every decomposable PCS $S$, we can associate a triple

$$(\mathrm{Skel}\, S, \mathrm{Dns}\, S, \mathbf{\Phi}(S)),$$

which is called the *triple associated with* $S$. These triples can be abstractly characterized as follows:

**Definition 366.** A *triple* is defined as $(B, D, \Phi)$, where

(i) $B = (B; \vee, \wedge, ', 0, 1)$ is a boolean algebra;

(ii) $D$ is a meet-semilattice with unit;

(iii) $\Phi$ is a $\{0, 1\}$-isotone mapping from $B$ into $\mathrm{Con}\, D$.

Observe that the triple associated with a decomposable PCS $S$ is a triple as defined in Definition 366.

The triple associated with a decomposable PCS $S$ determines the algebra.

$\Diamond$ **Theorem 367.** *Two decomposable PCS-s are isomorphic iff their associated triples are isomorphic.*

Next we characterize the triples associated with decomposable PCS-s.

$\diamond$ **Theorem 368.** *Let* $(B, D, \Phi)$ *be a triple. Then there exists a decomposable PCS* $S$ *such that* $(B, D, \Phi) \cong (\operatorname{Skel} S, \operatorname{Dns} S, \Phi(S))$.

This result is best possible:

**Corollary 369.** *The class of decomposable PCS-s is the largest class of algebras that have triple representations.*

Unfortunately, the congruences $a\Phi(L)$ are only semilattice congruences, even if $L$ is a decomposable modular p-algebra.

From Theorem 368, we see that every decomposable PCS or decomposable p-algebra $L$ can be constructed from a triple $(B, D, \Phi)$. We have to work with pairs $(a, d/a'\Phi)$, where $d/a'\Phi$ is a congruence class on $D$. This contrasts with the situation in which $L$ can be obtained from a triple $(B, D, \varphi)$, where $\varphi$ is the map $B \to \operatorname{Fil} D$ (Stone algebras, distributive p-algebras or relatively pseudocomplemented (semi)lattices). The advantage of the second construction is that we are working with pairs of elements $(a, d) \in B \times D$ only (see Exercises 6.17–6.31).

Further work in this area was done by T. Katriňák [460], [461] and T. Katriňák and P. Mederly [466], [467]. For a review of this field as of 1980, see T. Katriňák [462].

### Applications

The triple decomposition of a PCS or a p-algebra can also be considered as a method for the study of decomposable PCS's. There exist characterizations of homomorphisms, subalgebras, direct products, congruences, and so on, of PCS's in terms of the associated triples. The main result in the paper T. Katriňák [463] is the abstract characterization of the congruence-lattice $\operatorname{Con} S$ of an arbitrary PCS $S$. Since the join-semilattice $C$ of compact elements of $\operatorname{Con} S$ is a decomposable join-PCS, it is enough to characterize the associated triple of $C$.

The work on triple representation has been an inspiration for similar constructions in other areas. See J. Schmidt [632] for semigroups, T. S. Blyth and J. Varlet [77] for MS-algebras, T. Katriňák and J. Guričan [465] for semilattices.

### 1.9  $\diamond$ Identities and quasi-identities in submodule lattices by Gábor Czédli

A ring $R$ is an associative ring with unit 1, and all modules satisfy $1x = x$. For a (left) module $M$ over a ring $R$, let $\operatorname{Sub} M$ denote the lattice of all submodules of $M$. Clearly, $\operatorname{Sub} M$ and $\operatorname{Con} M$ are isomorphic lattices and they are modular and algebraic. Let $R\text{-}\mathbf{Mod}$ denote the class of all $R$-modules.

We are interested in identities that hold in the submodule lattices of all $R$-modules or, equivalently, in the lattice variety $\mathbf{Var}(\{\operatorname{Sub} M \mid M \in R\text{-}\mathbf{Mod}\})$. Let

$$\mathbf{L}(R) = \mathbf{S}\left(\{\operatorname{Sub} M \mid M \in R\text{-}\mathbf{Mod}\}\right)$$

denote the class of lattices that can be embedded in some $\operatorname{Sub} M$ with $M \in R\text{-}\mathbf{Mod}$. This class is a quasivariety, see M. Makkai and G. McNulty [509] or G. Czédli [110]. Hence

$$\mathbf{Var}(\{\operatorname{Sub} M \mid M \in R\text{-}\mathbf{Mod}\}) = \mathbf{H}\,\mathbf{L}(R).$$

For $m, n \in \mathbb{N}_0$, define the ring *divisibility condition* $D(m, n)$ as the property: there is an $x \in R$ such that $m \cdot x = n \cdot 1$.

The next two theorems are from G. Hutchinson and G. Czédli [414].

$\Diamond$ **Theorem 370.** *Let $R$ be a ring, let $(m, n) \in \mathbb{N}_0 \times \mathbb{N}$, and let $\lambda$ be a lattice identity. Consider the pair $(m_\lambda, n_\lambda) \in \mathbb{N}_0 \times \mathbb{N}$ and the lattice identity $\Delta(m, n)$ constructed in [414]. Then*

- *$\lambda$ holds in $\mathbf{H}\,\mathbf{L}(R)$ iff $D(m_\lambda, n_\lambda)$ holds in $R$.*

- *$\Delta(m, n)$ holds in $\mathbf{H}\,\mathbf{L}(R)$ iff $D(m, n)$ holds in $R$.*

Identities similar to $\Delta(0, n)$ and $\Delta(m, 1)$ were previously given by C. Herrmann and A. P. Huhn [392]. It follows from Theorem 370 that $\mathbf{H}\,\mathbf{L}(R)$ depends only on the divisibility conditions that hold in $R$. However, we do not need all of them. Let $P$ denote the set of prime numbers. By the *spectrum* of a ring $R$ we mean the function $\operatorname{Spec}_R \colon \{0\} \cup P \to \mathbb{N}_0 \cup \{\omega\}$ defined by

$$0 \mapsto \min\{n \in \mathbb{N} \mid D(0, n) \text{ holds in } R\}, \text{ the } characteristic \text{ of } R,$$
$$p \mapsto \min\{n \in \mathbb{N}_0 \mid D(p^{n+1}, p^n) \text{ holds in } R\} \quad (\text{for } p \in P).$$

(In the first case, $\min \emptyset$ is 0, but it is $\omega$ in the second case.)

$\Diamond$ **Theorem 371.** *For any pair of rings $R$ and $S$, the inclusion $\mathbf{H}\,\mathbf{L}(R) \subseteq \mathbf{H}\,\mathbf{L}(S)$ is equivalent to $\operatorname{Spec}_R \leq \operatorname{Spec}_S$, that is, $\operatorname{Spec}_R(0)$ divides $\operatorname{Spec}_S(0)$ and $\operatorname{Spec}_R(0) \leq \operatorname{Spec}_S(x)$ for all $x \in P$.*

The possible spectra are also determined in [414]. Hence there are continuously many lattice varieties $\mathbf{H}\,\mathbf{L}(R)$, and the order they form is completely understood.

Although the original proof of the following duality theorem uses abelian category theory, Theorem 370 offers an easier approach, see G. Czédli and G. Takách [119].

G. Hutchinson [410]–[413] published a number of deep results in this field, we mention two more.

$\Diamond$ **Theorem 372.** *For any ring $R$, the class $\mathbf{H}\mathbf{L}(R)$ is a selfdual variety of lattices.*

In order to get information about the quasivarieties $\mathbf{L}(R)$, recall that $R$-**Mod** is an abelian category. Consider another ring $S$. A functor

$$F\colon R\text{-}\mathbf{Mod} \to S\text{-}\mathbf{Mod}$$

is called an *embedding functor*, if it sends distinct morphisms to distinct morphisms.

For $2 \leq n \in \mathbb{N}$, a sequence

$$A_0 \xrightarrow{f_1} A_1 \xrightarrow{f_2} A_2 \xrightarrow{f_3} \cdots \xrightarrow{f_n} A_n$$

in the category $R$-**Mod** is called an *exact sequence* if the image

$$\mathrm{Im}(f_i) = f_i(A_{i-1})$$

equals the kernel

$$\mathrm{Ker}(f_{i+1}) = \{\, x \in A_i \mid f_{i+1}(x) = 0 \,\}$$

for $i = 1, \ldots, n-1$. A functor $F\colon R\text{-}\mathbf{Mod} \to S\text{-}\mathbf{Mod}$ is called *an exact functor* if it sends exact sequences to exact sequences.

$\Diamond$ **Theorem 373.** *For any two rings $R$ and $S$, the inclusion $\mathbf{L}(R) \subseteq \mathbf{L}(S)$ holds iff there exists an exact embedding functor $F\colon R\text{-}\mathbf{Mod} \to S\text{-}\mathbf{Mod}$.*

G. Hutchinson [413] reduces the study of the $\mathbf{L}(R)$ quasivarieties to the case when $R$ has a prime power characteristic char $R = p^k$. Let

$$\mathfrak{Q}(n) = \big(\{\, \mathbf{L}(R) \mid \text{char } R = n \,\}; \subseteq\big),$$
$$\mathfrak{H}(n) = \big(\{\, \mathbf{H}\mathbf{L}(R) \mid \text{char } R = n \,\}; \subseteq\big).$$

It is shown in G. Hutchinson and G. Czédli [414] that $\mathfrak{H}(n)$ is a singleton for every $n \in \mathbb{N}$. If $p$ is a prime, then $\mathfrak{Q}(p)$ is a singleton by G. Hutchinson [410]. However, G. Czédli and G. Hutchinson [114] proved that, for $2 \leq k \in \mathbb{N}$, the boolean algebra of all subsets of a denumerably infinite set can be order-embedded into $\mathfrak{Q}(p^k)$.

Some related posthumous works of George Hutchinson are available from the web sites of G. Czédli, F. Wehrung, and C. Herrmann:

- http://www.math.u-szeged.hu/∼czedli/

- http://www.math.unicaen.fr/∼wehrung/

- http://www.mathematik.tu-darmstadt.de:8080/∼herrmann/

When searching for **submodule and lattice**, *MathSciNet* returns 673 papers.

**Exercises**

1.1. Show that a lattice $L$ is modular iff it satisfies the identity

$$(x \vee (y \wedge z)) \wedge (y \vee z) = (x \wedge (y \vee z)) \vee (y \wedge z).$$

1.2. Let
$$p = (x \vee y) \wedge (y \vee z) \wedge (z \vee x),$$
$$d_0 = a \vee b \vee c,$$

and define, for all $n \geq 0$,

$$d_{n+1} = p(a \wedge d_n, b \wedge d_n, c \wedge d_n).$$

Show that the identity $d_1 = d_2$ is equivalent to modularity but $d_2 = d_3$ is not (B. Wolk).

1.3. A finite lattice $L$ is modular iff it does not contain a pentagon $\{0, a, b, c, i\}$ satisfying $a \succ b$.

1.4. Can the numbers of covering pairs in Exercise 1.3 be increased?

1.5. Let $L$ be a lattice and $a, b \in L$. Then $\varphi_b \colon x \mapsto x \wedge b$ maps $[a, a \vee b]$ into $[a \wedge b, b]$, and $\psi_a \colon x \mapsto x \vee a$ maps $[a \wedge b, b]$ into $[a, a \vee b]$. Show that $\varphi_b \psi_a \varphi_b = \varphi_b$ and $\psi_a \varphi_b \psi_a = \psi_a$.

1.6. Using the notation of Exercise 1.5, prove that

$$\{\, x \in [a \wedge b, b] \mid \varphi_b \psi_a(x) = x \,\}$$

is isomorphic to

$$\{\, x \in [a, a \vee b] \mid \psi_a \varphi_b(x) = x \,\}.$$

(Exercises 1.5 and 1.6 are due to W. Schwan [634]–[636].)

1.7. Does Corollary 349 characterize the modularity of a lattice?

1.8. Find a nonmodular lattice satisfying the Upper Covering Condition and the Lower Covering Condition.

1.9. In a planar lattice $M$, a *covering square* is a sublattice $S = \mathsf{B}_i^2$ with elements $\{o, a, b, i\}$ such that $o \prec a$, $o \prec b$, $a \prec i$, $b \prec i$. A *covering diamond* is a sublattice $S = \mathsf{M}_3$ with elements $\{o, a, b, c, i\}$ such that $o \prec a$, $o \prec b$, $o \prec c$, $a \prec i$, $b \prec i$, $c \prec i$. Since $S$ is planar, it has exactly one internal element.
A planar modular lattice $M$ is called *slim* if it has no covering diamond. Prove that a slim planar modular lattice $M$ is distributive.

1.10. Start with a planar modular lattice $M$; if it has a covering diamond, remove its internal element. Proceeding thus, we obtain a sublattice $D$ without covering diamonds.
Prove that $D$ is a distributive lattice and it is a cover-preserving sublattice of $M$.

1.11. Show that we can construct every planar modular lattice as follows. Start with a planar distributive lattice $D$. Do the following in a finite number of times: add an internal element to a covering square.

1.12. Show that the Kuroš-Ore Theorem fails in the first lattice of Figure 77.

1.13. Prove that the representation

$$a = x_0 \vee \cdots \vee x_{i-1} \vee y_j \vee x_{i+1} \vee \cdots \vee x_{n-1}$$

is always irredundant in the Kuroš-Ore Theorem.

1.14. Using the notation of the Kuroš-Ore Theorem, show that there is a permutation $\pi$ of $\{0, \ldots, m-1\}$ such that

$$a = x_0 \vee \cdots \vee x_{i-1} \vee y_{\pi(i)} \vee x_{i+1} \vee \cdots \vee x_{n-1}$$

holds for all $0 \leq i < n$ (R. P. Dilworth [156]).

1.15. Can the Kuroš-Ore Theorem be sharpened so as to require that all $x_i$ can be simultaneously replaced by $y_j$? (No, inspect the third lattice in Figure 85.)

1.16. Does Theorem 352 characterize modularity?

1.17. Prove Theorem 353 directly, that is, without reference to Free$_M(3)$.

1.18. Show that, for finite modular lattices, it is sufficient to consider projectivities of prime intervals. Simplify Theorems 352 and 353 for prime intervals.

1.19. Extend Exercise 1.18 to locally finite lattices.

1.20. Show that Exercise 1.18 does not generalize to nonmodular lattices.

1.21. Show that to define independence in modular lattices, it is sufficient to assume that $\bigvee X \wedge \bigvee Y = \varnothing$ for finite $X, Y \subseteq I$ with $X \cap Y = \varnothing$.

1.22. Let $L$ be a complete lattice and let $I$ be a set of compact elements of $L$. Then $I$ is independent iff $X \mapsto \bigvee X$ is an isomorphism between the boolean lattice of all subsets of $I$ and the complete sublattice generated by $I$.

1.23. Let $L$ be a complemented modular lattice and let

$$a_0 = 0 < a_1 < \cdots < a_n = 1$$

be elements of $L$. Let $b_1 = a_1$, let $b_2$ be a relative complement of $a_1$ in $[0, a_2]$, ..., let $b_n$ be a relative complement of $a_{n-1}$ in $[0, a_n]$. Then $B = \{b_1, \ldots, b_n\}$ is an independent set.

1.24. Prove the converse of Exercise 1.23.

1.25. Show that Theorem 363 characterizes modularity.

*1.26. Let $L$ be a modular lattice and let $C$, $C'$, and $C''$ be chains in $L$. The sublattice generated by $C \cup C' \cup C''$ is distributive iff

$$a \wedge (b \vee c) = (a \wedge b) \vee (a \wedge c),$$

for all $a \in C$, $b \in C'$, and $c \in C''$ (B. Jónsson [434]).

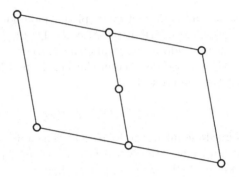

Figure 76. $d_2 = d_3$ fails

1.27. Let **K** be a variety of lattices and let $P$ be an order such that Free$_\mathbf{K}$ $P$ exists in the sense of Definition 66. Let **A** be the class of all *finite* lattices $L$ such that $L$ is generated by some homomorphic image of $P$ (in the sense defined in the proof of Theorem 69). Let us assume that there is an $L \in \mathbf{A}$ such that $|L| \geq |L'|$ for every $L' \in \mathbf{A}$. Is it true that Free$_\mathbf{K}$ $P$ exists and is finite, in fact, is it true that Free$_\mathbf{K}$ $P = L$?

1.28. Prove that in a modular lattice $L$, the sublattice generated by

$$[a \wedge b, a] \cup [a \wedge b, b]$$

equals the interval $[a \wedge b, a \vee b]$, for every $a, b \in L$, iff $L$ is distributive.

1.29. Find equations to replace "$a \wedge (b \vee c) = (a \wedge b) \vee (a \wedge c)$" in Theorem 362.

1.30. Find a direct proof of the statement that, in a modular lattice, every distributive element is neutral.

1.31. Show that, in a distributive lattice, if $[a, b] \approx [c, d]$ and $[c, d]$ is a subinterval of $[a, b]$, then $[a, b] = [c, d]$.

1.32. Show that the identity $d_2 = d_3$ of Exercise 1.2 fails in the lattice of Figure 76 (B. Wolk).

1.33. Show that a finite modular lattice is dismantlable iff it has breadth two or less (D. Kelly and I. Rival [469]).

1.34. Verify that if $L$ is a finite modular lattice satisfying

$$|L| \leq \frac{1}{3}(5 \operatorname{len}(L) + 7),$$

then $L$ has an $m$ element sublattice for all $m \leq |L|$. (I. Rival [615]. Hint: Use Exercise I.6.37.)

1.35. Does the fact that $\varphi_b$ of Theorem 348 is one-to-one, for all $a, b \in L$, characterizes the modularity of $L$?

## 2.   Semimodular Lattices

### 2.1   The basic definition

A lattice $L$ is called *semimodular* if it satisfies the *Upper Covering Condition*, that is, for all $a, b \in L$,

$$a \prec b \quad \text{implies that} \quad a \vee c \prec b \vee c \text{ or } a \vee c = b \vee c$$

(that is, $a \vee c \preceq b \vee c$). Examples of semimodular lattices include modular lattices and some important lattices from geometry (see Section 3). Two small examples are given in Figure 77. The simplest way to prove that a finite lattice is semimodular is to verify condition (ii) of Theorem 375.

Let $L$ be a lattice. Recall that, for the element $a \in L$, the length of a longest maximal chain in id$(a)$ is denoted by height$(a)$, provided that there is a finite longest maximal chain. Otherwise, put height$(a) = \infty$.

We first show that, for semimodular lattices of finite length, height$(a)$ is the length of *any* maximal chain in $[0, a]$, see O. Ore [563].

**Theorem 374 (The Jordan-Hölder Chain Condition).** *Let $L$ be a lattice of finite length and let $C$ and $C'$ be maximal chains in $L$. If $L$ is semimodular, then $C$ and $C'$ are of the same length.*

*Proof.* Let $C = \{a_0, \ldots, a_n\}$ be a maximal chain of length $n$:

$$0 = a_0 < \cdots < a_n = 1.$$

We prove that any other maximal chain is of length $n$, by induction on $n$. If $n \leq 1$, the statement is trivial. Let us assume that the statement holds for length $< n$. Let $C'$ be

$$0 = b_0 < b_1 < \cdots < b_m = 1,$$

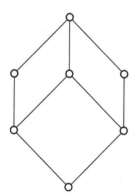
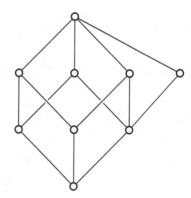

Figure 77. Two semimodular lattices

another maximal chain in $L$. If $a_1 = b_1$, then, in the semimodular lattice $\mathrm{fil}(a_1)$, the maximal chain $C - \{a_0\}$ is of length $n-1$, so the maximal chain $C' - \{b_0\}$ has to be of length $n-1$, therefore, $n = m$. If $a_1 \neq b_1$ (see Figure 78), then let $C''$ be a maximal chain in $\mathrm{fil}(a_1 \vee b_1)$ and let $k$ be the length of $C''$. Because of semimodularity, $a_1 \vee b_1 \succ a_1$ and $a_1 \vee b_1 \succ b_1$. Hence $C'' \cup \{a_1\}$ is a maximal chain of length $k+1$ and $C - \{a_0\}$ is a maximal chain of length $n-1$ in $\mathrm{fil}(a_1)$; thus $k+1 = n-1$. Similarly, $k+1 = m-1$, hence $n = m$. $\square$

Necessary and sufficient conditions for the validity of the Jordan-Hölder Chain Condition in a finite order $P$ is given in O. Ore [563]; see also S. Mac Lane [516]. To state Ore's condition, define a *cell* $C$ in $P$ as a subset

$$C = \{o, a_1, \ldots, a_n, b_1, \ldots, b_m, i\}$$

of $P$, where $\{o, a_1, \ldots, a_n, i\}$ and $\{o, b_1, \ldots, b_m, i\}$ are maximal chains in the interval $[o, i]$ of $P$ and

$$\sup\{a_k, b_j\} = i,$$
$$\inf\{a_k, b_j\} = o$$

for all $k = 1, \ldots, n$, $j = 1, \ldots, m$. Ore's condition is: *For every cell $C$, the equality $n = m$ holds ($n$ and $m$ depend on $C$).* This is obviously satisfied in semimodular lattices, where $n = m = 1$ for every cell. This result has been rediscovered (and published!) half a dozen times since 1943.

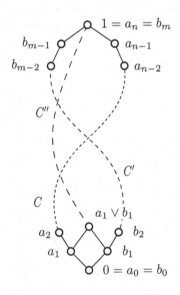

Figure 78. Proving the Jordan-Hölder Chain Condition

## 2.2   Equivalent formulations

Theorem 375 states the most important equivalent formulations of semimodularity:

**Theorem 375.** *Let $L$ be a lattice of finite length. The following conditions on $L$ are equivalent for all $a, b, c \in L$:*

  (i)  *$L$ is semimodular.*

 (ii)  *If $a \neq b$, and $a$ and $b$ cover $a \wedge b$, then $a \vee b$ covers $a$ and $b$.*

(iii)  *If $a \leq b$ and $C$ is a maximal chain in $[a, b]$, then $\{ x \vee c \mid x \in C \}$ is a maximal chain in $[a \vee c, b \vee c]$.*

(iv)  *$\operatorname{height}(a) + \operatorname{height}(b) \geq \operatorname{height}(a \wedge b) + \operatorname{height}(a \vee b)$.*

*Remark.* The equivalence of (i)–(iii), and that they imply (iv), can be found in G. Birkhoff [61]; see also G. Birkhoff [71].

*Proof.*
    (ii) *implies* (i). Let $a \prec b$. If $c \leq a$ or $a \vee c \geq b$, then $b \vee c \succeq a \vee c$. If $c \nleq a$ and $a \vee c \ngeq b$, then $b \wedge (a \vee c) = a$. Let

$$a = a_0 < a_1 < \cdots < a_n = a \vee c$$

be a maximal chain in $[a, a \vee c]$. Since $b$ and $a_1$ cover $a$ and $b \neq a_1$, the element $b \vee a_1$ covers $a_1$. An easy induction shows that $b \vee a_i$ covers $a_i$ for all $i = 1, 2, \ldots, n$. Thus $b \vee a_n$ covers $a_n$, that is, $b \vee c$ covers $a \vee c$.
    (i) *implies* (iii). Obvious, since if $x \prec y$ in $C$, then $c \vee x \preceq c \vee y$.
    (iii) *implies* (iv). Condition (iii) obviously implies semimodularity and so, by Theorem 374, we can assume that the Jordan-Hölder Chain Condition holds. Let $C$ be a maximal chain in $[a \wedge b, b]$. By the Jordan-Hölder Chain Condition, the length of $C$ is $\operatorname{height}(b) - \operatorname{height}(a \wedge b)$. By (iii),

$$D = \{ a \vee x \mid x \in C \}$$

is a maximal chain in $[a, a \vee b]$. The length of $D$ is at most the length of $C$, that is,

$$\operatorname{height}(b) - \operatorname{height}(a \wedge b);$$

on the other hand, the length of $D$ is

$$\operatorname{height}(a \vee b) - \operatorname{height}(a),$$

by the Jordan-Hölder Chain Condition, and so

$$\operatorname{height}(b) - \operatorname{height}(a \wedge b) \geq \operatorname{height}(a \vee b) - \operatorname{height}(a),$$

which was to be proved.

(iv) *implies* (ii). Assume that $a$ and $b$ cover $a \wedge b$. Then $a < a \vee b$. Rearranging the inequality in (iv), we get that

$$\text{height}(a \vee b) - \text{height}(a) \leq \text{height}(b) - \text{height}(a \wedge b) = 1,$$

and therefore, $a \vee b$ covers $a$. Similarly, $a$ covers $b$, verifying (ii).   □

Using Theorem 375, it is easy to check that some constructions yield semimodular lattices. We give one example. Take a semimodular lattice $L$ of length $n$ and pick a $k$ with $0 < k < n$. Let $L_k$ be the chopped lattice of all $x \in L$ with $\text{height}(x) \leq k$ along with 1. It is easy to check that $L_k$ is a semimodular lattice of length $k + 1$. The result of this construction with $L = \mathsf{C}_2^4$ and $k = 2$ is shown in Figure 79.

Another application of Theorem 375 is the following statement.

**Corollary 376.** *Let $L$ be a lattice of finite length. The following conditions on $L$ are equivalent:*

(i) *$L$ is modular.*
(ii) *$L$ satisfies the Upper and the Lower Covering Conditions.*
(iii) *$\text{height}(a) + \text{height}(b) = \text{height}(a \wedge b) + \text{height}(a \vee b)$ for all $a, b \in L$.*

*Proof.* We know, from Section 1.3, that (i) implies (ii). Assuming (ii), Theorem 375 and its dual (to be more precise, the dual of Theorem 375(iii)) yield condition (iii). Now assume (iii). If $L$ is not modular, then $L$ contains a pentagon $\{o, a, b, c, i\}$. Thus

$$\text{height}(i) = \text{height}(b) + \text{height}(c) - \text{height}(o),$$
$$\text{height}(i) = \text{height}(a) + \text{height}(c) - \text{height}(o),$$

implying that $\text{height}(a) = \text{height}(b)$, a contradiction.   □

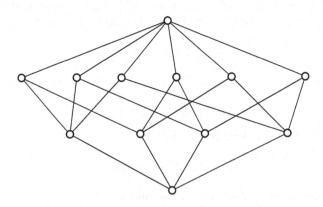

Figure 79. The construction illustrated

## 2.3   The Jordan-Hölder Theorem

A classical theorem of R. Dedekind [149] (see also R. Dedekind [150]) states that the factors of any chief series (maximal chain of normal subgroups) of a finite group are invariant. C. Jordan, O. Hölder, and H. Wielandt generalized this result to the factors of any composition series (maximal chain of subnormal subgroups of a group, where a subgroup $H$ of a group $G$ is *subnormal in $G$* if there exists a finite maximal chain $H = H_0 \prec H_1 \prec \cdots \prec H_n = G$ such that $H_i$ is normal in $H_{i+1}$ for each $i < n$).

Here is a formulation of this result for semimodular lattices (G. Grätzer and J. B. Nation [324]):

**Theorem 377 (Jordan-Hölder Theorem).** *Let $C$ and $D$ be two maximal chains in a finite length semimodular lattice, say*

$$C: 0 = c_0 \prec c_1 \prec \cdots \prec c_n = 1,$$
$$D: 0 = d_0 \prec d_1 \prec \cdots \prec d_n = 1.$$

*Then there is a permutation $\pi$ of the set $\{1, \ldots, n\}$ with the following property: there exists a (prime) interval $\mathfrak{p}_i$ such that $[c_{i-1}, c_i]$ is up-perspective to $\mathfrak{p}_i$ and $\mathfrak{p}_i$ is down-perspective to $[d_{\pi(i)-1}, d_{\pi(i)}]$ for all $1 \leq i \leq n$.*

## Remark

Recall that the lattice of subnormal subgroups of a finite group is lower semimodular (see H. J. Zassenhaus [742]; originally published in 1928 by O. Schreier and in 1934 by H. J. Zassenhaus), so the dual of this theorem applies to yield the full Jordan-Hölder Theorem for groups.

*Proof.* By induction on $\mathrm{len}(L)$. The statement is obvious for $\mathrm{len}(L) \leq 2$, so let $\mathrm{len}(L) > 2$.

Let $k$ be the largest integer with $c_1 \not\leq d_k$; note that $k < n$. If $k = 0$, then $c_1 = d_1$ and the statement follows by the induction hypothesis. So we can assume that $k > 0$.

Let $e_j = c_1 \vee d_j$ for all $0 \leq j \leq n$. Note that $e_0 = c_1$ and $e_k = e_{k+1} = d_{k+1}$, and indeed $e_j = d_j$ for $j \geq k + 1$. Now

$$c_1 = e_0 \prec e_1 \prec \cdots \prec e_k = e_{k+1} \prec e_{k+2} \prec \cdots \prec e_n = 1$$

is a maximal chain in the interval $[c_1, 1]$. By induction, there is an bijection

$$\sigma: \{2, \ldots, n\} \to \{1, \ldots, k, k+2, \ldots, n\}$$

such that, for all $i > 1$, the interval $[c_{i-1}, c_i]$ is up-perspective to some prime interval $\mathfrak{p}_i$ in $L$, which in turn is down-perspective to $[e_{\sigma(i)-1}, e_{\sigma(i)}]$. The interval $[e_{j-1}, e_j]$ is down-perspective to $[d_{j-1}, d_j]$, for all $j \leq k$, while the equality $[e_{j-1}, e_j] = [d_{j-1}, d_j]$ holds for all $j > k + 1$. Meanwhile, $[0, c_1]$ is up-perspective to $[d_k, d_{k+1}]$. So we may take $\pi$ to be the permutation with $\pi(i) = \sigma(i)$ for all $i \neq 1$, and $\pi(1) = k + 1$. $\qquad \square$

This result has an interesting consequence.

**Corollary 378.** *Let $L$ be a semimodular lattice and $C$ be a finite maximal chain in $L$. Then $C$ is a congruence-determining sublattice of $L$.*

*Proof.* We know that a congruence is determined by the prime intervals it collapses and every prime interval is contained in a maximal chain.

Let $\alpha$ be a congruence on $L$, let $C$ be a maximal chain. Let $\beta$ be the congruence of $L$ generated by $\alpha \rceil C$, the congruence $\alpha$ restricted to the chain $C$. Then $\beta \leq \alpha$. Also, if $[a, b]$ is a prime interval, then $[a, b]$ is projective to some prime interval $[c, d]$ in $C$, by Theorem 377. So if $a \equiv b \pmod{\alpha}$, then $c \equiv d \pmod{\alpha}$, whence $a \equiv b \pmod{\beta}$. Thus $\alpha \leq \beta$. We conclude that $\alpha = \beta$, verifying that $C$ is a congruence-determining sublattice of $L$. $\qquad \square$

In G. Czédli and E. T. Schmidt [117], there is an interesting addition to Theorem 377:

$\diamond$ **Theorem 379.** *The permutation $\pi$ in Theorem 377 is unique.*

## 2.4   Independence of atoms

Independence in semimodular lattices is easy to handle for atoms.

**Theorem 380.** *Let $I = \{a_1, \ldots, a_n\}$ be a set of $n$ atoms of a semimodular lattice. Then the following conditions on $I$ are equivalent:*

  (i) *$I$ is independent.*
  (ii) *$(a_1 \vee \cdots \vee a_i) \wedge a_{i+1} = 0$   for all $i = 1, 2, \ldots, n-1$.*
  (iii) *height$(a_1 \vee \cdots \vee a_n) = n$.*

*Proof.*
  (i) is equivalent to (ii) by Theorem 360.
  (ii) implies (iii). By induction on $i$, we prove that

$$\text{height}(a_1 \vee \cdots \vee a_i) = i.$$

This is true for $i = 1$. If height$(a_1 \vee \cdots \vee a_i) = i$, then by (ii) and semimodularity,

$$a_1 \vee \cdots \vee a_i \prec a_1 \vee \cdots \vee a_i \vee a_{i+1},$$

and so

$$\text{height}(a_1 \vee \cdots \vee a_i \vee a_{i+1}) = \text{height}(a_1 \vee \cdots \vee a_i) + 1 = i + 1.$$

(iii) implies (ii). It is obvious by semimodularity that, for atoms $b_1, \ldots, b_k$, the inequality

$$\text{height}(b_1 \vee \cdots \vee b_k) \leq k$$

holds. Assume now that (ii) fails. Then $a_{i+1} \leq a_1 \vee \cdots \vee a_i$. Using the last displayed inequality, we get that

$$\text{height}(a_1 \vee \cdots \vee a_n) = \text{height}(a_1 \vee \cdots \vee a_i \vee a_{i+2} \vee \cdots \vee a_n) \leq n - 1,$$

a contradiction. This verifies that (ii) holds.   □

Let $A$ be a set of atoms. Then $G \subseteq A$ *spans* $A$ if for every $a \in A$, there is a finite $G_1 \subseteq G$ such that $a \leq \bigvee G_1$. The following generalizes a result familiar from any first course on vector spaces.

**Theorem 381.** *Let $L$ be a semimodular lattice and let $A \subseteq \text{Atom}(L)$. Let $I$ be an independent subset of $A$, and let $G \supseteq I$ span $A$. Then there is an independent subset $J$ of $A$ such that $G \supseteq J \supseteq I$ and $J$ spans $A$.*

*Proof.* Let $\mathfrak{X}$ be the set of all independent subsets $X$ of $A$ with $I \subseteq X \subseteq G$. If $\mathcal{C} \subseteq \mathfrak{X}$ and $\mathcal{C}$ is a chain, then $\bigcup \mathcal{C}$ is again independent, since independence is tested with the finite subsets of $\bigcup \mathcal{C}$ and every finite subset of $\bigcup \mathcal{C}$ is also a finite subset of some $X \in \mathcal{C}$. Hence, by Zorn's Lemma (see Section II.1.4), there is a maximal independent subset $J$ of $A$ with $I \subseteq J \subseteq G$. We wish to show that $J$ spans $A$. It is sufficient to show that $J$ spans $G$. Let $g \in G$. If $g \in J$, there is nothing to prove. If $g \in G - J$, then $J \cup \{g\}$ is not independent, and so there is a finite $J_1 \subseteq J$ such that $J_1 \cup \{g\}$ is not independent, that is, by Theorem 380(iii) and by semimodularity,

$$\text{height}(\bigvee(J_1 \cup \{g\})) < |J_1| + 1.$$

Since $\text{height}(\bigvee J_1) = |J_1|$, we obtain that $\bigvee(J_1 \cup \{g\}) = \bigvee J_1$, that is, $g \leq \bigvee J_1$, proving that $A$ is spanned by $J$.   □

## 2.5 M-symmetry

The definition of semimodularity was given for arbitrary lattices but it is obvious that it is not very useful for lattices without many prime intervals since lattices without prime intervals are always semimodular. Various attempts have been made to rectify this situation, that is, to come up with a definition agreeing with semimodularity for lattices of finite length and that also selects an interesting class of lattices which are not of finite length.

We start with the definition of L. R. Wilcox [726] and S. Maeda [522].

**Definition 382.** *Let $L$ be a lattice. A pair of elements $(a, b)$ of $L$ is called modular (see Figure 80), in notation, $a \, M \, b$, if*

$$x \leq b \quad \text{implies that} \quad x \vee (a \wedge b) = (x \vee a) \wedge b.$$

The lattice $L$ is called M-*symmetric* if $a\,M\,b$ implies that $b\,M\,a$ for every $a, b \in L$.

Now it is obvious that $L$ is modular iff $a\,M\,b$ for all $a, b \in L$. In the proof of $[a \wedge b, b] \cong [a, a \vee b]$ in the Isomorphism Theorem, we only used that $a\,M\,b$ in $L$ and $b\,M\,a$ in $L^\delta$, hence

**Corollary 383.** *Let $L$ be a lattice and $a, b \in L$. If $a\,M\,b$ in $L$ and $b\,M\,a$ in $L^\delta$, then $[a \wedge b, b] \cong [a, a \vee b]$. This isomorphism is given by the map $\psi_a$, whose inverse is the map $\varphi_b$.*

*Remark.* The notation $\varphi_b$ and $\psi_a$ are from Section 1.2, see Figure 72.

If only $a\,M\,b$ is assumed, half of this conclusion still holds.

**Lemma 384.** *Let $L$ be a lattice and let $a, b \in L$. The following conditions are equivalent:*

   (i) *$a\,M\,b$;*
   (ii) *$\varphi_b$ is onto;*
   (iii) *$\psi_a$ is one-to-one.*

*Proof.*
   (i) *implies* (ii). If $x \in [a \wedge b, b]$, then $x \le b$, and so

$$\varphi_b(a \vee x) = (a \vee x) \wedge b = x \vee (a \wedge b) = x.$$

   (ii) *implies* (iii). Let $x, y \in [a \wedge b, b]$, and let $\psi_a(x) = \psi_a(y)$ with $x \ne y$. Then $\psi_a(x \vee y) = \psi_a(x) = \psi_a(y)$ and $x$ or $y < x \vee y$, say $x < x \vee y$. There exists a $z \in [a, a \vee b]$ such that $\varphi_b(z) = x$. Since $z \ge x$ and $z \ge a$, it follows that $z \ge x \vee a = \psi_a(x) = \psi_a(x \vee y)$. Thus $z \ge x \vee y$ and so $\varphi_b(z) \ge x \vee y > x$, a contradiction.

   (iii) *implies* (i). Let $x \le b$. Then

$$\psi_a(x \vee (a \wedge b)) = x \vee (a \wedge b) \vee a \quad = x \vee a,$$
$$\psi_a((x \vee a) \wedge b) = ((x \vee a) \wedge b) \vee a = x \vee a;$$

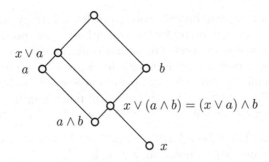

Figure 80. Modular elements

hence, we obtain that $a \, M \, b$, using that $\psi_a$ is one-to-one. To verify that

$$((x \vee a) \wedge b) \vee a = x \vee a,$$

observe that $x \vee a \geq$ and $x \vee a \geq (x \vee a) \wedge b$, hence $x \vee a \geq ((x \vee a) \wedge b) \vee a$. Moreover, if $t \geq (x \vee a) \wedge b$ and $t \geq a$, then $t \geq (x \vee a) \wedge b \geq x$, since $b \geq x$, and so $t \geq a \vee x$.                                                                               $\square$

The following result is implicit in L. R. Wilcox [726].

**Theorem 385.** *Let $L$ be a lattice of finite length. Then the lattice $L$ is semimodular iff it is M-symmetric.*

*Proof.* Let $L$ be a semimodular lattice of finite length. We shall prove that $a \, M \, b$ iff

(1)          $\mathrm{height}(b) - \mathrm{height}(a \wedge b) = \mathrm{height}(a \vee b) - \mathrm{height}(a),$

from which M-symmetry trivially follows. The length of $C$ is $\mathrm{height}(b) - \mathrm{height}(a \wedge b)$ and the length of $D$ is $\mathrm{height}(a \vee b) - \mathrm{height}(a)$. So if $a \, M \, b$, then $\psi_a$ is one-to-one; moreover, $|C| = |D|$ and (1) holds. Conversely, if $a \, M \, b$ fails, then $\psi_a$ is not one-to-one, and $C$ can be chosen so as to include $x, y \in [a \wedge b, b]$ with $\psi_a(x) = \psi_a(y)$. Then $|D| < |C|$ and we obtain (1).

To prove the converse, we do not have to assume that $L$ is of finite length. So let $L$ be an M-symmetric lattice, let $a, b, c \in L$, and let $b \succ a$. If $b \vee c = a \vee c$, then we have nothing to prove. If $b \vee c > a \vee c$, then put $d = a \vee c$ so $b \wedge d = a$ and $b \vee d = b \vee c$. We have to prove that $b \vee d \succ d$. Indeed, let $b \vee d > x \geq d$. Then $x \not\geq b$ and so $b \wedge x = a$ and $b \vee x = b \vee d$. Since $b \succ b \wedge x$, it follows that $\varphi_b$ as a map from $[x, x \vee b]$ into $[x \wedge b, b]$ is an onto map and so, by Lemma 384, we obtain that $x \, M \, b$. By M-symmetry, $b \, M \, x$, which means, by definition, that

$$y \vee (b \wedge x) = (y \vee b) \wedge x$$

for every $y \leq x$. Let $y = d$; then we obtain that

$$d = d \vee (b \wedge x) = (d \vee b) \wedge x = x,$$

that is, $b \vee c \succ a \vee c$.                                                                               $\square$

Examples of M-symmetric lattices not of finite length include the lattice of all closed subspaces of a Banach space and the projection lattice of a von Neumann algebra.

M-symmetry is one approach to extend the notion of semimodularity to lattices without many prime intervals. Another approach was given by S. Mac Lane [515] (see Exercises 2.22–2.24) and refined by R. P. Dilworth [154]. These and other approaches can be found in R. Croisot [108]. For a survey of these approaches and their interdependence, see M. Stern [667].

## 2.6   ◊ Consistency
## by Manfred Stern

Let $L$ be a lattice of finite length. A join-irreducible element $j$ of $L$ is *consistent* if, for every element $x$ in $L$, the element $x \vee j$ is join-irreducible also in the interval $[x, 1]$. The lattice $L$ is *consistent* if all the join-irreducible elements in $L$ are consistent.

Of course, consistency is a direct consequence of the Isomorphism Theorem for Modular Lattices (Theorem 348). Geometric lattices of finite length, see Section 3, are also trivially consistent. The *hexagon* (the smallest nonmodular semimodular lattice), see Figure 81, is not consistent.

The following result of P. Crawley [104] is a nice characterization of consistent lattices:

◊ **Theorem 386.** *In a lattice $L$ of finite length, consistency is equivalent to the following condition:*

*For all $x, y \in L$, if the interval $[x \wedge y, y]$ has exactly one dual atom, then the interval $[x, x \vee y]$ has exactly one dual atom.*

In fact, Crawley proved this equivalence for a large class of lattices that are not necessarily of finite length (see also P. Crawley and R. P. Dilworth [107]).

For the next result, we need one more class of lattices. Let $L$ be a lattice of finite length. A join-irreducible element $j$ is *strong* if $j \leq x \vee j_*$ implies that $j \leq x$ for all $x \in L$, where $j_*$ denotes the unique lower cover of $j$. (The element $j$ in the hexagon of Figure 81 is not strong.) A lattice of finite length is *strong* if all its join-irreducible elements are strong.

P. Crawley [104], U. Faigle [175], and K. Reuter [606] characterized consistency for semimodular lattices; a short summary of these results is given in M. Stern [664].

◊ **Theorem 387.** *For a finite lattice $L$, the following conditions are equivalent:*

(i) *$L$ is an S-glued sum of geometric lattices;*

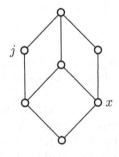

Figure 81. The hexagon

(ii) *L is semimodular and consistent;*

(iii) *L is semimodular and strong;*

(iv) *L is semimodular and has the replacement property described in the Kuroš-Ore Theorem (Theorem 351).*

*S*-glued sums are defined in Section IV.2.3.

To illustrate this result, we take an example from M. Stern [663]. We glue together the two geometric lattices $[0, b]$ and $[a, 1]$ of the lattice of Figure 82 to obtain the consistent semimodular lattice of Figure 82.

Other characterizations of consistent semimodular lattices are given in U. Faigle, G. Richter, and M. Stern [176] and T. Abe [1].

There are many interesting combinatorial results for semimodular lattices of finite length, see, for instance, H. H. Crapo [101], B. Korte, L. Lovász, R. Schrader [481], J. P. S. Kung [487], [488], and R. P. Stanley [662].

We only consider one more topic: the covering graph, Cov *L*, introduced in Section I.1.5.

G. Birkhoff [70], raised the question: "For which lattices *L* is there no information lost when passing from the diagram to the covering graph?"

An order *P* is called *graded* if an integer-valued function *h* can be defined on *P* satisfying

$$x \leq y \text{ and } h(x) + 1 = h(y) \quad \text{iff} \quad x \prec y$$

for all $x, y \in P$. By Exercise 2.16, every semimodular lattice of finite length is graded.

Birkhoff's question is answered by M. Stern [665] for a wide class of lattices.

◇ **Theorem 388.** *Let L and M be graded lattices of finite length with isomorphic covering graphs. Then L is both strong and dually strong iff M is both*

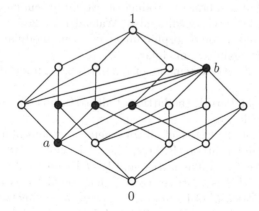

Figure 82. A consistent semimodular lattice obtained by gluing

*strong and dually strong. Moreover, if this condition is satisfied, then there are sublattices $A$ and $B$ of $L$ such that $L \cong A \times B$ and $M \cong A^\delta \times B$.*

This result contains as special cases two earlier contributions to Birkhoff's question, J. Jakubík [423] (for $L$ and $M$ modular) and D. Duffus and I. Rival [168] (for $L$ and $M$ atomistic and dually atomistic). The lattice of Figure 82 shows a nonmodular semimodular lattice that is both strong and dually strong but neither atomistic nor dually atomistic.

**Exercises**

2.1. Show that a lattice $L$ is semimodular iff $x \succ x \wedge y$ implies that $x \vee y \succ y$.

2.2. Modify the proof of the Jordan-Hölder Chain Condition. Assume only that $C'$ is a chain and $n < m$, and derive a contradiction. What conclusion can be drawn from this proof?

2.3. Let $L$ be a semimodular lattice. Let $A, B, C \subseteq \text{Atom}(L)$. Show that if $A$ spans $B$ and $B$ spans $C$, then $A$ spans $C$.

2.4. Show that (i)–(iii) of Lemma 384 are equivalent to

  (iv) $\varphi_b \psi_a$ is the identity map.

2.5. Let $L$ be a semimodular lattice and let $a \in L$. Prove that if $p$ and $q$ are atoms of $L$ and $a < a \vee q \le a \vee p$, then $a \vee p = a \vee q$ (*Steinitz-Mac Lane Exchange Axiom*).

2.6. Let $L$ be a lattice. Show that all sublattices of $L$ are semimodular iff $L$ is modular.

2.7. Show that direct products and convex sublattices of semimodular lattices are again semimodular.

2.8. Show that a subdirect product of two finite semimodular lattices is semimodular (G. Czédli and A. Walendziak [120]).

2.9. Prove that a homomorphic image of a semimodular lattice of finite length is again semimodular.

2.10. Investigate the statements of Exercises 2.6–2.9 for M-symmetric lattices.

2.11. Show that Part $A$, the lattice of all partitions on a set $A$, is a semimodular lattice.

2.12. Show that the lattice of all congruence relations of a semilattice is a semimodular lattice (R. Freese and J. B. Nation [188]).

2.13. Let $L$ be a modular lattice and let $I$ be an ideal of $L$. Then $L' = (L - I) \cup \{0\}$ is a lattice. Under what conditions is $L'$ semimodular?

2.14. Is the lattice $L'$ of Exercise 2.13 always M-symmetric?

2.15. Let $L$ be a lattice. Prove that $\text{Sub } L$ is semimodular iff $L$ is a chain (K. M. Koh [477]).

2.16. Show that an order $P$ is graded iff every interval of $P$ is of finite length and satisfies the Jordan-Hölder Chain Condition.

2.17. Let $L$ be a lattice. Show that $\operatorname{Sub} L$ is graded iff the dual of $\operatorname{Sub} L$ is semimodular, which, in turn, is equivalent to the condition that $L$ has no sublattice isomorphic to $C_2 \times C_3$ (H. Lakser [496]).

2.18. Prove that a lattice of finite length is semimodular iff it is graded and satisfies equation (1) on page 337 (G. Birkhoff [61]).

2.19. The *Infinite Jordan-Hölder Chain Condition* holds for a lattice $L$ if, for all maximal chains $C$ and $C'$ of $L$, the equality $|C| = |C'|$ holds. Show that this fails for the lattice $L = A \times B$, where $A$ is the $[0, 1]$ real interval and $B$ is the $[0, 1]$ rational interval.

2.20. Show that the Infinite Jordan-Hölder Chain Condition holds for the Boolean lattice of all subsets of a set $A$ iff $A$ is finite.

2.21. A lattice $L$ is said to satisfy the *Mac Lane Condition* if, for all elements $a, b, c$ of $L$ satisfying $a \parallel b$ and $a \wedge b < c \leq a$, there exists a $d \in L$ satisfying $a \wedge (b \vee d) = c$. Show that the Mac Lane Condition implies the Upper Covering Condition.

2.22. Prove that a lattice of finite length is semimodular iff it satisfies the Mac Lane Condition. (Exercises 2.21 and 2.22 are from S. Mac Lane [515].)

2.23. Find M-symmetric lattices that fail to satisfy the Mac Lane Condition.

2.24. Find lattices satisfying the Mac Lane Condition that are not M-symmetric.

2.25. Verify that from every planar semimodular lattice, we can obtain a slim one (defined in Exercise 1.9) by dropping the inner element in covering diamonds. And conversely, from slim, planar, semimodular lattices we can build up all planar semimodular lattices by reversing the process. (Use Figure 83 to visualize the process.)

2.26. Let $L$ and $K$ be finite lattices. Let $\varphi \colon L \to K$ be a join-homomorphism. We call $\varphi$ *cover-preserving* if it preserves the relation $\prec$; equivalently:

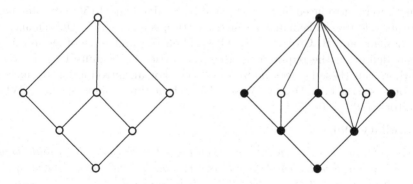

Figure 83. Two planar semimodular lattices

$x \prec y$ implies that $\varphi(x) \prec \varphi(y)$, for all $x$, $y \in L$, provided that $\varphi(x) \neq \varphi(y)$. Let $L$ be a slim semimodular lattice. Prove that there exists a planar distributive lattice $D$ and a cover-preserving join-homomorphism $\varphi$ of $D$ onto $L$.

2.27.  A planar semimodular lattice can be obtained from the direct product $A \times B$ of two finite chains $A$ and $B$ in the following three steps:

(a)  keep removing doubly irreducible elements from the boundaries, to obtain the lattice $D$.

(b)  Apply a cover-preserving join-homomorphism to $D$.

(c)  Add doubly-irreducible elements to the interiors of covering squares.

(See G. Grätzer and E. Knapp [291]–[295]. This was generalized in G. Czédli and E. T. Schmidt [116]. For a survey of recent developments in the field of finite semimodular lattices, see G. Czédli and E. T. Schmidt [118].)

# 3.  Geometric Lattices

## 3.1  Definition and basic properties

Just as most lattices arising out of algebraic examples are algebraic (see Section I.3.15), most lattices arising out of geometry are geometric in the following sense:

**Definition 389.** A lattice $L$ is called *geometric* if $L$ is semimodular, $L$ is algebraic, and the compact elements of $L$ are exactly the finite joins of atoms of $L$.

Equivalently, the lattice $L$ is complete, $L$ is *atomistic*, all atoms are compact, and $L$ is semimodular.

For lattices of finite length, this concept was introduced by G. Birkhoff [63], influenced by H. Whitney [725]. Geometric lattices with no restriction on length were introduced and investigated by S. Mac Lane [515] under the name *exchange lattices*. The name, geometric lattice, is due to M. L. Dubreil-Jacotin, L. Lesieur, and R. Croisot [167]. G. Birkhoff [71] and some others retain in their definition the requirement that the lattice be of finite length. Many authors call these lattices *matroid lattices*. For an introduction to matroid lattices, see the book D. J. A. Welsh [717] and the three volume series, N. White (editor) [719], [720], [721].

**Corollary 390.**

(i) *Let $L$ be a geometric lattice. Then the set $F$ of elements of finite height is an ideal of $L$. The lattice $F$ is semimodular and every element of $F$ is a finite join of atoms. The lattice $L$ is isomorphic to $\mathrm{Id}\,F$, the lattice of all ideals of $F$.*

(ii) *An interval of a geometric lattice is again a geometric lattice.*

*Proof.*

(i) Let $a, b \in F$, let $c \in L$ with $c \leq a$. Then $\mathrm{height}(c) \leq \mathrm{height}(a)$; it follows that $c \in F$. By Theorem 375,

$$\mathrm{height}(a \vee b) \leq \mathrm{height}(a) + \mathrm{height}(b) - \mathrm{height}(a \wedge b),$$

and so $a \vee b \in F$. Thus $F$ is an ideal. The second statement of (i) is obvious. The third statement follows from the proof of Theorem 42.

(ii) An interval $[a, b]$ of an algebraic lattice $L$ is again algebraic. Moreover, by semimodularity,

$$\{\, a \vee p \mid p \in \mathrm{Atom}(L),\ p \nleq a,\ \text{and}\ p \leq b \,\} = \mathrm{Atom}([a, b]).$$

The rest is easy.                                                            □

Now we utilize the theory of independence.

**Lemma 391.** *Let $L$ be a geometric lattice. Every element $a$ of $L$ is a join of an independent set of atoms, in fact, of any maximal independent set of atoms of $L$ in* $\mathrm{id}(a)$.

*Proof.* Let $A$ be a maximal independent set of atoms of $L$ in $\mathrm{id}(a)$ ($A$ exists by Theorem 381 with $I = \varnothing$ and $G = \mathrm{Atom}(a)$). By Theorem 381 and the compactness of atoms, $A$ spans all the atoms in $\mathrm{id}(a)$. Since $L$ is atomistic, it follows that $a = \bigvee A$.                                                            □

We also obtain the result of G. Birkhoff [63] and S. Mac Lane [515].

**Theorem 392.** *Any geometric lattice $L$ is complemented; in fact, it is relatively complemented.*

*Proof.* Let $L$ be a geometric lattice. Let $a \in L$. Let $A$ be a maximal independent set in $\mathrm{Atom}(a)$ and let $K$ be the set of all atoms not contained in $\mathrm{id}(a)$. Then by Theorem 381, there is a maximal independent set $I$ in $\mathrm{Atom}(L)$ satisfying $A \subseteq I \subseteq A \cup K$. Set

$$b = \bigvee (I - A).$$

Since $\bigvee I = 1$ by Theorem 381, we obtain that $a \vee b = 1$. Let $c = a \wedge b$. If $c \neq 0$, then there is an atom $p \leq c$. Since $p \leq \bigvee I$ and $p \leq \bigvee (I - A)$, by the compactness of $p$, there exist finite $I_1 \subseteq I$ and $I_2 \subseteq I - A$ such that $p \leq \bigvee I_1$ and $p \leq \bigvee I_2$. By the definition of independence,

$$p \leq \bigvee I_1 \wedge \bigvee I_2 = \bigvee (I_1 \cap I_2) = \bigvee \varnothing = 0,$$

a contradiction. Hence $a \wedge b = 0$, proving that $b$ is a complement of $a$. Therefore, $L$ is complemented. The second statement follows from the first and from Corollary 390.                                                            □

## 3.2   Structure theorems

Our next task is to prove the structure theorem for geometric lattices due to F. Maeda [519] and in its sharper form due to U. Sasaki and S. Fujiwara [628].

**Theorem 393.** *Every geometric lattice is isomorphic to a direct product of directly indecomposable geometric lattices.*

This theorem is augmented by a characterization theorem of direct indecomposability.

**Theorem 394.** *A geometric lattice is directly indecomposable iff every pair of atoms is perspective.*

For the proof of Theorem 393, we need a lemma. (Recall that $a \sim b$ was introduced in Definition 269.)

**Lemma 395.** *Let $L$ be a geometric lattice and let $a, b \in L$. If $a \sim b$ and $a$ and $b$ both have finite height, then there exists $x \in L$ with finite height such that $a \wedge x = b \wedge x = 0$ and $a \vee x = b \vee x$.*

*Proof.* Let $a, b \in L$, let $a$ and $b$ have finite height, and let

$$a \vee y = b \vee y,$$
$$a \wedge y = b \wedge y = 0.$$

Observe that $a \leq b \vee y$ and $\operatorname{height}(a), \operatorname{height}(b) < \infty$, hence, by the compactness of $a$, there is an $x_1 \leq y$ with $\operatorname{height}(x_1) < \infty$ satisfying $a \leq b \vee x_1$. Similarly, choose $x_2 \leq y$ with $\operatorname{height}(x_2) < \infty$ satisfying $b \leq a \vee x_2$. Then $x = x_1 \vee x_2$ will satisfy the lemma.                                    □

*Proof of Theorem 393.* For $a, b \in L$, let us write $a \approx b$ and call $a$ and $b$ *projective* if $a = x_0 \sim x_1 \sim \cdots \sim x_n = b$ for some $x_1, \ldots, x_{n-1} \in L$.
   For $p \in \operatorname{Atom}(L)$, we define the projective representative of $p$:

$$\operatorname{ProjRep}(p) = \{\, q \mid q \in \operatorname{Atom}(L) \text{ and } p \approx q \,\}$$

and the projective closure,

$$\operatorname{ProjCl} = \{\, \operatorname{ProjRep}(p) \mid p \in \operatorname{Atom}(L) \,\}.$$

For $x = \operatorname{ProjRep}(p) \in \operatorname{ProjCl}$, we set

$$\operatorname{unit}(x) = \operatorname{unit}(p) = \bigvee \operatorname{ProjRep}(p).$$

We are going to show that

$$L \cong \prod (\, [0, \operatorname{unit}(x)] \mid x \in \operatorname{ProjCl}).$$

Every element $a$ of $L$ is a join of atoms, hence if we set

$$a_{\mathrm{ProjRep}(p)} = \bigvee(\mathrm{id}(a) \cap \mathrm{ProjRep}(p)),$$

then

$$a = \bigvee(\, a_{\mathrm{ProjRep}(p)} \mid \mathrm{ProjRep}(p) \in \mathrm{ProjCl}\,),$$

where $a_{\mathrm{ProjRep}(p)} \le \mathrm{unit}(p)$.
   To establish that

$$a \mapsto (a_{\mathrm{ProjRep}(p)} \mid \mathrm{ProjRep}(p) \in \mathrm{ProjCl})$$

is the required isomorphism, it is sufficient by Exercise 3.20 to show that every element $a \in L$ has a unique representation in the form

$$a = \bigvee(\, a_{\mathrm{ProjRep}(p)} \mid \mathrm{ProjRep}(p) \in \mathrm{ProjCl}\,)$$

with $a_{\mathrm{ProjRep}(p)} \le \mathrm{unit}(p)$. Indeed, let $a$ have two such distinct representations:

$$a = \bigvee(\, x_{\mathrm{ProjRep}(p)} \mid \mathrm{ProjRep}(p) \in \mathrm{ProjCl}\,)$$
$$= \bigvee(\, y_{\mathrm{ProjRep}(p)} \mid \mathrm{ProjRep}(p) \in \mathrm{ProjCl}\,),$$

where

$$x_{\mathrm{ProjRep}(p)}, y_{\mathrm{ProjRep}(p)} \le \mathrm{unit}(p),$$
$$x_{\mathrm{ProjRep}(q)} \ne y_{\mathrm{ProjRep}(q)},$$

for at least one atom $q$. By taking the join of the two representations, we can assume that

$$x_{\mathrm{ProjRep}(p)} \le y_{\mathrm{ProjRep}(p)},$$

and so

$$x_{\mathrm{ProjRep}(q)} < y_{\mathrm{ProjRep}(q)}.$$

We now need the following easy statement:

   *Let $A \cup \{p\} \subseteq \mathrm{Atom}(L)$ and let $p \le \bigvee A$. Then $p \sim r$ for some $r \in A$.*

Indeed, $p \le \bigvee A_1$, for some finite set $A_1 \subseteq A$, by the compactness of $p$. Choose $A_1$ minimal. Therefore, $p \sim r$, for every $r \in A_1$, because if $x = \bigvee(A_1 - \{r\})$, then $r \vee x = \bigvee A_1$. Since $p \not\le x$ (because $A_1$ is minimal), it follows that

$$\mathrm{height}(p \vee x) = \mathrm{height}(x) + 1 = \mathrm{height}\Big(\bigvee A_1\Big)$$

and so $p \vee x = \bigvee A_1$. By minimality, the set $A_1$ is independent, hence $r \wedge x = 0$ and $p \not\leq x$, and so $p \wedge x = 0$. Thus $p \sim r$.

It follows from this statement that there is an atom $t \in \mathrm{ProjRep}(q)$ such that $t \leq y_{\mathrm{ProjRep}(q)}$ and $t \not\leq x_{\mathrm{ProjRep}(q)}$. It also follows that if we set

$$a_1 = x_{\mathrm{ProjRep}(q)},$$
$$a_2 = \bigvee (x_{\mathrm{ProjRep}(p)} \mid p \not\approx q),$$

then $a = a_1 \vee a_2$ and $a_1 \wedge a_2 = y_{\mathrm{ProjRep}(q)} \wedge a_2 = 0$. So we obtain elements $t, a_1$, and $a_2$ satisfying

$$t \wedge a_1 = 0,$$
$$a_1 \wedge a_2 = 0,$$
$$t \leq a_1 \vee a_2$$

(see Figure 84). Let $x_1$ be a relative complement of $t \vee a_1$ in $[a_1, a_1 \vee a_2]$ and $x_2$ a relative complement of $a_2 \wedge x_1$ in $[0, a_2]$. We claim that

$$t \vee x_1 = x_2 \vee x_1,$$
$$t \wedge x_1 = x_2 \wedge x_1 = 0,$$

and therefore, $t \sim x_2$. Indeed,

$$t \vee x_1 = t \vee a_1 \vee x_1 = a_1 \vee a_2,$$
$$x_2 \vee x_1 = x_2 \vee (a_2 \wedge x_1) \vee x_1 = a_2 \vee x_1 = a_1 \vee a_2,$$
$$t \wedge x_1 = t \wedge (t \vee a_1) \wedge x_1 = t \wedge a_1 = 0,$$
$$x_2 \wedge x_1 = a_2 \wedge x_2 \wedge x_1 = x_2 \wedge (a_2 \wedge x_1) = 0.$$

Now observe that $x_1 \prec t \vee x_1 = x_1 \vee x_2$. Therefore, for every atom $u \leq x_2$, the equality $x_1 \vee u = t \vee x_1$ holds and, of course, $x_1 \wedge u = 0$. Hence $t \sim u$. Now recall that $t \in \mathrm{ProjRep}(q)$, so $t \sim u$ implies that $u \in \mathrm{ProjRep}(q)$. But

$$u \leq a_2 = \bigvee (x_{\mathrm{ProjRep}(p)} \mid p \not\approx q),$$

hence $u \in \mathrm{ProjRep}(p)$, for some $p \neq q$, a contradiction.

Finally, we have to show that $[0, \mathrm{unit}(p)]$ is directly indecomposable. If it is not, then, by Theorem 275, it has a pair $a, b$ of complemented neutral elements, $a, b \notin \{0, \mathrm{unit}(p)\}$. So we can choose atoms $u, v$ of $[0, \mathrm{unit}(p)]$ such that $u \leq a$ and $v \leq b$. Since $u \approx v$ in $L$, this also holds in $[0, \mathrm{unit}(p)]$, since $[0, \mathrm{unit}(p)]$ is a direct factor. Taking the homomorphism $\varphi \colon x \mapsto x \wedge a$ of $[0, \mathrm{unit}(p)]$ onto $[0, a]$, we obtain that $\varphi(u) \approx \varphi(v)$. But $u = \varphi(u)$, $\varphi(v) = 0$, and $u \approx 0$ imply that $u = 0$, a contradiction. $\square$

It is clear, from the proof of Theorem 393, that indecomposability is equivalent with the projectivity of atoms. Therefore, to prove Theorem 394, it suffices to prove the following statement:

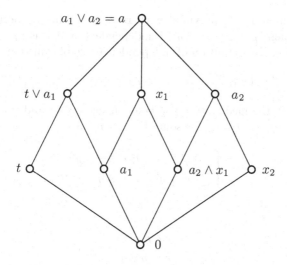

Figure 84. A step in the proof of Theorem 393

**Theorem 396.** *In a geometric lattice, perspectivity of atoms is transitive.*

*Proof.* Let $p, q, r \in \mathrm{Atom}(L)$. Let us assume that $p \sim q$ and $q \sim r$. We can choose, by Lemma 395, $x$ and $y$ such that

$$q \wedge y = r \wedge y = 0, \qquad q \vee y = r \vee h, \qquad \mathrm{height}(y) = n,$$
$$p \wedge x = q \wedge x = 0, \qquad p \vee x = q \vee x, \qquad \mathrm{height}(x) = m.$$

Choose $x$ so as to minimize $\mathrm{height}(x)$. Let $\{s_1, \ldots, s_m\}$ be an independent set in $\mathrm{Atom}(x)$, and $\{t_1, \ldots, t_n\}$ in $\mathrm{id}(y)$, and let

$$x_i = s_1 \vee \cdots \vee s_{i-1} \vee s_{i+1} \vee \cdots \vee s_m$$

for $1 \leq i \leq m$. We claim that $p \not\leq x_i \vee q$. Otherwise, $p \leq x_i \vee q$, which implies that $x_i \vee p \leq x_i \vee q$. However, $\mathrm{height}(x_i) = m - 1$ and therefore,

$$\mathrm{height}(x_i \vee p) = \mathrm{height}(x_i \vee q) = m,$$

implying that $x_i \vee p = x_i \vee q$ and then $x_i$ could replace $x$, contradicting the minimality of $\mathrm{height}(x)$. Therefore,

$$\mathrm{height}(p \vee q \vee x_i) = 2 + m - 1 = m + 1,$$
$$p \vee q \vee x_i = p \vee x = q \vee x,$$

for all $1 \leq i \leq m$. Now by Theorem 380, $\{s_1, \ldots, s_m, p\}$ is an independent set of atoms so by Theorem 381, we can choose a subset $\{t'_1, \ldots, t'_k\}$ of $\{t_1, \ldots, t_n\}$ such that

$$\{p, s_1, \ldots, s_m, t'_1, \ldots, t'_k\}$$

is independent and $k$ is maximal with respect to this property. Set $y' = t_1' \vee \cdots \vee t_k'$. Then $\{p, s_1, \ldots, s_m, y'\}$ is independent, and so is $\{q, s_1, \ldots, s_m, y'\}$. Therefore, using the definition of independence, we obtain that

$$y' = (x \vee y') \wedge (q \vee x_1 \vee y') \wedge \cdots \wedge (q \vee x_m \vee y').$$

Since $r \wedge y = 0$, the inequality $r \not\leq y$ holds; hence $r \not\leq y'$ and, therefore, either $r \not\leq x \vee y'$ or $r \not\leq q \vee x_i \vee y'$ for some $1 \leq i \leq m$. Set

$$a = \begin{cases} x \vee y', & \text{if } r \not\leq x \vee y'; \\ q \vee x_i \vee y', & \text{if } r \not\leq q \vee x_i \vee y'. \end{cases}$$

Then

$$r \wedge a = 0,$$
$$r \leq p \vee a,$$

the latter because $r \leq q \vee y \leq p \vee x \vee y = p \vee x \vee y' = p \vee q \vee x_i \vee y'$. Finally, since $r \not\leq a$, we obtain that $p \not\leq a$, that is,

$$p \wedge a = 0,$$

thus $a \leq a \vee r \leq a \vee p$ and $a \prec a \vee p$, and so

$$a \vee r = a \vee p,$$

proving that $p \sim r$. $\qquad \square$

If the geometric lattice $L$ is of finite length, then $L$ is said to be *finite dimensional*. In this case, Theorem 393 follows from Lemma 278. With some effort, the general case could be reduced to the finite dimensional case. However, without the present proof of Theorem 393, it would be more difficult to prove Theorem 394.

Theorem 279 could be used to prove the result that every finite dimensional geometric lattice is isomorphic to a direct product of simple geometric lattices. The factors we obtain from Theorems 393 and 394 are such that every pair of atoms is perspective. Now if $L$ is a geometric lattice of finite length in which every pair of atoms is perspective, then $L$ is simple. Indeed, if $\alpha$ is a congruence relation of $L$ and $\alpha \neq \mathbf{0}$, then there exist $a, b \in L$ with $a < b$, such that $a \equiv b \pmod{\alpha}$. Choose an atom $p$ such that $p \leq b$ and $p \not\leq a$. Then $p \equiv 0 \pmod{\alpha}$, since $p = p \wedge b \equiv p \wedge a = 0 \pmod{\alpha}$. Now let $q \in \text{Atom}(L)$. Since $p \sim q$, it follows that $p$ and $q$ have a common relative complement $x$ in some interval $[0, y]$. Then

$$[0, p] \overset{\text{up}}{\sim} [x, y] \overset{\text{dn}}{\sim} [0, q],$$

and so $q \equiv 0 \pmod{\alpha}$. Since height$(1) < \infty$, it follows that $1 = q_1 \vee \cdots \vee q_n$, for some finite set $\{q_1, \ldots, q_n\} \subseteq \mathrm{Atom}(L)$ satisfying $q_i \equiv 0 \pmod{\alpha}$, for all $i$ with $1 \leq i \leq n$, hence $1 \equiv 0 \pmod{\alpha}$. Thus $\alpha = 1$ and $L$ is simple.

Exactly the same proof yields that if $L$ is a geometric lattice in which any two atoms are perspective, then $L$ is subdirectly irreducible; the only atom $\beta$ of $\mathrm{Con}\, L$ contained in every $\alpha \neq 0$ is the congruence $\beta = \mathrm{con}(p, 0)$, where $p$ is any atom. Hence, by Theorem 392, a geometric lattice is subdirectly irreducible iff it is directly indecomposable. If $a \in L$ and height$(a) < \infty$, then $a \equiv 0 \pmod{\beta}$. In general, however, $L$ is not simple.

### 3.3 Geometries

By a *geometry* we mean a set $A$, certain subsets of which are called *subspaces* (or *flats*) such that some basic properties hold for the subspaces; we shall list these properties in the next definition.

**Definition 397.** A geometry $(A, ^-)$ is a set $A$ and a function $X \mapsto \overline{X}$ of $\mathrm{Pow}\, A$ into itself satisfying the following conditions:

(i) $(A, ^-)$ is a closure system on the set $A$ (see Definition 30).

(ii) $\overline{\varnothing} = \varnothing$, and $\overline{\{x\}} = \{x\}$ for any $x \in A$.

(iii) If $x \in \overline{X \cup \{y\}}$, but $x \notin \overline{X}$, then $y \in \overline{X \cup \{x\}}$.

(iv) If $x \in \overline{X}$, then $x \in \overline{X_1}$ for some finite $X_1 \subseteq X$.

A closure system is completely determined by the lattice $L$ of closed sets. In case of a geometry, we also have $\overline{\{x\}} = \{x\}$, meaning that the elements of $A$ can be identified as atoms of $L$, hence $L$ determines the geometry even if $L$ is known only up to isomorphism.

For geometries, a closed set is called a *subspace* and $\overline{X}$ is called the *subspace spanned by* $X$. Thus (iv) means that if $x$ belongs to the subspace spanned by the set $X$, then $x$ belongs to the subspace spanned by some finite subset $X_1$ of $X$. A closure system satisfying (iv) is usually called *algebraic*.

**Lemma 398.** *A lattice $L$ is algebraic iff $L$ is isomorphic to the lattice of closed sets of an algebraic closure system.*

*Proof.* Let $L$ be an algebraic lattice. By Theorem 42, we obtain the isomorphism $L \cong \mathrm{Id}\, F$, where $F$ is a join-semilattice with zero. Take $A = F$ and, for any $X \subseteq A$, define $\overline{X} = \mathrm{id}(X)$, the ideal generated by $X$. The verification of (i) and (iv) for $(A, ^-)$ is contained in the proof of Theorem 42. Then $\mathrm{Id}\, F$ is the set of closed sets and so $L \cong \mathrm{Id}\, F$.

Conversely, if $(A, ^-)$ is an algebraic closure system and $L$ is the lattice of closed sets, then it is easily seen that $X \in L$ is compact iff $X = \overline{X_1}$, for some finite $X_1 \subseteq X$, from which it follows immediately that $L$ is algebraic. $\square$

Note that Theorem 42 is a stronger form of Lemma 398.

Condition (iii) makes it possible to define the *dimension* of a subspace spanned by a finite set. Let $\dim(\varnothing) = 0$ and $\dim(\{a\}) = 1$. If $X$ is a subspace spanned by a finite set, $\dim(X) = n$, and $y \notin X$, then $\dim(\overline{X \cup \{y\}}) = n + 1$. (iii) implies that this defines dim uniquely.

Finally, (ii) states that $\varnothing$ is a subspace and so is every singleton $\{x\}$. If we drop this last condition, we get what is known as a *pregeometry*, from which a geometry can be obtained by identifying all pairs of elements, $x, y$ of $A$, for which $\overline{\{x\}} = \overline{\{y\}}$.

**Theorem 399.** *Let $(A, {}^-)$ be a geometry. Then $L = \mathrm{Clo}(A, {}^-)$ is a geometric lattice. Conversely, if $L$ is a geometric lattice, $A = \mathrm{Atom}(L)$, and $\overline{X}$ is the set of atoms spanned by $X$ (in the sense defined in Section 2.4), for any $X \subseteq A$, then $(A, {}^-)$ is a geometry and $L \cong \mathrm{Clo}(A, {}^-)$.*

*Proof.* Let $(A, {}^-)$ be a geometry. Then, by Lemma 398, $L = \mathrm{Clo}(A, {}^-)$ is algebraic. Since $X$ is compact iff $X = \overline{X_1}$, for some finite $X_1 \subseteq X$, we see that $X$ is compact iff $X$ is a finite join of atoms (by Definition 397(ii), $\{x\}$ is an atom for any $x \in A$).

It remains to show that $L$ is semimodular. Let $X, Y \in L$, let $Y = \overline{X \cup \{x\}}$, and let $x \notin X$. We claim that $X \prec Y$. Indeed, if $Z \in L$ and $X \subset Z \subseteq Y$, then there is a $z \in Z - X$ and $z \in Z \subseteq Y = \overline{X \cup \{x\}}$, and so $x \in \overline{X \cup \{z\}} \subseteq Z$, by Definition 397(iii). Thus $Y = \overline{X \cup \{x\}} \subseteq Z$. We conclude that $Y = Z$, proving that $X \prec Y$. Now let $U \in L$. Then

$$Y \vee U = \overline{Y \cup U} = \overline{X \cup U \cup \{x\}},$$
$$X \vee U = \overline{X \cup U},$$

hence either $x \in \overline{X \cup U}$ and so $X \vee U = Y \vee U$ or $x \notin \overline{X \cup U}$ and in this case $X \vee U \prec Y \vee U$.

Conversely, let $L$ be a geometric lattice, let $A = \mathrm{Atom}(L)$, and let $\overline{X}$ be the set of atoms spanned by $X$ for any $X \subseteq A$. It is immediate that $(A, {}^-)$ is a closure system. Definition 397(ii) and (iv) are also clear. To verify Definition 397(iii), let $x \in \overline{X \cup \{y\}}$ and $x \notin \overline{X}$. Since $\{y\}$ is an atom, by semimodularity,

$$\overline{X \cup \{y\}} = \overline{X} \vee \{y\} \succ \overline{X};$$

thus

$$X \subset \overline{X \cup \{x\}} \subseteq \overline{X \cup \{y\}}$$

implies that

$$\overline{X \cup \{x\}} = \overline{X \cup \{y\}};$$

so $y \in \overline{X \cup \{x\}}$.

Now let $\varphi$ be the map $X \mapsto \bigvee X$, for all $X \subseteq A$, and let $X \in \mathrm{Clo}(A, {}^-)$. Then $\varphi$ maps $\mathrm{Clo}(A, {}^-)$ into $L$. Since every element of $L$ is a join of atoms,

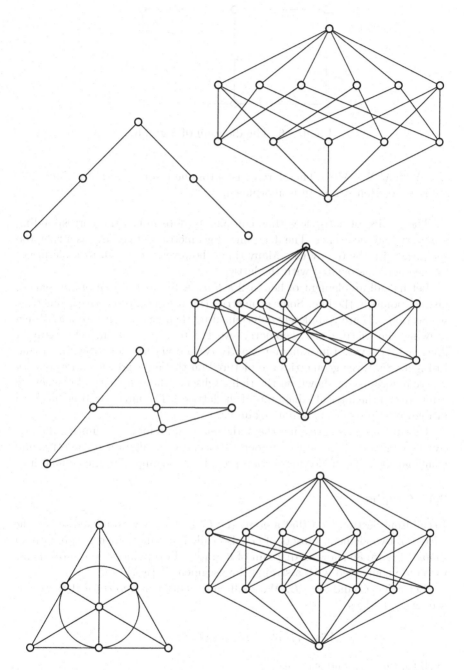

Figure 85. Three geometries and the corresponding lattices

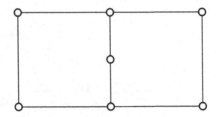

Figure 86. The diagram of a graph

$X \subseteq Y$ iff $\bigvee X \leq \bigvee Y$. Therefore, $\varphi$ is onto, one-to-one and both $\varphi$ and $\varphi^{-1}$ are isotone. Hence $\varphi$ is an isomorphism.                    $\square$

The lattice of subspaces thus completely determines the corresponding geometry and vice versa. The diagrams of geometric lattices are, as a rule, too complicated to be of any use. Many times, however, we can draw a picture of the geometry associated with the lattice.

Let us call an element of height 2 a *line*. A line is the join of any pair of distinct points in the line. Some geometries (for instance, projective geometries, see Section 5.3) are completely determined by their lines, some are not. When drawing the picture of a geometry, we try to represent lines by straight lines or by curves. Lines having exactly two points are not drawn as a rule. In Figure 85, three geometries are pictured on the left, and the corresponding lattice diagrams are shown on the right. Observe the simplicity of geometries compared to the intricate diagrams of the lattices. The intervals $[0, b]$ and $[a, 1]$ in Figure 82 provides two more examples.

Closure systems satisfying the additional requirements Definition 397(ii)–(iv) can be found in large numbers, especially in algebra, geometry, and combinatorics. The following section provides an example from combinatorics.

### 3.4   Graphs

Recall from Section I.1.4 that a *graph* $(G; E)$ is a set $G$ with a fixed set $E$ (the *edges*) of two-element subsets of $G$ (also called, an unoriented graph without loops). Figure 86 shows a diagram of a graph. Two points $a$ and $b$ (elements of $G$) are connected with a straight line segment if $\{a, b\} \in E$.

Let $a, b \in G$ and $A \subseteq E$. We shall call $a$ and $b$ *A-connected* if there is a sequence of edges

$$\{x_0, x_1\}, \{x_1, x_2\}, \dots, \{x_{n-1}, x_n\} \in A$$

such that $a = x_0$ and $x_n = b$.

Now we define a geometry, called the *edge geometry*, on $E$: for $\{a, b\} \in E$ and $A \subseteq E$, let $\{a, b\} \in \overline{A}$ if $a$ and $b$ are *A-connected*.

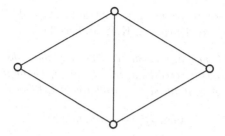

Figure 87. A small graph

**Theorem 400.** $(E, ^-)$ *is a geometry.*

*Proof.* All the properties of a geometry obviously hold except perhaps condition Definition 397(iii). Let $a$ and $b$ be $A$-connected; then there is an $A$-connecting sequence

$$\{x_0, x_1\}, \ldots, \{x_{n-1}, x_n\}$$

such that $a = x_0, x_n = b$ and for which $n$ is minimal. We claim that no edge can occur twice in this sequence. Indeed, let $0 \le i < j \le n-1$ and $\{x_i, x_{i+1}\} = \{x_j, x_{j+1}\}$. Now if $x_i = x_j$ and $x_{i+1} = x_{j+1}$, then, by dropping $\{x_{i+1}, x_{i+2}\}, \ldots, \{x_j, x_{j+1}\}$ from the sequence, we get a shorter $A$-connecting sequence. If $x_i = x_{j+1}$ and $x_{i+1} = x_j$, then we can drop all of $\{x_i, x_{i+1}\}, \ldots, \{x_j, x_{j+1}\}$.

Now we prove Definition 397(iii). Let $x \in \overline{X \cup \{y\}}$ but $x \notin \overline{X}$, where $x = \{a, b\}$ and $y = \{c, d\}$. Let $e_0, \ldots, e_{n-1} \in X \cup \{y\}$ be a shortest sequence connecting $a$ and $b$. Since $x \notin \overline{X}$, one of the $e_0, \ldots, e_{n-1}$ must be $y$. By the observation above, exactly one of $e_0, \ldots, e_{n-1}$, say $e_i$, equals $y$. But then $e_{i+1}, \ldots, e_{n-1}, x, e_0, \ldots, e_{i-1}$ will connect $c$ and $d$, hence $y \in \overline{X \cup \{x\}}$. $\qquad\square$

For instance, take the graph of Figure 87. Then in Figure 85 the first diagram on the left shows the edge geometry and on the right shows the lattice diagram, called the *edge lattice* associated with it.

## 3.5 Whitney numbers

To conclude this section we discuss two results of a combinatorial nature.

Let $L$ be a finite geometric lattice. Let $\mathrm{Lev}(i)$ be the set of elements of $L$ of height $i$ and let us define the *Whitney number*:

$$W_i = |\mathrm{Lev}(i)|.$$

Observe that each $\mathrm{Lev}(i)$ is an antichain of $L$.

An antichain $E$ of $L$ is a *maximum sized antichain* if $|X| \le |E|$ for every antichain $X$ of $L$. The following two theorems establish the size of a maximum

sized antichain for some classes of geometric lattices. The first is credited to
R. P. Dilworth and L. H. Harper in K. A. Baker [36].

**Theorem 401.** *Let the finite geometric lattice $L$ have the property that every
element $x$ of height $i$ is covered by $a_i$ and covers $b_i$ elements ($a_i$ and $b_i$ depend
only on $i$). Then a maximum sized antichain of $L$ has*

$$\max(W_i \mid 0 \le i \le \text{len}(L))$$

*elements.*

*Proof.* Let $n = \text{len}(L)$ and let $\max(W_i \mid 0 \le i \le n) = W_k$; then $\text{Lev}(k)$ is an
antichain with $W_k$ elements. We have to prove that if $S$ is an antichain, then
$|S| \le W_k$.

For $x \in L$, let $s(x)$ denote the number of maximal chains going through $x$.
It is obvious that if $\text{height}(x) = i$, then $s(x) = b_1 \cdots b_i \cdot a_i \cdots a_{n-1}$, hence $s(x)$
depends only on $i$. Since every maximal chain has exactly one element of
order $i$, we obtain that $s(x) \cdot W_i = s$, where $s$ is the number of maximal chains
in $L$. Using $W_i \le W_k$, we obtain that

$$s(x) = \frac{s}{W_i} \ge \frac{s}{W_k}.$$

Finally, a maximal chain goes through at most one element of the antichain $S$. Therefore,

$$s \ge \sum(s(x) \mid x \in S) \ge |S| \cdot \frac{s}{W_k},$$

that is, $\mid S \mid \le W_k$. □

Theorem 401, as well as the next result, was motivated by *Sperner's Lemma*
(E. Sperner [661]), stating the conclusions of Theorem 401 for $\text{Pow}\,X$.

A finite geometric *lattice* $L$ of length $n$ is *unimodal* if for some integer $k$
the inequalities

$$W_1 \le \cdots \le W_{k-1} \le W_k \ge W_{k+1} \ge \cdots \ge W_{n-1}$$

hold.

If $X$ and $Y$ are subsets of $L$, we say that there is a *matching* between $X$
and $Y$ if there is a one-to-one map $\varphi \colon X \mapsto Y$ (or $\varphi \colon Y \mapsto X$, if $|X| \ge |Y|$))
such that $x$ is comparable with $\varphi(x)$ for all $x \in X$. The following is G.-C. Rota's
approach to the problem of finding maximum sized antichains:

**Theorem 402.** *Let $L$ be a finite geometric lattice of length $n$. If $L$ is unimodal
and there is a matching between $\text{Lev}(i)$ and $\text{Lev}(i+1)$ for all $i < n$, then a
maximum sized antichain of $L$ has $W_k$ elements.*

*Proof.* For a subset $S$ of $L$, let

$$t(S) = \max(\text{height}(x) - \text{height}(y) \mid x, y \in S)$$

be the *thickness* of $S$. Let $S$ be an antichain of $L$. We shall prove $|S| \leq W_k$, by induction on $t(S)$.

If $t(S) = 0$, then $S \subseteq \text{Lev}(i)$ for some $i$, hence $|S| \leq W_i \leq W_k$. Now let $t(S) > 0$ and assume the inequality for antichains with a smaller thickness. Since $t(S) > 0$, there is an $x \in S$ with $\text{height}(x) \neq k$, say $\text{height}(x) < k$. Let $i = \min(\text{height}(x) \mid x \in S)$, and set $S = S_0 \cup S_1$, where $S_1 = S \cap \text{Lev}(i)$ and $S_0 = S - S_1$. By assumption, $i < k$, and so $|\text{Lev}(i)| \leq |\text{Lev}(i+1)|$, hence there is a matching $\varphi \colon \text{Lev}(i) \to \text{Lev}(i+1)$. Define

$$S' = S_0 \cup \varphi(S_1).$$

Observe that $S_0$ and $\varphi(S_1)$ are disjoint, because if $x \in S_0$ and $x \in \varphi(S_1)$, that is, if $x = \varphi(y)$ with $y \in S_1$, then $x, y \in S$ with $x \neq y$, and $x$ and $y$ are comparable, contradicting that $S$ is an antichain. Thus $|S'| = |S|$. Moreover, $S'$ is an antichain; indeed, if $x, y \in S'$ with $x \neq y$, and $x, y$ are comparable, then $x \in S_0$ and $y \in \varphi(S_1)$, that is, $y = \varphi(z)$ for some $z \in S_1$. Since $\text{height}(y) = i + 1$, the inequality $\text{height}(y) \leq \text{height}(x)$ holds, therefore, $x < y$ is impossible. Thus $y < x$, contradicting that $z \leq y$ and $z \parallel x$.

Thus $S'$ is an antichain, $t(S') = t(S) - 1$, and so

$$|S| = |S'| \leq W_k. \qquad \square$$

**Exercises**

3.1. Let $F$ be an atomistic semimodular lattice. Show that $\text{Id}\, F$ is a geometric lattice.

3.2. Is the converse of Exercise 3.1 true?

3.3. Prove that a lattice $L$ is geometric iff $L$ is atomic, relatively complemented, continuous, and semimodular (G. Birkhoff [70] and F. Maeda [519]).

3.4. Show that in Exercise 3.3 "semimodular" can be replaced by the condition that if $a$ and $b$ cover $a \wedge b$ and $a \neq b$, then $a \vee b$ covers $a$ and $b$.

3.5. Prove that in Exercise 3.3, "semimodularity" can be replaced by "M-symmetry" (F. and S. Maeda [521]).

3.6. Let $F$ be a semilattice and let $\text{Con}\, F$ be the lattice of all congruence relations of $F$. Prove that $\text{Con}\, F$ is geometric iff $\text{Con}\, F$ is atomistic.

3.7. Let $G$ be an abelian group. Let $L$ denote the lattice of all subgroups $H$ with the property that $G/H$ (the quotient group) has no element of finite order excepting zero. Prove that $L$ is a geometric lattice.

3.8. Consider the concepts of $p$-independence and $p$-basis in fields of characteristic $p$ (see, for instance, B. L. van der Waerden [693]). Can these be viewed as independence in a suitable geometric lattice?

3.9. Let $L$ be a geometric lattice. Show that $a \sim b$ in $L$ iff

$$a \vee x = b \vee x,$$
$$a \wedge x = b \wedge x$$

for some $x \in L$.

3.10. Let $a$ and $b$ be atoms in the geometric lattice $L$. Show that $a \sim b$ iff there exists a finite independent set of atoms $I$ such that $b \in I$ and, for all $I_1 \subseteq I$, the inequality $a \leq \bigvee I_1$ holds iff $I_1 = I$.

3.11. Under the conditions of Exercise 3.10, show that if $a \sim b$, then

$$[0, a] \stackrel{\text{up}}{\sim} [x, 1] \stackrel{\text{dn}}{\sim} [0, b]$$

for some $x \in L$.

3.12. Find a finite, simple, relatively complemented lattice where the relation of perspectivity on atoms is not transitive. (F. Wehrung.) (*Hint*: use the lattice in Figure 88.)

3.13. Prove that, in a geometric lattice, $a \sim b$ iff $[0, a] \sim [x, y] \sim [0, b]$ for some interval $[x, y]$.

3.14. Relate $a \approx b$ and $[0, a] \approx [0, b]$ in a general lattice, in a geometric lattice, and in a modular geometric lattice.

3.15. Let $L$ be a modular geometric lattice. Show that $L$ is simple iff $L$ is a directly indecomposable lattice of finite length.

3.16. Let $I$ and $J$ be maximal independent sets of atoms of a geometric lattice. Prove that $|I| = |J|$.

3.17. Show that in a geometric lattice $L$, for every element $a$, there is a smallest element $\text{unit}(a)$ of the center of $L$, satisfying $a \leq \text{unit}(a)$.

3.18. Prove that, in a geometric lattice, $a \sim b$ iff $\text{unit}(a) = \text{unit}(b)$ (with $\text{unit}(x)$ as in Exercise 3.17).

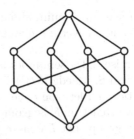

Figure 88. A lattice for Exercise 3.12

3.19. A lattice $L$ is called *left-complemented* if $L$ is bounded and, for all $a, b \in L$, there exists $b_1 \leq b$ such that

$$a \vee b_1 = a \vee b,$$
$$a \wedge b_1 = 0,$$

and $a \, M \, b_1$ (L. R. Wilcox [727]). Is every geometric lattice left-complemented?

3.20. Let $L$ be a complete lattice and let $X \subseteq L$. Then

$$L \cong \prod (\mathrm{id}(x) \mid x \in X)$$

iff every element $a \in L$ has one and only one expression of the form

$$a = \bigvee (a_x \mid x \in X),$$

where $a_x \leq x$ for all $x \in X$.

3.21. Define the isomorphism of two geometries. Prove that two geometries are isomorphic iff the associated geometric lattices are isomorphic.

3.22. Define a *circuit* of a graph $(G; E)$ as a set of edges

$$\{x_0, x_1\}, \ldots, \{x_{n-1}, x_n\}, \{x_n, x_0\}.$$

Define the subspaces of the geometry on the edges of the graph in terms of the circuits. What role is played by the minimal circuits?

3.23. Draw the lattice associated with the graph of Figure 86.

3.24. Does the lattice associated with a graph determine the graph (up to isomorphism)?

3.25. Generalize Theorem 401 to orders.

3.26. Prove Sperner's Lemma using Theorem 401.

3.27. Prove Sperner's Lemma using Theorem 402.

3.28. Does Theorem 401 or Theorem 402 apply to finite projective geometries?

3.29. Consider the graph $G_n$ of Figure 89. Let $L_n$ be the edge lattice of $G_n$. If $A \in L_n$ and $\{a, b\} \in A$, then $A$ is determined by

$$\{i \mid \{a, i\} \text{ and } \{b, i\} \in A\}.$$

If $\{a, b\} \notin A$, then $A$ is determined by

$$\{i \mid \{a, i\} \text{ or } \{b, i\} \in A\}$$

and a choice function selecting one of $\{a, i\}$ and $\{b, i\}$. Conclude from this the equality

$$W_k = 2^k \binom{n}{k} + \binom{n}{k-1}.$$

3.30. Show that in $L_{10}$ (as defined in Exercise 3.29) there is an antichain contained in $\mathrm{Lev}(6) \cup \mathrm{Lev}(7)$ that has more elements than any $\mathrm{Lev}(k)$. (Exercises 3.29 and 3.30 are due to R. P. Dilworth and C. Greene [162].)

*    *    *

Let $P$ be a finite order. The *Möbius function* an integer-valued function defined on $P \times P$ by the formulas:

$$\mu(x,y) = \begin{cases} 1, & \text{for } x = y \in P; \\ 0, & \text{if } x \not\le y; \\ -\sum(\mu(x,z) \mid x \le z < y), & \text{if } x < y. \end{cases}$$

(See A. F. Möbius [532] and L. Weisner [716].)

3.31. Establish the Möbius inversion formula: if $f$ and $g$ are real-valued functions on a finite order $P$ and

$$g(x) = \sum(f(y) \mid y \le x),$$

for every $x \in P$, then

$$f(x) = \sum(g(y)\mu(y,x) \mid y \le x).$$

3.32. Let $P$ and $Q$ be finite orders with Möbius functions $\mu_P$ and $\mu_Q$, respectively. Then the Möbius function $\mu_{P \times Q}$ of $P \times Q$ satisfies

$$\mu_{P \times Q}((x,y),(u,v)) = \mu_P(x,u)\mu_Q(y,v),$$

for $x, u \in P$ and $y, v \in Q$.

3.33. Show that the Möbius function $\mu$ of the boolean lattice $B$ is given by

$$\mu(x,y) = (-1)^{\mathrm{height}(y)-\mathrm{height}(x)},$$

for $x \le y$, where $\mathrm{height}(x)$ is the height of $x$ in $B$.

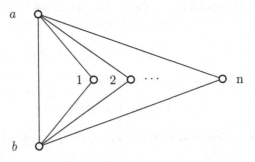

Figure 89. A graph for Exercise 3.29

3.34. Let $\mu$ be the Möbius function of a finite lattice $L$ and let $x, y, z \in L$.
(i) Let us assume that $x \leq y$ and $y$ is not the join of elements covering $x$; then $\mu(x, y) = 0$ (P. Hall [370]).
(ii) If $x \leq z \leq y$, then

$$\sum(\mu(x, t) \mid z \vee t = y) = \begin{cases} \mu(x, y), & \text{if } z = x; \\ 0, & \text{if } z \neq x. \end{cases}$$

(See L. Weisner [716].)

3.35. Show that the Möbius function $\mu$ of a finite distributive lattice $L$ is given by:

$$\mu(x, y) = \begin{cases} 0, & \text{if } y \text{ is not the join of elements covering } x; \\ (-1)^n, & \text{if } y \text{ is the join of } n \text{ distinct elements covering } x. \end{cases}$$

(Hint: Use Exercises 3.33 and 3.34(i).)

3.36. Let $\mu$ be the Möbius function of a finite geometric lattice $L$. Let $x, y \in L$ with $x \leq y$. Then $\mu(x, y) \neq 0$; $\mu(x, y)$ is positive if $\text{height}(y) - \text{height}(x)$ is even; $\mu(x, y)$ is negative, if $\text{height}(y) - \text{height}(x)$ is odd (G.-C. Rota [619]). (Hint: Apply Exercise 3.34(ii).)

# 4. Partition Lattices

## 4.1 Basic properties

As in Section I.1.2, a *partition* of a nonempty set $A$ is a set $\pi$ of nonempty pairwise disjoint subsets of $A$ whose union is $A$. The elements of $\pi$ are called the *blocks* of $\pi$. A singleton as a block is called *trivial*. If the elements $a$ and $b$ of $A$ belong to the same block, we write $a \equiv b \pmod{\pi}$ or $a \, \pi \, b$. For $a \in A$, we denote by $a/\pi$ the block $\{ b \in A \mid a \, \pi \, b \}$.

Let Part $A$ denote the set of all partitions of $A$ ordered by

$$\pi_0 \leq \pi_1 \quad \text{if} \quad x \equiv y \pmod{\pi_0} \text{ implies that } x \equiv y \pmod{\pi_1}.$$

The set Part $A$ with this ordering (which corresponds to set inclusion for the corresponding equivalence relations, see Section I.1.2) forms a complete lattice, called the *partition lattice* (or *equivalence lattice*) of $A$.

We draw a picture of a partition by drawing the boundary lines of the blocks. Then $\pi_0 \leq \pi_1$ if the boundary lines of the partition $\pi_1$ are also boundary lines of the partition $\pi_0$ (but $\pi_0$ may have some more boundary lines). Equivalently, the blocks of the partition $\pi_1$ are unions of blocks of the partition $\pi_0$, see Figure 90.

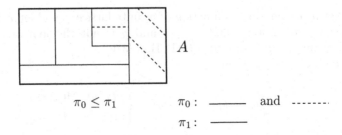

$$\pi_0 \leq \pi_1 \qquad\qquad \pi_0 : \quad\underline{\phantom{xxxx}}\quad \text{and} \quad \text{-------}$$
$$\pi_1 : \quad\underline{\phantom{xxxx}}$$

Figure 90. The ordering of partitions

**Lemma 403.**

(i) Part $A$ *is a complete lattice and*

$$x \equiv y \pmod{\bigwedge(\pi_i \mid i \in I)}$$

*iff* $x \equiv y \pmod{\pi_i}$ *for all* $i \in I$;

$$x \equiv y \pmod{\bigvee(\pi_i \mid i \in I)}$$

*iff there are* $i_0, \ldots, i_n \in I$ *and* $x_0, \ldots, x_{n+1} \in A$, *for some natural number* $n$, *such that* $x = x_0$, $y = x_{n+1}$, *and* $x_j \equiv x_{j+1} \pmod{\pi_{i_j}}$ *for all* $0 \leq j \leq n$.

*The zero,* $\mathbf{0}$, *of* Part $A$ *has only trivial blocks and the unit,* $\mathbf{1}$, *has only one block, namely,* $A$.

(ii) *The atoms of* Part $A$ *are the partitions with exactly one nontrivial block and this block has exactly two elements. The dual atoms of* Part $A$ *are the partitions with exactly two blocks.*

(iii) *The covering* $\pi_0 \prec \pi_1$ *holds in* Part $A$ *iff* $\pi_1$ *is the result of replacing two blocks of* $\pi_0$ *by their union.*

(iv) *In* Part $A$, *the filter* fil$(\pi)$ *is isomorphic to the partition lattice of the set* $\pi$.

(v) *In* Part $A$, *the ideal* id$(\pi)$ *is isomorphic to the direct product of all* Part $X$, *where* $X$ *ranges over the nontrivial blocks of* $\pi$.

*Proof.*
   (i) This is implicit in the proofs of Theorems 12 and 37.
   (ii) and (iii) are trivial.
   (iv) If $\pi \leq \xi$, then each block of $\xi$ is a union of blocks of $\pi$. Thus $\xi$ defines a partition of the blocks of $\pi$, that is, of $\pi$. This sets up the required isomorphism.

(v) If $\xi \leq \pi$, then for each nontrivial block $X$ of $\pi$, the partition $\xi$ of $A$ defines a partition $\xi_X$ of $X$. Since a block of $\xi$ that intersects $X$ is completely within $X$, the map

$$\xi \mapsto (\xi_X \mid X \text{ is a nontrivial block of } \pi)$$

sets up the required isomorphism.                                                $\square$

From Lemma 403 we easily conclude the following result of O. Ore [562].

**Theorem 404.** *The lattice* Part $A$ *is simple and geometric.*

*Proof.* By Lemma 403(i), Part $A$ is complete. It follows from Lemma 403(ii) that every element is a join of atoms and, by the formula for join, the atoms are compact. If $\pi_0$ and $\pi_1$ satisfy the condition of Lemma 403(iii), then so do $\pi_0 \vee \xi$ and $\pi_1 \vee \xi$ unless $\pi_0 \vee \xi = \pi_1 \vee \xi$. Therefore, Part $A$ is semimodular. This shows that Part $A$ is geometric.

Let $\pi$ and $\xi$ be atoms in Part $A$. By Lemma 403(ii), $\pi$ and $\xi$ have only one nontrivial block each, say $\{a, b\}$ and $\{c, d\}$, respectively. Let $\tau$ be a partition with two blocks as follows: if $\{a, b\} \cap \{c, d\} = \varnothing$, the blocks of $\tau$ are $A - \{a, c\}$ and $\{a, c\}$; if $\{a, b\} \cap \{c, d\} = \{e\}$, then the blocks of $\tau$ are $A - \{e\}$ and $\{e\}$. In either case, $\tau$ is a complement of $\pi$ and $\xi$ and so $\pi \sim \xi$. By the discussion in Section 3.2, the lattice Part $A$ is simple, if $A$ is finite.

Let $A$ be infinite and let $\alpha \neq \mathbf{0}$ be a congruence relation of Part $A$. Choose an atom $\xi \leq \mathbf{0}$. So $\xi \equiv \mathbf{0} \pmod{\alpha}$. As proved in the previous paragraph, all atoms of Part $A$ are congruent modulo $\alpha$ to the zero of Part $A$. Since $A$ is infinite, we can split $A$ up into two disjoint sets $A_0$ and $A_1$ of the same cardinality. Let $\varphi$ be a one-to-one map of $A_0$ onto $A_1$.

Now we define some partitions of $A$. Let $\pi$ have the two blocks, $A_0$ and $A_1$. Let $\xi_i$ have exactly one nontrivial block, $A_i$ for $i = 0, 1$. The blocks of $\tau$ are all the sets of the form $\{x, \varphi(x)\}$, where $x \in A_0$.

Then these partitions form a sublattice of Part $A$; a diagram of this sublattice is given in Figure 91.

Let $\varrho$ be any atom $\leq \tau$. Then $\varrho \equiv \mathbf{0} \pmod{\alpha}$, hence $\pi \equiv \pi \vee \varrho \pmod{\alpha}$. Since $\pi \prec \mathbf{1}$, we obtain that $\pi \vee \varrho = \mathbf{1}$ and so $\mathbf{1} \equiv \pi \pmod{\alpha}$. Taking the meets of both sides with $\tau$ and joining with $\xi_i$, we get $\xi_i \equiv \mathbf{1} \pmod{\alpha}$ for $i = 0, 1$, hence $\mathbf{0} = \xi_0 \wedge \xi_1 \equiv \mathbf{1} \pmod{\alpha}$. Thus $\alpha = \mathbf{1}$ and Part $A$ is simple.     $\square$

Call a partition $\pi$ *finite* if all blocks of $\pi$ are finite and only finitely many blocks are nontrivial. Then it is clear that $\pi$ is finite iff it is a finite join of atoms (recall Lemma 403(ii)). Since Part $A$ is geometric, this is, furthermore, equivalent to $\pi$ being compact.

Let Part$_{\text{fin}}$ $A$ be the set of all finite partitions. This is obviously an ideal of Part $A$. By Corollary 390 we obtain:

**Corollary 405.** Part$_{\text{fin}}$ $A$ *is an ideal of* Part $A$ *and the following isomorphism holds:*

$$\text{Id Part}_{\text{fin}} A \cong \text{Part } A.$$

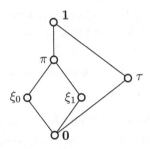

Figure 91. Partitions in the proof of Theorem 404

## 4.2  Type 3 representations

The importance of partition lattices lies in the fact that they are as universal for lattices as permutation groups are for groups. P. M. Whitman [724] proved that every lattice $L$ can be embedded in some partition lattice. One drawback of this result is that the joins of the partitions representing elements of $L$ have to be computed by Lemma 403(i), which can be rather complicated. This drawback of partition joins is eliminated in B. Jónsson [432]. To state Whitman's theorem as improved by Jónsson, we need some concepts.

A *representation* of a lattice $L$ is an embedding $\alpha$ of $L$ into a partition lattice, Part $A$; the embedding $\alpha$ is called *of type* 3 if

$$\alpha(a \vee b) = \alpha(a) \circ \alpha(b) \circ \alpha(a) \circ \alpha(b)$$

for all $a, b \in L$. In other words, type 3 means that for the partitions $\alpha(a)$, for $a \in L$, the join formula in Lemma 403(i) always holds with $n \leq 3$.

The number 3 appears arbitrary but we shall see in Theorem 406 that 3 cannot be reduced to 2 unless $L$ is modular.

**Theorem 406.** *Every lattice has a type 3 representation.*

*Proof.* A *meet-representation* $\mu$ of a lattice $L$ is an embedding of $L$ as a meet-semilattice into some partition lattice Part $A$; in other words, the map $\mu$ is a one-to-one map that preserves meets but not necessarily joins. We start out by describing a simple way of constructing meet-representations of a lattice.

Let $A$ be a set and let $L$ be a lattice with zero. A *distance function* $\delta$ maps the set $A^2$ onto the lattice $L$ and it satisfies the following rules:

(i) $\delta$ is *symmetric*, that is, $\delta(x, y) = \delta(y, x)$ for all $x, y \in A$;

(ii) $\delta$ is *normalized*, that is, $\delta(x, x) = 0$ for all $x \in A$;

(iii) $\delta$ satisfies the *triangle inequality*, that is,

$$\delta(x, y) \vee \delta(y, z) \geq \delta(x, z),$$

for all $x, y, z \in A$.

For instance, if $L$ is a lattice with zero and $A = L$, then defining $\delta(x, y) = x \vee y$, for all $x \neq y$, and $\delta(x, x) = 0$ yields a distance function.

With a distance function $\delta$, we can associate a meet-representation $\mu$ as follows: for $a \in L$ and $x, y \in A$, let

$$x \equiv y \pmod{\mu(a)} \quad \text{iff} \quad \delta(x, y) \leq a.$$

Indeed, $\mu(a)$ is a partition. Also

$$x \equiv y \pmod{\mu(a) \wedge \mu(b)} \quad \text{iff} \quad x \equiv y \pmod{\mu(a)} \text{ and } x \equiv y \pmod{\mu(b)},$$

which is equivalent to

$$\delta(x, y) \leq a \quad \text{and} \quad \delta(x, y) \leq b;$$

we obtained that $\delta(x, y) \leq a \wedge b$, which is the same as $x \equiv y \pmod{\mu(a \wedge b)}$. Thus

$$\mu(a) \wedge \mu(b) = \mu(a \wedge b),$$

that is, $\mu$ is a meet-homomorphism. To show that $\mu$ is one-to-one, take $a, b \in L$ with $a \neq b$; we can assume that $a \not\leq b$. Since $\delta$ is onto, there are $x, y \in A$ such that $\delta(x, y) = a$. Then $x \equiv y \pmod{\mu(a)}$. Since $a \not\leq b$, the congruence $x \equiv y \pmod{\mu(a)}$ fails to hold, and so $\mu(a) \neq \mu(b)$.

With a distance function $\delta \colon A^2 \to L$ and with a quadruple $(x, y, a, b)$ with $x, y \in A$ and $a, b \in L$ satisfying $\delta(x, y) \leq a \vee b$, we associate a new set

$$A^* = A \cup \{z_1, z_2, z_3\}$$

with three distinct elements $z_1, z_2, z_3$ not in $A$ and a new distance function $\delta^* \colon (A^*)^2 \to L$ defined as follows (see Figure 92): $\delta^*$ is symmetric, normalized,

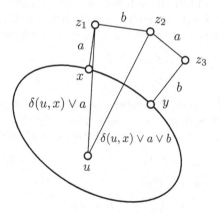

Figure 92. Constructing a type 3 representation

and $\delta^*(u, v) = \delta(u, v)$ for all $u, v \in A$. Moreover,

$$\delta^*(z_1, z_2) = b,$$
$$\delta^*(z_2, z_3) = a,$$
$$\delta^*(z_1, z_3) = a \vee b;$$

and for $u \in A$,

$$\delta^*(u, z_1) = \delta(u, x) \vee a,$$
$$\delta^*(u, z_2) = \delta(u, x) \vee a \vee b,$$
$$\delta^*(u, z_3) = \delta(u, y) \vee b.$$

It is easy to verify that $\delta^*$ is a distance function by virtue of $\delta(x, y) \le a \vee b$.

Now let $\{ (x_\gamma, y_\gamma, a_\gamma, b_\gamma) \mid \gamma < \chi \}$ be a well-ordered sequence of all quadruples satisfying $x_\gamma, y_\gamma \in A$, $a_\gamma, b_\gamma \in L$, and $\delta(x_\gamma, y_\gamma) \le a_\gamma \vee b_\gamma$. We define

$$A_0 = A^*,$$
$$\delta_0 = \delta^*;$$

this construction is performed with the quadruple $(x_0, y_0, a_0, b_0)$ for all $\gamma < \chi$;

$$A_\gamma = \left( \bigcup ( A_\xi \mid \xi < \gamma ) \right)^*,$$
$$\delta_\gamma = \left( \bigcup ( \delta_\xi \mid \xi < \gamma ) \right)^*,$$

this construction is performed with the quadruple $(x_\gamma, y_\gamma, a_\gamma, b_\gamma)$. Finally,

$$A^+ = \bigcup ( A_\gamma \mid \gamma < \chi ),$$
$$\delta^+ = \bigcup ( \delta_\gamma \mid \gamma < \chi ).$$

Summarizing, we obtain:

For every set $A$ and distance function $\delta$, there exist a set $A^+ \supseteq A$ and a distance function $\delta^+$ on $A^+$ extending $\delta$ such that, for all $x, y \in A$ and $a, b \in A$, if $\delta(x, y) \le a \vee b$, then there exist $z_1, z_2, z_3 \in A^+$ such that

$$\delta^+(x, z_1) = \delta^+(z_2, z_3) = a,$$
$$\delta^+(z_1, z_2) = \delta^+(z_3, y) = b.$$

Now to prove Theorem 406, embed the lattice $L$ into a lattice $K$ with zero. Let $A_0 = K$ and define $\delta_0$ as above: $\delta_0(x, y) = x \vee y$, if $x \ne y$, and $\delta_0(x, x) = 0$. Again, we define, inductively, sets and distance functions:

$$A_{n+1} = (A_n)^+ \qquad \text{for } n < \omega,$$
$$\delta_{n+1} = (\delta_n)^+ \qquad \text{for } n < \omega;$$
$$A = \bigcup ( A_n \mid n < \omega ),$$
$$\delta = \bigcup ( \delta_n \mid n < \omega ).$$

Obviously, $\delta\colon A^2 \to K$ is a distance function. Let $\mu$ be the meet-representation associated with $\delta$. If $x \equiv y \pmod{\mu(a \vee b)}$, that is, if $\delta(x, y) \leq a \vee b$, then $x, y \in A_n$, for some $n < \omega$, hence there exists $z_1, z_2, z_3 \in A$ (in fact, $z_1, z_2, z_3 \in A_{n+1}$) such that

$$\delta(x, z_1) = \delta(z_2, z_3) = a,$$
$$\delta(z_1, z_2) = \delta(z_3, y) = b,$$

and so

$$x \equiv z_1 \pmod{\mu(a)}, \qquad z_1 \equiv z_2 \pmod{\mu(b)},$$
$$z_2 \equiv z_3 \pmod{\mu(a)}, \qquad z_3 \equiv y \pmod{\mu(b)}.$$

Thus $x \equiv y \pmod{\mu(a) \vee \mu(b)}$, proving that $\mu(a \vee b) = \mu(a) \vee \mu(b)$. Therefore, $\mu$ is a representation, in fact, a type 3 representation of $K$ and hence also of $L$. □

Finally, we prove P. M. Whitman's result—see [724]—that every representation of a lattice yields a representation by subgroups.

**Corollary 407.** *Every lattice can be embedded in the lattice of all subgroups of some group.*

*Proof.* Let $\mu\colon L \to \operatorname{Part} A$ be a representation of $L$. Let $G$ be the group of all permutations of $A$. To an element $a \in L$, we assign the subgroup $H_a$ generated by all transpositions $(xy)$ such that $x \equiv y \pmod{\mu(a)}$. (The *transposition* $(xy)$ is the permutation interchanging $x$ and $y$ and leaving all other elements of $A$ fixed.) It is easy to see that $(xy) \in H_a$ iff $x \equiv y \pmod{\mu(a)}$, hence $H_a$ determines $\mu(a)$. Thus $a \mapsto H_a$ is one-to-one. Clearly, $H_a \wedge H_b = H_{a \wedge b}$. A transposition $(xy)$ belongs to $H_a \vee H_b$ iff

$$(xy) = (x_1 y_1)(x_2 y_2) \cdots (x_n y_n),$$

where $x_i \equiv y_i \pmod{\mu(a)}$ or $x_i \equiv y_i \pmod{\mu(b)}$, which is equivalent to

$$x \equiv y \pmod{\mu(a) \vee \mu(b)}.$$

Thus $H_a \vee H_b = H_{a \vee b}$. □

## 4.3   Type 2 representations

Let us call a representation $\mu\colon L \to \operatorname{Part} A$ *of type 2* if

$$\alpha(a \vee b) = \alpha(a) \circ \alpha(b) \circ \alpha(a)$$

for $a, b \in L$.

What kind of lattices have type 2 representations? This question was answered in B. Jónsson [432].

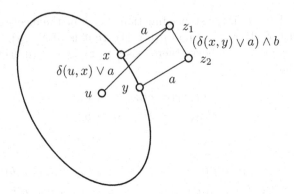

Figure 93. The "triangle" $\{u, z_1, z_2\}$

**Theorem 408.** *A lattice $L$ has a representation of type 2 iff $L$ is modular.*

*Proof.* Let $L$ have a representation $\mu\colon L \to$ Part $A$ of type 2. Let $a, b, c \in L$ and $a \geq c$. Since $(a \wedge b) \vee c \leq a \wedge (b \vee c)$ in any lattice, we wish to show that

$$a \wedge (b \vee c) \leq (a \wedge b) \vee c.$$

So let $x, y \in A$ and let $x \equiv y \pmod{\mu(a \wedge (b \vee c))}$, that is,

$$x \equiv y \pmod{\mu(a)},$$
$$x \equiv y \pmod{\mu(b \vee c)}.$$

Since $\mu$ is a type 2 representation, there exist elements $z_1$ and $z_2$ such that

$$x \equiv z_1 \pmod{\mu(c)},$$
$$z_1 \equiv z_2 \pmod{\mu(b)},$$
$$z_2 \equiv y \pmod{\mu(c)}.$$

Using $c \leq a$, we obtain that $z_1 \equiv x \pmod{\mu(a)}$, $x \equiv y \pmod{\mu(a)}$, and $y \equiv z_2$ $\pmod{\mu(a)}$; thus $z_1 \equiv z_2 \pmod{\mu(a)}$. Also, $z_1 \equiv z_2 \pmod{\mu(b)}$, hence

$$z_1 \equiv z_2 \pmod{\mu(a \wedge b)}.$$

Thus $x \equiv y \pmod{\mu((a \wedge b) \vee c)}$, implying that $\mu(a \wedge (b \vee c)) \leq \mu((a \wedge b) \vee c)$. Therefore, $a \wedge (b \vee c) \leq (a \wedge b) \vee c$, as claimed.

Now to prove the converse, let $L$ be a modular lattice and let the lattice $K$ be the lattice $L$ with a zero adjoined. Then $K$ is again modular.

For every distance function $\delta\colon A^2 \to K$ and $(x, y, a, b)$ satisfying $\delta(x, y) \leq a \vee b$, we construct a set $A^* = A \cup \{z_1, z_2\}$ having two more elements than $A$ and a distance function $\delta^*$ defined as follows (see Figure 93):

$\delta^*$ is symmetric, normalized, and $\delta^*(x, y) = \delta(x, y)$ for all $x, y \in A$; moreover,

$$\delta^*(z_1, z_2) = (\delta(x, y) \vee a) \wedge b;$$
$$\delta^*(u, z_1) = \delta(u, x) \vee a \qquad \text{for } u \in A;$$
$$\delta^*(u, z_2) = \delta(u, y) \vee at \qquad \text{for } u \in A.$$

The triangle inequality is again trivial, except for the "triangle" $\{u, z_1, z_2\}$, see Figure 93.

We have to verify the inequality

$$\delta^*(u, z_1) \vee \delta^*(z_1, z_2) \geq \delta^*(u, z_2).$$

Indeed,

$$\delta^*(u, z_1) \vee \delta^*(z_1, z_2) = \delta(u, x) \vee a \vee ((\delta(x, y) \vee a) \wedge b)$$

(by modularity)

$$= \delta(u, x) \vee ((\delta(x, y) \vee a) \wedge (a \vee b))$$

(by absorption, using $\delta(x, y) \leq a \vee b$)

$$= \delta(u, x) \vee \delta(x, y) \vee a$$

(by triangle inequality)

$$\geq \delta(u, y) \vee a = \delta^*(u, z_2).$$

Now we proceed the same way as the proof of Theorem 406.  □

## 4.4   Type 1 representations

Under what conditions can we improve the "2" of "type 2" to "1"? Call a representation $\mu \colon L \to \operatorname{Part} A$ *of type 1* if

$$\alpha(a \vee b) = \alpha(a) \circ \alpha(b)$$

for $a, b \in L$.

This is equivalent to the statement that $\mu(a)$ and $\mu(b)$ are *permutable*; for the definition, see Exercise II.1.39. Lattices having a type 1 representation satisfy a special identity that is not a consequence of the modular identity.

**Definition 409.** Let $x_0, x_1, x_2, y_0, y_1, y_2$ be variables. We define some terms:

$$z_{ij} = (x_i \vee x_j) \wedge (y_i \vee y_j) \quad \text{for } 0 \leq i < j < 3;$$
$$z = z_{01} \wedge (z_{02} \vee z_{12}).$$

The *arguesian identity* is

$$(x_0 \vee y_0) \wedge (x_1 \vee y_1) \wedge (x_2 \vee y_2) \leq ((z \vee x_1) \wedge x_0) \vee ((z \vee y_1) \wedge y_0).$$

A lattice satisfying this identity is called *arguesian*.

*Remark.* This identity is a lattice theoretic form of Desargues' Theorem; see B. Jónsson [432]; see also M. Schützenberger [633]. To understand the meaning of the identity, the reader should read the discussion of Desargues' Theorem and its connection with the arguesian identity in Section 5.5.

Do all modular lattices have type 1 representations? B. Jónsson [432] proved that this is not the case.

**Theorem 410.** *Any lattice having a type 1 representation is arguesian.*

*Proof.* Let $\mu\colon L \to \text{Part } A$ be a type 1 representation of the lattice $L$. We shall show that $\mu(L)$ is arguesian (this is sufficient to prove since $L \cong \mu(L)$). Let $a_0, a_1, a_2, b_0, b_1, b_2 \in \mu(L)$; thus $a_0, a_1, a_2, b_0, b_1, b_2$ are partitions of the set $A$. If $x, y \in A$ satisfy

$$x \equiv y \pmod{(a_0 \vee b_0) \wedge (a_1 \vee b_1) \wedge (a_2 \vee b_2)},$$

then there exist $u_0, u_1, u_2 \in A$ (see Figure 94) such that

$$x \equiv u_i \pmod{a_i},$$
$$u_i \equiv y \pmod{b_i}$$

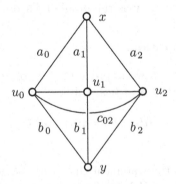

Figure 94. Proving the arguesian identity

for $i = 1, 2, 3$. Let
$$c_{ij} = (a_i \vee a_j) \wedge (b_i \vee b_j),$$
for $0 \leq i < j < 3$. Then $u_i \equiv u_j \pmod{c_{ij}}$ for all such $i$ and $j$. Set

$$c = c_{01} \wedge (c_{02} \vee c_{12}).$$

Then $u_0 \equiv u_1 \pmod{c}$. So

$$x \equiv u_0 \pmod{(c \vee a_1) \wedge a_0},$$
$$u_0 \equiv y \pmod{(c \vee b_1) \wedge b_0},$$

therefore
$$x \equiv y \pmod{((c \vee a_1) \wedge a_0) \vee ((c \vee b_1) \wedge b_0)},$$
proving that

$$(a_0 \vee b_0) \wedge (a_1 \vee b_1) \wedge (a_2 \vee b_2) \leq ((c \vee a_1) \wedge a_0) \vee ((c \vee b_1) \wedge b_0),$$

so $\mu(L)$ is arguesian. $\qquad\square$

Every arguesian lattice is modular. Indeed, to see this, it is sufficient to show that $\mathsf{N}_5$ is not arguesian. If we take $\mathsf{N}_5$, and substitute

$$x_0 = y_1 = c,$$
$$x_1 = x_2 = y_0 = b,$$
$$y_2 = a,$$

then $z = a$, however, the identity demands that $a \leq b$, which is false.

It follows from the results of Section 5.9 that there are modular lattices that are not arguesian.

Lattices having type 1 representations are characterized with an infinite set $\Sigma$ of identities in B. Jónsson [437], based on some ideas of R. C. Lyndon [506] and [508]. M. Haiman [368] proved no finite subset of $\Sigma$ would do, in fact, there is no first-order characterization. Haiman utilized Herrmann's S-glued sum construction, see Section IV.2.3.

By B. Jónsson [436], every arguesian lattice of length $\leq 4$ has a type 1 representation.

If a lattice $L$ can be embedded in the lattice of all normal subgroups of a group, then $L$ has a type 1 representation. The converse is false by B. Jónsson [433]; the counterexample can be chosen to have length 5.

### 4.5 ◇ Type 2 and 3 congruence lattices in algebras

Let $\mathfrak{A} = (A; F)$ be a universal algebra (see Section I.1.9) and let $\operatorname{Con} \mathfrak{A}$ be the congruence lattice of the algebra $\mathfrak{A}$. Of course, $\operatorname{Con} \mathfrak{A}$ is a sublattice of $\operatorname{Part} A$, so we can define $\operatorname{Con} \mathfrak{A}$ as type 3 in the obvious way:

Con $\mathfrak{A}$ is of type 3 if

$$\alpha \vee \beta = \alpha \circ \beta \circ \alpha \circ \beta$$

for $\alpha, \beta \in \mathrm{Con}\,\mathfrak{A}$.

In Section I.3.17, we discuss the result of G. Grätzer and E. T. Schmidt [338] (Theorem 50): *Every algebraic lattice L is isomorphic to the congruence lattice of a universal algebra* $\mathfrak{A}$. It is logical to ask whether this result can be combined with the existence of type 3 representations (Theorem 406).

Indeed, it can.

◇ **Theorem 411.** *Every algebraic lattice L is isomorphic to a type 3 congruence lattice of a universal algebra* $\mathfrak{A}$.

It should be clear now how to define type 2 congruence lattices of an algebra, and then we can combine Theorem 50 with Theorem 408:

◇ **Theorem 412.** *Every modular algebraic lattice L is isomorphic to a type 2 congruence lattice of a universal algebra* $\mathfrak{A}$.

These two theorems are from G. Grätzer and E. T. Schmidt [338]; see also G. Grätzer and W. A. Lampe, Appendix 7 of G. Grätzer [263].

### 4.6  ◇ Sublattices of finite partition lattices

Ever since P. M. Whitman [724] proved that every lattice $L$ can be embedded in some partition lattice, the problem has been unresolved as to whether every finite lattice can be embedded into a finite partition lattice.

A number of papers were published providing partial answers. Finally, P. Pudlák and J. Tůma [599] solved the problem:

◇ **Theorem 413.** *Let L be a finite lattice. Then there is a finite set X such that L can be embedded in the partition lattice* Part $X$.

Going back at least to G. Grätzer and E. T. Schmidt [338], a harder problem was raised: For every finite lattice $L$, is there a finite algebra $\mathfrak{A}$ such that $L$ is isomorphic to Con $\mathfrak{A}$?

P. Pálfy and P. Pudlák [576] show that every finite lattice is isomorphic to the congruence lattice of a finite algebra iff every finite lattice is isomorphic to an interval in the subgroup lattice of a finite group. This result points out that the universal algebraic problem is basically a group theoretical problem. A number of papers have been published on this group problem, here are two: J. W. Snow [656], P. Hegedűs and P. Pálfy [383].

An embedding $\varphi$ of a finite lattice $L$ into a finite partition lattice Part $X$ is called *cover-preserving* if $a \prec b$ in $L$ implies that $\varphi(a) \prec \varphi(b)$ in Part $X$. M. Wild [730] gives necessary and sufficient conditions for a finite modular lattice to have a cover-preserving embedding into a finite partition lattice; in particular, such a lattice must be 2-distributive, as defined in Exercise VI.3.18.

## 4.7  ◇ Generating partition lattices

A lattice $L$ is $(1 + 1 + 2)$-*generated* if it has a generating set $\{a_1, a_2, b, c\}$ such that $a_1 < a_2$ and $\{a_i, b, c\}$ is an antichain for $i = 1, 2$. More generally, if $L$ has a four-element generating set, then it is called *four-generated*.

H. Strietz [670] and L. Zádori [741] proved that all finite partition lattices are four-generated.

◇ **Theorem 414.** *If* $3 \leq |A| < \infty$, *then* Part $A$ *is four-generated. If* $7 \leq |A| < \infty$, *then* Part $A$ *is* $(1 + 1 + 2)$-*generated.*

For an infinite set $A$, the lattice Part $A$ is not finitely generated; however, the following result of G. Czédli [111]] holds:

◇ **Theorem 415.** *If* $|A| = \aleph_0$, *then* Part $A$ *has a* $(1+1+2)$-*generated sublattice* $S$ *such that* $\text{Part}_{\text{fin}} A \subseteq S$.

Let $X$ be a subset of a complete lattice $L$. The set $X$ *generates $L$ as a complete lattice* if no proper complete sublattice of $L$ includes $X$. An infinite cardinal $\mathfrak{m}$ is *inaccessible* if it satisfies the following three conditions:

(i) $\aleph_0 < \mathfrak{m}$;
(ii) $\mathfrak{n} < \mathfrak{m}$ implies that $2^{\mathfrak{n}} < \mathfrak{m}$;
(iii) if $I$ is a set of cardinals each $< \mathfrak{m}$ and $|I| < m$, then $\sup I < \mathfrak{m}$.

By a result of K. Kuratowski, see for instance A. Levy [502], ZFC has a model without inaccessible cardinals. For such models, G. Czédli [112] proves an analogue of Theorem 414 for complete lattices.

◇ **Theorem 416.** *If* $|A| \geq 7$ *and there is no inaccessible cardinal $\mathfrak{m}$ such that* $\mathfrak{m} \leq |A|$, *then* Part $A$, *as a complete lattice, is* $(1 + 1 + 2)$-*generated.*

## Exercises

4.1. Compute $|\text{Part } A|$ for $|A| \leq 5$.

4.2. Show that Part $A$ is modular iff $|A| \leq 3$.

4.3. Let $A \subset B$. Find embeddings of Part $A$ into Part $B$.

4.4. Let $A_i$, for $i \in I$, be pairwise disjoint sets and let $\alpha_i$ be a partition of $A_i$ for all $i \in I$. Define a partition $\alpha$ on $A = \bigcup(A_i \mid i \in I)$ by

$$u \equiv v \pmod{\alpha} \quad \text{if} \quad u, v \in A_i \text{ for some } i \in I \text{ and } u \equiv v \pmod{\alpha_i}.$$

Show that this defines an embedding of $P = \prod(\text{Part } A_i \mid i \in I)$ into Part $A$.

4.5. With $A_i$, for $i \in I$, and $A$, $P$ as in Exercise 4.4, is there a $\{0, 1\}$-embedding of $P$ into Part $A$?

4.6. Is there an embedding of $\prod(\text{Part } A_i \mid i \in I)$ into $\text{Part}\prod(A_i \mid i \in I)$?

4.7. Let $A$ be a set and let $B = A \cup \{z\}$ with $z \notin A$. For $X \subseteq A$, define a partition $\alpha(X)$ on $B$ by

$$x \equiv y \pmod{\alpha(X)} \quad \text{if} \quad x, y \in X \cup \{z\}.$$

Show that $X \mapsto \alpha(X)$ is an embedding of Pow $A$ into $\text{Part } B$.

4.8. Using Exercise 4.7, show that every distributive lattice has an embedding into a partition lattice.

4.9. Show that every finite distributive lattice can be embedded in a finite partition lattice.

4.10. Let $\mathfrak{A}$ be a finite (universal) algebra. Verify that the congruence lattice of $\mathfrak{A}$ can be embedded in a finite partition lattice.

4.11. Following O. Ore [562], for $\alpha, \beta \in \text{Part } A$, we define a graph $G(\alpha, \beta)$ on the set $\alpha \cup \beta$; an edge $\{X, Y\}$ of $G(\alpha, \beta)$ is a block $X$ of $\alpha$ and a block $Y$ of $\beta$ such that $X \cap Y \neq \varnothing$. Show that the blocks of $\alpha \vee \beta$ are maximal connected subgraphs of $G(\alpha, \beta)$.

4.12. Let $\alpha, \beta \in \text{Part } A$. Prove that the modular equation,

$$\alpha \wedge (\beta \vee \gamma) = (\alpha \wedge \beta) \vee \gamma,$$

holds for all $\gamma \leq \alpha \in \text{Part } A$ iff $G(\alpha, \beta)$ is a *tree*, that is, a graph without cycles (O. Ore [562]).

4.13. Let $\beta, \gamma \in \text{Part } A$. Show that

$$\alpha \wedge (\beta \vee \gamma) = (\alpha \wedge \beta) \vee (\alpha \wedge \gamma)$$

holds for all $\alpha \in \text{Part } A$ iff $\beta$ and $\gamma$ are permutable (O. Ore [562]).

4.14. Let $\alpha, \gamma \in \text{Part } A$. Prove that

$$\alpha \vee (\beta \wedge \gamma) = (\alpha \vee \beta) \wedge (\alpha \vee \gamma)$$

holds for all $\beta \in \text{Part } A$ iff $\alpha$ is the zero of Part $A$, or $\alpha \geq \gamma$, or $\gamma$ has only one nontrivial block and $\gamma \geq \alpha$ (O. Ore [562]).

4.15. Show that the zero and the unit are the only distributive elements of Part $A$.

4.16. Let $D$ be a proper distributive ideal of Part $A$. Show that every element of $D$ is majorized by another element of $D$ with only one nontrivial block. Also show that $\bigvee D$ is a distributive element of Part $A$.

4.17. Show that Part $A$ has no proper distributive ideal.

4.18. Use Theorem 272 and Exercise 4.17 to show that Part $A$ is simple (O. Ore [562]).

4.19. Define type $n$ representations. Find a representation that is not of type $n$ for any $n < \omega$.

4.20. Find a representation that is not associated with a distance function.

*4.21. Disprove the converse of Theorem 410 (B. Jónsson).

# 5. Complemented Modular Lattices

## 5.1 Congruences

In complemented (more generally, in sectionally complemented) modular lattices, congruences are very nice.

**Theorem 417.** *Let $L$ be a sectionally complemented modular lattice. Then all congruences of $L$ are neutral.*

*Proof.* Every congruence $\alpha$ of $L$ is standard (because $L$ is sectionally complemented by Lemma 99), which means that $0/\alpha$ is a standard element of $\operatorname{Id} L$. Now $L$ is modular, thus by Lemma 59, so is $\operatorname{Id} L$, and thus $\operatorname{Id} L$ is weakly modular. By Theorem 256, the sets of distributive, standard, and neutral elements in $\operatorname{Id} L$ are the same. In particular, $0/\alpha$ is a neutral element of $\operatorname{Id} L$, thus a neutral ideal of $L$. (Alternatively, utilize Lemma 271 and Theorem 272.) $\square$

**Corollary 418.** *The following statements are equivalent, for any ideal $I$ in a sectionally complemented modular lattice $L$:*

- (i) *$I$ is distributive;*
- (ii) *$I$ is standard;*
- (iii) *$I$ is neutral;*
- (iv) *$I$ is subperspectivity closed;*
- (v) *$I$ is perspectivity closed.*

By Lemma 99, every complemented modular lattice is sectionally complemented. In particular, we obtain

**Corollary 419.** *Let $L$ be a complemented modular lattice. Then every congruence of $L$ is neutral.*

## 5.2 Modular geometric lattices

The key to the investigation of complemented modular lattices is provided by modular geometric lattices and projective spaces. We begin with some properties of modular geometric lattices.

**Lemma 420.** *Let $L$ be a modular geometric lattice and let $(A,^-)$ be the associated geometry.*

- (i) *Let $p, q \in A$ and $p \neq q$. Then $p \sim q$ iff there exists an $r \in A$ such that $r \neq p, q$ and $r \leq p \vee q$ (thus $\{0, p, q, r, p \vee q\}$ is a diamond in $L$).*

- (ii) *Let $X \subseteq A$. Then $X$ is a subspace iff $p, q \in X$ implies that $r \in X$ for all $r \leq p \vee q$.*

*Proof.*

(i) Let $x$ be a complement of $p$ and $q$ and define $r = (p \vee q) \wedge x$. Then

$$p \vee r = p \vee ((p \vee q) \wedge x)$$

(by modularity and $p \leq p \vee q$)

$$= (p \vee x) \wedge (p \vee q) = 1 \wedge (p \vee q) = p \vee q.$$

Similarly, $q \vee r = p \vee q$. Obviously, $p \wedge r = q \wedge r = 0$. Hence

$$\text{height}(r) = \text{height}(p \vee q) + \text{height}(0) - \text{height}(p) = 1.$$

Thus $r \in A$. Since $r \neq 0$ and $r \wedge p = r \wedge q = 0$, clearly, $r \neq p, q$.

(ii) Let $X \subseteq A$ satisfy the condition that $r \leq p \vee q$ and $p, q \in X$ imply that $r \in X$. We prove by induction on $n$ that

$$r \in A, \ r \leq p_1 \vee \cdots \vee p_n, \ \text{and} \ p_1, \ldots, p_n \in X \quad \text{imply that} \quad r \in X.$$

This is obvious for $n = 1$ and, by assumption for $n = 2$. Let us assume this statement for $n - 1$ and let $r, p_1, \ldots, p_n$ be given satisfying $r \leq p_1 \vee \cdots \vee p_n$. We can assume that $\{p_1, \ldots, p_n\}$ is independent, since $r \leq p_{i_1} \vee \cdots \vee p_{i_k}$ otherwise, for some $\{i_1, \ldots, i_k\} \subset \{1, \ldots, n\}$, and so $r \in X$, by induction. Let $a = p_2 \vee \cdots \vee p_n$; then $r \vee a \leq p_1 \vee a$ and $\text{height}(r \vee a) = \text{height}(p_1 \vee a) = n$, hence $r \vee a = p_1 \vee a$. If $r \leq a$, then $r \in X$ by the induction hypothesis, so we can assume that $r \not\leq a$. Thus $r \wedge a = p_1 \wedge a = 0$. By the proof of (i), $t = (p_1 \vee r) \wedge a$ is an atom and, by the induction hypothesis, $t \in X$. Since $t, p_1 \in X$ and $r \leq t \vee p_1$, we conclude that $r \in X$. This completes the proof of the "if" part of (ii); the "only if" part is trivial.  □

**Lemma 421.** *Under the conditions of Lemma 420, the geometry $(A, \bar{\ })$ satisfies the following property (see Figure 95):*

*The Pasch Axiom. If $p, q, r, x, y$ are points, $x \leq p \vee q$, $y \leq q \vee r$, and $x \neq y$, then there is a point $z$ such that $z \leq (p \vee r) \wedge (x \vee y)$.*

*Remark.* Apart from degenerate cases, the Pasch Axiom requires that if a line $x \vee y$ intersects two sides, $p \vee q$ and $q \vee r$, of the triangle $\{p, q, r\}$, then it intersects the third side, $p \vee r$. If $x \neq p, q$ and $y \neq q, r$, then we shall also have $z \neq p, r$.

*Proof.* If $|\{p, q, r, x, y\}| < 5$, then we can always choose an element

$$z \in \{p, q, r, x, y\}$$

satisfying

$$z \leq (p \vee r) \wedge (x \vee y).$$

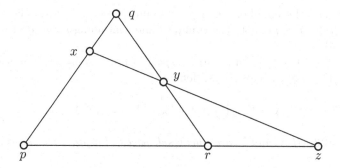

Figure 95. The Pasch Axiom

If $r \leq p \vee q$, again one can choose $z = x$. So let $|\{p, q, r, x, y\}| = 5$, and let $r \not\leq p \vee q$. Then

$$\text{height}(p \vee r) = \text{height}(x \vee y) = 2$$

and $\text{height}(p \vee q \vee r) = 3$.

Thus

$$\text{height}((p \vee r) \wedge (x \vee y)) = \text{height}(p \vee r) + \text{height}(x \vee y) - \text{height}(p \vee r \vee x \vee y).$$

Now observe that $p \vee q \vee x \vee y = p \vee q \vee r$, hence

$$\text{height}((p \vee r) \wedge (x \vee y)) = 2 + 2 - 3 = 1,$$

and so $z = (p \vee r) \wedge (x \vee y)$ is the required atom.     $\square$

## 5.3   Projective spaces

We can see, from Lemmas 420 and 421, that many important properties of a geometry associated with a modular geometric lattice are formulated in terms of points and lines. This pattern is followed in the next definition:

**Definition 422.** Let $A$ be a set and let $L$ be a collection of subsets of $A$. Then $(A, L)$ is called a *projective space* if the following properties hold:

(i) Every $l \in L$ has at least two elements.

(ii) For every pair of distinct $p, q \in A$, there is exactly one $l \in L$ satisfying $p, q \in l$.

(iii) For $p, q, r, x, y \in A$ and $l_1, l_2 \in L$ satisfying $p, q, x \in l_1$ and $q, r, y \in l_2$, there exist $z \in A$ and $l_3, l_4 \in L$ satisfying $p, r, z \in l_3$ and $x, y, z \in l_4$.

Let us call the members of $A$ *points*, and those of $L$, *lines*. For $p, q \in A$ with $p \neq q$, let $p + q$ denote the (unique) line containing $p$ and $q$; if $p = q$, set $p + q = \{p\}$.

A set $X \subseteq A$ is called a *linear subspace* if $p, q \in X$ imply that $p + q \subseteq X$. If $X$ and $Y$ are linear subspaces, define

$$X + Y = \bigcup( x + y \mid x \in X \text{ and } y \in Y ).$$

**Lemma 423.** *If $X$ and $Y$ are linear subspaces of a projective space, then so is $X + Y$.*

*Proof.* Take the points $p, q, r$ such that $r \in p + q$ and $p, q \in X + Y$. We wish to show that $r \in X + Y$. Now if $p, q \in X$ or $p, q \in Y$, then $r \in X \cup Y \subseteq X + Y$, since $X$ and $Y$ are linear subspaces. If $p \in X$ and $q \in Y$ (or the other way around), then $r \in X + Y$, by the definition of $X + Y$. Therefore, we can assume that $p \notin X \cup Y$, and so there exist $p_X \in X$ and $p_Y \in Y$ such that $p \in p_X + p_Y$.

Now we distinguish two cases. Firstly, if $q \in X \cup Y$, say $q \in X$, then consider the lines $q + p_X, q + p, p + p_X, r + p_Y$ (see the first diagram in Figure 96). We can assume that the points $q, p, p_X, p_Y, r$ are all distinct, and so $r + p_Y$ has a point in common with $q + p$ and $p + p_Y$; hence, by Definition 422(iii), there is a $t \in X$ such that $t \in q + p_X$ and $t \in r + p_Y$. If $t = p_Y$, then $p \in X$, a contradiction. Hence $t \neq p_Y$ and so $r \in t + p_Y \in X + Y$.

Secondly, let $q \notin X \cup Y$. Again we apply Definition 422(iii), this time to the lines $p + p_X, p + q, p_X + q, p_Y + r$ (see the lower half of Figure 96). Hence there is a point $t \in p_X + q$ such that $r \in t + p_Y$. By the first case, $t \in X + Y$. Again by the first case, $r \in X + Y$. $\qquad\square$

Now we come to the crucial step:

**Theorem 424.** *The linear subspaces of a projective space form a modular geometric lattice.*

*Proof.* Since the intersection of any number of linear subspaces is a linear subspace again, we have a closure system $(A, ^-)$. For $X \subseteq A$, the closure $\overline{X}$ can be described as follows: set

$$X_0 = X,$$

$$\ldots$$

$$X_n = X_{n-1} + X_{n-1},$$

$$\ldots\ldots$$

Then

$$\overline{X} = \bigcup( X_i \mid i = 0, 1, 2, \ldots ).$$

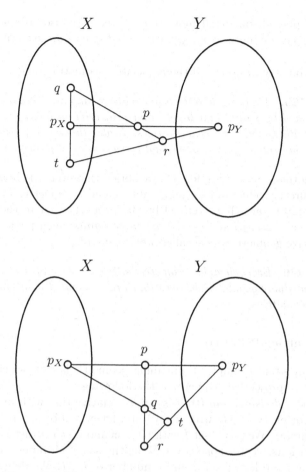

Figure 96. The two cases in the proof of Lemma 423

It follows immediately that $(A,^-)$ is an algebraic closure system, and so the linear subspaces form an algebraic lattice and, for the linear subspaces $X$ and $Y$, the formula $X \vee Y = \overline{X \cup Y}$ holds.

If $X, Y, Z$ are linear subspaces and $X \supseteq Z$, then $X \wedge (Y \vee Z) \supseteq (X \wedge Y) \vee Z$ obviously holds.

Now let $p \in X \wedge (Y \vee Z)$, that is, $p \in X$ and $p \in Y \vee Z$. Since $p \in Y \vee Z = Y + Z$, there exist $p_Y \in Y$ and $p_Z \in Z$ such that $p \in p_Y + p_Z$. From $X \supseteq Z$, it follows that $p$ and $p_Z \in X$. If $p = p_Z$, then $p \in Z$ and so $p \in (X \wedge Y) \vee Z$. If $p \neq p_Z$, then $p_Y \in p + p_Z \subseteq X$. Thus $p_Y \in X \wedge Y$ and $p_Z \in Z$; therefore, $p \in (X \wedge Y) \vee Z$. This proves that the lattice is modular.

Definition 397(i), (ii), (iv) have been verified or are obvious and (iii) follows from modularity.                                                                □

We have proved somewhat more than stated. We also obtained that the linear subspaces define a geometry whose subspaces are exactly the linear subspaces.

Call a geometry *projective* if the associated geometric lattice is modular.

**Theorem 425.** *There is a one-to-one correspondence between projective spaces—defined by points and lines—and projective geometries—defined as geometries with modular subspace lattices. Under this correspondence, linear subspaces of projective spaces correspond to subspaces of projective geometries.*

Applying the results of Section 3.2, we obtain that every modular geometric lattice is a direct product of indecomposable modular geometric lattices, which by Lemma 420(i), are characterized by the property that in the associated projective space *every line has at least three points*; such projective spaces (and projective geometries) are called *nondegenerate*.

**Corollary 426.** *Every modular geometric lattice can be represented as a direct product of modular geometric lattices that are associated with nondegenerate projective geometries.*

## 5.4   The lattice $\mathrm{PG}(D, \mathfrak{m})$

The most important example of a modular geometric lattice associated with a nondegenerate projective geometry is the following.

Let $D$ be a *division ring* (that is, an associative ring with unit, in which $a^{-1}$ exists for every $a \in D$ with $a \neq 0$) and let $\mathfrak{m} > 0$ be a cardinal number (finite or infinite). We take a set $I$ with $|I| = \mathfrak{m}$ and we construct $V = V(D, \mathfrak{m})$, an $\mathfrak{m}$-dimensional vector space over $D$, as the set of all functions $f \colon I \to D$ such that $f(i) = 0$ for all but a finite number of $i \in I$. We define $h = f + g$ and $k = af$ $(a \in D)$ by $h(i) = f(i) + g(i)$ and $k(i) = af(i)$, respectively.

A *submodule* $U$ of $V$ is a nonempty subset of $V$ such that $f, g \in U$ imply that $f + ag \in U$ for all $a \in D$. The submodules of $V$ form a lattice denoted by $\mathrm{PG}(D, \mathfrak{m})$. It is easily seen that $\mathrm{PG}(D, \mathfrak{m})$ is a modular geometric lattice. The atoms of $\mathrm{PG}(D, \mathfrak{m})$ are exactly the submodules $\{\, af \mid a \in D \,\}$ where $f$ is any element of $V$ that is not the identically zero function, $f_0$.

The submodules $\{\, af \mid a \in D \,\}$ $(f \neq f_0)$ form the points of the associated projective geometry. If $\{\, af \mid a \in D \,\}$ and $\{\, ag \mid a \in D \,\}$ are distinct points, then the set of points of the line through these two points is

$$\{\, a(xf + yg) \mid a \in D \,\}, \quad x, y \in D, \ xy \neq 0.$$

The fubction $f$ $(\neq f_0)$ will be called a *representative* of the point $\{\, af \mid a \in D \,\}$.

This geometry is nondegenerate: if $f$ and $g$ are representatives of two distinct points, then $f - g$ represents a third point of the line.

## 5.5 Desargues' Theorem

The projective geometry $PG(D, \mathfrak{m})$ is typical of projective geometries in which Desargues' Theorem holds.

In a projective geometry, let us call a set of points $X$ *collinear* if $X \subseteq l$ for some line $l$. A triple $(a_0, a_1, a_2)$ of noncollinear points is a *triangle*. The triangles $(a_0, a_1, a_2)$ and $(b_0, b_1, b_2)$ are *perspective with respect to the point $p$* if $a_i \neq b_i$, $a_i + a_j \neq b_i + b_j$, for all $0 \leq i, j < 3$, and the points $p, a_i, b_i$ are collinear for $i = 0, 1, 2$. They are *perspective with respect to a line $l$* if $c_{01}, c_{12}, c_{20} \subseteq l$, where $c_{ij}$ is the intersection of $a_i + a_j$ and $b_i + b_j$. The classical *Desargues' Theorem* states that if two triangles are perspective with respect to a point, then they are perspective with respect to a line.

Now we shall prove that the arguesian identity—of the last section—is a lattice theoretic formulation of Desargues' Theorem. See B. Jónsson [432], [433], [438] and M. Schützenberger [633].

**Theorem 427.** *Let $L$ be a modular geometric lattice. Then $L$ satisfies the arguesian identity iff Desargues' Theorem holds in the associated projective geometry.*

The proof of Theorem 427 will follow easily from Lemmas 428–430.

**Lemma 428.** *Let $L$ be a modular geometric lattice. Then the arguesian identity holds for the atoms of $L$ iff Desargues' Theorem holds in the associated projective geometry.*

*Proof.* Let us substitute the atoms $a_0, a_1, b_0, b_1, b_2$ in the arguesian identity. Let us form the $c_{ij}$, for all $0 \leq i < j \leq 2$, and let

$$c = c_{01} \wedge (c_{02} \vee c_{12}),$$
$$p = (a_0 \vee b_0) \wedge (a_1 \vee b_1) \wedge (a_2 \vee b_2);$$

see Figure 97. We have to verify that

$$p \leq ((c \vee a_1) \wedge a_0) \vee ((c \vee b_1) \wedge b_0).$$

Let us assume first that $a_0, a_1, a_2$ and $b_0, b_1, b_2$ are not collinear, that $a_i \neq b_i$, for $i = 0, 1, 2$, and that the triangles $(a_0, a_1, a_2)$ and $(b_0, b_1, b_2)$ are perspective with respect to the point $p$. Then the $c_{ij}$ are all distinct atoms and they are collinear iff $c = c_{01} \wedge (c_{12} \vee c_{02})$ is an atom, in fact, $c = c_{01}$; otherwise, $c = 0$. If $c = c_{01}$, then $((c \vee a_1) \wedge a_0) \vee ((c \vee b_1) \wedge b_0)$ is the line $a_0 \vee b_0$, and so

$$p \leq ((c \vee a_1) \wedge a_0) \vee ((c \vee b_1) \wedge b_0);$$

otherwise, $c = 0$, $(c \vee a_1) \wedge a_0 = (c \vee b_1) \wedge b_0 = 0$, and $p \not\leq 0$. Thus under the conditions stated, the identity holds iff the geometry satisfies Desargues' Theorem.

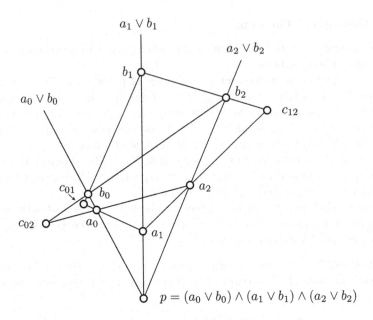

Figure 97. Arguesian identity

Now observe that if the above conditions are not satisfied, then

$$p \le ((c \vee a_1) \wedge a_0) \vee ((c \vee b_1) \wedge a_1)$$

always holds. There are several cases to consider ($p = 0$, $p$ is a line, $a_0, a_1, a_2$ are collinear, $a_1, b_1, b_2$ are collinear, $a_i = b_i$, for some $i$, and so on), but all are trivial.                                                                    □

**Lemma 429.** *Let $L$ be a modular geometric lattice. Let $p$ be an $n$-ary term in which each variable $x_i$, for all $0 \le i < n$, occurs at most once. Then, for an atom $a$ and elements $b_0, \ldots, b_{n-1} \in L$,*

$$a \le p(b_0, \ldots, b_{n-1})$$

*holds iff there exist $a_0, \ldots, a_{n-1} \in L$ such that*

   (i) $a \le p(a_0, \ldots, a_{n-1})$;
   (ii) $a_i \le b_i$ for all $0 \le i < n$;
  (iii) $a_i$ is an atom, if $b_i \ne 0$.

*Proof.* We use induction on the rank of $p$. Let $p = x_i$. Then $a \le b_i$, and so we can take $a_i = a$. Now let us assume that the statement holds for

$$p_0 = p_0(x_0, \ldots, x_{n-1}),$$
$$p_1 = p_1(y_0, \ldots, y_{m-1}),$$

where $\{x_0, \ldots, x_{n-1}\} \cap \{y_0, \ldots, y_{m-1}\} = \varnothing$.

First, let $p = p_0 \wedge p_1$. Now if

$$a \leq p(b_0, \ldots, b_{n-1}, b'_0, \ldots, b'_{m-1}) = p_0(b_0, \ldots, b_{n-1}) \wedge p_1(b'_0, \ldots, b'_{m-1}),$$

then

$$a \leq p_0(b_0, \ldots, b_{n-1}),$$
$$a \leq p_1(b'_0, \ldots, b'_{m-1}).$$

By the induction hypothesis, we can choose $a_i \leq b_i$ and $a'_j \leq b'_j$ such that $a_i$ is an atom if $b_i \neq 0$, and $a'_j$ is an atom if $b'_j \neq 0$, and

$$a \leq p_0(a_0, \ldots, a_{n-1}) \wedge p_1(a'_0, \ldots, a'_{m-1}) = p(a_0, \ldots, a_{n-1}, a'_0, \ldots, a'_{m-1}).$$

(Observe that if $x_i = y_j$, then we would not know how to choose $a_i = a'_j$.)

Second, let $p = p_0 \vee p_1$. Then

$$a \leq p_0(b_0, \ldots, b_{n-1}) \vee p_1(b'_0, \ldots, b'_{m-1}).$$

If $p_0(b_0, \ldots, b_{n-1}) = 0$, then $a \leq p_1(b'_0, \ldots, b'_{m-1})$, hence we can choose the element $a'_i$ by the induction hypothesis. We choose $a_i$ as an arbitrary atom $\leq b_i$, if $b_i \neq 0$, and let $a_i = 0$, if $b_i = 0$. If $p_1(b'_0, \ldots, b'_{m-1}) = 0$, we proceed similarly. Now if

$$p_0(b_0, \ldots, b_{n-1}) \neq 0,$$
$$p_1(b'_0, \ldots, b'_{m-1}) \neq 0,$$

then Lemma 423 and Theorem 425 imply that there exist atoms

$$a' \leq p_0(b_0, \ldots, b_{n-1}),$$
$$a'' \leq p_1(b'_0, \ldots, b'_{m-1})$$

such that $a \leq a' \vee a''$. Applying the induction hypothesis twice, we obtain the elements $a_0, \ldots, a_{n-1}, a'_0, \ldots, a'_{m-1}$ such that the inequalities $a_i \leq b_i$ hold for all $0 \leq i < n$; the inequalities $a'_i \leq b'_i$ hold for all $0 \leq i < m$; the element $a_i$ is an atom if $b_i \neq 0$; the element $a'_i$ is an atom if $b'_i \neq 0$; and

$$a' \leq p_0(a_0, \ldots, a_{n-1}),$$
$$a'' \leq p_1(a'_0, \ldots, a_{m-1}).$$

Thus

$$a \leq a' \vee a'' \leq p_0(a_0, \ldots, a_{n-1}) \vee p_1(a'_0, \ldots, a'_{m-1}),$$

completing the induction. $\qquad\square$

G. Grätzer and H. Lakser [302] prove that some identities it is sufficient to test with atoms and zero.

**Lemma 430.** *Let us assume that the lattice identity $\varepsilon$ can be written in the form $p \leq q$, where $p$ and $q$ are lattice terms and each variable occurs in $p$ at most once. If $\varepsilon$ holds for the atoms and the zero of a modular geometric lattice $L$, then $\varepsilon$ holds for $L$.*

*Proof.* Let $b_0, \ldots, b_{n-1} \in L$; we want to verify that

$$p(b_0, \ldots, b_{n-1}) \leq q(b_0, \ldots, b_{n-1}).$$

It is sufficient to show that $a \leq q(b_0, \ldots, b_{n-1})$ for every atom $a$ satisfying $a \leq p(b_0, \ldots, b_{n-1})$.

So let $a$ be an atom and $a \leq p(b_0, \ldots, b_{n-1})$. By Lemma 429, we can choose $a_i \leq b_i$, for $0 \leq i < n$, such that the elements $a_i$ are atoms or zero and

$$a \leq p(a_0, \ldots, a_{n-1}).$$

By assumption, $\varepsilon$ holds for atoms and zero, hence

$$p(a_0, \ldots, a_{n-1}) \leq q(a_0, \ldots, a_{n-1}).$$

Using these two inequalities and the fact that $q$ is isotone, we obtain that

$$a \leq p(a_0, \ldots, a_{n-1}) \leq q(a_0, \ldots, a_{n-1}) \leq q(b_0, \ldots, b_{n-1}). \qquad \square$$

*Proof of Theorem 427.* If Desargues' Theorem fails in the associated projective geometry, then by Lemma 428, the arguesian identity $\varepsilon$ fails in $L$. Conversely, if Desargues' Theorem holds in the associated projective geometry, then by Lemma 428, $\varepsilon$ holds for the atoms of $L$. A trivial step extends this to the statement that $\varepsilon$ holds for the atoms and the zero of $L$. Observe that $\varepsilon$ is of the form required in Lemma 430, hence $\varepsilon$ holds in $L$. $\qquad \square$

**Corollary 431.** *Desargues' Theorem holds in the projective space associated with the lattice $PG(D, \mathfrak{m})$.*

*Proof.* With a submodule $U$, we associate an equivalence relation $\alpha(U)$ on the vector space $V = V(D, \mathfrak{m})$, defined by $f \, \alpha(U) \, g$ if $f - g \in U$. Then obviously $U \mapsto \alpha(U)$ is a type 1 representation of the lattice $PG(D, \mathfrak{m})$, hence by Theorem 410, the lattice $PG(D, \mathfrak{m})$ is arguesian. Thus by Theorem 427, Desargues' Theorem holds in the projective space associated with $PG(D, \mathfrak{m})$. $\qquad \square$

## 5.6    Arguesian lattices

The next result is a well-known theorem of geometry; it is the classical proof of Desargues' Theorem in a three-dimensional space.

**Theorem 432.** *Let $L$ be a modular geometric lattice. Let us assume that the projective geometry associated with $L$ is nondegenerate. If the length of $L$ is at least 4, then $L$ is arguesian.*

*Proof.* Let the triangles $(a_0, a_1, a_2)$ and $(b_0, b_1, b_2)$ be perspective with respect to the point $p$. A *plane* $\alpha$ is an element of height 3. Clearly, $\alpha = a_0 \vee a_1 \vee a_2$ and $\beta = b_0 \vee b_1 \vee b_2$ are planes.

Let us assume that $\alpha \neq \beta$. Since

$$\alpha \vee \beta = p \vee \alpha = p \vee \beta,$$
$$\text{height}(\alpha \vee \beta) = 4,$$

it follows that

$$\text{height}(\alpha \wedge \beta) = \text{height}(\alpha) + \text{height}(\beta) - \text{height}(\alpha \vee \beta) = 3 + 3 - 4 = 2;$$

thus $\alpha \wedge \beta$ is a line. The lines $a_i \vee a_j$ and $b_i \vee b_j$ are in the plane $p \vee a_i \vee a_j$, hence

$$\text{height}(c_{ij}) = \text{height}(a_i \vee a_j) + \text{height}(b_i \vee b_j) - \text{height}(p \vee a_i \vee a_j) = 1,$$

and so $c_{ij}$ is a point. Since $a_i \vee a_j$ is in the plane $\alpha$ and $b_i \vee b_j$ is in $\beta$, the point $c_{ij}$ is in $\alpha \wedge \beta$. Therefore, $c_{01}, c_{02}, c_{12}$ are collinear.

Now let us assume that $\alpha = \beta$. Since the length of $L$ is at least 4, $\alpha$ is not the unit element. Take a $\pi \in L$ with $\pi \succ \alpha$. Let $l$ be a relative complement of $\alpha$ in $[p, \pi]$. Then

$$\text{height}(l) = \text{height}(\pi) + \text{height}(p) - \text{height}(\alpha) = 4 + 1 - 3 = 2,$$

so $l$ is a line. The projective geometry associated with $L$ is nondegenerate, hence we can choose two distinct points $p'$ and $p''$ on $l$ with $p' \neq p$ and $p'' \neq p$.

Now define

$$d_i = (p' \vee a_i) \wedge (p'' \vee b_i)$$

for $i = 0, 1, 2$. It is easily seen that $(d_0, d_1, d_2)$ is a triangle and the plane $\delta = d_0 \vee d_1 \vee d_2 \neq \alpha, \beta$. Furthermore, $(a_0, a_1, a_2)$ and $(d_0, d_1, d_2)$ are perspective with respect to $p'$, and $(b_0, b_1, b_2)$ and $(d_0, d_1, d_2)$ are perspective with respect to $p''$. Thus we can apply the first case to conclude that $(a_0 \vee a_1) \wedge (d_0 \vee d_1)$ and also $(b_0 \vee b_1) \wedge (d_0 \vee d_1)$ are in $l' = \alpha \wedge \delta = \beta \wedge \delta$; but this implies that $(a_0 \vee a_1) \wedge (b_0 \vee b_1) \in l'$, hence $c_{01} \in l'$. Similarly, $c_{02}$ and $c_{12} \in l'$. This shows that the two triangles are perspective with respect to the line $l'$. $\square$

An equivalent form of the arguesian identity that formally resembles Desargues' Theorem can be found in G. Grätzer, B. Jónsson, and H. Lakser [283]. Applying this characterization, it is proved that if $\{o, a, b, c, i\}$ is a diamond in a modular lattice, then the arguesian identity holds in $[o, a]$; this is also used in G. Grätzer and J. Sichler [355].

By B. Jónsson [436], every arguesian lattice of length $\leq 4$ has a type 1 representation.

### 5.7   The Coordinatization Theorem of Projective Geometry

Before we come to the classical Coordinatization Theorem, we prove one more lemma.

**Lemma 433.** *A projective space satisfying Desargues' Theorem also satisfies the converse (or "dual") statement:*

*If two triangles are perspective with respect to a line, then they are perspective with respect to a point.*

*Proof.* Let us use the notation of Figure 97. We now assume that $c_{01}, c_{02}, c_{12}$ are on a line $l$ and we do not have the point $p$. Define $p$ as the intersection of the lines $a_1 + b_1$ and $a_2 + b_2$. To show the two triangles perspective with respect to $p$, we have to verify that $a_0, b_0, p$ are collinear.

Consider the triangles $(a_2, b_2, c_{02})$ and $(a_1, b_1, c_{01})$; they are perspective with respect to the point $c_{12}$. By Desargues' Theorem, the intersections of the corresponding sides, $a_0, b_0, p$ are collinear.                    □

Now we are ready to state the classical Coordinatization Theorem of Projective Geometry.

**Theorem 434.** *Let $L$ be a directly irreducible arguesian geometric lattice of length at least three. Then there exist a division ring $D$, unique up to isomorphism, and a unique cardinal number $\mathfrak{m}$ such that $L \cong \mathrm{PG}(D, \mathfrak{m})$.*

*Remark.* For lattices of finite length, this was already included in O. Veblen and W. H. Young [690] (of course, in a form appropriate for projective spaces). Some authors claim that this proof is not complete and that no complete proof appeared until J. von Neumann [552], [553]. The condition of finite dimensionality was eliminated in O. Frink [205].

The method described below is "von Staudt's algebra of throws". A more modern (maybe less intuitive) proof can be found in R. Baer [35]; see also E. Artin [33].

*Sketch of proof.* A complete proof of this result is too long and of not enough interest for a lattice theory book. However, the mere statement that there is such an isomorphism $\varphi \colon L \to \mathrm{PG}(D, \mathfrak{m})$ is insufficient for workers in lattice theory. So we choose a middle course: we are going to construct $D$ and $\varphi$ but will not provide detailed proofs.

Let $l$ be an arbitrary line in $L$; we fix three distinct points of $l$ and call them $0, 1, \infty$. We set $D = l - \{\infty\}$, and define addition and multiplication on $D$.

Let $a, b \in D$. Fix two distinct points $p$ and $q$ not in $L$ such that $0, p, q$ are collinear (see Figure 98). Then form the points

$$r = (a \vee p) \wedge (q \vee \infty),$$
$$s = (p \vee \infty) \wedge (b \vee q),$$

and set

$$a + b = (r \vee s) \wedge l.$$

It is easy to see that $a + b$ does not depend on the choice of $p$ and $q$. Indeed, let $p'$ and $q'$ be another pair of distinct points not in $l$ such that $0, p', q'$ are collinear and let us form $r'$ and $s'$, as before. Then $(p, q, r)$ and $(p', q', r')$ and also $(p, q, s)$ and $(p', q', s')$ are perspective with respect to the line $l$; so by Lemma 433, they are perspective with respect to a point, in fact, the same point $u$, that is

$$u, p, p'; \quad u, q, q'; \quad u, r, r'; \quad u, s, s'$$

are all collinear. Hence $(r, s, q)$ and $(r', s', q')$ are perspective with respect to $u$; therefore, by Desargues' Theorem, they are perspective with respect to a line $l'$. But

$$(s \vee q) \wedge (s' \vee q') = b \in l,$$
$$(r \vee q) \wedge (r' \vee q') = \infty \in l,$$

hence $l = l'$. Therefore, $(r \vee s) \wedge (r' \vee s') \in l$, that is,

$$(r \vee s) \wedge l = (r' \vee s') \wedge l,$$

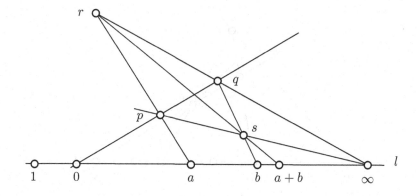

Figure 98. Constructing the division ring: addition

showing that $a + b$ is independent of the choice of $p$ and $q$.

To define $ab$, choose two distinct points $p$ and $q$ not in $l$ such that $p, q, \infty$ are collinear (see Figure 99). Then we form the points

$$r = (1 \vee p) \wedge (q \vee b),$$
$$s = (0 \vee r) \wedge (p \vee a),$$

and we define

$$ab = (q \vee s) \wedge l.$$

Then $D$ is a division ring with 0 as a zero and 1 as the unit element.

Now we choose a maximal independent set $I$ of atoms of $L$. Let $|I| = \mathfrak{m}$ and we consider the lattice $PG(D, \mathfrak{m})$, where each atom is represented by a function $f \colon I \to D$, which is 0 at all but a finite number of elements; we wish to associate with each $r$ of $L$ such a function $d_r \colon I \to D$.

Fix $p, t \in I$ with $p \neq t$, and set $l = p \vee t$. We choose a point $t_1$ on $l$, $t_1 \neq p$ and $t_1 \neq t$. We define a division ring $D$ on $l$ with $p$ as a zero, $t_1$ as the unit, and $t$ as infinity.

Now, for every $q \in I$ with $q \neq p$ and $q \neq t$, fix a point $q_1$ of the line $p \vee q$ with $q_1 \neq p$ and $q_1 \neq q$. The map

$$\psi_q \colon x \mapsto (x \vee ((t \vee q) \wedge (t_1 \vee q_1))) \wedge l$$

is a bijection between $p \vee q$ and $l$ and it fixes $p$ and takes $q_1$ into $p_1$.

We define $d_r \colon I \to D$ for every atom $r$ of $L$. First, let

$$r \not\leq \bigvee (I - \{p\}).$$

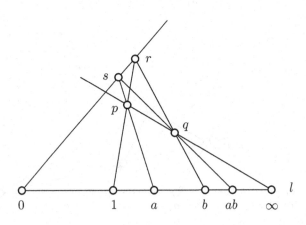

Figure 99. Constructing the division ring: multiplication

Let $r \leq \bigvee I_1$, where $I_1$ is a minimal (finite) subset of $I$ satisfying this inequality. By assumption, $p \in I_1$. We define

$$d_r(x) = \begin{cases} 1 \text{ (the unit of } D), & \text{for } x = p; \\ 0 \text{ (the zero of } D), & \text{for } x \notin I_1; \\ \psi_q((r \vee \bigvee(I_1 - \{x\})) \wedge (p \vee x)) \in D, & \text{for } x \in I_1 - \{p\}. \end{cases}$$

Second, if

$$r \leq \bigvee(I - \{p\}),$$

then $r \leq \bigvee I_1$, where $I_1$ is a minimal (finite) subset of $I - \{p\}$ with this property. By assumption, $p \notin I_1$. Let $s$ be any point on $p \vee r$ with $s \neq p$ and $s \neq r$. Then $s \nleq \bigvee(I - \{p\})$, hence $d_s$ has already been defined. Now we define

$$d_r(x) = \begin{cases} d_s(x), & \text{for } x \neq p; \\ 0, & \text{for } x = p. \end{cases}$$

The claim is then that $r \mapsto d_r$ maps the atoms of $L$ onto the set of functions representing the atoms of $\mathrm{PG}(D, \mathfrak{m})$ in such a manner that collinearity is preserved. Hence this map can be extended to an isomorphism $\varphi$ of $L$ and $\mathrm{PG}(D, \mathfrak{m})$. $\qquad\square$

Combining Theorems 432 and 434, we obtain

**Corollary 435.** *Let $L$ be a directly indecomposable modular geometric lattice of length at least 4. Then $L \cong \mathrm{PG}(D, \mathfrak{m})$, with a suitable division ring $D$ and cardinal number $\mathfrak{m}$.*

## 5.8  Frink's Embedding Theorem

We start with two statements to prepare for Frink's Embedding Theorem.

**Lemma 436.** *Let $L$ be a modular lattice with zero and let $p, a, b$ elements of $L$ satisfying the following conditions:*

(i) $p \leq a \vee b$;
(ii) $p$ is an atom;
(iii) $p \nleq a, b$.

*If there exists a sectional complement of $a \wedge b$ in $[0, b]$, then there are atoms $x \leq a$ and $y \leq b$ such that $p \leq x \vee y$.*

*Remark.* Figure 100 shows the "configuration" of $p, a, b$. The diagram is the sublattice generated by $p, a, b$ provided that $a \wedge b = 0$. To get the most general sublattice generated by $p, a, b$, take the appropriate quotient lattice of $\mathrm{Free}_{\mathsf{M}}(3)$, see Figure 20.

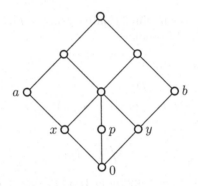

Figure 100. The configuration of $p, a, b$

*Proof.* As $p$ is an atom and $p \not\le a, b$, we get that $p \wedge a = p \wedge b = 0$.

Assume first that $a \wedge b = 0$ and set $x = a \wedge (b \vee p)$, $y = b \wedge (a \vee p)$, see Figure 100. If $x = 0$, then from $p \wedge b = 0$ and Theorem 348, it follows that $p \wedge (a \vee b) = 0$, a contradiction; so $x \ne 0$. Since $p$ is an atom, $a \wedge b = 0$, and $L$ is both upper and lower semimodular (see Corollary 376), it follows that $x$ is an atom. Similarly, $y$ is an atom. Now by modularity,

$$x \vee b = (a \vee b) \wedge (b \vee p) = b \vee p,$$

since $b \le b \vee p$. Again by modularity,

$$x \vee y = (x \vee b) \wedge (a \vee p) = (b \vee p) \wedge (a \vee p) \ge p,$$

since $x \le a \vee p$, which concludes this case.

Now we settle the general case. Let $b'$ be a sectional complement of $a \wedge b$ in $[0, b]$. Observe that $p \le a \vee b = a \vee (a \wedge b) \vee b' = a \vee b'$ and $p \not\le a, b'$. Since $a \wedge b' = 0$, we can apply the first case, and we find atoms $x \le a$ and $y \le b'$ (thus $y \le b$) such that $p \le x \vee y$.  □

The following result is the crucial step in the usual proof of Frink's Embedding Theorem, as stated in C. Herrmann and M. V. Semenova [393].

**Theorem 437.** *Let $L$ be a sectionally complemented lattice, let $M$ be a modular lattice with zero, and let $\varepsilon \colon L \to M$ be a $\{0\}$-lattice homomorphism. Denote by $F$ the set of all (possibly empty) finite joins of atoms in $M$. Then the map*

$$\varphi(a) = \{\, x \in F \mid x \le \varepsilon(a) \,\}, \quad \text{for each } a \in L,$$

*defines a $\{0\}$-lattice homomorphism $\varphi \colon L \to \operatorname{Id} F$. Furthermore, if $\varepsilon(c)$ majorizes an atom of $M$ whenever $c \in L - \{0\}$, then the map $\varphi$ is an embedding.*

*Proof.* By Theorem 361, $F$ is an ideal of $M$. It is obvious that $\varphi$ is a zero-preserving meet-homomorphism. To show that it is a join-homomorphism, let $a, b \in L$; we prove that $\varphi(a) \vee \varphi(b)$ majorizes $\varphi(a \vee b)$. It suffices to prove that $p \in \varphi(a \vee b)$ implies that $p \in \varphi(a) \vee \varphi(b)$ for all $a, b \in L$ and every atom $p$ of $M$.

The assumption $p \in \varphi(a \vee b)$ implies that $p \leq \varepsilon(a \vee b)$, which is equivalent to $p \leq \varepsilon(a) \vee \varepsilon(b)$. If either $p \leq \varepsilon(a)$ or $p \leq \varepsilon(b)$ then

$$p \in \varphi(a) \cup \varphi(b) \subseteq \varphi(a) \vee \varphi(b),$$

as desired.

Now assume that $p \not\leq \varepsilon(a), \varepsilon(b)$. If $b'$ is a sectional complement of $a \wedge b$ in $[0, b]$ in $L$, then $\varepsilon(b')$ is a sectional complement of $\varepsilon(a) \wedge \varepsilon(b)$ in $[0, \varepsilon(b)]$ in $M$. Therefore, by Lemma 436, there are atoms $x \leq \varepsilon(a)$ and $y \leq \varepsilon(b)$ of $M$ such that $p \leq x \vee y$. From $x \in \varphi(a)$ and $y \in \varphi(b)$, it follows that $p \in \varphi(a) \vee \varphi(b)$. So $\varphi$ is a $\{0\}$-lattice homomorphism.

Now assume that $\varepsilon(c)$ majorizes an atom of $M$, whenever $c \in L - \{0\}$. Let $a, b \in L$ with $a \not\leq b$. Let $c$ be a sectional complement of $a \wedge b$ in $[0, a]$; observe that $c > 0$. By assumption, there exists an atom $p \leq \varepsilon(c)$. Then $p \leq \varepsilon(a)$ and

$$p \wedge \varepsilon(b) = p \wedge \varepsilon(a) \wedge \varepsilon(b) = p \wedge \varepsilon(a \wedge b) = 0,$$

thus $\varepsilon(a) \not\leq \varepsilon(b)$. Therefore, $\varphi$ is an embedding. $\qquad\square$

The following result is close to the original result in O. Frink [205].

**Theorem 438 (Frink's Embedding Theorem).** *Every complemented modular lattice $L$ can be embedded into a modular geometric lattice $\overline{L}$. This embedding can be chosen to be a $\{0, 1\}$-embedding and $\overline{L}$ can be chosen to satisfy all the identities satisfied by $L$.*

*Remark.* The last observation was made by B. Jónsson [433].

*Proof.* Let $F$ be the dual of $\operatorname{Fil} L$ of $L$, let $A = \operatorname{Atom}(F)$, and let $J$ denote the set of joins of all finite subsets of $A$. It follows from the dual of Lemma 59 that every identity that holds in $L$ also holds in $F$. In particular, $F$ is modular, thus by Theorem 361, $J$ is an ideal of $F$, and it also satisfies every identity satisfied by $L$; hence so does $\overline{L} = \operatorname{Id} J$. Since $A = \operatorname{Atom}(J)$ and every element of $J$ is a finite join of atoms, it follows that $\overline{L}$ is atomistic; the are atoms the principal ideals determined by the elements of $A$. Since $\overline{L}$ is an algebraic lattice, it is geometric.

An application of Theorem 437, with $F$ in place of $M$ and $\varepsilon\colon a \mapsto \operatorname{fil}(a)$ the canonical embedding of $L$ into the dual of its filter lattice, yields a $\{0\}$-lattice homomorphism $\varphi\colon L \to \overline{L}$. Trivially, $\varphi$ preserves the unit. In order to prove that $\varphi$ is an embedding, we need to prove that for each $c \in L - \{0\}$, there exists a maximal proper filter $P$ of $L$ such that $c \in P$. This is an immediate consequence of Zorn's Lemma. $\qquad\square$

We can combine the previous results to obtain the following embedding theorems:

**Corollary 439.** *Every complemented modular lattice can be embedded in a direct product of lattices* $\mathrm{PG}(D_i, \mathfrak{m}_i)$ *and subspace lattices of projective planes that do not satisfy Desargues' Theorem.*

*Proof.* This is immediate from Theorem 438 and Corollary 426.    □

**Corollary 440.** *A complemented modular lattice $L$ can be embedded in a direct product of lattices* $\mathrm{PG}(D_i, \mathfrak{m}_i)$ *iff $L$ is arguesian.*

*Proof.* If $L$ has such an embedding, then it is arguesian by Corollary 431. If $L$ is arguesian, then, by Theorem 438, the lattice $L$ can be embedded in an arguesian geometric lattice. The direct factorization of Corollary 426 gives us arguesian directly indecomposable geometric lattices, which, by Theorem 434, are of the form $\mathrm{PG}(D_i, \mathfrak{m}_i)$.    □

Summarizing these results, we obtain a result of B. Jónsson [439]:

**Theorem 441.** *Let $L$ be a complemented modular lattice. The following conditions on $L$ are equivalent:*

(i) *$L$ can be embedded in a direct product of lattices* $\mathrm{PG}(D_i, \mathfrak{m}_i)$.
(ii) *$L$ can be embedded in an arguesian geometric lattice.*
(iii) *$L$ can be embedded in the lattice of all subgroups of an abelian group.*
(iv) *$L$ has a representation of type 1.*
(v) *$L$ is arguesian.*

*Proof.* (i) is equivalent to (v) by Corollary 440. The equivalence of (i) and (ii) was proved in the proof of Corollary 440. (i) implies (iii), since the elements of $L$ are represented by subgroups of the additive group of $\prod D_i$. (iii) implies (iv) was observed in the proof of Corollary 431. (iv) implies (v) by Theorem 410.    □

### 5.9    A weaker version of the arguesian identity

Theorem 441 explains the significance of the arguesian identity for complemented modular lattices . It is interesting that a weaker version of the arguesian identity holds for all complemented modular lattices, see G. Grätzer and H. Lakser [302].

Let

$$\varepsilon \colon p(x_1, \ldots, x_6) \leq q(x_1, \ldots, x_6)$$

be the arguesian identity. Choose three new variables, $x, y, z$, and define

$$u = (x \vee y) \wedge (y \vee z) \wedge (z \vee x),$$
$$v = (x \wedge y) \vee (y \wedge z) \vee (z \wedge x),$$
$$\overline{x}_i = x_i \wedge ((u \wedge x) \vee v)$$

for $i = 1, 2, \ldots, 6$.

**Theorem 442.** *The identity*

$$\bar{\varepsilon} \colon p(\bar{x}_1, \ldots, \bar{x}_6) \vee v \le q(\bar{x}_1, \ldots, \bar{x}_6) \vee v$$

*holds in any complemented modular lattice.*

*Proof.* Consider the following lattice property:

(P)    If $\{0, a, b, c, i\}$ is a diamond, then $\mathrm{id}(a)$ is an arguesian lattice.

We claim that if a lattice has property (P), then it satisfies the identity $\bar{\varepsilon}$.
Let $t, r, s \in L$, and let us consider the substitution $x = t$, $y = r$, $z = s$.
There are two cases to consider:
   Case 1.

$$(t \vee r) \wedge (r \vee s) \wedge (s \vee t) = (t \wedge r) \vee (r \wedge s) \vee (s \wedge t) = \alpha.$$

Then $\bar{x}_i = x_i \wedge \alpha$, and so $\bar{\varepsilon}$ becomes

$$p(x_1 \wedge \alpha, \ldots, x_6 \wedge \alpha) \vee \alpha \le q(x_1 \wedge \alpha, \ldots, x_6 \wedge \alpha) \vee \alpha.$$

But

$$p(x_1 \wedge \alpha, \ldots, x_6 \wedge \alpha) \le p(\alpha, \ldots, \alpha) = \alpha,$$

and similarly for $q$, so $\bar{\varepsilon}$ becomes $\alpha \le \alpha$, a triviality.
   Case 2.

$$i = (t \vee r) \wedge (r \vee s) \wedge s \vee t) \neq (t \wedge r) \vee (r \wedge s) \vee (s \wedge t) = o.$$

Then it follows from the diagram of $\mathrm{Free}_M(3)$ (see Figure 20), that $o$ and $i$ along with

$$a = (i \wedge t) \vee o,$$
$$b = (i \wedge r) \vee o,$$
$$c = (i \wedge s) \vee o$$

form a diamond. Thus property (P) applies and $\mathrm{id}(a)$ is arguesian. Since

$$x_1 \wedge a, \ldots, x_6 \wedge a \in \mathrm{id}(a),$$

we obtain that

$$p(x_1 \wedge a, \ldots, x_6 \wedge a) \le q(x_1 \wedge a, \ldots, x_6 \wedge a);$$

joining both sides by $o$ we get that $\bar{\varepsilon}$ holds in $L$.

Next we claim that property (P) holds for the two types of lattices listed in Corollary 439. Naturally, (P) holds for a geometry $PG(D, \mathfrak{m})$ because it is arguesian. Let $L$ be the subspace lattice of a projective plane. If $\{o, a, b, c, i\}$ is a diamond, then $a$ is either a point or a line. In either case, $\mathrm{id}(a)$ is arguesian.

By Corollary 439, every complemented modular lattice $K$ can be embedded in a direct product of lattices having property (P) and therefore satisfying $\bar{\varepsilon}$. Thus $K$ also satisfies $\bar{\varepsilon}$.                                    □

It is easy to construct a modular lattice in which $\bar{\varepsilon}$ fails. Let $L$ be the subspace lattice of a projective plane in which Desargues' Theorem fails. By our discussion in Lemma 428, there are seven atoms of $L$, namely,

$$a_0, a_1, a_2, b_0, b_1, b_2, d$$

such that

$$p(a_0, a_1, a_2, b_0, b_1, b_2) = d,$$
$$q(a_0, a_1, a_2, b_0, b_1, b_2) = 0.$$

Let $e$ be a complement of $d$; then $e$ is a dual atom of $L$. Let $K = L \cup \{b, c, i\}$ such that with $o = e$ and $a = 1$ (the unit of $L$); moreover, $\{o, a, b, c, i\}$ is a diamond. It is immediate that $K$ is a modular lattice. Substitute

$$x_1 = a_0, x_1 = a_1, x_2 = a_2, x_3 = b_0, x_4 = b_1, x_5 = b_2, x = a \ (= 1), y = b, z = c.$$

Then

$$u = i, v = 0, (u \wedge x) \vee o = a \ (= 1),$$

hence $x_i = \overline{x}_i$ for $i = 1, 2, \ldots, 6$. Thus $\bar{\varepsilon}$ yields $d \vee o \leq 0 \vee o$, that is, $1 \leq o$, which is not true. We conclude:

**Corollary 443 (R. P. Dilworth and M. Hall [163]).** *There exists a modular lattice that cannot be embedded in a complemented modular lattice.*

C. Herrmann and A. P. Huhn [392] prove that the lattice of all subgroups of $(\mathbb{Z}/4\mathbb{Z})^3$ (where $\mathbb{Z}$ is the additive group of integers) cannot be embedded in a complemented modular lattice.

## 5.10   Projective planes

The modular lattice $K$, which we constructed for Corollary 443, is of length 4. The next result shows that this is best possible.

**Theorem 444.** *Every modular lattice $B$ of length at most 3 can be embedded in a complemented modular lattice $C$. The embedding can always be constructed to be a $\{0, 1\}$-embedding and $C$ can be chosen to have the length of $B$.*

We shall prove Theorem 444 using the concept of a projective plane.

**Definition 445.** A *projective plane* $(A, L)$ is a set $A$ (of points) and a collection $L$ of subsets of $A$ (the lines) satisfying the following conditions:

(i) Every line $l \in L$ has at least two elements.

(ii) For every pair of points $p_0, p_1 \in A$ with $p_0 \neq p_1$, there is exactly one line $l \in L$ satisfying $p_0, p_1 \in l$.

(iii) For every pair of distinct lines $l_0$ and $l_1$, there is exactly one point $p \in A$ satisfying $p \in l_0$ and $p \in l_1$.

A *partial plane* satisfies (i)–(iii) with "at most one" replacing "exactly one" in (ii) and (iii). Observe that the first two of the three properties already define a partial plane.

It is easily seen that a projective plane is a projective space.

**Lemma 446.** *Every partial plane can be embedded in a projective plane, that is, for every partial plane $(A, L)$, there is a projective plane $(A', L')$ such that $A \subseteq A'$ and, for every $l \in L$, there is an $l' \in L'$ satisfying $l \subseteq l'$.*

*Proof.* Let $(A, L)$ be a partial plane. Define

$$A^+ = A \cup \{\, p(l, m) \mid l, m \in L \text{ and } l \cap m = \varnothing \,\},$$
$$L^+ = \{\, l \cup \{\, p(l, m) \mid m \in L,\ l \cap m = \varnothing \,\} \mid l \in L \,\}$$
$$\cup \{\, \{p, q\} \mid p, q \in A,\ p \neq q,\ p, q \in l \text{ for no } l \in L \,\},$$

where $p(l, m) = p(m, l)$, and the new points defined are distinct from each other and from the points in $A$.

Now define

$$A_0 = A, \qquad\qquad L_0 = L,$$
$$A_{n+1} = (A_n)^+, \qquad L_{n+1} = (L_n)^+,$$

for $n < \omega$, and

$$A' = \bigcup (\, A_n \mid n < \omega \,).$$

Lines are defined as follows: let $l_n$ be a line in $(A_n, L_n)$; then we obtain the lines $l_{k+1}$ for all $k \geq n$:

$$l_{k+1} = l_k \cup \{\, p(l_k, m) \mid m \in L_k,\ l_k \cap m = \varnothing \,\}.$$

$L'$ is the collection of all sets of the form $\bigcup(\, l_k \mid k \geq n \,)$. It is obvious that $(A', L')$ satisfies the requirements. $\qquad\square$

*Proof of Theorem 444.* If the length of the modular lattice $B$ is less than 3, the statement is obvious. So let $B$ be of length 3. We define a partial plane $(A, L)$ as follows: the points of $(A, L)$ are of two kinds, namely, the atoms of $B$ and for every $a \in B$ with height$(a) = 2$, which majorizes exactly one atom, a new point $p(a)$. For every $a \in B$ with height$(a) = 2$, majorizing at least two atoms, we define a line

$$\{ p \mid p \in \mathrm{Atom}(B), \ p \leq a \};$$

if $a \in B$ with height$(a) = 2$ and $a$ majorizes exactly one atom $p$, then we define the line $\{p, p(a)\}$. It is clear that $(A, L)$ is a partial plane. Embed $(A, L)$ into a projective plane $(A', L')$ by Lemma 446 and let $C$ be the subspace lattice of $(A', L')$.                                    □

### 5.11 ◊ Coordinatizing sectionally complemented modular lattices by Friedrich Wehrung

A *coordinatization theorem* could be loosely defined (see F. Wehrung [710]) as "*a statement that expresses a class of geometric objects in algebraic terms. Hence it is a path from* **synthetic** *geometry to* **analytic** *geometry*". The classical Coordinatization Theorem of Projective Geometry is an example of such a theorem; it is an immediate consequence of Theorems 393 and 434.

In order to state this theorem, we need some notation. We denote by $\mathrm{Sub}\, V$ the lattice of all subspaces of a vector space $V$ (over a division ring). For a cardinal number $\mathsf{m} \geq 3$, let $\mathsf{M}_\mathsf{m}$ denote the lattice of length 2 with $\mathsf{m}$ atoms.

**Theorem 447.** *Every modular geometric lattice is isomorphic to a direct product $\prod(L_i \mid i \in I)$, where each $L_i$ has one of the following forms:*

(i) $\mathsf{M}_\mathsf{m}$, *for a cardinal number $\mathsf{m} \geq 3$;*

(ii) $\mathrm{Sub}\, V$, *for a vector space $V$ (over some division ring) of dimension at least 3;*

(iii) *a nonarguesian projective plane.*

At first sight it is not so obvious to see how to extend this result to (non-geometric) complemented modular lattices. In the proof of Theorem 434, the atoms play a special role, and they also make it possible to follow the proof by pictures. However, many complemented modular lattices do not have any atoms.

The answer, that originates in J. von Neumann [552], [553], comes from *ring theory*. All our rings will be associative, but not necessarily unital.

**Definition 448.** A ring $R$ is *regular* if for each element $x \in R$ there exists an element $y \in R$ such that $xyx = x$. If in addition, $R$ is unital and $y$ can always be taken a unit of $R$, we say that $R$ is *unit-regular*.

We set $\operatorname{Lr} R = \{\, xR \mid x \in R \,\}$, the set of all principal right ideals of a ring $R$.

While $\operatorname{Lr} R$ is not necessarily a lattice under set inclusion (take $R = \mathbb{Z}[\sqrt{-5}]$), the case where $R$ is regular is special. The (nontrivial) proof of the non-unital case is contained in K. D. Fryer and I. Halperin [209, Section 3.2].

$\Diamond$ **Theorem 449.** *The lattice* $\operatorname{Lr} R$ *is a sectionally complemented sublattice of the lattice* $\operatorname{Sub} R_R$ *of all right ideals of* $R$ *for any regular ring* $R$.

As $\operatorname{Sub} R_R$ is arguesian, a noteworthy corollary of Theorem 449 is that $\operatorname{Lr} R$ *is arguesian.* Moreover, as $\operatorname{Sub} R_R$ is obviously an algebraic lattice, it follows that $\operatorname{Lr} R$ is exactly the $\{\vee, 0\}$-semilattice (which, here, turns out to be a *lattice with zero*) of all compact members of $\operatorname{Sub} R_R$.

The relation with Theorem 447 (especially part (ii) of that theorem) is as follows. The ring $R = \operatorname{End} V$ of all endomorphisms of a vector space $V$ (over an arbitrary division ring) is easily seen to be regular. Furthermore, the assignment that to any element of $R$ associates its *image* defines an *isomorphism* from $\operatorname{Lr} R$ onto $\operatorname{Sub} V$: in formula, $\operatorname{Lr} \operatorname{End} V \cong \operatorname{Sub} V$.

**Definition 450.** A lattice is *coordinatizable* if it is isomorphic to $\operatorname{Lr} R$ for some regular ring $R$.

In particular, it follows from Theorem 447 that *Every coordinatizable lattice is sectionally complemented and arguesian.* J. von Neumann [552], [553] considers only the case where both the lattice and the ring have a unit, however, it is not hard to verify that a *regular* ring $R$ is unital iff the lattice $\operatorname{Lr} R$ has a largest element. For a cardinal number $\mathfrak{m} \geq 3$, the lattice $\mathsf{M}_\mathfrak{m}$ is coordinatizable iff either $\mathfrak{m}$ is infinite or $\mathfrak{m} - 1$ is a prime power; in particular, the least non-coordinatizable such lattice is $\mathsf{M}_7$. By Theorem 449, a geometric lattice is coordinatizable iff it is a product of lattices of the form $\operatorname{Sub} V$, for vector spaces $V$, and coordinatizable $\mathsf{M}_\mathfrak{m}$-s.

This suggests that coordinatizability is a property of lattices with "enough geometry". The latter intuition is best captured by the following definition. An independent finite sequence $(a_i \mid i < n)$ in a lattice $L$ with zero is *homogeneous* if the elements $a_i$ are pairwise perspective. An element $x \in L$ is *large* if $\operatorname{con}_L(0, x) = \mathbf{1}$. An independent sequence $(a_i \mid 0 \leq i < n)$ together with a sequence $(c_i \mid 1 \leq i < n)$ form

- an *$n$-frame* if $a_0 \vee c_i = a_i \vee c_i$ while $a_0 \wedge c_i = a_i \wedge c_i = 0$ for $1 \leq i < n$;
- a *large $n$-frame* if it is an $n$-frame and $a_0$ is large;
- a *spanning $n$-frame* if it is an $n$-frame, $L$ has a unit, and $1 = \bigvee_{i<n} a_i$.

In particular, spanning implies large, while the converse fails (even in presence of a unit).

The most decisive landmark on coordinatization of complemented modular lattices, the *von Neumann Coordinatization Theorem*, is due to J. von Neumann [552], [553], and it states the following.

◇ **Theorem 451 (von Neumann Coordinatization Theorem).** *Let* $L$ *be a complemented modular lattice. If* $L$ *has a spanning n-frame for some* $n \geq 4$, *then* $L \cong \mathrm{Lr}\, R$ *for some regular ring* $R$. *Furthermore,* $R$ *is unique up to isomorphism.*

This result has been extended and the proof simplified by many authors, in particular, by I. Halperin and co-authors. The most powerful known extension of Theorem 451 is due to B. Jónsson [439].

◇ **Theorem 452.** *Let* $L$ *be a complemented modular lattice. If* $L$ *is arguesian with a large 3-frame, then* $L \cong \mathrm{Lr}\, R$ *for some regular ring* $R$. *Furthermore,* $R$ *is unique up to isomorphism.*

In particular, if $L$ has a large 4-frame, then it is arguesian, thus by Theorem 452, $L \cong \mathrm{Lr}\, R$ for a unique regular ring $R$.

While a fully detailed proof of Theorem 434 is very long, it is, conceptually, relatively simple, as it can be followed by pictures. While the proof of von Neumann's Theorem (Theorem 451) does not have a much higher logical complexity than the one of Theorem 434, it is often perceived as being much harder to follow, because it cannot be so easily visualized. This problem is overcome in C. Herrmann [391], by providing a much easier proof of Jónsson's Theorem *assuming Theorem 434.*

The basic idea is the following. Start with a complemented arguesian lattice $L$ with a large 3-frame. Using Frink's Embedding Theorem (Theorem 438) together with Theorem 447, we may assume that $L$ is a sublattice, with the same bounds, of the subgroup lattice $\mathrm{Sub}\, A$ of an Abelian group $A$. Then construct directly a regular subring $R$ of the ring $\mathrm{End}\, A$ of all endomorphisms of $A$ such that $L \cong \mathrm{Lr}\, R$. It turns out that there is a very simple formula describing $R$, attributed in C. Herrmann [391] to L. Giudici [232]: namely, $R$ *is the subring of* $\mathrm{End}\, A$ *generated by the idempotent endomorphisms* $f$ *of* $A$ *such that both the kernel and the image of* $f$ *belong to* $L$.

B. Jónsson [441, Introduction] observes that there are infinitely many lattices of the form $\mathsf{M}_n$ (with $3 \leq n < \omega$) which are not coordinatizable (take $\mathsf{M}_{4k+7}$), however, any ultraproduct of those lattices, with respect to a non-principal ultrafilter, is isomorphic to $\mathsf{M}_\mathfrak{m}$ for some infinite $\mathfrak{m}$, and thus it is coordinatizable. It follows that *The class of all non-coordinatizable lattices is not first-order axiomatizable.* Jónsson asks in [439] a similar question for the class of all coordinatizable lattices. The answer turns out to be also negative, see F. Wehrung [710].

◇ **Theorem 453.** *There are countable, 2-distributive, complemented modular lattices* $K$ *and* $L$ *with a spanning* $\mathsf{M}_3$, *such that* $K$ *is an elementary submodel of* $L$, *the lattice* $L$ *is coordinatizable, but the lattice* $K$ *is not coordinatizable. In particular, the class of all coordinatizable lattices is not first-order definable.*

Another question is whether the unit of the lattice is really required in the statement of Theorem 452 (indeed, notice that the existence of a large

3-frame does not require a unit). This question is raised, and partly answered, in B. Jónsson [442].

◊ **Theorem 454.** *Let L be a sectionally complemented arguesian lattice with a large 3-frame. If L has a countable cofinal subset, then L is coordinatizable.*

Due to a counterexample in B. Jónsson [442], the uniqueness statement about the coordinatizing ring in Theorem 454 no longer holds. Without the countable cofinal sequence assumption, a weaker coordinatization theorem, involving *locally projective modules over regular rings*, is obtained in that paper. Again, C. Herrmann [391] simplifies greatly the proof of that result, assuming Theorem 434. The question whether full coordinatization could be achieved (without the countable cofinal sequence assumption) is finally answered, in the negative, in F. Wehrung [715].

◊ **Theorem 455.** *There exists a non-coordinatizable sectionally complemented arguesian lattice L with a large 4-frame. Furthermore, L has $\aleph_1$ elements, and it is an ideal in a coordinatizable complemented modular lattice.*

The proof of Theorem 455 includes the following ingredients:

(i) An infinite combinatorial statement, due to P. Gillibert [227], [228], stating that every *tree* (it suffices to do it for the chain $\omega_1$) has a suitable "norm-covering" (called, in P. Gillibert and F. Wehrung [230], an $\aleph_1$-*lifter*).

(ii) Tools from F. Wehrung [714] (in particular, the *Banaschewski functions*), together with more infinite combinatorics, that make it possible to find a *diagram counterexample* (indexed by the chain $\omega_1$) to Jónsson's question.

(iii) Tools from P. Gillibert and F. Wehrung [230], called *larders*, see Section 5.5.

## 5.12  ◊ The dimension monoid of a lattice
by Friedrich Wehrung

For a lattice $L$ and a commutative monoid $M$, an *M-valued dimension function* on $L$ is a function $D$, defined on all pairs $(x, y) \in L \times L$ such that $x \leq y$, taking its values in $M$, satisfying the following properties:

(D0) $D(x, x) = 0$ for all $x \in L$;

(D1) $D(x, z) = D(x, y) + D(y, z)$, for all $x \leq y \leq z$ in $L$;

(D2) $D(x, x \vee y) = D(x \wedge y, y)$ for all $x, y \in L$.

Here are some examples of dimension functions: for an arbitrary lattice $L$, set $M = \mathrm{Con}_c\, L$ (the addition on $M$ is the join) and $D(x, y) = \mathrm{con}_L(x, y)$; for a modular lattice $L$ of finite length, set $M = \mathbb{Z}^+ = \{0, 1, 2, \dots\}$, endowed with its natural addition, and define $D(x, y)$, for $x \leq y$, as the length of the interval $[x, y]$.

The "most general" dimension function on a lattice $L$ takes its value in the *dimension monoid* of $L$, denoted by $\mathrm{Dim}\, L$. Formally, $\mathrm{Dim}\, L$ is the commutative monoid defined by the generators $\dim(x, y)$, for $x \leq y$ in $L$, and the relations (D0)–(D2) above applied to $D = \dim$. This construction is extensively studied in F. Wehrung [701]. What is denoted in the present book by $\dim(x, y)$ (for consistency with the notation system used) is denoted there by $\Delta(x, y)$.

Here are a few highlights about the dimension monoid construct.

Because of the congruence lattice example above, the dimension monoid of a lattice $L$ is a precursor of the congruence lattice of $L$. The kernel of the canonical monoid homomorphism $\rho \colon \mathrm{Dim}\, L \twoheadrightarrow \mathrm{Con_c}\, L$ can be characterized as follows. For any elements $x$ and $y$ in a commutative monoid $M$, we set

$x \leq y$ iff there exists an element $z \in M$ satisfying $x + z = y$;

$x \propto y$ iff there exists an element $n \in \mathbb{Z}^+$ satisfying $x \leq ny$.

Then $\rho(\xi) \leq \rho(\eta)$ (in $\mathrm{Con_c}\, L$) iff $\xi \propto \eta$ (in $\mathrm{Dim}\, L$) for all $\xi, \eta \in \mathrm{Dim}\, L$.

The domain of the dim function can be extended to all pairs of elements of $L$, by setting $\dim(x, y) = \dim(x \wedge y, x \vee y)$ for arbitrary $x, y \in L$. Then

$$\dim(x, y) = \dim(y, x),$$
$$\dim(x, z) \leq \dim(x, y) + \dim(y, z)$$

for all $x, y, z \in L$: hence the extended dim map is a $(\mathrm{Dim}\, L)$-valued *distance* on the lattice $L$. Because of (D2), this distance gives equal value to projective intervals.

A commutative monoid $M$ is *conical* if $x + y = 0$ implies that $x = y = 0$ for all $x, y \in M$. The monoid $\mathrm{Dim}\, L$ is easily seen to be conical for any lattice $L$. It is still an open problem whether $\mathrm{Dim}\, L$ is always a *refinement monoid*, that is, a commutative monoid in which $a_0 + a_1 = b_0 + b_1$ implies the existence of elements $c_{i,j}$, for $i, j < 2$, such that $a_i = c_{i,0} + c_{i,1}$ and $b_i = c_{0,i} + c_{1,i}$ for all $i < 2$. However, for two important classes of lattices the answer to this question is known:

◇ **Theorem 456.** *Let $L$ be a lattice. If either $L$ is modular or $L$ has no infinite bounded chains, then $\mathrm{Dim}\, L$ is a refinement monoid.*

For a generalized Boolean lattice $B$, denote by $\mathbb{Z}^+[B]$ the commutative monoid freely generated by a copy of $B$ with relations $a + b = (a \wedge b) + (a \vee b)$ for all $a, b \in B$. This monoid is easily seen to be cancellative, and in fact, a subdirect power of $\mathbb{Z}^+$. (The notation $\mathrm{BR}\, L$ was introduced in Section II.4.1.)

◇ **Theorem 457.** *Let $L$ be a distributive lattice and set $B = \mathrm{BR}\, L$. Then $\mathrm{Dim}\, L \cong \mathbb{Z}^+[B]$, with $\dim(x, y) = y - x$ for all $x \leq y$ in $L$.*

The dimension monoid can also be fully described for modular lattices of finite length:

◇ **Theorem 458.** *Let $L$ be a modular lattice without infinite chains and denote by $P$ the set of all projectivity classes of prime intervals of $L$. Then $\mathrm{Dim}\, L$ is isomorphic to the monoid of all families $(x_\pi \mid \pi \in P)$ of nonnegative integers such that $\{\, \pi \in P \mid x_\pi \neq 0 \,\}$ is finite; that is, the free commutative monoid on $P$. For $x \leq y$ in $L$ and $\pi \in P$, the dimension $\dim(x, y)_\pi$ is the number of occurrences of an element of $\pi$ in any maximal chain from $x$ to $y$.*

Using the notation of Section III.1.4, $P = \mathrm{PrInt}(L)/\Leftrightarrow$.

In particular, Theorem 458 together with standard results about von Neumann regular rings (see K. R. Goodearl [238]) makes it possible in P. Gillibert [229] to evaluate further *critical points* between some modular varieties of lattices. It is also used in F. Wehrung [708] to prove that the distributive $\{\vee, 0, 1\}$-semilattice $S_{\omega_1}$ (of cardinality $\aleph_1$) constructed there is not isomorphic to $\mathrm{Con}_c L$ for any *modular* lattice $L$ which is *locally finite* (or, more generally, in which every finite subset generates a sublattice of finite length). Recall that by contrast, every distributive $\{\vee, 0\}$-semilattice of cardinality at most $\aleph_1$ is isomorphic to $\mathrm{Con}_c L$ for some lattice $L$ which can be either be

- locally finite and relatively complemented with zero, see G. Grätzer, H. Lakser, and F. Wehrung [320], or

- sectionally complemented and modular, see F. Wehrung [703].

Hence there is no "best of two worlds" for these results.

A commutative monoid is *primitive* if it is defined by a set of generators $\Sigma$ with relations of the form $x + y = y$ for certain pairs $(x, y) \in \Sigma \times \Sigma$. This definition is equivalent to the one introduced in R. S. Pierce [581]. In particular, every primitive monoid is a conical refinement monoid.

◇ **Theorem 459.** *The following statements hold, for any lattice $L$ without infinite bounded chains:*

(i) *The monoid $\mathrm{Dim}\, L$ is primitive.*

(ii) *If $L$ is simple, then $\mathrm{Dim}\, L$ is isomorphic to $\mathbb{Z}^+$ if $L$ is modular, and to the semilattice $\{0, 1\}$, otherwise.*

F. Wehrung [705] introduces more efficient computation techniques of the dimension monoid of a finite lattice. These techniques involve refinements of the join-dependency relation $D$, see Section IV.4.4.

So far, the most fruitful aspects of the dimension monoid have been developed in relation to the *nonstable K-theory* of (von Neumann) regular rings (see Section 5.11). Denote by $\mathrm{FP}(R)$ the class of all finitely generated projective right modules over a unital, regular ring $R$, and denote by $[X]$

the isomorphism class of a member $X$ of $\mathrm{FP}(R)$. Isomorphism classes can be added *via* the rule

$$[X] + [Y] = [X \oplus Y] \quad \text{for } X, Y \in \mathrm{FP}(R).$$

The *nonstable K-theory* of $R$ is encoded by the commutative monoid

$$\mathrm{Vn}\, R = \{ [X] \mid X \in \mathrm{FP}\, R \},$$

endowed with the addition defined above. Then $\mathrm{Vn}\, R$ is always a conical refinement monoid. Furthermore, the usual $K_0$ (preordered) group of $R$ is just the universal (Grothendieck) group of the monoid $\mathrm{Vn}\, R$. For example, if $R$ is a field (or a matrix ring over a field), then $\mathrm{Vn}\, R \cong \mathbb{Z}^+$ while $K_0(R) \cong \mathbb{Z}$.

◇ **Theorem 460.** *Let $R$ be a unital regular ring which is either isomorphic to $\mathrm{M}_2(R')$ (the ring of all $2 \times 2$ matrices over $R'$) for some ring $R'$, or it is unit-regular. Then $\mathrm{Vn}\, R \cong \mathrm{Dim}\, \mathrm{Lr}\, R$.*

In particular, as $\mathrm{Vn}\, R \cong \mathrm{Vn}(\mathrm{M}_2(R))$, Theorem 460 states that the non-stable K-theory of regular rings and the dimension theory of the associated lattices are practically interchangeable.

A lattice $L$ with zero is *normal* if for any $x, y \in L$, if $x$ and $y$ are projective and $x \wedge y = 0$, then $x$ and $y$ are perspective. An example of a non-normal modular ortholattice, pointed out by C. Herrmann, is given in [701, Section 10.4].

We define a lattice $L$ *conditionally $\aleph_0$-meet-continuous* if every increasing majorized sequence $(b_n \mid n < \omega)$ of elements of $L$ has a join and

$$a \wedge \bigvee_n b_n = \bigvee_n (a \wedge b_n)$$

for all $a \in L$. By using Jónsson's Coordinatization Theorem (see Theorem 452), the following consequence can be derived (recall that the symbol $\sim$ denotes the relation of perspectivity, see Definition 269):

◇ **Theorem 461.** *Let $L$ be a sectionally complemented modular lattice. If either $L$ or its dual is conditionally $\aleph_0$-meet-continuous, then $L$ is normal. Furthermore, for all $x, y \in L$, $\dim(0, x) = \dim(0, y)$ iff there are $x_0, x_1, y_0, y_1 \in L$ such that $x_0 \wedge x_1 = y_0 \wedge y_1 = 0$, $x = x_0 \vee x_1$, $y = y_0 \vee y_1$, and $x_i \sim y_i$ for $i < 2$.*

The dimension theory of *complete* (not only countably complete) lattices, subjected to additional conditions, can be pushed further, yielding in certain cases a complete description, presented in K. R. Goodearl and F. Wehrung [239]. Here is a very rough outline of how this works. For an ordinal $\gamma$, each of the commutative monoids

$$\mathbb{Z}_\gamma = \mathbb{Z}^+ \cup \{ \aleph_\alpha \mid \alpha \leq \gamma \},$$
$$\mathbb{R}_\gamma = \mathbb{R}^+ \cup \{ \aleph_\alpha \mid \alpha \leq \gamma \},$$
$$\mathbf{2}_\gamma = \{0\} \cup \{ \aleph_\alpha \mid \alpha \leq \gamma \},$$

endowed with the interval topology, is a compact Hausdorff topological monoid. Now denote by $\mathbf{C}(\Omega, M)$ the additive monoid of all continuous maps from a topological space $\Omega$ to a topological monoid $M$. A partial monoid $M$ is a *continuous dimension scale* if it is isomorphic to a down-set (endowed with componentwise addition) of a product

$$\mathbf{C}(\Omega_{\mathrm{I}}, \mathbb{Z}_\gamma) \times \mathbf{C}(\Omega_{\mathrm{II}}, \mathbb{R}_\gamma) \times \mathbf{C}(\Omega_{\mathrm{III}}, \mathbf{2}_\gamma),$$

for zero-dimensional compact Hausdorff spaces $\Omega_{\mathrm{I}}$, $\Omega_{\mathrm{II}}$, $\Omega_{\mathrm{III}}$. If $M$ is a monoid (not only a partial monoid), we say that $M$ is a *total* continuous dimension scale. The property, for a partial commutative monoid, of being a continuous dimension scale can also be expressed by a list of axioms, some of them similar to completeness (for lattices).

$\Diamond$ **Theorem 462.** *Let $L$ be a sectionally complemented modular lattice in which every directed majorized subset $X$ has a join and*

$$a \wedge \bigvee X = \bigvee (a \wedge x \mid x \in X)$$

*for $a \in L$. Then $\operatorname{Dim} L$ is a total continuous dimension scale. Furthermore, every total continuous dimension scale can be represented this way.*

In K. R. Goodearl and F. Wehrung [239], similar results are obtained in abstract measure theory (unrestrictedly additive positive measures on complete Boolean algebras), but also nonstable K-theory of either self-injective regular rings, von Neumann algebras (W*-algebras), or their abstract generalizations, the AW*-algebras.

To conclude this section, we mention an application of the dimension monoid to coordinatization of sectionally complemented modular lattices (see Section 5.11), obtained in F. Wehrung [710].

$\Diamond$ **Theorem 463.** *Let $L$ be a complemented arguesian lattice. Then $L$ has a large 3-frame iff there exists an independent sequence $(a_0, a_1, a_2, b)$ in $L$ such that*

(i) $a_0 \vee a_1 \vee a_2 \vee b = 1$;
(ii) $a_i \sim a_j$ *for all distinct $i$, $j < 3$;*
(iii) $b \lesssim a_0 \vee a_1$.

Hence, for a complemented arguesian lattice, Jónsson's sufficient condition of having a large 3-frame (see Theorem 452) can be expressed by a single first-order sentence.

## 5.13   $\Diamond$ Dilworth's Covering Theorem
by Joseph P. S. Kung

In a finite distributive lattice $L$, the equality $|\operatorname{Ji} L| = |\operatorname{Mi} L|$ holds for the set of join-irreducible and the set of meet-irreducible elements, see the comment following Corollary 112. Since the modular law is a weaker form of the distributive law, a natural conjecture, from "the middle 1930's", claimed that this equality holds for finite modular lattices as well. This conjecture, and more, was proved by R. P. Dilworth [159] in 1954.

If $L$ is a finite lattice, let

$$\mathrm{Cov}_k\, L = \{\, x \in L \mid x \text{ covers exactly } k \text{ elements}\,\},$$

$$\mathrm{Cov}^k\, L = \{\, x \in L \mid x \text{ is covered by exactly } k \text{ elements}\,\}.$$

Note that $\mathrm{Cov}_0\, L = \{0\}$, $\mathrm{Cov}^0\, L = \{1\}$, $\mathrm{Cov}_1\, L = \mathrm{Ji}\, L$, and $\mathrm{Cov}^1\, L = \mathrm{Mi}\, L$.

◇ **Theorem 464 (Dilworth's Covering Theorem).** *Let $k$ be a nonnegative integer and $L$ be a finite modular lattice. Then $|\mathrm{Cov}_k\, L| = |\mathrm{Cov}^k\, L|$.*

We sketch Dilworth's proof. For an element $x$ in a finite lattice, let $x^\dagger$ be the join of all elements covering $x$, and dually, let $x_\dagger$ be the meet of all elements covering $x$. (If $x$ is join-irreducible, then $x_\dagger = x_*$, as introduced in Section I.6.3.) If $L$ is modular, then the intervals $[x, x^\dagger]$ and $[x_\dagger, x]$ are *complemented* modular lattices.

We begin by showing that the theorem holds for finite complemented modular lattices. This can be done independently or deduced from Corollary 426.

Next, by Möbius inversion (see Exercise 3.31), we show that the equation

$$\sum_{x \in L} \mu(0, x)|\, \mathrm{Cov}_k([x, 1])| = \begin{cases} 1, & \text{if } k \text{ equals the number of atoms;} \\ 0, & \text{otherwise,} \end{cases}$$

and its dual analog, with $|\mathrm{Cov}^k[x, 1]|$ replacing $|\mathrm{Cov}_k[x, 1]|$, hold. From these equations, we conclude that

$$\sum_{x \in L} \mu(0, x)(|\, \mathrm{Cov}_k[x, 1]| - |\, \mathrm{Cov}^k[x, 1]|) = 0.$$

We can now finish the proof by induction down the lattice $L$, starting with the interval $[1_\dagger, 1]$.

Another proof of Theorem 464, not using Möbius inversion, can be found in B. Ganter and I. Rival [218]; see Exercise 5.26.

Dilworth's theorem is one of many results saying that in a finite modular lattice, the number of elements satisfying a property and the number of elements satisfying the dual of that property are equal. A further example is the following result.

◇ **Theorem 465.** *Let $k$ be a nonnegative integer and $L$ be a finite modular lattice. Then the number of elements $x$ such that $\mathrm{height}(x^\dagger) - \mathrm{height}(x) = k$ equals the number of elements $y$ such that $\mathrm{height}(y) - \mathrm{height}(y_\dagger) = k$.*

In case $k = 1$, I. Rival conjectured a matching version of Theorem 464, namely that there exists a bijection

$$\varphi \colon \{0\} \cup \operatorname{Ji} L \to \{1\} \cup \operatorname{Mi} L$$

such that $j \leq \varphi(j)$ for all $j \in \{0\} \cup \operatorname{Ji} L$. This was proved by J. P. S. Kung [487, 488] in the following general form.

$\Diamond$ **Theorem 466.** *There are bijective functions $\varphi$ and $\psi$ from the set*

$$\bigcup(\operatorname{Cov}_i L \mid 0 \leq i \leq k)$$

*of elements covering $k$ or fewer elements to the set*

$$\bigcup(\operatorname{Cov}^i L \mid 0 \leq i \leq k)$$

*of elements covered by $k$ or fewer elements such that $x \leq \varphi(x)$ and $x \not\leq \psi(x)$ for every $x$. In particular, there is a bijection*

$$\varphi \colon \{0\} \cup \operatorname{Ji} L \to \{1\} \cup \operatorname{Mi} L$$

*such that $j \leq \varphi(j)$ for every $j \in \{0\} \cup \operatorname{Ji} L$.*

## Exercises

5.1. Find a geometry $(A, ^-)$ in which $X$ is a subspace iff $r \leq p_0 \vee p_1 \vee p_2$ and $p_0, p_1, p_2 \in X$ imply that $r \in X$, but $(A, ^-)$ is not a projective geometry.

5.2. Why do we need $x \neq y$ in the Pasch Axiom? Phrase the Pasch Axiom so that this assumption can be dropped.

5.3. For points $p$ and $q$ of a projective geometry, define $p \equiv q$ iff there is a third point $r \leq p \vee q$. Show that this relation is an equivalence relation.

5.4. Derive Corollary 426 from Exercise 5.3.

5.5. Prove by direct computation that the projective geometry associated with $\operatorname{PG}(D, \mathfrak{m})$ satisfies Desargues' Theorem.

5.6. Show that the Pasch Axiom is equivalent to the following implication holding for all points:

If $a_0 \leq a_1 \vee a_2$ and $a_1 \wedge (a_0 \vee a_2) \leq a_3 \vee a_4$, then

$$
\begin{aligned}
a_0 \leq \,& (a_3 \wedge (a_4 \vee (a_1 \wedge (a_0 \vee a_2)))) \\
& \vee ((a_0 \vee a_3) \wedge ((a_1 \vee a_0) \wedge a_2)) \\
& \vee ((a_3 \vee (a_1 \wedge (a_0 \vee a_2))) \wedge a_4).
\end{aligned}
$$

5.7. Prove that the implication of Exercise 5.6 holds for all subspaces of a projective space.

*5.8. Prove that the algebra $D$ constructed in the proof of Theorem 434 is a division ring.

5.9. Let $K$ and $L$ be modular geometric lattices and let $\varphi$ be a one-to-one map from $\text{Atom}(K)$ onto $\text{Atom}(L)$. Prove that $\varphi$ can be extended to an isomorphism iff the statement

$$a \leq b \vee c \quad \text{is equivalent to} \quad \varphi(a) \leq \varphi(b) \vee \varphi(c)$$

holds for all atoms $a, b, c \in K$.

*5.10. In the proof of Theorem 434, verify that, for atoms $r, s, t \in L$ with $r \leq s \vee t$ iff $ad_r + bd_s + cd_t = 0$ holds for some $a, b, c \in D$.

5.11. Combine Exercises 5.9 and 5.10 to prove $L \cong \text{PG}(D, \mathfrak{m})$ in Theorem 434.

5.12. Show that the identity $\overline{\varepsilon}$ is not selfdual and that the dual of $\overline{\varepsilon}$ could also be used in Theorem 442.

5.13. The projective plane constructed in Lemma 446 is the *free projective plane* generated by the partial plane. What does "free" mean for projective planes?

5.14. Investigate the lattice theoretic properties of the embedding of the subspace lattice of $(A, L)$ into the subspace lattice of $(A^+, L^+)$ in the proof of Lemma 446.

5.15. Why could Lemma 446 not be extended to provide embeddings of modular lattices of length 4?

5.16. Let $L$ be a modular lattice of length 3 or 4. Let $a$ be the join of all atoms of $L$ and $b$ the meet of all dual atoms of $L$. Consider the situations shown in Figure 101. Show that they (along with their duals) suggest a complete classification of all modular lattices of length 3 or 4 with the exception of distributive lattices and of complemented modular lattices (B. Jónsson [436]).

5.17. Develop a duality theory for projective planes. The dual of $(A, L)$ is $(L, A^{\delta})$, where $a^{\delta} = \{\, l \mid a \in l \,\}$ for all $a \in A$.

5.18. Show that $|A| = |L|$, for a finite projective plane $(A, L)$.

5.19. What is the "dual" of Desargues' Theorem?

5.20. Find a finite modular lattice with an atom $p$ and elements $a, b$ such that $p \wedge a = p \wedge b = 0$, $p \leq a \vee b$, but there are no atoms $x \leq a$ and $y \leq b$ such that $p \leq x \vee y$.

5.21. On the Euclidean plane, fix a line $l$ (see Figure 102). Let $u$ be a line and let $\alpha$ be the angle determined by $u$ and $l$. A *refracted line* is defined as follows: if $0 \leq \alpha \leq \pi/2$, then form the line $u_r$ with angle $\alpha/2$; below $l$, the refracted line is $u$, above $l$, it is $u_r$. Otherwise, $u$ is the refracted line.

Define a projective space as the Euclidean plane with a line at infinity; the lines are the refracted lines with a point at infinity and the line at

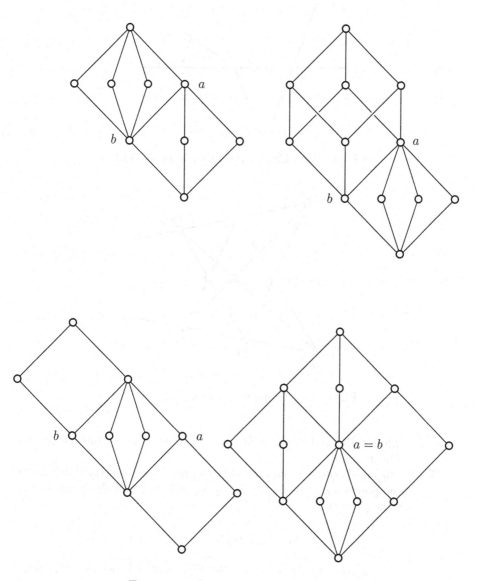

Figure 101. Lattices for Exercise 5.16

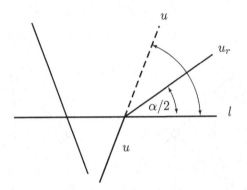

Figure 102. Configuration for Exercise 5.21

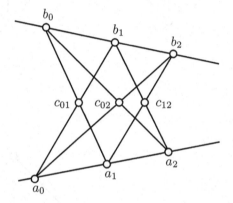

Figure 103. Configuration for Exercise 5.22

infinity. Prove that this defines a projective plane in which Desargues'
Theorem fails.

*5.22. *Pappus' Theorem* is said to hold in a projective geometry if, whenever
the points $a_0, a_1, a_2$ and $b_0, b_1, b_2$ are collinear and all six are in the
same plane, and

$$c_{ij} = (a_i \vee b_j) \wedge (a_j \vee b_i) \qquad (0 \le i < j < 3),$$

then $c_{01}, c_{02}, c_{12}$ are collinear. (See Figure 103.) Prove that if Pappus'
Theorem holds in a projective space, then the division ring constructed
in Theorem 434 is commutative.

*5.23. Prove the converse of Exercise 5.22.

*5.24. Use the Coordinatization Theorem to show that a finite projective
geometry of length 4 or more and its dual are isomorphic.

5.25. Prove that in a finite complemented modular lattice, the number of $k$-element subsets of atoms whose join is the unit is equal to the number of $k$-element subsets of dual atoms whose meet is the zero.

5.26. In a finite modular lattice, let $L_n$ (resp., $U_n$) denote the set of all $(n+1)$-element subsets $\{p, q_1, q_2, \ldots, q_n\}$ such that every $q_i$ covers $p$ (resp., $p$ covers every $q_i$). Show that $|L_n| = |U_n|$ for every positive integer $n$. (Hint: use Exercise 5.25.)

# Varieties of Lattices

## 1. Characterizations of Varieties

### 1.1 Basic definitions and results

In this section, we shall discuss the basic properties of varieties of lattices. Of the four characterizations and descriptions given, three apply to arbitrary varieties of universal algebras; the fourth is valid only for those varieties of universal algebras that are *congruence distributive* (that is, the congruence lattice of any algebra in the variety is distributive). For the sake of simplicity, all these results are stated and proved only for lattices.

For a class $\mathbf{K}$ of lattices, let $\mathrm{Iden}(\mathbf{K})$ denote the set of all identities holding in all lattices of $\mathbf{K}$; if $\mathbf{K} = \{L\}$, we write $\mathrm{Iden}(L)$ for $\mathrm{Iden}(\mathbf{K})$. (The same convention will be used for all "operators" as well; if $\mathbf{X}$ is an "operator", that is, $\mathbf{X}$ is any one of $\mathbf{H}, \mathbf{S}, \mathbf{P}, \mathbf{Var}, \mathbf{P_s}, \mathbf{P_u}, \mathbf{Si}, \mathbf{I}$ and the class $\mathbf{K} = \{L\}$, then we write $\mathbf{X}(L)$ for $\mathbf{X}(\mathbf{K})$.) For a class $\Sigma$ of identities, let $\mathbf{Mod}(\Sigma)$ denote the class of all "lattice models of $\Sigma$"—lattices in which all the identities of $\Sigma$ hold. By definition, the class $\mathbf{K}$ is a *variety of lattices* if $\mathbf{K} = \mathbf{Mod}(\Sigma)$, for some set $\Sigma$ of identities, or equivalently, if

$$\mathbf{K} = \mathbf{Mod}(\mathrm{Iden}(\mathbf{K})).$$

Let $\varepsilon\colon p = q$ be an identity where $p$ and $q$ are $n$-ary lattice terms. Let $\mathbf{K}$ be a variety and let $\mathrm{Free}_{\mathbf{K}}(n)$ be the free lattice over $\mathbf{K}$ with $n$ generators, $u_0, \dots, u_{n-1}$; then $\mathrm{Free}_{\mathbf{K}}(n)$ exists by Corollary 70. Let $\mathrm{Free}_{\mathbf{K}}(\aleph_0)$ denote the lattice over $\mathbf{K}$ with $\aleph_0$ free generators, $u_0, u_1, \dots, u_n, \dots$.

G. Grätzer, *Lattice Theory: Foundation*, DOI 10.1007/978-3-0348-0018-1_6,
© Springer Basel AG 2011

Let us suppose that $\varepsilon$ holds for the generators of $\mathrm{Free}_{\mathbf{K}}(n)$, that is,

$$p(u_0, \ldots, u_{n-1}) = q(u_0, \ldots, u_{n-1}).$$

Let $L$ be any lattice in $\mathbf{K}$ and let $a_0, \ldots, a_{n-1} \in L$. Then the map, $u_i \mapsto a_i$ for $i = 0, 1, \ldots, n-1$, can be extended to a homomorphism $\varphi$ of $\mathrm{Free}_{\mathbf{K}}(n)$ into $L$, and so

$$\begin{aligned}
p(a_0, \ldots, a_{n-1}) &= p(\varphi(u_0), \ldots, \varphi(u_{n-1})) = \varphi(p(u_0, \ldots, u_{n-1})) \\
&= \varphi(q(u_0, \ldots, u_{n-1})) = q(\varphi(u_0), \ldots, \varphi(u_{n-1})) \\
&= q(a_0, \ldots, a_{n-1}),
\end{aligned}$$

that is, $\varepsilon$ holds in $L$. In particular,

$$\varepsilon \in \mathrm{Iden}(\mathbf{K}) \quad \text{iff} \quad \varepsilon \in \mathrm{Iden}(\mathrm{Free}_{\mathbf{K}}(n)).$$

If we make no restrictions on the arity of $p$ and $q$, we obtain

$$\varepsilon \in \mathrm{Iden}(\mathbf{K}) \quad \text{iff} \quad \varepsilon \in \mathrm{Iden}(\mathrm{Free}_{\mathbf{K}}(\aleph_0)).$$

It follows that $\mathbf{K}$ is completely determined by $\mathrm{Free}_{\mathbf{K}}(\aleph_0)$.

**Theorem 467.**

(i) *There is a one-to-one correspondence between varieties of lattices and (up to isomorphism) free lattices with $\aleph_0$ generators, wherein $\mathbf{K}$ corresponds to $\mathrm{Free}_{\mathbf{K}}(\aleph_0)$.*

(ii) *A lattice $L$ is a free lattice with $\aleph_0$ generators over some variety $\mathbf{K}$ of lattices iff $L$ has a countable generating set $U$ such that every map $U \to L$ can be extended to an endomorphism of $L$.*

*Remark.* (i) sets up the correspondence between varieties and countably generated free lattices, while (ii) describes which lattices occur in this correspondence. It follows from the results of Chapter VII that the $U$ in (ii) is uniquely determined as the set of doubly irreducible elements of $L$.

*Proof.* (i) has already been proved. The "only if" part of (ii) is trivial. To prove the "if" part, let $L$ and $U$ be given as in (ii). We wish to construct a variety $\mathbf{K}$ such that $L$ is free over $\mathbf{K}$. Let $\mathbf{K}$ be the class of all lattices $A$ such that any map $U \to A$ can be extended to a homomorphism of $L$ into $A$. Let $\Sigma$ be the set of all identities $\varepsilon \colon p = q$ such that

$$p(u_0, u_1, \ldots) = q(u_0, u_1, \ldots)$$

holds in $L$ for all $u_0, u_1, \ldots \in U$.

We claim that $\mathbf{K} = \mathbf{Mod}(\Sigma)$. Indeed, arguing as at the beginning of this section, we obtain that $\mathbf{K} \subseteq \mathbf{Mod}(\Sigma)$. Conversely, let $A \in \mathbf{Mod}(\Sigma)$ and let $\alpha \colon U \to A$ be a map. If $\varepsilon \colon p = q$ is an identity and

$$p(u_0, u_1, \ldots) = q(u_0, u_1, \ldots),$$

for $u_0, u_1, \ldots \in U$ (with all $u_i$ distinct), then $\varepsilon \in \Sigma$. Since $A \in \mathbf{Mod}(\Sigma)$, we obtain that $\varepsilon$ holds in $A$ and therefore,

$$p(\alpha(u_0), \alpha(u_1), \ldots) = q(\alpha(u_0), \alpha(u_1), \ldots).$$

Thus (see Exercise I.5.45) $\alpha$ can be extended to a homomorphism. We conclude that $L \in \mathbf{K}$.

This proves that $\mathbf{K} = \mathbf{Mod}(\Sigma)$, so $\mathbf{K}$ is a variety. The lattice $L$ is free over $\mathbf{K}$, by the definition of $\mathbf{K}$.  $\square$

## 1.2  Fully invariant congruences

Let $u_0, u_1, \ldots$ be the generators of $\mathrm{Free}(\aleph_0)$ (the free lattice on $\aleph_0$ generators, see Section I.5.1) and let $v_0, v_1, \ldots$ be the generators of $\mathrm{Free}_{\mathbf{K}}(\aleph_0)$. Then the map $u_i \mapsto v_i$, for $i = 0, 1, \ldots$, extends to a homomorphism $\alpha = \alpha_{\mathbf{K}}$. Let us define the congruence $\boldsymbol{\alpha}$ as the kernel of $\alpha$. Since by Theorem 467, the variety $\mathbf{K}$ is determined by $\mathrm{Free}_{\mathbf{K}}(\aleph_0)$ and the lattice $\mathrm{Free}_{\mathbf{K}}(\aleph_0)$ is determined by $\boldsymbol{\alpha}$, we conclude that $\mathbf{K}$ is determined by $\boldsymbol{\alpha}$. If we can ascertain which congruences arise this way, we shall have another description of varieties.

Call a congruence relation $\boldsymbol{\beta}$ of a lattice $L$ *fully invariant* if $a \equiv b \pmod{\boldsymbol{\beta}}$ implies that $\gamma(a) \equiv \gamma(b) \pmod{\boldsymbol{\beta}}$ for all $a, b \in L$ and for all endomorphisms $\gamma$ of $L$.

B. H. Neumann [549] connects fully invariant congruence relations and varieties.

**Theorem 468.** *There is a one-to-one correspondence between varieties of lattices and fully invariant congruence relations of* $\mathrm{Free}(\aleph_0)$.

*Proof.* Let $\mathbf{K}$ be a variety and let the map $\alpha$ $(= \alpha_{\mathbf{K}})$, the congruence $\boldsymbol{\alpha}$ $(= \boldsymbol{\alpha}_{\mathbf{K}})$, and the elements $u_i, v_i$ be as described above. We show that $\boldsymbol{\alpha}$ is fully invariant. Let $\gamma$ be an endomorphism of $\mathrm{Free}(\aleph_0)$, let $a, b \in \mathrm{Free}(\aleph_0)$, and let $a \equiv b \pmod{\boldsymbol{\alpha}}$.

Let $\beta$ be the endomorphism of $\mathrm{Free}_{\mathbf{K}}(\aleph_0)$ extending the map $v_i \mapsto \alpha\gamma(u_i)$ for $i = 0, 1, \ldots$. Since $a \in \mathrm{Free}(\aleph_0)$, it follows that $a = p(u_0, \ldots, u_{n-1})$ for some integer $n$ and for some $n$-ary term $p$. We compute:

$$
\begin{aligned}
\beta\alpha(a) &= \beta\alpha(p(u_0, \ldots, u_{n-1})) = p(\beta\alpha(u_0), \ldots, \beta\alpha(u_{n-1})) \\
&= p(\beta(v_0), \ldots, \beta(v_{n-1})) = p(\alpha\gamma(u_0), \ldots, \alpha\gamma(u_{n-1})) \\
&= \alpha\gamma(p(u_0, \ldots, u_{n-1})) = \alpha\gamma(a),
\end{aligned}
$$

and similarly for $b$. Thus

$$\alpha\gamma(a) = \beta\alpha(a) = \beta\alpha(b) = \alpha\gamma(b),$$

and therefore, $\gamma(a) \equiv \gamma(b) \pmod{\boldsymbol{\alpha}}$.

Now let $\boldsymbol{\alpha}$ be a fully invariant congruence relation of $\mathrm{Free}(\aleph_0)$. We shall show that $\mathrm{Free}(\aleph_0)/\boldsymbol{\alpha}$ satisfies the condition of Theorem 467(ii). If

$$u_i \equiv u_j \pmod{\boldsymbol{\alpha}},$$

for some $i \neq j$, then it is easily shown that $\boldsymbol{\alpha} = \mathbf{1}$ and so $\mathrm{Free}(\aleph_0)/\boldsymbol{\alpha}$ is the one-element lattice. So let us assume that

$$u_i \not\equiv u_j \pmod{\boldsymbol{\alpha}},$$

for all $i \neq j$. We set

$$U = \{\, u_i/\boldsymbol{\alpha} \mid i = 0, 1, \dots \,\} \subseteq \mathrm{Free}(\aleph_0)/\boldsymbol{\alpha},$$

and consider a map $\gamma\colon U \to \mathrm{Free}(\aleph_0)/\boldsymbol{\alpha}$. In $\mathrm{Free}(\aleph_0)$, choose the elements $a_0, a_1, \dots$ so that

$$\gamma(u_i/\boldsymbol{\alpha}) = a_i/\boldsymbol{\alpha}, \quad i = 0, 1, \dots.$$

Then the map $u_i \mapsto a_i$, for $i = 0, 1, \dots$, can be extended to an endomorphism $\delta$ of $\mathrm{Free}(\aleph_0)$. Since $\boldsymbol{\alpha} = \mathrm{Ker}(\alpha)$ is fully invariant, it follows that $\mathrm{Ker}(\gamma) \subseteq \mathrm{Ker}(\delta\alpha)$ and so $\alpha(x) \mapsto \delta\alpha(x)$ extends $\gamma$ to an endomorphism of $\mathrm{Free}(\aleph_0)/\boldsymbol{\alpha}$. $\square$

## 1.3   Formulas for Var(K)

In Section I.4.2, we have already proved that a variety is closed under the formation of homomorphic images, sublattices, and direct products. The converse, which is due to G. Birkhoff [62], is the third description of varieties.

**Theorem 469.** *A class* **K** *of lattices is a variety iff* **K** *is closed under the formation of homomorphic images, sublattices, and direct products.*

*Remark.* The direct product of an empty family of lattices is the one-element lattice and, therefore, if **K** is closed under the formation of direct products, then **K** is not the empty class. Observe also, that if **K** is closed under the formation of homomorphic images, then **K** is closed under the formation of isomorphic copies.

*Proof.* Let **K** be closed under the formation of homomorphic images, sublattices, and direct products. If **K** consists of one-element lattices only, then **K** can be defined by the identity: $x_0 = x_1$ and so **K** is equational. Now we can assume that **K** contains a lattice of more than one element. Therefore, we conclude, just as in Section I.5.2, that $\mathrm{Free}_{\mathbf{K}}(\mathfrak{m})$ exists for any cardinal $\mathfrak{m}$.

Let $L \in \mathbf{Mod}(\text{Iden}(\mathbf{K}))$, let $\mathfrak{m}$ be a cardinal with $|L| + \aleph_0 = \mathfrak{m}$, and take a free lattice $\text{Free}_{\mathbf{K}}(\mathfrak{m})$ in $\mathbf{K}$. We denote by $U$ the set of free generators of $\text{Free}_{\mathbf{K}}(\mathfrak{m})$. Since $|L| + \aleph_0 = |U|$, there is a map $\alpha$ of $U$ onto $L$. Let

$$p(u_0, u_1, \ldots, u_{n-1}) = q(u_0, u_1, \ldots, u_{n-1})$$

hold in $\text{Free}_{\mathbf{K}}(\mathfrak{m})$ with $u_0, u_1, \ldots, u_{n-1} \in U$. Without loss of generality, we can assume that the $u_i$ are all distinct. Therefore, $p = q \in \text{Iden}(\text{Free}_{\mathbf{K}}(\mathfrak{m}))$ (as argued at the beginning of this section) and so $p = q \in \text{Iden}(\mathbf{K})$, by the freeness of $\text{Free}_{\mathbf{K}}(\mathfrak{m})$. Because $L \in \mathbf{Mod}(\text{Iden}(\mathbf{K}))$, it follows that $p = q \in \text{Iden}(L)$; in particular,

$$p(\alpha(u_0), \alpha(u_1), \ldots, \alpha(u_{n-1})) = q(\alpha(u_0), \alpha(u_1), \ldots, \alpha(u_{n-1})).$$

This shows that $\alpha$ satisfies the hypothesis of Exercise I.5.45 and can, therefore, be extended to a homomorphism $\beta$ of $\text{Free}_{\mathbf{K}}(\mathfrak{m})$ onto $L$. Thus $L$ is a homomorphic image of a member of $\mathbf{K}$, and so $L \in \mathbf{K}$.

The converse was proved in Lemma 59. □

To obtain a slightly different version of this result, see [A. Tarski [674]], we introduce some notation. For a class $\mathbf{K}$ of lattices, let $\mathbf{I}(\mathbf{K})$, $\mathbf{H}(\mathbf{K})$, $\mathbf{S}(\mathbf{K})$, and $\mathbf{P}(\mathbf{K})$ denote the class of all isomorphic copies, homomorphic images, sublattices, and direct products of members of $\mathbf{K}$, respectively. Note that $\mathbf{IH} = \mathbf{HI} = \mathbf{H}$, by definition.

**Corollary 470.** *Let $\mathbf{K}$ be a class of lattices. Then $\mathbf{HSP}(\mathbf{K})$ is the smallest variety containing $\mathbf{K}$.*

*Proof.* We start out by observing three formulas for any class $\mathbf{K}$ of lattices:

   (i) $\mathbf{SH}(\mathbf{K}) \subseteq \mathbf{HS}(\mathbf{K})$;
  (ii) $\mathbf{PH}(\mathbf{K}) \subseteq \mathbf{HP}(\mathbf{K})$;
 (iii) $\mathbf{PS}(\mathbf{K}) \subseteq \mathbf{SP}(\mathbf{K})$.

To prove (i), let $L \in \mathbf{SH}(\mathbf{K})$. Then there is an $A \in \mathbf{K}$ and a homomorphism $\alpha$ of $A$ onto a lattice $B$ containing $L$ as a sublattice. Set

$$C = \{\, x \in A \mid \alpha(x) \in L \,\}.$$

Then $C \in \mathbf{S}(\mathbf{K})$ and $L \in \mathbf{H}(C)$. Hence $L \in \mathbf{HS}(\mathbf{K})$, proving (i).

To prove (ii), let $L \in \mathbf{PH}(\mathbf{K})$. Then there exist lattices $A_i \in \mathbf{K}$ and homomorphisms $\alpha_i$ of $A_i$ onto $B_i$, for all $i \in I$, such that

$$L = \prod(\, B_i \mid i \in I \,).$$

It is clear that $L$ is a homomorphic image of

$$\prod(\, A_i \mid i \in I \,) \in \mathbf{P}(\mathbf{K}),$$

proving (ii).

The proof of (iii) follows that of (ii) with $A_i \leq B_i$ for all $i \in I$.

To show that $\mathbf{HSP}(\mathbf{K}) = \mathbf{K}_1$ is a variety using Theorem 469, we have to verify three formulas:

$$\mathbf{H}(\mathbf{K}_1) \subseteq \mathbf{K}_1,$$
$$\mathbf{S}(\mathbf{K}_1) \subseteq \mathbf{K}_1,$$
$$\mathbf{P}(\mathbf{K}_1) \subseteq \mathbf{K}_1.$$

Indeed (using (i)–(iii) and that $\mathbf{XX}(\mathbf{K}_2) = \mathbf{X}(\mathbf{K}_2)$ for all $\mathbf{X} \in \{\mathbf{H}, \mathbf{S}, \mathbf{P}\}$ and any class $\mathbf{K}_2$):

$$\mathbf{H}(\mathbf{K}_1) = \mathbf{HHSP}(\mathbf{K}) = \mathbf{HSP}(\mathbf{K}) = \mathbf{K}_1,$$
$$\mathbf{S}(\mathbf{K}_1) = \mathbf{SHSP}(\mathbf{K}) \subseteq \mathbf{HSSP}(\mathbf{K}) = \mathbf{HSP}(\mathbf{K}) = \mathbf{K}_1,$$
$$\mathbf{P}(\mathbf{K}_1) = \mathbf{PHSP}(\mathbf{K}) \subseteq \mathbf{HPSP}(\mathbf{K}) \subseteq \mathbf{HSPP}(\mathbf{K}) = \mathbf{HSP}(\mathbf{K}) = \mathbf{K}_1.$$

If $\mathbf{K}_2$ is any variety containing $\mathbf{K}$, then

$$\mathbf{K}_2 = \mathbf{HSP}(\mathbf{K}_2) \supseteq \mathbf{HSP}(\mathbf{K}) = \mathbf{K}_1,$$

so indeed $\mathbf{HSP}(\mathbf{K})$ is the smallest variety containing $\mathbf{K}$. $\qquad\square$

Let $\mathbf{Var}(\mathbf{K})$ denote the smallest variety containing $\mathbf{K}$. Then Corollary 470 can be summarized in the formula

$$\mathbf{Var} = \mathbf{HSP}.$$

For a class $\mathbf{K}$ of lattices, let $\mathbf{P_s}(\mathbf{K})$ denote the class of all lattices that are isomorphic to a subdirect product of members of $\mathbf{K}$. The following variant of Corollary 470—S. R. Kogalovskiĭ [476]—is not especially useful, but the construction used in the proof has found some applications.

**Corollary 471.** *A class $\mathbf{K}$ of lattices is a variety iff $\mathbf{K}$ is closed under the formation of homomorphic images and subdirect products. Moreover, for every class $\mathbf{K}$ of lattices,*
$$\mathbf{Var}(\mathbf{K}) = \mathbf{HP_s}(\mathbf{K}).$$

*Remark.* In other words, the equality

$$\mathbf{Var} = \mathbf{HP_s}$$

holds.

*Proof.* We leave it to the reader to verify that both statements follow readily from the following inequality:

(iv) $\mathbf{S}(\mathbf{K}) \subseteq \mathbf{HP_s}(\mathbf{K}).$

To verify (iv), let $L \in \mathbf{S(K)}$, that is, let $L$ be a sublattice of some $A \in \mathbf{K}$. Now take $A^I$ with $|I| = \aleph_0$ and form a sublattice $B \subseteq A^I$ defined as follows: for $f \in A^I$, let $f \in B$ if $f(i) = a$, for some $a \in L$ and for all but finitely many $i \in I$; a map $\varphi \colon B \to L$ is defined by mapping $f$ to this $a$. It is trivial that $B \in \mathbf{P_s}(A)$ and $L \in \mathbf{H}(B)$, concluding that $L \in \mathbf{HP_s(K)}$.    $\square$

## 1.4   Jónsson's Lemma

We have obtained two formulas for **Var**, namely,

$$\mathbf{Var} = \mathbf{HSP} = \mathbf{HP_s}.$$

Their usefulness is, however, somewhat limited when applied to describing members of varieties of lattices. A great improvement of these formulas is possible for lattices and we shall proceed to develop it.

Let $L_i$, for $i \in I$, be lattices, let $I \neq \varnothing$, and let $\mathcal{D}$ be a filter of the lattice Pow $I$ of all subsets of $I$. For $f, g \in \prod(L_i \mid i \in I)$, we set

$$\mathrm{Equ}(f, g) = \{\, i \in I \mid f(i) = g(i)\,\}.$$

(Equ in $\mathrm{Equ}(f, g)$ stands for "equal".)

We introduce a congruence relation $\boldsymbol{\alpha} = \boldsymbol{\alpha}(\mathcal{D})$ on $L = \prod(L_i \mid i \in I)$. For $f, g \in L$, let

$$f \equiv g \pmod{\boldsymbol{\alpha}} \quad \text{iff} \quad \mathrm{Equ}(f, g) \in \mathcal{D}.$$

(View members of $\mathcal{D}$ as "sets of measure 1". Then $f \equiv g \pmod{\boldsymbol{\alpha}}$ iff $f$ and $g$ are equal on a set of measure 1.) The relation $\boldsymbol{\alpha}$ is obviously reflexive and symmetric. Let $f, g, h \in L$ satisfy

$$f \equiv g \pmod{\boldsymbol{\alpha}},$$
$$g \equiv h \pmod{\boldsymbol{\alpha}}.$$

Then

$$\mathrm{Equ}(f, g), \mathrm{Equ}(g, h) \in \mathcal{D},$$

therefore,

$$\mathrm{Equ}(f, h) \supseteq \mathrm{Equ}(f, g) \cap \mathrm{Equ}(g, h) \in \mathcal{D}.$$

Thus $\mathrm{Equ}(f, h) \in \mathcal{D}$ and we conclude that $f \equiv h \pmod{\boldsymbol{\alpha}}$. If $f, g, h \in L$ and $f \equiv g \pmod{\boldsymbol{\alpha}}$, then

$$\mathrm{Equ}(f \vee h, g \vee h) \supseteq \mathrm{Equ}(f, g) \in \mathcal{D}.$$

It follows that

$$\mathrm{Equ}(f \vee h, g \vee h) \in \mathcal{D},$$

and so
$$f \vee h \equiv g \vee h \pmod{\alpha}.$$
Similarly,
$$f \wedge h \equiv g \wedge h \pmod{\alpha}.$$
Thus $\alpha$ is a congruence relation and we can form the lattice $L/\alpha$, denoted by
$$\prod_{\mathcal{D}}(L_i \mid i \in I),$$
called a *reduced product* of the family $(L_i \mid i \in I)$ of lattices. When $\mathcal{D}$ is a prime filter, then $\prod_{\mathcal{D}}(L_i \mid i \in I)$ is called an *ultraproduct* (also called, *prime product*), see J. Łoś [504]. For a class **K** of lattices, $\mathbf{P_u(K)}$ will denote the class of all lattices that are isomorphic to an ultraproduct of members of **K**.

We start out by proving some elementary properties of these constructions:

**Lemma 472.** *Let $I$ be a nonempty set and let $\mathcal{D}$ be a filter of* Pow $I$. *Let $J \in \mathcal{D}$ with $J \neq \varnothing$ and define $\mathcal{E} = \{X \cap J \mid X \in \mathcal{D}\}$. Then $\mathcal{E}$ is a filter of* Pow $J$ *and if $\mathcal{D}$ is prime, so is $\mathcal{E}$. Furthermore, for every family $(L_i \mid i \in I)$ of lattices,*
$$\prod_{\mathcal{D}}(L_i \mid i \in I) \cong \prod_{\mathcal{E}}(L_i \mid i \in J).$$

*Proof.* The statements concerning $\mathcal{E}$ are trivial. Let $\pi$ denote the homomorphism $f \mapsto f_J$, the restriction of $f$ to $J$. Let
$$\varphi \colon \prod(L_i \mid i \in I) \to \prod_{\mathcal{D}}(L_i \mid i \in I),$$
$$\psi \colon \prod(L_i \mid i \in J) \to \prod_{\mathcal{E}}(L_i \mid i \in J)$$
be the natural homomorphisms and let $\alpha = \alpha(\mathcal{D})$ be the kernel of $\varphi$. Let $\beta$ be the kernel of $\psi\pi$. Then $f \equiv g \pmod{\beta}$ iff $\mathrm{Equ}(f_J, g_J) \in \mathcal{E}$, or equivalently, if $J \cap \mathrm{Equ}(f, g) \in \mathcal{D}$. Since $J \in \mathcal{D}$, this is equivalent to $\mathrm{Equ}(f, g) \in \mathcal{D}$. Thus $\alpha = \beta$ and the isomorphism follows. $\square$

**Corollary 473.** *Let $I$ be a nonempty set and let $\mathcal{D}$ be a filter of* Pow $I$. *If $\mathcal{D}$ is principal, $\mathcal{D} = \mathrm{fil}(J)$, then for any family $(L_i \mid i \in I)$ of lattices,*
$$\prod_{\mathcal{D}}(L_i \mid i \in I) \cong \prod(L_i \mid i \in J);$$
*in particular, if $\mathcal{D}$ is principal and prime, $\mathcal{D} = \mathrm{fil}(\{j\})$, then*
$$\prod_{\mathcal{D}}(L_i \mid i \in I) \cong L_j.$$

**Lemma 474.** *Let $L_0, \ldots, L_{n-1}$ be finite lattices. Let $I \neq \varnothing$ and let $(L_i \mid i \in I)$ be a family of lattices with $L_i \in \{L_0, \ldots, L_{n-1}\}$ for all $i \in I$. Let $\mathcal{D}$ be prime over $I$. Then there is a $j$ with $0 \leq j < n$, such that*
$$\prod_{\mathcal{D}}(L_i \mid i \in I) \cong L_j.$$

*Proof.* We can assume that the lattices $L_0, \ldots, L_{n-1}$ are pairwise nonisomorphic. Thus if we define

$$I_j = \{\, i \in I \mid L_i = L_j \,\},$$

for $0 \le j < n$, then $I_0, \ldots, I_{n-1}$ are pairwise disjoint and $I_0 \cup \cdots \cup I_{n-1} = I$. Since $\mathcal{D}$ is prime, there is exactly one $j$ with $0 \le j < n$, such that $I_j \in \mathcal{D}$. Applying Corollary 473 to $I_j$, we obtain that

$$\prod_{\mathcal{D}}(L_i \mid i \in I) \cong \prod_{\mathcal{E}}(L_i \mid i \in I_j),$$

where $\mathcal{E} = \{\, X \cap I_j \mid X \in \mathcal{D} \,\}$ is prime over $I_j$. For an element $a$ in $L_j$, let

$$f_a \in \prod(L_i \mid i \in I_j)$$

be defined by $f_a(i) = a$ for all $i \in I$. Then $\mathrm{Equ}(f_a, f_b) = \varnothing \notin \mathcal{E}$, for all $a, b \in L_j$ with $a \ne b$, and therefore, $f_a \not\equiv f_b \pmod{\boldsymbol{\alpha}(\mathcal{E})}$. Thus

$$\alpha \colon a \mapsto f_a/\boldsymbol{\alpha}(\mathcal{E})$$

embeds $L_j$ in $\prod_{\mathcal{E}}(L_i \mid i \in I_j)$. In fact, $\alpha$ is an isomorphism. To verify this, it is sufficient to prove that $\alpha$ is onto. So let $f \in \prod(L_i \mid i \in I_j)$. Then, for each $a \in L$, we define

$$I_{j,a} = \{\, i \in I_j \mid f(i) = a \,\}.$$

Since the $I_{j,a}$ are pairwise disjoint and $\bigcup(I_{j,a} \mid a \in L_j) = I_j$, we conclude that there exists exactly one $a \in L_j$ such that $I_{j,a} \in \mathcal{E}$. Now $\mathrm{Equ}(f, f_a) = I_{j,a} \in \mathcal{E}$, thus $f \equiv f_a \pmod{\boldsymbol{\alpha}(\mathcal{E})}$ and $\alpha$ is onto. Therefore,

$$\prod_{\mathcal{E}}(L_i \mid i \in I_j) \cong L_j. \qquad \square$$

Now we are ready to state the formula of B. Jónsson [444] for **Var**:

**Theorem 475 (Jónsson's Lemma).** *For a class* **K** *of lattices,*

$$\mathbf{Var}(\mathbf{K}) = \mathbf{P_s}\mathbf{H}\mathbf{S}\mathbf{P_u}(\mathbf{K}).$$

*Proof.* By Theorem 221 (Birkhoff's Subdirect Product Representation Theorem), it is sufficient to prove that if $A$ is a subdirectly irreducible lattice in the variety $\mathbf{Var}(\mathbf{K})$, then the lattice $A$ is in $\mathbf{HSP_u}(\mathbf{K})$. By Corollary 470, the lattice $A$ is in $\mathbf{HSP}(\mathbf{K})$, and therefore, there exist $A_i \in \mathbf{K}$, for $i \in I$, a sublattice $B$ of $\prod(A_i \mid i \in I)$, and a congruence relation $\beta$ on $B$ such that $B/\beta \cong A$.

We claim that there is a filter $\mathcal{D}$ prime over $I$ such that $\boldsymbol{\alpha}(\mathcal{D})$ restricted to $B$ is contained in $\beta$. Indeed, if we have such a prime filter $\mathcal{D}$, then by the

Second Isomorphism Theorem (Lemma 220), $B/\beta$ is a homomorphic image of $B/\alpha(\mathcal{D})]B$, which is, in turn, a sublattice of

$$\prod( A_i \mid i \in I )/\alpha(\mathcal{D}) = \prod\nolimits_{\mathcal{D}}(A_i \mid i \in I).$$

Thus $A \cong B/\beta \in \mathbf{HSP_u(K)}$, as required.

For $J \subseteq I$, set $\alpha_J = \alpha(\mathrm{fil}(J))$, that is, for all $f, g \in \prod( A_i \mid i \in I )$,

$$f \equiv g \pmod{\alpha_J} \quad \text{iff} \quad \mathrm{Equ}(f, g) \supseteq J.$$

Observe that $\alpha(\mathcal{D}) = \bigcup(\alpha_J \mid J \in \mathcal{D})$; consequently, to find $\mathcal{D}$, we have to look for it in

$$\mathcal{E} = \{ J \mid J \subseteq I \text{ and } \alpha_J]B \le \beta \}.$$

This set $\mathcal{E}$ has the following three properties:

(i) $I \in \mathcal{E}$ and $\varnothing \notin \mathcal{E}$;
(ii) $J_0 \in \mathcal{E}$ and $J_0 \subseteq J_1 \subseteq I$ imply that $J_1 \in \mathcal{E}$;
(iii) $M, N \subseteq I$ and $M \cup N \in \mathcal{E}$ imply that $M \in \mathcal{E}$ or $N \in \mathcal{E}$.

Obviously, $\alpha_I = 0$ and $\alpha_\varnothing = 1$, so (i) is trivial. If $J_0 \subseteq J_1$, then $\alpha_{J_0} \ge \alpha_{J_1}$, proving (ii). To prove (iii), let $M$ and $N \subseteq I$. It is trivial that

$$\alpha_{M \cup N} = \alpha_M \wedge \alpha_N,$$
$$(\alpha_{M \cup N})]B = (\alpha_M)]B \wedge (\alpha_N)]B.$$

Now let $M \cup N \in \mathcal{E}$. Then $(\alpha_{M \cup N})]B \le \beta$, that is,

$$(\alpha_M)]B \wedge (\alpha_N)]B \le \beta.$$

Since $B/\beta$ is subdirectly irreducible, the congruence $\beta$ is meet-irreducible in $\mathrm{Con}\, B$. So using that $\mathrm{Con}\, B$ is distributive, we conclude that $(\alpha_M)]B \le \beta$ or $(\alpha_N)]B \le \beta$, that is, $M$ or $N \in \mathcal{E}$, proving (iii).

Now let $\mathcal{D}$ be a filter of $\mathrm{Pow}\, I$ maximal with respect to the property $\mathcal{D} \subseteq \mathcal{E}$. We show that $\mathcal{D}$ is prime. By (i), $\varnothing \notin \mathcal{D}$, so $\mathcal{D}$ is proper. If $\mathcal{D}$ is proper but not prime, then there exists a $J \subseteq I$ such that $J \notin \mathcal{D}$ and $I - J \notin \mathcal{D}$. If $J \cap J' \in \mathcal{E}$, for every $J' \in \mathcal{D}$, then by (ii), the filter $\mathcal{D}$ and the set $J$ would generate a filter contained in $\mathcal{E}$, contradicting $J \notin \mathcal{D}$ and the maximality of $\mathcal{D}$. Thus there exists a $J_0 \in \mathcal{D}$ such that $J \cap J_0 \notin \mathcal{E}$. Similarly, there exists a $J_1 \in \mathcal{D}$ such that $(I - J) \cap J_1 \notin \mathcal{E}$. Then

$$J_0 \cap J_1 = (J \cap (J_0 \cap J_1)) \cup ((I - J) \cap (J_0 \cap J_1)),$$

contradicting (iii). Thus $\mathcal{D}$ is prime.    $\square$

Let $\mathbf{Si(K)}$ be the class of subdirectly irreducible lattices in $\mathbf{K}$. An equivalent form of Theorem 475 is the following:

**Corollary 476.** *For a class* **K** *of lattices,*

$$\mathbf{Si}\,\mathbf{Var}(\mathbf{K}) \subseteq \mathbf{HSP_u}(\mathbf{K}).$$

We shall illustrate the power of Theorem 475 with two simple applications from B. Jónsson [444].

If **K** is a finite set of finite lattices, then, by Lemma 474, $\mathbf{P_u}(\mathbf{K})$ is, up to isomorphic copies, the same as **K**, so we conclude:

**Corollary 477.** *Let* **K** *be a finite set of finite lattices. Then*

$$\mathbf{Si}\,\mathbf{Var}(\mathbf{K}) \subseteq \mathbf{HS}(\mathbf{K}).$$

Observe that $\mathbf{HS}(\mathbf{K})$ is, up to isomorphic copies, a finite set of finite lattices.

**Corollary 478.** *Let A and B be finite nonisomorphic subdirectly irreducible lattices. If* $|A| \leq |B|$, *then there exists an identity* $\varepsilon$ *holding in A but not holding in B.*

*Proof.* Note that $B \notin \mathbf{HS}(A)$ since $|B| \geq |A|$ and $B$ is not isomorphic to $A$. Hence, by Corollary 477, $B \notin \mathbf{Var}(A)$ and so some identity holding in $A$ must fail in $B$. $\qquad\square$

### Exercises

1.1. Show that $\mathbf{K} \mapsto \mathrm{Iden}(\mathbf{K})$ and $\Sigma \mapsto \mathbf{Mod}(\Sigma)$ set up a Galois connection.

1.2. Prove that **K** is a variety iff $\mathbf{K} = \mathbf{Mod}(\mathrm{Iden}(\mathbf{K}))$. For a set $\Sigma$ of identities, $\Sigma = \mathrm{Iden}(\mathbf{K})$, for some class **K** of lattices, iff $\Sigma = \mathrm{Iden}(\mathbf{Mod}(\Sigma))$.

1.3. Characterize sets of identities $\Sigma$ that are of the form $\mathrm{Iden}(\mathbf{K})$ for some class **K** of lattices.

1.4. Find properties of varieties **K** of lattices satisfying

$$\mathrm{Iden}(\mathbf{K}) = \mathrm{Iden}(\mathrm{Free}_{\mathbf{K}}(n))$$

for some integer $n$.

1.5. Let $L$ be a lattice and let $U$ be a generating set of $L$. Show that if $L$ is free over **K** with $U$ as a free generating set, then the same holds over $\mathbf{Var}(\mathbf{K})$.

1.6. Reprove Theorem 467(ii) using Exercise 1.5 and Theorem 469.

1.7. Prove that the elements of a free generating set are doubly irreducible (B. Jónsson [446]).

1.8. Let **K** be a variety of lattices. Let $\varphi$ be the natural homomorphism of Free($\aleph_0$) onto Free$_{\mathbf{K}}$($\aleph_0$) and let $\alpha$ be the kernel of $\varphi$. We assign to $\alpha$ a variety $\mathbf{K}_1$ over which Free($\aleph_0$)/$\alpha$ is free. Prove that $\mathbf{K} = \mathbf{K}_1$.

1.9. Let $\alpha$ be a fully invariant congruence relation of Free($\aleph_0$). We assign to $\alpha$ a variety **K** and to **K** a fully invariant congruence relation $\alpha_1$ of Free($\aleph_0$) as in the proof of Theorem 468. Prove that $\alpha = \alpha_1$.

1.10. Let $L$ be a lattice. Show that the fully invariant congruence relations of $L$ form a complete sublattice of Con $L$.

1.11. Prove that the lattice of fully invariant congruence relations of a lattice $L$ is a distributive algebraic lattice. Characterize the compact elements.

1.12. A congruence relation $\alpha$ of a lattice $L$ is called *invariant* if $a \equiv b$ (mod $\alpha$) implies that $\varphi(a) \equiv \varphi(b)$ (mod $\alpha$) for every automorphism $\varphi$ of $L$. Do the invariant congruences form a lattice; if so, is this lattice distributive or algebraic?

1.13. Let $(A; F)$ and $(A; G)$ be algebras and $F \subseteq G$. Show that the congruence lattice of $(A; G)$ is a complete sublattice of the congruence lattice of $(A; F)$.

1.14. Derive Exercises 1.10–1.12 from Exercise 1.13.

1.15. A variety **K** of lattices is *generated by a lattice A* if $\mathbf{K} = \mathbf{Var}(A)$. Show that every variety of lattices is generated by a lattice.

1.16. A variety **K** of lattices is *locally finite* if every finitely generated member of **K** is finite. Prove that a variety generated by a finite lattice is locally finite. (Do not use Theorem 475 or any of its consequences. This result is true even in classes of algebras for which Theorem 475 fails.)

1.17. Is the converse of the statement of Exercise 1.16 true?

1.18. Prove that $\mathbf{P_sH(K)} \subseteq \mathbf{HP_s(K)}$ for every class **K** of lattices.

1.19. For a class **K** of lattices, let $\mathbf{P_r(K)}$ denote the class of reduced products (up to isomorphism) of members of **K**. Prove that

$$\mathbf{P_r(K)} \subseteq \mathbf{P_sP_u(K)}$$

(G. Grätzer and H. Lakser [301]).

*       *       *

An *implication* (or *quasi-identity*) is a sentence (all variables are universally quantified):

$$p_0(x_0, \ldots, x_{n-1}) = q_0(x_0, \ldots, x_{n-1})$$
$$\text{and } p_{m-1}(x_0, \ldots, x_{n-1}) = q_{m-1}(x_0, \ldots, x_{n-1})$$
$$\text{imply that}$$
$$p_m(x_0, \ldots, x_{n-1}) = q_m(x_0, \ldots, x_{n-1}).$$

The form of modularity, as stated in Lemma 63, is an implication. The meet-semidistributive law of Chapter VII is an implication:

$$x \wedge y = x \wedge z \text{ implies that } x \wedge y = x \wedge (y \vee z).$$

Also, any identity is an implication.

An *implicational class* (or *quasivariety*) is the class **K** of all lattices satisfying a set of implications. For a class **K**, the class of all isomorphic copies of members of **K** is denoted **I(K)**.

1.20. For a class **K** of lattices (algebras), prove that

$$\mathbf{ISP_r(K)}$$

is the smallest implicational class containing **K** (A. I. Mal'cev [525] and G. Grätzer and H. Lakser [301]).

1.21. Let $\mathcal{D}$ be a prime filter over $I$. An *ultrapower* $L_\mathcal{D}^I$ of a lattice $L$ is an ultraproduct $\prod_\mathcal{D}(L_i \mid i \in I)$ such that $L_i = L$ for all $i \in I$. Prove that $L$ can be embedded into $L_\mathcal{D}^I$.

1.22. Prove that every lattice $L$ can be embedded into an ultraproduct of all finitely generated sublattices of $L$. (Hint: let $I$ be the set of all nonempty finite subsets of $L$. For $J \in I$, let $L_J$ be the sublattice generated by $J$. Choose a $\mathcal{D}$ prime over $I$ such that

$$\{ K \in I \mid K \supseteq J \} \in \mathcal{D}$$

for all $J \in I$.)

$$* \quad * \quad *$$

Let us define (first-order) formulas:

(i) $p = q$ is a formula for the $n$-ary terms $p$ and $q$;

(ii) if $\Phi$ is a formula, so is $\neg\Phi$ (read: not $\Phi$);

(iii) if $\Phi_0$ and $\Phi_1$ are formulas, then so is $\Phi_0 \vee \Phi_1$ (read: $\Phi_0$ or $\Phi_1$);

(iv) if $\Phi$ is a formula, so is $(\exists x_k)\Phi$ (read: there exists an $x_k$ such that $\Phi$).

Then the formula $\neg((\neg\Phi_0) \vee (\neg\Phi_1))$ is interpreted as $\Phi_0 \wedge \Phi_1$ (read: $\Phi_0$ and $\Phi_1$), the formula $\neg((\exists x_k)\neg\Phi)$ is interpreted as $(x_k)\Phi$ (or $(\forall x_k)\Phi$, read: for all $x_k$, $\Phi$), and so on.

1.23. Define inductively what it means for a formula $\Phi$ to be satisfied by certain elements of a lattice $L$.

1.24. Define precisely a *free variable* $x$ which is not "bound" by the quantifier $\exists x$. A *sentence* is a formula without free variables. Find sentences expressing that the lattice $L$ has at most or exactly $n$ elements.

1.25. For a finite lattice $K$, construct a sentence $\Phi$ such that $\Phi$ holds for a lattice $L$ iff $L$ has a sublattice isomorphic to $K$.

*1.26. Prove that a sentence $\Phi$ holds for an ultraproduct $\prod_{\mathcal{D}}(L_i \mid i \in I)$ iff $\{\, i \mid \Phi$ holds in $L_i \,\} \in \mathcal{D}$ (J. Łoś [504]).

1.27. Prove Lemma 474 using Exercise 1.26.

1.28. A *model* of a set $\Sigma$ of sentences is a lattice $L$ in which all sentences $\Phi \in \Sigma$ are satisfied. Prove the *Compactness Theorem*: Let $\Sigma$ be a set of sentences. If every finite subset of $\Sigma$ has a model, then $\Sigma$ has a model. (Hint: for every nonempty finite $\Omega \subseteq \Sigma$, choose a model $L_\Omega$. Let $I$ be the set of all finite nonempty subsets of $\Sigma$ and choose a $\mathcal{D}$ prime over $I$ containing all sets of the form $\{\, X \in I \mid X \supseteq J \,\}$, where $J \in I$. Then $L = \prod_{\mathcal{D}}(L_\Omega \mid \Omega \in I)$ is a model of $\Sigma$.)

1.29. Let $\Sigma$ and $\Sigma_1$ be sets of sentences. Then $\Sigma$ *implies* $\Sigma_1$ iff every model of $\Sigma$ is also a model of $\Sigma_1$. Also, $\Sigma$ *is equivalent to* $\Sigma_1$ iff $\Sigma$ implies $\Sigma_1$ and $\Sigma_1$ implies $\Sigma$. Prove that if $\Sigma$ is equivalent to $\{\Phi\}$, then there is a finite $\Sigma_1 \subseteq \Sigma$ that is equivalent to $\{\Phi\}$. (Use the Compactness Theorem.)

$$* \qquad * \qquad *$$

1.30. Call a lattice $L$ *finitely subdirectly irreducible* if $\omega$ is meet-irreducible in $\operatorname{Con} L$. Prove that Corollary 476 holds also for the finitely subdirectly irreducible members of $\mathbf{Var}(\mathbf{K})$.

1.31. Let $\mathbf{M}_4$ be the subspace lattice of a projective line with four points. Show that $\mathbf{M}_4 = \mathbf{Var}(\mathbf{M}_4) \supset \mathbf{M}_3 = \mathbf{Var}(\mathbf{M}_3)$ and $\mathbf{M}_4 \supset \mathbf{K} \supseteq \mathbf{M}_3$ implies that $\mathbf{K} = \mathbf{M}_3$ for every variety $\mathbf{K}$.

1.32. Find a variety $\mathbf{N}$ of lattices such that $\mathbf{N} \supset \mathbf{N}_5 = \mathbf{Var}(\mathbf{N}_5)$, $\mathbf{N} \not\supseteq \mathbf{M}_3$, and $\mathbf{N} \supset \mathbf{K} \supseteq \mathbf{N}_5$ implies that $\mathbf{K} = \mathbf{N}_5$ for every variety $\mathbf{K}$.

1.33. Show that $\mathbf{Var}(\mathbf{K}) = \mathbf{SHPS}(\mathbf{K})$, if $\mathbf{K}$ is one of the classes of lattices listed below:

   (i) $\mathbf{K} = \{L\}$, where $L$ is a finite lattice;

   (ii) $\mathbf{K} = \{\mathsf{M}_{\aleph_0}\}$, where $\mathsf{M}_{\aleph_0}$ is the subspace lattice of a projective line with countably many points;

   (iii) for each prime number $p$, we choose a field $F_p$ of characteristic $p$ and $\mathbf{K} = \{\, \mathsf{L}_p \mid p$ is a prime number $\}$, where $\mathsf{L}_p$ is the lattice of subspaces of the projective plane coordinatized by $F_p$.

1.34. Let $\mathbf{K}$ be a variety of lattices. Prove that

$$\mathrm{Free}_{\mathbf{K}}(\mathfrak{m}) \in \mathbf{ISP}(\mathrm{Free}_{\mathbf{K}}(\aleph_0))$$

for any cardinal $\mathfrak{m}$.

1.35. Prove the Compactness Theorem for any type of universal algebras. Find applications of this result to lattices that go beyond the Compactness Theorem of Exercise 1.28. (Hint: use a type with $\vee$ and $\wedge$ and infinitely many constants.)

## 2.   The Lattice of Varieties of Lattices

### 2.1   Basic properties

Let $\mathbf{K}_0$ and $\mathbf{K}_1$ be varieties of lattices. Then $\mathbf{K}_0 \cap \mathbf{K}_1$ is again a variety and

$$\mathbf{K}_0 \cap \mathbf{K}_1 = \mathbf{Mod}(\mathrm{Iden}(\mathbf{K}_0) \cup \mathrm{Iden}(\mathbf{K}_1)).$$

There is also a smallest variety $\mathbf{K}_0 \vee \mathbf{K}_1$ containing both $\mathbf{K}_0$ and $\mathbf{K}_1$, namely,

$$\mathbf{K}_0 \vee \mathbf{K}_1 = \mathbf{Mod}(\mathrm{Iden}(\mathbf{K}_0) \cap \mathrm{Iden}(\mathbf{K}_1)).$$

All varieties of lattices form a lattice $\mathbf{\Lambda}$ with the lattice operations $\mathbf{K}_0 \vee \mathbf{K}_1$ and $\mathbf{K}_0 \wedge \mathbf{K}_1 = \mathbf{K}_0 \cap \mathbf{K}_1$. (Axiomatic set theory does not permit the formation of a set whose elements are classes. Thus formally, one cannot form the lattice of varieties. It is easy to get around this difficulty. For instance, replace a variety $\mathbf{K}$ by $\mathrm{Iden}(\mathbf{K})$, which is a subset of the countable set of all lattice identities. Then we form the lattice of all subsets of the form $\mathrm{Iden}(\mathbf{K})$ of the set of all lattice identities.)

The zero of $\mathbf{\Lambda}$ is $\mathbf{T}$ (the trivial variety) consisting of all one-element lattices. Let $\mathbf{K}$ be a variety of lattices different from $\mathbf{T}$. Then there is a lattice $L$ in $\mathbf{K}$ with $|L| > 1$. Therefore, $L$ has $\mathsf{C}_2$ as a sublattice and so $\mathsf{C}_2 \in \mathbf{K}$. By Corollary 120, $\mathbf{Var}(\mathsf{C}_2) = \mathbf{D}$, the class of all distributive lattices. Thus $\mathbf{K} \supseteq \mathbf{D}$. We have just verified that $\mathbf{D}$ is the only atom of $\mathbf{\Lambda}$ and every nonzero member contains $\mathbf{D}$, as illustrated in Figure 105.

Now let $\mathbf{K}$ be a variety of lattices properly containing $\mathbf{D}$. Then there is a nondistributive lattice $L$ in $\mathbf{K}$. By Theorem 101, either $\mathsf{N}_5$ or $\mathsf{M}_3$ is a sublattice of $L$, hence either $\mathsf{N}_5$ or $\mathsf{M}_3$ is in $K$. Set

$$\mathbf{N}_5 = \mathbf{Var}(\mathsf{N}_5),$$
$$\mathbf{M}_3 = \mathbf{Var}(\mathsf{M}_3).$$

We have just proved that $\mathbf{K} \supseteq \mathbf{N}_5$ or $\mathbf{K} \supseteq \mathbf{M}_3$. Thus $\mathbf{D}$ is covered by exactly two varieties, $\mathbf{N}_5$ and $\mathbf{M}_3$, and every variety properly containing $\mathbf{D}$ contains $\mathbf{N}_5$ or $\mathbf{M}_3$, see Figure 104.

The correspondence, set up in Theorem 468, between varieties of lattices and fully invariant congruences of $\mathrm{Free}(\aleph_0)$ is an isomorphism between $\mathbf{\Lambda}^{\delta}$ and the lattice of fully invariant congruences of $\mathrm{Free}(\aleph_0)$.

**Theorem 479.** *The lattice of varieties of lattices, $\mathbf{\Lambda}$, is distributive and dually algebraic. The dually compact elements are exactly those varieties that can be defined by finitely many identities.*

*Proof.* Observe that an identity $p = q$ corresponds to forming the smallest fully invariant congruence under which $p(u_0, u_1, \ldots) \equiv q(u_0, u_1, \ldots)$. Thus the compact fully invariant congruence relations correspond to finite sets of identities. $\qquad\square$

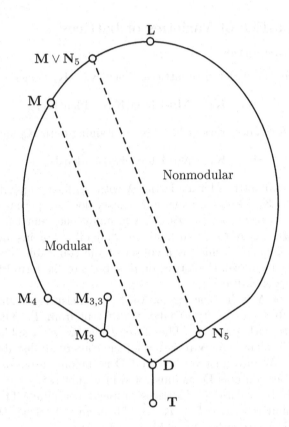

Figure 104. The lattice $\Lambda$ of varieties

The distributivity of the lattice $\Lambda$ can also be derived from the following formula (see B. Jónsson [444]).

**Theorem 480.** *Let $\mathbf{K}_0$ and $\mathbf{K}_1$ be varieties of lattices. Then*

$$\mathbf{Si}(\mathbf{K}_0 \vee \mathbf{K}_1) = \mathbf{Si}(\mathbf{K}_0) \cup \mathbf{Si}(\mathbf{K}_1).$$

*Proof.* Apply Corollary 476 to $\mathbf{K} = \mathbf{K}_0 \cup \mathbf{K}_1$: if $L \in \mathbf{K}_0 \vee \mathbf{K}_1$ and $L$ is subdirectly irreducible, then $L \in \mathbf{HSP_u}(\mathbf{K}_0 \cup \mathbf{K}_1)$, that is $L \in \mathbf{HS}(L')$, where $L'$ is an ultraproduct of lattices from $\mathbf{K}_0 \cup \mathbf{K}_1$, specifically, $L' = \prod_{\mathcal{D}}(L_i \mid i \in I)$, where $L_i \in \mathbf{K}_0 \cup \mathbf{K}_1$. Set

$$I_j = \{\, i \in I \mid L_i \in \mathbf{K}_j \,\}$$

for $j = 0, 1$. Then $I_0 \cup I_1 = I \in \mathcal{D}$, hence $I_0$ or $I_1 \in \mathcal{D}$. If $I_j \in \mathcal{D}$, then by Lemma 472, $L' \in \mathbf{P_u}(\mathbf{K}_j) = \mathbf{K}_j$. Thus we conclude that $L' \in \mathbf{K}_0$ or $L' \in \mathbf{K}_1$. $\qquad\square$

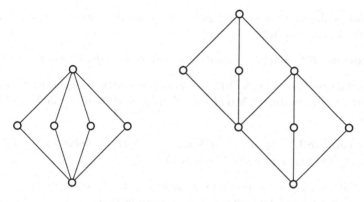

Figure 105. The lattices $M_4$ and $M_{3,3}$

Thus $\mathbf{K} \mapsto \mathbf{Si}(\mathbf{K})$ is analogous to the set representation of distributive lattices. (It is easy to turn this into a set representation, see Exercise 2.2.) In particular, we should note that if $\mathbf{Si}(\mathbf{K_0}) \subset \mathbf{Si}(\mathbf{K_1})$ and up to isomorphism $\mathbf{Si}(\mathbf{K_1})$ has only one more member, then $\mathbf{K_0} \prec \mathbf{K_1}$. To illustrate this, consider the lattices $M_4$ and $M_{3,3}$ of Figure 105, and the varieties $\mathbf{M_4}$ and $\mathbf{M_{3,3}}$ they generate, respectively. By Corollary 477, $\mathbf{Si}(\mathbf{M_4}) \subseteq \mathbf{HS}(\mathbf{M_4})$ and $\mathbf{Si}(\mathbf{M_{3,3}}) \subseteq \mathbf{HS}(\mathbf{M_{3,3}})$. So, up to isomorphic copies,

$$\mathbf{Si}(\mathbf{M_4}) = \{C_2, M_3, M_4\},$$
$$\mathbf{Si}(\mathbf{M_{3,3}}) = \{C_2, M_3, M_{3,3}\}.$$

Thus $\mathbf{M_3} \prec \mathbf{M_4}$ and $\mathbf{M_3} \prec \mathbf{M_{3,3}}$ in $\mathbf{\Lambda}$, see Figure 104. Also, $\mathbf{M_3} = \mathbf{M_4} \wedge \mathbf{M_{3,3}}$. All the varieties considered above are of finite height in $\mathbf{\Lambda}$.

**Lemma 481.** *The collection of varieties of lattices that can be generated by a single finite lattice is an ideal in $\mathbf{\Lambda}$. This ideal contains only elements of finite height.*

*Proof.* If the variety $\mathbf{K}$ is generated by a finite lattice $L$, then, by Corollary 477, up to isomorphism, $\mathbf{Si}(\mathbf{K})$ is a finite set of finite lattices. Thus if $\mathbf{K_0} \subset \mathbf{K}$, then, up to isomorphism, $\mathbf{Si}(\mathbf{K_0})$ must be a subset of this finite set, hence there are only finitely many such $\mathbf{K_0}$. All the statements of this corollary now follow immediately.                                                     □

## 2.2   ◇ Varieties of finite height

Lemma 481 was known by 1964. From about the same time originates the question, is the converse of the second statement of the lemma true?

In 1995, J. B. Nation [544] provided a answer in the negative: the variety generated by lattice $J$ of Figure 106. This diagram is a "spherical" diagram, it should be drawn on a sphere: pairs of black-filled elements with the same label should be identified.

◇ **Theorem 482.** **Var**($J$) *cannot be generated by a finite lattice.*

The lattice $F$ of Figure 107 is shown also with a "spherical" diagram. The lattice $F$ is finite, so **Var**($F$) is of finite height in the lattice of lattice varieties.

◇ **Theorem 483.** *The covering* **Var**($F$) $\prec$ **Var**($J$) *holds in the lattice* $\Lambda$. *Therefore,* **Var**($J$) *is of finite height in* $\Lambda$.

So Nation's lattice $J$ provides a variety, **Var**($J$), of finite height in the lattice $\Lambda$ that cannot be generated by a finite lattice.

### 2.3   Join-irreducible varieties

From the observations made above, it is clear that the join-irreducible elements of $\Lambda$ are connected with varieties generated by a single subdirectly irreducible lattice. The following result states a number of connections of this type (R. N. McKenzie [512]).

**Theorem 484.** *Let* $\mathbf{K} \in \Lambda$.

(i) *If* fil($\mathbf{K}$) *is a* completely prime filter (*that is,* $\bigvee(\mathbf{K}_i \mid i \in I) \in$ fil($\mathbf{K}$) *implies that* $\mathbf{K}_i \in$ fil($\mathbf{K}$) *for some* $i \in I$), *then* $\mathbf{K}$ *can be generated by a finite subdirectly irreducible lattice.*

(ii) *If* $\mathbf{K}$ *can be generated by a finite subdirectly irreducible lattice, then* $\mathbf{K}$ *is completely join-irreducible.*

(iii) *If* $\mathbf{K}$ *is completely join-irreducible, then* $\mathbf{K}$ *can be generated by a subdirectly irreducible lattice.*

(iv) *If* $\mathbf{K}$ *can be generated by a subdirectly irreducible lattice, then* $\mathbf{K}$ *is join-irreducible.*

*Proof.* (i) Let $\mathbf{K}_n$ denote the variety generated by the partition lattice on an $n$-element set. By Lemma 403 and Corollary 405, it follows that

$$\bigvee(\mathbf{K}_n \mid n = 1, 2, 3, \dots) = \mathbf{L} \supseteq \mathbf{K}.$$

Since fil($\mathbf{K}$) is completely prime, we conclude that $\mathbf{K}_n \supseteq \mathbf{K}$ for some integer $n$. Thus by Lemma 481, $\mathbf{K}$ can be generated by finitely many finite subdirectly irreducible lattices. Since $\mathbf{K}$ is join-irreducible, $\mathbf{K}$ can be generated by a single finite subdirectly irreducible lattice.

(ii) If $\mathbf{K}$ is generated by the finite subdirectly irreducible lattice $L$ and $\mathbf{K} = \mathbf{K}_0 \vee \mathbf{K}_1$, then by Theorem 480, $L \in \mathbf{Si}(\mathbf{K}_0)$ or $L \in \mathbf{Si}(\mathbf{K}_1)$, implying

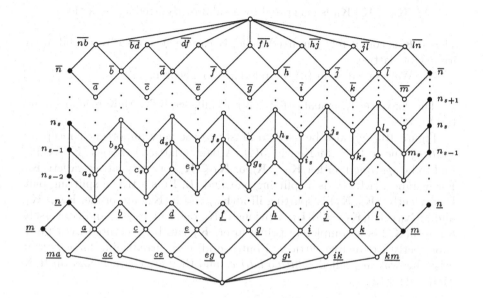

Figure 106. The lattice $J$

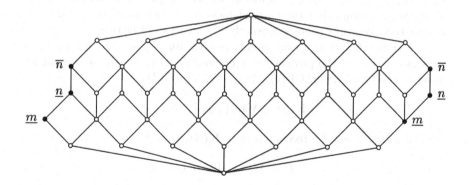

Figure 107. The lattice $F$

that $\mathbf{K} = \mathbf{K}_0$ or $\mathbf{K} = \mathbf{K}_1$. Thus $\mathbf{K}$ is join-irreducible. By Lemma 481, $\mathbf{K}$ is of finite height, hence $\mathbf{K}$ is completely join-irreducible.

(iii) Any variety $\mathbf{K}$ is of the form

$$\bigvee (\, \mathbf{K}_0 \subseteq \mathbf{K} \mid \mathbf{K}_0 \text{ is generated by a subdirectly irreducible lattice}\,).$$

Thus if $\mathbf{K}$ is completely join-irreducible, $\mathbf{K}$ can be generated by a subdirectly irreducible lattice.

(iv) We proceed as in (ii), by reference to Theorem 480.    $\square$

The converse statements of (i)–(iv) all fail, see R. N. McKenzie [512] and the Exercises.

Figure 104 can also be used to illustrate the very important concept of *splitting* due to R. N. McKenzie [512]. A pair of lattice varieties $(\mathbf{K}_0, \mathbf{K}_1)$ is said to be *splitting* if either $\mathbf{K}_2 \subseteq \mathbf{K}_0$ or $\mathbf{K}_1 \subseteq \mathbf{K}_2$ for every lattice variety $\mathbf{K}_2$. For instance, $(\mathbf{M}, \mathbf{N}_5)$ is a splitting pair and $(\mathbf{L}, \mathbf{T})$ is a trivial splitting pair. Equivalently, $(\mathbf{K}_0, \mathbf{K}_1)$ is splitting iff $\mathrm{id}(\mathbf{K}_0)$ and $\mathrm{fil}(\mathbf{K}_1)$ are prime, $\mathbf{K}_0 \not\supseteq \mathbf{K}_1$, and $\mathrm{id}(\mathbf{K}_0) \cup \mathrm{fil}(\mathbf{K}_1) = \mathrm{id}(\mathbf{L})$. Obviously, $\mathbf{K}_0$ determines $\mathbf{K}_1$, and conversely. Since $\mathrm{fil}(\mathbf{K}_1)$ is a completely prime filter, $\mathbf{K}_1$ can be generated by a finite subdirectly irreducible lattice. Finite subdirectly irreducible lattices that arise this way are called *splitting lattices* and they are characterized in R. N. McKenzie [512].

## 2.4    $2^{\aleph_0}$ lattice varieties

How big is the lattice $\mathbf{\Lambda}$? Since there are $\aleph_0$ identities and $2^{\aleph_0}$ sets of identities, it follows that there are at most $2^{\aleph_0}$ varieties. Now we shall show, by construction, that there are exactly $2^{\aleph_0}$ varieties of lattices.

Let $\Pi$ be the set of prime numbers and, for any $p \in \Pi$, let $L_p$ be the subspace lattice of the projective plane coordinatized by the $p$-element field (that is, the Galois field, $\mathrm{GF}(p)$). For a subset $S \subseteq \Pi$, set

$$\mathbf{K}(S) = \mathbf{Var}(\{\, L_i \mid i \in S \,\}).$$

We claim that $S$ can be recovered from $\mathbf{K}(S)$, in fact,

$$L_p \in \mathbf{K}(S) \quad \text{iff} \quad p \in S.$$

Obviously, if $p \in S$, then $L_p \in \mathbf{K}(S)$. Now let $L_p \in \mathbf{K}(S)$ and $p \notin S$. Since $L_p$ is subdirectly irreducible, we can apply Corollary 476 to obtain that

$$L_p \in \mathbf{HSP}_{\mathrm{u}}(\{\, L_i \mid i \in S \,\}),$$

that is, $L_p \in \mathbf{HS}(L)$, where we define the lattice $L$ as $\prod_{\mathcal{D}}(L_i \mid i \in I)$ with $\mathcal{D}$ prime over $I$, and $L_i$, for each $i \in I$, is an $L_j$ with $j \in S$.

Each $L_i$ is a complemented modular lattice of length 3, in which any pair of atoms is perspective. Since these properties can be expressed by (first-order) sentences, by Exercise 1.26, the lattice $L$ has the same property. By the results of Section V.5, the lattice $L$ is the subspace lattice of a nondegenerate projective plane.

Each $L_i$ satisfies Desargues' Theorem and, by Theorem 427, this property can be expressed by a sentence. Thus $L$ satisfies Desargues' Theorem and, by the Coordinatization Theorem, $L$ can be coordinatized by a division ring $\mathcal{D}$. We assumed that $p \notin S$, so $L_i$, for each $i \in I$, is coordinatized by a division ring not of characteristic $p$. This again can be expressed by a formula, hence $\mathcal{D}$ is not of characteristic $p$.

The lattice $L_p$ belongs to $\mathbf{HS}(L)$, that is, the lattice $L$ has a sublattice $L'$ such that $L_p$ is a homomorphic image of $L'$. Since both $L$ and $L_p$ are of length 3 and a proper homomorphic image of a modular lattice of length 3 is of length less than 3, clearly we can assume that $L_p = L'$, that is, $L_p$ is a sublattice of $L$. This is a contradiction: we can introduce $x + y$ for points $x$ and $y$ in $L_p$ using only elements of $L_p$; thus for points $x$ and $y$ of $L_p$, the line $x + y$ in $L_p$ is the same as in $L$. But we have $p \cdot x = 0$ in $L_p$, while $p \cdot x \neq 0$, for all $x \neq 0$ in $L$, which is impossible.

Thus there are at least as many varieties of lattices as there are subsets of $\Pi$, which number $2^{\aleph_0}$.

**Theorem 485.** *There are $2^{\aleph_0}$ varieties of (modular) lattices.*

This result was proved by K. A. Baker [37] (on whose example the above discussion was based), R. N. McKenzie [512] (without modularity), and R. Wille [734].

## 2.5  ◇ The covers of the variety generated by the pentagon

In the lattice of all lattice varieties, the variety $\mathbf{N_5}$ is, of course, covered by $\mathbf{N_5} \vee \mathbf{M_3}$. Exercise 2.1 lists fifteen subdirectly irreducible lattices of R. N. McKenzie [512], each generating a variety covering the variety $\mathbf{N_5}$ in the lattice of all lattice varieties. The converse was proved in B. Jónsson and I. Rival [451].

◇ **Theorem 486.** *Let $\mathbf{V} \not\subseteq \mathbf{M}$ be a variety of lattices covering $\mathbf{N_5}$ in $\Lambda$. Then either $\mathbf{V} = \mathbf{N_5} \vee \mathbf{M_3}$ or $\mathbf{V}$ is generated by one of the fifteen lattices listed in Exercise 2.1.*

These fifteen lattices are subdirectly irreducible, so we have fifteen join-irreducible covers of $\mathbf{N_5}$ on $\Lambda$.

The modular covers of $\mathbf{M_3}$ in $\Lambda$ are discussed in Section 3.3.

## 2.6  ◇ Products of varieties

Let **V** and **W** be classes of lattices. The *product* **V** ∘ **W** of **V** and **W** consists of all those lattices $L$ for which there exists a congruence $\alpha$ such that all blocks of $\alpha$ are in **V** and $L/\alpha$ is in **W**. Note that the product operation is nonassociative, even for lattice varieties.

The corresponding concept for varieties of groups was extensively studied in H. Neumann [551] and generalized to universal algebras in A. I. Mal'cev [510].

The first lattice theoretic result on products is in V. B. Lender [501]: nontrivial prevarieties (classes closed under **I**, **S**, and **P**) of lattices under product form a free groupoid.

Clearly, the interval doubling construction (see Section IV.1.2) is very important for products of lattice varieties. Indeed, by Lemma 297, if **V** is a lattice variety and $L \in \mathbf{V}$, then $L[I] \in \mathbf{D} \circ \mathbf{V}$. A. Day [136] proves the following technical lemma:

◇ **Lemma 487.** *Let* **K** *be the smallest class of lattices with the property that* $\mathsf{C}_2 \in \mathbf{K}$ *and* **K** *is closed under the interval doubling construction. Then* $\mathbf{Var}(\mathbf{K}) = \mathbf{L}$.

As it turns out, the class **K** of Lemma 487 consists exactly of the *bounded lattices* discussed in Section IV.4.4.

Following G. Grätzer and D. Kelly [285], let us define the classes $\mathbf{D}^0 = \mathbf{T}$, $\mathbf{D}^1 = \mathbf{D}$, and $\mathbf{D}^{n+1} = \mathbf{D}^n \circ \mathbf{D}$ for all $n > 0$. For a variety **V**, the *dimension* of **V** is the largest integer $n$ with $\mathbf{D}^n \subseteq \mathbf{V}$. It follows from Lemma 487 that every variety $\mathbf{V} \neq \mathbf{L}$ has a dimension, $\dim(\mathbf{V})$. By [285],

$$\dim(\mathbf{V} \circ \mathbf{W}) \geq \dim(\mathbf{V}) + \dim(\mathbf{W});$$

we conclude that every variety $\mathbf{V} \neq \mathbf{L}$ is a product of ∘-indecomposable varieties.

As another consequence of Lemma 487, A. Day [137] showed that **L** and **T** are the only idempotent elements of the groupoid of nontrivial prevarieties; in other words, $\mathbf{X} \circ \mathbf{X} = \mathbf{X}$ only has trivial solutions. This was generalized in G. Grätzer and D. Kelly [284]: $\mathbf{X} \circ \mathbf{Y} = \mathbf{X} \vee \mathbf{Y}$ only has trivial solutions.

Products of varieties have been studied in G. Grätzer and D. Kelly [284]–[287] and E. Fried and G. Grätzer [197], [198]. Here is one more result from these papers (G. Grätzer and D. Kelly [286]):

◇ **Theorem 488.** *The class* **D** ∘ **D** *is a variety with continuumly many subvarieties.*

The deepest result of this field is in T. Harrison [374]:

◇ **Theorem 489.** *If* **V** *is any nonmodular variety of lattices,* **W** *is any nontrivial variety of modular lattices, and* **V** ∘ **W** *is a variety, then* $\mathbf{V} = \mathbf{L}$.

Let $\mathbf{V}$ and $\mathbf{W}$ be lattice varieties. In view of Lemma 21, it was natural for by R. N. McKenzie to conjecture that a lattice $K$ belongs to the variety $\mathbf{H}(\mathbf{V} \circ \mathbf{W})$ (the variety generated by $\mathbf{V} \circ \mathbf{W}$) iff there is a tolerance relation $\tau$ on $K$ satisfying:

(i) all $\tau$-blocks of $K$ are in $\mathbf{V}$;
(ii) $K/\tau$ is in $\mathbf{W}$.

In E. Fried and G. Grätzer [198] the following result is proved:

$\diamond$ **Theorem 490.** *The lattice $K$ of Figure 108 is in* $\mathbf{H}(\mathbf{M}_3 \circ \mathbf{D})$. *However, there is no tolerance relation $\tau$ on $K$ satisfying the conditions* (i) *and* (ii) *with the varieties* $\mathbf{V} = \mathbf{M}_3$ *and* $\mathbf{W} = \mathbf{D}$.

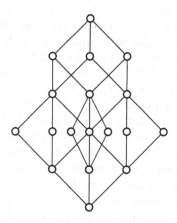

Figure 108. The lattice $K$ for Theorem 490

## 2.7   $\diamond$ Lattices of equational and quasi-equational theories
## by Kira Adaricheva

Many of the results of Sections 1 and 2 hold for algebras, in general. For instance, the proof of Theorem 468 verifies that if $\mathbf{V}$ is a variety of algebras and $\mathrm{Free}_{\mathbf{V}}(\aleph_0)$ is a free algebra in $\mathbf{V}$ with countably many generators, then the lattice of subvarieties of $\mathbf{V}$ is dually isomorphic to the lattice of fully invariant congruence relations of $\mathrm{Free}_{\mathbf{V}}(\aleph_0)$. Equivalently, the lattice of fully invariant congruence relations of $\mathrm{Free}_{\mathbf{V}}(\aleph_0)$ is isomorphic to the lattice $\mathrm{LTh}(\mathbf{V})$ of equational theories extending $\mathrm{Iden}(\mathbf{V})$.

Quasi-identities were introduced preceding Exercise 1.20. In particular, every identity is a quasi-identity, so one can consider the richer structure of all *quasi-equational* theories extending $\mathrm{Iden}(\mathbf{V})$. We denote this lattice by $\mathrm{QTh}(\mathbf{V})$.

G. Birkhoff [69] and A. I. Mal'cev [526] independently posed the following problem:

*Which lattices can be represented as lattices of equational theories, or as lattices of quasi-equational theories?*

It follows from the Compactness Theorem (Exercise 1.28) that $\mathrm{LTh}(\mathbf{V})$ is an algebraic lattice with a compact unit, moreover, the compact elements are represented by finitely axiomatizable equational theories. In 1983, R. N. McKenzie [514] showed that every $\mathrm{LTh}(\mathbf{V})$ is isomorphic to the lattice of congruence relations of a groupoid with a left unit and a right zero. Based on this, W. A. Lampe [498] established the first restricting condition on the structure of lattices of the form $\mathrm{LTh}(\mathbf{V})$. Nevertheless, the restriction is imposed only at the top of the lattice, and adjoining a new top element may turn a forbidden lattice into a representable one. Indeed, as shown by D. Pigozzi and G. Tardos [582], for every algebraic lattice $L$, the sum $L + C_1$ is representable as $\mathrm{LTh}(\mathbf{V})$, for some variety $\mathbf{V}$. In 2008, A. Nurakunov [555] showed that a lattice is representable as a lattice of an equational theory $\mathrm{LTh}(\mathbf{V})$ iff it is isomorphic to the lattice of congruence relations of some special monoid enriched by two unary operations.

Unlike their equational counterparts, the lattices of quasi-equational theories were initially studied in their dual form, called Q-lattices, namely, as lattices of subquasivarieties of some given variety (or quasivariety).

The study of Q-lattices started with V. A. Gorbunov [240] and V. A. Gorbunov and V. I. Tumanov [244]. Besides the most general properties of Q-lattices such as the join-semidistributive law $(\mathrm{SD}_\vee)$, the dual algebraicity and the atomicity, they discovered an important class of Q-lattices. These are lattices of the form $\mathrm{S_p}\, A$, where $A$ is an algebraic lattice and $\mathrm{S_p}\, A$ stands for the lattice of subsets of $A$ closed under arbitrary meets and joins of nonempty chains (see also Section VII.2.7). Such a description mimics the description of an arbitrary quasivariety as a class closed with respect to forming subdirect products and direct limits.

In V. A. Gorbunov and V. I. Tumanov [245], the construction $\mathrm{S_p}\, A$ was generalized: it was shown that any Q-lattice can be represented as $\mathrm{S_p}(A, \varepsilon)$, where $A$ is an algebraic lattice with a preorder $\varepsilon$ on $A$, and $\mathrm{S_p}(A, \varepsilon)$ consists of elements $X$ of $\mathrm{S_p}\, A$ which are $\varepsilon$-*hereditary*, that is, $a\,\varepsilon\,b$ and $a \in X$ imply that $b \in X$. In V. A. Gorbunov [241], it was proved that any Q-lattice is a complete sublattice of a suitable $\mathrm{S_p}\, A$.

In K. V. Adaricheva and V. A. Gorbunov [18], following the lead of W. Dziobiak [170], the concept of an *equaclosure operator* (equational closure operator) on a complete lattice $L$ is introduced as a closure operator $h$ satisfying the following four conditions: (1) $h(0) = 0$; (2) if $h(a) = h(b)$, then $h(a) = h(a \wedge b)$; (3) $h(a) \wedge (b \vee c) = (h(a) \wedge b) \vee (h(a) \wedge c)$; (4) every $h$-closed element can be represented as a meet of dually compact elements of $L$. This concept captured the closure of a given subclass by formation of homomorphic images and became very important for the characterization problem of Q-lattices; see also R. Freese, K. Kearnes, and J. B. Nation [187].

Finite atomistic lattices that can be represented as lattices of quasivarieties were characterized in K. V. Adaricheva, W. Dziobiak, and V. A. Gorbunov [16]. This result was extended to algebraic atomistic lattices in K. V. Adaricheva, W. Dziobiak, and V. A. Gorbunov [17]:

◇ **Theorem 491.** *The following conditions on an algebraic atomistic lattice L are equivalent:*

(i) *L is a Q-lattice.*

(ii) *L is isomorphic to the lattice of meet subsemilattices of some algebraic lattice in which any proper ideal satisfies both chain conditions.*

(iii) *L is a dually algebraic lattice admitting an equaclosure operator.*

The proof uses a number of earlier results, including some from K. V. Adaricheva [13]. The book V. A. Gorbunov [243] gives a rather complete theory of Q-lattices as of 1998.

In two papers, K. V. Adaricheva and J. B. Nation [24] took an approach that, in some sense, unified the study of Q-lattices with the approach to lattices of equational theories. Namely, they represent lattices of the form $QTh(\mathbf{V})$ as congruence lattices, $Con(S, \mathcal{F})$, of semilattices with operators. In this model, a semilattice $S$ stands for the semilattice of compact congruence relations on $Free_{\mathbf{V}}(\aleph_0)$, while the operators $\mathcal{F}$ are natural extensions to $S$ of endomorphisms of $Free_{\mathbf{V}}(\aleph_0)$.

This approach established that a lattice of the form $Con(S, \mathcal{F})$ has a natural operator that mimics the behavior of the equaclosure operator on a Q-lattice. It was named *an equa-interior operator*. Also, a new operator was discovered that acts in conjunction with the equa-interior operator on a lattice $Con(S, \mathcal{F})$. As a result, more properties of equa-interior operators were found. The new operator helped explain why Q-lattices are *bi-atomic*.

The second paper in [24] proves a partial converse: every lattice of the form $Con(S, \mathcal{F})$, where the operators $\mathcal{F}$ form a group and $S$ has a unit, is representable as $QTh(\mathbf{W})$ for some quasivariety $\mathbf{W}$. In particular, one of the most attractive conjectures of the theory of Q-lattices, namely, that every finite Q-lattice is lower bounded, was disproved. In J. B. Nation [545], some modification of this argument was used to show that, for any semilattice $S$ with operators, the lattice $Con(S, \mathcal{F})$ is isomorphic to the lattice of implicational theories in the language that may not contain equality.

Great progress has been made in understanding the general structure of $QTh(\mathbf{V})$ and its dual lattice $L_q(\mathbf{V})$ of subquasivarieties of $\mathbf{V}$; however, a large number of (quasi)varieties $\mathbf{V}$ are known for which $L_q(\mathbf{V})$ is highly complex. The concept of a Q-universal quasivariety was introduced in M. Sapir [627]; it indicates the extreme complexity of a quasivariety, from the perspective of its subquasivariety lattice. A quasivariety **V** of finite type is *Q-universal*, if for every other quasivariety **W** of finite type, $L_q(\mathbf{W})$ is a homomorphic image of a sublattice of $L_q(\mathbf{V})$.

In M. Adams and W. Dziobiak [7], and V. A. Gorbunov [242], different sufficient conditions of Q-universality were established. These conditions also yield that the lattice of ideals Id Free($\aleph_0$) of a free lattice of countable rank is embeddable into $L_q(\mathbf{V})$. It remains an open problem whether $L_q(\mathbf{V})$ for a Q-universal quasivariety **V** always has Id Free($\aleph_0$) as a sublattice.

There is a surprising connection between Q-universality and another measurement of the complexity of a quasivariety, taken from the point of view of categories. If the category of all directed graphs is isomorphic to a subcategory of **V** in such a way that finite graphs correspond to finite algebras from **V**, then **V** is called *finite-to-finite universal*. In M. Adams and W. Dziobiak [8], it was shown that if a quasivariety of finite signature is finite-to-finite universal, then it is Q-universal. A similar result was obtained in V. Koubek and J. Sichler [484]. The description of Q-universal quasivarieties of directed graphs is still unknown, but considerable progress toward the description is made in A. V. Kravchenko [485] and S. V. Sizyĭ [650].

The whole issue of Studia Logica in 2004 is devoted to the theory of quasivarieties. M. E. Adams, K. V. Adaricheva, W. Dziobiak, A. V. Kravchenko [5] lists a number of open problems in the field.

Several problems proposed in this article have already been solved. To find out which, in *MathSciNet*, search for `Author adar*` and `Author adams`, then **Search**, and you find this article; click on `From References`, and it lists the papers referencing the survey.

## 2.8  ◇ Modified Priestley dualities as natural dualities by Brian A. Davey and Miroslav Haviar

The theory of *natural dualities* is a general theory for quasi-varieties of algebras that generalizes 'classical' dualities such as Stone duality for boolean algebras, Pontryagin duality for abelian groups, and Priestley duality for distributive lattices.

The theory was first developed in the early 1980s by B. A. Davey and H. Werner [133], and its rapid development over the next 30 years is covered in the survey papers by B. A. Davey [123] and by H. A. Priestley [595], and in the monographs by D. M. Clark and B. A. Davey [91] and by J. G. Pitkethly and B. A. Davey [583].

Every quasivariety of the form $\mathbf{A} = \mathbf{ISP}(M)$, where $M$ is a finite lattice-based algebra, has a natural duality. In the case that $M$ is distributive-lattice based, it is possible to use the restricted Priestley duality and the natural duality for **A** simultaneously. In tandem, these dualities can provide an extremely powerful tool for the study of **A**, see Clark and Davey [91, Chapter 7]. As well as being a natural area of application of natural duality theory, distributive-lattice-based algebras in general, and distributive lattices in particular, have provided deep insights into the general theory. Important examples have been Heyting algebras, particularly the finite Heyting chains, and Kleene algebras; but here we concentrate on the three-element bounded distributive lattice

$$3 = (\{0, d, 1\}; \vee, \wedge, 0, 1),$$

which was seminal in developments that led to the solution of the *Full versus Strong Problem*, one of the most tantalizing problems in the theory of natural dualities.

For a natural-duality viewpoint, Priestley duality for the class **D** of bounded distributive lattices is obtained via homsets based on the two-element chain 2 and uses the fact that $\mathbf{D} = \mathbf{ISP}(2)$. By using the fact that $\mathbf{D} = \mathbf{ISP}(3)$, B. A. Davey, M. Haviar, and H. A. Priestley [127] introduced the following modified Priestley duality for **D** as a natural duality based on 3. Let $f, g$ be the non-identity endomorphisms of 3 (see Figure 109) and let

$$\underset{\sim}{3} = (\{0, d, 1\}; f, g, \mathcal{T}),$$

where $\mathcal{T}$ is the discrete topology. The topological structure $\underset{\sim}{3}$ is called an *alter ego* of the lattice 3. Let $\mathbf{X} = \mathbf{IS_c P^+}(\underset{\sim}{3})$ be the class of all isomorphic

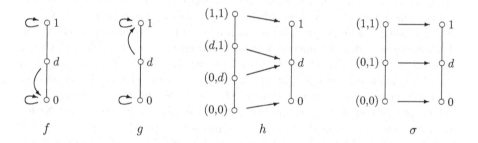

Figure 109. The (partial) operations $f$, $g$, $h$ and $\sigma$ on 3

copies of closed substructures of nonzero powers of $\underset{\sim}{3}$. One can set up natural hom-functors $D: \mathbf{D} \to \mathbf{X}$ and $E: \mathbf{X} \to \mathbf{D}$ as follows:

(On objects:)      $D: A \mapsto \mathbf{D}(A, 3) \leqslant \underset{\sim}{3}^A$    and    $E: X \mapsto \mathbf{X}(X, \underset{\sim}{3}) \leqslant 3^X$;

(On morphisms:)    $D: u \mapsto - \circ u$    and    $E: \varphi \mapsto - \circ \varphi.$

The functors $D$ and $E$ define a dual adjunction between **D** and **X** with unit and co-unit the evaluation maps

$$e_A: A \to ED(A) \quad \text{and} \quad \varepsilon_X: X \to DE(X),$$

defined by $e_A(a)(x) = x(a)$, for all $a \in A$ and $x \in \mathbf{D}(A, 3)$, and $\varepsilon_X(\alpha)(x) = \alpha(x)$ for all $x \in X$ and $\alpha \in \mathbf{X}(X, \underset{\sim}{3})$. Since every $A \in \mathbf{D}$ is isomorphic to its *double dual* $ED(A)$ via the map $e_A$ [127], we say that $\underset{\sim}{3}$ *yields a* (*natural*) *duality for* $\mathbf{D}$ *based on* 3.

Davey, Haviar, and Priestley [127] showed that such a modified Priestley duality for $\mathbf{D}$, in which the order is replaced by endomorphisms, can be based on any finite non-boolean distributive lattice $M$. They also showed that, while the order relation cannot be removed in the boolean case, it can at least be replaced by any finitary relation on $M$, which itself, like the order on 2, forms a non-boolean lattice.

B. A. Davey and M. Haviar [124], studied the enrichment of $\underset{\sim}{3}$ given by

$$\underset{\sim}{3}_\sigma = (\{0, d, 1\}; f, g, \sigma, \mathfrak{T}),$$

and B. A. Davey, M. Haviar, and R. Willard [128] explored deeply the enrichments $\underset{\sim}{3}_\sigma$ and

$$\underset{\sim}{3}_h = (\{0, d, 1\}; f, g, h, \mathfrak{T}).$$

(The binary partial operations $h$ and $\sigma$ are given in Figure 109.) If in the above scheme for the modified Priestley duality for $\mathbf{D}$ based on 3, the alter ego $\underset{\sim}{3}$ of 3 is replaced with the alter ego $\underset{\sim}{3}_\sigma$, then not only the map $e_A \colon A \to E\tilde{D}(A)$ is an isomorphism, for all $A \in \mathbf{D}$, establishing a duality between $\mathbf{D} = \mathbf{ISP}(3)$ and $\mathbf{X}_\sigma = \mathbf{IS_c P^+}(\underset{\sim}{3}_\sigma)$, but moreover the map $\varepsilon_X \colon X \to DE(X)$ is an isomorphism, for all $X \in \mathbf{X}_\sigma$, establishing a *full duality* between $\mathbf{D}$ and $\mathbf{X}_\sigma$. In general, such a scheme provides us with a canonical way of constructing, via hom-functors, a dual adjunction between a category of algebras $\mathbf{A} = \mathbf{ISP}(M)$, generated by a finite algebra $M$, and a category $\mathbf{X} = \mathbf{IS_c P^+}(\underset{\sim}{M})$ of structured topological spaces, generated by the alter ego $\underset{\sim}{M}$ of the algebra $M$. (It should be noted that for some finite algebras $M$ there is no choice of alter ego $\underset{\sim}{M}$ for which the resulting dual adjunction yields a duality between $\mathbf{A}$ and $\mathbf{X}$; for example, the two-element implication algebra $I = (\{0, 1\}; \to)$, see [91, Chapter 10].) If $\underset{\sim}{M}$ yields a full duality between $\mathbf{A} = \mathbf{ISP}(M)$ and $\mathbf{X} = \mathbf{IS_c P^+}(\underset{\sim}{M})$ and the alter ego $\underset{\sim}{M}$ is injective in $\mathbf{X}$, then the full duality is said to be a *strong duality*. If the hom-functors $D, E$ are restricted to the categories $\mathbf{A}_{\text{fin}}$ and $\mathbf{X}_{\text{fin}}$ of finite members of $\mathbf{A}$ and $\mathbf{X}$ only, then the concepts of a *finite-level* duality, full duality or strong duality are obtained.

The properties of the modified Priestley dualities for $\mathbf{D}$ based on 3 given by the alter egos $\underset{\sim}{3}$, $\underset{\sim}{3}_h$, and $\underset{\sim}{3}_\sigma$ are summarized in the following theorem.

◇ **Theorem 492.** *Let* $\underset{\sim}{3}$, $\underset{\sim}{3}_h$, *and* $\underset{\sim}{3}_\sigma$ *be the alter egos of* 3 *defined above.*

(i) $\underset{\sim}{3}$ *yields a duality on* $\mathbf{D}$ (Davey, Haviar, Priestley [127]).

(ii) $\underset{\sim}{3}_h$ *yields a full duality, which is not strong, on the category* $\mathbf{D}_{\text{fin}}$ *and yields a duality, which is not full, on the category* $\mathbf{D}$ (Davey, Haviar, Willard [128]).

(iii) $\underset{\sim}{3}_\sigma$ *yields a strong duality for* **D** (Davey, Haviar [124]).

(iv) *Every full duality on* **D** *based on* 3 *is strong* (Davey, Haviar, Willard [128]).

Let $R = (\{0, a, b, 1\}; t, \vee, \wedge, 0, 1)$ be the four-element chain with $0 < a < b < 1$ enriched with the ternary discriminator function $t$. Let $u$ be the partial endomorphism of $R$ with domain $\{0, a, 1\}$ given by $u(a) = b$. D. M. Clark, B. A. Davey, and R. Willard [92] showed that $R$ provides a negative solution to the *Full versus Strong Problem*, which dates back to the beginnings of the theory of natural dualities and asks: *Is every full duality strong?*

$\diamond$ **Theorem 493.** *The alter ego* $\underset{\sim}{R}_\perp = (\{0, a, b, 1\}; \mathrm{graph}(u), \mathfrak{T})$ *yields a full but not strong duality on* **ISP**$(R)$ (Clark, Davey, Willard [92]).

In general, a finite algebra $M$ admits essentially only one finite-level strong duality, but can admit many different finite-level full dualities. The alter egos $\underset{\sim}{M}$ yielding the finite-level full dualities for **ISP**$_{\mathrm{fin}}(M)$ form a doubly algebraic lattice $\mathbf{F}(M)$ introduced and studied in B. A. Davey, J. G. Pitkethly, and R. Willard [129]. The following theorem summarizes results in this direction.

$\diamond$ **Theorem 494.**

(i) $|\mathbf{F}(M)| = 1$ *for any finite semilattice, abelian group or relative Stone Heyting algebra* $M$ (Davey, Haviar, Niven [125]).

(ii) $\mathbf{F}(M)$ *is finite for any finite quasi-primal algebra* $M$; *in particular, for the algebra* $R$ *defined above,* $|\mathbf{F}(R)| = 17$ (Davey, Pitkethly, Willard [129]).

(iii) *The lattice* $\mathbf{F}(3)$ *is non-modular and has size* $2^{\aleph_0}$ (Davey, Haviar, and Pitkethly [126]).

### Exercises

2.1. Consider the lattices of Figures 110 and their duals (fifteen in all). Show that each one generates a variety covering $\mathbf{N}_5$ (R. N. McKenzie [512]).

2.2. Let $A$ be a fixed countable set. For a variety $\mathbf{K}$, let $\mathbf{Si}_A(\mathbf{K})$ denote the set of all subdirectly irreducible lattices $L$ in $\mathbf{K}$ satisfying $L \subseteq A$. Prove that $\mathbf{K} \mapsto \mathbf{Si}_A(\mathbf{K})$ is a set representation of the lattice of all varieties of lattices. (Hint: $\mathrm{Free}_{\mathbf{K}}(\aleph_0)$ can be recovered from $\mathbf{Si}_A(\mathbf{K})$.)

2.3. Does the representation of Exercise 2.2 preserve infinite joins and meets of varieties?

2.4. Prove that a variety **K** of lattices can be generated by finitely many finite lattices iff **K** can be generated by a single finite lattice, which in turn, is equivalent to the statement that all subdirectly irreducible lattices in **K** are finite and there are only finitely many nonisomorphic subdirectly irreducible lattices.

2.5. Show that **L** can be generated by a subdirectly irreducible lattice and **L** is completely join-reducible.

2.6. Prove that the variety generated by the lattice of Figure 111 is join-irreducible but it cannot be generated by a single subdirectly irreducible lattice (R. N. McKenzie [512]).

2.7. Let $\mathbf{K}_0$ and $\mathbf{K}_1$ be the varieties of lattices generated by all projective planes satisfying Desargues' Theorem and Pappus' Theorem, respectively. Prove that $\mathbf{K}_0$ and $\mathbf{K}_1$ have the same finite members (K. A. Baker [37]).

2.8. Find a variety of lattices that is not generated by its finite members (K. A. Baker [37] and R. Wille [734]).

2.9. Find a sublattice of $\boldsymbol{\Lambda}$ isomorphic to the lattice of all subsets of a countable set (K. A. Baker [37]).

2.10. Let **K** be a variety of lattices. If $\mathbf{K} \neq \mathbf{L}$, then $\mathbf{K} \prec \mathbf{K}_0$ for some variety $\mathbf{K}_0$ (B. Jónsson [444]). (Hint: Start with $\mathbf{K} \vee \mathbf{Var}(L)$, where $L$ is a finite lattice not in **K**.)

2.11. Prove that **L** is join-irreducible (B. Jónsson [444]). (Hint: If $\mathbf{K} \subset \mathbf{L}$, then the partition lattice on an infinite set does not belong to **K**.)

2.12. There is no variety of lattices covered by **L** (B. Jónsson [444]).

2.13. Show that every proper interval of $\boldsymbol{\Lambda}$ contains a prime interval.

2.14. Let $\mathbf{K}_0$ and $\mathbf{K}_1$ be varieties of lattices. If $\mathbf{K}_0 \subset \mathbf{K}_1$ and $\mathbf{K}_0$ can be defined by finitely many identities, then there exists a variety **K** satisfying $\mathbf{K}_0 \prec \mathbf{K} \subseteq \mathbf{K}_1$. (Hint: use Theorem 479.)

2.15. Prove that, in Exercise 2.14, we cannot require that $\mathbf{K}_0 \subseteq \mathbf{K} \prec \mathbf{K}_1$.

*2.16. Prove that $\mathbf{N}_5$ contains all lattices of width 2 (O. T. Nelson, Jr. [546]).

# 3.  Finding Equational Bases

## 3.1   UDE-s and identities

An *equational basis* $\Sigma$ of a class **K** is a set of identities such that

$$\mathbf{Var}(\mathbf{K}) = \mathbf{Mod}(\Sigma).$$

This $\Sigma$ is of special interest if it is *irredundant*, that is, $\mathbf{Var}(\mathbf{K}) \neq \mathbf{Mod}(\Sigma_1)$, for every $\Sigma_1 \subset \Sigma$, or if it is *finite*. There is not much one can say about the problem of finding equational bases in general. However, if **K** has some special properties, then there is hope for some meaningful results.

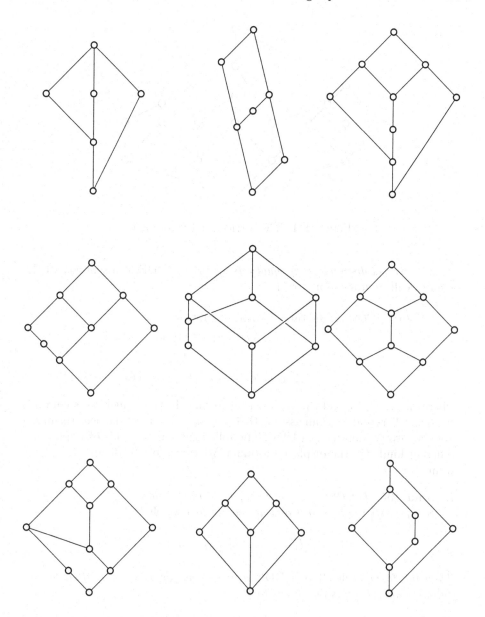

Figure 110. The lattices for Exercise 2.1

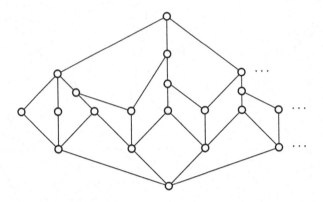

Figure 111. The lattice for Exercise 2.6

A *universal disjunction of equations*, or briefly, UDE, is a sentence of the form (recall, $\vee$ stands for "or")

$$(\forall x_0) \cdots (\forall x_{n-1})(f_0(x_0, \ldots, x_{n-1}) = g_0(x_0, \ldots, x_{n-1})$$
$$\vee f_1(x_0, \ldots, x_{n-1}) = g_1(x_0, \ldots, x_{n-1})$$
$$\cdots$$
$$\vee f_{m-1}(x_0, \ldots, x_{n-1}) = g_{m-1}(x_0, \ldots, x_{n-1})),$$

where $f_0, \ldots, f_{m-1}$ and $g_0, \ldots, g_{m-1}$ are terms. In the sequel, we shall omit the quantifiers and we shall assume that $f_i \leq g_i$ holds in any lattice. Examples abound: every identity is a UDE. The following lemma yields examples of a different kind. (In the displayed formula, $\bigvee$ stands for the disjunction of the terms.)

**Lemma 495.** *Let* $P = \{a_0, \ldots, a_{n-1}\}$ *be a finite order.*
*Set* $g_i = \bigvee(x_j \mid a_j \leq a_i)$ *and define the formula* $\Phi(P)$:

$$\bigvee(g_i = g_i \vee g_k \mid a_k \not\leq a_i).$$

*Then the formula* $\Phi(P)$ *is a UDE which holds for a lattice* $L$ *iff* $L$ *has no subset isomorphic (as an order) to* $P$.

*Proof.* If $Q = \{b_0, \ldots, b_{n-1}\} \subseteq L$ is isomorphic to $P$ and the map defined by $b_i \mapsto a_i$, for $0 \leq i < n$, is an isomorphism, then setting $x_i = b_i$, for $0 \leq i < n$, we find that $g_i(b_0, \ldots, b_{n-1}) = b_i$. Thus if $a_k \not\leq a_i$, then $g_k \not\leq g_i$, and so $g_i \neq g_i \vee g_k$. Since all the terms of $\Phi(P)$ fail, $\Phi(P)$ itself fails in $L$.

Conversely, if $\Phi(P)$ fails in $L$, then there are elements $b_0, \ldots, b_{n-1} \in L$ such that $g_k(b_0, \ldots, b_{n-1}) \not\leq g_i(b_0, \ldots, b_{n-1})$, whenever $a_k \not\leq a_i$. Since $a_k \leq a_i$ obviously implies that $g_k(b_0, \ldots, b_{n-1}) \leq g_i(b_0, \ldots, b_{n-1})$ by the definition of the term $g_i$, we conclude that $a_i \mapsto g_i(b_0, \ldots, b_{n-1})$ is an isomorphism of $P$ with

$$\{g_0(b_0, \ldots, b_{n-1}), \ldots, g_{n-1}(b_0, \ldots, b_{n-1})\}. \qquad \square$$

Another important UDE is $\Phi(n)$:

$$\bigvee ( x_i = x_j \mid 0 \leq i < j \leq n )$$

which holds for $L$ iff $|L| \leq n$. (To follow the inequality convention introduced above, we should replace "$x_i = x_j$" by "$x_i \wedge x_j = x_i \vee x_j$".)

Now let $\Phi$ be an arbitrary UDE:

$$\bigvee ( f_i(x_0, \ldots, x_{n-1}) = g_i(x_0, \ldots, x_{n-1}) \mid 0 \leq i < m )$$

and consider the following statement:

$S(\Phi)$: for any integer $k$, the term

$$\begin{aligned}
r_k = p_{2m}(x, \; &p_k(g_0(x_0, \ldots, x_{n-1}), \quad y_0^0, \ldots, \quad y_{k-1}^0), \\
&p_k(f_0(x_0, \ldots, x_{n-1}), \quad y_0^0, \ldots, \quad y_{k-1}^0), \\
&\quad \ldots \\
&p_k(g_{m-1}(x_0, \ldots, x_{n-1}), y_0^{m-1}, \ldots, y_{k-1}^{m-1}), \\
&p_k(f_{m-1}(x_0, \ldots, x_{n-1}), y_0^{m-1}, \ldots, y_{k-1}^{m-1}))
\end{aligned}$$

does not depend on $x$, where $p_k$ is the term introduced in Lemma 229. In the term $r_k$, the variables are $x, x_0, \ldots, x_{n-1}$ and $y_j^i$ for all $0 \leq i < m$ and $0 \leq j < k$.

We claim that the statement $S(\Phi)$ holds in a subdirectly irreducible lattice $L$ iff the UDE $\Phi$ does.

Let the UDE $\Phi$ hold in $L$. Substituting $x_i = a_i$, for all $0 \leq i < n$, we obtain the elements $f_i(a_0, \ldots, a_{n-1})$ and $g_i(a_0, \ldots, a_{n-1})$ for all $0 \leq i < m$. It follows from $\Phi$ that $f_i(a_0, \ldots, a_{n-1}) = g_i(a_0, \ldots, a_{n-1})$ for some $i$. Then two successive elements substituted in $p_{2m}$ agree and so, by a trivial property of $p_{2m}$ (see Exercise III.1.8), $r_k$ does not depend on $x$, verifying the statement $S(\Phi)$.

Now let the UDE $\Phi$ fail in $L$. Then there exist $a_0, \ldots, a_{n-1} \in L$ such that

$$f_i(a_0, \ldots, a_{n-1}) < g_i(a_0, \ldots, a_{n-1})$$

for $0 \leq i < m$. Since $L$ is subdirectly irreducible, by Corollary 232, there are $a, b \in L$ satisfying $b < a$ and

$$[f_i(a_0, \ldots, a_{n-1}), g_i(a_0, \ldots, a_{n-1})] \Rightarrow [b, a].$$

Thus for a suitable integer $k$ and elements $c_0^i, \ldots, c_{k-1}^i \in L$ with $0 \leq i < m$,

$$p_k(g_i(a_0, \ldots, a_{n-1}), c_0^i, \ldots, c_{k-1}^i) = a,$$
$$p_k(f_i(a_0, \ldots, a_{n-1}), c_0^i, \ldots, c_{k-1}^i) = b.$$

Hence $r_k$ with $x_i = a_i$, for all $0 \le i < n$, and $y_j^i = c_j^i$, for all $0 \le i < m$ and $0 \le j < k$, takes the form

$$r_k = p_{2k}(x, a, b, \dots, a, b).$$

But it is evident that $r_k = x$ on the interval $[b, a]$ (see Exercise III.1.9) and so $r_k$ is dependent on $x$. Thus the statement $S(\Phi)$ fails in $L$.

The statement that $r_k$ does not depend on $x$ is can be expressed with the identity $\varepsilon_k$:

$$r_k(x, \dots) = r_k(z, \dots),$$

where $z$ is a variable distinct from $x, x_i, y_j^i$. Set

$$\Sigma(\Phi) = \{\, \varepsilon_k \mid k = 1, 2, \dots \,\}.$$

Now we are ready to state K. A. Baker's solution to our problem, see [38]:

**Theorem 496.** *Let $\Phi$ be a UDE. Then $\Sigma(\Phi)$ is an equational basis for* **Mod**$(\Phi)$. *For a set $\Omega$ of UDE-s, the set of identities*

$$\bigcup(\,\Sigma(\Phi) \mid \Phi \in \Omega\,)$$

*is an equational basis for* **Mod**$(\Omega)$.

*Proof.* Let $\mathbf{K}_0$ be the variety defined by the set of identities

$$\bigcup(\,\Sigma(\Phi) \mid \Phi \in \Omega\,)$$

and let $\mathbf{K}_1$ be the variety generated by the class of lattices **Mod**$(\Omega)$.

If $L \in \mathbf{Si}(\mathbf{K}_0)$, then by the discussion above, we obtain that

$$L \in \mathbf{Mod}(\Omega) \subseteq \mathbf{K}_1.$$

Hence $\mathbf{K}_0 \subseteq \mathbf{K}_1$. Conversely, if $L \in \mathbf{Si}(\mathbf{K}_1)$, then by Corollary 476,

$$L \in \mathbf{HSP_u}(\mathbf{Mod}(\Omega)).$$

But $\Omega$ is preserved under ultraproducts (see Exercise 1.26) and obviously, by the formation of sublattices and homomorphic images. Hence $L \in \mathbf{Mod}(\Omega)$ and so, by the discussion above,

$$L \in \mathbf{Mod}(\bigcup(\,\Sigma(\Phi) \mid \Phi \in \Omega\,)) = \mathbf{K}_0.$$

Thus $\mathbf{K}_1 \subseteq \mathbf{K}_0$ proving that $\mathbf{K}_0 = \mathbf{K}_1$. ☐

## 3.2 Bounded sequences of intervals

We can recast this proof using the following concept: a set or sequence of intervals of a lattice $L$ is $(k\text{-})bounded$ if there is a proper interval $[a, b]$ such that each interval in the set is congruence projective (in $k$ steps) to $[a, b]$. Corollary 232 (slightly sharpened) then reads:

**Lemma 497.**

(i) *Any finite set of proper intervals is bounded in a subdirectly irreducible lattice.*

(ii) *Let $\varphi$ be a homomorphism of the lattice $L$ onto $L'$, let*

$$[a_0, b_0], \ldots, [a_{n-1}, b_{n-1}]$$

*be intervals of $L$, and let*

$$[\varphi(a_0), \varphi(b_0)], \ldots, [\varphi(a_{n-1}), \varphi(b_{n-1})]$$

*be $k$-bounded. Then*

$$[a_0, b_0], \ldots, [a_{n-1}, b_{n-1}]$$

*is $(k + 2)$-bounded.*

Now for a UDE $\Phi$ defined as $\bigvee f_i = g_i$, let $\alpha_k = \alpha_k(\Phi)$ denote the sentence stating that $[f_0, g_0], \ldots, [f_{m-1}, g_{m-1}]$ is not $k$-bounded for any substitution. We can write this simply by requiring that if $[f_i, g_i] \Rightarrow [u, v]$, for all $i$, then $u = v$. Then observe:

**Lemma 498.**

(i) $\Phi$ *implies $\alpha_k$ and $\alpha_k$ implies $\alpha_t$ for every $t \leq k$.*

(ii) *If $\alpha_k$, for all $k = 1, 2, \ldots$, hold in a subdirectly irreducible lattice, then so does $\Phi$.*

(iii) $\alpha_k$ *is preserved under the formation of direct products and sublattices.*

(iv) *For any set $\Omega$ of UDE-s, the equality*

$$\mathbf{Var}(\mathbf{Mod}(\Omega)) = \mathbf{Mod}(\{\, \alpha_k(\Phi) \mid \Phi \in \Omega \text{ and } k = 1, 2, \ldots \,\})$$

*holds.*

*Proof.*
(i) and (iii) are trivial.
(ii) restates Lemma 497(i).
(iv) follows from Corollary 476 and the previous statements. □

The reader should have no difficulty relating the sentence $\alpha_k$ to the identity $\varepsilon_k$.

We can use these ideas to give a simple proof of the following result of R. N. McKenzie [511]:

**Theorem 499.** *Any finite lattice has a finite equational basis.*

*Remark.* The present proof is due to C. Herrmann [386], which is based on K. A. Baker [41].

Let $L$ have $n$ elements. If $\Sigma$ is a finite equational basis for $\mathbf{Mod}(\Phi(n))$ (the class of at most $n$ element lattices), then we can easily find a finite equational basis for $L$: let $A$ be a finite set of (up to isomorphism) all finite lattices $N$ satisfying $N \notin \mathbf{Var}(L)$ and $|N| \leq n$; for each $N \in A$, choose an identity $\varepsilon_N$ holding in $L$ but not in $N$; then $\Sigma \cup \{\, \varepsilon_N \mid N \in A \,\}$ is a finite equational basis for $L$.

In the class $\mathbf{K} = \mathbf{Mod}(\Phi(n))$, any lattice has two properties: (i) it is defined by an at most $n^2$-termed UDE; (ii) every bounded set of at most $n^2$ intervals is $n^2$-bounded (because there are at most $n^2$ intervals in the lattice). Thus the following lemma completes the proof of Theorem 499:

**Lemma 500.** *Let $\mathbf{K}$ be a class of lattices and let $m$ be an integer with the following two properties:*

(i) $\mathbf{K} = \mathbf{Mod}(\Sigma)$, *where $\Sigma$ is a finite set of at most $m$-termed UDE-s.*

(ii) *There is an integer $r$ with the property that every bounded set of $m$ intervals is $r$-bounded in every subdirectly irreducible lattice $L \in \mathbf{K}$.*

*Then $\mathbf{K}$ has a finite equational basis.*

*Proof.* Let $\varrho_{m,r}$ denote the sentence expressing that any set of $m$ intervals that is $(r + 1)$-bounded is $r$-bounded. A trivial induction shows that $\varrho_{m,r}$ implies that any bounded set of $m$ intervals is $r$-bounded.

Let $L \in \mathbf{Var}(\mathbf{K})$. Then $L$ is a subdirect product of subdirectly irreducible lattices $L_i$, for $i \in I$, where each $L_i \in \mathbf{K}$ by Corollary 476, since UDE-s are preserved under the formation of ultra products, sublattices, and homomorphic images. Let $c \neq d$ and $[a_j, b_j] \Rightarrow [c, d]$ in $L$ for all $0 \leq j < m$. Then $\varphi_i(c) \neq \varphi_i(d)$ and $[a_j, b_j] \Rightarrow [\varphi_i(c), \varphi_i(d)]$ in $L_i$ for all $0 \leq j < m$ and for some $i \in I$, where $\varphi_i$ is the projection of $L$ onto $L_i$. By the second hypothesis, $\varrho_{m,r}$ holds in $L_i$ and so, by Lemma 497(ii), the set $\{\, [a_j, b_j] \mid 0 \leq j < m \,\}$ is $(r + 2)$-bounded. Thus the sentence $\varrho_{m,r+2}$ holds in any $L \in \mathbf{Var}(\mathbf{K})$.

By Lemma 498(iv), the variety $\mathbf{Var}(\mathbf{K})$ of lattices is defined by

$$\{\, \alpha_k(\Phi) \mid \Phi \in \Omega \text{ and } k = 1, 2, \dots \,\}.$$

Hence this set of sentences implies that $\varrho_{m,r+2}$ holds and so, by the Compactness Theorem (see Exercise 1.29), finitely many $\alpha_k(\Phi)$ imply that $\varrho_{m,r+2}$

holds. Let
$$A = \{\, \alpha_k(\Phi) \mid \Phi \in \Omega \text{ and } k = 1, 2, \ldots, t \,\}$$
imply that $\varrho_{m,r+2}$ holds and let us further assume that $t \geq r + 2$. We claim
that $A$ defines $\mathbf{Var(K)}$. Indeed, $A$ holds in $\mathbf{Var(K)}$ by Lemma 498, hence the
containment $\mathbf{Mod}(A) \supseteq \mathbf{Var(K)}$. Conversely, $A$ includes $\alpha_{r+2}(\Phi)$, for every
$\Phi \in \Omega$, and $A$ implies the sentence $\varrho_{m,r+2}$. But $\alpha_{r+2}(\Phi)$ and $\varrho_{m,r+2}$ imply
$\alpha_i(\Phi)$ for every $i$; thus $A$ implies $\alpha_i(\Phi)$, for every $i$ and any $\Phi \in \Omega$, proving
that $\mathbf{Mod}(A) \subseteq \mathbf{Var(K)}$.

We have proved that $A$ is equivalent to $\mathrm{Iden}(\mathbf{K})$ and $A$ is finite, hence,
by the Compactness Theorem, $A$ is equivalent to some finite $\Sigma \subseteq \mathrm{Iden}(\mathbf{K})$.
Thus $\Sigma$ is a finite equational basis for $\mathbf{K}$. $\qquad\square$

The proof of Theorem 499, as exhibited above, does not give a finite
equational basis; it only proves that there is one. In R. N. McKenzie [511] and
in K. A. Baker [41], more complicated arguments are presented that actually
construct a finite equational basis.

Baker's method of finding equational bases is exploited in great detail in
K. A. Baker [41]; for further results and applications, see K. A. Baker [39] and
[40]. C. Herrmann [386] and M. Makkai [524] present further variants on the
same theme; see also G. Grätzer and H. Lakser [299].

If the varieties $\mathbf{K}_0$ and $\mathbf{K}_1$ have finite equational bases, it does not follow
that $\mathbf{K}_0 \vee \mathbf{K}_1$ has a finite equational basis. For lattices this negative result
has been verified by K. A. Baker (unpublished) and B. Jónsson [447]. Some
positive results are also given in B. Jónsson [447]; for instance $\mathbf{M} \vee \mathbf{N}_5$ has
finite equational bases.

## 3.3   The modular varieties covering $\mathbf{M}_3$

In some cases, a finite equational basis for a finite lattice can be found using
the following method. Let $L$ be a finite lattice. Let us assume that we have
constructed the lattices $L_0, L_1, \ldots, L_{n-1}$ such that

  (i) $\mathbf{Var}(L)$ is covered by each $\mathbf{Var}(L_i)$;
 (ii) if $\mathbf{K}$ is a variety of lattices and $\mathbf{Var}(L) \subset \mathbf{K}$, then $L_i \in \mathbf{K}$ for some $i$.

Find lattice identities $\varepsilon_0, \ldots, \varepsilon_{n-1}$ that hold in $L$ the identity $\varepsilon_i$ fails in $L_i$ for
all $1 \leq i \leq n$. Then $\{\varepsilon_0, \ldots, \varepsilon_{n-1}\}$ is a finite equational basis for $L$. We shall
illustrate this method with an example.

**Theorem 501.** *In the lattice of all varieties of lattices, $\mathbf{M}_3$ is covered by $\mathbf{M}_4$*
*and $\mathbf{M}_{3,3}$. If $\mathbf{K}$ is any variety of modular lattices and $\mathbf{M}_3 \subset \mathbf{K}$, then $\mathbf{M}_4 \subseteq \mathbf{K}$*
*or $\mathbf{M}_{3,3} \subseteq \mathbf{K}$.*

*Remark.* This result was proved in G. Grätzer [253] under the additional
hypothesis that $\mathbf{K}$ is generated by a finite lattice. This hypothesis was removed
in B. Jónsson [445], where the following corollary was also stated.

This result has been extended in B. Jónsson [445] and D. X. Hong [399]. Related questions are considered in C. Herrmann [386] and in the papers of R. Freese.

**Corollary 502.** *An equational basis for* $\mathbf{M}_3$ *is given by the modular identity and the identity*

$$x \wedge (y \vee z) \wedge (z \vee w) \wedge (w \vee y) \leq (x \wedge y) \vee (x \wedge z) \vee (x \wedge w).$$

By the discussion above, to prove this corollary, it is sufficient to see that this identity holds in $\mathbf{M}_3$ but fails in $\mathbf{M}_4$ and $\mathbf{M}_{3,3}$; this is left to the reader.

The proof of Theorem 501 is based on the following two lemmas.

**Lemma 503.** *Let* $L$ *be a modular lattice and let* $\{o, a, b, c, i\}$ *be a diamond in* $L$. *Let* $a \leq x^a < i$ *and set (see Figure 112)*

$$x^b = b \vee (x^a \wedge c),$$
$$x^c = c \vee (x^a \wedge b),$$
$$o_1 = (x^a \wedge b) \vee (x^a \wedge c).$$

*Then* $\{o_1, x^a, x^b, x^c, i\}$ *is a diamond.*

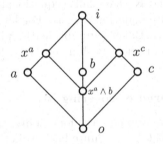

Figure 112. Illustrating Lemma 503

*Proof.* Since

$$a \leq x^a < i,$$
$$b \leq x^b < i,$$
$$c \leq x^c < i,$$

the equalities

$$x^a \vee x^b = x^a \vee x^c = x^b \vee x^c = i$$

obviously hold. Also,

$$x^a \wedge x^b = x^a \wedge (b \vee (x^a \wedge c)) = (x^a \wedge b) \vee (x^a \wedge c) = o_1,$$

and similarly, $x^a \wedge x^c = o_1$; finally,

$$
\begin{aligned}
x^b \wedge x^c &= (b \vee (x^a \wedge c)) \wedge (c \vee (x^a \wedge b)) \\
&= ((b \vee (x^a \wedge c)) \wedge c) \vee (x^a \wedge b) \\
&= (b \wedge c) \vee (x^a \wedge c) \vee (x^a \wedge b) = o \vee o_1 = o_1. \qquad \square
\end{aligned}
$$

**Lemma 504.** *Let $L$ be a subdirectly irreducible modular lattice which has no sublattice of which $\mathsf{M}_{3,3}$ is a homomorphic image. Let $[a, b] \overset{n}{\approx} [c, d]$ in $L$ such that $[a, b]$ has no proper subinterval projective to a subinterval of $[c, d]$ in fewer than $n$ steps. Then $n \leq 3$.*

*Proof.* Let $[y_0, x_0] \overset{4}{\approx} [y_4, x_4]$ be such that no proper subinterval of $[y_0, x_0]$ is projective to a proper subinterval of $[y_4, x_4]$ in fewer than four steps. By duality, we can assume that

$$
[y_0, x_0] \overset{\text{up}}{\sim} [y_1, x_1] \overset{\text{dn}}{\sim} [y_2, x_2] \overset{\text{up}}{\sim} [y_3, x_3] \overset{\text{dn}}{\sim} [y_4, x_4].
$$

By (the proof of) Theorem 352, we can assume that this is a normal sequence. Let $[q_i, p_i]$ be a proper subinterval of $[y_i, x_i]$, for all $0 \leq i \leq 4$, such that

$$
[q_0, p_0] \overset{\text{up}}{\sim} [q_1, p_1] \overset{\text{dn}}{\sim} [q_2, p_2] \overset{\text{up}}{\sim} [q_3, p_3] \overset{\text{dn}}{\sim} [q_4, p_4].
$$

We claim that the set $X_i = \{p_{i-1}, q_{i-1}, p_i, q_i, p_{i+1}, q_{i+1}\}$ cannot generate a distributive sublattice for $i = 1, 2,$ or $3$. Indeed, if one does, for instance, $X_1$ generates a distributive sublattice, then

$$
[q_0, p_0] \overset{\text{dn}}{\sim} [q_0 \wedge q_2, p_0 \wedge p_2] \overset{\text{up}}{\sim} [q_3, p_3] \overset{\text{dn}}{\sim} [q_4, p_4],
$$

by Theorem 353, contrary to the hypothesis. Thus the sublattice generated by $\{x_{i-1}, y_{i-1}, x_i, y_i, x_{i+1}, y_{i+1}\}$ is isomorphic to a homomorphic image of the third lattice of Figure 75 not collapsing $\mathsf{M}_3$ or to the dual of such a lattice (see Figure 113).

    Now we claim that $o_1 \vee i_2 = i_1$, using the notation of Figure 113. Let $o_1 \vee i_2 \neq i_1$. Since $i_1 > o_1$ and $x_1 = i_1 > i_2$, we must have $i_1 > o_1 \vee i_2 = x^c$. Thus by Lemma 503, there are elements $x^b$ and $x^a$ such that

$$
A_1 = \{x^b \wedge x^c, x^a, x^b, x^c, i_1\}
$$

is a diamond and

$$
\begin{aligned}
b_1 &\leq x^b < i_1, \\
a_1 &\leq x^a < i_1.
\end{aligned}
$$

Since $[b_1, i_1] \overset{\text{dn}}{\sim} [a_2, i_2]$, there is an element $y^a$ satisfying $[x^b, i_1] \overset{\text{dn}}{\sim} [y^a, i_2]$. Again, by Lemma 503, we find the elements $y^b$ and $y^c$ satisfying

$$
\begin{aligned}
b_2 &\leq y^b < i_2, \\
c_2 &\leq y^c < i_1,
\end{aligned}
$$

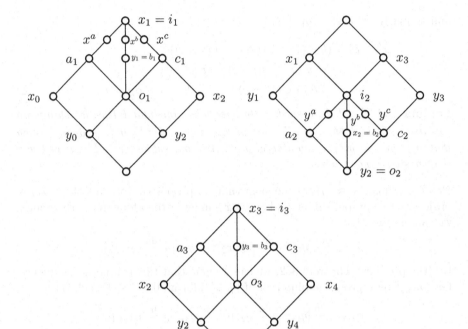

Figure 113. Proving Lemma 503

and

$$A_2 = \{y^a \wedge y^b, y^a, y^b, y^c, i_2\}$$

is a diamond. Thus $[x^b, i_1] \overset{\text{dn}}{\sim} [y^a, i_2]$ and since $i_2 \leq x^c < i_1$, the perspectivity $[x^b, x^c] \wedge x^c \overset{\text{dn}}{\sim} [y^a, i_2]$ also holds. This last relation means that

$$A_1 \cup A_2 \cong \mathsf{M}_{3,3}$$

or if $x^c \neq i_2$ (and $x^b \wedge x^c \neq y^a$), then

$$(A_1 \cup A_2)/\mathrm{con}(x^c, i_2) \cong \mathsf{M}_{3,3},$$

a contradiction.

Therefore, $o_1 \vee i_2 = i_1$. By duality, $o_1 \wedge i_2 = o_2$, that is

$$[o_1, i_1] \overset{\text{dn}}{\sim} [o_2, i_2],$$

and similarly,

$$[o_3, i_2] \overset{\text{up}}{\sim} [o_3, i_3].$$

By normality, $o_2 = o_1 \wedge o_3$, and by definition, $i_2 = i_1 \wedge i_3$. Thus the set

$$\{o_1, o_2, o_3, i_1, i_2, i_3\}$$

is contained in the sublattice generated by $\{o_1, i_1, o_3, i_3\}$, which is distributive by Theorem 363. By Theorem 353, we conclude that

$$[o_1, i_1] \overset{\text{up}}{\sim} [o_1 \vee o_3, i_1 \vee i_3] \overset{\text{dn}}{\sim} [o_3, i_3],$$

which trivially implies that

$$[y_0, x_0] \overset{\text{up}}{\sim} [y_1 \vee o_3, x_1 \vee o_3] \overset{\text{dn}}{\sim} [y_4, x_4],$$

contrary to the hypothesis.                                        $\square$

*Proof of Theorem 501.* Let $\mathbf{K}$ be a variety of modular lattices. Let us assume that $\mathbf{K} \supseteq \mathbf{M}_3$ and $\mathbf{M}_4, \mathbf{M}_{3,3} \notin \mathbf{K}$. In order to show that $\mathbf{K} = \mathbf{M}_3$, it is sufficient to verify that if $L \in \mathbf{Si}(\mathbf{K})$, then $L \leq \mathbf{M}_3$.

Assume, to the contrary, that $L \in \mathbf{Si}(\mathbf{K})$, but $L$ is not a sublattice of $\mathbf{M}_3$. Since $\mathbf{M}_4, \mathbf{M}_{3,3} \notin \mathbf{K}$, we must have $\mathbf{M}_4, \mathbf{M}_{3,3} \notin \mathbf{HS}(L)$.

Obviously, $|L| > 2$. If $L$ is of length 2, then $L$ must have $\mathbf{M}_4$ as a sublattice, a contradiction. Thus $L$ has a chain, $c_0 < c_1 < c_2 < c_3$, of length 3. Since $L$ is subdirectly irreducible, $\mathrm{con}(c_0, c_1) \wedge \mathrm{con}(c_1, c_2) \neq \mathbf{0}$ and so, by applying Theorem 230 twice, we obtain a proper interval $[y, x]$ such that $[c_1, c_2]$ and $[c_0, c_1]$ are c-projective to $[y, x]$. By Theorem 352 and by the symmetry of projectivity, there is a proper subinterval $[b, a]$ of $[c_1, c_2]$ and a subinterval $[d, c]$ of $[c_0, c_1]$ such that $[b, a] \overset{n}{\approx} [d, c]$. Choose these $a, b, c, d$ so that $n$ is minimal in $[b, a] \overset{n}{\approx} [d, c]$. Then $a, b, c, d$, and $L$ satisfy the conditions of Lemma 504 and therefore, $n \leq 3$. We claim that $n = 3$. Indeed, in any lattice $L$, we cannot have $a > b \geq c > d$ and $a/b \overset{n}{\approx} c/d$ with $n \leq 2$. (Proof. If

$$[b, a] \overset{\text{up}}{\sim} [q, p] \overset{\text{dn}}{\sim} [d, c],$$

then

$$d = c \wedge q \geq c \wedge b = c.$$

If

$$[b, a] \overset{\text{dn}}{\sim} [q, p] \overset{\text{up}}{\sim} [d, c],$$

then

$$a = b \vee p \leq b \vee c = b.$$

The case $n = 1$ is trivial.)

Now let

$$[b, a] = [y_0, x_0] \sim [y_1, x_1] \sim [y_2, x_2] \sim [y_3, x_3] = [d, c].$$

If $[b, a] \overset{\text{up}}{\sim} [y_1, x_1]$, then $x_1 > y_1 \geq c > d$ and $[y_1, x_1] \overset{2}{\approx} [d, c]$, which is impossible as noted in the previous paragraph. Thus

$$[y_0, x_0] \overset{\text{dn}}{\sim} [y_1, x_1] \overset{\text{up}}{\sim} [y_2, x_2] \overset{\text{dn}}{\sim} [y_3, x_3].$$

Applying the same arguments to $\text{con}(c_3, c_2) \wedge \text{con}(a, b)$, we obtain that

$$[u_0, z_0] \overset{\text{dn}}{\sim} [u_1, z_1] \overset{\text{up}}{\sim} [u_2, z_2] \overset{\text{dn}}{\sim} [u_3, z_3],$$

where $[u_0, z_0]$ and $[u_3, z_3]$ are proper subintervals of $[c_2, c_3]$ and $[b, a]$, respectively. By making the trivial replacements

$$
\begin{array}{lll}
[y_0, x_0] & \text{by} & [u_3, z_3], \\
[y_1, x_1] & \text{by} & [x_1 \wedge u_3, x_1 \wedge z_3], \\
[y_2, x_2] & \text{by} & [y_2 \vee (x_1 \wedge u_3), y_2 \vee (x_1 \wedge z_3)], \\
[y_3, x_3] & \text{by} & x_3 \wedge [x_3 \wedge (y_2 \vee (x_1 \wedge u_3)), (y_2 \vee (x_1 \wedge z_3))],
\end{array}
$$

we can assume that $[y_0, x_0]$ equals $[y_3, x_3]$. Thus by Theorem 353, there are diamonds

$$A_j = \{o_j, a_j, b_j, c_j, i_j\},$$

$j = 0, 1$, such that $[b, a] \overset{\text{dn}}{\sim} [a_0, i_0]$ and $[b, a] \overset{\text{up}}{\sim} [o_1, a_1]$. We conclude that $[o_1, a_1] \overset{\text{dn}}{\sim} [a_0, i_0]$ and so $A_0 \cup A_1$ is a sublattice of which $M_{3,3}$ is a homomorphic image, a contradiction.    $\square$

## Exercises

3.1. A *universal sentence* $\Psi$ is a sentence of the form

$$(x_0) \cdots (x_{n-1})\Phi,$$

and only $x_0, \ldots, x_{n-1}$ can occur as variables in $\Phi$. Show that every universal sentence in which no negation or implication occurs is equivalent to a finite set of UDE-s.

3.2. Show that a UDE is preserved under the formation of sublattices and homomorphic images.

3.3. Are all UDE-s preserved under direct products?

3.4. In the proof of Theorem 499, the statement is used that in an $n$-element lattice every bounded set of proper intervals is $n^2$-bounded. Can $n^2$ be improved in this statement?

3.5. Write out the formulas $\alpha_k$ and $\varrho_{m,r}$ to prove formally that these are indeed (first-order) formulas.

3.6. An algebra $(A; \vee, \wedge)$ is called a *weakly associative lattice*[1] (or WA lattice) if the following are satisfied: the two binary operations satisfy the idempotent, commutative, and absorption identities, and

$$x \leq z \text{ and } y \leq z \quad \text{imply that} \quad x \vee y \leq z$$

and its dual hold, where $a \leq b$ means that $a = a \wedge b$ or, equivalently, $a \vee b = b$ (E. Fried [193] and H. L. Skala [651]). Show that all the results of this section hold for WA lattices (K. A. Baker [41]).

*3.7. Take the three-element weakly associative lattice $\mathsf{T} = \{0, 1, 2\}$ defined by $0 \leq 1 \leq 2 \leq 0$. Find a finite equational basis for $\mathsf{T}$ (E. Fried and G. Grätzer [194]).

*3.8. If $L$ is a modular lattice of length $n$, then any bounded set of proper intervals is $k$-bounded, where $k \leq \left[\frac{3n}{2}\right] + 2$ and $[x]$ stands for the largest integer $\leq x$ (C. Herrmann [386]).

3.9. Let $\mathbf{M}^n$ denote the class of modular lattices of length at most $n$. Prove that $\mathbf{M}^n$ has a finite equational basis. (For $n = 2$, this is due to B. Jónsson [445]; for $n = 3$, this is due to D. X. Hong [399]. For general $n$, this is due to K. A. Baker; a reference to this fact and a proof of this result based on Exercise 3.8 is due to C. Herrmann [386].)

3.10. Show that a finite equational basis for $\mathbf{M}^2$, for the notation see Exercise 3.9, is given by the modular identity and

$$(x \wedge (y \vee (z \wedge u))) \vee (z \wedge u) \leq y \vee (x \wedge z) \vee (x \wedge u)$$

(B. Jónsson [445]). (Hint: Use Lemma 504 and the reasoning in the proof of Theorem 501.)

3.11. Let $L$ be a subdirectly irreducible lattice of length at most 3. Show that any bounded set of proper intervals is, in fact, 5-bounded (C. Herrmann [386]).

3.12. Let $\mathbf{L}_3$ denote the class of all lattices of length at most 3. Prove that $\mathbf{L}_3$ has a finite equational basis (C. Herrmann [386]).

3.13. Show that Lemma 503 can be derived from Theorem 353 (and it is implicit in the same).

3.14. Using the notation of Lemma 503, project $x^a$ into the intervals: $[o, a], [o, b], [o, c]$, obtaining the elements: $x_a, x_b, x_c$, respectively. Show that $\{o, x_a, x_b, x_c, o_1\}$ is a diamond.

*3.15. Let $L$ be a subdirectly irreducible lattice of width greater than 4. Then either the third lattice of Figure 85 or one of the eight lattices of Figures 114 is a homomorphic image of a sublattice of $L$ (R. Freese [181]).

---

[1] Also called a *trellis*. *MathSciNet* has about 50 references for this topic.

3.16. Let **M**(4) denote the class of all modular lattices of width 4. Prove that **M**(4) has a finite equational basis (R. Freese [181]). (Hint: use Exercise 3.15.)

3.17. Show that to prove Theorem 501 for a variety generated by a finite lattice, it is sufficient to analyze projectivities of prime intervals. To what extent would this simplify the proof?

3.18. Let $L$ be a modular lattice and let $n$ be a positive integer. After A. P. Huhn [403], the lattice $L$ is called $n$-*distributive* if the following identity holds:

$$x \wedge \bigvee_{0 \le i \le n} y_i = \bigvee_{0 \le j \le n} x \wedge \bigvee_{\substack{0 \le i \le n \\ i \ne j}} y_i.$$

The following form of the identity is easier to visualize:

$$x \wedge \bigvee_i y_i = \bigvee_j (x \wedge \bigvee_{i \ne j} y_i).$$

Prove that $L$ is $n$-distributive iff it satisfies the following identity in the variables $x_0, x_1, \ldots, x_{n+1}$:

$$\bigwedge_j (\bigvee_{i \ne j} x_i) = \bigvee_k (\bigwedge_{j \ne k} (\bigvee_{i \ne j,k} x_i))$$

(G. M. Bergman [59] and A. P. Huhn [403], [404]).

*3.19. For a positive integer $n$, we define a partial lattice $\mathsf{P}_n$ as follows: $\mathsf{P}_n = \mathsf{B}_{n+1} \cup \{w\}$. For $x, y \in \mathsf{B}_{n+1}$, we define $x \vee y$ and $x \wedge y$ as in $\mathsf{B}_{n+1}$. Define

$$w \vee 0 = 0 \vee w = w,$$
$$w \wedge 1 = 1 \wedge w = w.$$

For $x \in \mathsf{B}_{n+1} - \{1\}$, set

$$w \wedge x = x \wedge w = 0.$$

If $d$ is a dual atom of $\mathsf{B}_{n+1}$ or if $d = 1$,

$$w \vee d = d \vee w = 1.$$

The joins $w \vee x$ and $x \vee w$ are not defined for any $x \in \mathsf{B}_{n+1}$ satisfying $0 < \text{height}(x) < n$.

Prove that a modular lattice $L$ is $n$-distributive iff $L$ does not contain $\mathsf{P}_n$ as a partial sublattice (A. P. Huhn [404]).

3.20. Prove that every lattice of breadth at most $n$ is $n$-distributive. Find a counterexample to the converse for $n = 2$.

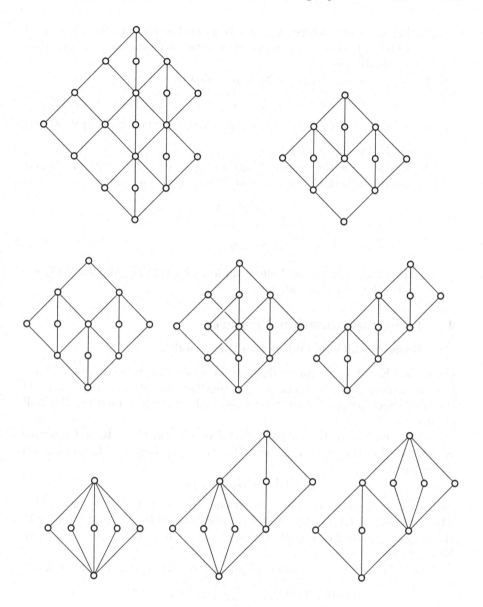

Figure 114. Eight lattices for Exercise 3.15

3.21. For a positive integer $k$ and a division ring $D$, form the lattice $L = PG(D, k)$. How is $k$ determined by the smallest integer $n$ such that $L$ is $n$-distributive?

*3.22. An equational basis for $\mathbf{N}_5$ is provided by

$$x \wedge (y \vee u) \wedge (y \vee v) \leq (x \wedge (y \vee (u \wedge v))) \vee (x \wedge u) \vee (x \wedge v),$$
$$x \wedge (y \vee (u \wedge (x \vee v))) = (x \wedge (y \vee (u \wedge x))) \vee (x \wedge ((x \wedge y) \vee (u \wedge v)))$$

(R. N. McKenzie [512]).

*3.23. Consider the three-element algebra $M = (\{0, 1, 2\}; \cdot)$ with one binary operation such that $0$ is a zero $(0 \cdot x = x \cdot 0 = 0)$ and

$$0 = 1 \cdot 1,$$
$$1 = 1 \cdot 2,$$
$$2 = 2 \cdot 1 = 2 \cdot 2.$$

Prove that $M$ has no finite equational basis (V. L. Murskiĭ [536]; see also R. C. Lyndon [507]).

## 4.   The Amalgamation Property

### 4.1   Basic definitions and elementary results

For a class $\mathbf{K}$ of lattices (or of algebras, in general), it is important to know how the lattices in the class can be "put together" to obtain a larger lattice of the class. Such properties are known as amalgamation properties. We shall discuss two.

A $\mathbb{V}$-*formation in* $\mathbf{K}$ is a pair of lattices $B_0$ and $B_1$ in $\mathbf{K}$ and a lattice $A \in \mathbf{K}$, a sublattice of both $B_0$ and $B_1$. More precisely, a $\mathbb{V}$-formation is a quintuple

$$(A, B_0, B_1, \varphi_0, \varphi_1)$$

such that $A, B_0, B_1 \in \mathbf{K}$ and $\varphi_i$ is an embedding of $A$ and $B_i$ for $i = 0, 1$. The $\mathbb{V}$-formation $(A, B_0, B_1, \varphi_0, \varphi_1)$ is *amalgamated by* $(\psi_0, \psi_1, C)$ if $C \in \mathbf{K}$, the map $\psi_i$ is an embedding of $B_i$ into $C$, for $i = 0, 1$, and $\psi_0 \varphi_0 = \psi_1 \varphi_1$ (see Figure 115).

The $\mathbb{V}$-formation is *strongly amalgamated by* $(\psi_0, \psi_1, C)$ if, in addition,

$$\psi_0(B_0) \cap \psi_1(B_1) = \psi_0 \varphi_0(A) \; (= \psi_1 \varphi_1(A)).$$

A class $\mathbf{K}$ is said to have the *(Strong) Amalgamation Property* if every $\mathbb{V}$-formation can be (strongly) amalgamated.

The class $\mathbf{L}$ has the Strong Amalgamation Property. To see this, take a $\mathbb{V}$-formation $(A, B_0, B_1, \varphi_0, \varphi_1)$; we can assume that $B_0 \cap B_1 = A$ and that $A$ is a sublattice of $B_0$ and $B_1$. On the set $P = B_0 \cup B_1$, we define an ordering $\leq$ as follows:

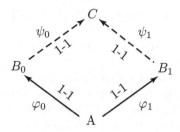

Figure 115. Amalgamation

(i) let $a, b \in B_i$; then $a \leq b$ in $P$ if $a \leq b$ in $B_i$ for $i = 0, 1$;

(ii) let $a \in B_i$ and $b \in B_j$ with $i \neq j$; then $a \leq b$ in $P$ if $a \leq c$ in $B_i$ and $c \leq b$ in $B_j$ for some $c \in A$ and $i, j \in \{0, 1\}$.

It is easy to check that $P$ is an order. Then

$$a \vee b = \sup\{a, b\},$$
$$a \wedge b = \inf\{a, b\}$$

(the $P^{\mathrm{max}}$ of Definition 87) turns $P$ into a partial lattice (see Lemma 88) of which $A, B_0, B_1$ are sublattices. Thus by Theorem 84, $P$ can be embedded into a lattice, proving the Strong Amalgamation Property for **L**.

The class **D** does not have the Strong Amalgamation Property. Indeed, take the distributive lattices $B_0 = B_1 = \mathsf{B}_2$ and $A = \mathsf{C}_3 = \{0, a, 1\}$, and let the maps $\varphi_0 = \varphi_1 = \varphi$ be given by

$$\varphi(x) = \begin{cases} (0, 0) & \text{for } x = 0; \\ (1, 0) & \text{for } x = a; \\ (1, 1) & \text{for } x = 1. \end{cases}$$

Let $(\psi_0, \psi_1, C)$ strongly amalgamate $(A, B_0, B_1, \varphi_0, \varphi_1)$ with $C \in \mathbf{D}$. Then $\psi_0(B_0) \cap \psi_1(B_1) = A$ and so $\psi_0(0, 1) \neq \psi_1(0, 1)$. Both of these elements are relative complements of $\psi_0(1, 0)$ $(= \psi_1(1, 0))$ in the interval $[\psi_0(0, 0), \psi_0(1, 1)]$ of $C$, contradicting $C \in \mathbf{D}$ and Corollary 103.

However, the class **D** has the Amalgamation Property. This we shall prove shortly, after some general remarks.

Let **K** be a variety of lattices (algebras, in general) and let

$$Q = (A, B_0, B_1, \varphi_0, \varphi_1)$$

be a $\mathbb{V}$-formation in **K**. We assume that $B_0 \cap B_1 = \varnothing$.

Let $F$ be the lattice freely generated in **K** by the partial lattice $B_0 \cup B_1$. We define in $F$ the congruence

$$\alpha = \bigvee (\mathrm{con}(\varphi_0(a), \varphi_1(a)) \mid a \in A).$$

Obviously, $\alpha$ is the smallest congruence relation of $F$ such that

$$\alpha_i \colon x \mapsto x/\alpha, \quad i = 0, 1,$$

is a homomorphism of $B_i$ into $F/\alpha$ and $\alpha_0 \varphi_0 = \alpha_1 \varphi_1$. Thus $(\alpha_0, \alpha_1, F/\alpha)$ amalgamates $Q$ iff $\alpha_0$ and $\alpha_1$ are one-to-one. Therefore, if $\alpha_0$ and $\alpha_1$ are one-to-one, then $Q$ can be amalgamated. Conversely, if $(\psi_0, \psi_1, C)$ amalgamates $Q$ with $C \in \mathbf{K}$, then map $B_0 \cup B_1$ into $C$ defined by

$$x \mapsto \psi_i(x), \quad \text{for } x \in B_i, \ i = 0, 1.$$

extends to a homomorphism $\beta$ of $F$ into $C$ and obviously $\mathrm{Ker}(\beta) \geq \alpha$, and $\beta(x) = \psi_i(x)$ for all $x \in B_i$. Thus $\alpha_i$ followed by the natural homomorphism of $F/\alpha$ into $C$ equals $\psi_i$, which is one-to-one by assumption. So $\alpha_i$ is one-to-one for $i = 0, 1$. We have proved (several of these results are from G. Grätzer [261]):

**Theorem 505.** *Let $\mathbf{K}$ be a variety, let*

$$Q = (A, B_0, B_1, \varphi_0, \varphi_1)$$

*be a $\mathbb{V}$-formation in $\mathbf{K}$, and let $F$ and $\alpha$ be as constructed above. Then $Q$ can be amalgamated in $\mathbf{K}$ iff, for $i = 0$ or $1$ and $x, y \in B_i$,*

$$x \equiv y \pmod{\alpha} \quad \text{implies that} \quad x = y.$$

**Corollary 506.** *A variety $\mathbf{K}$ has the Amalgamation Property iff, for every $\mathbb{V}$-formation $(A, B_0, B_1, \varphi_0, \varphi_1)$ in $\mathbf{K}$ and $x, y \in B_0$ with $x \neq y$, there exist a lattice $C \in \mathbf{K}$ and homomorphisms $\psi_i \colon B_i \to C$ such that $\psi_0 \varphi_0 = \psi_1 \varphi_1$ and $\psi_0(x) \neq \psi_0(y)$.*

*Proof.* In order to prove the $\alpha_i$ (introduced in the proof of Theorem 505) one-to-one, it is sufficient to have homomorphisms $\psi_i$ separating a pair of distinct elements. The necessity is obvious. $\quad\square$

**Corollary 507.** *Let $\mathbf{K}$ be a variety and let*

$$Q = (A, B_0, B_1, \varphi_0, \varphi_1)$$

*be a $\mathbb{V}$-formation in $\mathbf{K}$. If $Q$ cannot be amalgamated in $\mathbf{K}$, then there are finitely generated subalgebras $A'$ of $A$ and $B_i'$ of $B_i$, for $i = 0, 1$, such that*

$$Q' = (A', B_0', B_1', \varphi_0', \varphi_1')$$

*cannot be amalgamated in $\mathbf{K}$, where $\varphi_i'$ is the restriction of $\varphi_i$ to $A'$ for $i = 0, 1$.*

*Proof.* Let $F$ and $\alpha$ be given as in Theorem 505. If $Q$ cannot be amalgamated in $\mathbf{K}$, then there are $i \in \{0,1\}$ and $x, y \in B_i$ such that $x \neq y$ and $x \equiv y$ (mod $\alpha$). By Theorem 230 and Lemma 233, we can select finite subsets $A^*$ of $A$ and $B_i^*$ of $B_i$, for $i = 0, 1$, such that in computing $x \equiv y$ (mod $\alpha$), we use only elements of $A^* \cup B_0^* \cup B_1^*$. Thus we can set $A' = \mathrm{sub}(A^*)$ and $B_i' = \mathrm{sub}(B_i^*)$ for $i = 0, 1$. A trivial application of Theorem 505 shows that $Q'$ cannot be amalgamated in $\mathbf{K}$.                                         □

Call a $\mathbb{V}$-formation $(A, B_0, B_1, \varphi_0, \varphi_1)$ *finitely generated* (resp., *finite*) if $A, B_0, B_1$ are finitely generated (resp., finite).

**Corollary 508.** *A variety $\mathbf{K}$ has the Amalgamation Property iff every finitely generated $\mathbb{V}$-formation in $\mathbf{K}$ can be amalgamated in $\mathbf{K}$.*

A variety $\mathbf{K}$ is called *locally finite* if every finitely generated member of $\mathbf{K}$ is finite. It is easily seen that $\mathbf{K}$ is locally finite iff $\mathrm{Free}_{\mathbf{K}}(n)$ is finite for all $n < \omega$. This always holds if $\mathbf{K}$ is generated by a single finite lattice $L$. Let $\mathbf{K}_{\mathrm{fin}}$ denote the class of finite members of $\mathbf{K}$.

**Corollary 509.** *Let $\mathbf{K}$ be a locally finite variety. Then $\mathbf{K}$ has the Amalgamation Property iff all finite $\mathbb{V}$-formations in $\mathbf{K}$ can be amalgamated in $\mathbf{K}$, or equivalently, iff $\mathbf{K}_{\mathrm{fin}}$ has the Amalgamation Property.*

The last equivalence follows from the observation that if

$$Q = (A, B_0, B_1, \varphi_0, \varphi_1)$$

is finite and can be amalgamated in $\mathbf{K}$, then some quotient of $\mathrm{Free}_{\mathbf{K}} Q$ will amalgamate $Q$; but $\mathrm{Free}_{\mathbf{K}} Q = \mathrm{Free}_{\mathbf{K}}(B_0 \cup B_1)$ is finite since $\mathbf{K}$ is locally finite.

Now we return to the class $\mathbf{D}$. By Theorem 126, $|\mathrm{Free}_{\mathbf{D}}(n)| < 2^{2^n}$ and so $\mathbf{D}$ is locally finite.

**Corollary 510.** *The class $\mathbf{D}$ has the Amalgamation Property.*

*Proof.* Let

$$Q = (A, B_0, B_1, \varphi_0, \varphi_1)$$

be a finite $\mathbb{V}$-formation in $\mathbf{D}$ and let $x, y \in B_0$ with $x \neq y$. Set $C = \mathsf{C}_2$ and let $\psi_0$ be a homomorphism of $B_0$ into $\mathsf{C}_2$ such that $\psi_0(x) \neq \psi_0(y)$ (see Corollary 109). Let $P$ be the ideal kernel of $\psi_0 \varphi_0$. If $P = A$, define the map $\psi_1 \colon B_1 \to \mathsf{C}_2$ by $\psi_1(x) = 0$ for all $x \in B_1$. If $P = \varnothing$, define $\psi_1 \colon B_1 \to \mathsf{C}_2$ by $\psi_1(x) = 1$ for all $x \in B_1$. If $P \neq A$ and $P \neq \varnothing$, then $P = \mathrm{id}(a)$, where the element $a \neq 1$ is meet-irreducible in $A$.

Thus there is a unique $b \in A$ such that $b \succ a$. By Corollary 111, there is a meet-irreducible element $p$ in $B_1$ such that $a \leq p$ and $b \not\leq p$. We then define the map $\psi_1 \colon B_1 \to C_2$ by

$$\psi_1(x) = \begin{cases} 0, & \text{if } x \le p; \\ 1, & \text{if } x \nleq p. \end{cases}$$

It is obvious that $(\psi_0, \psi_1, C_2)$ satisfies the conditions of Corollary 506, thus $Q$ can be amalgamated in **D**.    □

To further illustrate the usefulness of Corollary 506, for a class **K** of lattices (or algebras, in general) define $A \in \mathbf{K}$ to be *injective* in **K** if, for every $B, C \in \mathbf{K}$ with $B \le C$, any homomorphism of $B$ into $A$ can be extended to a homomorphism of $C$ into $A$.

The following result of R. S. Pierce [579] connects injectivity with the Amalgamation Property.

**Corollary 511.** *Let* **K** *be a variety. If any member of* **K** *can be embedded in an injective member of* **K***, then* **K** *has the Amalgamation Property.*

*Proof.* We apply Corollary 506. Let $C$ be an injective member of **K** into which $B_0$ can be embedded; let $\psi_0$ be this embedding. Then $\psi_0(x) \ne \psi_0(y)$. Set $A' = \varphi_1(A)$. Then $\psi_0 \varphi_1^{-1}$ is a homomorphism (in fact, an embedding) of $A'$ into $C$, thus this homomorphism can be extended to a homomorphism $\psi_1$ into $B_1$ into $C$. Obviously $\psi_0 \varphi_0 = \psi_1 \varphi_1$.    □

A further application of Corollary 507 will be given at the end of this section.

A typical application of the Amalgamation Property is the sublattice theorem of free products of lattices discussed in Section VII.1.8.

### 4.2    Lattice varieties with the Amalgamation Property

The Amalgamation Property for lattices and orders was discussed in 1956 in B. Jónsson [435]; it was observed for boolean algebras in 1965 in A. Daigneault [121]. The early development of the Amalgamation Property was described in 1965 in B. Jónsson [443]. It was B. Jónsson's work more than any other influence that convinced the algebraists of the importance of this property.

There were only three lattice varieties known to have the Amalgamation Property: **T**, **D**, and **L**. The major open problem was: are there any others.

The solution came in two steps.

In January of 1971, B. Jónsson published an abstract in the Notices of the American Mathematical Society **18** (1971), p. 400, stating that the class **M** of modular lattices does not have the Amalgamation Property, solving a long standing problem of lattice theory. In fact, he stated more: Any equational class **K** of modular lattices having the Amalgamation Property must satisfy the arguesian identity. This was followed by an announcement by G. Grätzer and H. Lakser (see volume **18**, p. 618) stating that every member of such a variety $\mathbf{K}$ can be embedded into the subspace lattice of an infinite dimensional projective geometry. The two manuscripts were combined in G. Grätzer, B. Jónsson, and H. Lakser [283], proving that there are no other modular varieties with the Amalgamation Property.

More than a decade later, in the paper A. Day and J. Ježek [141] (utilizing the work of V. Slavík [653], see Section IV.2), the nonmodular case was handled, yielding the result:

$\Diamond$ **Theorem 512.** *There are only three lattice varieties with the Amalgamation Property:* $\mathbf{T}$, $\mathbf{D}$, *and* $\mathbf{L}$.

This result is far too technical to prove in this book. But we are going to discuss two interesting special cases. The first result is due to A. Day, S. D. Comer, and S. Fajtlowicz, as quoted in G. Grätzer, B. Jónsson, and H. Lakser [283].

**Theorem 513.** *Let* $\mathbf{K}$ *be a variety generated by a finite lattice. If* $\mathbf{K} \supset \mathbf{D}$, *then* $\mathbf{K}$ *does not have the Amalgamation Property.*

*Proof.* If $\mathbf{K}$ is generated by a finite lattice $L$, then by Corollary 477, $\mathbf{Si}(\mathbf{K}) \subseteq \mathbf{HS}(L)$. Thus no subdirectly irreducible lattice in $\mathbf{K}$ has more than $n = |L|$ elements.

Since $\mathbf{K} \supset \mathbf{D}$, there is a nondistributive lattice in $\mathbf{K}$ and thus $\mathsf{N}_5$ or $\mathsf{M}_3 \in \mathbf{K}$, by Theorem 101.

If $\mathsf{N}_5 \in \mathbf{K}$ (we use the notation $\mathsf{N}_5 = \{o, a, b, c, i\}$ as in Figure 24), then consider the $\mathbb{V}$-formation $(\mathsf{C}_2, \mathsf{N}_5, \mathsf{N}_5, \varphi_0, \varphi_1)$, where $\mathsf{C}_2 = \{0, 1\}$ and

$$\varphi_0(0) = o, \qquad \varphi_0(1) = i,$$
$$\varphi_1(0) = b, \qquad \varphi_1(1) = a.$$

Let $(\psi_0, \psi_1, A)$ amalgamate this $\mathbb{V}$-formation. Then $A$ will have the lattice $A_1$ of Figure 116 as a sublattice. In fact, $A_1$ is the union of the two images of $\mathsf{N}_5$ in $A$. Observe that $A_1$ is again subdirectly irreducible and $|A_1| = 8$. Then we take the $\mathbb{V}$-formation $(\mathsf{C}_2, \mathsf{N}_5, A_1, \varphi_0, \varphi_1)$ with

$$\varphi_0(0) = o, \qquad \varphi_0(1) = i,$$
$$\varphi_1(0) = b, \qquad \varphi_1(1) = a.$$

The union of the images of $\mathsf{N}_5$ and $A_1$ will form a sublattice $A_2$, which is again subdirectly irreducible and $|A_2| = 11$. Proceeding by induction, we obtain the subdirectly irreducible lattice $A_k \in \mathbf{K}$ with $|A_k| = 5 + 3k$. Choosing $k$ so that $5 + 3k > n$, we get a contradiction.

If $\mathsf{M}_3 \in \mathbf{K}$ (using the notation $\mathsf{M}_3 = \{o, a, b, c, i\}$ as in Figure 24), then we take the $\mathbb{V}$-formation $(\mathsf{C}_2, \mathsf{M}_3, \mathsf{M}_3, \varphi_0, \varphi_1)$, where

$$\varphi_0(0) = a, \qquad \varphi_0(1) = i,$$
$$\varphi_1(0) = o, \qquad \varphi_1(1) = c.$$

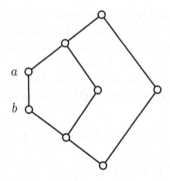

Figure 116. Proving Theorem 513

Then the union of the images is again a sublattice of the amalgam, thus obtaining $\mathsf{M}_{3,3} \in \mathbf{K}$. Proceeding the same way, we obtain a sequence of simple modular lattices of increasing size, leading to a contradiction as above. $\quad\square$

**Theorem 514.** $\mathbf{M}$ *does not have the Amalgamation Property.*

*Remark.* Theorem 512 states that the same result holds for all modular varieties $\mathbf{K} \supset \mathbf{D}$. The proof of this more general result is quite technical, but the basic idea is illustrated in the special case $\mathbf{K} = \mathbf{M}$.

Let us assume that $\mathbf{M}$ has the Amalgamation Property; under this assumption, we prove a few embedding theorems.

**1.** *Any modular lattice $A$ can be embedded in a bounded modular lattice $B$ that has a five-element chain.*

*Proof.* This is trivial. If $A$ has no five-element chain, we add new zeros and ones until it has a five-element chain. If $A$ is not bounded, we add bounds. $\quad\square$

**2.** *Any bounded modular lattice $A$ can be embedded in a modular lattice $B$ having the same bounds and having the property that, for every $a \in A$ with $a \neq 0,1$, the lattice $B$ has a diamond $\{0, a, b, c, 1\}$.*

*Proof.* To prove this, let $A - \{0,1\} = \{ a_\gamma \mid \gamma < \alpha \}$ and we define an increasing sequence of modular lattices $A_\gamma$ with the same bounds 0 and 1 for all $\gamma < \alpha$. Set $\overline{A}_0 = A$. If the $A_\gamma$, for $\gamma < \delta$ (and $\delta < \alpha$), have already been defined, set

$$\overline{A}_\delta = \bigcup ( A_\gamma \mid \gamma < \delta ).$$

Let $\mathsf{C}_3 = \{0,1,2\}$ and consider the $\mathbb{V}$-formation $Q_\delta = (\mathsf{C}_3, \mathsf{M}_3, \overline{A}_\delta, \varphi_0, \varphi_1)$ defined by

$$\varphi_0(0) = o, \qquad \varphi_0(1) = a, \qquad \varphi_0(2) = i,$$
$$\varphi_1(0) = 0, \qquad \varphi_1(1) = a_\delta, \qquad \varphi_1(2) = 1.$$

Let $(\psi_0, \psi_1, A)$ amalgamate $Q_\delta$. By forming the interval $[\psi_0\varphi_0(0), \psi_0\varphi_0(2)]$ in the amalgamated lattice, we obtain $A_\delta$. It is obvious that $A_\delta$ with $a = a_\delta$ has the property required of $B$. We define $B = \bigcup(A_\gamma \mid \gamma < \alpha)$, which obviously has the required properties. $\qquad\square$

**3.** *Any bounded modular lattice $A$ can be embedded in a modular lattice $B$ having the same bounds as $A$ and having the following properties:*

  (i) *For every $a \in B$ with $a \neq 0, 1$, there is a diamond $\{0, a, b, c, 1\}$ in $B$.*
  (ii) *$B$ is a simple, complemented, modular lattice.*

*Proof.* Indeed, set $A_0 = A$, and if $A_n$ is defined, then define $A_{n+1}$ as the lattice constructed from $A_n$ in the second embedding theorem. Then

$$B = \bigcup(A_n \mid n < \omega)$$

obviously satisfies (i).

Let $\alpha$ be a congruence relation of $B$ with $\mathbf{0} \neq \alpha$. Since $B$ is a complemented modular lattice, $B$ is relatively complemented; hence there is an $a \in B$ such that $a \neq 0$ and $a \equiv 0 \pmod{\alpha}$. If $a \neq 1$, then, by (i), there is a diamond $\{0, a, b, c, 1\}$. Since $\mathsf{M}_3$ is simple, we obtain that $0 \equiv 1 \pmod{\alpha}$, that is, $\alpha = \mathbf{1}$. If $a = 1$, again $\alpha = \mathbf{1}$. Thus $B$ is simple. $\qquad\square$

*Proof of Theorem 514.* Now we are ready to prove the theorem.

Let $A \in \mathbf{M}$. By the embeddings 1–3, we can embed $A$ into a simple complemented modular lattice $B$ having a five-element chain. By Theorem 438, $B$ can be $\{0, 1\}$-embedded into a modular geometric lattice $C$. The lattice $C$ is directly indecomposable because it has a simple $\{0, 1\}$-sublattice (namely, $B$). The lattice $C$ has a five-element chain because $B$ has one. Thus Corollary 435 applies and $C \cong \mathrm{PG}(D, \mathfrak{m})$. By Lemma 428 and Corollary 431, the arguesian identity holds in $C$. Since $A \leq C$, the arguesian identity holds in $A$.

The lattice $A$ is an arbitrary member of $\mathbf{M}$ and the arguesian identity does not hold in $\mathbf{M}$; this is a contradiction. $\qquad\square$

## 4.3 The class Amal(K)

The Amalgamation Property fails for most varieties of lattices; so it seems reasonable to ask to what extent it holds in general. The answer is rather surprising.

Following G. Grätzer and H. Lakser [298], for a class $\mathbf{K}$ of lattices (or algebras, in general), let $\mathbf{Amal(K)}$ be the class of all those lattices $A \in \mathbf{K}$ for which all $\mathbb{V}$-formations in $\mathbf{K}$ of the form $(A, \ldots)$ can be amalgamated in $\mathbf{K}$. Obviously, $\mathbf{K}$ has the Amalgamation Property iff $\mathbf{Amal(K)} = \mathbf{K}$.

It is easy to see that $\mathsf{C}_1 \in \mathbf{Amal(K)} \neq \varnothing$ for every variety $\mathbf{K}$ of lattices. To prove this, take the $\mathbb{V}$-formation $(\mathsf{C}_1, B_0, B_1, \varphi_0, \varphi_1)$. We can amalgamate the V-formation with $(\psi_0, \psi_1, B_0 \times B_1)$, where for $x \in B_0$, we define $\psi_0(x) = (x, \varphi_1(0))$ and symmetrically.

Call a subclass $\mathbf{K}_1$ of the class $\mathbf{K}$ *cofinal in* $\mathbf{K}$ if for all $A \in \mathbf{K}$, there is an extension $B \in \mathbf{K}_1$.

M. Yasuhara [740] proves that $\mathbf{Amal}(\mathbf{K})$ is a large subclass for a variety $\mathbf{K}$.

**Theorem 515.** *For a variety* $\mathbf{K}$*, the class* $\mathbf{Amal}(\mathbf{K})$ *is cofinal in* $\mathbf{K}$*.*

The proof of this result will be given in the next two lemmas. Some of the ideas of Lemma 516 originate in A. Robinson [616], [617]. The present form of Lemma 516 is a slight variant of a lemma of M. Yasuhara and is due to M. Makkai.

For a class $\mathbf{K}$, let $\mathbf{Ec}(\mathbf{K})$ stand for the class of all $A \in \mathbf{K}$ having the following property:

> For any extension $B \in \mathbf{K}$ of $A$ and for any finite $X \subseteq A$ and $Y \subseteq B$, there is an embedding $\varphi \colon [X \cup Y] \to A$ fixing $X$ (that is, $\varphi(x) = x$ for all $x \in X$).

**Lemma 516.** *For a variety* $\mathbf{K}$*, the subclass* $\mathbf{Ec}(\mathbf{K})$ *is cofinal in* $\mathbf{K}$*.*

*Proof.* To facilitate the proof, we introduce some concepts. For $A \in \mathbf{K}$, define $(\varphi, X, C)$ to be a *triple over* $A$ *in* $\mathbf{K}$ if the following three conditions are satisfied:

(i) $X$ is a finite subset of $A$;
(ii) $C$ is a finitely generated member of $\mathbf{K}$;
(iii) $\varphi \colon X \to C$ is a homomorphism of the partial sublattice $X$ of $A$ into $C$.

The last condition means that if $x \vee y = z$ or $x \wedge y = z$, for $x, y, z \in X$, then $\varphi(x) \vee \varphi(y) = \varphi(z)$ or $\varphi(x) \wedge \varphi(y) = \varphi(z)$ in $C$, respectively.

The triple $(\varphi, X, C)$ *is realized over* $A$ *by* $B$ if the following three conditions are satisfied:

(i) $B \in \mathbf{K}$;
(ii) $B$ is an extension of $A$;
(iii) there is an embedding $\psi$ of $C$ into $B$ such that $\psi\varphi(x) = x$ for all $x \in X$.

Now let $(\varphi_\gamma, X_\gamma, C_\gamma)$, for $\gamma < \alpha$, list all the triples over $A$. We define two increasing sequences of lattices as follows:

$$\overline{A}_0 = A;$$

if $A_\gamma$ has been defined for $\gamma < \delta < \alpha$, then set

$$\overline{A}_\delta = \bigcup ( A_\gamma \mid \gamma < \delta ).$$

Regard $(\varphi_\delta, X_\delta, C_\delta)$ as a triple over $\overline{A}_\delta$. If it is realized in $\mathbf{K}$, define $A_\delta$ as any member of $\mathbf{K}$ realizing it. If it cannot be realized in $\mathbf{K}$, define $A_\delta = \overline{A}_\delta$. Then set

$$A^{(1)} = \bigcup(A_\gamma \mid \gamma < \alpha),$$
$$A^{(n+1)} = (A^{(n)})^{(1)},$$
$$A^* = \bigcup(A^{(n)} \mid n < \omega).$$

Since $\mathbf{K}$ is a variety, it follows that $A^* \in \mathbf{K}$. We claim that $A^* \in \mathbf{Ec}(K)$. Indeed, let $B \in \mathbf{K}$ be an extension of $A^*$, and choose finite subsets $X \subseteq A^*$ and $Y \subseteq B$. Define $C = \mathrm{sub}(X \cup Y)$ and let $\varphi\colon X \to A$ be the identity map. Then $(\varphi, X, C)$ is a triple over $A^*$ that can be realized by $B$. Since $X \subseteq A^*$ and $X$ is finite, we obtain that $X \subseteq A^{(n)}$ for some $n < \omega$. Hence $(\varphi, X, C)$ can be regarded as a triple over $A^{(n)}$. Thus $(\varphi, X, C)$ occurs as some $(\varphi_\delta, X_\delta, C_\delta)$ in the list of all triples over $A^{(n)}$.

In the $\delta$-th step of the construction of $A^{(n+1)} = (A^{(n)})^{(1)}$, we view $(\varphi, X, C)$ as a triple over $(\overline{A^{(n)}})_\delta$; we observe that $(\varphi, X, C)$ can be realized by $B$, and so $(A^{(n)})_\delta$ realizes $(\varphi, X, C)$, that is, there is an embedding $\psi\colon C \to (A^{(n)})_\delta$ such that $\psi\varphi(x) = x$ for all $x \in X$. However, $\varphi(x) = x$, for all $x \in X$, and therefore, $\psi\varphi(x) = \psi(x) = x$ for all $x \in X$. Thus $\psi$ is an embedding of $\mathrm{sub}(X \cup Y)$ into $A^*$ keeping $X$ fixed, proving that $A^* \in \mathbf{Ec}(K)$. $\quad\square$

**Lemma 517.** *For any variety $\mathbf{K}$, the subclass $\mathbf{Ec}(\mathbf{K}) \subseteq \mathbf{Amal}(\mathbf{K})$.*

*Proof.* Let $A \in \mathbf{Ec}(\mathbf{K})$ and consider a $\mathbb{V}$-formation $Q = (A, B_0, B_1, \varphi_0, \varphi_1)$ in $\mathbf{K}$. We can assume that $A \leq B_0, B_1$ and $\varphi_0 = \varphi_1$ is the identity map on $A$. We wish to show that $Q$ can be amalgamated in $\mathbf{K}$. Assume, to the contrary, that $Q$ cannot be amalgamated in $\mathbf{K}$. Then, by Corollary 507, there exist finite subsets $X \subseteq A$, $Y_0 \subseteq B_0$, and $Y_1 \subseteq B_1$ such that

$$Q' = (\mathrm{sub}(X), \mathrm{sub}(X \cup Y_0), \mathrm{sub}(X \cup Y_1), \varphi_0', \varphi_1')$$

cannot be amalgamated, where $\varphi_i'$ is the restriction of $\varphi_i$ to $\mathrm{sub}(X \cup Y_i)$ for $i = 0, 1$. Since $A \in \mathbf{Ec}(K)$, there exist embeddings $\psi_i\colon \mathrm{sub}(X \cup Y_i) \to A$ keeping $X$ fixed for $i = 0, 1$. Thus $\psi_i\varphi_i(x) = x$, for $x \in X$ and $i = 0, 1$, and so $\psi_i\varphi_i(a) = a$ for all $a \in \mathrm{sub}(X)$. This shows that $(\psi_0, \psi_1, A)$ amalgamates $Q'$, which is a contradiction. $\quad\square$

The existentially complete algebras of M. Yasuhara [740] are different from members of $\mathbf{Ec}(\mathbf{K})$. Thus Lemma 516 is slightly stronger and Lemma 517 is slightly weaker than the corresponding results in M. Yasuhara [740].

Finite members of $\mathbf{Amal}(\mathbf{M}_3)$, $\mathbf{Amal}(\mathbf{M}_4)$, and so on, are described in E. Fried, G. Grätzer, and H. Lakser [200].

G. Grätzer, B. Jónsson, and H. Lakser [283] states that if **K** is a variety of modular lattices and **K** is not arguesian, then $C_2 \notin \mathbf{Amal(K)}$. This was strengthened in G. Grätzer and H. Lakser [302].

The amalgamation class of the variety generated by the pentagon was described in B. Jónsson [452].

For a variety **V**, is $\mathbf{Amal(V)}$ an elementary class? C. Bergman [57] proved that the answer is negative for the variety **V** generated by a suitable finite modular lattice. See also P. Ouwehand and H. Rose [565].

## Exercises

4.1. Show that the class of all groups has the Strong Amalgamation Property.

4.2. Investigate which varieties of pseudocomplemented distributive lattices have which Amalgamation Property (G. Grätzer and H. Lakser [298]).

4.3. Show that $\mathbf{L_{fin}}$ has the Strong Amalgamation Property.

4.4. Define the $\mathbf{StAmal(K)}$ as the analogue of $\mathbf{Amal(K)}$ for the Strong Amalgamation Property. Determine $\mathbf{StAmal(D)}$.

4.5. Show that "$C \in \mathbf{K}$" can be changed to "$C \in \mathbf{Si(K)}$" in Corollary 506.

4.6. Did we use the Axiom of Choice in verifying Corollary 509?

4.7. Let $L$ be injective in **K**. If $A$ is an extension of $L$ in **K**, then $L$ is a *retract* of $A$, that is, there is a homomorphism $\varphi \colon A \to L$ such that $\varphi(x) = x$ for all $x \in L$.

4.8. Show that any retract of a complete boolean lattice is a complete boolean lattice.

4.9. A distributive lattice $L$ is injective in **D** iff $L$ is a complete boolean lattice (P. R. Halmos [371]). (Hint: use Exercises 4.7, 4.8, and Corollary 132.)

4.10. Prove Corollary 510 using Corollary 511.

4.11. Find further examples of the phenomenon observed twice in the proof of Theorem 513, namely, that for some special $\vee$-formation, if $(\psi_0, \psi_1, C)$ amalgamates $(A, B_0, B_1, \varphi_0, \varphi_1)$, then $\psi_0(B_0) \cup \psi_1(B_1)$ is a uniquely determined sublattice of $C$. (This topic is discussed in V. Slavík [653], [654] and later in E. Fried and G. Grätzer [195]–[199]— in these five papers this special amalgamation was called *pasting*—and E. Fried, G. Grätzer, and E. T. Schmidt [202]. The main result of E. Fried and G. Grätzer [195] and [196] is that **M** is closed under pasting.)

4.12. Show that there are $2^{\aleph_0}$ varieties of modular lattices not having the Amalgamation Property.

4.13. Let **K** be a variety. Then, for every $A \in \mathbf{K}$, we construct an extension $B \in \mathbf{Amal(K)}$. Give an upper bound for $|B|$.

*        *        *

The following exercises are from G. Grätzer, B. Jónsson, and H. Lakser [494]. Exercise 4.14 is implicit in B. H. Neumann and H. Neumann [550].

4.14. Let **K** be a variety having the Amalgamation Property. Let $A, B \in \mathbf{K}$ with $A \leq B$, and let $\alpha$ be an automorphism of $A$. Show that there exist an extension $C \in \mathbf{K}$ of $B$ and an automorphism $\beta$ of $C$ extending $\alpha$.

4.15. Let **K** be a variety of lattices. Then $\mathsf{C}_2 \in \mathbf{Amal(K)}$ iff there is a $D \in \mathbf{Amal(K)}$ with $|D| > 1$.

4.16. Let **K** be a variety of lattices. Prove that if $\operatorname{Id} L \in \mathbf{Amal(K)}$, then $L \in \mathbf{Amal(K)}$.

4.17. Let $\mathbf{K} = \mathbf{Var}(L)$, where $L$ is a finite subdirectly irreducible lattice. Then $L \in \mathbf{Amal(K)}$.

# Chapter

# VII

# Free Products

## 1.  Free Products of Lattices

### 1.1  Introduction

The formation of a free product of a family of lattices is one of the most fundamental constructions of lattice theory. This specializes to the construction of free lattices, which form a class of lattices that is probably the closest rival of the class of distributive lattices in the richness of its structure. Also, free products provide a very useful tool for the construction of pathological lattices.

Most of the results are based on the Structure Theorem for Free Products, Theorem 528. Since this theorem is based on a large number of long inductive definitions, we shall present first a short intuitive description of its contents. We ask the reader to follow it up with a careful reading of the theorem, bearing in mind that the final result is very simple and most efficient in use.

Let $A$ and $B$ be lattices and let $L$ be a free product of $A$ and $B$, in symbols, $L = A * B$; for the present discussion, this should mean that $A$ and $B \leq L$, the set $A \cup B$ generates $L$, and $L$ is the "most general" lattice with these properties, as in Section I.5.1. Figure 117 provides two small examples: the diagrams of $C_2 * C_1$ and $C_3 * C_1$.

Since $L$ is generated by $A \cup B$, every element of $L$ can be written in the form

$$p(a_0, \ldots, a_{n-1}, b_0, \ldots, b_{m-1}), \quad a_0, \ldots, a_{n-1} \in A, \ b_0, \ldots, b_{m-1} \in B,$$

G. Grätzer, *Lattice Theory: Foundation*, DOI 10.1007/978-3-0348-0018-1_7,
© Springer Basel AG 2011

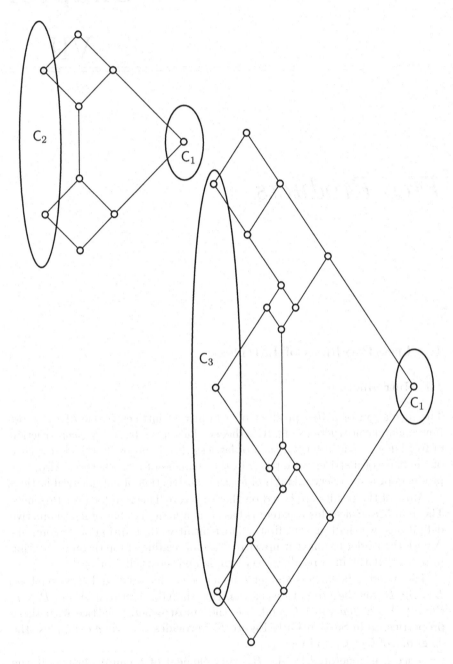

Figure 117. The lattices $C_2 * C_1$ and $C_3 * C_1$

where $p$ is an $(n+m)$-ary lattice term. For instance, if $A$ and $B$ are as given in Figure 118, then

$$a_1,$$

$$b_2,$$

$$a_3 \vee b_1,$$

$$(a_3 \wedge b_2) \vee (a_0 \wedge b_0),$$

$$((a_5 \wedge b_0) \vee a_3) \wedge ((a_2 \wedge b_0) \vee a_2 \vee b_3)$$

are examples of elements of $L$. Observe, however, that expressions of this sort that appear to be very different may, in fact, denote the same element of $L$. For instance,

$$a_5 \vee (b_3 \wedge ((a_3 \vee b_3) \wedge (a_4 \vee b_3))) = a_5 \vee b_3.$$

Theorem 528 will help the reader to construct more complicated examples.

The problem is, then, how to decide whether two expressions represent the same element of $L$. The key fact is that if

$$p = p(a_0, \ldots, a_{n-1}, b_0, \ldots, b_{m-1}) \le a$$

for some $a \in A$, then there is a smallest element of $A$ with this property. In fact, this smallest element (called the *upper cover* of $p$ in $A$) is easy to compute knowing the element $p$ and the lattices $A$ and $B$.

So our plan is the following: define formally the free product, expressions of the form $p(a_0, \ldots, a_{n-1}, b_0, \ldots, b_{m-1})$, upper and (dually) lower covers, present the algorithm deciding whether

$$p(a_0, \ldots, a_{n-1}, b_0, \ldots, b_{m-1}) \le q(a_0, \ldots, a_{n-1}, b_0, \ldots, b_{m-1}),$$

and, finally, prove that this algorithm works.

*Remark.* The idea of the upper and lower cover of an element in a free product, and how to compute them, goes back to R. P. Dilworth [155]. It becomes more explicit in C. C. Chen and G. Grätzer [88], in which the free product of a lattice $L$ with $\mathrm{Free}(\aleph_0)$ is considered.

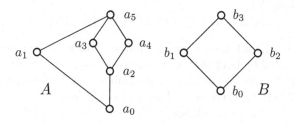

Figure 118. Two sample lattices

## 1.2   The basic definitions

In this section, let $\mathcal{L} = (L_i \mid i \in I)$ be a fixed family of lattices with $I \neq \varnothing$; we assume that $L_i$ and $L_j$ are disjoint for all $i, j \in I$ with $i \neq j$. We set

$$\bigcup \mathcal{L} = \bigcup ( L_i \mid i \in I )$$

and we consider $\bigcup \mathcal{L}$ an order under the natural ordering:

For $a, b \in \bigcup \mathcal{L}$, $a \leq b$ if $a, b \in L_i$ and $a \leq b$ in $L_i$ for some $i \in I$.

A free product $L$ of the $L_i$, for $i \in I$, is a lattice

$$\text{Free} \bigcup \mathcal{L} \quad (= \text{Free}_{\mathbf{L}} \bigcup \mathcal{L})$$

freely generated by $\bigcup \mathcal{L}$ in the sense of Definition 66. Or, more generally, stated for any variety $\mathbf{K}$ of lattices:

**Definition 518.** Let $\mathbf{K}$ be a variety of lattices and let $L \in \mathbf{K}$. Let $L_i \in \mathbf{K}$ for $i \in I$. The lattice $L \in \mathbf{K}$ is a *free $\mathbf{K}$-product of the lattices $L_i$, for $i \in I$*, if the following conditions are satisfied:

(i) Each $L_i \leq L$ and, for all $i, j \in I$ with $i \neq j$, the sublattices $L_i$ and $L_j$ are disjoint.

(ii) The lattice $L$ is generated by $\bigcup ( L_i \mid i \in I )$.

(iii) For any lattice $A \in \mathbf{K}$ and for any family of homomorphisms $\varphi_i \colon L_i \to A$, for $i \in I$, there exists a homomorphism $\varphi \colon L \to A$ such that the homomorphism $\varphi$ on $L_i$ agrees with the homomorphism $\varphi_i$ for all $i \in I$.

If $\mathbf{K} = \mathbf{L}$, the lattice $L$ is called a *free product*.

The next definition is a slight adaptation of Definition 52.

**Definition 519.** Let $X$ be an arbitrary set. The set $\text{Term}(X)$ of terms in $X$ is the smallest set satisfying (i) and (ii):

(i) $X \subseteq \text{Term}(X)$.
(ii) If $p, q \in \text{Term}(X)$, then $(p \vee q), (p \wedge q) \in \text{Term}(X)$.

The reader should keep in mind that a term is a sequence of symbols (no longer denoted by bold characters and symbols) and equality means formal equality (that is, equality as strings of symbols). As before, parentheses will be dropped whenever there is no danger of confusion.

In what follows, we shall deal with terms in $\bigcup \mathcal{L}$. Let $a, b, c \in L_i$, and let $a \vee b = c$ in $L_i$. Observe that $a \vee b$—which stands for $(a \vee b)$—and $c$ are distinct as terms in $\bigcup \mathcal{L}$.

## 1.3   Covers

It will be convenient to use a lattice $A$ with two new bounds:

**Definition 520.** For a lattice $A$, we define $A^b = A \cup \{0^b, 1^b\}$, where $0^b, 1^b \notin A$; we order $A^b$ by the two rules:

(i) $0^b < x < 1^b$ for all $x \in A$.
(ii) $x \leq y$ in $A^b$ if $x \leq y$ in $A$ for all $x, y \in A$.

Thus $A^b$ is a bounded lattice. Note, however, that $A^b \neq A$ even if $A$ is itself bounded. It is important to observe that $0^b$ is meet-irreducible and $1^b$ is join-irreducible. Thus if $a \vee b = 1^b$ in $A^b$, then either $a$ or $b$ is $1^b$, and dually. This will be quite important in subsequent computations.

**Definition 521.** Let $p \in \mathrm{Term}(\bigcup \mathcal{L})$ and let $i \in I$. The *upper $i$-cover of $p$*, in notation, $p^{(i)}$, is an element of $L_i^b$ defined as follows:

(i) An element $a \in \bigcup \mathcal{L}$ belongs to $L_j$ for exactly one $j$. Define $a^{(i)} = a$ if $i = j$, and $a^{(i)} = 1^b$ if $i \neq j$.

(ii) If $p^{(i)}$ and $q^{(i)}$ are known, we compute $(p \vee q)^{(i)}$ and $(p \wedge q)^{(i)}$ as follows:

$$(p \vee q)^{(i)} = p^{(i)} \vee q^{(i)},$$
$$(p \wedge q)^{(i)} = p^{(i)} \wedge q^{(i)},$$

where $\vee$ and $\wedge$ on the right side of these equations are to be taken in $L_i^b$.

The definition of the *lower $i$-cover* of $p$, in notation, $p_{(i)}$, is analogous, with $0^b$ replacing $1^b$ in (i). An upper cover or a lower cover is *proper* if it is not $0^b$ or $1^b$. Observe that $0^b$ never occurs as an upper cover, nor $1^b$ as a lower cover.

**Corollary 522.** *For any $p \in \mathrm{Term}(\bigcup \mathcal{L})$ and $i \in I$, the inequality*

$$p_{(i)} \leq p^{(i)},$$

*holds in $L_i^b$. If $p_{(i)}$ and $p^{(j)}$ are both proper, then $i = j$.*

*Proof.* If $p \in \bigcup \mathcal{L}$, then $p = p_{(i)} = p^{(i)}$, for all $p \in L_i$, and $p_{(i)} = 0^b < 1^b = p^{(i)}$, for all $p \notin L_i$, so the first statement is true. If the first statement holds for $p$ and $q$, then by Definition 520(ii),

$$(p \vee q)_{(i)} = p_{(i)} \vee q_{(i)} \leq p^{(i)} \vee q^{(i)} = (p \vee q)^{(i)},$$

and so the first statement holds for $p \vee q$, and similarly for $p \wedge q$.

To prove the second statement, it is sufficient to verify that if $p_{(i)}$ is proper, then $p^{(j)}$ is not proper for any $j \neq i$. This is obvious for $p \in \bigcup \mathcal{L}$, by

Definition 521(i). Now assume that $p = q \vee r$ and let $p_{(i)}$ be proper; assume inductively that the result holds for $q$ and $r$. Either $q_{(i)}$ or $r_{(i)}$ must be proper, hence $q^{(j)} = 1^b$ or $r^{(j)} = 1^b$, ensuring that $p^{(j)} = q^{(j)} \vee r^{(j)} = 1^b$.

Finally, if $p = q \wedge r$ and $p_{(i)}$ is proper, then both $q_{(i)}$ and $r_{(i)}$ are proper, hence $q^{(j)} = r^{(j)} = 1^b$, and so $p^{(j)} = 1^b$. $\qquad\square$

### 1.4   The algorithm

We introduce a preordering on $\mathrm{Term}(\bigcup \mathcal{L})$, which for typographic convenience we will denote by $\subseteq$.

**Definition 523.** For $p, q \in \mathrm{Term}(\bigcup \mathcal{L})$, set $p \subseteq q$ if one of the following rules applies:

| | | | |
|---|---|---|---|
| (C) | $p^{(i)} \leq q_{(i)},$ | in $L_i^b$ for some $i \in I$; | |
| ($_\vee$W) | $p = p_0 \vee p_1,$ | where $p_0 \subseteq q$ | and $p_1 \subseteq q$; |
| ($_\wedge$W) | $p = p_0 \wedge p_1,$ | where $p_0 \subseteq q$ | or $p_1 \subseteq q$; |
| (W$_\vee$) | $q = q_0 \vee q_1,$ | where $p \subseteq q_0$ | or $p \subseteq q_1$; |
| (W$_\wedge$) | $q = q_0 \wedge q_1,$ | where $p \subseteq q_0$ | and $p \subseteq q_1$. |

*Remark.* In $L_i^b$, we denote the two new bounds by $0^b$ and $1^b$ as opposed to $0_i^b$ and $1_i^b$. This will create no ambiguity.

*Remark.* In (C), C stands for Cover; in the other four conditions, W stands for P. M. Whitman. Each Whitman condition assumes a $\vee$ or $\wedge$ on the left or right of the $\subseteq$ in $p \subseteq q$; at most two of these conditions may be applicable to a particular $p \subseteq q$. Note also, that if $p, q \in \bigcup \mathcal{L}$, then only (C) can apply.

Definition 523 essentially gives the algorithm we have been looking for. For $p, q \in \mathrm{Term}(\bigcup \mathcal{L})$, we will show that $p$ and $q$ represent the same element of the free product iff $p \subseteq q$ and $q \subseteq p$. We shall show this by actually exhibiting the free product as the set of equivalence classes of $\mathrm{Term}(\bigcup \mathcal{L})$ under this relation. To be able to do this, we have to establish a number of properties of the relation $\subseteq$. All the proofs are by induction on $\mathrm{rank}(p)$, see Section I.4.1.

### 1.5   Computing the algorithm

**Lemma 524.** *Let $p, q, r \in \mathrm{Term}(\bigcup \mathcal{L})$ and let $i \in I$. Then*

(i) $p \subseteq p$.
(ii) $p \subseteq q$ *implies that* $p_{(i)} \leq q_{(i)}$ *and* $p^{(i)} \leq q^{(i)}$.
(iii) $p \subseteq q$ *and* $q \subseteq r$ *imply that* $p \subseteq r$.

*Proof.*
(i) Proof by induction on $\mathrm{rank}(p)$. If $p \in \bigcup \mathcal{L}$, then $p \in L_i$ for some $i \in I$. Hence $p = p_{(i)} = p^{(i)}$ by Definition 521(i), and so $p \subseteq p$, by (C).

Let $p = q \vee r$. Then $q \subseteq q \vee r$ and $r \subseteq q \vee r$ by $(W_\vee)$, hence $q \vee r \subseteq q \vee r$ by $(\vee W)$, that is, $p \subseteq p$. If $p = q \wedge r$, we proceed similarly.

(ii) If $p \subseteq q$ by (C), then $p^{(j)} \leq q_{(j)}$ for some $j \in I$. We conclude that $p_{(j)} \leq q_{(j)}$ and $p^{(j)} \leq q^{(j)}$ by Corollary 522 and that $p^{(j)}$ and $q_{(j)}$ are proper by the observation immediately preceding this corollary. Thus again by Corollary 522, $p_{(i)}$ and $q^{(i)}$ are not proper, for all $i \neq j$, hence $p_{(i)} = 0^b$ and $q^{(i)} = 1^b$; so $p_{(i)} \leq q_{(i)}$ and $p^{(i)} \leq q^{(i)}$ in $L_i^b$.

Now we induct on $\mathrm{rank}(p) + \mathrm{rank}(q)$. First, let $\mathrm{rank}(p) + \mathrm{rank}(q) = 2$; then $p, q \in \bigcup \mathcal{L}$, so only (C) can yield $p \subseteq q$. In this case, and also in the induction step if (C) is applied, we obtain the result as in the previous paragraph.

Then let $\mathrm{rank}(p) + \mathrm{rank}(q) > 2$. Then $p$ or $q \notin \bigcup \mathcal{L}$, say, $p \notin \bigcup \mathcal{L}$. Now if $p = p_0 \wedge p_1$ and $(\wedge W)$ applies, then $p_0 \subseteq q$ or $p_1 \subseteq q$, say $p_0 \subseteq q$. Then $(p_0)_{(i)} \leq q_{(i)}$ and $(p_0)^{(i)} \leq q^{(i)}$, hence

$$p_{(i)} = (p_0)_{(i)} \wedge (p_1)_{(i)} \leq (p_0)_{(i)} \leq q_{(i)},$$

and similarly for upper covers with "and" replaced by "or". If $p = p_0 \vee p_1$, we proceed dually.

(iii) If $p \subseteq q$ by (C), then $p^{(i)} \leq q_{(i)}$ for some $i \in I$. By (ii), $q_{(i)} \leq r_{(i)}$, hence $p^{(i)} \leq r_{(i)}$ and so by (C), $p \subseteq r$. This takes care of the base of the induction, $p, q, r \in \bigcup \mathcal{L}$, since then only (C) applies. We induct on the sum of the ranks of $p, q, r$.

If $p \subseteq q$ by (C), then $p \subseteq r$ has already been proved.

If $p \subseteq q$ follows from $(\vee W)$, then $p = p_0 \vee p_1, p_0 \subseteq q, p_1 \subseteq q$, and so by induction, $p_0 \subseteq r$ and $p_1 \subseteq r$, implying that $p_0 \vee p_1 = p \subseteq r$ by $(\vee W)$.

If $p \subseteq q$ follows from $(\wedge W)$, then $p = p_0 \wedge p_1$ and $p_0 \subseteq q$ or $p_1 \subseteq q$. Thus by the induction hypotheses, $p_0 \subseteq r$ or $p_1 \subseteq r$, and so by $(\wedge W)$, $p_0 \wedge p_1 = p \subseteq r$.

If $q \subseteq r$ follows from (C), $(W_\vee)$, or $(W_\wedge)$, we can proceed dually (that is, by interchanging $\vee$ and $\wedge$).

The only cases that remain are where $p \subseteq q$ arises by the rule $(W_\vee)$ or $(W_\wedge)$, and $q \subseteq r$ by $(\vee W)$ or $(\wedge W)$. Note that if the first inequality arises from $(W_\vee)$, then $q$ has the form $q_1 \vee q_2$, so that the latter inequality can only arise from $(\wedge W)$; while we have the dual situation if the first inequality arises from $(W_\wedge)$. By duality, it suffices to consider the former case. Then $p \subseteq q_0$ and $p \subseteq q_1$, and $q_0 \subseteq r$ or $q_1 \subseteq r$. Hence $p \subseteq q_i \subseteq r$, for $i = 0$ or for $i = 1$, hence the induction hypotheses yield that $p \subseteq r$.      $\square$

## 1.6   Representing the free product

By Lemma 524, the relation $\subseteq$ is a preordering and so we can define:

$$p \equiv q \quad \text{if} \quad p \subseteq q \text{ and } q \subseteq p \qquad\qquad \text{for } p, q \in \mathrm{Term}(\textstyle\bigcup \mathcal{L}).$$

$$\mathrm{rep}(p) = \{\, q \in \mathrm{Term}(\textstyle\bigcup \mathcal{L}) \mid p \equiv q \,\} \qquad \text{for } p \in \mathrm{Term}(\textstyle\bigcup \mathcal{L})).$$

$$\mathrm{Rep}(\textstyle\bigcup \mathcal{L}) = \{\, \mathrm{rep}(p) \mid p \in \mathrm{Term}(\textstyle\bigcup \mathcal{L}) \,\}.$$

$$\mathrm{rep}(p) \leq \mathrm{rep}(q) \quad \text{if} \quad p \subseteq q.$$

In other words, we split $\mathrm{Term}(\bigcup \mathcal{L})$ into blocks under $\equiv$ and $\mathrm{Rep}(\bigcup \mathcal{L})$ is the set of blocks, which we order by $\leq$, see Section I.1.2.

**Lemma 525.** $\mathrm{Rep}(\bigcup \mathcal{L})$ *is a lattice, in fact,*

$$\mathrm{rep}(p) \vee \mathrm{rep}(q) = \mathrm{rep}(p \vee q),$$
$$\mathrm{rep}(p) \wedge \mathrm{rep}(q) = \mathrm{rep}(p \wedge q)$$

*for $p, q \in \mathrm{Term}(\bigcup \mathcal{L})$.*

*Now let $a, b, c, d \in L_i$ for some $i \in I$. Then*

$$a \wedge b = c,$$
$$a \vee b = d$$

*in $L_i$ is equivalent to*

$$\mathrm{rep}(a) \wedge \mathrm{rep}(b) = \mathrm{rep}(c),$$
$$\mathrm{rep}(a) \vee \mathrm{rep}(b) = \mathrm{rep}(d)$$

*in $\mathrm{Rep}(\bigcup \mathcal{L})$. If $x, y \in \bigcup \mathcal{L}$ with $x \neq y$, then $\mathrm{rep}(x) \neq \mathrm{rep}(y)$.*

*Proof.* Let $p, q \in \mathrm{Term}(\bigcup \mathcal{L})$. Clearly, $p \wedge q \subseteq p$ and $p \wedge q \subseteq q$ follow from $p \subseteq p$, $q \subseteq q$, and $(\wedge W)$. If $r \subseteq p$ and $r \subseteq q$, then $r \subseteq p \wedge q$ by $(W_\wedge)$; this argument and its dual give the first statement.

If $a \wedge b = c$ in some $L_i$, then $(a \wedge b) \in \mathrm{Term}(\bigcup \mathcal{L})$ satisfies

$$(a \wedge b)_{(i)} = (a \wedge b)^{(i)} = c.$$

Hence from $(C)$ we get both $(a \wedge b) \subseteq c$ and $c \subseteq (a \wedge b)$, that is, $(a \wedge b) \equiv c$. By the result of the preceding paragraph, this means that $\mathrm{rep}(a) \wedge \mathrm{rep}(b) = \mathrm{rep}(c)$, as desired. The corresponding statement for $\vee$ follows by duality.

Finally, if $\mathrm{rep}(x) = \mathrm{rep}(y)$, for some $x, y \in \bigcup \mathcal{L}$, then $x \subseteq y$. Since only $(C)$ applies, $x^{(i)} \leq y_{(i)}$ for some $i \in I$. Thus $x^{(i)}$ and $y_{(i)}$ are proper and so $x = x^{(i)}$ and $y = y_{(i)}$. We conclude that $x \leq y$; similarly, $y \leq x$. Thus $x = y$. $\square$

By Lemma 525,
$$a \mapsto \mathrm{rep}(a), \quad \text{for } a \in L_i,$$
is an embedding of $L_i$ into $\mathrm{Rep}(\bigcup \mathcal{L})$ for any $i \in I$. Therefore, by identifying the element $a \in L_i$ with $\mathrm{rep}(a)$, we get that each $L_i \leq \mathrm{Rep}(\bigcup \mathcal{L})$ and, therefore, $\bigcup \mathcal{L} \subseteq \mathrm{Rep}(\bigcup \mathcal{L})$. It is also obvious that the ordering induced by $\mathrm{Rep}(\bigcup \mathcal{L})$ on $\bigcup \mathcal{L}$ agrees with the original ordering.

**Theorem 526.** $\mathrm{Rep}(\bigcup \mathcal{L})$ *is a free product of the $L_i$ for $i \in I$.*

*Proof.* Conditions (i) and (ii) of Definition 518 have already been observed. To get condition (iii), let $A$ be a lattice and let the lattice homomorphisms $\varphi_i \colon L_i \to A$ be given for all $i \in I$. We define inductively a map

$$\psi \colon \mathrm{Term}(\bigcup \mathcal{L}) \to A$$

as follows: for $p \in \bigcup \mathcal{L}$, there is exactly one $i \in I$ with $p \in L_i$; set $\psi(p) = \varphi_i(p)$; if $p = p_0 \vee p_1$ or $p = p_0 \wedge p_1$, and if $\psi(p_0)$ and $\psi(p_1)$ have already been defined, then set $\psi(p) = \psi(p_0) \vee \psi(p_1)$ or $\psi(p) = \psi(p_0) \wedge \psi(p_1)$, respectively. Now we prove:

**Lemma 527.** *Let $p, q \in \mathrm{Term}(\bigcup \mathcal{L})$ and $i \in I$.*

(i) *If $p_{(i)}$ is proper, then $\psi(p_{(i)}) \leq \psi(p)$.*
(ii) *If $p^{(i)}$ is proper, then $\psi(p) \leq \psi(p^{(i)})$.*
(iii) *$p \subseteq q$ implies that $\psi(p) \leq \psi(q)$.*
(iv) *$p \equiv q$ implies that $\psi(p) = \psi(q)$.*

*Proof.*
   (i) If $p \in \bigcup \mathcal{L}$ and $p_{(i)}$ is proper, then $p \in L_i$. Hence $p = p_{(i)}$ and so the inequality $\psi(p_{(i)}) \leq \psi(p)$ is obvious.
   So assume that $p \notin \bigcup \mathcal{L}$, and let us induct on $\mathrm{rank}(p)$.
   If $p = q \vee r$, then $q_{(i)}$ or $r_{(i)}$ is proper. If both $q_{(i)}$ and $r_{(i)}$ are proper, the calculation is straightforward. If, say, $q_{(i)}$ is proper and $r_{(i)} = 0^b$, then

$$\psi(p_{(i)}) = \psi(q_{(i)} \vee r_{(i)}) = \psi(q_{(i)}) \leq \psi(q) \leq \psi(p).$$

If $p = q \wedge r$, then $p_{(i)} = q_{(i)} \wedge r_{(i)}$, so $q_{(i)}$ and $r_{(i)}$ are proper. Therefore, $\psi(q_{(i)}) \leq \psi(q)$ and $\psi(r_{(i)}) \leq \psi(r)$, by induction; hence

$$\psi(p_{(i)}) = \psi(q_{(i)}) \wedge \psi(r_{(i)}) \leq \psi(q) \wedge \psi(r) = \psi(q \wedge r) = \psi(p).$$

   (ii) This follows by duality from (i).
   (iii) If $p \subseteq q$ follows from (C), then $p^{(i)} \leq q_{(i)}$ for some $i \in I$. Thus $p^{(i)}$ and $q_{(i)}$ are proper. Therefore, $\psi(p) \leq \psi(p^{(i)})$ by (ii), $\psi(p^{(i)}) \leq \psi(q_{(i)})$, because $p^{(i)}$ and $q_{(i)} \in \bigcup \mathcal{L}$, and $\psi(q_{(i)}) \leq \psi(q)$ by (i), implying that the inequality $\psi(p) \leq \psi(q)$ holds.

This takes care of $p, q \in \bigcup \mathcal{L}$ and of the first case in the induction step.

If $p \subseteq q$ follows from $(\wedge W)$, then $p = p_0 \wedge p_1$, where $p_0 \subseteq q$ or $p_1 \subseteq q$. Hence $\psi(p_0) \le \psi(q)$ or $\psi(p_1) \le \psi(q)$, therefore, $\psi(p) = \psi(p_0) \wedge \psi(p_1) \le \psi(q)$.

If $p \subseteq q$ follows from $(\vee W)$, $(W_\wedge)$, or $(W_\vee)$, the proof is analogous to the proof in the last case.

(iv) follows from (iii). $\qquad\qquad\qquad\qquad\qquad\qquad\qquad\qquad\qquad\qquad\square$

Now take a $p \in \mathrm{Term}(\bigcup \mathcal{L})$ and define

$$\varphi(\mathrm{rep}(p)) = \psi(p).$$

The map $\varphi$ is well-defined by Lemma 527(iv). Since

$$\begin{aligned}
\varphi(\mathrm{rep}(p) \vee \mathrm{rep}(q)) &= \varphi(\mathrm{rep}(p \vee q)) \\
&= \psi(p \vee q) = \psi(p) \vee \psi(q) = \varphi(\mathrm{rep}(p)) \vee \varphi(\mathrm{rep}(q)),
\end{aligned}$$

and similarly for $\wedge$, we conclude that $\varphi$ is a homomorphism. Finally, for $p \in L_i$, for some $i \in I$,

$$\varphi(\mathrm{rep}(p)) = \psi(p) = \varphi_i(p),$$

by the definition of $\psi$; hence $\varphi$ restricted to $L_i$ agrees with $\varphi_i$. $\qquad\qquad\square$

Lemma 524(ii) implies that for $p, q \in \mathrm{Term}(\bigcup \mathcal{L})$, if $p \equiv q$, then $p_{(i)} = q_{(i)}$ and $p^{(i)} = q^{(i)}$ for all $i \in I$. Hence we can define

$$\begin{aligned}
(\mathrm{rep}(p))_{(i)} &= p_{(i)}, \\
(\mathrm{rep}(p))^{(i)} &= p^{(i)}.
\end{aligned}$$

## 1.7    The Structure Theorem for Free Products

Recall that $\mathcal{L} = (L_i \mid i \in I)$ is a family of lattices with $I \ne \varnothing$; we assume that $L_i$ and $L_j$ are disjoint for all $i, j \in I$ with $i \ne j$.

All our results will now be summarized (G. Grätzer, H. Lakser, and C. R. Platt [311]):

**Theorem 528 (The Structure Theorem of Free Products).** *Let $L$ be a free product of $\mathcal{L} = (L_i \mid i \in I)$. For every $a \in L$ and $i \in I$, if some element of $L_i$ is majorized by $a$, then there is a largest one with this property, namely, $a_{(i)}$; dually, we get the element $a^{(i)}$.*

*Let $a = p(a_0, \dots, a_{n-1})$, where $p$ is an $n$-ary term and*

$$a_0, \dots, a_{n-1} \in \bigcup \mathcal{L}.$$

*Then $a_{(i)}$ can be computed by the algorithm given in Definition 521. Dually, $a^{(i)}$ can be computed.*

*For some $n, m < \omega$, let $p$ be an $n$-ary term, let $q$ be an $m$-ary term, let*

$$a_0, \ldots, a_{n-1}, b_0, \ldots, b_{m-1} \in \bigcup \mathcal{L}.$$

*Then we can decide whether*

$$p(a_0, \ldots, a_{n-1}) \leq q(b_0, \ldots, b_{m-1})$$

*using the algorithm of Definition 523.*

The idea of the proof of Theorem 528 goes back to P. M. Whitman [722] and R. P. Dilworth [155].

We should comment on the use of the term "algorithm" in Theorem 528 and elsewhere in this section. (C) of Definition 523 brings in covers and so does the procedure described in Definition 521. This procedure is an algorithm insofar as the structure of $L_i$ is described in an effective way; such procedures are commonly called "relative algorithms" or an "algorithm using an oracle". In this case, we only need an "oracle" for computation in the lattice $L_i$ for $i \in I$.

Thus if we consider the free product of finitely many finite lattices, then we really deal with an algorithm. In the general case, algorithm should be interpreted intuitively and not be given the precise meaning assigned to it in mathematical logic.

The existence of covers is not special to the variety $\mathbf{L}$ of all lattices, as was observed in B. Jónsson [446]. To prove this for lower covers, let $L$ be a free $\mathbf{K}$-product of $L_i$ for $i \in I$. Fix an $i \in I$, and for $j \in I$, define $\psi_j \colon L_j \to L_i^b$ as follows: $\psi_i$ is the identity map on $L_i$; $\psi_j(a) = 0^b$ for all $j \neq i$ and all $a \in L_j$. By Exercise I.4.14, the bounded lattice $L_i^b$ belongs to $\mathbf{K}$ and so, by the definition of free $\mathbf{K}$-product, there is a homomorphism $\psi \colon L \to L_i^b$ extending the $\psi_j$ for all $j \in I$.

From the construction of $a_{(i)}$ we can verify inductively that $a_{(i)} = \psi(a)$ for all $a \in L$. Now if $b \in L_i$ is majorized by $a$, then $\psi(b) \leq \psi(a)$, that is, $b \leq a_{(i)}$; conversely, if $b \leq a_{(i)} \leq a$, then $b \leq a$

Therefore, if there is $b \in L_i$ majorized by $a$, then $a_{(i)}$ is the largest such element. Otherwise, $\psi(a) = 0^b$; indeed, if $\psi(a) \neq 0^b$, then $b = \psi(a) \in L_i$, a contradiction in view of $b = \psi(a) \leq a$. (We use the fact that $\psi(a) \leq a$ for all $a \in L$, since this holds for all $a \in \bigcup \mathcal{L}$.)

Free products (in fact, free $\mathbf{K}$-products) of two lattices satisfy a special condition (G. Grätzer, H. Lakser, and C. R. Platt [311]):

**Theorem 529 (The Splitting Theorem).** *Let the lattice $L$ be a free product of the lattices $L_0$ and $L_1$, in notation, $L = L_0 * L_1$. For every $a \in L$, if $a^{(0)}$ is not proper, then $a_{(1)}$ is proper and conversely. Thus*

$$L = \mathrm{id}(L_0) \cup \mathrm{fil}(L_1),$$

*where the union is a disjoint union. In other words, for every element $a$ of $L$, either $a$ is majorized by some element of $L_0$ or $a$ majorizes an element of $L_1$.*

*Proof.* $\mathrm{id}(L_0) \cup \mathrm{fil}(L_1) \leq L$ contains a generating set, namely, $L_0 \cup L_1$. Hence $\mathrm{id}(L_0) \cup \mathrm{fil}(L_1) = L$. Now Theorem 528 yields the statement. $\qquad\square$

**Corollary 530.** *The free product $L_0 * L_1$ can be represented as a disjoint union of four convex sublattices: the convex sublattices generated by $L_0$ and $L_1$, $\mathrm{id}(L_0) \cap \mathrm{id}(L_1)$, and $\mathrm{fil}(L_0) \cap \mathrm{fil}(L_1)$.*

P. M. Whitman [722] observed that for $a \in \mathrm{Free}(n)$, if $a \geq x_1$ does not hold, then $a \leq x_2 \vee \cdots \vee x_n$ holds; proof by induction on $\mathrm{rank}(a)$ using Theorem 540. This is a precursor of the Splitting Theorem.

The next result is a trivial application of Theorem 528.

**Corollary 531.** *Let $K_i \leq L_i$, for $i \in I$, and let $L$ be a free product of the $L_i$ for $i \in I$. Let $K \leq L$ be the sublattice generated by $\bigcup( K_i \mid i \in I )$. Then $K$ is a free product of the $K_i$ for $i \in I$.*

*Proof.* This is obvious since $p \subseteq q$ iff it follows from Definition 523, for $p, q \in \mathrm{Term}(\bigcup( K_i \mid i \in I ))$, and in applying Definition 523, we use only elements of the $K_i$ for $i \in I$. Thus by the proof of Theorem 526, $K$ is a free product of the $K_i$ for $i \in I$. $\qquad\square$

The next result is a special case of a result of B. Jónsson [440] and [443] (see Figure 119):

**Theorem 532.** *Let $\mathbf{K}$ be a variety of lattices with the Amalgamation Property. Let us assume that $A, B \in \mathbf{K}$ and let $C$ be a free $\mathbf{K}$-product of $A$ and $B$. Let $A_1 \leq A$, let $B_1 \leq B$, and let $C_1 \leq C$ be the sublattice generated by $A_1 \cup B_1$ in $C$. Then $C_1$ is a free $\mathbf{K}$-product of $A_1$ and $B_1$.*

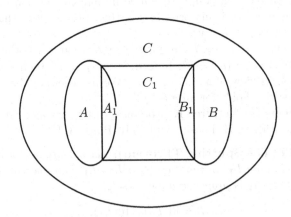

Figure 119. Illustrating Theorem 532

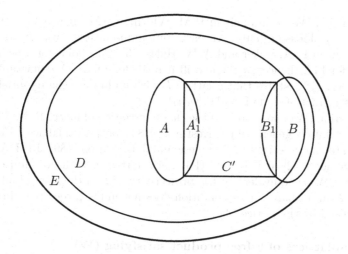

Figure 120. Proving Theorem 532

*Proof.* Let $C'$ be a free **K**-product of $A_1$ and $B_1$. Thus there exists a homomorphism $\chi$ of $C'$ into $C_1$ such that $\chi$ is the identity on $A_1$ and $B_1$. Since $A_1 \leq A$ and $A_1 \leq C'$, by the Amalgamation Property, there is a lattice $D$ in **K** containing $A$ and $C'$ as sublattices; clearly, $A_1 \subseteq A \cap C'$. Similarly, $B_1 \leq D$ and $B$, and thus there exists a lattice $E$ in **K** containing $B$ and $D$ as sublattices; clearly, $B_1 \subseteq B \cap D$ (see Figure 120). Since $C$ is a free product of $A$ and $B$, there exists a homomorphism $\varphi_1$ of $C$ into $E$ such that $\varphi_1$ is the identity on $A$ and $B$.

Let $\varphi$ be the restriction of $\varphi_1$ to $C_1$. Then $\varphi$ maps $C_1$ into $C'$, the map $\chi\varphi$ is the identity on $A_1$ and $B_1$, and is thus the identity on $C_1$. Similarly, $\varphi\chi$ is the identity on $C'$, so $\varphi$ is an isomorphism between $C_1$ and $C'$. □

There are three important properties that we shall prove below for free lattices:

(W) $\quad x \wedge y \leq u \vee v \quad$ implies that $\quad [x \wedge y, u \vee v] \cap \{x, y, u, v\} \neq \varnothing.$

(SD$_\vee$) $\quad u = x \vee y = x \vee z \quad$ implies that $\quad u = x \vee (y \wedge z).$

(SD$_\wedge$) $\quad u = x \wedge y = x \wedge z \quad$ implies that $\quad u = x \wedge (y \vee z).$

(W) is called the *Whitman condition*, (SD$_\vee$) the *join-semidistributive law*, and (SD$_\wedge$) the *meet-semidistributive law*.

We can rephrase the Whitman condition:

$x \wedge y \leq u \vee v$ implies that $x \leq u \vee v$ or $y \leq u \vee v$ or $x \wedge y \leq u$ or $x \wedge y \leq v$.

Observe, that a free product need not satisfy (W); indeed, $x \wedge y \leq u \vee v$ may hold in one of the factors or on account of (C).

Condition (W) is implicit in P. M. Whitman [722]; it was first explicitly stated in B. Jónsson [440]. An easy, but illuminating, property of (W) is pointed out in K. A. Baker and A. W. Hales [43]: (W) holds for a lattice $L$ iff it holds for $\operatorname{Id} L$ (hence, by duality, iff it holds for $\operatorname{Fil} L$). A remarkable result of R. Freese [182] states that a lattice $L$ with no infinite chains satisfies (W) iff $L$ is a retract of some $\operatorname{Fil}(\operatorname{Id}\operatorname{Free}(n))$.

Condition (W) plays a role in the characterization of finite (R. N. McKenzie [512]) and finitely generated (A. Kostinsky [483]) projective lattices. Projective lattices are characterized in R. Freese and J. B. Nation [189]. In B. A. Davey and B. Sands [132], it is proved that, for a lattice $L$ with no infinite chains, condition (W) is equivalent to the projectivity of $L$ in the class of all lattices with no infinite chains. This condition does not imply projectivity in general, as witnessed by the lattice $\mathsf{M}_3$.

## 1.8    Sublattices of a free product satisfying (W)

Let us say that a *subset* $A$ of a lattice $L$ *satisfies* (W) if (W) holds in $L$ for all $x, y, z, u \in A$. The next result (G. Grätzer and H. Lakser [303]) shows that many subsets of a free product satisfy (W).

**Theorem 533.** *Let the lattice $L$ be a free product of $\mathcal{L} = (L_i \mid i \in I)$. Let $A_i$ be a subset of $L_i^b$ satisfying (W) for each $i \in I$. Let $A$ be a subset of $L$ satisfying*

$$A_{(i)} = \{ a_{(i)} \mid a \in A \} \subseteq A_i,$$
$$A^{(i)} = \{ a^{(i)} \mid a \in A \} \subseteq A_i$$

*for all $i \in I$. Then $A$ satisfies (W) in $L$.*

*Proof.* Let $x, y, u, v \in A$, let $x = \operatorname{rep}(p), y = \operatorname{rep}(q), u = \operatorname{rep}(r), v = \operatorname{rep}(s)$, where $p, q, r, s \in \operatorname{Term}(\bigcup \mathcal{L})$. Let $x \wedge y \le u \vee v$. Then $p \wedge q \subseteq r \vee s$. This relation might arise under Definition 523 via $(_\wedge W)$, $(W_\vee)$, or (C). In either of the first two cases, we immediately get the desired case of (W).

If (C) applies, then $(p \wedge q)^{(i)} \le (r \vee s)_{(i)}$, for some $i \in I$, hence the inequality

$$p^{(i)} \wedge q^{(i)} \le r_{(i)} \vee s_{(i)}$$

holds in $A_i$. Therefore, since $A_i$ satisfies (W), one of the following holds:

$$p^{(i)} \le r_{(i)} \vee s_{(i)},$$
$$q^{(i)} \le r_{(i)} \vee s_{(i)},$$
$$p^{(i)} \wedge q^{(i)} \le r_{(i)},$$
$$p^{(i)} \wedge q^{(i)} \le s_{(i)}.$$

Again, by (C), we conclude that one of the following holds:

$$p \subseteq r \vee s,$$
$$q \subseteq r \vee s,$$
$$p \wedge q \subseteq r,$$
$$p \wedge q \subseteq s.$$

So again (W) holds in $A$.                                                    □

Observe how strong (W) really is. For instance, if (W) holds in a lattice $A$, then $A$ has no doubly reducible element.

**Corollary 534.**

(i) *A free product of lattices satisfies* (W) *iff each lattice satisfies* (W).
(ii) *A free product* (*as lattices*) *of any family of chains satisfies* (W).
(iii) *Any free lattice satisfies* (W).

## 1.9   Minimal representations

Let the lattice $L$ be a free product of $\mathcal{L} = (L_i \mid i \in I)$. Naturally, every $a \in L$ has infinitely many representations of the form

$$a = p(a_0, \dots, a_{n-1}) \in \mathrm{Term}(\bigcup \mathcal{L}) \quad \text{for } a_0, \dots, a_{n-1} \in \bigcup \mathcal{L}.$$

If $\mathrm{rank}(p)$ is minimal, we call $p$ a *minimal representation of* $a$ and we call $p$ a *minimal term*. Using the notation of Theorem 528, the next result (H. Lakser [491]) tells us how to recognize a minimal representation:

**Theorem 535.** *Let* $p \in \mathrm{Term}(\bigcup \mathcal{L})$. *Then* $p$ *is a minimal term iff one of the following three conditions holds:*

(a) $p \in \bigcup \mathcal{L}$.

(b) $p = p_0 \vee \cdots \vee p_{n-1}$ *with* $n > 1$, *where no* $p_j$ *is a join of more than one term and conditions* (i)–(v) *below hold.*

   (i) *Each* $p_j$ *is minimal for* $0 \le j < n$.
   (ii) $p_j \not\subseteq p_0 \vee \cdots \vee p_{j-1} \vee p_{j+1} \vee \cdots \vee p_{n-1}$ *for all* $0 \le j < n$.
   (iii) *If* $0 \le j < n$ *with* $\mathrm{rank}(p_j) > 1$ *and* $i \in I$, *then* $p_j^{(i)} \not\subseteq p_{(i)}$ *in* $L_i^b$.
   (iv) *If* $p_j = p_j' \wedge p_j''$, *where* $0 \le j < n$ *and* $p_j', p_j'' \in \mathrm{Term}(\bigcup \mathcal{L})$, *then* $p_j' \not\subseteq p$ *and* $p_j'' \not\subseteq p$.
   (v) *If* $p_j, p_k \in L_i$, *where* $0 \le j \le k < n$ *and* $i \in I$, *then* $j = k$.

(c) *The dual of* (b) *holds.*

*Proof.* We follow our convention of dropping parentheses in multiple meets and joins; so that an expression $p_0 \vee \cdots \vee p_{n-1}$ for $n > 1$ represents any of a number of possible bracketed expressions.

Let $p \in \mathrm{Term}(\bigcup \mathcal{L})$ and $p \notin \bigcup \mathcal{L}$. We can assume that

$$p = p_0 \vee \cdots \vee p_{n-1}$$

with $n > 1$. If one of conditions (i)-(v) fails, consider the terms

$$q_{(i)} = p_0 \vee \cdots \vee p_{j-1} \vee p_j' \vee p_{j+1} \vee \cdots \vee p_{n-1},$$

$$\text{where } p_j' \equiv p_j \text{ and } \mathrm{rank}(p_j') < \mathrm{rank}(p_j),$$

$$q_{(ii)} = p_0 \vee \cdots \vee p_{j-1} \vee p_{j+1} \vee \cdots \vee p_{n-1},$$

$$q_{(iii)} = p_0 \vee \cdots \vee p_{j-1} \vee p_j^{(i)} \vee p_{j+1} \vee \cdots \vee p_{n-1},$$

$$q_{(iv)} = p_0 \vee \cdots \vee p_{j-1} \vee p_j' \vee p_{j+1} \vee \cdots \vee p_{n-1},$$

$$q_{(v)} = p_0 \vee \cdots \vee p_{j-1} \vee p_{j+1} \vee \cdots \vee p_{k-1} \vee p_{k+1} \vee \cdots \vee p_{n-1} \vee a,$$

where in the last formula $a \in L_i$ and $a = p_j \vee p_k$ in $L_i$. It is obvious that, if condition (y) fails ($i \leq y \leq v$), then $p \equiv q_{(y)}$ and $\mathrm{rank}(q_{(y)}) < \mathrm{rank}(p)$, contradicting the minimality of $p$.

From this, and the dual observations in case (c), we see the "only if" direction of our theorem.

For the converse, let us assume that

(1) $p, q \in \mathrm{Term}(\bigcup \mathcal{L})$;
(2) $p \equiv q$;
(3) $p$ satisfies (i)–(v);
(4) $q$ is minimal.

We show that $\mathrm{rank}(p) = \mathrm{rank}(q)$. This gives us an algorithm to reduce any term to a minimal one and, at the same time, verifies the converse. Indeed, if $p$ were not minimal, there would be a term $q$ satisfying (2), (4), and $\mathrm{rank}(p) > \mathrm{rank}(q)$, contradicting the above statement.

So let the terms $p$ and $q$ be given as specified, let $p = p_0 \vee \cdots \vee p_{n-1}$ with $n > 1$.

Firstly, we claim that $\mathrm{rank}(q) > 1$. Indeed, if $\mathrm{rank}(q) = 1$, then $q \in L_i$ for some $i \in I$. Thus by Lemma 524(ii), $p_{(i)} = p^{(i)} = q$.

It follows that all $p_j^{(i)}$ are proper, hence that if any $p_j$ has rank 1, that is, lies in $\bigcup \mathcal{L}$, it lies in $L_i$. But by (v), at most one of these terms can lie in $L_i$; so at least one $p_j$ has rank $> 1$. But since $p_j^{(i)} \leq p^{(i)} = q = p_{(i)}$, this contradicts condition (iii).

Secondly, we claim that the term $q$ is not of the form $q_0 \wedge q_1$. Indeed, let $q = q_0 \wedge q_1$. Consider $q \subseteq p$. If (C) applies, then $q^{(i)} \leq p_{(i)} = q_{(i)}$, and, therefore, $q_{(i)} = q^{(i)} \equiv q$, which, since $\mathrm{rank}(q) > 1$, contradicts the minimality of $q$.

If $(\wedge W)$ applies, then $q_0 \subseteq p$ or $q_1 \subseteq p$. Obviously, $p \subseteq q_i$, thus if say $q_0 \subseteq p$, then $p \equiv q_0$, contradicting the minimality of $q$.

Finally, if $(W_\vee)$ applies, then $q \subseteq p_0 \vee \cdots \vee p_{n-2}$ or $q \subseteq p_{n-1}$. The first possibility yields that $p_{n-1} \subseteq p_0 \vee \cdots \vee p_{n-2}$, while the second gives $p_{n-1} \equiv p$ (and therefore, $p_0 \subseteq p_1 \vee \cdots \vee p_{n-1}$), both contradicting that the term $p$ satisfies (ii).

Thus the term $q$ is of the form

$$q = q_0 \vee \cdots \vee q_{m-1}, \quad \text{for } m > 1,$$

where no $q_i$ is the join of two terms.

Next we show that there are functions

$$f \colon \{0, 1, \ldots, n-1\} \to \{0, 1, \ldots, m-1\},$$
$$g \colon \{0, 1, \ldots, m-1\} \to \{0, 1, \ldots, n-1\}$$

satisfying the following conditions:

($\alpha$) $g(f(j)) = j$, for all $0 \leq j < n$, and $f(g(j)) = j$ for all $0 \leq j < m$.

($\beta$) If $0 \leq j < n$ and $\operatorname{rank}(p_j) > 1$, then $q_{f(j)} \equiv p_j$ and $\operatorname{rank}(q_{f(j)}) = \operatorname{rank}(p_j)$, and similarly for any $0 \leq j < m$ with $\operatorname{rank}(q_i) > 1$.

($\gamma$) If $p_j \in L_i$, for $0 \leq j < n$ and $i \in I$, then $q_{f(j)} \in L_i$, and similarly for $q_j \in L_i$.

Let $0 \leq j < n$ and $\operatorname{rank}(p_j) > 1$. Then $p_j \subseteq q$. If ($\gamma$) is applicable, then the inequality $p_j^{(i)} \leq q_{(i)}$ holds for some $i \in I$; since $q_{(i)} = p_{(i)}$, we obtain $p_j^{(i)} \leq p_{(i)}$, contradicting condition (iii) for $p$. Condition $(\wedge W)$ is not applicable either because it would contradict that $p$ satisfies (iv). Hence only $(W_\vee)$ is applicable. Therefore, $p_j \subseteq q_0 \vee \cdots \vee q_{m-2}$ or $p_j \subseteq q_{m-1}$. Continuing this argument, we conclude that $p_j \subseteq q_{f(j)}$ for some $0 \leq f(j) < m$. If $q_{f(j)} \in L_i$, for some $i \in I$, then $p_j \subseteq q_{f(j)}$ implies that $p_j^{(i)} \leq q_{f(j)}$ in $L_i$, thus $p_j^{(i)} \leq q_{f(j)} \leq q_{(i)} = p_{(i)}$, contradicting that condition (iii) holds for $p$. Therefore, $\operatorname{rank}(q_{f(j)}) > 1$.

Since $q$ is minimal, it satisfies (i)–(v), hence, reversing the roles of $p$ and $q$, we can similarly define the function $g$ on those indices $j$ with $\operatorname{rank}(q_j) > 1$. Thus

$$p_j \subseteq q_{f(j)} \subseteq p_{g(f(j))},$$

and so by (ii), $g(f(j)) = j$ and $p_j \equiv q_{f(j)}$. Similarly, we obtain $f(g(j)) = j$ and $q_j \equiv p_{g(j)}$.

Now let $0 \leq j < n$ and $\operatorname{rank}(p_j) = 1$. Then $p_j \in L_i$ for some $i \in I$. In view of $p_j \subseteq q$, we can assume that $q_{(i)}$ is proper. Since $q_{(i)} = (q_0)_{(i)} \vee \cdots \vee (q_{m-1})_{(i)}$, some $(q_j)_{(i)}$ must be proper. By renumbering $q_0, \ldots, q_{m-1}$, we get that $(q_k)_{(i)}$ is proper iff $0 \leq k < t$, where $t \leq m$. Thus

$$q_{(i)} = (q_0)_{(i)} \vee \cdots \vee (q_{t-1})_{(i)}.$$

If $\operatorname{rank}(q_s) > 1$, for all $0 \le s < t$, then $\operatorname{rank}(p_{g(s)}) > 1$ and $(p_{g(s)})_{(i)} = (q_s)_{(i)}$. Therefore, $g(s) \ne j$, for all $0 \le s < t$, and so

$$p_j \le (q_0)_{(i)} \vee \cdots \vee (q_{t-1})_{(i)} \le (p_0 \vee \cdots \vee p_{j-1} \vee p_{j+1} \vee \cdots \vee p_{n-1})_{(i)}.$$

Thus $p_j \subseteq p_0 \vee \cdots \vee p_{j-1} \vee p_{j+1} \vee \cdots \vee p_{n-1}$, contradicting the fact that (ii) holds for $p$.

Consequently, we can choose $0 \le f(j) < n$ such that $\operatorname{rank}(q_{f(j)}) = 1$ and $(q_{f(j)})_{(i)}$ is proper, that is, such that $q_{f(j)} \in L_i$. By (v), the choice of $f(j)$ is unique. Similarly, we define $g(j)$ for all $0 \le j < m$ and $\operatorname{rank}(q_j) = 1$. It is obvious that $(\alpha)$, $(\beta)$, and $(\gamma)$ are satisfied.

Now $(\alpha)$ implies that $f$ and $g$ are one-to-one, hence $n = m$. Since $\operatorname{rank}(p_j) = \operatorname{rank}(q_{f(j)})$, we conclude that $\operatorname{rank}(p) = \operatorname{rank}(q)$.

This completes the proof that $p$ is minimal in case (b). The dual argument applies in case (c), while the conclusion is immediate in case (a).     □

Compare the minimal representation with the canonical representation of Exercise 1.17.

### 1.10   Sublattices of a free product satisfying (SD$_\vee$)

A *subset* $A$ of a lattice $L$ *satisfies* (SD$_\vee$) if (SD$_\vee$) holds for all $x, y, z \in A$. The following result (G. Grätzer and H. Lakser [303]) establishes for (SD$_\vee$) what was done for (W) in Theorem 533.

**Theorem 536.** *Let the lattice $L$ be a free product of $\mathcal{L} = (L_i \mid i \in I)$. Let $A_i$ be a subset of $L_i^b$ satisfying* (SD$_\vee$) *for each $i \in I$. Let $A$ be a subset of $L$ such that $A_{(i)} \subseteq A_i$ for all $i \in I$. Then $A$ satisfies* (SD$_\vee$).

*Proof.* Let

$$x = \operatorname{rep}(p), \ y = \operatorname{rep}(q), \ z = \operatorname{rep}(s) \in A, \ x \vee y = x \vee z,$$

and let

$$u = u_0 \vee \cdots \vee u_{n-1}, \quad \text{for } n \ge 1,$$

be a representation of $p \vee q \equiv p \vee s$ satisfying (i)–(v) of Theorem 535(b). We show that, for each $j$ with $0 \le j < n$ and $\operatorname{rank}(u_j) > 1$, either $u_j \subseteq p$ or $u_j \subseteq q$ holds. Consider $u_j \subseteq p \vee q$. If (C) applies, then $u_j^{(i)} \le (p \vee q)_{(i)} = u_{(i)}$, contradicting Theorem 535(b)(iii). If ($\wedge$W) applies, that is, if $u_j = u_j' \wedge u_j''$ and $u_j' \subseteq p \vee q$, we get a contradiction with Theorem 535(b)(iv). Thus only (W$_\vee$) can apply, yielding $u_j \subseteq p$ or $u_j \subseteq q$. Similarly, $u_j \subseteq p$ or $u_j \subseteq s$, and so $u_j \subseteq p \vee (q \wedge s)$.

On the other hand, assume that $\operatorname{rank}(u_j) = 1$; it follows that $u_j \le u_{(i)}$ for some $i \in I$. Now $u_{(i)} = p_{(i)} \vee q_{(i)}$ and $u_{(i)} = p_{(i)} \vee s_{(i)}$ in $L_i^b$ and $p_{(i)}, q_{(i)}, s_{(i)} \in A_i$, and so

$$u_j \le u_{(i)} = p_{(i)} \vee (q_{(i)} \wedge s_{(i)})$$

by $(\mathrm{SD}_\vee)$. Thus $u_j \subseteq p \vee (q \wedge s)$ by (C). Since $u_j \subseteq p \vee (q \wedge s)$ is proved for all $j$ with $0 \leq j < n$, we conclude that

$$u_0 \vee \cdots \vee u_{n-1} \subseteq p \vee (q \wedge s),$$

and so $u \equiv x \vee (y \wedge z)$, as claimed. $\qquad\qquad\qquad\qquad\qquad\qquad$ $\square$

By dualizing Theorem 536, we can obtain the analogous result for $(\mathrm{SD}_\wedge)$.

## 1.11   The Common Refinement Property

We now prove that the *Common Refinement Property* holds for free products (G. Grätzer and J. Sichler [355]).

**Theorem 537 (The Common Refinement Property for Free Products).**
*Let the lattice $L$ be a free product of the lattices $A_0$ and $A_1$ and also of the lattices $B_0$ and $B_1$. Then $L$ is a free product of the lattices*

$$(A_i \cap B_j \mid i,\ j = 0, 1,\ A_i \cap B_j \neq \varnothing).$$

*Proof.* We shall write $a_{A_i}$ for the lower cover of $a$ in $A_i$, and $a_{B_j}$ for its lower cover in $B_j$. Let $a \in A_0$. We claim that

$$a_{B_0} \in (A_0 \cap B_0) \cup \{0^b\}.$$

Indeed, since $L$ is generated by $B_0 \cup B_1$,

$$a = p(b_{0,0}, b_{0,1}, \ldots, b_{1,0}, b_{1,1}, \ldots),$$

where $b_{0,0}, b_{0,1}, \ldots \in B_0$ and $b_{1,0}, b_{1,1}, \ldots \in B_1$. Computing the lower $A_0$-covers and observing that $a_{A_0} = a$, we obtain that

$$a = p((b_{0,0})_{A_0}, (b_{0,1})_{A_0}, \ldots, (b_{1,0})_{A_0}, (b_{1,1})_{A_0}, \ldots).$$

Forming lower $B_0$-covers in the original expression for $a$, we get that

$$a_{B_0} = p(b_{0,0}, b_{0,1}, \ldots, 0^b, 0^b, \ldots),$$

and from the previous formula

$$a_{B_0} = (a_{A_0})_{B_0} = p(((b_{0,0})_{A_0})_{B_0}, \ldots, ((b_{1,0})_{A_0})_{B_0}, \ldots).$$

But $b_{1,m} \geq (b_{1,m})_{A_0}$ and so

$$0^b = (b_{1,m})_{B_0} \geq ((b_{1,m})_{A_0})_{B_0},$$

hence $((b_{1,m})_{A_0})_{B_0} = 0^b$. Thus

$$a_{B_0} = p(b_{0,0}, b_{0,1}, \ldots, 0^b, 0^b, \ldots)$$
$$= p(((b_{0,0})_{A_0})_{B_0}, ((b_{0,1})_{A_0})_{B_0}, \ldots, 0^b, 0^b, \ldots).$$

Since $p$ is isotone and $b_{0,m} \geq (b_{0,m})_{A_0} \geq ((b_{0,m})_{A_0})_{B_0}$, we obtain that

$$a_{B_0} = p(b_{0,0}, b_{0,1}, \ldots, 0^b, 0^b, \ldots) \geq p((b_{0,0})_{A_0}, (b_{0,1})_{A_0}, \ldots, 0^b, 0^b, \ldots)$$
$$\geq p(((b_{0,0})_{A_0})_{B_0}, ((b_{0,1})_{A_0})_{B_0}, \ldots, 0^b, 0^b, \ldots) = a_{B_0},$$

and so

$$a_{B_0} = p((b_{0,0})_{A_0}, (b_{0,1})_{A_0}, \ldots, 0^b, 0^b, \ldots) \in A_0 \cup \{0^b\}.$$

By definition, $a_{B_0} \in B_0 \cup \{0^b\}$, hence $a_{B_0} \in (A_0 \cap B_0) \cup \{0^b\}$. Similarly, $a_{B_j} \in (A_i \cap B_j) \cup \{0^b\}$ for all $a \in A_i$ and $i, j \in \{0, 1\}$. It follows immediately, that

$$a = p((b_{0,0})_{A_0}, (b_{0,1})_{A_0}, \ldots, (b_{1,0})_{A_0}, (b_{1,1})_{A_0}, \ldots)$$
$$\in \text{sub}((A_0 \cap B_0) \cup (A_0 \cap B_1) \cup \{0^b\})$$

for $a \in A_0$.

Now a simple induction on the rank of a term proves that for all $a \in A_0$ and for the term $p$ of smallest rank representing $a$ in the form

$$a = p(a_0, \ldots, a_{n-1}), \quad a_0, \ldots, a_{n-1} \in (A_0 \cap B_0) \cup (A_0 \cap B_1) \cup \{0^b\},$$

no $a_i$ is $0^b$. We conclude that

$$A_0 \subseteq \text{sub}((A_0 \cap B_0) \cup (A_0 \cap B_1)).$$

Thus

$$L = \text{sub}(A_0 \cup A_1) = \text{sub}\left(\bigcup (A_i \cap B_j \mid i, j = 0, 1)\right).$$

Applying Corollary 531 twice, we get that

$$A_0 = (A_0 \cap B_0) * (A_0 \cap B_1), \qquad A_1 = (A_1 \cap B_0) * (A_1 \cap B_1),$$
$$B_0 = (A_0 \cap B_0) * (A_1 \cap B_0), \qquad B_1 = (A_0 \cap B_1) * (A_1 \cap B_1),$$

hence

$$L = (A_0 \cap B_0) * (A_0 \cap B_1) * (A_1 \cap B_0) * (A_1 \cap B_1)$$

(to be more precise, drop all $A_i \cap B_j = \varnothing$), and this is the common refinement of $A_0 * A_1 = B_0 * B_1$.     $\square$

The Structure Theorem for Free Products has its limitations. Despite many attempts, no one has succeeded in finding a proof of the Common Refinement Property using the Structure Theorem.

A very special case of the Common Refinement Property for free products was proved in A. Kostinsky [482]. For any variety **K** of lattices, the Common Refinement Property holds for free **K**-products by G. Grätzer and J. Sichler [355]. B. Jónsson and E. Nelson [450] verify the same result for *regular varieties* of algebras, that is, varieties defined by identities in which the same variables occur on both sides. In G. Grätzer and J. Sichler [355], examples are exhibited of lattices that cannot be represented as free products of freely indecomposable lattices; the same result holds also relative to any variety **K** of lattices **K** ≠ **T**.

Let

$$g(L) = \min\{\, |X| \mid X \text{ generates } L \,\}.$$

By G. Grätzer and J. Sichler [354], the formula

$$g(A * B) = g(A) + g(B)$$

holds; the same formula holds in any variety **K** ≠ **T** of lattices.

### 1.12  ◇ Bounded and amalgamated free products

If we start with a family $\mathcal{L} = (L_i \mid i \in I)$ of bounded lattices, then we can define the $\{0,1\}$-*free product* of $\mathcal{L}$ in a natural way, see Definition 197 and the discussion following it. There are advantages to this approach in that the $i$-th covers, $p_{(i)}$ and $p^{(i)}$ (see Definition 521), belong to $L_i$, we do not have to introduce "fictional" bounds.

The Structure Theorem for Free Products, Theorem 528, carries over with minor changes. Interestingly, the new version contains the old one. If $\mathcal{L} = (L_i \mid i \in I)$ is a family of lattices, then $\mathcal{L}^b = (L_i^b \mid i \in I)$ is a family of bounded lattices; form the $\{0,1\}$-free product $K$ of $\mathcal{L}^b$, and take the sublattice $L$ generated by $\bigcup(L_i \mid i \in I)$. It is easy to see that $L$ is a free product of $\mathcal{L}$. So the two approaches are, in a sense, equivalent.

The Common Refinement Property for Free Products, Theorem 537, as we pointed out in the last section, is not proved using the Structure Theorem. Problem VI.2 in the first edition of this book raised the question whether the common refinement property holds for $\{0,1\}$-free products.

This was verified in G. Grätzer and A. P. Huhn [279] (see also [280], [281]). The basic concept in this paper is *amalgamated free products*.

Let $Q$, $A_0$, $A_1$ be lattices ($Q = \varnothing$ is allowed). Let $Q$ be a sublattice of both $A_0$ and $A_1$ and let $A_0 \cap A_1 = Q$. Then $A_0 \cup A_1$ is a partial lattice in a natural way. The lattice freely generated by this partial lattice will be called the *free product of $A_0$ and $A_1$ amalgamated over $Q$,* or the *$Q$-free product of $A_0$ and $A_1$*; it will be denoted by $A_0 *_Q A_1$.

### ◇ Theorem 538 (The Common Refinement Property for Amalgamated Free Products).

*Let the lattice $L$ be an amalgamated free product of the lattices $A_0$ and $A_1$*

*over $Q$ and also of the lattices $B_0$ and $B_1$ over $Q$. If $Q$ is finite, then $L$ is an amalgamated free product of the lattices*

$$(A_i \cap B_j \mid i, j = 0, 1, \ A_i \cap B_j \neq \varnothing)$$

*over $Q$.*

The special case $Q = \{0, 1\}$ yields the common refinement property for $\{0, 1\}$-free products.

### 1.13  ◇ Distributive free products

We cannot say much about the free product of distributive lattices beyond what can be said about free products of lattices in general; the free lattice, Free($n$), is such a lattice. However, we can consider the free product of distributive lattices in the variety of distributive lattices, see Section II.5.5.

A solution to the word problem of free $\{0, 1\}$-distributive products is given in G. Grätzer and H. Lakser [297]. In the same paper, it is proved that if $\mathfrak{m}$ is a regular cardinal and $L_i$, for $i \in I$, are bounded distributive lattices satisfying $|C| < \mathfrak{m}$ for any chain $C$ in any $L_i$, then the same holds in the free $\{0, 1\}$-distributive product. In particular, if all $L_i$ satisfy the Countable Chain Condition, so does the free $\{0, 1\}$-distributive product; see also Section 12 of G. Grätzer [257]. M. E. Adams and D. Kelly [10] extended this result to free products of lattices. See also B. Jónsson [446]. The analogous problem for $a$-disjoint sets is considered in H. Lakser [495] and M. E. Adams and D. Kelly [9]. An interesting description of free $\{0, 1\}$-distributive products can be found in R. W. Quackenbush [601]; see also B. A. Davey [122].

### Exercises

1.1. Prove that $C_2 * C_1$ is indeed the first lattice of Figure 117.

1.2. Show that $C_3 * C_1$ is the second lattice of Figure 117.

1.3. Find an infinite descending chain in $C_4 * C_1$. (Hint: Let $C_4 = \{a_0, a_1, a_2, a_3\}$ with $a_0 < a_1 < a_2 < a_3$ and $C_1 = \{b\}$ be the two chains. Define

$$c_1 = ((a_2 \wedge (a_1 \vee b)) \vee (a_3 \wedge b)) \wedge ((a_3 \wedge (a_0 \vee b)) \vee a_1)$$

and, for all $n > 1$,

$$c_n = ((c_{n-1} \wedge a_2) \vee (a_3 \wedge b)) \wedge ((c_{n-1} \wedge (a_0 \vee b)) \vee a_1).$$

Then $c_1 > c_2 > c_3 > \cdots$.)

1.4. Construct an infinite descending chain in $C_2 * C_2$. (Hint: Let $\{a_0, a_1\}$ with $a_0 < a_1$ and $\{b_0, b_1\}$ with $b_0 < b_1$ be the two chains. Define

$$c_1 = ((a_1 \wedge (a_0 \vee b_1)) \vee b_0) \wedge ((b_1 \wedge (a_1 \vee b_0)) \vee a_0)$$

and, for all $n > 1$,

$$c_n = ((c_{n-1} \wedge a_1) \vee b_0) \wedge ((c_{n-1} \wedge b_1) \vee a_0).)$$

1.5. $A * B$ is finite iff $A$ or $B$ is the one-element lattice and the other is a chain of not more than three elements (Ju. I. Sorkin [658]).

1.6. Let $a_1 < a_2$ and $b_1 < b_2$ be two chains. Introduce the following notation (see Figure 121):

$$A_1 = a_2,$$
$$B_1 = b_2,$$
$$A_1' = a_1,$$
$$B_1' = b_1,$$

and for all $n > 1$,

$$A_n = a_2 \wedge (a_1 \vee B_{n-1}),$$
$$B_n = b_2 \wedge (b_1 \vee A_{n-1}),$$
$$C_n = a_1 \vee B_n,$$
$$D_n = b_1 \vee A_n,$$
$$P_n = A_n \vee B_n,$$
$$Q_n = C_n \wedge D_n,$$
$$M_1 = a_1 \vee b_1,$$
$$M_2 = (a_2 \wedge b_2) \vee a_1 \vee b_1,$$
$$V_1 = b_2 \wedge ((a_2 \wedge b_2) \vee a_1 \vee b_1),$$
$$V_2 = (a_2 \wedge b_2) \vee (b_2 \wedge (a_1 \vee b_1)),$$
$$V_3 = b_2 \wedge (a_1 \vee b_1),$$
$$W_1 = a_2 \wedge ((a_2 \wedge b_2) \vee a_1 \vee b_1),$$
$$W_2 = (a_2 \wedge b_2) \vee (a_2 \wedge (a_1 \vee b_1)),$$
$$W_3 = a_2 \wedge (a_1 \vee b_1).$$

Show that the partial operation $'$ on the generators extends to an antiautomorphism $'$ of $C_2 * C_2$ of order 2; and that the elements listed above and their images under $'$ are distinct, and are all the elements of $C_2 * C_2$ as shown on Figure 121 (H. L. Rolf [618]).

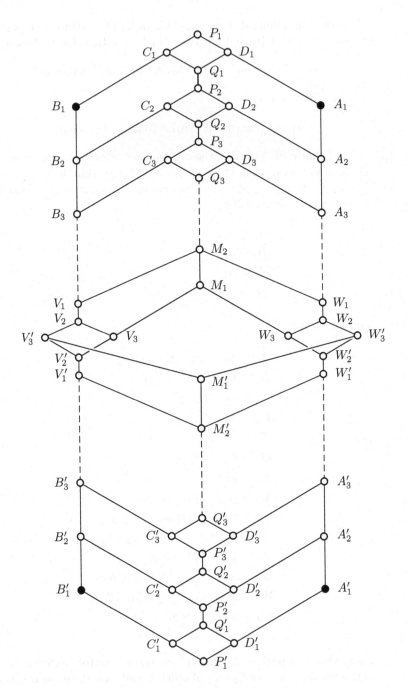

Figure 121. The lattice $C_2 * C_2$

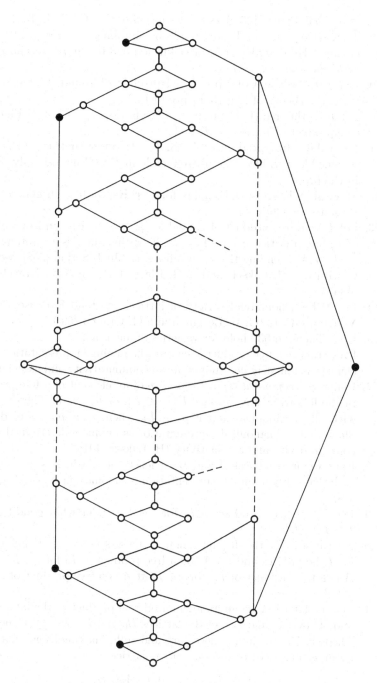

Figure 122. The lattice $C_4 * C_1$

1.7. Show that Figure 122 gives a description of $C_4 * C_1$ (H. L. Rolf [618]).

1.8. To what extent can Theorem 528 be simplified for free products of chains? (Hint: replace (C) in Definition 523 by "$p, q \in L_i$ and $p \leq q$ in $L_i$ for some $i \in I$".)

1.9. Define the concept of a free $\{0,1\}$-product of bounded lattices. Develop the theory of free $\{0,1\}$-products.

1.10. Let $L$ be the free $\{0,1\}$-product of the lattices $L_i^b$ for $i \in I$. Find the free product of $L_i$, for $i \in I$, in $L$.

1.11. Let $L$ be a free product of $L_i$ for $i \in I$. Show that $a \in L_i$ is join-irreducible in $L$ iff $a$ is join-irreducible in $L_i$. (Hint: use only covers in the proof.)

1.12. Show that Exercise 1.11 holds for free **K**-products in general (see B. Jónsson [446]).

1.13. Let $L$ be a free product of $\mathcal{L}$, and let $\varphi_i \colon L_i \to A$ be an isotone map of $L_i$ into a lattice $A$ for all $i \in I$. Show that there is an isotone map $\varphi \colon L \to A$ extending the $\varphi_i$ for all $i \in I$. (Ju. I. Sorkin [658]; see also G. Grätzer, H. Lakser, and C. R. Platt [311] and A. V. Kravchenko [486].)

1.14. Let **V** be a nontrivial variety of lattices. Extend Exercise 1.13 to **V**-free products (G. M. Bergman and G. Grätzer [60]).

1.15. Does Theorem 532 hold for varieties of algebras?

1.16. Prove that in a free product of chains, the minimal representation of an element is uniquely determined up to commutativity and associativity.

*1.17. Define the *canonical representation* $p$ of an element of a free product in the following way: $p \in \bigcup(L_i \mid i \in I)$; or $p$ is as in Theorem 535 with all $p_i$ canonical and if $p_j \in L_k$, then $p_j = p_{(k)}$; and dually. Show that the canonical representation is uniquely determined up to commutativity and associativity (H. Lakser [491]).

1.18. Find a pair of lattices $A$ and $B$, at least one of which is not a chain, such that every minimal representation is canonical in $A*B$ (H. Lakser [491]).

1.19. Let $L$ be a free product of $L_i$ for $i \in I$. Under what conditions is $L - L_i \leq L$?

1.20. Let the set $I$ be the disjoint union of the sets $I_j$ for $j \in J$. Let $L_i$, for $i \in I$, be lattices and let $A_j$ be a free product of $L_i$ for $i \in I_j$. Then $A$ is a free product of $L_i$, for $i \in I$, iff $A$ is a free product of $A_j$ for $j \in J$.

*1.21. Let the lattice $L$ be an amalgamated free product of the lattices $A_0$ and $A_1$ over $Q$ and also of the lattices $B_0$ and $B_1$ over $Q$. Generalize Theorem 538 by finding necessary and sufficient conditions that $L$ be an amalgamated free product of the lattices

$$(A_i \cap B_j \mid i, \, j = 0, 1, \ A_i \cap B_j \neq \varnothing)$$

over $Q$ (G. Grätzer and A. P. Huhn [279]).

*1.22. For a lattice $K$, let $g(K)$ denote the cardinality of the smallest generating set of $K$. Let $L$ be the free product of the lattices $A$ and $B$. Prove that $g(L) = g(A) + g(B)$, as asserted at the end of Section 1.11 (G. Grätzer and J. Sichler [354]).

*1.23. Generalize Exercise 1.22 to amalgamated free products (G. Grätzer and A. P. Huhn [280]).

## 2.   The Structure of Free Lattices

In this section, we describe the structure of free lattices as it follows from the results in Section 1. For a much deeper treatment of this topic, we refer the reader to the book R. Freese, J. Ježek, and J. B. Nation [186].

### 2.1   The structure theorem

By comparing the definition of a free lattice Free(m) (see Section I.5.1) with the definition of a free product (see Section 1.2), we observe:

*A lattice is free iff it is a free product of a family of one-element lattices.*

Thus the results of the previous section can be specialized to describe the structure of free lattices.

**Definition 539.** For a set $X$ and $p, q \in \mathrm{Term}(X)$, set $p \subseteq q$ if it follows from $x \subseteq x$, for all $x \in X$, and from $(\wedge W)$, $(\vee W)$, $(W_\wedge)$, and $(W_\vee)$.

Again, for $p, q \in \mathrm{Term}(X)$, we set

$$p \equiv q \quad \text{iff} \quad p \subseteq q \text{ and } q \subseteq p;$$
$$\mathrm{rep}(p) = \{\, q \in \mathrm{Term}(X) \mid p \equiv q \,\};$$
$$\mathrm{Rep}(X) = \{\, \mathrm{rep}(p) \mid p \in \mathrm{Term}(X) \,\};$$
$$\mathrm{rep}(p) \leq \mathrm{rep}(q) \quad \text{iff} \quad p \subseteq q.$$

Now we can state the celebrated result of P. M. Whitman [722]:

**Theorem 540.** $\mathrm{Rep}(X)$ *is a free lattice on* $|X|$ *generators. In other words, Definition 539 provides an algorithm to decide whether* $a \leq b$ *in the free lattice, where* $a$ *and* $b$ *are represented by the terms* $p$ *and* $q$, *respectively.*

*Proof.* Compare Definition 539 with Definition 523. To make the comparison, we have to set $X = I$ and $L_i = \{i\}$ for all $i \in I$. It is sufficient to show that if $p \subseteq q$ by (C), then $p \subseteq q$ by Definition 539. We prove this by induction. This is obvious if $p, q \in X$. Now let $\mathrm{rank}(p) + \mathrm{rank}(q) > 2$, and let $p^{(i)} \leq q_{(i)}$, for some $i \in I$, that is, $p^{(i)} = q_{(i)} = i$. If $p = p_0 \wedge p_1$, then $(p_0 \wedge p_1)^{(i)} = p_0^{(i)} \wedge p_1^{(i)}$, hence $p_0^{(i)} = i$ or $p_1^{(i)} = i$. Thus $p_0^{(i)} \leq q_{(i)}$ or $p_1^{(i)} \leq q_{(i)}$, hence by the induction hypothesis, $p_0 \subseteq q$ or $p_1 \subseteq q$, and so by $(\wedge W)$, $p \subseteq q$. We proceed similarly in the other three cases.  $\square$

Theorem 540 is called the solution to the word problem. Exercise 2.2 gives a solution to the word problem for Free $P$, the lattice freely generated by an order $P$.

From Theorem 535, we learned that minimal terms differ only in some $L_i$ component. If all $|L_i| = 1$, they have to be identical.

P. M. Whitman [722] characterized minimal polynomials:

**Theorem 541.** *A minimal representation of an element is unique up to commutativity and associativity. Let $p \in \mathrm{Term}(X)$. Then $p$ is minimal iff $p \in X$, or if $p = p_0 \vee \cdots \vee p_{n-1}$ with $n > 1$, where no $p_j$ is a join of more than one term and conditions (i)–(iii) below hold, or the dual case.*

(i)  *Each $p_j$ is minimal for all $0 \le j < n$.*

(ii)  *The formula $p_j \not\le p_0 \vee \cdots \vee p_{j-1} \vee p_{j+1} \vee \cdots \vee p_{n-1}$ holds for all $j$ with $0 \le j < n$.*

(iii)  *If $p_j = p_j' \wedge p_j''$, where $0 \le j < n$ and $p_j', p_j'' \in \mathrm{Term}(X)$, then $p_j', p_j'' \not\le p$.*

*Proof.* In the special case considered, Theorem 535(v) is made superfluous by Theorem 535(ii). Also, Theorem 535(iii) does not apply since $p_j^{(i)}, p_{(i)} \in L_i$, hence $p_j^{(i)} = p_{(i)}$. The remaining conditions of Theorem 535 are identical with (i)–(iii) above.    □

A minimal representation of an element in a free lattice is called a *canonical form* of the element in the literature.

## 2.2   ◇ The word problem for modular lattices

The solution to the word problem for $\mathrm{Free}_{\mathbf{M}}(3)$ is trivial, since $\mathrm{Free}_{\mathbf{M}}(3)$ is finite, see Figure 20. The solution to the word problem for $\mathrm{Free}_{\mathbf{M}}(n)$, for $n > 3$, and for various $\mathrm{Free}_{\mathbf{M}'}(n)$, where $\mathbf{M}'$ is a subvariety of $\mathbf{M}$ (for instance, the variety of arguesian lattices or the variety generated by subspace lattices of rational vector spaces) has been one of the most active research fields for 60 years, with dozens of references in *MathSciNet*.

The most outstanding result in this field is in R. Freese [183]:

◇ **Theorem 542.** *The word problem for the free modular lattice on five generators is unsolvable.*

C. Herrmann [390] improved "five" to "four" in this result.

## 2.3   Applications

Now we shall study in some detail the structure of $\mathrm{Free}(3)$. Let $x_0, x_1, x_2$ be the free generators of $\mathrm{Free}(3)$. The zero and unit of $\mathrm{Free}(3)$ are $x_0 \wedge x_1 \wedge x_2$

and $x_0 \vee x_1 \vee x_2$, respectively. By Lemma 73, $x_0 \vee x_1, x_0 \vee x_2, x_1 \vee x_2$ generate a sublattice isomorphic to $C_2^3$. Since

$$\text{Free}(3) = \{x_0\} * \{x_1 \wedge x_2, x_1, x_2, x_1 \vee x_2\}$$

(see Exercise 1.20), the Splitting Theorem (Theorem 529) gives that, for every element $x \in \text{Free}(3)$, either $x \geq x_0$ or $x \leq x_1 \vee x_2$.

Similarly, $x \geq x_1$ or $x \leq x_0 \vee x_2$; and $x \geq x_2$ or $x \leq x_0 \vee x_1$. From this, we can infer that

$$x_0 \vee x_1 \prec x_0 \vee x_1 \vee x_2,$$
$$(x_0 \vee x_1) \wedge (x_0 \vee x_2) \prec x_0 \vee x_1;$$

these and the symmetric cases give the nine coverings at the top of Figure 123. Let us prove the second covering; if

$$(x_0 \vee x_1) \wedge (x_0 \vee x_2) < t \leq x_0 \vee x_1,$$

then $t \not\leq x_0 \vee x_2$, hence $t \geq x_1$; similarly, $t \geq x_0$, hence $t \geq x_0 \vee x_1$, proving that $t = x_0 \vee x_1$.

If $t > x_0$, then $t \not\leq x_0$, and so $t \geq x_1 \wedge x_2$; hence

$$x_0 \prec x_0 \vee (x_1 \wedge x_2) = x_0 \vee a;$$

by symmetry and duality, this accounts for six coverings in the middle of the diagram. The other six are typified by ($a$ and $b$ are defined in Figure 123)

$$x_0 \vee a \succ (x_0 \vee a) \wedge b,$$

which is again a trivial consequence of the Splitting Theorem.

Again, easy applications of the Splitting Theorem yield that all elements of rank 4 or more lie in one of the intervals:

$$[x_0 \vee a, (x_0 \vee x_1) \wedge (x_0 \vee x_2)],$$

its two symmetric intervals, the three dual intervals, and $[a, b]$. Thus to complete the verification of Figure 123, we have to show that

$$[x_0 \vee a, (x_0 \vee x_1) \wedge (x_0 \vee x_2)] = [x_0 \vee a, x_0 \vee b] \cup \{(x_0 \vee x_1) \wedge (x_0 \vee x_2)\}.$$

The trick (due to P. Gumm) is to verify first, by induction on $\text{rank}(x)$, that if $x \in [x_0 \vee a, (x_0 \vee x_1) \wedge (x_0 \vee x_2)]$, then $x$ is comparable with $x_0 \vee b$. Condition (W), and therefore, Theorem 544, is needed in this step. Then

$$x_0 \vee b \prec (x_0 \vee x_1) \wedge (x_0 \vee x_2)$$

follows easily. The details are left to the reader.

A word of warning about Figure 123: it gives a "generalized diagram" of Free(3) but it does not contain all the elements or even all the orderings of the elements shown. For instance, $a < x_0 \vee a$ is not shown nor is the element $(x_0 \wedge b) \vee (x_1 \wedge b)$ (the element $u_0$ on page 498), an element of $[a, b]$. Nevertheless, Figure 123 can be very useful in visualizing Free(3).

We obtain some additional coverings in Free(3) from a result of R. A. Dean [146]:

**Theorem 543.** *For any*

$$c \in [x_0 \vee (x_1 \wedge x_2), (x_0 \vee x_1) \wedge (x_0 \vee x_2)],$$

*the covering*

$$c \wedge (x_1 \vee x_2) \prec (c \wedge (x_1 \vee x_2)) \vee x_0$$

*holds.*

*Proof.* These two elements are obviously distinct, since they belong to disjoint intervals. If

$$c \wedge (x_1 \vee x_2) \leq t \leq (c \wedge (x_1 \vee x_2)) \vee x_0$$

and $t \geq x_0$, then obviously $t = (c \wedge (x_1 \vee x_2)) \vee x_0$. Otherwise, $t \leq x_1 \vee x_2$, hence

$$t \leq ((c \wedge (x_1 \vee x_2)) \vee x_0) \wedge (x_1 \vee x_2) \leq c \wedge (x_1 \vee x_2),$$

and so $t = c \wedge (x_1 \vee x_2)$.     □

Many of the observations made about Free(3) hold for all $\text{Free}_\mathbf{K}(3)$ for any variety $\mathbf{K} \neq \mathbf{T}$ of lattices. An interesting example is

$$x_0 \wedge b \prec x_0 \prec x_0 \vee a.$$

This even holds for $\mathbf{K} = \mathbf{D}$, where $a = b$.

Apart from the Splitting Theorem, the most important properties of a free lattice are given in the next result, which is due to P. M. Whitman [722] and B. Jónsson [440]:

**Theorem 544.** *Conditions* (W), $(SD_\wedge)$, *and* $(SD_\vee)$ *hold in any free lattice.*

*Proof.* If $|A_i| = 1$, then $A_i^b = \mathbf{C}_3$, which satisfies all three conditions. So choose the lattices $A_i = L_i^b$, and the result follows from Theorems 533 and 536.     □

We illustrate the power of these conditions by two results.

**Theorem 545 (B. Jónsson [440]).** *A sublattice $A$ of finite length of a free lattice (of any lattice satisfying $(SD_\wedge)$) is finite.*

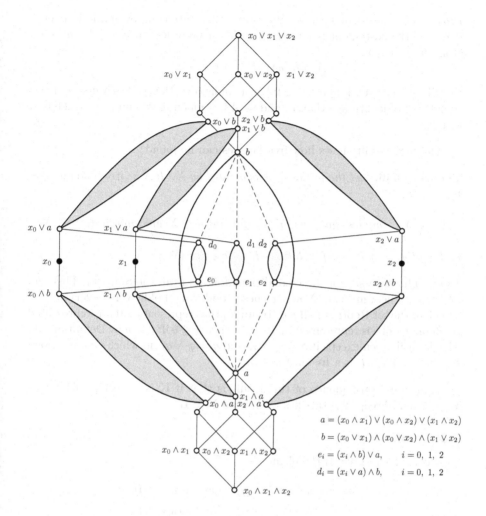

Figure 123. The free lattice on three generators, Free(3)

*Proof.* Let $A$ be of length $n$. We prove the statement by induction on $n$. If $n = 1$, the statement is trivial. Now let $A$ be of length $n$ and let $a$ be an atom of $A$. Then

$$A = \mathrm{fil}(a) \cup \{ x \mid a \wedge x = 0 \}.$$

By (SD$_\wedge$), the set $x\{ x \mid a \wedge x = 0 \}$ is a sublattice. Hence $A$ is a union of two sublattices of length less than $n$, and so $A$ is a union of two finite sets. Thus $A$ is finite.    □

The next result shows how free lattices can be found.

**Theorem 546.** *Let the lattice $L$ be generated by $X$. If $X$ is irredundant, that is,*

$$x \leq \bigvee Y \quad \textit{implies that} \quad x \in Y \qquad \textit{for any } x \in X \textit{ and any finite } Y \subseteq X,$$

*and dually, then $L$ is a free lattice iff $L$ satisfies* (W).

*Proof.* The necessity of these conditions follows from the Splitting Theorem and from Theorem 544. Now let these conditions hold. With every $a \in L$, associate the set $\mathrm{Rep}(a)$ of all $p \in \mathrm{Term}(X)$ that represent $a$. If $a$ is represented by $\mathrm{Rep}(p)$ and $b$ is represented by $\mathrm{Rep}(q)$, then $a \leq b$ iff $p \subseteq q$ by Definition 539. This is easily seen except if $p = p_0 \wedge p_1$ and $q = q_0 \vee q_1$, in which case it follows from (W). Thus $L$ is a free lattice.    □

A curious consequence of this result is that if $\mathbf{K}$ is a variety of lattices, $\mathbf{K} \neq \mathbf{T}$, and $\mathrm{Free}_{\mathbf{K}}(\aleph_0)$ satisfies (W), then $\mathbf{K} = \mathbf{L}$.

## 2.4 Sublattices

Now consider in $\mathrm{Free}(3)$ the elements

$$u_0 = (x_0 \wedge (x_1 \vee x_2)) \vee (x_1 \wedge (x_0 \vee x_2)),$$
$$u_1 = (x_0 \wedge (x_1 \vee x_2)) \vee (x_2 \wedge (x_0 \vee x_1)),$$
$$u_2 = (x_0 \vee (x_1 \wedge x_2)) \wedge (x_1 \vee (x_0 \wedge x_2)),$$
$$u_3 = (x_0 \vee (x_1 \wedge x_2)) \wedge (x_2 \vee (x_0 \wedge x_1)).$$

It is an easy computation to show that $U = \{u_0, u_1, u_2, u_3\}$ is irredundant. Hence $\mathrm{Free}(3) \geq \mathrm{Free}(4)$. Now define $y_0 = u_0, a_0 = u_1, b_0 = u_2, c_0 = u_3$. If $y_n, a_n, b_n, c_n$ are defined, construct $y_{n+1}, a_{n+1}, b_{n+1}, c_{n+1}$ from $a_n, b_n, c_n$, just as the $u_i$ were constructed from $x_0, x_1, x_2$.

We claim that $Y = \{y_0, y_1, y_2, \ldots\}$ is an irredundant set. To see this, we prove by induction that $y_0, \ldots, y_n, a_n, b_n, c_n$ generate a free lattice on $n + 4$ generators. This is true for $n = 0$. Assume it for $n$. Then

$$\mathrm{sub}(y_0, \ldots, y_n, y_{n+1}, a_{n+1}, b_{n+1}, c_{n+1})$$

is the sublattice generated by $\mathrm{sub}(y_0, \ldots, y_n)$ and $\mathrm{sub}(y_{n+1}, a_{n+1}, b_{n+1}, c_{n+1})$ in the lattice

$$\mathrm{sub}(y_0, \ldots, y_n, a_n, b_n, c_n) = \mathrm{sub}(y_0, \ldots, y_n) * \mathrm{sub}(a_n, b_n, c_n).$$

Thus by Theorem 532,

$$\mathrm{sub}(y_0, \ldots, y_n, y_{n+1}, a_{n+1}, b_{n+1}, c_{n+1})$$
$$\cong \mathrm{sub}(y_0, \ldots, y_n) * \mathrm{sub}(y_{n+1}, a_{n+1}, b_{n+1}, c_{n+1})$$
$$\cong \mathrm{Free}(n+1) * \mathrm{Free}(4) \cong \mathrm{Free}(n+5).$$

This proves the statement. Thus $Y$ is irredundant and so, by Theorem 546, the isomorphism $\mathrm{sub}(Y) \cong \mathrm{Free}(\aleph_0)$ holds. So we have proved a result of P. M. Whitman [722].

**Theorem 547.** $\mathrm{Free}(3)$ *contains* $\mathrm{Free}(\aleph_0)$ *as a sublattice.*

A much stronger version of this result was proved in S. T. Tschantz [681]:

◇ **Theorem 548.** *Every infinite interval in a free lattice contains a sublattice isomorphic to* $\mathrm{Free}(3)$.

A free generator of a free lattice is doubly irreducible, and, of course, no other element is such. Hence we obtain:

**Theorem 549.** *There is a one-to-one correspondence between automorphisms of a free lattice and permutations of its free generating set.*

This is used in proving the following result of F. Galvin and B. Jónsson [217].

**Theorem 550.** *Every chain of a free lattice is countable.*

*Proof.* Let $X$ be the free generating set of $L$. Assume that $X$ is not countable. Let $X_0$ be a countable subset of $L$, let $L_0 = \mathrm{sub}(X_0)$, and let $C$ be a chain in $L$.

For $a, b \in L$, let $a \equiv b$ mean that $\varphi(a) = b$ for some automorphism $\varphi$ of $L$. Obviously, $\equiv$ is an equivalence relation. Now if $a \in L$, then $a \in \mathrm{sub}(Y)$ for some finite $Y \subseteq X$. Take a permutation $\pi$ of $X$ with $\pi(Y) \subseteq X_0$, and let $\varphi$ be the automorphism of $L$ extending $\pi$. Then $\varphi(a) \in L_0$. Hence there are no more blocks than $|L_0| = \aleph_0$. Therefore, the proof will be complete if we show that $a \not\equiv b$ for every $a, b \in C$ with $a \neq b$. Indeed, let $a < b$ and let $\varphi(a) = b$ for some automorphism $\varphi$. Then $a \in \mathrm{sub}(Y)$ for some finite $Y \subseteq X$. There is a permutation $\varrho$ of $X$ such that $\varrho(x) = \varphi(x)$, for $x \in Y$, and $\varrho(x) = x$ for $x \in X - (Y \cup \varphi(Y))$. Let $\psi$ be the automorphism of $L$ extending $\varrho$. Then $\psi(a) = b$ and $\psi$ is of some finite order $n$ (as a group element). Thus

$$a < \psi(a) < \psi^2(a) < \cdots < \psi^n(a) = a,$$

a contradiction. □

A somewhat unexpected dividend of the study of the structure of free lattices is the following result of R. P. Dilworth [155] and R. A. Dean [143]:

**Theorem 551.** *There exists a three-generated partial lattice with infinitely many maximal elements.*

*Proof.* Let $x_0, x_1, x_2$ be the free generators of Free(3) and define $y_0 = x_0$,

$$y_n = x_0 \vee (x_1 \wedge (x_2 \vee (x_0 \wedge (x_1 \vee (x_2 \wedge y_{n-1}))))),$$
$$y_{-n} = x_0 \wedge (x_1 \vee (x_2 \wedge (x_0 \vee (x_1 \wedge (x_2 \vee y_{-n+1}))))),$$
$$a_n = x_1 \vee (y_n \wedge (y_{-n} \vee x_2)).$$

It is easy to check that $\mathrm{sub}(a_1, a_2, \ldots) \cong \mathrm{Free}(\aleph_0)$.

For a term $p$, define the *component subset* of $p$, denoted by $\mathrm{Komp}(p)$, as follows:

$$\mathrm{Komp}(x_i) = \{x_i\};$$
$$\mathrm{Komp}(p_0 \wedge p_1) = \mathrm{Komp}(p_0) \cup \mathrm{Komp}(p_1) \cup \{p_0 \wedge p_1\},$$

and similarly for $p_0 \vee p_1$.

If $a \in \mathrm{Free}(3)$, let $a = \mathrm{rep}(p)$, where $p$ is the canonical term representing $a$. Define

$$\mathrm{Komp}(a) = \{\, \mathrm{rep}(q) \mid q \in \mathrm{Komp}(p) \,\}.$$

Then $\mathrm{Komp}(a) \subseteq \mathrm{Free}(3)$. Regard $\mathrm{Komp}(a)$ as a partial sublattice of Free(3). Then $\mathrm{Komp}(a)$ becomes a partial lattice. $\mathrm{Komp}(a)$ is generated by

$$\{x_0, x_1, x_2\} \cap \mathrm{Komp}(a).$$

Now define
$$A = \bigcup (\mathrm{Komp}(a_n) \mid n = 1, 2, \ldots).$$

All the elements of $A$ are easily enumerated, and we observe that $a_1, a_2, \ldots$ are maximal elements of $A$. $\qquad\square$

This theorem, in turn, found an application in Ju. I. Sorkin [659] and R. A. Dean [143].

**Corollary 552.** *Let $L$ be a countable lattice. Then $L$ can be embedded in a three-generated lattice.*

*Proof.* Let $L = \{b_0, b_1, b_2, b_3, \ldots\}$ and, with the partial lattice $A$ of Theorem 551, form the set $B = L \cup A$ (we assume that $L$ and $A$ are disjoint). We define an ordering $\leq$ on $B$:

(i)  for $u, v \in L$, let $u \leq v$ in $B$ if $u \leq v$ in $L$;
(ii) for $u, v \in A$, let $u \leq v$ in $B$ if $u \leq v$ in $A$;

(iii) let $a_{2n+1} \leq b_n$ and $a_{2n+2} \leq b_n$;
(iv) for $u \in A$ and $v \in L$, let $u \leq v$ if there are $v_0, \ldots, v_{n-1} \in L$ such that

   (a) $v_0 \wedge \cdots \wedge v_{n-1} \leq v$ in $L$;
   (b) for each $i$ with $0 \leq i < n$, there is a $b_j$ such that $b_j \leq v_i$ in $L$ and $u \leq a_{2j+1}$, $u \leq a_{2j+2}$ in $A$;

(v) $v \leq u$ never holds in $B$ for $u \in A$ and $v \in L$.

Now it is easy to check that $B$ is a partial lattice, and the joins

$$b_n = a_{2n+1} \vee a_{2n+2} \quad \text{for } n = 0, 1, \ldots$$

are valid in $B$. Since $A$ is three-generated, so is $B$. The partial lattice $B$ can be embedded in a lattice $C$ and obviously, $\mathrm{sub}(B)$ in $C$ is three-generated and contains $L$ as a sublattice. $\qquad\square$

Corollary 552 suggests that there are very many three-generated lattices. Indeed, in P. Crawley and R. A. Dean [106], it is proved that there are $2^{\aleph_0}$ three-generated lattices. Call a lattice $L$ $\mathfrak{m}$-*universal* if $|L| = \mathfrak{m}$ and every lattice of cardinality at most $\mathfrak{m}$ is isomorphic to a sublattice of $L$. Thus there are no $\aleph_0$-universal lattices. Assuming the Generalized Continuum Hypothesis, B. Jónsson [435] proves that $\aleph_\alpha$-universal lattices exist for each $\alpha > 0$.

## 2.5  ◇ More covers

We have already found infinitely many covers in Free(3), but we have hardly scratched the surface. A number of papers have been published on covers in Free(3). Maybe the best result is in A. Day [138]:

◇ **Theorem 553.** *Finitely generated free lattices are relatively atomic.*

Based on Day's work, R. Freese and J. B. Nation [191] (see also R. Freese, J. Ježek, and J. B. Nation [186]) describe all covers. Here are some more of their results.

◇ **Theorem 554.** *A chain of covers in a free lattice can have length at most 4. Chains of covers of length 3 and 4 in Free(n) occur only in the connected component of 0 or of 1. On the other hand, Free(n) has infinitely many chains of covers of length 2 for all $n \geq 3$.*

However, the analogue of Theorem 553 does not hold for chains of covers of length 2: there are infinite intervals in Free(3) that do not contain any chain of covers of length 2.

The bottom of Free(3), see Figure 123, is a finite interval. It turns out this and its dual are the only finite intervals of length 4 in Free(3).

## 2.6    ◇ Finite sublattices and transferable lattices

What are the sublattices of the free lattices? As opposed to groups, where the subgroup of a free group is free, this problem for lattices seems very difficult. There are a number of early papers on the subject, in 1959–1962: P. Crawley and R. A. Dean [106], R. A. Dean [145], B. Jónsson [440] and B. Jónsson and J. E. Kiefer [448]. *MathSciNet* lists several dozen papers on this topic.

The conjecture that finite sublattices of a free lattice can be characterized by (W), (SD$_\wedge$), and (SD$_\vee$) (see these conditions in Section 1.4) is attributed to B. Jónsson; it first appeared in print in B. Jónsson and J. E. Kiefer [448]. About 20 years of work, first by Jónsson and then by J. B. Nation, resolved this problem (see J. B. Nation [539]):

◇ **Theorem 555.** *Finite sublattices of a free lattice are characterized by* (W), (SD$_\wedge$), *and* (SD$_\vee$).

A seemingly unrelated topic started in 1965: I was invited by J. C. Abbott and G. Birkhoff to give a lecture on my forthcoming book on universal algebra (G. Grätzer [254]) at a conference at the United States Naval Academy in May of 1966 (this lecture was published in G. Grätzer [256]). In preparation for this lecture, I gave a series of talks at McMaster University in December 1965 and January 1966, where I introduced the concept of a transferable lattice.

Recall from Section IV.1.2 that a lattice $K$ is *transferable* if, for every lattice $L$, whenever $K$ has an embedding $\varphi$ into Id $L$, then $K$ has an embedding $\psi$ into $L$. The lattice $K$ is called *sharply transferable*, if, in addition, this $\psi$ can always be chosen to satisfy

$$\psi(a) \in \varphi(a) \text{ but } \psi(a) \notin \varphi(b), \text{ for any } b < a.$$

The problem of characterizing (sharply) transferable lattices (and several related questions) was raised in my lectures and in G. Grätzer [256], as Problem 14 in G. Grätzer [257] and as Problems I.22–I.29 in G. Grätzer [262].

The crucial concept in dealing with transferability was introduced by H. S. Gaskill. Let $K$ be a finite lattice, $p \in K$, and $J \subseteq K$. We call $(p, J)$ a *pair* if $p \notin J$, $2 \le |J|$, $p \le \bigvee J$, and $p \nleq j$ for all $j \in J$. A pair $(p, J)$ is *minimal* if whenever $(p, J')$ is also a pair and for every $a \in J'$ there exists a $b \in J$ with $a \le b$, then $J \subseteq J'$. (Minimal pairs are the most economical way of describing the join-structure of a finite lattice.) A finite lattice $K$ satisfies the condition (T$_\vee$) if a rank function $r$ can be defined on Ji $K$ (with integer values) such that if $(p, J)$ is a minimal pair of $K$ and $q \in J$, then $r(p) < r(q)$. Condition (T$_\wedge$) is the dual of (T$_\vee$).

For finite lattices, transferability and sharp transferability are completely described in the following result:

◇ **Theorem 556.** *For a finite lattice $K$, the following conditions are equivalent:*

(i)  $K$ *is transferable*
(ii)  $K$ *is sharply transferable.*
(iii)  $K$ *satisfies* $(T_\vee)$, $(T_\wedge)$, *and* $(W)$.
(iv)  $K$ *can be embedded in* Free(3).

This result was discovered over about a decade as a collaborative effort involving H. S. Gaskill, C. R. Platt, B. Sands, and myself (others making a contribution to this field include K. A. Baker, A. Day, A. W. Hales, E. Nelson, T. Tan, and A. G. Waterman); see G. Grätzer [256], H. S. Gaskill [220], H. S. Gaskill, G. Grätzer, and C. R. Platt [222], H. S. Gaskill and C. R. Platt [221], G. Grätzer, C. R. Platt, B. Sands [326], and G. Grätzer and C. R. Platt [325].

The equivalence of the first four of the following five conditions combines the previous two theorems:

◇ **Theorem 557.** *For a finite lattice $K$, the following conditions are equivalent:*

(i)  $K$ *is transferable*
(ii)  $K$ *is sharply transferable.*
(iii)  $K$ *satisfies* $(SD_\vee)$, $(SD_\wedge)$, *and* $(W)$.
(iv)  $K$ *can be embedded in* Free(3).
(v)  $K$ *is projective.*

The equivalence of conditions (iv) and (v) is due to B. Jónsson and R. N. McKenzie, see R. N. McKenzie [512]; in fact, this equivalence holds for finitely generated lattices, see A. Kostinsky [483].

B. Jónsson and J. B. Nation [449] attempted to bring closer (and in a sense, unify) Theorem 555 and an earlier version of Theorem 556.

Note that the majority of the problems I.22–I.29 in G. Grätzer [262] concerning transferable lattices are still open.

## 2.7  ◇ Semidistributive lattices
### by Kira Adaricheva

We saw the importance of the semidistributive laws $(SD_\vee)$ and $(SD_\wedge)$ earlier in this chapter. Unlike other lattice laws, such as the distributive law and the modular law, these laws cannot be expressed by lattice identities. As a result, the classes of all lattices satisfying either of the semidistributive laws, or both of them, are not closed under homomorphic images. They are examples of *implicational classes*, or *quasivarieties*, and the laws $(SD_\vee)$ and $(SD_\wedge)$ are examples of *quasi-identities*.

We denote by $\mathbf{SD_\vee}$ the quasivariety of join-semidistributive lattices.

Varieties of lattices contained in $\mathbf{SD_\vee}$ are described in B. Jónsson and I. Rival [451], see also P. Jipsen and H. Rose [427]. These varieties are characterized by not containing six finite lattices, $M_3$ among them.

Semidistributive laws play an important role as properties of congruence lattices of algebras in a variety. We saw Jónsson's Lemma (Theorem 475) as an example of the strong connection between the structure of a variety and the structure of the congruence lattices of algebras in the variety: the argument of Jónsson's Lemma works not only in the variety of all lattices, but in any variety with congruence distributive algebras. In D. Hobby and R. N. McKenzie [395], both laws (SD$_\vee$) and (SD$_\wedge$) were studied in congruence lattices of varieties. K. Kearnes and E. Kiss [468] devote to these properties a chapter of the book, proving, among other things, that congruence lattices of all algebras of a variety **V** are join-semidistributive iff the congruence lattices of algebras in **V** are meet-semidistributive and satisfy some nontrivial identity.

The quasivariety **SD**$_\vee$ attracts a great deal of interest, because of the abundance of important join-semidistributive lattices.

The following result (K. V. Adaricheva, V. Gorbunov and V. Tumanov [21]; see also V. Gorbunov [243]) establishes one of the basic properties of the class **SD**$_\vee$:

◊ **Theorem 558.** *The quasivariety* **SD**$_\vee$ *is generated by its finite members.*

In particular, this implies that the universal theory of **SD**$_\vee$ is decidable.

An important subclass of **SD**$_\vee$ is the class of *lower bounded lattices*. The term is inherited from the concept of a *lower bounded homomorphism*, introduced in R. N. McKenzie [512]. The most general definition of a lower bounded lattice was suggested in K. V. Adaricheva and V. Gorbunov [19]: a lattice $L$ is *lower bounded* if, for any homomorphism $\varphi$: Free $X \to L$ with $X$ finite and for any $a \in L$, the filter $\varphi^{-1}(\mathrm{fil}(a))$ is either empty or has a least element.

There are the dual notions of an *upper bounded homomorphism* and an *upper bounded lattice*, as well as the conjunction of the two: *bounded homomorphisms* and *bounded lattices*. The latter form a subclass in the class of semidistributive lattices.

A *splitting lattice* is a finite and subdirectly irreducible bounded lattice, see Section VI.2.3.

Finite lower bounded lattices were intensively studied, producing various descriptions, see B. Jónsson and J. B. Nation [449], A. Day [138], and P. Pudlák and J. Tůma [598], among others; see also Section IV.4.4.

None of these descriptions can be extended to arbitrary lower bounded lattices, as examples in K. V. Adaricheva and V. A. Gorbunov [19] show.

The description of lower bounded lattices remains an open problem. In F. Wehrung [709], some infinitary lattice identities define a proper subclass of **SD**$_\vee$ that includes all lower bounded lattices, while K. V. Adaricheva and J. B. Nation [23] give the description of the lower bounded lattices of a finite rank.

Examples of join-semidistributive lattices include subquasivariety lattices, as well as related lattices of the form $S_p A$, for an upper-continuous lattice $A$, see the discussion in Section VI.2.7. The lattice $S_p A$ is the lattice of subsets of $A$ closed under arbitrary meets and joins of arbitrary nonempty chains. The survey K. V. Adaricheva [14] examines various aspects of these lattices.

In combinatorics, a *convex geometry* is a closure system $(\text{Pow } X, \phi)$ with a finite set $X$ and a closure operator $\phi$ satisfying *the anti-exchange axiom*:

If $A$ is closed, $x \neq y \in X - A$, and $x \in \phi(A \cup y)$, then $y \notin \phi(A \cup x)$.

Convex geometries provide an important source of join-semidistributive lattices.

The structure of open sets of such a closure system, that is, the complements of closed sets, is called an *anti-matroid*. This points to a relation with *matroids*, which are defined as closure systems $(\text{Pow } X, \phi)$, where the closure operator satisfies *the exchange axiom*:

If $A$ is closed, $x \neq y \in X - A$, and $x \in \phi(A \cup y)$, then $y \in \phi(A \cup x)$.

B. Dietrich [152] surveys the relationship between matroids and anti-matroids.

The lattices of closed sets of matroids are geometric lattices, and in many important cases they satisfy the modular law. The description of lattices of closed sets of convex geometries was rather elusive, since they reappeared in disguised form in various studies, see a short but informative survey, B. Monjardet [533]. The first connections between combinatorial and lattice properties of convex geometries was established in P. H. Edelman and R. Jamison [171]. The following result first appeared in explicit form in V. Duquenne [169]; see also K. V. Adaricheva, V. A. Gorbunov, and V. I. Tumanov [21].

◇ **Theorem 559.** *A finite lattice is isomorphic to a lattice of closed sets of some finite convex geometry iff it is join-semidistributive and lower semimodular.*

In general, convex geometries are not always join-semidistributive, but in many important types of convex geometries this law holds; these are, for example, convex subsets of orders, as shown by M. K. Bennett and G. Birkhoff [74] or convex bodies of vector spaces, see [21].

It seems that the juxtaposition between the exchange and the anti-exchange axioms of underlying closure systems explains the difference between the modular law and the join-semidistributive law, which are both generalizations of the distributive law. Indeed, it is easy to observe that, $\mathbf{M} \cap \mathbf{SD}_\vee = \mathbf{D}$, that is, the former two laws can both hold only in distributive lattices.

The article [21] made a considerable effort to show deeper connections between convex geometries and join-semidistributive lattices. We note that in general the convex geometries are not atomistic, but the atomistic ones play an essential role.

◇ **Theorem 560.** *Any finite join-semidistributive lattice $L$ can be embedded into the lattice of closed sets of some finite, atomistic, convex geometry.*

This result was sharpened in K. V. Adaricheva and J. B. Nation [22], where it was shown how to embed $L$ into the largest convex geometry on the set $\mathrm{Ji}\,L$ of join-irreducible elements of $L$.

It remains an open problem whether there exists a special type of *finite* convex geometry that contains all finite join-semidistributive lattices as sublattices (see Problem 3 in [21]). When the restriction on finiteness of the convex geometry is removed, such convex geometries exist (K. V. Adaricheva, V. A. Gorbunov, and V. I. Tumanov [21]):

◇ **Theorem 561.** *Every finite join-semidistributive lattice can be embedded into an atomistic and join-semidistributive convex geometry of the form $S_p(A)$ for some algebraic and dually algebraic lattice $A$.*

A number of papers establish embedding results into various types of convex geometries such as biatomic convex geometries in K. V. Adaricheva and F. Wehrung [25], convex subsets of orders in M. Semenova and F. Wehrung [641] and M. Semenova and A. Zamojska-Dzienio [643], convex subsets of vector spaces in M. Semenova and F. Wehrung [642], suborders of an order in M. Semenova [640], or algebraic subsets of an algebraic lattice in K. V. Adaricheva [15].

G. M. Bergman [59] investigates various convexity lattices in $\mathbf{R}^n$, such as convex sets containing the origin, open bounded convex sets, compact and relatively convex subsets of a fixed set $S \subseteq \mathbf{R}^n$. Various aspects of convex geometries of relatively convex sets of points configurations in $\mathbf{R}^n$ are studied in K. Adaricheva and M. Wild [26].

**Exercises**

2.1. Regard Term$(X)$ (see Definition 539) as an algebra with the binary operations $\vee$ and $\wedge$. Define the concept of a congruence relation $\alpha$ on Term$(X)$ and the corresponding *quotient algebra* Term$(X)/\alpha$. Prove that $\equiv$ is a congruence relation on Term$(X)$ and the corresponding quotient algebra is isomorphic to the lattice Rep$(X)$.

2.2. Verify the solution to the word problem for the free lattice Free $P$ over the order $P$. Show that an algorithm is provided for deciding whether $a \leq b$ in Free $P$ by the rules:

(i) let $a, b \in P$, then $a \subseteq b$ iff $a \leq b$ in $P$;

(ii) otherwise, $a \subseteq b$ follows from (i) and the rules $(\wedge\mathrm{W})$, $(\vee\mathrm{W})$, $(\mathrm{W}_\wedge)$, and $(\mathrm{W}_\vee)$

(R. P. Dilworth [155]).

2.3. Let $L$ be a free product of $\mathcal{L} = (L_i \mid i \in I)$. Show that $L$ is completely freely generated by $\bigcup \mathcal{L}$ iff all the $L_i$ are chains.

2.4. Prove that $C_2^n$ is a sublattice of a free lattice iff $n \leq 3$.

2.5. Let $L$ be a modular lattice. Verify that if $L$ is a sublattice of a free lattice, then $L$ is distributive.

2.6. Let the lattice $L$ be of length $n$. Show that if $L$ is a sublattice of a free lattice, then $|L| \leq 2^n$.

2.7. Let the lattice $L$ be generated by a finite set $X$ and let us assume that $x \nleq \bigvee(X - \{x\})$ for some $x \in X$. Then

$$a \wedge \bigvee(X - \{x\}) \prec (a \wedge \bigvee(X - \{x\})) \vee x$$

for every $a \geq x$.

2.8. Let $\mathfrak{m} \geq \aleph_0$. Show that there is no covering in Free($\mathfrak{m}$).

2.9. Let $\mathfrak{m} \geq \aleph_0$ and let $a, b \in$ Free($\mathfrak{m}$) with $a < b$. Show that $[a, b]$ has a sublattice isomorphic to Free($\mathfrak{m}$).

2.10. Let $f(n)$ be the maximum cardinality of a sublattice of length $n$ of a free lattice. Then $(\sqrt{2})^n \leq f(n)$ (B. Jónsson and J. E. Kiefer [448]).

2.11. A *linear decomposition* $A_i$, for $i \in I$, of a lattice $A$ consists of a chain $I$, sublattices $A_i$ of $A$, for $i \in I$, such that if $i, j \in I$ with $i < j$, then $a < b$ in $A$, for all $a \in A_i$, $b \in A_j$, and $A = \bigcup(A_i \mid i \in I)$. Show that if all $A_i$ are sublattices of a free lattice and $I$ is countable, then $A$ is also a sublattice of a free lattice.

2.12. A distributive lattice $D$ is a sublattice of a free lattice iff $D$ has a linear decomposition $A_i$, for $i \in I$, such that $|I| \leq \aleph_0$ and each $A_i$ is either $C_1$, or $C_2^3$, or of the form $C_2 \times C$ where $C$ is a countable chain (F. Galvin and B. Jónsson [217]).

2.13. Let $A$ be an algebra with the following properties: an ordering $\leq$ is defined on $A$; $A$ has a generating set $X$ such that every permutation on $X$ can be extended to an isotone automorphism of $A$. Prove that every chain in $A$ is countable.

2.14. Enumerate in canonical form all the elements of $A$ in the proof of Theorem 551.

2.15. Prove that $B$ in the proof of Corollary 552 is a partial lattice.

2.16. Is there a general lemma about "gluing together" two partial lattices that contains the construction of Corollary 552?

2.17. Use component subsets to prove that if $a, b \in$ Free($\aleph_0$) with $a \neq b$, then there is a homomorphism $\varphi$ of Free($\aleph_0$) onto a finite lattice such that $\varphi(a) \neq \varphi(b)$ (R. A. Dean [143]).

2.18. Prove that every finite lattice can be embedded in a finite three-generated lattice.

2.19. Using the notation of Figure 123, let $y_i \in$ Free(3) and $e_i \leq y_i \leq d_i$ for $i = 0, 1, 2$. Prove that sub($y_0, y_1, y_2$) is a proper sublattice of Free(3) isomorphic to Free(3).

*2.20. Compute Free$_{\mathbf{N}_5}$(3). (Hint: it has 99 elements and a rather complicated structure, see A. G. Waterman [697].)

# 3.  Reduced Free Products

## 3.1  Basic definitions

In this section, let $\mathcal{L} = (L_i \mid i \in I)$, be a fixed family of *bounded lattices* and let $L$ be a free $\{0, 1\}$-product of $\mathcal{L}$ (see Definition 197 and the discussion following it; see also Section 1.12).

As we shall see, a pair of elements $x, y$ is *complementary in $L$* (that is, they satisfy $x \vee y = 1$ and $x \wedge y = 0$) if either they belong to some $L_i$ and they are complementary in $L_i$ or if there exist elements $x_0, x_1, y_0, y_1$ in some $L_i$ such that $x_0 \leq x \leq y_0$, $x_1 \leq y \leq y_1$, and $\{x_0, x_1\}$, $\{y_0, y_1\}$ are complementary in $L_i$. We shall describe a construction in which there are many more complements than in the free $\{0, 1\}$-product, but in which we can still keep track of the complements. We call this construction the *reduced free product*. Several applications will be given in this section and the next.

**Definition 562.** A *C-relation* C on $\mathcal{L} = (L_i \mid i \in I)$ is a set of two-element subsets of $\bigcup \mathcal{L}$ such that if $\{a, b\} \in C$, then there exist $i, j \in I$ with $i \neq j$, satisfying $a \in L_i - \{0_i, 1_i\}$ and $b \in L_j - \{0_j, 1_j\}$.

**Definition 563.** Let $C$ be a *C-relation* on $\mathcal{L} = (L_i \mid i \in I)$. A lattice $L$ is a *C-reduced free product of $\mathcal{L}$* if the following conditions hold:

(i)  $L = \mathrm{sub}(\bigcup \mathcal{L})$ and the lattice $L_i$ is a $\{0, 1\}$-sublattice of $L$ for all $i \in I$.

(ii)  If $\{a, b\} \in C$, then $a, b$ is a complementary pair in $L$.

(iii)  Let $\varphi_i$ be a $\{0, 1\}$-homomorphism of $L_i$ into a bounded lattice $A$, for all $i \in I$, let $\{a, b\} \in C$, where $a \in L_i$ and $b \in L_j$ with $i \neq j$, and let $\varphi_i(a), \varphi_j(b)$ be complementary in $A$; then there is a homomorphism $\varphi$ of $L$ into $A$ extending the $\varphi_i$ for all $i \in I$.

## 3.2  The structure theorem

Using the technique of Section I.5.2, we can verify the existence of a $C$-reduced free product by finding a lattice satisfying Definition 563(i) and (ii). Such a lattice is

$$\{0, 1\} \cup \bigcup (L_i - \{0_i, 1_i\} \mid i \in I)$$

with the obvious ordering. The uniqueness of $C$-reduced free products can be established as in Section I.5.1. The next result shall give a description of a $C$-reduced free product based on the Structure Theorem for Free Products. This description is then used to describe the complementary pairs in a $C$-reduced free product.

Theorems 565 and 566 appeared in their present form in G. Grätzer [259]; earlier versions can be found in C. C. Chen and G. Grätzer [88] and G. Grätzer [258]. The germ of the idea can be traced back to R. P. Dilworth [155].

**Definition 564.** Let $\mathcal{L} = (L_i \mid i \in I)$ be a family of pairwise disjoint bounded lattices. We define a subset $S$ of $\mathrm{Term}(\bigcup \mathcal{L})$:

For $p \in \mathrm{Term}(\bigcup \mathcal{L})$, let $p \in S$ be defined by induction on the rank of $p$:

(i) For $p \in \bigcup \mathcal{L}$, let $p \in S$ if $p \in L_i - \{0_i, 1_i\}$ for some $i \in I$.

(ii) For $p = q \wedge r$, let $p \in S$ if $q, r \in S$ and the following two conditions hold:

    (ii$_1$) $p \subseteq 0_i$ for no $i \in I$;
    (ii$_2$) $p \subseteq x \wedge y$ for no $\{x, y\} \in C$.

(iii) For $p = q \vee r$, let $p \in S$ if $q, r \in S$ and the following two conditions hold:

    (iii$_1$) $1_i \subseteq p$ for no $i \in I$;
    (iii$_2$) $x \vee y \subseteq p$ for no $\{x, y\} \in C$.

Now we set

$$L = \{0, 1\} \cup \{\, \mathrm{rep}(p) \mid p \in S \,\},$$

and order $L$ by

$$0 < \mathrm{rep}(p) < 1 \quad \text{for } p \in S,$$
$$\mathrm{rep}(p) \leq \mathrm{rep}(q) \quad \text{if } \quad p \subseteq q.$$

If we identify $a \in L_i$ with $\mathrm{rep}(a)$, then we get the setup we need:

**Theorem 565.** *The set $L$ with the ordering $\leq$ is a lattice and it is a $C$-reduced free product of $\mathcal{L} = (L_i \mid i \in I)$.*

*Proof.* The set $L$ is obviously an order. To show that $L$ is a lattice, we have to find the meet and the join of $\mathrm{rep}(p)$ and $\mathrm{rep}(q)$ in $L$ for $p, q \in S$. We claim that

$$\mathrm{rep}(p) \wedge \mathrm{rep}(q) = \begin{cases} \mathrm{rep}(p \wedge q), & \text{if } p \wedge q \in S; \\ 0, & \text{otherwise.} \end{cases}$$

To verify this claim, it is sufficient to prove that, for every $u \in \mathrm{Term}(\bigcup \mathcal{L})$, if $u \subseteq 0_i$, for some $i \in I$, or $u \subseteq x \wedge y$, for some $\{x, y\} \in C$, then $u \notin S$. We prove this by induction on the rank of $u$.

If $u \in \bigcup \mathcal{L}$ and $u \subseteq 0_i$, then $u = 0_i \notin S$. If $u \in \bigcup \mathcal{L}$ and $u \subseteq x \wedge y$, for some $\{x, y\} \in C$, then $u \in L_i$, and $x \in L_k$ and $y \in L_n$ with $k \neq n$, and $u \leq x, y$; these imply that $i = k$ and $i = n$, a contradiction.

If $u = u_0 \wedge u_1$, and $u_0$ or $u_1 \notin S$, then $u \notin S$ by Definition 564(ii); if $u_0, u_1 \in S$, then $u \notin S$ by Definition 564(ii$_1$) or Definition 564(ii$_2$). Finally, if $u = u_0 \vee u_1$, then $u_0 \subseteq 0_i$ or $u_1 \subseteq 0_i$ in the first case, and $u_0 \subseteq x \wedge y$ or $u_i \subseteq x \wedge y$ in the second case, and so $u_0$ or $u_1 \notin S$ implying $u \notin S$ by Definition 564(iii).

Dually,

$$\mathrm{rep}(p) \vee \mathrm{rep}(q) = \begin{cases} \mathrm{rep}(p \vee q), & \text{if } p \vee q \in S; \\ 1, & \text{otherwise.} \end{cases}$$

Now it is obvious that $a \mapsto \mathrm{rep}(a)$ is a $\{0,1\}$-embedding of $L_i$ into $L$. So after the identification, Definition 563(i) becomes obvious. Definition 563(ii) is clear in view of Definition 564(ii$_1$), (ii$_2$), and our description of meet and join in $L$.

Let $K$ be the free product of $\mathcal{L}$ as constructed in Section 1.2.

Then $L - \{0,1\} \subseteq K$. We define a congruence $\boldsymbol{\alpha}$ on $K$:

$$\boldsymbol{\alpha} = \bigvee(\,\mathrm{con}(0_i, x) \mid i \in I,\ x \leq 0_i\,) \vee \bigvee(\,\mathrm{con}(x, 1_i) \mid i \in I,\ x \geq 1_i\,)$$

$$\vee \bigvee(\,\mathrm{con}(x, u \wedge v) \mid x \leq u \wedge v,\ \{u,v\} \in C\,)$$

$$\vee \bigvee(\,\mathrm{con}(u \vee v, x) \mid x \geq u \vee v,\ \{u,v\} \in C\,).$$

In other words, $\boldsymbol{\alpha}$ is the smallest congruence relation under which all $0_i$ and $u \wedge v$, for $u, v \in C$, are in the block which is the zero of $K/\boldsymbol{\alpha}$ and dually. We claim that

$$K/\boldsymbol{\alpha} \cong L.$$

To see this, it is sufficient to prove that every block modulo $\boldsymbol{\alpha}$, except the two extremal ones, contains one and only one element of $S$.

Let $\varepsilon_i$ be the identity map of $L_i$ into $L$. Then there is a map $\varphi$ extending the $\varepsilon_i$, for all $i \in I$, into a homomorphism of $K$ into $L$.

Observe, that $\varphi(\mathrm{rep}(p)) = \mathrm{rep}(p)$ for all $p \in S$. Indeed, $p = p(a_0, \ldots, a_{n-1})$, where $a_0, \ldots, a_{n-1} \in S \cap \bigcup \mathcal{L}$; hence

$$\varphi(\mathrm{rep}(p)) = \mathrm{rep}(\varphi(p)) = \mathrm{rep}(\varphi(p(a_0, \ldots, a_{n-1})))$$
$$= \mathrm{rep}(\varphi(p(a_0)), \ldots, \varphi(a_{n-1}))) = \mathrm{rep}(p(a_0, \ldots, a_{n-1})) = \mathrm{rep}(p),$$

since $\varphi(a_0) = a_0, \ldots, \varphi(a_{n-1}) = a_{n-1}$.

Let $\boldsymbol{\varphi}$ be the congruence kernel of $\varphi$. Since $L$ satisfies Definition 563(i) and (ii), it follows that $\boldsymbol{\alpha} \leq \boldsymbol{\varphi}$. Now if $p, q \in S$, and $\varphi(\mathrm{rep}(p)) = \varphi(\mathrm{rep}(q))$, then $\mathrm{rep}(p) = \mathrm{rep}(q)$. In other words, $\mathrm{rep}(p) \equiv \mathrm{rep}(q) \pmod{\boldsymbol{\varphi}}$ implies that $\mathrm{rep}(p) = \mathrm{rep}(q)$. Therefore, the same holds for $\boldsymbol{\alpha}$. This proves that there is at most one $\mathrm{rep}(p)$ in the nonextremal blocks of $\boldsymbol{\alpha}$. To show "at least one", take a $p \in \mathrm{Term}(\bigcup \mathcal{L})$ such that

$$\mathrm{rep}(p) \not\equiv 0_i \pmod{\boldsymbol{\alpha}},$$
$$\mathrm{rep}(p) \not\equiv 1_i \pmod{\boldsymbol{\alpha}}$$

for any $i \in I$; we prove that there exists a $q \in S$ such that

$$\mathrm{rep}(p) \equiv \mathrm{rep}(q) \pmod{\boldsymbol{\alpha}}.$$

Let $p \in L_i$ for some $i \in I$. Then, by assumption, $p \neq 0_i$ and $1_i$; hence we can take $q = p$. Let

$$p = p_0 \wedge p_1,$$
$$\mathrm{rep}(p_0) \equiv \mathrm{rep}(q_0) \pmod{\alpha},$$
$$\mathrm{rep}(p_1) \equiv \mathrm{rep}(q_1) \pmod{\alpha},$$

where $q_0, q_1 \in S$. If $q_0 \wedge q_1 \in S$, take $q = q_0 \wedge q_1$. Otherwise, by Definition 564(ii), $q_0 \wedge q_1 \equiv 0_i \pmod{\alpha}$, hence $p \equiv 0_i \pmod{\alpha}$, contrary to our assumption. The dual argument completes the proof.

Thus we have proved that $K/\alpha \cong L$.

Now we are ready to verify Definition 563(iii). For each $i \in I$, let $\varphi_i$ be a $\{0,1\}$-homomorphism of $L_i$ into the bounded lattice $A$. Since $K$ is the free product of the $L_i$, for $i \in I$, there is a homomorphism $\psi$ of $K$ into $A$ extending the $\varphi_i$ for all $i \in I$. Let $\psi$ be the congruence kernel of $\psi$. It obviously follows from the definition of $\alpha$ that $\alpha \leq \psi$. Therefore, by the Second Isomorphism Theorem, $x/\alpha \mapsto \psi(x)$ is a homomorphism of $K/\alpha$ into $A$. Combining this with the isomorphism $L \cong K/\alpha$ as described above, we get a $\{0,1\}$-homomorphism $\varphi$ of $L$ into $A$ extending the $\varphi_i$ for all $i \in I$. □

## 3.3   Getting ready for applications

**Theorem 566.** *Let $L$ be a $\mathcal{C}$-reduced free product of $\mathcal{L} = (L_i \mid i \in I)$. Let $a, b$ be a complementary pair in $L$. Then there exist $a_0, b_0$ and $a_1, b_1$ satisfying*

$$a_0 \leq a \leq a_1,$$
$$b_0 \leq b \leq b_1,$$

*such that either $\{a_0, b_0\}, \{a_1, b_1\} \in C$ or $a_0, b_0$ and $a_1, b_1$ are complementary pairs in $L_i$, for some $i \in I$, and conversely.*

*Proof.* The converse is, of course, obvious. Indeed, in either case, by Definition 563, $a_0, b_0$ and $a_1, b_1$ are complementary in $L$, hence

$$a \wedge b \leq a_1 \wedge b_1 = 0,$$
$$a \vee b \geq a_0 \vee b_0 = 1,$$

and so $a$ and $b$ are complementary in $L$.

Now we prove the main part of the theorem. We take $p, q \in S$ such that $a = \mathrm{rep}(p)$ and $b = \mathrm{rep}(q)$ are complementary in $L$. Then $p \wedge q$ violates Definition 564(ii$_1$) or (ii$_2$) and $p \vee q$ violates Definition 564(iii$_1$) or (iii$_2$).

The four cases will be handled separately.

*Case 1. $p \wedge q$ violates Definition 564(ii$_1$) and $p \vee q$ violates (iii$_1$).* Hence, for some $i, j \in I$,

$$p \wedge q \subseteq 0_i,$$
$$1_j \subseteq p \vee q.$$

Thus in the free product $K$ of the $L_i$, for $i \in I$,

$$(p \wedge q)^{(i)} = 0_i,$$
$$(p \vee q)_{(j)} = 1_j.$$

Note that $q^{(i)}$ is proper, because otherwise $p^{(i)} = 0_i$, that is, $p \subseteq 0_i$, contradicting $p \in S$. Similarly, $q_{(j)}$ is proper. This is a contradiction unless $i = j$, in which case, we can put $a_0 = p_{(i)}$, $b_0 = q_{(i)}$, $a_1 = p^{(i)}$, $b_1 = q^{(i)}$ and these obviously satisfy the requirements of the theorem.

*Case 2. $p \wedge q$ violates Definition 564(ii$_1$) and $p \vee q$ violates (iii$_2$).* Hence there exist $i \in I$ and $\{x, y\} \in C$ such that

$$p \wedge q \subseteq 0_i,$$
$$x \vee y \subseteq p \vee q.$$

Let $x \in L_j$ and $y \in L_k$ with $j, k \in I$ and $j \neq k$. Then $i \neq j$ or $i \neq k$; let us assume that $i \neq j$. Since

$$(p \wedge q)^{(i)} = 0_i,$$
$$p^{(i)} \wedge q^{(i)} = 0_i,$$

we conclude, just as in Case 1, that $p^{(i)}$ and $q^{(i)}$ are proper. From $i \neq j$, we conclude that $p_{(j)}$ and $q_{(j)}$ are not proper, that is, $p_{(j)} = q_{(j)} = 0^b$. Thus

$$(p \vee q)_{(j)} = p_{(j)} \vee q_{(j)} = 0^b,$$

contradicting that $p \vee q \supseteq x \in L_j$. Case 2 cannot occur.

*Case 3. $p \wedge q$ violates Definition 564(ii$_2$) and $p \vee q$ violates (iii$_1$).* This leads to a contradiction just as Case 2 does.

*Case 4. $p \wedge q$ violates Definition 564(ii$_2$) and $p \vee q$ violates (iii$_2$).* Then there exist $\{a_0, b_0\} \in C$ and $\{a_1, b_1\} \in C$ such that $a_0 \in L_i$, $b_0 \in L_j$, $a_1 \in L_k$, $b_1 \in L_n$ and $i, j, k, n \in I$ with $i \neq j$ and $k \neq n$, and

$$a_0 \vee b_0 \subseteq p \vee q,$$
$$p \wedge q \subseteq a_1 \wedge b_1.$$

We conclude, as above, that

$$a_0 \leq p_{(i)} \vee q_{(i)},$$
$$b_0 \leq p_{(j)} \vee q_{(j)},$$
$$p^{(k)} \wedge q^{(k)} \leq a_1,$$
$$p^{(n)} \wedge q^{(n)} \leq b_1.$$

Therefore, $p_{(i)}$ or $q_{(i)}$ is proper, $p_{(j)}$ or $q_{(j)}$ is proper, $p^{(k)}$ or $q^{(k)}$ is proper, and $p^{(n)}$ or $q^{(n)}$ is proper.

We cannot have both $p_{(i)}$ and $q_{(i)}$ proper, because then neither $p^{(k)}$ nor $q^{(k)}$ could be proper unless $i = k$; and similarly, $i = n$, contradicting $k \neq n$.

We can assume that $p_{(i)}$ is proper and $q_{(i)} = 0^b$, and so $p_{(i)} = (p \vee q)_{(i)} \geq a_0$. We cannot have $p_{(j)}$ proper, because then $q_{(j)} = 0^b$ and $p_{(j)} \geq b_0$; and therefore, $p \supseteq a_0 \vee b_0$. Now $k \neq i$ and $k \neq j$ yields a contradiction (neither $p^{(k)}$ nor $q^{(k)}$ could be proper), hence $k = i$ or $k = j$. Let $k = i$ (the case $k = j$ is similar). Then $q^{(i)}$ is not proper, since $i \neq j$, hence $p^{(i)} = (p \wedge q)^{(i)} \leq a_1$ is proper. But $i \neq n$, hence $p^{(n)}$ is not proper and so $q^{(n)}$ must be proper. Since $q_{(j)}$ is proper, we conclude that $n = j$ and $q^{(j)} = (p \wedge q)^{(j)} \leq b_1$. To sum up, we have obtained that

$$a_0 \leq p_{(i)}, \qquad p^{(i)} \leq a_1,$$
$$b_0 \leq q_{(i)}, \qquad q^{(i)} \leq b_1.$$

Hence,

$$a_0 \subseteq p \subseteq a_1,$$
$$b_0 \subseteq q \subseteq b_1,$$

as required. $\qquad\qquad\qquad\qquad\qquad\qquad\qquad\qquad\qquad\qquad\qquad\quad$ $\square$

Let us say that a bounded lattice $A$ has *no comparable complements* if $A$ contains no pentagon $\mathsf{N}_5 = \{0, a, b, c, 1\}$. Let $\mathrm{Comp}(A)$ stand for the set of complementary pairs in $A$.

A $C$-relation $C$ is said to have *no comparable complements* if

$$\{a_0, b_0\},\ \{a_1, b_1\} \in C,\ a_0 \leq a_1 \text{ and } b_0 \leq b_1 \text{ imply that } a_0 = a_1 \text{ and } b_0 = b_1.$$

The following result is immediate from Theorem 566.

**Corollary 567.** *Let $\mathcal{L} = (L_i \mid i \in I)$ be a family of bounded lattices with no comparable complements and let $C$ be a $C$-relation on $\mathcal{L}$ with no comparable complements. Let $L$ be a $C$-reduced free product of the $\mathcal{L}$. Then*

$$\mathrm{Comp}(L) = C \cup \bigcup (\,\mathrm{Comp}(L_i) \mid i \in I\,),$$

*and $L$ has no comparable complements.*

The $C$-reduced free product construction has an interesting generalization, the $\mathcal{R}$-reduced free product construction of M. E. Adams and J. Sichler [11] and [12]. $\mathcal{R}$-reduced free products extend $C$-reduced free products in two important ways: (i) An $\mathcal{R}$-reduction is not necessarily determined by a $C$-relation. (ii) An $\mathcal{R}$-reduction can be done in many lattice varieties not only in the variety of all lattices.

## 3.4  Embedding into uniquely complemented lattices

Now we are ready for our first application from C. C. Chen and G. Grätzer [88].

**Theorem 568.** *Let $K$ be a bounded lattice in which every element has at most one complement (an* at most uniquely complemented lattice*). Then $K$ has a $\{0,1\}$-embedding into a* uniquely complemented lattice *$L$ (that is, into a lattice $L$ in which every element has exactly one complement).*

*Proof.* If $K$ is complemented, then set $L = K$. Otherwise, let $K = K_0$. We define, by induction, the lattice $K_n$. If $K_{n-1}$ is defined, let $I_{n-1}$ be the set of noncomplemented elements of $K_{n-1}$. For $i \in I_{n-1}$, let $K_i = \{a_i\}^b$. Define the $\mathcal{C}$-relation $R_{n-1}$ on the family $\{K_{n-1}\} \cup (K_i \mid i \in I_{n-1})$ by the rule

$$\{a,b\} \in R_{n-1} \quad \text{if} \quad \{a,b\} = \{i, a_i\} \text{ for some } i \in I_{n-1}.$$

Let $K_n$ be the $\mathcal{C}$-reduced free product with respect to the relation $R_{n-1}$. Since

$$K = K_0 \subseteq K_1 \subseteq K_2 \subseteq \cdots$$

and all these containments are $\{0,1\}$-embeddings, we can form

$$L = \bigcup (K_i \mid i \in I).$$

Now observe that $K_0$ is a lattice with no comparable complements. By induction, if this is known for $K_{n-1}$, then it is true for $K_n$ since $R_{n-1}$ has no comparable complements and so Corollary 567 applies. Therefore, again by Corollary 567,

$$\mathrm{Comp}(K_n) = \mathrm{Comp}(K_{n-1}) \cup R_{n-1}.$$

Thus $K_n$ is at most uniquely complemented and every element of $K_{n-1}$ has a complement in $K_n$. It is now obvious that $L$ is uniquely complemented.    $\square$

Every lattice $K$ can be embedded in a lattice in which every element has at most one complement, namely into $K^b$. As a special case of Theorem 568, we get the celebrated result of R. P. Dilworth [155]:

**Corollary 569.** *Every lattice can be embedded into a uniquely complemented lattice.*

See G. Grätzer [270] for an elementary exposition of the background of this result,

A number of alternative proofs of Corollary 569 have been published:

(i) The original proof of R. P. Dilworth [155].

(ii) P. Crawley and R. P. Dilworth [107].

(iii) M. E. Adams and J. Sichler [11] and [12], which extend the result to a large number of lattice varieties.

(iv) G. Grätzer and H. Lakser [309] utilizing Dean's Lemma (Theorem 575).

The result was generalized to $\mathfrak{m}$-complete lattices in J. Harding [373].

For a graph $G$ (or, more precisely, $(G; E)$), let $(L_a \mid a \in G)$ be a family of lattices, where $L_a = \{0_a, a, 1_a\}$ is a three-element lattice. Define the $\mathcal{C}$-relation $C$ on $(L_a \mid a \in G)$ by

$$\{x, y\} \in C \quad \text{if} \quad \{x, y\} \in E.$$

Let $\operatorname{Lat} G$ (the lattice representation of $G$) denote the $\mathcal{C}$-reduced free product of $(L_a \mid a \in G)$. Some examples are given in Figure 124.

For a bounded lattice $L$, let $\operatorname{End}_{\{0,1\}} L$ denote the *monoid* (that is, semigroup with identity) of all $\{0, 1\}$-endomorphisms of $L$. For a graph $G$, we denote by $\operatorname{End} G$ the monoid of endomorphisms of $G$; a map $\varphi \colon G \to G$ is an *endomorphism* if $(a, b) \in E$ implies that $(\varphi(a), \varphi(b)) \in E$.

Observe that $G \subseteq \operatorname{Lat} G$ and, in fact, $G$ generates $\operatorname{Lat} G$ as a $\{0, 1\}$-sublattice (that is, $G \cup \{0, 1\}$ generates $\operatorname{Lat} G$). Therefore, every $\varphi \in \operatorname{End} G$ has at most one extension $\overline{\varphi}$ to a $\{0, 1\}$-endomorphism of $\operatorname{Lat} G$.

**Corollary 570.** *Every endomorphism $\varphi$ of a graph $G$ has exactly one extension $\overline{\varphi}$ to a $\{0, 1\}$-endomorphism of $\operatorname{Lat} G$. If $\varphi$ is onto, so is $\overline{\varphi}$.*

*Proof.* This is clear from Definition 563(iii) with $A = \operatorname{Lat} G$.     $\square$

For an integer $n \geq 2$, an *$n$-cycle* of a graph $G$ is an $n$-tuple of elements $(a_0, \ldots, a_{n-1})$ with $\{a_0, a_1\}, \ldots, \{a_{n-2}, a_{n-1}\}, \{a_{n-1}, a_0\} \in E$.

**Theorem 571.** *Let $G$ be a graph satisfying the property that every element of $G$ is contained in some cycle of odd length. Then $\operatorname{End} G$ is isomorphic with the monoid $\operatorname{End}_{\{0,1\}}(\operatorname{Lat} G)$.*

*Proof.* It is obvious that the lattices and the $\mathcal{C}$-relation used in forming $\operatorname{Lat} G$ satisfy the hypotheses of Corollary 567. Therefore,

$$\operatorname{Comp}(\operatorname{Lat} G) = E \cup \{0, 1\}.$$

By our assumption, every element of $G$ is the endpoint of an edge, and $G$ is recognized in $\operatorname{Lat} G$ as the set of complemented elements, other than 0 and 1. Since a $\{0, 1\}$-endomorphism $\psi$ takes a complementary pair into a complementary pair, we conclude that $\psi(G) \subseteq G \cup \{0, 1\}$. Let $\psi(g) \in \{0, 1\}$ for some $g \in G$; for instance, $\psi(g) = 0$. By assumption, there is a cycle of odd

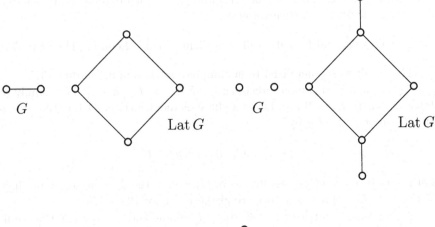

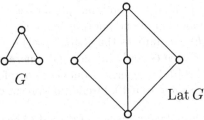

Figure 124. Constructing lattices from graphs

length $(g_0, \ldots, g_{2n})$ with $g = g_0$. This means that $\{g_0, g_1\}, \ldots, \{g_{2n-1}, g_{2n}\}$, and $\{g_{2n}, g_0\}$ are complementary pairs. Thus

$$\psi(g) = \psi(g_0) = 0,$$
$$\psi(g_1) = 1, \ \psi(g_2) = 0, \ \psi(g_3) = 1, \ldots, \psi(g_{2n-1}) = 1, \ \psi(g_{2n}) = 0,$$
$$\psi(g_0) = 1,$$

a contradiction. This shows that every $\{0, 1\}$-endomorphism $\psi$ is the unique extension of a map $\varphi$ of $G$ into itself; this $\varphi$ is obviously a graph endomorphism. Thus the map $\varphi \mapsto \overline{\varphi}$ is the required isomorphism of End $G$ with $\text{End}_{\{0,1\}}$ Lat $G$. □

As is shown in the Exercises, every monoid is the endomorphism semigroup of a graph in which every element lies on a cycle of odd length. Thus we conclude a result of G. Grätzer and J. Sichler [352]:

**Theorem 572.** *Every monoid can be represented as the $\{0, 1\}$-endomorphism semigroup of a bounded lattice.*

Theorem 572 is one of a large body of results representing monoids as endomorphism monoids of various types of algebras. All these results are based on P. Vopěnka, A. Pultr, and Z. Hedrlín [692] proving the existence of rigid relations and on Z. Hedrlín and A. Pultr [381] proving the representation for graphs. See also Z. Hedrlín and A. Pultr [382] for the case of algebras with two unary operations; Z. Hedrlín and J. Lambek [380] for the case of semigroups and for an alternate proof of the existence of rigid graphs. The result on triangle connected graphs is a special case of a result of P. Hell [385].

## 3.5  ◇ Dean's Lemma

An alternative to reduced free products is Dean's Lemma. To state it, we need a few definitions that are variants of definitions covered earlier in Chapters I and VII.

Let $P$ be an order, and let $\mathcal{J}$ and $\mathcal{M}$ be sets of two-element subsets of $P$ such that $\sup X$ exists in $P$, for each $X \in \mathcal{J}$, and, for each $X \in \mathcal{M}$, $\inf X$ exists in $P$. ($\mathcal{J}$ is a subset of the "sup-table" of $P$, and $\mathcal{M}$ is a subset of the "inf-table" of $P$.) Define a $\mathcal{J}$-ideal $I$ as a down-set of $P$ with the property that if $X \subseteq I$ and $X \in \mathcal{J}$, then $\sup X \in I$. Define $\mathcal{M}$-filters dually.

If $L$ is a lattice and the map $\varphi \colon P \to L$ is isotone and satisfies

$$\varphi(\sup\{x, y\}) = \varphi(x) \vee \varphi(y),$$

for each $\{x, y\} \in \mathcal{J}$ and

$$\varphi(\inf\{x, y\}) = \varphi(x) \wedge \varphi(y),$$

for each $\{x, y\} \in \mathcal{M}$, we say that $\varphi$ is a $(\mathcal{J}, \mathcal{M})$-*morphism*. If, in addition, $\varphi$ is an order embedding, then we say that $\varphi$ is a $(\mathcal{J}, \mathcal{M})$-*embedding*.

A lattice $\mathrm{Free}(P; \mathcal{J}, \mathcal{M})$ along with a $(\mathcal{J}, \mathcal{M})$-embedding

$$\eta \colon P \to \mathrm{Free}(P; \mathcal{J}, \mathcal{M})$$

is said to be a *free lattice generated by $P$ and preserving the sups in $\mathcal{J}$ and the infs in $\mathcal{M}$* if $\mathrm{Free}(P; \mathcal{J}, \mathcal{M})$ is generated as a lattice by the set $\eta(P)$ and the *universal mapping property* holds: let $L$ be a lattice and let $\varphi \colon P \to L$ be a $(\mathcal{J}, \mathcal{M})$-morphism; then there is a lattice homomorphism $\varphi' \colon \mathrm{Free}(P; \mathcal{J}, \mathcal{M}) \to L$ satisfying $\varphi = \varphi'\eta$.

The results of Section I.5 carry over to this setup, in particular, the lattice $\mathrm{Free}(P; \mathcal{J}, \mathcal{M})$ is unique up to isomorphism.

Now we modify Definition 539.

By recursion on $\mathrm{rank}(p)$, we associate with each $p \in \mathrm{Term}(P)$ a $\mathcal{J}$-ideal $\underline{p}$ of $P$, the *lower cover* of $p$, and an $\mathcal{M}$-filter $\overline{p}$ of $P$, the *upper cover* of $p$, as follows:

**Definition 573.**

(i) If $p \in P$, then $\underline{p} = {\downarrow}p$, the down-set generated by $p$, and $\overline{p} = {\uparrow}p$, the up-set generated by $p$.

(ii) $\underline{p \vee q} = \underline{p} \vee \underline{q}$, the join in the lattice of $\mathcal{J}$-ideals, and $\overline{p \vee q} = \overline{p} \wedge \overline{q} = \overline{p} \cap \overline{q}$.

(iii) $\overline{p \wedge q} = \overline{p} \wedge \overline{q} = \overline{p} \cap \overline{q}$, and $\overline{p \wedge q} = \overline{p} \vee \overline{q}$, the join in the lattice of $\mathcal{M}$-filters.

We now define a binary relation $\subseteq$ on $\mathrm{Term}(P)$.

**Definition 574.** Let $p, q \in \mathrm{Term}(P)$. Then $p \subseteq q$ if it follows from $(\vee\mathrm{W})$, $(\wedge\mathrm{W})$, $(\mathrm{W}_\vee)$, $(\mathrm{W}_\wedge)$ (see Definition 523) and the additional rule:

$$(\mathrm{C_H}) \qquad\qquad \overline{p} \cap \underline{q} \neq \varnothing.$$

Again, for all $p, q \in \mathrm{Term}(P)$, we set

$$p \equiv q \quad \text{iff} \quad p \subseteq q \text{ and } q \subseteq p;$$
$$\mathrm{rep}(p) = \{\, q \in \mathrm{Term}(P) \mid p \equiv q \,\};$$
$$\mathrm{rep}(P) = \{\, \mathrm{rep}(p) \mid p \in \mathrm{Term}(P) \,\};$$
$$\mathrm{rep}(p) \leq \mathrm{rep}(q) \quad \text{iff} \quad p \subseteq q.$$

Now we can state Dean's Lemma (R. A. Dean [147]):

$\Diamond$ **Theorem 575.** $\mathrm{Rep}(P)$ *is a* $\mathrm{Free}(P; \mathcal{J}, \mathcal{M})$.

The reader may find surprising that the complexity introduced in this section has not much influence in carrying out the proof as given in Section 1.

For the missing details the reader may want to consult H. Lakser [497]; in this paper, Section 4 compares the method of proof outlined here with Dean's.

### 3.6   Some applications of Dean's Lemma

We provide an alternative proof of Theorem 568, based on G. Grätzer and H. Lakser [309].

Let $K$ be a bounded lattice. Let $a \in K - \{0, 1\}$, and let $u$ be an element not in $K$. Let $0 < u < 1$ extend the partial ordering $\leq$ of $K$ to $Q = K \cup \{u\}$.

We extend the lattice operations $\vee$ and $\wedge$ of $K$ to $Q$ as *commutative partial meet and join operations*. For $x \leq y$ in $Q$, define $x \vee y = y$ and $x \wedge y = x$. In addition, let $a \vee u = 1$ and $a \wedge u = 0$, see Figure 125.

A subset $I$ of $Q$ is an *ideal* if it is a down-set and it is closed under the joins defined. *Filters* are defined dually. For ideals $I$ and $J$ of $Q$, the meet is given by $I \cap J$, while the join is described by the rule:

$$(1) \quad I \vee J = \begin{cases} I \vee_K J, & \text{if } I, J \subseteq K \text{ and } I \vee_K J \subset K; \\ ((I \cap K) \vee_K (J \cap K)) \cup \{u\}, & \text{if } u \in I \cup J \text{ and} \\ & \qquad a \notin (I \cap K) \vee_K (J \cap K); \\ Q, & \text{otherwise.} \end{cases}$$

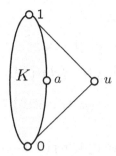

Figure 125. The partial lattice $Q$

The following statement is easy to prove:

**Lemma 576.** *A finitely generated ideal of $Q$ is either principal or of the form*

$$\mathrm{id}(x) \vee \mathrm{id}(u) = \mathrm{id}(x) \cup \{u\}, \quad \text{with } x \in K, \ 0 < x, \text{ and } a \nleq x.$$

We now discuss $\mathrm{Free}\,Q$, the lattice freely generated by $Q$ and preserving the partial joins and meets of $Q$.

It follows from Lemma 576 that, given any term $A \in \mathrm{Term}(Q)$, there are uniquely defined elements $A_*$ and $A^*$ of $K$ with $\underline{A} \cap K = \mathrm{id}(A_*)_K$ and $\overline{A} \cap K = \mathrm{fil}(A^*)_K$. So we have:

**Lemma 577.** *For $x \in K$, the inequality $x \leq A$ holds iff $x \leq A_*$. If $A \leq B$, then $A_* \leq B_*$.*

The most important properties of $A_*$ and of $u \leq A$ are summarized as follows:

**Theorem 578.** *The following statements hold:*

(i) $u \leq u$. *If $x \in K$, then $u \leq x$ iff $x = 1$.*
(ii) $u_* = 0$. *If $x \in K$, then $x_* = x$.*
(iii) $u \leq A \wedge B$ *iff $u \leq A$ and $u \leq B$.*
(iv) $(A \wedge B)_* = A_* \wedge B_*$.
(v) $u \leq A \vee B$ *iff either $u \leq A$, or $u \leq B$, or $A_* \vee B_* = 1$.*
(vi)

$$(A \vee B)_* = \begin{cases} 1, & \text{if } a \leq A_* \vee B_* \text{ and either } u \leq A \text{ or } u \leq B; \\ A_* \vee B_*, & \text{otherwise.} \end{cases}$$

We leave the easy proof to the reader.
Now we want to describe the complements in $\mathrm{Free}\,Q$.

**Theorem 579.**

(i) *The only complement of $u$ in* $\operatorname{Free} Q$ *is $a$.*

(ii) *Let $K$ contain no spanning* $\mathsf{N}_5$. *Let* $\operatorname{rep}(A)$, $\operatorname{rep}(B)$ *be complementary in* $\operatorname{Free} Q$. *Then either*

$$\{\operatorname{rep}(A), \operatorname{rep}(B)\} \subseteq K$$

*or*

$$\{\operatorname{rep}(A), \operatorname{rep}(B)\} = \{u, a\}.$$

*Proof of* (i). Let $A \in \operatorname{Term}(Q)$ be such that $\operatorname{rep}(A)$ is a complement of $u$ in $\operatorname{Free} Q$, that is,

$$A \wedge u \equiv 0 \quad \text{and} \quad A \vee u \equiv 1.$$

By Theorem 578(vi),

$$1 = (A \vee u)_* = \begin{cases} 1, & \text{if } a \leq A_* \vee u_* = A_*; \\ A_*, & \text{otherwise.} \end{cases}$$

So either $a \leq A_*$ or $1 = A_*$; in either case, $a \leq A_*$. Dually, $a \geq A^*$. Thus

$$A \leq A^* \leq a \leq A_* \leq A,$$

and so $A \equiv a$. □

*Proof of* (ii). We have, by assumption,

$$A \wedge B \equiv 0 \quad \text{and} \quad A \vee B \equiv 1.$$

By Theorem 578(ii) and (iv),

$$(2) \qquad\qquad\qquad A_* \wedge B_* = 0,$$

and, dually,

$$(3) \qquad\qquad\qquad A^* \vee B^* = 1.$$

Since $u \leq A \vee B$, we conclude, by Theorem 578(v), that one of the following conditions holds:

(a) $A_* \vee B_* = 1$,   (b) $u \leq A$,   (c) $u \leq B$.

Dually, since $u \geq A \wedge B$, one of the following conditions holds:

($\alpha$) $A^* \wedge B^* = 0$,   ($\beta$) $u \geq A$,   ($\chi$) $u \geq B$.

*First case:* (a) holds.

If ($\alpha$) holds, then

$$A^* \vee B^* = 1 = A_* \vee B_*,$$
$$A_* \wedge B_* = 0 = A^* \wedge B^*.$$

Since $A_* \leq A^*$ and $B_* \leq B^*$, and since $K$ contains no spanning $\mathsf{N}_5$, we conclude that $A^* = A_*$ and $B^* = B_*$, that is, that $(A)$, $(B) \in K$.

If ($\beta$) holds, then $0 = u_* = A_*$, and so, by (a), $1 = B_*$.

Thus $B^* \leq 1 = B_*$, that is, $B \equiv 1$. Then $A \equiv 0$, and so

$$\{\mathrm{rep}(A), \mathrm{rep}(B)\} = \{0, 1\}.$$

Similarly, if ($\gamma$) holds, the same conclusion holds.

Thus in this case, $\{\mathrm{rep}(A), \mathrm{rep}(B)\} \subseteq K$.

*Second case:* ($\alpha$) holds. By duality, we get that $\{\mathrm{rep}(A), \mathrm{rep}(B)\} \subseteq K$.

*Third case:* One of (b) or (c) holds, and one of ($\beta$) or ($\gamma$) holds.

If (b) and ($\beta$) hold, then $A \equiv u$, and, by Statement (i) of our theorem, then $B \equiv a$, that is $\{\mathrm{rep}(A), \mathrm{rep}(B)\} = \{u, a\}$.

If (b) and ($\gamma$) hold, then $B \leq A$, and so $A \equiv 1$ and $B \equiv 0$.

The two remaining cases are similar to the two immediately above, with the roles of $A$ and $B$ reversed.    $\square$

Now we reprove Theorem 568. So let $K$ be a bounded, at most uniquely complemented lattice. Clearly, $K$ contains no spanning $\mathsf{N}_5$. If $K$ is uniquely complemented, there is nothing to do. If not, pick an $a \in K$ that has no complement, define $Q = K \cup \{u\}$, and form $L_1 = \mathrm{Free}\,Q$. By Theorem 579, the lattice $L_1$ is an at most uniquely complemented $\{0, 1\}$-extension of $K$, and the element $a$ has a complement in $L_1$, namely, $u$. By transfinite induction, we obtain an at most uniquely complemented $\{0, 1\}$-extension $\overline{L}$ of $K$ in which every element of $K$ has a complement. Repeating this construction $\omega$-times, we obtain the lattice $L$ of this theorem.

Let $\mathfrak{m}$ be a cardinal number. A lattice $K$ is called (at most) $\mathfrak{m}$-complemented, if $K$ is bounded and every $x \in K - \{0, 1\}$ has (at most) $\mathfrak{m}$ complements.

**Theorem 580.** *Let $K$ be an at most $\mathfrak{m}$-complemented lattice with no spanning $\mathsf{N}_5$. Then $K$ has a $\{0, 1\}$-embedding into an $\mathfrak{m}$-complemented lattice $L$.*

*Proof.* Follow the idea of the re-proof of Theorem 214.    $\square$

Much stronger results are proved in G. Grätzer and H. Lakser [307]. Here is an example:

$\Diamond$ **Theorem 581.** *Let $K$ be a lattice, let $[a, b]$ be an interval in $K$. If the interval $[a, b]$ in the lattice $K$ is at most uniquely relatively complemented, then the lattice $K$ has an extension $L$ such that the interval $[a, b]$ of $L$ is uniquely relatively complemented.*

Such a result cannot be proved using the techniques of this section.

**Exercises**

The following exercises should provide the reader with the necessary background in category theory (theory of concrete categories, graph theory) necessary for Theorem 572 and for the results of the next section. The present sequence of exercises is based on a set of lecture notes of J. Sichler.

Let $A$ be a nonempty set; in what follows, we shall deal with relational systems $(A; R)$, $(A; R_0, \ldots, R_{n-1})$, $(A; R_0, \ldots, R_n, \ldots)$, where $R$ and the $R_i$ are binary relations. Given two systems of the same type, say

$$(A; R_0, \ldots, R_n, \ldots),$$
$$(B; S_0, \ldots, S_n, \ldots),$$

a map $\varphi \colon A \to B$ is called a *homomorphism* if

$$(a, b) \in R_i \quad \text{implies that} \quad (\varphi(a), \varphi(b)) \in S_i \quad \text{for all } i = 0, 1, \ldots.$$

An *endomorphism* of $(A; R_0, \ldots)$ is a homomorphism of $(A; R_0, \ldots)$ into itself. The endomorphism monoid $\mathrm{End}(A; R_0, \ldots)$ is defined as before, see Sections II.1.6 and 3.4. The relational system $(A; R_0, \ldots)$ is *rigid* if the only endomorphism is the identity map.

3.1. Let $(A; \leq)$ be a well-ordered set with unit. Let $A_c$ consist of all $a \in A$ that are cofinal with $\omega$, that is, for which there is a sequence

$$a_1 < \cdots < a_n < \cdots$$

such that

$$a = \bigvee (a_i \mid i = 1, 2, \ldots).$$

For each $a \in A_c$, fix such a sequence $(a_1, a_2, \ldots)$. Let $R_0$ be the relation $<$, and let $(x, y) \in R_i$ for all $i = 1, 2, \ldots$, if $y \in A_c$ and $x = y_i$ (that is, $x$ is the $i$-th member of the sequence associated with $y$). Let $\varphi$ be an endomorphism of $(A; R_0, R_1, \ldots)$. Prove that $x \leq \varphi(x)$ for all $x \in A$.

3.2. Let $a < \varphi(a)$ for some $a \in A$. Set $a_1 = \varphi(a)$, $a_2 = \varphi(a_1)$, $\ldots$, and

$$b = \bigvee (a_n \mid n = 1, 2, \ldots).$$

Prove that $\varphi(b) = b$.

3.3. Prove that $\varphi(b_n) = b_n$ for all $n = 1, 2, \ldots$ Conclude that the relational system $(A; R_0, R_1, \ldots)$ is rigid.

3.4. Given a system $(A; R_0, R_1, \ldots)$, we construct a new one $(B; S_0, S_1, \ldots)$ as follows: $B = A \times N$ ($N = \{0, 1, 2, \ldots\}$),

$$((x, n), (y, m)) \in S_0 \quad \text{if} \quad x = y \text{ and } m = n + 1, \text{ or}$$
$$x = y \text{ and } n = 0, \ m = 2;$$
$$((x, n), (y, m)) \in S_1 \quad \text{if} \quad n = m \text{ and } (x, y) \in R_n.$$

Prove that
$$\text{End}(A; R_0, R_1, \ldots) \cong \text{End}(B; S_0, S_1)$$

and if $A$ is infinite, then $|A| = |B|$.

3.5. Given a system $(A; R_0, R_1)$, we construct a new one $(B; S)$ as follows: $B = A \times \{0, 1, 2, 3, 4\}$,

$$((a, i), (b, j)) \in S \quad \text{if} \quad a = b \text{ and } j = i + 1, \text{ or}$$
$$a = b \text{ and } i = 0, \ j = 4, \text{ or}$$
$$i = 0, \ j = 2, \text{ and } (a, b) \in R_0, \text{ or}$$
$$i = 2, \ j = 4, \text{ and } (a, b) \in R_1.$$

Prove that $\text{End}(A; R_0, R_1) \cong \text{End}(B; S)$ and if $A$ is infinite, then $|A| = |B|$.

3.6. Prove that for each infinite cardinal $\mathfrak{m}$, there exists a rigid $(A; R)$ with $|A| = \mathfrak{m}$.

3.7. Prove the statement of Exercise 3.6 for finite cardinals.

3.8. Let $(A; (R_i \mid i \in I))$ be a system without any restriction on $|I|$. Let $(I; R)$ be a connected rigid graph. We define a new system $(B; R_0, R_1)$ as follows: $B = A \times I$,

$$((a, i), (b, j)) \in R_0 \quad \text{if} \quad a = b \text{ and } (i, j) \in R,$$
$$((a, i), (b, j)) \in R_1 \quad \text{if} \quad i = j \text{ and } (a, b) \in R_i.$$

Prove that $\text{End}(A; (R_i \mid i \in I)) \cong \text{End}(B; R_0, R_1)$.

3.9. Let $M$ be a monoid. For every $a \in M$, define

$$R_a = \{ (x, ax) \mid x \in M \}.$$

Show that $\text{End}(M; (R_a \mid a \in M)) \cong M$.

3.10. For any monoid $M$, find a system $(A; R)$ such that $\text{End}(A; R) \cong M$.

3.11. For a system $(A; R)$ and $B \subseteq A$, we construct a new system $(A; R, S_B)$ as follows: $(x, y) \in S_B$ if $x \neq y$ and $x, y \in B$ or $x, y \notin B$. Show that if $(A; R)$ is rigid, then the new systems are *mutually rigid*, that is, if $\varphi$ is a one-to-one homomorphism of one system to another, then the two systems are the same and $\varphi$ is the identity map.

3.12. Prove that there are $2^{\mathfrak{m}}$ mutually rigid systems of cardinality $\mathfrak{m}$, where $\mathfrak{m}$ is any infinite cardinal.

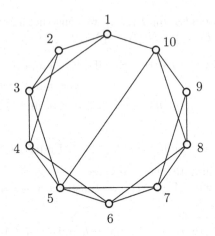

Figure 126. The graph for Exercise 3.13

3.13. Consider the graph $(G; E)$ of Figure 126. Let $(A; R)$ be a system satisfying $(a, a) \in R$ for no $a \in A$. We construct a new graph $(H; F)$ by "replacing each $(a, b) \in R$ by a copy of $(G; E)$ with the pair $(4, 8)$ signifying $(a, b)$". Formally,

$$H = (R \times \{1, 2, 3, 5, 6, 7, 9, 10\}) \cup A,$$

and, for each $e = (a, b) \in R$,

$$(e, 1), (e, 2), (e, 3), a, (e, 5), (e, 6), (e, 7), b, (e, 9), (e, 10)$$

should form a subgraph isomorphic with $(G; E)$. Discuss the connections between $\mathrm{End}(A; R)$ and $\mathrm{End}(H; F)$.

3.14. A *triangle* $\{a, b, c\}$ in a graph $(G; E)$ is a set of three elements of $G$ such that

$$\{a, b\}, \{b, c\}, \{c, a\} \in E.$$

A graph $(G; E)$ is *triangle connected* if, for every $a, b \in G$ with $a \neq b$, there is a sequence

$$T_0, \ldots, T_{n-1}$$

of triangles such that $a \in T_0$, $b \in T_{n-1}$, and $T_i \cap T_{i+1} \neq \emptyset$ for all $i = 0, \ldots, n-2$. Prove that for each infinite cardinal $\mathfrak{m}$, there are $2^{\mathfrak{m}}$ mutually rigid triangle connected graphs.

3.15. Prove that, for each infinite cardinal $\mathfrak{m}$, there are $2^{\mathfrak{m}}$ mutually rigid, connected graphs in which every element lies on a cycle of length 7.

3.16. Let $M$ be a monoid and let $\mathfrak{m}$ be an infinite cardinal satisfying $|M| \leq \mathfrak{m}$. Prove that there are $2^{\mathfrak{m}}$ pairwise nonisomorphic graphs $(G; E)$ of cardinality $\mathfrak{m}$ satisfying $\mathrm{End}(G; E) \cong M$.

* * *

3.17. Prove that the $L$ of Theorem 565 is a $\mathcal{C}$-reduced free product of the $L_i$, for $i \in I$, by verifying directly Definition 563(iii). (Hint: argue as in the proof of Theorem 526.)

* * *

Exercises 3.18–3.21 are based on C. C. Chen and G. Grätzer [88].

3.18. A *bi-uniquely complemented lattice* $L$ is a bounded lattice in which every element $x \neq 0, 1$ has exactly two complements. Define the concept of a *free bi-uniquely complemented lattice* and prove the existence and uniqueness (up to isomorphism) of a free bi-uniquely complemented lattice on $\mathfrak{m}$ generators.

3.19. Let $F_0$ and $F_1$ be free bi-uniquely complemented lattices and let $X_0$ and $X_1$ be the free generating sets of $F_0$ and $F_1$, respectively. Let $\alpha$ be a one-to-one map of $X_0$ onto $X_1$. Let $|X_0| = |X_1| = \aleph_0$. Show that there are $2^{\aleph_0}$ isomorphisms of $F_0$ and $F_1$ extending $\alpha$.

3.20. In a bounded lattice $L$, the *complementation is transitive* if whenever $x, y$ and $y, z$ are complementary pairs, then either $x = z$ or $x, z$ is also a complementary pair. Prove that every lattice can be embedded into a bi-uniquely complemented lattice with transitive complementation.

3.21. Let $L$ be a bounded lattice. Under what conditions is there a $\{0, 1\}$-embedding of $L$ into a bi-uniquely complemented lattice with transitive complementation?

* * *

3.22. Show that a relatively complemented, uniquely complemented lattice is boolean. (Hint: use Exercise I.6.4.)

3.23. Prove that a modular uniquely complemented lattice is boolean.

3.24. Show that an atomic uniquely complemented lattice is boolean(T. Ogasawara and U. Sasaki [556]). (Hint: the map $a \mapsto \mathrm{Atom}(a)$ is a set representation.)

3.25. Let $L$ be a uniquely complemented lattice and let $a \succ b$ in $L$. Prove that $a \wedge b'$ is an atom, where $b'$ is the complement of $b$ (see H.-J. Bandelt and R. Padmanabhan [49]). (Hint: if $a \wedge b' = 0$, then $b'$ has two complements; if $0 < x < a \wedge b'$, then $b \vee a'$ has two complements.)

3.26. Show that if a lattice $L$ is relatively atomic and uniquely complemented, then it is boolean. (This includes the result of V. N. Saliĭ [624] that every uniquely complemented, algebraic lattice is boolean.)

3.27. A *lattice with complementation* $(L; \vee, \wedge,', 0, 1)$ is a bounded lattice in which $a'$ is a complement of $L$ for each $a \in L$. An *endomorphism* $\varphi$ is a lattice endomorphism satisfying $(\varphi(a))' = \varphi(a')$ for all $a \in L$. Prove that, for every monoid $M$, there is a lattice with complementation whose endomorphisms semigroup is isomorphic to $M$ (G. Grätzer and J. Sichler [352]).

3.28. In Theorem 572 and in Exercise 3.27, how many pairwise nonisomorphic lattices of a given cardinality can be constructed satisfying the requirements?

3.29. Generalize the definition of Free($P; \mathcal{J}, \mathcal{M}$) to allow $\mathcal{J}$ to contain arbitrary nonempty finite subsets of $P$ such that $\sup X$ exists in $P$ and the dual for $\mathcal{M}$. Show that this does not affect the existence and description of Free($P; \mathcal{J}, \mathcal{M}$) (R. A. Dean [147]).

## 4. Hopfian Lattices

### 4.1 Basic definitions

A lattice $L$ is *hopfian* if every onto endomorphism is an automorphism. Equivalently, a lattice $L$ is hopfian if $L \cong L/\alpha$ implies that $\alpha = 0$, that is, $L$ is not isomorphic to a proper quotient of itself. The hopfian property is similarly defined for groups, rings, and algebras, in general. The first definition of the hopfian property also applies for graphs; one should note, however, that for a graph, a one-to-one and onto endomorphism need not be an automorphism; for graphs, an automorphism has to be defined as an invertible endomorphism.

Obviously, every finite lattice (algebra, graph) is hopfian. In fact, the hopfian property is a generalization of one of the most important characteristics of finiteness.

Closest to finite lattices are *finitely presented lattices*, that is, lattices of the form Free $Q$ (see Section I.5.1), where $Q$ is a finite partial lattice. (It is easily seen that these are the lattices that can be described by finitely many generators and finitely many relations, so this concept agrees with "finitely presented" as is used for groups and semigroups.)

An important property of finitely presented lattices is proved in T. Evans [173].

**Theorem 582.** *Every finitely presented lattice is hopfian.*

The proof of Theorem 582 is contained in the following three lemmas.

**Lemma 583.** *A finitely presented lattice can be represented as a subdirect product of finite lattices.*

*Proof.* It is sufficient to prove that if $Q$ is a finite partial lattice, $a, b$ are elements of Free $Q$ with $a \neq b$, then there exists a finite lattice $K$ and a homomorphism $\varphi \colon \text{Free } Q \to K$ such that $\varphi(a) \neq \varphi(b)$. Just as in Section 2.4, we can define the component subsets of $a$ and $b$ and set

$$Q' = Q \cup \text{Komp}(a) \cup \text{Komp}(b).$$

Regarding $Q'$ as a partial lattice, obviously $\operatorname{Free} Q = \operatorname{Free} Q'$. By (the proof of) Theorem 84, $Q'$ can be embedded into some finite lattice $K$. This embedding extends to a homomorphism $\varphi$ of $\operatorname{Free} Q' = \operatorname{Free} Q$ into $K$. Finally, $\varphi(a) \neq \varphi(b)$, since $\varphi$ is one-to-one on $Q'$. $\qquad\square$

**Lemma 584.** *Let $L$ be a finitely generated lattice and let $\boldsymbol{\alpha}$ be a congruence relation of $L$. If $L/\boldsymbol{\alpha}$ is finite, then there exists a fully invariant congruence relation $\boldsymbol{\varphi}$ of $L$ such that $\boldsymbol{\varphi} \le \boldsymbol{\alpha}$ and $L/\boldsymbol{\varphi}$ is finite.*

*Proof.* Set $\mathbf{K} = \mathbf{Var}(L/\boldsymbol{\alpha})$. We define

$$\boldsymbol{\varphi} = \bigvee \big(\operatorname{con}(p(a_0, \ldots, a_{n-1}), q(a_0, \ldots, a_{n-1})) \mid p = q \in \operatorname{Iden}(\mathbf{K}),$$

$$a_0, \ldots, a_{n-1} \in L \big).$$

It is easily seen that $\boldsymbol{\varphi}$ is fully invariant. Since the identities used to construct $\boldsymbol{\varphi}$ all hold in $L/\boldsymbol{\alpha}$, we conclude that $\boldsymbol{\varphi} \le \boldsymbol{\alpha}$. Finally, $L/\boldsymbol{\varphi}$ is finitely generated and $L/\boldsymbol{\varphi} \in \mathbf{K}$, which is a locally finite class, hence $L/\boldsymbol{\varphi}$ is finite. $\qquad\square$

A *subdirect representation* of a lattice $L$ is associated with a family $\boldsymbol{\alpha}_i$, for $i \in I$, of congruence relations of $L$ such that

$$\bigwedge (\boldsymbol{\alpha}_i \mid i \in I) = \mathbf{0}.$$

If the $\boldsymbol{\alpha}_i$, for all $i \in I$, are fully invariant, we call this subdirect representation *fully invariant*.

**Lemma 585.** *Let the lattice $L$ have a representation as a fully invariant subdirect product of hopfian lattices. Then $L$ is hopfian.*

*Proof.* Let $(\boldsymbol{\alpha}_i \mid i \in I)$ be a family of fully invariant congruence relations of $L$ satisfying $\bigwedge (\boldsymbol{\alpha}_i \mid i \in I) = \mathbf{0}$. Let $\varphi$ be an onto endomorphism of $L$. Since $\boldsymbol{\alpha}_i$ is fully invariant, for all $i \in I$, we can define

$$\varphi_i \colon a/\boldsymbol{\alpha}_i \mapsto \varphi(a)/\boldsymbol{\alpha}_i,$$

which is an endomorphism of $L/\boldsymbol{\alpha}_i$. Obviously, all $\varphi_i$ are onto, hence they all are automorphisms. Now if $a, b \in L$ and $\varphi(a) = \varphi(b)$, then

$$\varphi_i(a/\boldsymbol{\alpha}_i) = \varphi_i(b/\boldsymbol{\alpha}_i),$$

hence $a/\boldsymbol{\alpha}_i = b/\boldsymbol{\alpha}_i$ for all $i \in I$. Since $\bigwedge (\boldsymbol{\alpha}_i \mid i \in I) = \mathbf{0}$, we conclude that $\varphi$ is one-to-one. $\qquad\square$

R. Wille [736] exhibits a finitely generated lattice that is not hopfian. An earlier example of a finitely generated lattice that is not finitely presented (or presentable, to be more precise) is $\operatorname{Free}_M(4)$ (see T. Evans and D. X. Hong [174]); of course, $\operatorname{Free}_M(4)$ is hopfian.

## 4.2    Free product of hopfian lattices

The free product of two finitely presented lattices is again finitely presented. What about hopfian lattices? A negative answer was obtained in G. Grätzer and J. Sichler [353].

**Theorem 586.** *The free product of two hopfian lattices is not necessarily hopfian.*

The proof of Theorem 586 is based on the construction of the lattice Lat $G$ from the graph $G$ introduced in Section 3.4 and a lattice theoretic result (Theorem 587) which is due to J. Sichler [645] and H. Lakser [493]. In fact, the result we shall prove will be stronger than Theorem 586 and it will be based on J. Sichler [646].

For graphs $G_i$ ($= (G_i; E_i)$), for $i \in I$, we form the lattices Lat $G_i$ (with bounds $0_i$ and $1_i$) and the free product $L$ of the Lat $G_i$ for $i \in I$.

A *triangle* of $G_i$ is a three-element set $\{a, b, c\}$ such that

$$\{a, b\}, \{a, c\}, \{b, c\} \in E_i.$$

If $\{a, b, c\}$ is a triangle of $G_i$, then $\{0_i, a, b, c, 1_i\}$ is a diamond. We call this sublattice the *diamond associated with a triangle*.

**Theorem 587.** *Any diamond in $L$ is associated with a triangle of some $G_i$ for some $i \in I$.*

*Proof.* Let $M = \{o, a, b, c, i\}$ be a diamond in $L$. Let us first assume that $M \subseteq$ Lat $G_j$ for some $j \in I$.

  *Case 1.* $o = 0_j$ and $i = 1_j$. Then

$$\{a, b\}, \{a, c\}, \{b, c\} \in \mathrm{Comp}(\mathrm{Lat}\, G_j),$$

hence by Corollary 567,

$$\{a, b\}, \{a, c\}, \{b, c\} \in E_i,$$

and indeed, $M$ is associated with the triangle $\{a, b, c\}$ of $G_j$.

  *Case 2.* $o = 0_j$ and $i < 1_j$. Recall that

$$\mathrm{Lat}\, G_j - \{0_j, 1_j\} \subseteq \mathrm{Free}(G_j)$$

(the free lattice generated by the set $G_j$), hence $a, b, c, i \in \mathrm{Free}(G_j)$, and there is a congruence relation $\alpha$ on $\mathrm{Free}(G_j)$ such that $\mathrm{Free}(G_j)/\alpha \cong \mathrm{Lat}\, G_j$ and $x/\alpha \mapsto x$ under this isomorphism for all $x \in \mathrm{Lat}\, G_j - \{0_j, 1_j\}$. Since $i = a \vee b = a \vee c$ in $\mathrm{Free}(G_j)$, by $(\mathrm{SD}_\vee)$, $i = a \vee (b \wedge c)$. But $b \wedge c = 0_j$, that is, $(b \wedge c)/\alpha$ is the zero of $\mathrm{Free}(G_j)/\alpha$. Thus

$$i/\alpha = a/\alpha \vee (b \wedge c)/\alpha = a/\alpha,$$

and we conclude that $a = i$, a contradiction. Hence this case is not possible.

*Case 3.* $o > 0_j$ *and* $i = 1_j$. This is impossible; argue as in Case 2.

*Case 4.* $o > 0_j$ *and* $i < 1_j$. Then $M \subseteq \mathrm{Free}(G_j)$, which is impossible since there is no diamond in a free lattice.

Thus Theorem 587 is proved for sublattices of $\mathrm{Lat}\,G_j$. Now let $M \subseteq L$ and consider, for each $j \in J$,

$$M_{(j)} = \{\, x_{(j)} \mid x \in M \,\}.$$

Since $x \mapsto x_{(j)}$ is a homomorphism, $M_{(j)} \cong M$ or $M_{(j)}$ is a singleton. If $M_{(j)}$ is a singleton, for all $j \in J$, then Theorem 536 yields that $(\mathrm{SD}_\vee)$ holds in $M$, a contradiction. Hence there exists a $j \in I$ such that $M_{(j)}$ is a diamond and it is, therefore, associated with a triangle of $G_j$, in particular, $o_{(j)} = 0_j$ and $i_{(j)} = 1_j$.

By duality, there exists a $k \in I$ such that $M^{(k)}$ is a diamond; so $o^{(k)} = 0_k$ and $i^{(k)} = 1_k$. This yields immediately that $j = k$ and $o = 0_j$, $i = 1_j$. Since

$$a_{(j)} \vee b_{(j)} = 1_j,$$
$$a^{(j)} \wedge b^{(j)} = 0_j,$$
$$a_{(j)} \leq a^{(j)},$$
$$b_{(j)} \leq b^{(j)},$$

we conclude that $a_{(j)}$ and $a^{(j)}$ are complements of $b_{(j)}$ in $\mathrm{Lat}\,G_j$. But $\mathrm{Lat}\,G_j$ is a lattice with no comparable complements, hence $a_{(j)} = a^{(j)} = a \in \mathrm{Lat}\,G_j$. Similarly, $b, c \in \mathrm{Lat}\,G_j$. Hence $M \subseteq \mathrm{Lat}\,G_j$ and $M$ is associated with a triangle of $G_j$. $\qquad\square$

*Proof of Theorem 586.* Let $I$ be a set of at least two elements. For each $i \in I$, a lattice $L_i$ will be constructed such that the free product of all $L_i$, for $i \in I$, is not hopfian but the free product of $L_i$, for $i \in I'$, is hopfian for all $\varnothing \neq I' \subset I$. The special case $|I| = 2$ is Theorem 586.

Let $N = \{1, 2, 3, \ldots\}$, and we consider maps $\varphi \colon N \to I$. The map $\varphi$ is *eventually constant* if there exists an integer $n$ such that $\varphi(n) = \varphi(n+1) = \cdots$. Let $M(I)$ denote the set of all eventually constant maps of $N$ into $I$. By $M_n(I)$ we denote the set of all maps $\varphi \colon \{1, \ldots, n\} \to I$ and

$$M_\omega(I) = \bigcup (\, M_n(I) \mid n = 1, 2, \ldots ).$$

For $\varphi \in M(I)$, $n \in N$, and for $\varphi \in M_m(I)$, $n \leq m$, let $\varphi_n$ be the restriction of $\varphi$ to $\{1, \ldots, n\}$. Finally, for $\varphi \in M_n(I)$, we write $n = \mathrm{Dom}(\varphi)$ and $\mathrm{FV}(\varphi) = \varphi(n)$ (Dom for domain and FV for final value).

For a graph $G$ $(= (G; E))$ and $a, b \in G$, we say that $a$ and $b$ are *triangle connected* (see Exercise 3.14) if $a = b$ or there is a sequence $T_0, \ldots, T_n$ of triangles of $G$ such that $a \in T_0$, $b \in T_n$, and $T_i \cap T_{i+1} \neq \varnothing$ for all $i = 0, \ldots, n-1$.

Any graph can be decomposed into a disjoint union of triangle connected components. The graph $G$ is *triangle connected* if it has a single component.

Now choose a cardinal $\mathfrak{m} > |M(I)|$, a triangle connected graph $G_0$, and for each $\alpha \in M(I)$, choose a triangle connected graph $G_\alpha$ such that

$$|G_0| = |G_\alpha| = \mathfrak{m}$$

and the graphs $G_0$ and $G_\alpha$, for all $\alpha \in M(I)$, are pairwise disjoint and mutually rigid (Exercise 3.14). We fix the elements $a_0 \in G_0$ and $a_\alpha \in G_\alpha$ for all $\alpha \in M(I)$.

For every $n \in N$ and $\varphi \in M_n(I)$, we define a graph $G^\varphi$:

$$G^\varphi = \bigcup(\, G_\alpha \mid \alpha \in M(I) \text{ and } \alpha_n = \varphi\,),$$

$$E^\varphi = \bigcup(\, E_\alpha \mid \alpha \in M(I) \text{ and } \alpha_n = \varphi\,) \cup \{\, \{a_0, a_\alpha\} \mid \alpha \in M(I)\,\}.$$

In words, $G^\varphi$ is a disjoint union of all $G_\alpha$ such that $\alpha_n = \varphi$ and we add the edges connecting the distinguished element of $G_0$ with the distinguished element of the $G_\alpha$. Observe that $G^\varphi$ is connected and that $G_0$ and the $G_\alpha$ are the triangle connected components of $G^\varphi$.

Let $\chi$ be a homomorphism of $G^\varphi$ into $G^\psi$ for $\varphi \in M_n(I)$ and $\psi \in M_m(I)$. Then the triangle connected components of $G^\varphi$ have to be mapped into the triangle connected components of $G^\psi$. Thus if $\alpha \in M(I)$ with $\alpha_n = \varphi$ or $\alpha = 0$, then $\chi(G_\alpha) \subseteq G_\beta$, where $\beta \in M(I)$ and $\beta_m = \psi$ or $\beta = 0$. In view of the mutual rigidity of these graphs we must have $\alpha = \beta$ and $\chi$ is the identity on $G_\alpha$. So we conclude that if $\alpha \in M(I)$ and $\alpha_n = \varphi$, then $\alpha_m = \psi$; this is possible iff $m \leq n$ and $\alpha_m = \beta$. Thus there is a homomorphism of $G^\varphi$ into $G^\psi$ iff $\mathrm{Dom}(\varphi) \geq \mathrm{Dom}(\psi)$ and $\varphi_{\mathrm{Dom}(\psi)} = \psi$. Moreover, in this case, there is exactly one homomorphism which is the identity map on $G_0$ and on all $G_\alpha$ with $\alpha_n = \varphi$.

Now we are ready to define $L_i$ for all $i \in I$: let $L_i$ be a free product of all lattices $\mathrm{Lat}\, G^\varphi$ satisfying $\mathrm{FV}(\varphi) = i$.

Let $L$ be a free product of all $L_i$ for $i \in I$; in other words, $L$ is a free product of all $\mathrm{Lat}\, G^\varphi$ for $\varphi \in M_\omega(I)$. We verify that $L$ is not hopfian. We define a map $\beta$:

$\beta$ restricted to $G^\varphi$ is the homomorphism of $G^\varphi$ into $G^{\varphi_{n-1}}$, where $n = \mathrm{Dom}(\varphi) > 1$;

$\beta$ on $G^\varphi$ is the identity, if $\mathrm{Dom}(\varphi) = 1$.

Thus $\beta$ extends to a homomorphism $\gamma$ of $\mathrm{Lat}\, G^\varphi$ into $\mathrm{Lat}\, G^{\varphi_{n-1}}$, if $n = \mathrm{Dom}(\varphi) > 1$, and to the identity map $\gamma$ on $\mathrm{Lat}\, G^\varphi$, if $\mathrm{Dom}(\varphi) = 1$. Since $L$ is a free product, $\gamma$ extends to an endomorphism $\delta$ of $L$.

Observe that the image of $\beta$ covers all $G^\varphi$. Indeed, if $a \in G^\varphi$, it follows that $a \in G_\alpha$, for some $\alpha \in M(I)$ with $\alpha_n = \varphi$, where $n = \mathrm{Dom}(\varphi)$. Then

$\beta$ maps $a \in G_{\alpha}{}' \subseteq G^{\alpha_{n+1}}$ onto $a \in G_{\alpha} \subseteq G^{\varphi}$. Thus $\delta$ is an onto endomorphism of $L$. The map $\beta$ is not one-to-one since every element of any $G^{\varphi}$ with $\text{Dom}(\varphi) = 1$ is the image of two elements: of itself and of a suitable $G^{\psi}$ with $\text{Dom}(\psi) = 2$. Therefore, $\delta$ is not one-to-one. We have proved that $L$ is not hopfian.

Now let $\varnothing \neq I' \subset I$ and let $L'$ be the free product of the $L_i$ for $i \in I'$. We have to show that $L'$ is hopfian.

Observe that $L'$ is the free product of all $\text{Lat}\, G^{\varphi}$ satisfying $\text{FV}(\varphi) \in I'$. Let $\delta$ be an onto endomorphism of $L'$. We shall verify that $\delta$ is the identity map, implying that $L'$ is hopfian.

Assume to the contrary that $\delta$ is not the identity map. Then there is an $a \in G^{\varphi}$ with $\text{FV}(\varphi) \in I'$ and $\text{Dom}(\varphi) = n$ such that $a \neq \delta(a)$. Since $a$ is an element of a triangle, it is an atom of the associated diamond $M$. If $|\varphi(M_{\flat})| = 1$, then the bounds of $\text{Lat}\, G^{\varphi}$ are collapsed by $\delta$, hence all of $\text{Lat}\, G^{\varphi}$ is mapped by $\delta$ onto a single element. Otherwise, $\varphi(M) \cong M$, hence, by Theorem 587, $\delta(a) \in G^{\psi}$ for some $\psi \in M_{\omega}(I)$. Since $G^{\varphi}$ is connected, all of $G^{\varphi}$ is mapped by $\delta$ into $G^{\psi}$. Hence $\text{Dom}(\varphi) \geq \text{Dom}(\psi)$ and $\psi = \varphi_m$ where $m = \text{Dom}(\psi) < \text{Dom}(\varphi)$, since $\text{Dom}(\psi) = \text{Dom}(\varphi)$ implies that $\varphi = \psi$, contradicting the rigidity of $G^{\varphi}$.

In either case, we see that $G^{\varphi} \cap \delta(G^{\varphi})$ has at most one element. Let $i \in I - I'$ and consider the map $\beta \colon N \to I$ defined by $\beta_n = \varphi$ and $\beta(k) = i$ for all $k > n$. Then $\beta \in M(I)$ and $G_{\beta} \subseteq G^{\varphi}$. Let $G_{\beta}^* = G_{\beta} - (G^{\varphi} \cap \delta(G^{\varphi}))$. Since $\delta$ is onto, every $a \in G_{\beta}^*$ must be in the image of some $G^{\psi}$; but it cannot come from the homomorphism of $G^{\psi}$ into $G^{\varphi}$, since it would imply that $\text{Dom}(\psi) = m > n = \text{Dom}(\varphi)$ and $\beta_m = \psi$, contradicting the definition of $L'$. ($\psi(G)$ is not in $L'$ since $\text{FV}(\psi) = i \notin I'$). Thus $a \in G_{\beta}^*$ must come from a $G^{\psi}$ collapsed by $\delta$ onto $a$. However, $|G_{\beta}^*| = \mathfrak{m}$ and $|M(I)| < \mathfrak{m}$, a contradiction. Thus $\delta$ must be the identity map.     $\square$

## Exercises

4.1. Prove that $\text{Free}_{\mathbf{K}}(\mathfrak{m})$ is not hopfian for any nontrivial variety $\mathbf{K}$ and infinite cardinal $\mathfrak{m}$.

4.2. Prove that $\text{Free}_{\mathbf{K}}(n)$ is hopfian for every variety $\mathbf{K}$ of lattices and natural number $n$.

4.3. Find a bijective endomorphism of a graph that is not an automorphism.

4.4. Can the graph in Exercise 4.3 be chosen to be finite?

4.5. Define a finitely presented lattice as the "most free" lattice generated by $x_0, \ldots, x_{n-1}$ satisfying the "relations"

$$p_i(x_0, \ldots, x_{n-1}) = q_i(x_0, \ldots, x_{n-1}) \quad \text{for } i = 1, \ldots, m,$$

where $p_i$ and $q_i$ are $n$-ary terms. Show that this is equivalent to forming a Free $Q$ with a suitable finite partial lattice $Q$.

4.6. Let Free $Q \subseteq Q' \subseteq Q$, where $Q$ and $Q'$ are partial lattices. Show that Free $Q \cong$ Free $Q'$ provided that $Q'$ is generated by $Q$.

4.7. Formulate and prove the converse of Exercise 4.6.

4.8. For a lattice $L$ and variety $\mathbf{K}$ of lattices, we constructed, in the proof of Lemma 584, a congruence relation $\varphi = \varphi(\mathbf{K})$. Prove that $\varphi(\mathbf{K})$ is fully invariant.

4.9. Let $\varphi$ be a fully invariant congruence relation of a lattice $L$. Prove that $\varphi = \varphi(\mathbf{K})$ for some variety $\mathbf{K}$.

4.10. Let $L$ be a fully invariant subdirect product of the $L_i$ for $i \in I$. For an onto endomorphism $\varphi$, write down a formula for $\mathrm{Ker}(\varphi)$ from which Lemma 585 can be derived.

4.11. What is the analogue of Lemma 585 for graphs?

4.12. Show that the free product of two finitely presented lattices is finitely presented.

4.13. Prove that $\mathrm{Lat}(G_0 \cup G_1)$ is a free $\{0,1\}$-product of $\mathrm{Lat}\, G_0$ and $\mathrm{Lat}\, G_1$, where $G_0 \cup G_1$ is the disjoint union of the graphs $G_0$ and $G_1$.

4.14. Find two bounded hopfian lattices whose free $\{0,1\}$-product is not hopfian (G. Grätzer and J. Sichler [353]).

4.15. Let $G_0$ and $G_1$ be hopfian graphs. Prove that a free product of the lattices $\mathrm{Lat}\, G_0$ and $\mathrm{Lat}\, G_1$ is hopfian.

# *Afterword*

In case you are wondering how this book evolved, here is the story.

## Lattice Theory: First Concepts and Distributive Lattices

It all started in my formative years as a mathematician, 1955–1961, during my collaboration with E. T. Schmidt. A typical paper of ours started with a page or two of basic concepts and notation; how nice it would be to instead reference a standard book.

However, the time was not ripe for such a project. For instance, such a book would have to deal with uniquely complemented lattices. To accomplish this, one would have to reproduce the almost thirty pages of the famous argument of R. P. Dilworth embedding every lattice in a uniquely complemented lattice. Much more would have to be learned about free lattices and varieties of lattices before the project could be attempted.

In 1962, I wrote a proposal for a book on lattice theory that would survey the whole field in depth. Apart from doing some of the research necessary for the project, no writing was done. Then M. H. Stone enticed me to write a book on universal algebra, to be published in the D. Van Nostrand *University Series in Higher Mathematics* and I concentrated on this until the end of 1967.

Maybe because mathematicians in general (or I, in particular) are like hobbits (according to J. R. R. Tolkien [680]: "Hobbits delighted in such things if they were accurate: they liked to have books filled with things they already knew, set out fair and square with no contradictions.") or maybe because I felt that there was a real need for an in-depth book on lattice theory, I started in 1968 on this book. In the academic year 1968–1969, I gave a course on lattice theory and I wrote a set of lecture notes. The first two chapters of *First Concepts* are based on those notes.

This material was augmented by a chapter on pseudocomplemented distributive lattices and published under the title *Lattice Theory: First Concepts and Distributive Lattices* in 1971 [257]. (This book is still available; it was

G. Grätzer, *Lattice Theory: Foundation*, DOI 10.1007/978-3-0348-0018-1,
© Springer Basel AG 2011

reissued in 2008.) The *Introduction* of this book promised a companion volume, in preparation, on general lattices; I did not realize at the time that "the preparation" would require seven more years.

## Acknowledgements

The undergraduates and graduate students who took my course in 1968–69, and many of my colleagues who attended, helped by criticizing my lectures and by simplifying proofs. The lecture notes were also read by P. Burmeister, who offered many helpful remarks. A rewritten version of the lecture notes was read by R. Balbes, M. I. Gould, K. M. Koh, H. Lakser, S. M. Lee, P. Penner, C. R. Platt, and R. Padmanabhan, who coordinated the work of this group (just as he coordinated the work of the "readers" 42 years later).

I rewrote a substantial part of the manuscript as a result of the changes they suggested. I am very thankful to the whole group, and especially to C. R. Platt and R. Padmanabhan, for their untiring work.

I am grateful to K. D. Magill, Jr., who invited me to conduct a course on lattice theory at the State University of New York at Buffalo in the summer of 1970, and to the students of this class, especially J. H. Hoffman.

B. Jónsson read the manuscript for the publisher. The typing and retyping of the manuscript were done by Mrs. M. McTavish. The rest of the secretarial work was handled by Mrs. N. Buckingham.

The galley proofs were read by R. Antonius, J. A. Gerhard, K. M. Koh, W. A. Lampe, R. W. Quackenbush, I. Rival, and the group coordinator, R. Padmanabhan. E. Fried assisted me in compiling the Index. Thanks are also due to Professor N. S. Mendelsohn, who relieved me of all teaching and administrative duties to allow me time to conduct research, to supervise the research of graduate students, and to prepare the manuscript of this book.

## General Lattice Theory

A number of research breakthroughs in the 1960s and 1970s provided the material I needed to complete the project.

But then it became apparent that a complete revision of my plans was in order. While back in the late 1950s it seemed reasonable to try to give a complete picture of lattice theory, this became patently unfeasible in the 1970s. For instance, in 1958 there was one paper on Stone algebras; by 1974, there were more than fifty. A number of books have also appeared dealing with specialized aspects of lattice theory and with various applications.

Another change took place in the publishing field. For the second volume it became desirable to choose a publisher with a greater interest in monographs. The new arrangement made it necessary to produce a volume that does not depend on the previous publication. That is why most of the first two chapters of the new book were based on *Lattice Theory: First Concepts and Distributive Lattices*, thus making the new book, *General Lattice Theory*, self-contained.

The work on *General Lattice Theory* started in 1972 and then continued with an advanced course on lattice theory at the University of Manitoba in 1973–1974. The lecture notes of this course form the basis of most of Chapters III–VI.

### Acknowledgements

I am grateful to my students who took the course in 1973–1974 and to my colleagues who attended for their helpful criticisms and for many simplified proofs. In the proofreading of *General Lattice Theory* I was assisted by M. E. Adams, K. A. Baker, R. Beazer, J. Berman, B. A. Davey, J. A. Gerhard, M. I. Gould, D. Haley, D. Kelly, C. R. Platt, and G. H. Wenzel.

A great deal of organizational work was necessary in the distribution of manuscripts and the collation of corrections; this was faithfully carried out by R. Padmanabhan. M. E. Adams undertook the arduous task of getting the manuscript ready for the publisher.

I received help from various individuals in specific areas, including M. Doob (matroids), I. Rival (exercises on combinatorial topics), R. Venkataraman (partially ordered vector spaces), and B. Wolk (projective geometry).

Special thanks are due to Professor N. S. Mendelsohn for creating a very good environment for work. Mrs. M. McTavish did an excellent job of typing and retyping the manuscript. Finally, I would like to thank the members and the many visitors of my seminar who, over a period of eight years, have been lecturing an average of four hours a week, 52 weeks a year, in an attempt to teach me lattice theory. Without their help I could not even have tried.

## General Lattice Theory, Second Edition

In twenty years, tremendous progress has been made in Lattice Theory. Nevertheless, the change was in the superstructure not in the foundation. Accordingly, I decided to add appendices to record the change. In the first appendix: *Retrospective*, I briefly reviewed developments from the point of view of the first edition, specifically, the major results of 1978–1998 and solutions of the problems proposed in this book. It is remarkable how many difficult problems had been solved!

### Acknowledgements

An exceptional group of people wrote the other appendices: Brian A. Davey and Hilary A. Priestley on distributive lattices and duality, Friedrich Wehrung on continuous geometries, Marcus Greferath and Stefan E. Schmidt on projective lattice geometries, Peter Jipsen and Henry Rose on varieties, Ralph Freese on free lattices, Bernhard Ganter and Rudolf Wille on formal concept analysis;

Thomas E. Schmidt collaborated with me on congruence lattices. Many of these same people are responsible for definitive books on the same subjects.

## Lattice Theory: Foundation

When I started on this project, it did not take me very long to realize that what I attempted to accomplish in 1968–1978, I cannot even try in 2009. To lay the foundation, to survey the contemporary field, to pose research problems, would require more than one volume or more than one person. So I decided to cut back and concentrate in this volume on the foundation.

The foundation of a field does not change all that much over time. My plan was to revise, reorganize, and up-to-date the old chapters, add the foundation for congruence lattices of finite lattices (lattice constructions), and modernize notation.

Having started the writing in January of 2009, this project took about 22 months. Now it is time to start working on the companion volume, *Special Topics and Applications*.

### Acknowledgements

Early versions of Chapter I were read by K. A. Baker and R. W Quackenbush, who made numerous corrections.

F. Wehrung contributed many useful comments, suggestions for notational changes, and a new presentation of Frink's Embedding Theorem.

Throughout the writing, I received useful corrections from Ivan Chajda, Gábor Czédli, Joseph Kung, Luis Sequeira, and Manfred Stern. Many errors were corrected by Marcel Wild, Yuri Movsisyan, Jeffrey S. Olson, Chihiro Oshima, and Parameshwara Bhatta. Their work was coordinated by R. Padmanabhan. I am grateful to them all.

G. M. Bergman (and his seminar) submitted more than 1,000 comments on the first two chapters and on the first section of Chapter VII. George went far and beyond the call of duty, twice over.

And special thanks to Padmanabhan for 42 years of faithful help with the writing of this book and its predecessors.

This book was edited by Edwin Beschler; thorough editing is his trademark. Kati Fried did a professional job of compiling the index.

Dr. Thomas Hempfling, Executive Editor, Mathematics, of Birkhäuser Basel, guided this book, ably assisted by Sylvia Lotrovsky.

The final manuscript was read for the publisher by Gábor Czédli and Friedrich Wehrung; the manuscript turned out to be not so final after all.

### Notation

Not so long ago, a typical paper in lattice theory would have a formula such as

$$\Theta = \Theta(a, b) \in \Theta(L).$$

Congruences were denoted by $\Theta$, the principal congruence collapsing the elements $a$ and $b$ by $\Theta(a, b)$, and the congruence lattice by $\Theta(L)$. The symbol $\Theta$ was slightly overused ...

With the advent of LaTeX, operators entered the picture, so $\Theta(L)$ became Con $L$, $I(L)$ was changed to Id $L$, and $G(L)$ was replaced by Aut $L$. The new notation looks nicer and is mnemonic, easier to remember.

In this book, I try to carry out these changes systematically. I use operators and do not use parentheses whenever from one structure we construct a new structure (unless we need parentheses to avoid confusion as in $\text{Con}(A * B)$). I consulted widely about the notation changes. There was widespread (although not unanimous) acceptance to use Con $L$ for the congruence lattice, $\text{con}(a, b)$ for the principal congruence; Id $L$ for the ideal lattice and $\text{id}(X)$ for the ideal generated by $X$, and so on.

On the other hand, if we construct a set or produce a number, we use commands with arguments and parentheses, for instance, $\text{Atom}(L)$, $\text{width}(P)$.

The greatest resistance was to change $J(L)$; this denotates an order constructed from a lattice, so it should be with a mnemonic operator and no parentheses. I tried JIr $L$, JI $L$, Jir $L$, JI $L$ and some others, and ended up with Ji $L$. All choices elicited strongly opposing e-mails. I wonder if anybody will follow my lead.

In rewriting thousands of formulas, for instance,

$$(x_i \varphi \psi]$$

became

$$\text{id}(\psi \varphi(x_i))$$

I am sure I introduced far too many errors. Despite the extreme effort of so many to help me, many errors of this—and varied other—types remained in the book. For this, I apologize.

# Bibliography

[1] T. Abe, A characterization of relatively balanced lattices, *Algebra Universalis* **47** (2002), 45–50.

[2] M. E. Adams, The Frattini sublattice of a distributive lattice. *Algebra Universalis* **3** (1973), 216–228.

[3] ———, Maximal subalgebras of Heyting algebras, *Proc. Edin. Math. Soc. (2)*, **29** (1986), 359–365.

[4] ———, Uniquely complemented lattices. *The Dilworth theorems*, 79–84, Contemp. Mathematicians, *Birkhäuser Boston, Boston, MA*, 1990.

[5] M. E. Adams, K. V. Adaricheva, W. Dziobiak, and A. V. Kravchenko, Open questions related to the problem of Birkhoff and Maltsev. *Studia Logica* **78** (2004), 357–378.

[6] M. E. Adams and W. Dzobiak (eds), Special Issue of Studia Logica on Priestley duality, *Studia Logica* **56** (1986).

[7] ———, Q-universal quasivarieties of algebras. *Proc. Amer. Math. Soc.* **120** (1994), 1053–1059.

[8] ———, Finite-to-finite universal quasivarieties are Q-universal. *Algebra Universalis* **46** (2001), 253–283.

[9] M. E. Adams and D. Kelly, Homomorphic images of free distributive lattices that are not sublattices. *Algebra Universalis* **5** (1975), 143–144.

[10] ———, Chain conditions in free products of lattices. *Algebra Universalis* **7** (1977), 235–244.

[11] M. E. Adams and J. Sichler, Cover set lattices. *Canad. J. Math.* **32** (1980), 1177–1205.

[12] M. E. Adams and J. Sichler, Lattices with unique complementation. *Pacific. J. Math.* **92** (1981), 1–13.

[13] K. V. Adaricheva, The structure of finite lattices of subsemilattices. (Russian.) *Algebra i Logika* **30** (1991), 385–404, 507; translation in *Algebra and Logic* **30** (1991), 249–264.

[14] _____, Lattices of algebraic subsets. *Algebra Universalis* **52** (2004), 167–183.

[15] _____, On the prevariety of perfect lattices. *Algebra Universalis.* To appear.

[16] K. V. Adaricheva, W. Dziobiak, and V. A. Gorbunov, Finite atomistic lattices that can be represented as lattices of quasivarieties. *Fund. Math.* **142** (1993), 19–43.

[17] _____, Algebraic atomistic lattices of quasivarieties. *Algebra and Logic* **36** (1997), 213–225.

[18] K. V. Adaricheva and V. A. Gorbunov, Equaclosure operator and forbidden semidistributive lattices. (Russian.) *Sibirsk. Mat. Zh.* **30** (1989), 7–25; translation in *Siberian Math. J.* **30** (1989), 831–849.

[19] _____, On lower bounded lattices. *Algebra Universalis,* **46** (2001), 203–213.

[20] K. V. Adaricheva, V. A. Gorbunov, and M. V. Semenova, On continuous noncomplete lattices. The Viktor Aleksandrovich Gorbunov memorial issue. *Algebra Universalis* **46** (2001), 215–230.

[21] K. V. Adaricheva, V. A. Gorbunov, and V. I. Tumanov, Join-semidistributive lattices and convex geometries. *Adv. Math.* **173** (2003), 1–49.

[22] K. V. Adaricheva and J. B. Nation, Largest extension of a finite convex geometry. *Algebra Universalis* **52** (2004), 185–195.

[23] _____, Reflections on lower bounded lattices. *Algebra Universalis* **53** (2005), 307–330.

[24] _____, Lattices of quasi-equational theories as congruence lattices of semilattices with operators, Parts I and II. *Int. J. Alg. Comp.* To appear.

[25] K. V. Adaricheva and F. Wehrung, Embedding finite lattices into finite biatomic lattices. *Order* **20** (2003), 31–48.

[26] K. V. Adaricheva and M. Wild, Realization of abstract convex geometries by point configurations, *European J. Combin.* **31** (2010), 379–400.

[27] M. Aigner, Graphs and partial orderings. *Monatsh. Math.* **73** (1969), 385–396.

[28] J. W. Alexander, Ordered sets, complexes and the problem of compactification. *Proc. Natl. Acad. Sci. USA* **25** (1939), 296–298.

[29] F. W. Anderson and R. L. Blair, Representations of distributive lattices as lattices of functions. *Math. Ann.* **143** (1961), 187–211.

[30] J. Anderson and N. Kimura, The tensor product of semilattices, *Semigroup Forum* **16** (1968), 83–88.

[31] G. Ya. Areškin, On congruence relations in distributive lattices with zero. (Russian.) *Dokl. Akad. Nauk* **90** (1953), 485–486.

[32] _____, Free distributive lattices and free bicompact $T_0$-spaces. (Russian.) *Mat. Sb.* **33** (**75**) (1953), 133–156.

[33] E. Artin, Coordinates in affine geometry. *Rep. Math. Colloq.* **2** (1940), 15–20.

[34] S. P. Avann, Upper and lower complementation in a modular lattice. *Proc. Amer. Math. Soc.* **11** (1960), 17–22.

[35] R. Baer, Linear Algebra and Projective Geometry. *Academic Press, New York, N. Y.,* 1952.

[36] K. A. Baker, A generalization of Sperner's lemma. *J. Combin. Theory* **6** (1969), 224–225.

[37] _____, Equational classes of modular lattices. *Pacific J. Math.* **28** (1969), 9–15.

[38] _____, Equational axioms for classes of lattices. *Bull. Amer. Math. Soc.* **77** (1971), 97–102.

[39] _____, Primitive satisfaction and equational problems for lattices and other algebras. *Trans. Amer. Math. Soc.* **190** (1974), 125–150.

[40] _____, Equational axioms for classes of Heyting algebras. *Algebra Universalis* **6** (1976), 105–120.

[41] _____, Finite equational bases for finite algebras in a congruence-distributive equational class. *Advances in Math.* **24** (1977), 207–243.

[42] K. A. Baker, P. C. Fishburn, and F. S. Roberts, Partial orders of dimension 2, interval orders, and interval graphs. *Rand Corp. P-4376,* 1970.

[43] K. A. Baker and A. W. Hales, From a lattice to its ideal lattice. *Algebra Universalis* **4** (1974), 250–258.

[44] R. Balbes, Projective and injective distributive lattices. *Pacific J. Math.* **21** (1967), 405–420.

[45] R. Balbes and P. Dwinger, Distributive Lattices. *Univ. Missouri Press, Columbia, Miss.,* 1974. xiii+294 pp.

[46] R. Balbes and A. Horn, Stone lattices. *Duke Math. J.* **37** (1970), 537–545.

[47] _____, Injective and projective Heyting algebras. *Trans. Amer. Math. Soc.* **148** (1970), 549–559.

[48] B. Banaschewski and C. J. Mulvey, Stone-Čech compactification of locales. *Houston J. Math.* **6** (1980), 301–312.

[49] H.-J. Bandelt and R. Padmanabhan, A note on lattices with unique comparable complements. *Abh. Math. Sem. Univ. Hamburg* **48** (1979), 112–113.

[50] V. A. Baranskiĭ, On the independence of the automorphism group and the congruence lattice for lattices. *Abstracts of lectures of the 15th All-Soviet Algebraic Conference (Krasnojarsk, July 1979),* vol. 1, 11.

[51] _____, Independence of lattices of congruences and groups of automorphisms of lattices. (Russian.) *Izv. Vyssh. Uchebn. Zaved. Mat.* (1984), 12–17, 76; translation in *Soviet Math. (Iz. VUZ)* **28** (1984), 12–19.

[52] E. A. Behrens, Distributiv darstellbare Ringe. I. *Math. Z.* **73** (1960), 409–432.

[53] _____, Distributiv darstellbare Ringe. II. *Math. Z.* **76** (1961), 367–384.

[54] M. K. Bennett, Lattices and geometry. *Lattice theory and its applications (Darmstadt, 1991),* 27–50, Res. Exp. Math., **23**, Heldermann, Lemgo, 1995.

[55] L. Beran, On the pasting and cutting of lattices. *Atti Sem. Mat. Fis. Univ. Modena* **51** (2003), 85–98.

[56] C. Bergman, Amalgamation classes of some distributive varieties. *Algebra Universalis* **20** (1985), 143–165.

[57] _____, Non-axiomatizability of the amalgamation class of modular lattice varieties. *Order* **6** (1989), 49–58.

[58] G. M. Bergman, An Invitation to General Algebra and Universal Constructions. *Henry Helson, Berkeley, CA,* 1998. ii+398 pp. Available online: http://math.berkeley.edu/~gbergman/245

[59] _____, On lattices of convex sets in $\mathbb{R}^n$. *Algebra Universalis* **53** (2005), 357–395.

[60] G. M. Bergman and G. Grätzer, Isotone maps on lattices. Manuscript.

[61] G. Birkhoff, On the combination of subalgebras. *Proc. Cambridge Philos. Soc.* **29** (1933), 441–464.

[62] _____, On the structure of abstract algebras. *Proc. Cambridge Philos. Soc.* **31** (1935), 433–454.

[63] _____, Abstract linear dependence and lattices. *Amer. J. Math.* **57** (1935), 800–804.

[64] _____, Combinatorial relations in projective geometries. *Ann. of Math.* **36** (1935), 743–748.

[65] G. Birkhoff, Lattice Theory. First edition. *Amer. Math. Soc., Providence, R.I.,* 1940. v+155 pp.

[66] _____, Neutral elements in general lattices. *Bull. Amer. Math. Soc.* **46** (1940), 702–705.

[67] _____, Subdirect unions in universal algebra. *Bull. Amer. Math. Soc.* **50** (1944), 764–768.

[68] _____, On groups of automorphisms. (Spanish), *Rev. Un. Mat. Argentina* **11** (1946), 155–157.

[69] G. Birkhoff, Universal algebra. *Proc. First Canadian Math. Congress, Montreal, 1945,* 310–326. *University of Toronto Press, Toronto, 1946.*

[70] G. Birkhoff, Lattice Theory. Revised ed. American Mathematical Society Colloquium Publications, vol. 25. *American Mathematical Society, Providence, R.I.,* 1961. xiii+283 pp.

[71] G. Birkhoff, Lattice Theory. Corrected reprint of the 1967 third edition. American Mathematical Society Colloquium Publications, Vol. XXV. *American Mathematical Society, Providence, R.I.,* 1979. vi+418 pp.

[72] _____, Ordered sets in geometry. *Ordered sets (Banff, Alta., 1981),* 407–443, NATO Adv. Study Inst. Ser. C: Math. Phys. Sci., **83**, *Reidel, Dordrecht-Boston, Mass.,* 1982.

[73] _____, Lattices and their applications. *Lattice theory and its applications (Darmstadt, 1991),* 7–25, Res. Exp. Math., 23, *Heldermann, Lemgo,* 1995.

[74] G. Birkhoff and M. K. Bennett, The convexity lattice of a poset. *Order* **2** (1985), 223–242.

[75] G. Birkhoff and O. Frink, Representations of lattices by sets. *Trans. Amer. Math. Soc.* **64** (1948), 299–316.

[76] P. Blackburn, M. de Rijke, and Y. Venema, Modal Logic. *Cambridge Univ. Press, Cambridge, 2001.*

[77] T. S. Blyth and J. Varlet, Sur la construction de certaines MS-algèbres. *Port. Math.* **39** (1980), 489–496.

[78] K. P. Bogart, C. Greene, and J. P. S. Kung, The impact of the chain decomposition theorem on classical combinatorics. *The Dilworth theorems*, 19–29, Contemp. Mathematicians, *Birkhäuser Boston, Boston, MA, 1990.*

[79] J. R. Büchi, Representation of complete lattices by sets. *Port. Math.* **11** (1952), 151–167.

[80] L. Byrne, Two brief formulations of boolean algebra. *Bull. Amer. Math. Soc.* **52** (1946), 269–272.

[81] N. Caspard, B. Leclerc, and B. Monjardet, Finite Partially Ordered Sets: Concepts, Results, and Applications. (French.) Mathématiques & Applications (Berlin) [Mathematics and Applications], **60**. *Springer, Berlin, 2007.* xiv+340 pp.

[82] L. Carroll, Alice's Adventures in Wonderland. *McMillan and Co., London, 1865.*

[83] A. Chagrov and M. Zakharyaschev, Modal logic. Oxford Logic Guides **35**, *Oxford Univ. Press, Oxford, 1997.*

[84] I. Chajda, Algebraic Theory of Tolerance Relations. *Monograph series of Palacký University Olomouc, 1991, 117 pp.*

[85] I. Chajda, G. Eigenthaler, and H. Länger, Congruence classes in universal algebra. Research and Exposition in Mathematics, 26. *Heldermann Verlag, Lemgo, 2003.* x+217 pp.

[86] I. Chajda, R. Halaš, and J. Kühr, Semilattice structures. Research and Exposition in Mathematics, 30. *Heldermann Verlag, Lemgo, 2007.* vi+228 pp.

[87] I. Chajda and G. Czédli, A note on representation of lattices by tolerances. *J. Algebra* **148** (1992), 274–275.

[88]  C. C. Chen and G. Grätzer, On the construction of complemented lattices. *J. Algebra* **11** (1969), 56–63.

[89]  _____, Stone lattices I. Construction theorems. *Canad. J. Math.* **21** (1969), 884–894.

[90]  _____, Stone lattices II. Structure theorems. *Canad. J. Math.* **21** (1969), 895–903.

[91]  D. M. Clark and B. A. Davey, Natural dualities for the working algebraist. Cambridge Studies in Advanced Mathematics, **57**. *Cambridge University Press, Cambridge,* 1998; xii+356 pp.

[92]  D. M. Clark, B. A. Davey and R. Willard, Not every full duality is strong! *Algebra Universalis* **57** (2007), 375–381.

[93]  I. S. Cohen, Commutative rings with restricted minimum conditions. *Duke Math. J.* **17** (1950), 27–42.

[94]  P. M. Cohn, Free Rings and Their Relations. London Mathematical Society Monographs, No. 2. *Academic Press, London-New York,* 1971. xvi+346 pp.

[95]  P. M. Cohn, Free ideal rings and localization in general rings. New Mathematical Monographs, 3. *Cambridge University Press, Cambridge,* 2006. xxii+572 pp.

[96]  W. H. Cornish, *n*-normal lattices. *Proc. Amer. Math. Soc.* **45** (1974), 48–54.

[97]  _____, Pseudo-complemented modular semilattices. *J. Austral. Math. Soc.* **18** (1974), 239–251.

[98]  _____, On H. Priestley's dual of the category of bounded distributive lattices. *Mat. Vesnik.* **12 (27)** (1975), 329–332.

[99]  _____, Antimorphic Action. Categories of Algebraic Structures with Involutions or Anti-endomorphisms. R. & E. Research and Exposition in Mathematics, **12**. *Heldermann Verlag,* 1986.

[100]  M. Cotlar, A method of construction of structures and its application to topological spaces and abstract arithmetic. *Univ. Nac. Tucumán Rev. Ser. A.* **4** (1944), 105–157.

[101]  H. H. Crapo, Selectors: a theory of formal languages, semimodular lattices, and branching and shelling processes. *Adv. Math.* **54** (1984), 233–277.

546     Bibliography

[102]  H. H. Crapo and G.-C. Rota, Geometric lattices. *Trends in Lattice Theory*
       (*Sympos., U.S. Naval Academy, Annapolis, Md., 1966*), 127–172. *Van*
       *Nostrand Reinhold, New York*, 1970.

[103]  P. Crawley, The isomorphism theorem in compactly generated lattices.
       *Bull. Amer. Math. Soc.* **65** (1959), 377–379.

[104]  ――――, Decomposition theory for nonsemimodular lattices. *Trans. Amer.*
       *Math. Soc.* **99** (1961), 246–254.

[105]  ――――, Regular embeddings which preserve lattice structure. *Proc.*
       *Amer. Math. Soc.* **13** (1962), 748–752.

[106]  P. Crawley and R. A. Dean, Free lattices with infinite operations. *Trans.*
       *Amer. Math. Soc.* **92** (1959), 35–47.

[107]  P. Crawley and R. P. Dilworth, Algebraic Theory of Lattices. *Prentice-*
       *Hall, Englewood Cliffs, N.J.*, 1973.

[108]  R. Croisot, Contribution à l'étude des trellis semi-modulaires de longueur
       infinie. *Ann. Sci. École Norm. Sup.* (3) **68** (1951), 203–265.

[109]  G. Czédli, Factor lattices by tolerances. *Acta Sci. Math.* (*Szeged*) **44**
       (1982), 35–42.

[110]  ――――, Mal'cev conditions for Horn sentences with congruence per-
       mutability. *Acta Math. Hungar.* **44** (1984), 115–124.

[111]  ――――, Lattice generation of small equivalences of a countable set. *Order*
       **13** (1996), 11–16.

[112]  ――――, (1+1+2)-generated equivalence lattices. *J. Algebra* **221** (1999),
       439–462.

[113]  G. Czédli and G. Grätzer, Lattice tolerances and congruences. *Algebra*
       *Universalis.* To appear.

[114]  G. Czédli and G. Hutchinson, Submodule lattice quasivarieties and exact
       embedding functors for rings with prime power characteristic. *Algebra*
       *Universalis* **35** (1996), 425–445.

[115]  G. Czédli and Gy. Pollák, When do coalitions form a lattice? *Acta Sci.*
       *Math.* (*Szeged*) **60** (1995), 197–206.

[116]  G. Czédli and E. T. Schmidt, How to derive finite semimodular lattices
       from distributive lattices? *Acta Math. Hungar.* **121** (2008) 277–282.

[117]  ――――, The Jordan-Hölder theorem with uniqueness for groups and
       semimodular lattices. Manuscript.

[118] ———, Some results on semimodular lattices. *Contributions to General Algebra 19 (Proc. Olomouc Conf. 2010), Johannes Hein Verlag, Klagenfurt.* To appear.

[119] G. Czédli and G. Takách, On duality of submodule lattices. *Discuss. Math. Gen. Algebra Appl.* **20** (2000), 43–49.

[120] G. Czédli and A. Walendziak, Subdirect representation and semimodularity of weak congruence lattices. *Algebra Universalis* **44** (2000), 371–373.

[121] A. Daigneault, Products of polyadic algebras and of their representations. *Ph.D. Thesis, Princeton University, 1959.*

[122] B. A. Davey, Free products of bounded distributive lattices. *Algebra Universalis* **4** (1974), 106–107.

[123] ———, Duality theory on ten dollars a day. *Algebras and Orders (Montreal, PQ, 1991)*, 71–111, NATO Adv. Sci. Inst. Ser. C Math. Phys. Sci., **389**, *Kluwer Acad. Publ., Dordrecht, 1993.*

[124] B. A. Davey and M. Haviar, A schizophrenic operation which aids the efficient transfer of strong dualities. *Houston J. Math.* **26** (2000), 215–222.

[125] B. A. Davey, M. Haviar, and T. Niven, When is a full duality strong? *Houston J. Math.* **33** (2007), 1–22.

[126] B. A. Davey, M. Haviar, and J. G. Pitkethly, Using coloured ordered sets to study finite-level full dualities. *Algebra Universalis.* To appear.

[127] B. A. Davey, M. Haviar, and H. A. Priestley, Endoprimal distributive lattices are endodualisable. *Algebra Universalis* **34** (1995), 444–453.

[128] B. A. Davey, M. Haviar, and R. Willard, Full does not imply strong, does it? *Algebra Universalis* **54** (2005), 1–22.

[129] B. A. Davey, J. G. Pitkethly, and R. Willard, The lattice of alter egos. Manuscript.

[130] B. A. Davey, W. Poguntke, and I. Rival, A characterization of semi-distributivity. *Algebra Universalis* **5** (1975), 72–75.

[131] B. A. Davey and H. A. Priestley, Introduction to lattices and order. Second edition. *Cambridge University Press, New York, 2002.* xii+298 pp.

[132] B. A. Davey and B. Sands, An application of Whitman's condition to lattices with no infinite chains. *Algebra Universalis* **7** (1977), 171–178.

[133] B. A. Davey and H. Werner, Dualities and equivalences for varieties of algebras. *Contributions to Lattice Theory (Szeged, 1980)*, 101–275, Colloq. Math. Soc. János Bolyai, **33**, *North-Holland, Amsterdam, 1983.*

[134] A. C. Davis, A characterization of complete lattices. *Pacific J. Math.* **5** (1955), 311–319.

[135] A. Day, A simple solution of the word problem for lattices. *Canad. Math. Bull.* **13** (1970), 253–254.

[136] _____, Splitting lattices generate all lattices. *Algebra Universalis* **7** (1976), 163–170.

[137] _____, Idempotents in the groupoid of all **SP** classes of lattices. *Canad. Math. Bull.* **21** (1978), 499–501.

[138] _____, Characterizations of finite lattices that are bounded-homomorphic images of sublattices of free lattices. *Canad. J. Math.* **31** (1979), 69–78.

[139] A. Day and R. Freese, The role of gluing constructions in modular lattice theory. *The Dilworth theorems*, 251–260, Contemp. Mathematicians, *Birkhäuser Boston, Boston, MA, 1990.*

[140] A. Day and C. Herrmann, Gluings of modular lattices. *Order* **5** (1988), 85–101.

[141] A. Day and J. Ježek, The amalgamation property for varieties of lattices. *Trans. Amer. Math. Soc.* **286** (1984), 251–256.

[142] G. W. Day, Maximal chains in atomic boolean algebras. *Fund. Math.* **67** (1970), 293–296.

[143] R. A. Dean, Component subsets of the free lattice on $n$ generators. *Proc. Amer. Math. Soc.* **7** (1956), 220–226.

[144] _____, Completely free lattices generated by partially ordered sets. *Trans. Amer. Math. Soc.* **83** (1956), 238–249.

[145] _____, Sublattices of free lattices. 1961 *Proc. Sympos. Pure Math., Vol. II*, 31–42. *American Mathematical Society, Providence, R.I.*

[146] _____, Coverings in free lattices. *Bull. Amer. Math. Soc.* **67** (1961), 548–549.

[147] _____, Free lattices generated by partially ordered sets and preserving bounds. *Canad. J. Math.* **16** (1964), 136–148.

[148] R. A. Dean and R. H. Oehmke, Idempotent semigroups with distributive right congruence lattices. *Pacific J. Math.* **14** (1964), 1187–1209.

[149] R. Dedekind, Über die von drei Moduln erzeugte Dualgruppe. *Math. Ann.* **53** (1900), 371–403.

[150] R. Dedekind, Gesammelte mathematische Werke. Bände I–III. Herausgegeben von Robert Fricke, Emmy Noether und Öystein Ore. *Chelsea Publishing Co., New York,* 1968. Vol. I: iii+397 pp.; Vol. II: iv+442 pp.; Vol. III: iii+223–508 pp.

[151] A. H. Diamond and J. C. C. McKinsey, Algebras and their subalgebras. *Bull. Amer. Math. Soc.* **53** (1947), 959–962.

[152] B. Dietrich, Matroids and antimatroids—a survey. *Discrete Math.* **78** (1989), 223–23.

[153] R. P. Dilworth, Lattices with unique irreducible decompositions. *Ann. of Math.* (2) **41** (1940), 771–777.

[154] _____, The arithmetical theory of Birkhoff lattices. *Duke Math. J.* **8** (1941), 286–299.

[155] _____, Lattices with unique complements. *Trans. Amer. Math. Soc.* **57** (1945), 123–154.

[156] _____, Note on the Kurosch-Ore theorem. *Bull. Amer. Math. Soc.* **52** (1946), 659–663.

[157] _____, A decomposition theorem for partially ordered sets. *Ann. of Math.* (2) **51** (1950), 161–166.

[158] _____, The structure of relatively complemented lattices. *Ann. of Math.* (2) **51** (1950), 348–359.

[159] _____, Proof of a conjecture on finite modular lattices. *Ann. of Math.* (2) **60** (1954), 359–364.

[160] _____, Structure and decomposition theory of lattices. 1961 *Proc. Sympos. Pure Math., Vol. II,* 3–16. *American Mathematical Society, Providence, R.I.*

[161] R. P. Dilworth and P. Crawley, Decomposition theory for lattices without chain conditions. *Trans. Amer. Math. Soc.* **96** (1960), 1–22.

[162] R. P. Dilworth and C. Greene, A counterexample to the generalization of Sperner's theorem. *J. Combin. Theory* **10** (1971), 18–21.

[163] R. P. Dilworth and M. Hall, The embedding problem for modular lattices. *Ann. of Math.* (2) **45** (1944), 450–456.

[164] S. Z. Ditor, Cardinality questions concerning semilattices of finite breadth. *Discrete Math.* **48** (1984), 47–59.

[165] H. Dobbertin, Vaught's measures and their applications in lattice theory. *J. Pure Appl. Algebra* **43** (1986), 27–51.

[166] D. Dorninger and G. Eigenthaler, On compatible and order-preserving functions on lattices. *Universal algebra and applications* (*Warsaw, 1978*), 97–104, Banach Center Publ., 9, *PWN, Warsaw*, 1982.

[167] M. L. Dubreil-Jacotin, L. Lesieur, and R. Croisot, Leçons sur la théorie des treillis des structures algébriques ordonnées et des treillis géométriques. *Gauthier-Villars, Paris*, 1953. viii+385 pp.

[168] D. Duffus and I. Rival, Path length in the covering graph of a lattice. *Discrete Math.* **19** (1977), 139–158.

[169] V. Duquenne, On the core of finite lattices. *Discrete Math.* **88** (1991), 133–147.

[170] W. Dziobiak, On atoms in the lattice of quasivarieties. *Algebra Universalis* **24** (1987), 32–35.

[171] P. H. Edelman and R. Jamison, The theory of convex geometries. *Geom. Dedicata* **19** (1985), 247–274.

[172] P. Erdős, A. Hajnal, A. Máté, and R. Rado, Combinatorial Set Theory: Partition Relations for Cardinals. Studies in Logic and the Foundations of Mathematics, **106**. *North-Holland Publishing Co., Amsterdam*, 1984. 347 pp.

[173] T. Evans, Finitely presented loops, lattices, etc. are Hopfian. *J. London Math. Soc.* **44** (1969), 551–552.

[174] T. Evans and D. X. Hong, The free modular lattice on four generators is not finitely presentable. *Algebra Universalis* **2** (1972), 284–285.

[175] U. Faigle, Geometries on partially ordered sets. *J. Combin. Theory* Ser. B **28** (1980), 26–51.

[176] U. Faigle, G. Richter, and M. Stern, Geometric exchange properties in lattices of finite length, *Algebra Universalis* **19** (1984), 355–365.

[177] J. D. Farley, Functions on distributive lattices with the congruence substitution property: some problems of Grätzer from 1964. *Adv. Math.* **149** (2000), 193–213.

[178]  T. S. Fofanova, General lattice theory. (Russian.) *Ordered sets and lattices,* 79–152, *Univ. Komenského, Bratislava,* 1985. (AMS Translation, ser. 2, vol. 141 (1989).)

[179]  _____, General lattice theory. (Russian.) *Ordered sets and lattices, II,* 149–214, *Univ. Komenského, Bratislava,* 1988. (AMS Translation, ser. 2, vol. 152 (1992).)

[180]  G. A. Fraser, The semilattice tensor product of distributive semilattices, *Trans. Amer. Math. Soc..* **217** (1976), 183–194.

[181]  R. Freese, Varieties generated by modular lattices of width four. *Bull. Amer. Math. Soc.* **78** (1972), 447–450.

[182]  _____, Ideal lattices of lattices. *Pacific J. Math.* **57** (1975), 125–133.

[183]  _____, Free modular lattices. *Trans. Amer. Math. Soc.* **261** (1980), 81–91.

[184]  _____, Finitely presented lattices: canonical forms and the covering relation *Trans. Amer. Math. Soc.* **312** (1989), 841–860.

[185]  R. Freese, G. Grätzer, and E. T. Schmidt, On complete congruence lattices of complete modular lattices. *Internat. J. Algebra Comput.* **1** (1991), 147–160.

[186]  R. Freese, J. Ježek, and J. B. Nation, Free lattices. Mathematical Surveys and Monographs, **42**. *American Mathematical Society, Providence, RI,* 1995. viii+293 pp.

[187]  R. Freese, K. Kearnes, and J. B. Nation, Congruence lattices of congruence semidistributive algebras. *Lattice theory and its applications* (*Darmstadt, 1991*), 63–78, Res. Exp. Math., **23**, *Heldermann, Lemgo,* 1995.

[188]  R. Freese and J. B. Nation, Congruence lattices of semilattices. *Pacific J. Math.* **49** (1973), 51–58.

[189]  _____, Projective lattices. *Pacific J. Math.* **75** (1978), 93–106.

[190]  _____, Finitely presented lattices. *Proc. Amer. Math. Soc.* **77** (1979), 174–178.

[191]  _____, Covers in free lattices. *Trans. Amer. Math. Soc.* **288** (1985), 1–42.

[192]  P. A. Freĭdman, Rings with a distributive lattice of subrings. (Russian.) *Mat. Sb.* **73** (4) (1967), 513–534.

[193] E. Fried, Tournaments and nonassociative lattices. *Ann. Univ. Sci. Budapest. Eötvös Sect. Math.* **13** (1970), 151–164.

[194] E. Fried and G. Grätzer, A nonassociative extension of the class of distributive lattices. *Pacific J. Math.* **49** (1973), 59–78.

[195] _____, Pasting and modular lattices. *Proc. Amer. Math. Soc.* **106** (1989), 885–890.

[196] _____, Pasting infinite lattices. *J. Austral. Math. Soc.* Ser. A **47** (1989), 1–21.

[197] _____, Generalized congruences and products of lattice varieties. *Acta Sci. Math. (Szeged)* **54** (1990), 21–36.

[198] _____, Notes on tolerance relations of lattices: On a conjecture of R. N. McKenzie. *J. Pure Appl. Algebra* **68** (1990), 127–134.

[199] _____, The unique amalgamation property for lattices. *Ann. Univ. Sci. Budapest. Eötvös Sect. Math.* **33** (1990), 167–176.

[200] E. Fried, G. Grätzer, and H. Lakser, Amalgamation and weak injectives in the equational class of modular lattices $M_n$. Abstract. *Notices Amer. Math. Soc.* **18** (1971), 624.

[201] E. Fried, G. Grätzer, and H. Lakser, Projective geometries as cover preserving sublattices. *Algebra Universalis* **27** (1990), 270–278.

[202] E. Fried, G. Grätzer, and E. T. Schmidt, Multipasting of lattices. *Algebra Universalis* **30** (1993), 241–261.

[203] E. Fried and E. T. Schmidt, Standard sublattices. *Algebra Universalis* **5** (1975), 203–211.

[204] H. Friedman and D. Tamari, Problèmes d'associativité: une structure de treillis finis induite par une loi demi-associative. *J. Combin. Theory* **2** (1967), 215–242.

[205] O. Frink, Complemented modular lattices and projective spaces of infinite dimension. *Trans. Amer. Math. Soc.* **60** (1946), 452–467.

[206] _____, Pseudo-complements in semi-lattices. *Duke Math. J.* **29** (1962), 505–514.

[207] R. Frucht, On the construction of partially ordered systems with a given group of automorphisms. *Rev. Un. Mat. Argentina* **13** (1948), 12–18.

[208] _____, Lattices with a given abstract group of automorphisms. *Canad. J. Math.* **2** (1950), 417–419.

[209] K. D. Fryer and I. Halperin, The von Neumann coordinatization theorem for complemented modular lattices. *Acta Sci. Math. (Szeged)* **17** (1956), 203–249.

[210] L. Fuchs, Über die Ideale arithmetischer Ringe. *Comment. Math. Helv.* **23** (1949), 334–341.

[211] N. Funayama, On the completion by cuts of distributive lattices. *Proc. Imp. Acad. Tokyo* **20** (1944), 1–2.

[212] ———, Notes on lattice theory. IV. On partial (semi) lattices. *Bull. Yamagata Univ. Natur. Sci.* **2** (1953), 171–184.

[213] ———, Imbedding infinitely distributive lattices completely isomorphically into boolean algebras. *Nagoya Math. J.* **15** (1959), 71–81.

[214] N. Funayama and T. Nakayama, On the distributivity of a lattice of lattice-congruences. *Proc. Imp. Acad. Tokyo* **18** (1942), 553–554.

[215] H. Gaifman, Infinite boolean polynomials. I. *Fund. Math.* **54** (1964), 229–250.

[216] F. Galvin, A proof of Dilworth's Chain Decomposition Theorem. *Amer. Math. Monthly* **101** (1994), 352–353.

[217] F. Galvin and B. Jónsson, Distributive sublattices of a free lattice. *Canad. J. Math.* **13** (1961), 265–272.

[218] B. Ganter and I. Rival, Dilworth's covering theorem for modular lattices: a simple proof. *Algebra Universalis* **3** (1973), 348–350.

[219] B. Ganter and R. Wille, Formal concept analysis. Mathematical foundations. Translated from the 1996 German original by C. Franzke. *Springer-Verlag, Berlin*, 1999. x+284 pp.

[220] H. S. Gaskill, On transferable semilattices. *Algebra Universalis* **2** (1972), 303–316.

[221] H. S. Gaskill, G. Grätzer, and C. R. Platt, Sharply transferable lattices. *Canad. J. Math.* **27** (1975), 1247–1262.

[222] H. S. Gaskill and C. R. Platt, Sharp transferability and finite sublattices of a free lattice. *Canad. J. Math.* **27** (1975), 1036–1041.

[223] M. Gehrke and H. A. Priestley, Canonical extensions and completions of posets and lattices. *Rep. Math. Logic* **43** (2008), 133–152.

[224] A. Ghouilà-Houri, Caractérisation des graphes non orientés dont on peut orienter les arêtes de manière à obtenir le graphe d'une relation d'ordre. *C. R. Acad. Sci. Paris. Sér. A–B* **254** (1962), 1370–1371.

[225] G. Gierz, K. H. Hofmann, K. Keimel, J. D. Lawson, M. Mislove, and D. S. Scott, Continuous lattices and domains. Encyclopedia of Mathematics and its Applications, **93**. *Cambridge University Press, Cambridge, 2003.* xxxvi+591 pp.

[226] E. N. Gilbert, Lattice theoretic properties of frontal switching functions. *J. Mathematical Phys.* **33** (1954), 57–67.

[227] P. Gillibert, Points critiques de couples de variétés d'algèbres. Doctorat de l'Université de Caen, December 8, 2008. Available online: http://tel.archives-ouvertes.fr/tel-00345793

[228] P. Gillibert, Critical points of pairs of varieties of algebras. *Internat. J. Algebra Comput.* **19** (2009), 1–40.

[229] _____ , Critical points between varieties generated by subspace lattices of vector spaces. *J. Pure Appl. Algebra* **214** (2010), 1306–1318.

[230] P. Gillibert and F. Wehrung, From objects to diagrams for ranges of functors. Manuscript, 2010.

[231] P. C. Gilmore and A. J. Hoffman, A characterization of comparability graphs and of interval graphs. *Canad. J. Math.* **16** (1964), 539–548.

[232] L. Giudici, Dintorni del Teorema di Coordinatizzazione di von Neumann. *Tesi di Dottorato in Matematica, Università di Milano*, 1995. Available online at http://www.nohay.net/mat/tesi.1995/tesi.pdf

[233] V. Glivenko, Sur quelques points de la logique de M. Brouwer. *Bull. Acad. des Sci. de Belgique* **15** (1929), 183–188.

[234] M. M. Gluhov, On the problem of isomorphism of lattices. (Russian.) *Dokl. Akad. Nauk SSSR* **132** (1960), 254–256.

[235] R. Goldblatt, Varieties of complex algebras, *Annals Pure App. Logic* **44** (1989), 174–242.

[236] M. Goldstern and S. Shelah, Order polynomially complete lattices must be large. *Algebra Universalis* **39** (1998), 197–209.

[237] _____ , There are no infinite order polynomially complete lattices, after all. *Algebra Universalis* **42** (1999), 49–57.

[238] K. R. Goodearl, von Neumann Regular Rings. Second edition. *Robert E. Krieger Publishing Co., Inc., Malabar, FL*, 1991. xviii+412 pp.

[239] K. R. Goodearl and F. Wehrung, The complete dimension theory of partially ordered systems with equivalence and orthogonality. *Mem. Amer. Math. Soc.* **176** (2005), no. 831, vii+117 pp.

[240] V. A. Gorbunov, Lattices of quasivarieties. (Russian.) *Algebra i Logika* **15** (1976), 436–457, 487–488.

[241] _____, The structure of the lattices of quasivarieties. *Algebra Universalis* **32** (1994), 493–530.

[242] _____, Structure of lattices of varieties and lattices of quasivarieties: similarity and difference. II. (Russian.) *Algebra i Logika* **34** (1995), 369–397; translation in *Algebra and Logic* **34** (1995), 203–218.

[243] V. A. Gorbunov, Algebraic theory of quasivarieties. Translated from the Russian. Siberian School of Algebra and Logic. *Consultants Bureau, New York*, 1998. xii+298 pp.

[244] V. A. Gorbunov and V. I. Tumanov, A class of lattices of quasivarieties. *Algebra i Logika* **19** (1980), 59–80, 132–133.

[245] _____, Construction of lattices of quasivarieties. (Russian.) *Mathematical logic and the theory of algorithms*, 12–44, Trudy Inst. Mat., **2** *"Nauka"* *Sibirsk. Otdel., Novosibirsk*, 1982.

[246] G. Grätzer, Standard ideals. (Hungarian.) *Magyar Tud. Akad. Mat. Fiz. Oszt. Közl.* **9** (1959), 81–97.

[247] _____, A characterization of neutral elements in lattices. (Notes on Lattice Theory. I). *Magyar Tud. Akad. Mat. Fiz. Oszt. Közl.* **7** (1962), 191–192.

[248] _____, On boolean functions. (Notes on Lattice Theory. II). *Rev. Roumaine Math. Pures Appl.* **7** (1962), 693–697.

[249] _____, A generalization of Stone's representation theorem for boolean algebras. *Duke J. Math.* **30** (1963), 469–474.

[250] _____, On semi-discrete lattices whose congruence relations form a boolean algebra. *Acta Math. Acad. Sci. Hungar.* **14** (1963), 441–445.

[251] _____, Boolean functions on distributive lattices. *Acta Math. Acad. Sci. Hungar.* **15** (1964), 195–201.

[252] _____, On the family of certain subalgebras of a universal algebra. *Indag. Math.* **27** (1965), 790–802.

[253] _____, Equational classes of lattices. *Duke J. Math.* **33** (1966), 613–622.

[254] G. Grätzer, Universal Algebra. *D. Van Nostrand Co., Inc., Princeton, N.J.-Toronto, Ont.-London*, 1968. xvi+368 pp.

[255] G. Grätzer, Stone algebras form an equational class. (Notes on Lattice Theory. III.) *J. Austral. Math. Soc.* **9** (1969), 308–309.

[256] _____, Universal algebra. 1970 *Trends in Lattice Theory (Sympos., U.S. Naval Academy, Annapolis, Md., 1966)*, 173–210. *Van Nostrand Reinhold, New York.*

[257] G. Grätzer, Lattice Theory. First Concepts and Distributive Lattices. *W. H. Freeman and Co., San Francisco, Calif.*, 1971. xv+212 pp.

[258] _____, A reduced free product of lattices. *Fund. Math.* **73** (1971), 21–27.

[259] _____, Free products and reduced free products of lattices. *Proceedings of the University of Houston Lattice Theory Conference (Houston, Tex., 1973)*, 539–563. *Dept. Math., Univ. Houston, Houston, Tex.*, 1973.

[260] _____, A property of transferable lattices. *Proc. Amer. Math. Soc.* **43** (1974), 269–271.

[261] _____, A note on the Amalgamation Property. Abstract. *Notices Amer. Math. Soc.* **22** (1975), A–453.

[262] G. Grätzer, General Lattice Theory. Pure and Applied Mathematics **75**. *Academic Press, Inc. (Harcourt Brace Jovanovich, Publishers), New York-London;* Lehrbücher und Monographien aus dem Gebiete der Exakten Wissenschaften, Mathematische Reihe, Band 52. *Birkhäuser Verlag, Basel-Stuttgart*, 1978. xiii+381 pp.

[263] G. Grätzer, Universal Algebra. Revised reprint of the 1979 second edition. *Springer-Verlag, New York*, 2008. xx+586 pp. Softcover edition with a new epilogue, 2008.

[264] _____, Birkhoff's Representation Theorem is equivalent to the Axiom of Choice. *Algebra Universalis* **23** (1986), 58–60.

[265] _____, The Amalgamation Property in lattice theory. *C. R. Math. Rep. Acad. Sci. Canada* **9** (1987), 273–289.

[266] _____, The complete congruence lattice of a complete lattice. *Lattices, semigroups, and universal algebra (Lisbon, 1988)*, 81–87, *Plenum, New York*, 1990.

[267] _____, On the complete congruence lattice of a complete lattice with an application to universal algebra. *C. R. Math. Rep. Acad. Sci. Canada* **11** (1989), 105–108.

[268] _____, A "lattice theoretic" proof of the independence of the automorphism group, the congruence lattice, and subalgebra lattice of an infinitary algebra. *Algebra Universalis* **27** (1990), 466–473.

[269] G. Grätzer, General Lattice Theory. Second edition. New appendices by the author with B. A. Davey, R. Freese, B. Ganter, M. Greferath, P. Jipsen, H. A. Priestley, H. Rose, E. T. Schmidt, S. E. Schmidt, F. Wehrung, and R. Wille. *Birkhäuser Verlag, Basel*, 1998. xx+663 pp.
Reprint of the 1998 second edition. *Birkhäuser Verlag, Basel–Boston–Berlin*, 2003.

[270] _____, Two problems that shaped a century of lattice theory. *Notices Amer. Math. Soc.* **54** (2007), 696–707.

[271] G. Grätzer, The Congruences of a Finite Lattice, A Proof-by-Picture Approach. *Birkhäuser Boston, Inc., Boston, MA*, 2006. xxiii+281 pp.

[272] G. Grätzer and M. Greenberg, Lattice tensor products. I. Coordinatization. *Acta Math. Hungar.* **95 (4)** (2002), 265–283.

[273] _____, Lattice tensor products. II. Ideal lattices. *Acta Math. Hungar.* **97** (2002), 193–198.

[274] _____, Lattice tensor products. III. Congruences. *Acta Math. Hungar.* **98** (2003), 167–173.

[275] _____, Lattice tensor products. IV. Infinite lattices. *Acta Math. Hungar.* **103** (2004), 71–84.

[276] G. Grätzer, M. Greenberg, and E. T. Schmidt, Representing congruence lattices of lattices with partial unary operations as congruence lattices of lattices. II. Interval ordering. *J. Algebra.* **286** (2005), 307–324.

[277] G. Grätzer, D. S. Gunderson, and R. W. Quackenbush, A note on finite pseudocomplemented lattices. *Algebra Universalis* **61** (2009), 407–411.

[278] G. Grätzer and A. P. Huhn, A note on finitely presented lattices. *C. R. Math. Rep. Acad. Sci. Canada* **2** (1980), 291–296.

[279] _____, Amalgamated free product of lattices. I. The common refinement property. *Acta Sci. Math. (Szeged)* **44** (1982), 53–66.

[280] _____, Amalgamated free product of lattices. II. Generating sets. *Studia Sci. Math. Hungar.* **16** (1981), 141–148.

[281] _____, Amalgamated free product of lattices. III. Free generating sets. *Acta Sci. Math. (Szeged)* **47** (1984), 265–275.

[282] G. Grätzer, A. P. Huhn, and H. Lakser, On the structure of finitely presented lattices. *Canad. J. Math.* **33** (1981), 404–411.

[283] G. Grätzer, B. Jónsson, and H. Lakser, The Amalgamation Property in equational classes of modular lattices. *Pacific J. Math.* **45** (1973), 507–524.

[284] G. Grätzer and D. Kelly, On a special type of subdirectly irreducible lattice with an application to products of varieties. *C. R. Math. Rep. Acad. Sci. Canada* **2** (1980/81), 43–48.

[285] _____ , Products of lattice varieties. *Algebra Universalis* **21** (1985), 33–45.

[286] _____ , The lattice variety **D** ∘ **D**. *Acta Sci. Math. (Szeged)* **51** (1987), 73–80. Addendum. **52** (1988), 465.

[287] _____ , Subdirectly irreducible members of products of lattice varieties. *Proc. Amer. Math. Soc.* **102** (1988), 483–489.

[288] _____ , A new lattice construction. *Algebra Universalis* **53** (2005), 253–265.

[289] _____ , Which freely generated lattices contain $F(3)$? *Algebra Universalis* **59** (2008), 117–132.

[290] G. Grätzer, G. Klus, and A. Nguyen, On the algorithmic construction of the 1960 sectional complement. *Acta Sci. Math. (Szeged)* . To appear.

[291] G. Grätzer and E. Knapp, Notes on planar semimodular lattices. I. Construction. *Acta Sci. Math. (Szeged)* **73** (2007), 445–462.

[292] _____ , A note on planar semimodular lattices. *Algebra Universalis* **58** (2008), 497–499.

[293] _____ , Notes on planar semimodular lattices. II. Congruences. *Acta Sci. Math. (Szeged)* **74** (2008), 37–47.

[294] _____ , Notes on planar semimodular lattices. III. Congruences of rectangular lattices. *Acta Sci. Math. (Szeged)* **75** (2009), 29–48.

[295] _____ , Notes on planar semimodular lattices. IV. The size of a minimal congruence lattice representation with rectangular lattices. *Acta Sci. Math. (Szeged)* **75** (2009), 29–48.

[296] G. Grätzer and H. Lakser, Extension theorems on congruences of partial lattices. Abstract. *Notices Amer. Math. Soc.* **15** (1968), 732, 785.

[297] _____, Chain conditions in the distributive free product of lattices. *Trans. Amer. Math. Soc.* **144** (1969), 301–312.

[298] _____, The structure of pseudocomplemented distributive lattices. II. Congruence extension and amalgamation. *Trans. Amer. Math. Soc.* **156** (1971), 343–358.

[299] _____, Identities for equational classes generated by tournaments. Abstract. *Notices Amer. Math. Soc.* **18** (1971), 794.

[300] _____, The structure of pseudocomplemented distributive lattices. III. Injective and absolute subretracts. *Trans. Amer. Math. Soc.* **169** (1972), 475–487.

[301] _____, A note on the implicational class generated by a class of structures. *Canad. Math. Bull.* **16** (1973), 603–605.

[302] _____, Three remarks on the Arguesian identity. Abstract. *Notices Amer. Math. Soc.* **20** (1973), A-253, A-313.

[303] _____, Free lattice like sublattices of free products of lattices. *Proc. Amer. Math. Soc.* **44** (1974), 43–45.

[304] _____, Homomorphisms of distributive lattices as restrictions of congruences. *Canad. J. Math.* **38** (1986), 1122–1134.

[305] _____, On complete congruence lattices of complete lattices. *Trans. Amer. Math. Soc.* **327** (1991), 385–405.

[306] _____, On congruence lattices of m-complete lattices. *J. Austral. Math. Soc.* Ser. A **52** (1992), 57–87.

[307] _____, Notes on sectionally complemented lattices. I. Characterizing the 1960 sectional complement. *Acta Math. Hungar.* **108** (2005), 115–125.

[308] _____, Freely adjoining a relative complement to a lattice. *Algebra Universalis* **53** (2005), 189–210.

[309] _____, Freely adjoining a complement to a lattice. *Math. Slovaca* **56** (2006), 93–104.

[310] _____, Representing homomorphisms of congruence lattices as restrictions of congruences of isoform lattices. *Acta Sci. Math. (Szeged)* **75** (2009), 393–421.

[311] G. Grätzer, H. Lakser, and C. R. Platt, Free products of lattices. *Fund. Math.* **69** (1970), 233–240.

[312] G. Grätzer, H. Lakser, and R. W. Quackenbush, The structure of tensor products of semilattices with zero, *Trans. Amer. Math. Soc.* **267** (1981), 503–515.

[313] _____,Congruence-preserving extensions of congruence-finite lattices to isoform lattices. *Acta Sci. Math. (Szeged)* **75** (2009), 13–28.

[314] G. Grätzer, H. Lakser, and M. Roddy, Notes on sectionally complemented lattices. III. The general problem. *Acta Math. Hungar.* **108** (2005), 325–334.

[315] G. Grätzer, H. Lakser, and E. T. Schmidt, On a result of Birkhoff. *Period. Math. Hungar.* **30** (1995), 183–188.

[316] _____, Congruence representations of join-homomorphisms of distributive lattices: A short proof. *Math. Slovaca* **46** (1996), 363–369.

[317] _____, Isotone maps as maps of congruences. I. Abstract maps. *Acta Math. Hungar.* **75** (1997), 105–135.

[318] _____, Congruence lattices of finite semimodular lattices. *Canad. Math. Bull.* **41** (1998), 290–297.

[319] _____, Isotone maps as maps of congruences. II. Concrete maps. *Acta Math. Hungar.* **92** (2001), 253–258.

[320] G. Grätzer, H. Lakser, and F. Wehrung, Congruence amalgamation of lattices. *Acta Sci. Math. (Szeged)* **66** (2000), 3–22.

[321] G. Grätzer, H. Lakser, and B. Wolk, On the lattice of complete congruences of a complete lattice: On a result of K. Reuter and R. Wille. *Acta Sci. Math. (Szeged)* **55** (1991), 3–8.

[322] G. Grätzer and W. A. Lampe, On subalgebra lattices of universal algebras. *J. Algebra* **7** (1967), 263–270.

[323] G. Grätzer and R. N. McKenzie, Equational spectra and reduction of identities. Abstract. *Notices Amer. Math. Soc.* **14** (1967), 697.

[324] G. Grätzer and J. B. Nation, On the Jordan-Dedekind chain condition. *Algebra Universalis.* To appear.

[325] G. Grätzer and C. R. Platt, A characterization of sharply transferable lattices. *Canad. J. Math.* **32** (1980), 145–154.

[326] G. Grätzer, C. R. Platt, and B. Sands, Embedding lattices into lattices of ideals. *Pacific J. Math.* **85** (1979), 65–75.

[327] G. Grätzer and R. W. Quackenbush, On the variety generated by planar modular lattices. *Algebra Universalis* **63** (2010), 187–201.

[328] G. Grätzer, R. W. Quackenbush, and E. T. Schmidt, Congruence-preserving extensions of finite lattices to isoform lattices. *Acta Sci. Math. (Szeged)* **70** (2004), 473–494.

[329] G. Grätzer and M. Roddy, Notes on sectionally complemented lattices. IV. How far does the Atom Lemma go? *Acta Math. Hungar.* **117** (2007), 41–60.

[330] G. Grätzer and E. T. Schmidt, On the Jordan-Dedekind chain condition. *Acta Sci. Math. (Szeged)* **18** (1957), 52–56.

[331] _____, On a problem of M. H. Stone. *Acta Math. Acad. Sci. Hungar.* **8** (1957), 455–460.

[332] _____, Two notes on lattice-congruences. *Ann. Univ. Sci. Budapest. Eötvös Sect. Math.* **1** (1958), 83–87.

[333] _____, On ideal theory for lattices. *Acta Sci. Math. (Szeged)* **19** (1958), 82–92.

[334] _____, Ideals and congruence relations in lattices. *Acta. Math. Acad. Sci. Hungar.* **9** (1958), 137–175.

[335] _____, On the generalized boolean algebra generated by a distributive lattice. *Indag. Math.* **20** (1958), 547–553.

[336] _____, Standard ideals in lattices. *Acta Math. Acad. Sci. Hungar.* **12** (1961), 17–86.

[337] _____, On congruence lattices of lattices. *Acta Math. Acad. Sci. Hungar.* **13** (1962), 179–185.

[338] _____, Characterizations of congruence lattices of abstract algebras. *Acta Sci. Math. (Szeged)* **24** (1963), 34–59.

[339] _____, Algebraic lattices as congruence lattices: The $\mathfrak{m}$-complete case. *Lattice theory and its applications (Darmstadt, 1991)*, 91–101, Res. Exp. Math., **23**, Heldermann, Lemgo, 1995. viii+262 pp.

[340] _____, "Complete-simple" distributive lattices. *Proc. Amer. Math. Soc.* **119** (1993), 63–69.

[341] _____, Complete congruence lattices of complete distributive lattices. *J. Algebra* **171** (1995), 204–229.

[342] G. Grätzer and E. T. Schmidt, Do we need complete-simple distributive lattices? *Algebra Universalis* **33** (1995), 140–141.

[343] _____, The Strong Independence Theorem for automorphism groups and congruence lattices of finite lattices. *Beiträge Algebra Geom.* **36** (1995), 97–108.

[344] _____, A lattice construction and congruence-preserving extensions. *Acta Math. Hungar.* **66** (1995), 275–288.

[345] _____, Congruence-preserving extensions of finite lattices into sectionally complemented lattices. *Proc. Amer. Math. Soc.* **127** (1999), 1903–1915.

[346] _____, Congruence-preserving extensions of finite lattices to semimodular lattices. *Houston J. Math.* **27** (2001), 1–9.

[347] _____, Representing congruence lattices of lattices with partial unary operations as congruence lattices of lattices. I. Interval equivalence. *J. Algebra.* **269** (2003), 136–159.

[348] _____, On the Independence Theorem of related structures for modular (arguesian) lattices. *Studia Sci. Math. Hungar.* **40** (2003), 1–12.

[349] _____, Finite lattices with isoform congruences. *Tatra Mt. Math. Publ.* **27** (2003), 111–124.

[350] G. Grätzer, E. T. Schmidt, and K. Thomsen, Congruence lattices of uniform lattices. *Houston J. Math.* **29** (2003), 247–263.

[351] G. Grätzer, E. T. Schmidt, and D. Wang, A short proof of a theorem of Birkhoff. *Algebra Universalis* **37** (1997), 253–255.

[352] G. Grätzer and J. Sichler, On the endomorphism semigroup (and category) of bounded lattices. *Pacific J. Math.* **34** (1970), 639–647.

[353] _____, Free products of Hopfian lattices. *J. Austral. Math. Soc.* **17** (1974), 234–245.

[354] _____, On generating free products of lattices. *Proc. Amer. Math. Soc.* **46** (1974), 9–14.

[355] _____, Free decompositions of a lattice. *Canad. J. Math.* **27** (1975), 276–285.

[356] G. Grätzer and T. Wares, Notes on planar semimodular lattices. V. Cover-preserving embeddings of finite semimodular lattices into simple semimodular lattices. *Acta Sci. Math.* (*Szeged*) **76** (2010), 27–33.

[357] G. Grätzer and F. Wehrung, Proper congruence-preserving extensions of lattices. *Acta Math. Hungar.* **85** (1999), 175–185.

[358] _____, A new lattice construction: the box product, *J. Algebra.* **221** (1999), 315–344.

[359] _____, The $M_3[D]$ construction and $n$-modularity. *Algebra Universalis* **41** (1999), 87–114.

[360] _____, Tensor products and transferability of semilattices. *Canad. J. Math.* **51** (1999), 792–815.

[361] _____, Flat semilattices. *Colloq. Math.* **79** (1999), 185–191.

[362] _____, Tensor products of lattices with zero, revisited. *J. Pure Appl. Algebra* **147** (2000), 273–301.

[363] _____, The Strong Independence Theorem for automorphism groups and congruence lattices of arbitrary lattices. *Adv. in Appl. Math.* **24** (2000), 181–221.

[364] _____, A survey of tensor products and related constructions in two lectures. *Algebra Universalis* **45** (2001), 117–134.

[365] G. Grätzer and G. H. Wenzel, Tolerances, covering systems, and the Axiom of Choice. *Arch. Math. (Brno)* **25** (1989), 27–34.

[366] _____, Notes on tolerance relations of lattices. *Acta Sci. Math. (Szeged)* **54** (1990), 229–240.

[367] G. Grätzer and B. Wolk, Finite projective distributive lattices. *Canad. Math. Bull.* **13** (1970), 139–140.

[368] M. Haiman, Arguesian lattices which are not type-1. *Algebra Universalis* **28** (1991), 128–137.

[369] A. W. Hales, On the non-existence of free complete boolean algebras. *Fund. Math.* **54** (1964), 45–66.

[370] P. Hall, A contribution to the theory of groups of prime power order. *Proc. London Math. Soc. (2)* **36** (1934), 29–95.

[371] P. R. Halmos, Lectures on boolean algebras. Van Nostrand Mathematical Studies, No. 1. *D. Van Nostrand Co., Inc., Princeton, N.J.*, 1963. v+147 pp.

[372] W. Hanf, Representing real numbers in denumerable boolean algebras. *Fund. Math.* **91** (1976), 167–170.

[373] J. Harding, $\kappa$-complete uniquely complemented lattices. *Order* **25** (2008), 121–129.

[374] T. Harrison, A problem concerning the lattice varietal product. *Algebra Universalis* **25** (1988), 40–84.

[375] J. Hashimoto, Ideal theory for lattices. *Math. Japon.* **2** (1952), 149–186.

[376] J. Hashimoto and S. Kinugawa, On neutral elements in lattices. *Proc. Japan Acad.* **39** (1963), 162–163.

[377] M. Haviar and T. Katriňák, Lattices whose congruence lattice is relative Stone. *Acta Sci. Math. (Szeged)* **51** (1987), 81–91.

[378] _____ , Semi-discrete lattices with $(L_n)$-congruence lattices. *Contributions to general algebra, 7 (Vienna, 1990)*, 189–195, Hölder-Pichler-Tempsky, Vienna, 1991.

[379] M. Haviar and M. Ploščica, Affine complete Stone algebras. *Algebra Universalis* **34** (1995), 355–365.

[380] Z. Hedrlín and J. Lambek, How comprehensive is the category of semigroups? *J. Algebra* **11** (1969), 195–212.

[381] Z. Hedrlín and A. Pultr, Relations (graphs) with given finitely generated semigroups. *Monatsh. Math.* **68** (1964), 213–217.

[382] _____ , On full embeddings of categories of algebras. *Illinois J. Math.* **10** (1966), 392–406.

[383] P. Hegedűs and P. Pálfy, Finite modular congruence lattices. *Algebra Universalis* **54** (2005), 105–120.

[384] Z. Heleyová, Distributive lattices whose congruence lattice is Stone. *Acta Math. Univ. Comenian. (N.S.)* **64** (1995), 273–282.

[385] P. Hell, Full embeddings into some categories of graphs. *Algebra Universalis* **2** (1972), 129–141.

[386] C. Herrmann, Weak (projective) radius and finite equational bases for classes of lattices. *Algebra Universalis* **3** (1973), 51–58.

[387] _____ , On the equational theory of submodule lattices. *Proceedings of the University of Houston Lattice Theory Conference (Houston, Tex., 1973)*, 105–118. Dept. Math., Univ. Houston, Houston, Tex., 1973.

[388] _____ , S-verklebte Summen von Verbänden. *Math. Z.* **130** (1973), 225–274.

[389] _____ , Concerning M. M. Gluhov's paper on the word problem for free modular lattices. *Algebra Universalis* **5** (1975), 445.

[390] _____ , On the word problem for the modular lattice with four free generators. *Math. Ann.* **265** (1983), 513–527.

[391] _____ , Generators for complemented modular lattices and the von Neumann-Jónsson Coordinatization Theorems. *Algebra Universalis* **63** (2010), 45–64.

[392] C. Herrmann and A. P. Huhn, Zum Begriff der Charakteristik modularer Verbände. *Math. Z.* **144** (1975), 185–194.

[393] C. Herrmann and M. V. Semenova, Existence varieties of regular rings and complemented modular lattices. *J. Algebra* **314** (2007), 235–251.

[394] D. Higgs, Lattices isomorphic to their ideal lattices. *Algebra Universalis* **1** (1971), 71–72.

[395] D. Hobby and R. N. McKenzie, The structure of finite algebras. Contemporary Mathematics, **76**. *American Mathematical Society, Providence, RI*, 1988. xii+203 pp.

[396] M. Hochster, Prime ideal structure in commutative rings. *Trans. Amer. Math. Soc.* **142** (1969), 43–60.

[397] K. H. Hofmann and J. D. Lawson, The spectral theory of distributive continuous lattices. *Trans. Amer. Math. Soc.* **246** (1978), 285–310.

[398] K. H. Hofmann and A. L. Stralka, The algebraic theory of compact Lawson semilattices. Applications of Galois connections to compact semilattices. *Dissertationes Math.* **137** (1976), 58pp.

[399] D. X. Hong, Covering relations among lattice varieties. *Pacific J. Math.* **40** (1972), 575–603.

[400] A. Horn, A property of free boolean algebras. *Proc. Amer. Math. Soc.* **19** (1968), 142–143.

[401] S. Huang and D. Tamari, Problems of associativity: A simple proof for the lattice property of systems ordered by a semi-associative law. *J. Combin. Theory* Ser. A **13** (1972), 7–13.

[402] D. Huguet, La structure du treillis des polyèdres de paranthésages. *Algebra Universalis* **5** (1975), 82–87.

[403] A. P. Huhn, Weakly distributive lattices. (Hungarian.) *Ph.D. Thesis, University of Szeged*, 1972.

[404] A. P. Huhn, Schwach distributive Verbände. I. *Acta Sci. Math. (Szeged)* **33** (1972), 297–305.

[405] _____, On G. Grätzer's problem concerning automorphisms of a finitely presented lattice. *Algebra Universalis* **5** (1975), 65–71.

[406] _____, On the representation of distributive algebraic lattices, I. *Acta Sci. Math.(Szeged)* **45** (1983), 239–246.

[407] _____, On the representation of distributive algebraic lattices. II, *Acta Sci. Math. (Szeged)* **53** (1989), 3–10.

[408] _____, On the representation of distributive algebraic lattices. III. *Acta Sci. Math. (Szeged)* **53** (1989), 11–18.

[409] E. V. Huntington, Sets of independent postulates for the algebra of logic. *Trans. Amer. Math. Soc.* **5** (1904), 288–309.

[410] G. Hutchinson, On classes of lattices representable by modules. *Proceedings of the University of Houston Lattice Theory Conference (Houston, Tex., 1973)*, 69–94. Dept. Math., Univ. Houston, Houston, Tex., 1973.

[411] _____, A duality principle for lattices and categories of modules. *J. Pure Appl. Algebra* **10** (1977/78), 115–119.

[412] _____, Exact embedding functors between categories of modules. *J. Pure Appl. Algebra* **25** (1982), 107–111.

[413] _____, Exact embedding functors for module categories and submodule lattice quasivarieties. *J. Algebra* **219** (1999), 766–778.

[414] G. Hutchinson and G. Czédli: A test for identities satisfied in lattices of submodules. *Algebra Universalis* **8** (1978), 269–309.

[415] Iqbalunnisa, On neutral elements in a lattice. *J. Indian Math. Soc. (N. S.)* **28** (1964), 25–31.

[416] _____, On some problems of G. Grätzer and E. T. Schmidt. *Fund. Math.* **57** (1965), 181–185.

[417] _____, Normal, simple, and neutral congruences on lattices. *Illinois J. Math.* **10** (1966), 227–234.

[418] _____, On lattices whose lattices of congruences are Stone lattices. *Fund. Math.* **70** (1971), 315–318.

[419] J. R. Isbell, Atomless parts of spaces. *Math. Scand.* **31** (1972), 5–32.

[420] A. A. Iskander, Correspondence lattices of universal algebras. *Izv. Akad. Nauk SSSR Ser. Mat.* **29** (1965), 1357–1372.

[421] _____, On subalgebra lattices of universal algebras. *Proc. Amer. Math. Soc.* **32** (1972), 32–36.

[422] T. Iwamura, A lemma on directed sets (Japanese), *Zenkoku Shijo Sugaku Danwakai* **262** (1944), 107–111.

[423] J. Jakubík, On lattices whose graphs are isomorphic. (Russian.) *Czech. Math. J.* **4** (1954), 131–141.

[424] _____, Directly indecomposable direct factors of a lattice. *Math. Bohem.* **121**, no. 3 (1996), 281–292.

[425] M. F. Janowitz, The center of a complete relatively complemented lattice is a complete sublattice. *Proc. Amer. Math. Soc.* **18** (1967), 189–190.

[426] C. U. Jensen, On characterisations of Prüfer rings. *Math. Scand.* **13** (1963), 90–98.

[427] P. Jipsen and H. Rose, Varieties of Lattices. Lecture Notes in Mathematics, **1533**. *Springer-Verlag, Berlin,* 1992. x+162 pp.

[428] P. T. Johnstone, Tychonoff's Theorem without the axiom of choice. *Fund. Math.* **113** (1981), 21–35.

[429] _____, The point of pointless topology. *Bull. Amer. Math. Soc.* (N.S.) **8** (1983), 41–53.

[430] _____, Stone spaces. Reprint of the 1982 edition. Cambridge Studies in Advanced Mathematics, **3**. *Cambridge University Press, Cambridge,* 1986. xxii+370 pp.

[431] _____, The Art of Pointless Thinking: A Student's Guide to the Category of Locales. *Category Theory at Work (Bremen, 1990),* 85–107, Res. Exp. Math., **18**, *Heldermann, Berlin,* 1991.

[432] B. Jónsson, On the representation of lattices. *Math. Scand.* **1** (1953), 193–206.

[433] _____, Modular lattices and Desargues' theorem. *Math. Scand.* **2** (1954), 295–314.

[434] _____, Distributive sublattices of a modular lattice. *Proc. Amer. Math. Soc.* **6** (1955), 682–688.

[435] _____, Universal relational systems. *Math. Scand.* **4** (1956), 193–208.

[436] _____, Arguesian lattices identity of dimension $n \leq 4$. *Math. Scand.* **7** (1959), 133–145.

[437] B. Jónsson, Representation of modular lattices and of relation algebras. *Trans. Amer. Math. Soc.* **92** (1959), 449–464.

[438] ———, Lattice-theoretic approach to projective and affine geometry. 1959 *The axiomatic method. With special reference to geometry and physics. Proceedings of an International Symposium held at the Univ. of Calif., Berkeley, Dec. 26, 1957–Jan. 4, 1958 (edited by L. Henkin, P. Suppes and A. Tarski)*, 188–203. Studies in Logic and the Foundations of Mathematics, *North-Holland Publishing Co., Amsterdam.*

[439] ———, Representations of complemented modular lattices. *Trans. Amer. Math. Soc.* **97** (1960), 64–94.

[440] ———, Sublattices of a free lattice. *Canad. J. Math.* **13** (1961), 256–264.

[441] ———, Extensions of von Neumann's coordinatization theorem. 1961 *Proc. Sympos. Pure Math., Vol. II* pp. 65–70. *American Mathematical Society, Providence, R.I.*

[442] ———, Representations of relatively complemented modular lattices. *Trans. Amer. Math. Soc.* **103** (1962), 272–303.

[443] ———, Extensions of relational structures. 1965 *Theory of Models (Proc. Internat. Sympos. Berkeley, 1963)*, 146–157. *North-Holland, Amsterdam.*

[444] ———, Algebras whose congruence lattices are distributive. *Math. Scand.* **21** (1967), 110–121.

[445] ———, Equational classes of lattices. *Math. Scand.* **22** (1968), 187–196.

[446] ———, Relatively free products of lattices. *Algebra Universalis* **1** (1971), 362–373.

[447] ———, Sums of finitely based lattice varieties. *Adv. Math.* **14** (1974), 454–468.

[448] B. Jónsson and J. E. Kiefer, Finite sublattices of a free lattice. *Canad. J. Math.* **14** (1962), 487–497.

[449] B. Jónsson and J. B. Nation, A report on sublattices of a free lattice. *Contributions to universal algebra (Colloq., József Attila Univ., Szeged, 1975)*, 223–257. Colloq. Math. Soc. János Bolyai, **17**. *North-Holland, Amsterdam*, 1977.

[450] B. Jónsson and E. Nelson, Relatively free products in regular varieties. *Algebra Universalis* **4** (1974), 14–19.

[451] B. Jónsson and I. Rival, Lattice varieties covering the smallest nonmodular variety. *Pacific J. Math.* **82** (1979), 463–478.

[452] _____, Amalgamation in small varieties of lattices. *J. Pure Appl. Algebra* **68** (1990), 195–208.

[453] K. Kaarli, Polynomials and polynomial completeness. The concise handbook of algebra. Edited by A. V. Mikhalev and G. Pilz, 457–460. *Kluwer Academic Publishers, Dordrecht*, 2002. xvi+618 pp.

[454] K. Kaarli and V. Kuchmei, Order affine completeness of lattices with Boolean congruence lattices. *Czechoslovak Math. J.* **57** (2007), 1049–1065.

[455] K. Kaarli and A. F. Pixley, Polynomial Completeness in Algebraic Systems. *Chapman & Hall/CRC, Boca Raton, FL*, 2001. xvi+358 pp.

[456] K. Kaarli and K. Täht, Strictly locally order affine complete lattices. *Order* **10** (1993), 261–270.

[457] J. A. Kalman, A two axiom definition for lattices. *Rev. Roumaine Math. Pures Appl.* **13** (1968), 669–670.

[458] T. Katriňák, Die Kennzeichnung der distributiven pseudokomplementären Halbverbände. *J. Reine Angew. Math.* **241** (1970), 160–179.

[459] _____, Über eine Konstruktion der distributiven pseudokomplementären Verbände. *Math. Nachr.* **53** (1972), 85–99.

[460] _____, A new proof of the construction theorem for Stone algebras. *Proc. Amer. Math. Soc.* **40** (1973), 75–78.

[461] _____, Construction of modular double S-algebras. *Algebra Universalis* **8** (1978), 15–22.

[462] _____, p-algebras, *Contributions to lattice theory (Szeged, 1980)*, 549–573, Colloq. Math. Soc. János Bolyai, **33**, *North-Holland, Amsterdam*, 1983.

[463] _____, Congruence lattices of pseudocomplemented semilattices. *Semigroup Forum* **55** (1997), 1–23.

[464] _____, A new proof of the Glivenko-Frink theorem. *Bull. Soc. Roy. Sci. Liège* **50** (1981), 3–4.

[465] T. Katriňák and J. Guričan, Projective extensions of semilattices, *Algebra Universalis* **55** (2006), 45–55.

[466] T. Katriňák and P. Mederly, Construction of modular p-algebras. *Algebra Universalis* **4** (1974), 301–315.

[467]  T. Katriňák and P. Mederly, Construction of p-algebras. *Algebra Universalis* **17** (1983), 288–316.

[468]  K. Kearnes and E. Kiss, The shape of congruence lattices. Manuscript, 2006.

[469]  D. Kelly and I. Rival, Crowns, fences, and dismantlable lattices. *Canad. J. Math.* **26** (1974), 1257–1271.

[470]  ———, Planar lattices. *Canad. J. Math.* **27** (1975), 636–665.

[471]  M. Kindermann, Über die Äquivalenz von Ordnungspolynomvollständigkeit und Toleranzeinfachheit endlicher Verbände. *Contributions to General Algebra* (*Proc. Klagenfurt Conf., Klagenfurt, 1978*), 145–149, Heyn, Klagenfurt, 1979.

[472]  S. Kinugawa and J. Hashimoto, On relative maximal ideals in lattices. *Proc. Japan Acad.* **42** (1966), 1–4.

[473]  D. Kleitman, On Dedekind's problem: The number of monotone boolean functions. *Proc. Amer. Math. Soc.* **21** (1969), 677–682.

[474]  D. Kleitman and G. Markowski, On Dedekind's problem: the number of isotone boolean functions. II. *Trans. Amer. Math. Soc.* **213** (1975), 373–390.

[475]  B. Knaster, Un théorème sur les fonctions d'ensembles. *Ann. Soc. Polon. Math.* **6** (1928), 133–134.

[476]  S. R. Kogalovskiĭ, On a theorem of Birkhoff. (Russian.) *Uspehi Mat. Nauk.* **20** (1965), 206–207.

[477]  K. M. Koh, On sublattices of a lattice. *Nanta Math.* **6** (1973), 68–79.

[478]  M. Kolibiar, On the axiomatics of modular lattices. (Russian.) *Czechoslovak Math. J.* **6** (**81**) (1956), 381–386.

[479]  A. Komatu, On a characterization of join homomorphic transformation-lattice. *Proc. Imp. Acad. Tokyo* **19** (1943), 119–124.

[480]  A. D. Korshunov, Monotone boolean functions. (Russian.) *Uspekhi Mat. Nauk* **58** (2003), 89–162.

[481]  B. Korte, L. Lovász, and R. Schrader, Greedoids. Algorithms and Combinatorics, 4. *Springer-Verlag, Berlin,* 1991. viii+211 pp.

[482]  A. Kostinsky, Some problems for rings and lattices within the domain of general algebra. *Ph.D. Thesis, Univ. Calif., Berkeley,* 1971.

[483] _____, Projective lattices and bounded homomorphisms. *Pacific J. Math.* **40** (1972), 111–119.

[484] V. Koubek and J. Sichler, Almost ff-universality implies Q-universality. *Appl. Categ. Structures* **17** (2009), 419–434.

[485] A. V. Kravchenko, On lattice complexity of quasivarieties of graphs and endographs. (Russian.) *Algebra i Logika* **36** (1997), 273–281; translation in *Algebra and Logic* **36** (1997), 164–168.

[486] _____, On an article by Sorkin. *Algebra Universalis.* To appear.

[487] J. P. S. Kung, Matchings and Radon Transforms in lattices I. Consistent lattices. *Order* **2** (1985), 105–112.

[488] _____, Matchings and Radon Transforms in lattices II. Concordant sets. *Math. Proc. Camb. Phil. Soc.* **101** (1987), 221–231.

[489] K. Kuratowski, Sur une caractérisation des alephs. *Fund. Math.* **38** (1951), 14–17.

[490] A. G. Kuroš, Durchschnittsdarstellungen mit irreduziblen Komponenten in Ringen und in sogenannten Dualgruppen. *Mat. Sb.* **42** (1935), 613–616.

[491] H. Lakser, Normal and canonical representations in free products of lattices. *Canad. J. Math.* **22** (1970), 394–402.

[492] _____, The structure of pseudocomplemented distributive lattices. I. Subdirect decomposition. *Trans. Amer. Math. Soc.* **156** (1971), 335–342.

[493] _____, Simple sublattices of free products of lattices. Abstract. *Notices Amer. Math. Soc.* **19** (1972), A 509.

[494] _____, Principal congruences of pseudocomplemented distributive lattices. *Proc. Amer. Math. Soc.* **37** (1973), 32–36.

[495] _____, Disjointness conditions in free products of distributive lattices: An application of Ramsey's theorem. *Proceedings of the University of Houston Lattice Theory Conference (Houston, Tex., 1973)*, 156–168. Dept. Math., Univ. Houston, Houston, Tex., 1973.

[496] _____, A note on the lattice of sublattices of a finite lattice. *Nanta Math.* **6** (1973), 55–57.

[497] _____, Lattices freely generated by an order and preserving certain bounds. Manuscript.

[498] W. A. Lampe, A property of the lattice of equational theories. *Algebra Universalis* **23** (1986), 61–69.

[499] J. D. Lawson, Topological semilattices with small semilattices. *J. London Math. Soc.* **1** (1969), 719–724.

[500] K. B. Lee, Equational classes of distributive pseudo-complemented lattices. *Canad. J. Math.* **22** (1970), 881–891.

[501] V. B. Lender, The groupoid of prevarieties of lattices. (Russian.) *Sibirsk. Mat. Zh.* **16** (1975), 1214–1223, 1370.

[502] A. Levy, Basic Set Theory. *Springer-Verlag, Berlin-New York*, 1979.

[503] L. Libkin, Direct decompositions of atomistic algebraic lattices. *Algebra Universalis* **33** (1995), 127–135.

[504] J. Łoś, Quelques remarques, théorèmes et problèmes sur les classes définissables d'algèbres. *Mathematical Interpretation of Formal Systems*, 98–113. *North-Holland Publishing Co., Amsterdam*, 1955.

[505] H. F. J. Löwig, On the importance of the relation

$$[(A,B),(A,C)] < [(B,C),(C,A),(A,B)]$$

between three elements of a structure. *Ann. of Math.* (2) **44** (1943), 573–579.

[506] R. C. Lyndon, The representation of relational algebras. *Ann. of Math.* **51** (1950), 707–729.

[507] _____ , Identities in finite algebras. *Proc. Amer. Math. Soc.* **5** (1954), 8–9.

[508] _____ , The representation of relation algebras. II. *Ann. of Math.* **63** (1956), 294–307.

[509] M. Makkai and G. McNulty, Universal Horn axiom systems for lattices of submodules. *Algebra Universalis* **7** (1977), 25–31.

[510] A. I. Mal'cev, Multiplication of classes of algebraic systems. *Sibirsk. Mat. Ž.* **8** (1967), 346–365.

[511] R. N. McKenzie, Equational bases for lattice theories. *Math. Scand.* **27** (1970), 24–38.

[512] _____ , Equational bases and nonmodular lattice varieties. *Trans. Amer. Math. Soc.* **174** (1972), 1–43.

[513] _____ , Some unsolved problems between lattice theory and equational logic. *Proceedings of the University of Houston Lattice Theory Conference (Houston, Tex., 1973)*, 564–573. *Dept. Math., Univ. Houston, Houston, Tex.*, 1973.

[514] _____, Finite forbidden lattices. *Universal algebra and lattice theory* (*Puebla, 1982*), 176–205, Lecture Notes in Math., **1004**, *Springer-Verlag, Berlin,* 1983.

[515] S. Mac Lane, A lattice formulation for transcendence degrees and $p$-bases. *Duke Math. J.* **4** (1938), 455–468.

[516] _____, A conjecture of Ore on chains in partially ordered sets. *Bull. Amer. Math. Soc.* **49** (1943), 567–568.

[517] H. M. MacNeille, Partially ordered sets. *Trans. Amer. Math. Soc.* **42** (1937), 416–460.

[518] _____, Extension of a distributive lattice to a boolean ring. *Bull. Amer. Math. Soc.* **45** (1939), 452–455.

[519] F. Maeda, Lattice theoretic characterization of abstract geometries. *J. Sci. Hiroshima Univ. Ser. A-I Math.* **15** (1951), 87–96.

[520] F. Maeda, Kontinuierliche Geometrien. Die Grundlehren der mathematischen Wissenschaften in Einzeldarstellungen mit besonderer Berücksichtigung der Anwendungsgebiete, Bd. 95. *Springer-Verlag, Berlin-Göttingen-Heidelberg,* 1958. x+244 pp.

[521] F. Maeda and S. Maeda, Theory of symmetric lattices. Die Grundlehren der mathematischen Wissenschaften, Band 173. *Springer-Verlag, New York-Berlin,* 1970. xi+190 pp.

[522] S. Maeda, On the symmetry of the modular relation in atomic lattices. *J. Sci. Hiroshima Univ. Ser. A-I Math.* **29** (1965), 165–170.

[523] _____, Infinite distributivity in complete lattices. *Mem. Ehime. Univ. Sect. II Ser A.* **5.** (1966), 11–13.

[524] M. Makkai, A proof of Baker's finite-base theorem on equational classes generated by finite elements of congruence distributive varieties. *Algebra Universalis* **3** (1973), 174–181.

[525] A. I. Mal'cev, Several remarks on quasivarieties of algebraic systems. (Russian.) *Algebra i Logika Sem.* **5** (1966), 3–9.

[526] _____, Some borderline problems of algebra and logic. 1968 *Proc. Internat. Cong. Math.* (*Moscow, 1966*), 217–231. Izdat. *"Mir", Moscow.*

[527] W. McCune and R. Padmanabhan, Automated deduction in equational logic and cubic curves. Lecture Notes in Computer Science, 1095. Lecture Notes in Artificial Intelligence. *Springer-Verlag, Berlin,* 1996. x+231 pp.

[528]  P. Mederly, A characterization of modular pseudocomplemented semilattices. Colloq. Math. Soc. J. Bolyai, Lattice Theory, Szeged (Hungary), **14** (1974), 231–248.

[529]  K. Menger, New foundations of projective and affine geometry. *Ann. of Math.* (2) **37** (1936), 456–482.

[530]  G. Michler and R. Wille, Die primitiven Klassen arithmetischer Ringe. *Math. Z.* **113** (1970), 369–372.

[531]  F. Micol and G. Takách, Notes on pasting. *Acta Sci. Math.* (*Szeged*) **72** (2006), 3–14.

[532]  A. F. Möbius, Über eine besondere Art von Umkehrung der Reihen. *J. Reine Angew. Math.* **9** (1832), 105–123.

[533]  B. Monjardet, A use for frequently rediscovering a concept. *Order* **1** (1985), 415–417.

[534]  A. Monteiro, Axiomes indépendants pour les algèbres de Brouwer. *Rev. Un. Mat. Argentina* **17** (1955), 149–160.

[535]  A. Mostowski and A. Tarski, Boolesche Ringe mit geordneter Basis. *Fund. Math.* **32** (1939), 69–86.

[536]  V. L. Murskiĭ, The existence in the three-valued logic of a closed class with a finite basis having no finite complete system of identities. (Russian.) *Dokl. Akad. Nauk* **163** (1965), 815–818.

[537]  L. Nachbin, Une propriété caractéristique des algèbres booléiennes. *Portugal. Math.* **6** (1947), 115–118.

[538]  ———, On a characterization of the lattice of all ideals of a boolean ring. *Fund. Math.* **36** (1949), 137–142.

[539]  J. B. Nation, Finite sublattices of a free lattice. *Trans. Amer. Math. Soc.* **269** (1982), 311–337.

[540]  ———, Lattice varieties covering $V(L_1)$. *Algebra Universalis* **23** (1986), 132–166.

[541]  ———, An approach to lattice varieties of finite height. *Algebra Universalis* **27** (1990), 521–543.

[542]  J. B. Nation, Notes on Lattice Theory. 1991–2009, http://www.math.hawaii.edu/jb/books.html.

[543]  ———, Alan Day's doubling construction. *Algebra Universalis* **34** (1995), 24–34.

[544] _____, A counterexample to the finite height conjecture. *Order* **13** (1996), 1–9.

[545] _____, Lattices of theories in the languages without equality. Manuscript.
http://www.math.hawaii.edu/~jb/coopers.pdf

[546] O. T. Nelson, Jr., Subdirect decompositions of lattices of width two. *Pacific J. Math.* **24** (1968), 519–523.

[547] W. C. Nemitz, Implicative semi-lattices. *Trans. Amer. Math. Soc.* **117** (1965), 128–142.

[548] A. Nerode, Some Stone spaces and recursion theory. *Duke Math. J.* **26** (1959), 397–406.

[549] B. H. Neumann, Universal Algebra. Lecture Notes. *Courant Inst. of Math. Sci., New York University,* 1962.

[550] B. H. Neumann and H. Neumann, Extending partial endomorphisms of groups. *Proc. London Math. Soc.* (3) **2** (1952), 337–348.

[551] H. Neumann, Varieties of groups. *Springer-Verlag New York, Inc., New York,* 1967. x+192 pp.

[552] J. von Neumann, Lectures on Continuous Geometries. *Institute of Advanced Studies, Princeton, N.J.,* 1936.

[553] J. von Neumann, Continuous Geometry. Foreword by Israel Halperin. Princeton Mathematical Series, No. 25. *Princeton Univ. Press, Princeton, N.J.,* 1960.

[554] E. Noether, Abstrakter Aufbau der Idealtheorie in algebraischen Zahl- und Funktionenköpern. *Math. Ann.* **96** (1926), 26–61.

[555] A. M. Nurakunov, Equational theories as congruences of enriched monoids. *Algebra Universalis* **58** (2008), 357–372.

[556] T. Ogasawara and U. Sasaki, On a theorem in lattice theory. *J. Sci. Hiroshima Univ. Ser. A-I Math.* **14** (1949), 13.

[557] O. Ore, On the foundation of abstract algebra. I. *Ann. of Math.* **36** (1935), 406–437.

[558] _____, On the foundation of abstract algebra. II. *Ann. of Math.* **37** (1936), 265–292.

[559] _____, Structures and group theory. I. *Duke Math. J.* **3** (1937), 149–174.

[560] O. Ore, Structures and group theory. II. *Duke Math. J.* **4** (1938), 247–269.

[561] _____, Remarks on structures and group relations. *Vierteljschr. Natur-forsch. Ges. Zürich* **85** (1940), Beiblatt (Festschrift Rudolf Fueter), 1–4.

[562] _____, Theory of equivalence relations. *Duke Math. J.* **9** (1942), 573–627.

[563] _____, Chains in partially ordered sets. *Bull. Amer. Math. Soc.* **49** (1943), 558–566.

[564] _____, Galois connexions. *Trans. Amer. Math. Soc.* **55** (1944), 493–513.

[565] P. Ouwehand and H. Rose, Lattice varieties with non-elementary amalgamation classes. *Algebra Universalis* **55** (1999), 317–336.

[566] R. Padmanabhan, A note on Kalman's paper. *Rev. Roumaine Math. Pures Appl.* **13** (1968), 1149–1152.

[567] _____, Two identities for lattices. *Proc. Amer. Math. Soc.* **20** (1969), 409–412.

[568] _____, On identities defining lattices. *Algebra Universalis* **1** (1972), 359–361.

[569] _____, Two results on uniquely complemented lattices. *Proceedings of the Lattice Theory Conference (Ulm, 1975)*, pp. 146–147. Univ. Ulm, Ulm, 1975.

[570] _____, Equational theory of algebras with a majority polynomial. *Algebra Universalis* **7** (1977), 273–275.

[571] _____, A selfdual equational basis for boolean algebras. *Canad. Math. Bull.* **26** (1983), 9–12.

[572] R. Padmanabhan and W. McCune, Single identities for lattice theory and for weakly associative lattices. *Algebra Universalis* **36** (1996), 436–449.

[573] R. Padmanabhan, W. McCune, and R. Veroff, Yet another single law for lattices. *Algebra Universalis* **50** (2003), 165–169.

[574] R. Padmanabhan and S. Rudeanu, Axioms for lattices and boolean algebras. *World Scientific Publishing Co. Pte. Ltd., Hackensack, NJ*, 2008. xii+216 pp.

[575] P. P. Pálfy, On partial ordering of chief factors in solvable groups, *Manuscripta Math.* **55** (1986), 219–232.

[576] P. Pálfy and P. Pudlák, Congruence lattices of finite algebras and intervals in subgroup lattices of finite groups. *Algebra Universalis* **11** (1980), 22–27.

[577] D. Papert, Congruence relations in semilattices. *J. London Math. Soc.* **39** (1964), 723–729.

[578] M. A. Perles, On Dilworth's theorem in the infinite case. *Israel J. Math.* **1** (1963), 108–109.

[579] R. S. Pierce, Introduction to the Theory of Abstract Algebras. *Holt, Rinehart, and Winston, New York, N.Y.*, 1968.

[580] _____, Bases of countable boolean algebras. *J. Symbolic Logic* **38** (1973), 212–214.

[581] R. S. Pierce, Countable Boolean Algebras. *Handbook of Boolean Algebras, Vol. 3, 775–876, North-Holland, Amsterdam*, 1989.

[582] D. Pigozzi and G. Tardos, The representation of certain abstract lattices as lattices of subvarieties. Manuscript, 1999.

[583] J. G. Pitkethly and B. A. Davey, Dualisability: Unary Algebras and Beyond. Advances in Mathematics **9**, *Springer, New York*, 2005. xii+263 pp.

[584] C. R. Platt, Planar lattices and planar graphs. *J. Combin. Theory*, Ser. B **21** (1976), 30–39.

[585] M. Ploščica, Affine complete distributive lattices. *Order* **11** (1994), 385–390.

[586] _____, Separation properties in congruence lattices of lattices. *Colloq. Math.* **83** (2000), 71–84.

[587] _____, Non-representable distributive semilattices. *J. Pure Appl. Algebra* **212** (2008), 2503–2512.

[588] M. Ploščica and M. Haviar, Order polynomial completeness of lattices. *Algebra Universalis* **39** (1998), 217–219.

[589] _____, Congruence-preserving functions on distributive lattices. *Algebra Universalis* **59** (2008), 179–196.

[590] M. Ploščica, J. Tůma, and F. Wehrung, Congruence lattices of free lattices in non-distributive varieties. *Colloq. Math.* 76 (1998), 269–278.

[591] H. A. Priestley, Representation of distributive lattices by means of ordered Stone spaces. *Bull. London Math. Soc.* **2** (1970), 186–190.

[592] H. A. Priestley, Ordered topological spaces and the representation of distributive lattices. *Proc. London Math. Soc.* (3) **24** (1972), 507–530.

[593] _____, Stone lattices: A topological approach. *Fund. Math.* **84** (1974), 127–143.

[594] _____, The construction of spaces dual to pseudocomplemented distributive lattices. *Quart. J. Math. Oxford* (3) **26** (1975), 215–228.

[595] _____, Natural dualities. *Lattice theory and its applications (Darmstadt, 1991)*, 185–209, Res. Exp. Math., **23**, *Helderman, Lemgo, 1995*.

[596] P. Pudlák, A new proof of the congruence lattice representation theorem. *Algebra Universalis* **6** (1976), 269–275.

[597] _____, On congruence lattices of lattices. *Algebra Universalis* **20** (1985), 96–114.

[598] P. Pudlák and J. Tůma, Yeast graphs and fermentation of algebraic lattices. *Coll. Math Soc. János Bolyai* **14** (1976), 301–341.

[599] _____, Every finite lattice can be embedded in a finite partition lattice. *Algebra Universalis* **10** (1980), 74–95.

[600] A. Pultr, Frames. *Handbook of algebra, Vol. 3, 791–857, North-Holland, Amsterdam, 2003.*

[601] R. W. Quackenbush, Free products of bounded distributive lattices. *Algebra Universalis* **2** (1972), 393–394.

[602] _____, Near-boolean algebras. I. Combinatorial aspects. *Discrete Math.* **10** (1974), 301–308.

[603] _____, Non-modular varieties of semimodular lattices with a spanning $M_3$. Special volume on ordered sets and their applications (L'Arbresle, 1982). *Discrete Math.* **53** (1985), 193–205.

[604] G. N. Raney, Completely distributive complete lattices. *Proc. Amer. Math. Soc.* **3** (1952), 677–680.

[605] B. C. Rennie, The Theory of Lattices. *Forster and Jagg, Cambridge, England,* 1951. 51 pp.

[606] K. Reuter, The Kurosh-Ore exchange property. *Acta Math. Hungar.* **53** (1989), 119–127.

[607] K. Reuter and R. Wille, Complete congruence relations of concept lattices. *Acta Sci. Math. (Szeged)* **51** (1987), 319–327.

[608] I. Reznikoff, Chaînes de formules. *C. R. Acad. Sci. Paris* **256** (1963), 5021–5023.

[609] P. Ribenboim, Characterization of the sup-complement in a distributive lattice with least element. *Summa Brasil. Math.* **2** (1949), 43–49.

[610] B. Riečan, To the axiomatics of modular lattices. (Slovak.) *Acta Fac. Rerum Natur. Univ. Comenian. Math.* **2** (1958), 257–262.

[611] L. Rieger, A note on topological representations of distributive lattices. *Časopis Pěst. Mat. Fys.* **74** (1949), 55–61.

[612] I. Rival, Maximal sublattices of finite distributive lattices. *Proc. Amer. Math. Soc.* **37** (1973), 417–420.

[613] _____, Maximal sublattices of finite distributive lattices. II. *Proc. Amer. Math. Soc.* **44** (1974), 263–268.

[614] _____, Lattices with doubly irreducible elements. *Canad. Math. Bull.* **17** (1974), 91–95.

[615] _____, Sublattices of modular lattices of finite length. *Canad. Math. Bull.* **18** (1975), 95–98.

[616] A. Robinson, Infinite forcing in model theory. *Proceedings of the Second Scandinavian Logic Symposium (Oslo, 1970)*, 317–340. Studies in Logic and the Foundations of Mathematics, **63**, *North-Holland, Amsterdam*, 1971.

[617] _____, On the notion of algebraic closedness for noncommutative groups and fields. *J. Symbolic Logic* **36** (1971), 441–444.

[618] H. L. Rolf, The free lattice generated by a set of chains. *Pacific J. Math.* **8** (1958), 585–595.

[619] G.-C. Rota, On the foundations of combinatorial theory. I. Theory of Möbius functions. *Z. Wahrscheinlichkeitstheorie und Verw. Gebiete* **2** (1964), 340–368.

[620] S. Rudeanu, Axioms for Lattices and Boolean algebras. (Romanian.) Monografii Asupra Teoriei Algebrice A Mecanismelor Automate, *Editura Academiei Republicii Populare Romine, Bucharest*, 1963. 159 pp.

[621] S. Rudeanu, Boolean Functions and Equations. *North-Holland Publishing Co., Amsterdam-London; American Elsevier Publishing Co., Inc., New York*, 1974. xix+442 pp.

[622] P. Růžička, Free trees and the optimal bound in Wehrung's theorem. *Fund. Math.* **198** (2008), 217–228.

[623] P. Růžička, J. Tůma, and F. Wehrung, Distributive congruence lattices of congruence-permutable algebras, *J. Algebra* **311** (2007), 96–116.

[624] V. N. Saliĭ, A compactly generated lattice with unique complements is distributive. (Russian.) *Mat. Zametki* **12** (1972), 617–620.

[625] V. N. Saliĭ, Lattices with unique complements. Translated from the Russian by G. A. Kandall. Translations of Mathematical Monographs, **69**. *American Mathematical Society, Providence, RI*, 1988. x+113 pp.

[626] L. Santocanale, A duality for finite lattices. Manuscript, 2010.

[627] M. V. Sapir, The lattice of quasivarieties of semigroups. *Algebra Universalis* **21** (1985), 172–180.

[628] U. Sasaki and S. Fujiwara, The decomposition of matroid lattices. *J. Sci. Hiroshima Univ. Ser. A-I Math.* **15** (1952), 183–188.

[629] E. T. Schmidt, Every finite distributive lattice is the congruence lattice of some modular lattice. *Algebra Universalis* **4** (1974), 49–57.

[630] _____, The ideal lattice of a distributive lattice with 0 is the congruence lattice of a lattice. *Acta Sci. Math. (Szeged)* **43** (1981), 153–168.

[631] _____, Pasting and semimodular lattices. *Algebra Universalis* **27** (1990), 595–596.

[632] J. Schmidt, Quasi-decompositions, exact sequences, and triple sums of semigroups, I. General theory. *Contributions to universal algebra (Colloq., József Attila Univ., Szeged, 1975)*, 365–39. Colloq. Math. Soc. J. Bolyai, Vol. **17**, *North-Holland, Amsterdam*, 1977.

[633] M. Schützenberger, Sur certains axiomes de la théorie des structures. *C. R. Acad. Sci. Paris* **221** (1945), 218–220.

[634] W. Schwan, Perspektivitäten in allgemeinen Verbänden. *Math. Z.* **51** (1948), 126–134.

[635] _____, Ein allgemeiner Mengenisomorphiesatz der Theorie der Verbände. *Math. Z.* **51** (1948), 346–354.

[636] _____, Ein Homomorphiesatz der Theorie der Verbände. *Math. Z.* **52** (1949), 193–201.

[637] _____, Zusammensetzung von Schwesterperspektivitäten in Verbänden. *Math. Z.* **52** (1949), 150–167.

[638] D. S. Scott, Outline of a mathematical theory of computations, 169–176, *Proc. 4-th Annual Princeton Conf. Information Sc. and Systems, Princeton Univ. Press, Princeton*, 1970.

[639] _____, Continuous lattices, *Toposes, algebraic geometry and logic*, (*Conf., Dalhousie Univ., Halifax, N. S., 1971*), pp. 97–136, Lecture Notes in Math., Vol. 274, Springer, Berlin, 1972.

[640] M. Semenova, On lattices embeddable into lattices of suborders. (Russian.) *Algebra and Logic* **44** (2005), 483–511; translation in *Algebra and Logic* **44** (2005), 270–285.

[641] M. Semenova and F. Wehrung, Sublattices of lattices of order-convex sets, I. The main representation theorem. *J. Algebra* **277** (2004), 825–860.

[642] _____, Sublattices of lattices of convex subsets of vector spaces. (Russian.) Algebra Logika **43** (2004), 261–290; translation in *Algebra and Logic* **43** (2004), 145–161.

[643] M. Semenova and A. Zamojska-Dzienio, On lattices embeddable into lattices of order-convex sets. Case of trees. *Internat. J. Algebra Comput.* **17** (2007), 1667–1712.

[644] M. Sholander, Postulates for distributive lattices. *Canad. J. Math.* **3** (1951), 28–30.

[645] J. Sichler, Non-constant endomorphisms of lattices. *Proc. Amer. Math. Soc.* **34** (1972), 67–70.

[646] _____, Note on free products of Hopfian lattices. *Algebra Universalis* **5** (1975), 145–146.

[647] R. Sikorski, Boolean algebras. Second edition. Ergebnisse der Mathematik und ihrer Grenzgebiete, Band 25. *Academic Press, New York, N.Y.,* 1964. x+237 pp.

[648] H. L. Silcock, Generalized wreath products and the lattice of normal subgroups of a group. *Algebra Universalis* **7** (1977), 361–372.

[649] F. M. Sioson, Equational bases of boolean algebras. *J. Symbolic Logic* **29** (1964), 115–124.

[650] S. V. Sizyĭ, Quasivarieties of graphs. *Siberian Math. J.* **35** (1994), 783–794.

[651] H. L. Skala, Trellis theory. *Algebra Universalis* **1** (1971), 218–233.

[652] H. L. Skala, Trellis Theory. Memoirs Amer. Math. Soc. No. **121**. *American Mathematical Society, Providence, R.I.,* 1972. iii+42 pp.

[653] V. Slavík, A note on the amalgamation property in lattice varieties. *Comment. Math. Univ. Carolin.* **21** (1980), 473–478.

[654]  V. Slavík, The amalgamation property of varieties determined by primitive lattices. *Comment. Math. Univ. Carolin.* **21** (1980), 473–478.

[655]  _____, A note on the amalgamation property in lattice varieties. *Contributions to lattice theory (Szeged, 1980)*, 723–736, Colloq. Math. Soc. János Bolyai, **33**, *North-Holland, Amsterdam,* 1983.

[656]  J. W. Snow, Every finite lattice in $\mathcal{V}(M_3)$ is representable. *Algebra Universalis* **50** (2003), 75–81.

[657]  Ju. I. Sorkin, Independent systems of axioms defining a lattice. (Russian.) *Ukrain. Mat. Ž.* **3** (1951), 85–97.

[658]  _____, Free unions of lattices. (Russian.) *Mat. Sb.* **30** (72) (1952), 677–694.

[659]  _____, On the imbedding of latticoids in lattices. (Russian.) *Doklady Akad. Nauk SSSR. N. S.* **95** (1954), 931–934.

[660]  T. P. Speed, On Stone lattices. *J. Austral. Math. Soc.* **9** (1969), 293–307.

[661]  E. Sperner, Ein Satz über Untermengen einer endlichen Menge. *Math. Z.* **27** (1928), 544–548.

[662]  R. P. Stanley, Enumerative Combinatorics. Vol. I. The Wadsworth & Brooks/Cole Mathematics Series. *Wadsworth & Brooks/Cole Advanced Books & Software, Monterey, CA,* 1986.

[663]  M. Stern, Generalized matroid lattices. *Algebraic Methods in Graph Theory, Vol. I, II (Szeged, 1978)*, 727–748, Colloq. Math. Soc. J. Bolyai, **25**, *North-Holland, Amsterdam-New York,* 1981.

[664]  _____, The impact of Dilworth's work on the Kurosch-Ore Theorem. *The Dilworth Theorems*, 203–204, Contemp. Mathematicians, Birkhäuser Boston, Boston, MA, 1990.

[665]  _____, On the covering graph of balanced lattices. *Discrete Math.* **156** (1996), 311–316.

[666]  _____, Pentagons and hexagons in semimodular lattices. *Acta Sci. Math. (Szeged)* **17** (1998), 389–395.

[667]  M. Stern, Semimodular lattices. Theory and applications. Encyclopedia of Mathematics and its Applications, **73** *Cambridge University Press, Cambridge,* 1999, xiv+370. Paperback edition, 2009.

[668]  M. H. Stone, The theory of representations for boolean algebras. *Trans. Amer. Math. Soc.* **40** (1936), 37–111.

[669] ———, Topological representations of distributive lattices and Brouwerian logics. *Časopis Pěst. Mat.* **67** (1937), 1–25.

[670] H. Strietz, Finite partition lattices are four-generated. *Proceedings of the Lattice Theory Conference (Ulm, 1975)*, 257–259. *Univ. Ulm, Ulm*, 1975.

[671] K. Takeuchi, On maximal proper sublattices. *J. Math. Soc. Japan* **2** (1951), 228–230.

[672] D. Tamari, Monoides préordonnés et chaînes de Malcev. *Thèse, Université de Paris*, 1951.

[673] A. Tarski, Sur les classes d'ensembles closes par rapport à certaines opérations élémentaires. *Fund. Math.* **16** (1930), 181–304.

[674] ———, A remark on functionally free algebras. *Ann. of Math.* (2) **47** (1946), 163–165.

[675] ———, A lattice-theoretical fixpoint theorem and its applications. *Pacific J. Math.* **5** (1955), 285–309.

[676] ———, Equational logic and equational theories of algebras. 1968 *Contributions to Math. Logic (Colloquium, Hannover, 1966)*, 275–288. *North-Holland, Amsterdam*.

[677] S.-K. Teo, Representing finite lattices as complete congruence lattices of complete lattices. *Ann. Univ. Sci. Budapest. Eötvös Sect. Math.* **33** (1990), 177–182.

[678] ———, On the length of the congruence lattice of a lattice. *Period. Math. Hungar.* **21** (1990), 179–186.

[679] M. Tischendorf, The representation problem for algebraic distributive lattices, Ph.D. thesis, TH Darmstadt, 1992.

[680] J. R. R. Tolkien, The Lord of the Rings. *George Allen & Unwin, London*, 1954.

[681] S. T. Tschantz, Infinite intervals in free lattices. *Order* **6** (1990), 367–388.

[682] A. A. Tuganbaev, Distributive modules and rings. The concise handbook of algebra. 280–283. Edited by A. V. Mikhalev and G. Pilz, *Kluwer Academic Publishers, Dordrecht*, 2002. xvi+618 pp.

[683] J. Tůma, On the existence of simultaneous representations. *Acta Sci. Math. (Szeged)* **64** (1998), 357–371.

[684]  J. Tůma and F. Wehrung, Simultaneous representations of semilattices
       by lattices with permutable congruences. *Int. J. Algebra Comput.* **11 (2)**
       (2001), 217–246

[685]  _____ , A survey of recent results on congruence lattices of lattices.
       *Algebra Universalis* **48** (2002), 439–471.

[686]  H. Tverberg, On Dilworth's decomposition theorem for partially ordered
       sets. *J. Combin. Theory* **3** (1967) 305–306.

[687]  A. Urquhart, A topological representation theory for lattices. *Algebra
       Universalis* **8** (1978), 45–58.

[688]  J. C. Varlet, Contribution à l'étude des treillis pseudo-complémentés et
       des treillis de Stone. *Mém. Soc. Roy. Sci. Liège Coll.*, **8** (1963), 71 pp.

[689]  _____ , On the characterization of Stone lattices. *Acta Sci. Math.*
       (*Szeged*) **27** (1966), 81–84.

[690]  O. Veblen and W. H. Young, Projective Geometry. 2 volumes. *Ginn and
       Co., Boston,* 1910.

[691]  S. Vickers, Topology via Logic. Cambridge Tracts in Theoretical Com-
       puter Science, **5**. *Cambridge University Press, Cambridge,* 1989. xvi+200
       pp.

[692]  P. Vopěnka, A. Pultr, and Z. Hedrlín, A rigid relation exists on any set.
       *Comment. Math. Univ. Carolin.* **6** (1965), 149–155.

[693]  B. L. van der Waerden, Modern Algebra. Vol. I. Translated from the sec-
       ond revised German edition by Fred Blum. With revisions and additions
       by the author. *Frederick Ungar Publishing Co., New York, N.Y.,* 1949.
       xii+264 pp.

[694]  A. Walendziak, On direct decompositions of lattices. *Algebra Univer-
       salis* **37** (1997), 185–190.

[695]  Shih-chiang Wang, Notes on the permutability of congruence relations
       (Chinese), *Acta Math. Sinica* **3** (1953), 133–141.

[696]  M. Ward, A characterization of Dedekind structures. *Bull. Amer. Math.
       Soc.* **45** (1939), 448–451.

[697]  A. G. Waterman, The free lattice with 3 generators over $N_5$. *Portugal.
       Math.* **26** (1967), 285–288.

[698]  F. Wehrung, Treillis bi-locaux équationnellement compacts. *C. R. Acad.
       Sci. Paris Sér. I Math.* **318** (1994), 5–9.

[699] _____, Equational compactness of bi-frames and projection algebras. *Algebra Universalis* **33** (1995), 478–515.

[700] _____, Non-measurability properties of interpolation vector spaces. *Israel J. Math.* **103** (1998), 177–206.

[701] _____, The dimension monoid of a lattice. *Algebra Universalis* **40** (1998), 247–411.

[702] _____, A uniform refinement property of certain congruence lattices. *Proc. Amer. Math. Soc.* **127** (1999), 363–370.

[703] _____, Representation of algebraic distributive lattices with $\aleph_1$ compact elements as ideal lattices of regular rings. *Publ. Mat. (Barcelona)* **44** (2000), 419–435.

[704] _____, Join-semilattices with two-dimensional congruence amalgamation. *Colloq. Math.* **93** (2002), 209–235.

[705] _____, From join-irreducibles to dimension theory for lattices with chain conditions. *J. Algebra Appl.* **1** (2002), 215–242.

[706] _____, Direct decompositions of non-algebraic complete lattices. *Discrete Math.* **263** (2003), 311–321.

[707] _____, Forcing extensions of partial lattices. *J. Algebra* **262** (2003), 127–193.

[708] _____, Semilattices of finitely generated ideals of exchange rings with finite stable rank, *Trans. Amer. Math. Soc.* **356** (2004), 1957–1970.

[709] _____, Sublattices of complete lattices with continuity conditions. *Algebra Universalis* **53** (2005), 149–173.

[710] _____, Von Neumann coordinatization is not first-order. *J. Math. Log.* **6** (2006), 1–24.

[711] _____, A solution to Dilworth's congruence lattice problem. *Adv. Math.* **216** (2007), 610–625.

[712] _____, Poset representations of distributive semilattices. *Internat. J. Algebra Comput.* **18** (2008), 321–356.

[713] _____, Large semilattices of breadth three. *Fund. Math.* **208** (2010), 1–21.

[714] _____, Coordinatization of lattices by regular rings without unit and Banaschewski functions. *Algebra Universalis*. To appear.

[715] F. Wehrung, A non-coordinatizable sectionally complemented modular lattice with a large Jónsson four-frame. Manuscript, 2010.

[716] L. Weisner, Abstract theory of inversion of finite series. *Trans. Amer. Math. Soc.* **38** (1935), 474–484.

[717] D. J. A. Welsh, Matroid theory. L. M. S. Monographs, No. 8. *Academic Press [Harcourt Brace Jovanovich, Publishers], London-New York,* 1976. xi+433 pp.

[718] H. Werner and R. Wille, Charakterisierungen der primitiven Klassen arithmetischer Ringe. *Math. Z.* **115** (1970), 197–200.

[719] N. White (editor), Theory of matroids. Encyclopedia of Mathematics and its Applications, 26. *Cambridge University Press, Cambridge,* 1986. xviii+316 pp.

[720] N. White (editor), Combinatorial geometries. Encyclopedia of Mathematics and its Applications, 29. *Cambridge University Press, Cambridge,* 1987. xii+212 pp.

[721] N. White (editor), Matroid applications. Encyclopedia of Mathematics and its Applications, 40. *Cambridge University Press, Cambridge,* 1992. xii+363 pp.

[722] P. M. Whitman, Free lattices. *Ann. of Math.* (2) **42** (1941), 325–330.

[723] ———, Free lattices. II. *Ann. of Math.* (2) **43** (1942), 104–115.

[724] ———, Lattices, equivalence relations, and subgroups. *Bull. Amer. Math. Soc.* **2** (1946), 507–522.

[725] H. Whitney, On the abstract properties of linear dependence. *Amer. J. Math.* **57** (1935), 509–533.

[726] L. R. Wilcox, Modularity in the theory of lattices. *Ann. of Math.* **40** (1939), 490–505.

[727] ———, A note on complementation in lattices. *Bull. Amer. Math. Soc.* **48** (1942), 453–458.

[728] M. Wild, Join epimorphisms which preserve certain lattice identities. *Algebra Universalis* **27** (1990), 398–410.

[729] ———, Cover-preserving order embeddings into Boolean lattices. *Order* **9** (1992), 209–232.

[730] ———, Cover-preserving embedding of modular lattices into partition lattices. *Discrete Math.* **112** (1993), 207–244.

[731] _____, A theory of finite closure spaces based on implications. *Adv. Math.* **108** (1994), 118–139.

[732] _____, The minimal number of join irreducibles of a finite modular lattice. *Algebra Universalis* **35** (1996), 113–123.

[733] _____, Optimal implicational bases for finite modular lattices. *Quaest. Math.* **23** (2000), 153–161.

[734] R. Wille, Primitive subsets of lattices. *Algebra Universalis* **2** (1972), 95–98.

[735] _____, Jeder endlich erzeugte, modulare Verband endlicher Weite ist endlich. *Mat. Časopis Sloven. Akad. Vied.* **24** (1974), 77–80.

[736] _____, An example of a finitely generated non-Hopfian lattice. *Algebra Universalis* **5** (1975), 101–103.

[737] _____, Eine Charakterisierung endlicher, ordnungspolynomvollständiger Verbände. *Arch. Math. (Basel)* **28** (1977), 557–560.

[738] _____, Über endliche, ordnungsaffinvollständige Verbände. *Math. Z.* **155** (1977), 103–107.

[739] _____, Restructuring lattice theory: an approach based on hierarchies of concepts. *Ordered sets (Banff, Alta., 1981)*, 445–470, NATO Adv. Study Inst. Ser. C: Math. Phys. Sci., **83**, *Reidel, Dordrecht-Boston, Mass.*, 1982.

[740] M. Yasuhara, The Amalgamation Property, the Universal-Homogeneous Models, and the Generic Models. *Math. Scand.* **34** (1974), 5–36.

[741] L. Zádori, Generation of finite partition lattices. *Lectures in universal algebra (Szeged, 1983)*, 573–586, Colloq. Math. Soc. J. Bolyai, **43**, *North-Holland, Amsterdam*, 1986.

[742] H. J. Zassenhaus, The theory of groups. 2nd ed. *Chelsea Publishing Company, New York*, 1958. x+265 pp.

# Index

589

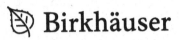

# The Congruences of a Finite Lattice
## A Proof-by-Picture Approach

### George Grätzer
University of Manitoba, Winnipeg, Canada

2006
XXII, 282 p., 110 illus.
With online files/update
Hardcover
ISBN 978-0-8176-3224-3

The congruences of a lattice form the congruence lattice. In the past half-century, the study of congruence lattices has become a large and important field with a great number of interesting and deep results and many open problems. This self-contained exposition by one of the leading experts in lattice theory, George Grätzer, presents the major results on congruence lattices of finite lattices featuring the author's signature "Proof-by-Picture" method and its conversion to transparencies.

Key features:
* Includes the latest findings from a pioneering researcher in the field
* Insightful discussion of techniques to construct "nice" finite lattices with given congruence lattices and "nice" congruence-preserving extensions
* Contains complete proofs, an extensive bibliography and index, and nearly 80 open problems
* Additional information provided by the author online at: http://www.maths.umanitoba.ca/homepages/gratzer.html/

The book is appropriate for a one-semester graduate course in lattice theory, yet is also designed as a practical reference for researchers studying lattices.

*The book is self-contained, with many detailed proofs presented that can be followed step-by-step. In addition to giving the full formal detals of the proofs, the author chooses a somehow more pedagogical way that he calls Proof-by-Picture, somehow related to the combinatorial (as opposed to algebraic) nature of many of the presented results. I believe that this book is a much-needed tool for any mathematician wishing a gentle introduction to the field of congruences representations of finite lattices, with emphasisi on the more 'geometric' aspects.* — Mathematical Reviews